Food Security and Nutrition

Food Security and Nutrition: Utilizing Undervalued Food Plants explores the potential of underutilized and indigenous food plants to enhance food and nutrition security amid global challenges, such as climate change, water scarcity, and population growth. It emphasizes the importance of dietary diversification to reduce dependency on major staple crops, highlighting the nutritional and health benefits of lesser-known crops, such as fonio, pigeonpea, finger millet, and a wide array of wild vegetables and mushrooms. The book reviews innovative farming techniques, like the organic medium–enclosed trough system and hydroponics, which improve crop yield and nutrient content while conserving resources. Additionally, it addresses the commercialization potential of indigenous fruits and wild herbal teas, emphasizing their market value and health benefits. The book also discusses the reduction of goitrogenic compounds in certain crops to mitigate health risks.

By examining the unique nutritional profiles and local adaptability of these crops, the book advocates for their broader use to build a more resilient and sustainable food system. Through a combination of scientific research, practical farming methods, and market strategies, the book aims to promote food diversity, improve health outcomes, and support sustainable agricultural practices.

Key Features

- Highlights the critical role of dietary diversification in reducing dependency on major staple crops, advocating for the inclusion of lesser-known, nutrient-rich plants to improve overall health and nutrition.
- Introduces cutting-edge farming methods, such as the organic medium–enclosed trough system and hydroponics.
- Explores the market value and health benefits of indigenous fruits and wild herbal teas, offering strategies for their commercialization to boost local economies and food security.
- Addresses the reduction of harmful compounds, and provides practical solutions to mitigate associated health risks and enhance the safety of food consumption.

Food Security and Nutrition

Utilizing Undervalued Food Plants

Edited by
Callistus Bvenura and Learnmore Kambizi

CRC Press
Taylor & Francis Group
Boca Raton London New York

CRC Press is an imprint of the
Taylor & Francis Group, an **informa** business

Designed cover image: Photographed by Ms. Senzeni Bvenura at the Bulawayo Vegetable Market in Zimbabwe on May 24, 2024. This image captures a vibrant vendor stall brimming with fresh and dried, underutilized indigenous foods.

First edition published 2025
by CRC Press
2385 NW Executive Center Drive, Suite 320, Boca Raton FL 33431

and by CRC Press
4 Park Square, Milton Park, Abingdon, Oxon, OX14 4RN

CRC Press is an imprint of Taylor & Francis Group, LLC

ISBN: 978-1-032-73655-6 (hbk)
ISBN: 978-1-032-74543-5 (pbk)
ISBN: 978-1-003-46976-6 (ebk)

DOI: 10.1201/9781003469766

Typeset in Times
by Apex CoVantage, LLC

Contents

Preface

In matters of music and food, I have often been accused of being too traditional – a label I embrace with great pride. Born and raised in Masase, in the Midlands province of Zimbabwe, I am what is commonly known in Zimbabwe as an "SRB" – a person with a "strong rural background." This upbringing has endowed me with extensive knowledge of both medicinal and food plants, which I have applied throughout my life. When I was growing up, our diet was predominantly plant-based, consisting mainly of indigenous and underutilized food plants. The rare occasions we ate meat were special, but it is the plant-based meals that evoke a profound sense of nostalgia, reminding me of evenings around a bonfire, listening to folklore passed down through generations. These gatherings fostered a strong sense of belonging and identity.

It is, therefore, no surprise that my career path has brought me back to my roots – both literally and figuratively. Reflecting on my childhood, I realize how rarely we fell ill. The adage "You are what you eat" rings true, validated by scientific research into these food plants, which are now recognized as superfoods, with diverse nutritional and health benefits. Their consumption is a significant step toward living a healthy, disease-free life. However, it is essential to acknowledge that physical activity and other factors also play crucial roles in overall well-being.

This book has been a long time coming. After the success of our first book, *Sustainable Uses and Prospects of Medicinal Plants*, my colleague Prof. Learnmore Kambizi and I decided to collaborate again with experts in related fields to compile this work. The response was overwhelming, prompting us to select a few contributions for this edition and reserve the rest for a second edition, God willing. Prof. Kambizi, with whom I share a totem and a similar upbringing, has made this collaboration one of the most fulfilling experiences of my career. The content of this book resonates deeply with us, especially the plants of choice, and I am confident it will resonate with you, the reader, regardless of your background.

This book, with its carefully selected chapters, is not the end but a call to action. To achieve food and nutrition security and eliminate hunger for posterity, humanity must work together, adopt the pragmatic steps suggested herein, and turn to traditional underutilized food plants for their environmental and health benefits. Incorporating these plants into our daily diets is a crucial step toward this goal.

We have enjoyed compiling this work and hope you will enjoy reading it even more. We are immensely grateful to each contributing author for their invaluable time and effort. This project would have been in vain without their contributions. We also extend our heartfelt thanks to our parents, families, and loved ones for their subliminal yet meaningful support throughout this journey.

Finally, we thank God for giving us the strength to see this project through to completion!

About the Editors

Dr. Callistus Bvenura graduated with a BSc in natural resources management and agriculture in 2006 from the Midlands State University, Zimbabwe, and completed his PhD in ethnobotany at the University of Fort Hare, South Africa, in 2014. His areas of specialization encompass interdisciplinary research at the intersection of plant biology, ethnobotany, food science, agriculture, and public health, with a focus on promoting sustainable food and nutrition security through the utilization of diverse plant resources. Callistus's research profile is comprehensive, including co-edited books, numerous peer-reviewed research articles, book chapters, and review papers in these focus areas, and is constantly presenting his research findings at local and international conferences. In addition, he has interests in product development, where he and his fellow colleagues developed Niselo, a probiotic sorghum beverage which has since then been commercialized in South Africa and the United Kingdom. He is also part of a team that developed broiler chicken feed using indigenous and underutilized grains and insects. Furthermore, he is a South African National Research Foundation–rated researcher and panel expert reviewer. Dr. Bvenura is also an editorial board member of the *Journal of Medicinal Plants and Economic Development* (JOMPED) and reviews for several local and international journals.

Prof. Learnmore Kambizi attained his BSc degree from Cuba in 1998 before enrolling for his master's and PhD degrees at the University of Fort Hare, which he completed in 2004. His PhD focused on ethnobotanical studies of plants used for the treatment of sexually transmitted infections in South Africa. He went on to pursue a postdoctoral fellowship within the fields of ethnobotany and phytomedicine. This discipline enabled him to realize his passion for sustainable utilization and conservation of medicinal plants on a local and international level. He worked as a lecturer at the University of Zimbabwe in 2005 before moving to the University of the Free State, South Africa, in 2007. In the same year, he was appointed by Walter Sisulu University as a senior lecturer until 2011. He moved to Cape Peninsula University of Technology, South Africa, where he is now a full professor of botany. In his academic journey to date, he has hosted four postdoctoral fellows and supervised five PhD, 16 master's, and numerous honors students. In addition, he has edited and published a book, published numerous articles in peer-reviewed journals, and presented at many national and international conferences. He recently edited a book titled *Sustainable uses and prospects of medicinal plants*. He is a South African National Research Foundation–rated researcher. Prof. Kambizi has also served as a panel member of the National Research Foundation of South Africa and has received many research grants to support postgraduate students and to assist communities. He also collaborates with traditional medical practitioners in collaboration with the South African Research Council as an advisor. In addition, Prof. Kambizi was appointed the first director of the African Centre for Herbal Research, based in Nigeria, a position he holds to date. He is a reviewer of many national and international journals.

About the Editors

Contributors

Oluwaseyi Moyode Aboyade
Department of Pharmaceutical Sciences
Tshwane University of Technology
South Africa

Oluwagbenga Oluwasola Adeogun
Department of Botany
University of Lagos
Nigeria

Jesulewami Precious Adeoye
Department of Plant Science
Olabisi Onabanjo University
Nigeria

Karimat Adeyinka Ahmed
Department of Plant Science
Olabisi Onabanjo University
Nigeria

Stephen Oluwaseun Amoo
Agricultural Research Council – Vegetables, Industrial, and Medicinal Plants
South Africa

David Adedayo Animasaun
Department of Plant Biology
University of Ilorin
Nigeria

Adeyemi Oladapo Aremu
Faculty of Natural and Agricultural Sciences
North-West University
South Africa

John Awungnjia Asong
Faculty of Natural and Agricultural Sciences
North-West University
South Africa

Yetunde Oyinkansade Babalola
Biomedicinal Research Centre
Forestry Research Institute of Nigeria
Nigeria

Moji Taibat Bakare-Odunola
Department of Pharmaceutical and Medicinal Chemistry
University of Ilorin
Nigeria

Samuel Oloruntoba Bamigboye
Department of Plant Science
Olabisi Onabanjo University
Nigeria

Oluwadamilola Mariam Baruwa
Department of Plant Science
Olabisi Onabanjo University
Nigeria

Evison Bhebhe
Engineering and Agriculture
University of Venda
South Africa

Callistus Bvenura
Department of Horticultural Sciences
Cape Peninsula University of Technology
South Africa

Sinorita Chauke
School of Biology and Environmental Sciences
University of Mpumalanga
South Africa

Ahmad Cheikhyoussef
Science and Technology Division, Multidisciplinary Research
University of Namibia
Namibia

Rumbidzai Lesley Chitsiku
National Biotechnology Authority
Zimbabwe

Rogerio Marcos Chiulele
Centre of Excellence in Agri-Food Systems and Nutrition
Mozambique

Babajide Charles Falemara
Research Coordinating Unit
Forestry Research Institute of Nigeria
Nigeria

Eastonce Tendayi Gwata
Department of Plant and Soil Sciences
University of Venda
South Africa

Lydia Ndinelao Horn
Zero Emission Research Initiative, Multidisciplinary Research
University of Namibia
Namibia

Liana Jank
Embrapa Beef Cattle
Brazil

Muhali Olaide Jimoh
Department of Plant Science
Olabisi Onabanjo University
Nigeria

MacDonald Bright Jumbo
International Crops Research Institute for Semi-Arid Tropics
Mali

Nailoke Pauline Kadhila
Department of Biochemistry
University of Namibia
Namibia

Learnmore Kambizi
Department of Horticultural Sciences
Cape Peninsula University of Technology
South Africa

David Roy Katerere
Department of Pharmaceutical Sciences
Tshwane University of Technology
South Africa

Mothibi Elizabeth Kola
School of Biology and Environmental Sciences
University of Mpumalanga
South Africa

Fausat Lola Kolawole
Department of Home Economics and Food Science
University of Ilorin
Nigeria

Ramar Krishnamurthy
C. G. Bhakta Institute of Biotechnology
Uka Tarsadia University
India

Charles Petrus Laubscher
Department of Horticultural Sciences
Cape Peninsula University of Technology
South Africa

Abdulazeez Opeyemi Lawal
Department of Plant Science
Olabisi Onabanjo University
Nigeria

Cyprian Mahuni
National Biotechnology Authority
Zimbabwe

Dorcus Maja
Faculty of Natural and Agricultural Sciences
North-West University
South Africa

Alen Manyevere
Department of Agronomy
University of Fort Hare
South Africa

Ruvimbo Jessy Mapaya
Faculty of Science and Technology
Midlands State University
Zimbabwe

Tsakani Maria Maphosa
Department of Plant and Soil Sciences
University of Venda
South Africa

Alfred Maroyi
Department of Botany
University of Fort Hare
South Africa

Rose Tafadzwa Masekesa
National Biotechnology Authority
Zimbabwe

Dipoo Charity Mashifane
Department of Plant and Soil Sciences
University of Venda
South Africa

Munyaradzi Mativavarira
National Biotechnology Authority
Zimbabwe

Salmina Ngoakoana Mokgehle
School of Agricultural Sciences
University of Mpumalanga
South Africa

Rorisang Michelle Moyo
National Biotechnology Authority
Zimbabwe

Shantel Moyo
National Biotechnology Authority
Zimbabwe

Semakaleng Mpai
Green Biotechnologies Research Centre of Excellence
University of Limpopo
South Africa

Esther Kathini Muema
Department of Soil Science
Stellenbosch University
South Africa

Faith Tariro Mukoko
National Biotechnology Authority
Zimbabwe

Olivia Mukondwa
Crop Breeding Institute
Matopos Research Station
Zimbabwe

Davis Mumbengegwi
Centre for Research Services
University of Namibia
Windhoek, Namibia

Lissette Mupanda
National Biotechnology Authority
Zimbabwe

Tatenda Clive Murashiki
National Biotechnology Authority
Zimbabwe

Hapson Mushoriwa
International Institute of Tropical Agriculture
Nigeria

Reward Muzerengwa
National Biotechnology Authority
Zimbabwe

Blessing Mwanalota
National Biotechnology Authority
Zimbabwe

James Lwambi Mwinga
Faculty of Natural and Agricultural Sciences
South Africa

Odeshnee Naicker
Department of Plant and Soil Sciences
University of Venda
South Africa

Felix Nchu
Department of Horticultural Sciences
Cape Peninsula University of Technology
South Africa

Prisca Nonceba Ncube
National Biotechnology Authority
Zimbabwe

Ashwell Rungano Ndhlala
Green Biotechnologies Research Centre of Excellence
University of Limpopo
South Africa

Peter Tshepiso Ndhlovu
School of Biology and Environmental Sciences
University of Mpumalanga
South Africa

Charity Ngirande
National Biotechnology Authority
Zimbabwe

Sihle Ngxabi
Department of Horticultural Sciences
Cape Peninsula University of Technology
South Africa

Keamogela Innocentia Nko
Faculty of Natural and Agricultural Sciences
North-West University
South Africa

Makomborero Nyoni
National Biotechnology Authority
Zimbabwe

Paul Oluwatimilehin Ogunbamowo
Biomedicinal Research Centre
Forestry Research Institute of Nigeria
Nigeria

Peter Avwerosuo Ohwofasa
Department of Pharmaceutical Sciences
Tshwane University of Technology
South Africa

Kehinde Stephen Olorunmaiye
Department of Plant Biology
University of Ilorin
Nigeria

Oluwatoyin Oladayo Onipe
Department of Biotechnology and Food Technology
University of Johannesburg
South Africa

Wilfred Otang-Mbeng
School of Biology and Environmental Sciences
University of Mpumalanga
South Africa

Stephen Oyedeji
C. G. Bhakta Institute of Biotechnology
Uka Tarsadia University
India

Rukayat Abiodun Oyegoke
Department of Biochemistry
University of Ilorin
Nigeria

Nikita Patel
C. G. Bhakta Institute of Biotechnology
Uka Tarsadia University
India

Iwanette du Preez-Bruwer
Centre for Research Services
University of Namibia
Windhoek, Namibia

Serifat Olatundun Salami
Department of Agricultural Education
Federal College of Education (Special)
Nigeria

Comfort Titilayomi Senjobi
Department of Plant Science
Olabisi Onabanjo University
Nigeria

Whitney Shingenge
Centre for Research Services
University of Namibia
Namibia

Fimanekeni Ndaitavela Shivute
Zero Emission Research Initiative, Multidisciplinary Research
University of Namibia
Namibia

Addmore Shonhai
Department of Biochemistry and Microbiology
University of Venda
South Africa

Melisa Nomalanga Sithole
National Biotechnology Authority
Zimbabwe

Avela Sogoni
Department of Horticultural Sciences
Cape Peninsula University of Technology
South Africa

Linda Ifeanyichukwu Sowunmi
Biomedicinal Research Centre
Forestry Research Institute of Nigeria
Nigeria

Elizabeth Vimbai Tavengana
National Biotechnology Authority
Zimbabwe

Fhatuwani Thovhogi
Department of Plant and Soil Sciences
University of Venda
South Africa

Henry Okwudili Udeh
Department of Applied Microbiology and Brewing
State University of Medical and Applied Sciences
Nigeria

Isabella Shaningika Etuhole Ueitele
Zero Emission Research Initiative, Multidisciplinary Research
University of Namibia
Namibia

Sonja Louise Venter
Agricultural Research Council – Vegetables, Industrial, and Medicinal Plants
South Africa

Hildegard Witbooi
Department of Agronomy
University of Fort Hare
South Africa

Janet Chiedza Zambuko
National Biotechnology Authority
Zimbabwe

Tariro Sarah Zimuto
National Biotechnology Authority
Zimbabwe

1 The Future of Undervalued Indigenous Crops

A Brief Overview

**Learnmore Kambizi and Callistus Bvenura*

Department of Horticulture, Faculty of Applied Sciences, Cape Peninsula University of Technology, Bellville 7535, South Africa.

*Corresponding author: kambiziL@cput.ac.za

There are no qualms that the growing global challenges related to food and nutrition security, such as climate change, population growth, and resource scarcity, demand a shift toward more diverse and resilient food sources. Historically, agriculture has relied heavily on a limited number of staple crops, leading to vulnerabilities in food systems (Bvenura & Afolayan, 2015). In fact, although close to 250,000 plant species are available for agriculture, about 7,500 of these species are in use today (FAO, 2004). Furthermore, out of these 7,500, only 150 supply the world's food needs (IDRC.CRDI, 2010). Disappointingly, out of the 150 species, about 12 supply close to 75% of the world's food needs, while only three, including maize, rice, and wheat, supply more than 50% of the world's calory needs (FAO, 2004). It is therefore not surprising that these crops are known as "mega-crops." Nevertheless, mega-crops are well-known for their high-maintenance requirements and the demands they place on growth and development. In the face of drought, these mega-crops are easily decimated, often leading to extreme food and nutrition insecurity. The UN's SDG 2, to feed 8.6 billion people by 2030 from the current 8.1 billion, and 9.8 billion by 2050, seems insurmountable if humanity continues to look to a few crops for salvation. Recent reports indicate that 29.3% of the global population (2.3 billion people) experienced moderate or severe food insecurity and nearly 828 million people (10.5%) suffered from hunger (Delgado et al., 2023). Another report indicates that the number of people facing food insecurity increased by 24 million in 2022 (FSIN and Global Network Against Food Crises, 2024). Unfortunately, Africa carries the heaviest burden of food and nutrition insecurity. In 2021, 20.2% of its population faced hunger, compared to 9.1% in Asia, 8.6% in Latin America and the Caribbean, 5.8% in Oceania, and less than 2.5% in North America and Europe (Delgado et al., 2023). Africa is rich in natural resources, including minerals, oil, natural gas, arable land, forests, and fresh water, yet faces the paradox of significant resource wealth coexisting with challenges like political instability, inadequate infrastructure, and insufficient investment in technology and education, which hinder effective utilization and limit economic prosperity and food security (Akpan, 2012).

Therefore, to address these challenges, it is crucial to explore the potential of indigenous food plants, which are often underutilized and undervalued, despite their nutritional and agricultural benefits. It is the main thrust of this book to highlight the importance of these crops and their potential to enhance food security and nutrition now and for posterity. But what are *undervalued food plants*?

Undervalued food plants, often referred to as "underutilized crops," or "neglected crops," or "orphan crops" and "hidden harvest," are species that have significant nutritional, agricultural, and economic potential but are not widely recognized or cultivated on a large scale (Bvenura & Afolayan, 2015). These plants are typically indigenous to specific regions and have been traditionally used by

DOI: 10.1201/9781003469766-1

local communities since time immemorial (Bvenura & Sivakumar, 2017). Despite their potential to enhance food security, nutrition, and biodiversity, they remain underexploited in modern agricultural systems.

Crops such as fonio (*Digitaria exilis*), pigeon pea (*Cajanus cajan*), finger millet (*Eleusine coracana*), sorghum (*Sorghum bicolor*), black nightshade (*Solanum nigrum*), jute mallow (*Corchorus olitorius*), cat's whiskers (*Cleome gynandra*), and indigenous fruits such as marula (*Sclerocarya birrea*) and baobab (*Adansonia digitata*), are some good examples of crops that have been traditionally used in local diets, especially in Africa, for centuries. A crop such as moringa (*Moringa oleifera*) is well-known as a superfood around the world, indicating its potential in the food matrix. Most of these crops are well-adapted to the local climate and soil conditions, yet they remain underexploited in modern agriculture. Fonio, for example, is one of the oldest cereals in West Africa, known for its drought resilience and quick growth cycle (Diarra & Dicko, 2014). However, it remains unknown throughout the continent, except in the few areas where it grows. If this crop is integrated into mainstream agriculture, it could easily play a huge role in ameliorating food and nutrition insecurity. Pigeon pea is valued for its high protein content and ability to fix nitrogen in the soil, improving soil fertility and benefiting subsequent crops (Jorrin et al., 2021). Cultivating this "supercrop" could prove beneficial for the environment while ensuring the amelioration of hunger and nutritional deficiencies. Nutritional, polyphenolic constituents, and biological activities of these plants are awash on numerous databases, underscoring the need for their inclusion in the diet for diversity and, perhaps, as staples in the future. Perhaps since these attributes are in the public domain and much work has been put into publicizing them over the past few decades, there is need to move forward and put this knowledge into good use.

Therefore, this book proposes several innovative ways of ameliorating food and nutrition insecurity using undervalued food plants.

One of the significant advantages of these indigenous crops is their adaptability to local conditions. They are often more resilient to drought, poor soils, and extreme weather compared to conventional staples. In this book, innovative farming techniques, such as the organic medium-enclosed trough (OMET) system, are proposed because of their ability to enhance growth, yield, and nutritional attributes. The OMET system reduces water and nutrient seepage, making it an efficient method for small-scale and commercial farming (Ferreira, 2013). Furthermore, hydroponics systems are promising technologies that can boost the production of underutilized indigenous crops. Hydroponics allows for year-round cultivation in controlled environments, saving water and overcoming land limitations (Kumar & Singh, 2024). This is especially important if humanity is to survive through severe weather patterns. These techniques can play a crucial role in modernizing the cultivation of underutilized crops and making them more competitive in the market.

Furthermore, breeding programs to enhance preferred qualities of these crops could prove beneficial to the overall goal. Antinutritional factors, including goitrogens that are prevalent especially in cruciferous vegetables and some members of the Fabaceae family, can be reduced through these breeding programs as well as other means proposed in this book.

Without a doubt, the market potential for undervalued indigenous fruits and crops is significant, driven by increasing consumer demand for unique and healthy food products. For example, marula and baobab fruits, as well as moringa, have shown promising market value, with substantial exports to the EU (ABioSA, 2020). Commercializing these crops can contribute to sustainable agriculture, biodiversity preservation, and economic opportunities for local communities. Product development and market research are essential to optimize formulations and enhance consumer acceptance, ensuring successful commercialization. This is why, in this book, underutilized herbal teas and wild mushrooms are also targets for improvement and commercialization for food and nutrition security, health, as well as medicine. From this perspective, a plant that can fit these attributes is a *superfood*.

However, despite their potential, several barriers hinder the wider adoption of underutilized indigenous crops. As such, each one in the following 25 strong chapters ends with a "conclusion and prospects" aspect. This is crucial in mapping the way forward and knowing what to expect in the next 15–20 years, including what can be done to reach humanity's intended goal of eradicating hunger as well as food nutrition insecurity. Underutilized indigenous crops hold immense potential to contribute to food security and nutrition in the face of global challenges. Their unique nutritional profiles, adaptability, and resilience make them valuable assets in building a more diverse and sustainable food system. To fully realize their potential, increased research, funding, and policy support are essential. By integrating these undervalued crops into mainstream agriculture, we can enhance food security, improve public health, and promote sustainable development.

Finally, we hope that you, the reader, will take pleasure in this remarkable work and join the fight for a hunger-free future, where food and nutrition are secure for all. In a world rich with plant biodiversity, boasting an estimated 250,000 agricultural plants, no one should go to bed hungry.

REFERENCES

ABioSA. (2020). Accessing international markets for Marula fruit & oil. *Technical Brief 1*. https://unctad.org/system/files/official-document/ditc-ted-05052020-BioTradeSSC-Marula.pdf

Akpan, U. (2012). Natural resources and economic development in Africa. *Journal of Sustainable Development in Africa*, *14*(8), 45–57.

Bvenura, C., & Afolayan, A. J. (2015). The role of wild vegetables in household food security in South Africa: A review. *Food Research International*, *76*, 1001–1011.

Bvenura, C., & Sivakumar, D. (2017). The role of wild fruits and vegetables in delivering a healthy and balanced diet. *Food Research International*, *99*, 15–30.

Delgado, C., Tschunkert, K., & Smith, D. (2023). *Food insecurity in Africa: Drivers and solutions*. SIPRI Research Policy Paper, SIPRI, Stockholm.

Diarra, B., & Dicko, M. H. (2014). Fonio (*Digitaria exilis* stapf) grain: A review of its nutritional properties, processing, and potential health benefits. *Food Reviews International*, *30*(3), 289–303.

FAO. (2004). *What is agrobiodiversity*. Author.

Ferreira, J. (2013, September 3). OMET farming better than hydroponics? *Farmers Weekly Publication*. www.farmersweekly.co.za/crops/field-crops/omet-farming-better-than-hydroponics/

FSIN and Global Network Against Food Crises. (2024). *GRFC 2024*. Author.

IDRC.CRDI. (2010). *Facts & figures on food and biodiversity*. IDRC.

Jorrin, B., Maluk, M., Atoliya, N., Kumar, S. C., Chalasani, D., Tkacz, A., Singh, P., Basu, A., Pullabhotla, S., Kumar, M., Mohanty, S. R., East, A. K., Ramachandran, V. K., James, E. K., Podile, A. R., Saxena, A. K., Rao, D., & Poole, P. S. (2021). Genomic diversity of pigeon pea (Cajanus cajan L. Millsp.) endosymbionts in India and selection of potential strains for use as agricultural inoculants. *Frontiers in Plant Science*, *12*, 680981.

Kumar, P., & Singh, J. (2024). Hydroponic: An eco-friendly future. In N. Kumar (Ed.), *Hydroponics and environmental bioremediation* (pp. 223–267). Springer Water; Springer.

2 The Contribution of Underutilized Fruits and Vegetables to Enhanced Food and Nutrition Security

A Review

**Peter Tshepiso Ndhlovu,[1] Salmina Ngoakoana Mokgehle,[2] Mothibi Elizabeth Kola,[1] Sinorita Chauke,[1] Babajide Charles Falemara,[3] and Wilfred Otang-Mbeng[1]*

[1]School of Biology and Environmental Sciences, Faculty of Agriculture and Natural Sciences, University of Mpumalanga, Private Bag X11283, Mbombela 1200, Mpumalanga Province, South Africa

[2]School of Agricultural Sciences, Faculty of Agriculture and Natural Sciences, University of Mpumalanga, Private Bag X11283, Mbombela 1200, South Africa

[3]Research Coordinating Unit, Forestry Research Institute of Nigeria, Ibadan, PMB 5054, Forest Hills, Jericho, Oyo State, Nigeria

*Corresponding author: Tshepiso.Ndhlovu@ump.ac.za

2.1 INTRODUCTION

Africa draws strength from its diverse plant species and ways of using them for food security (Ngugi, 2000). Yet this diversity of food is threatened by negligence, insufficient knowledge, and population growth, leading to habitat destruction. Most people living in rural areas live below the poverty datum line (FAO et al., 2017; FAO & UNIDO, 2008). A limited number of households can afford three meals a day, and most meals are unbalanced due to affordability (Willett et al., 2019). Indigenous people have used underutilized fruits and vegetables since ancient times. Fruits and vegetables grow and develop in response to a wide range of stimuli, especially the availability of CO_2, moisture, mineral nutrients, temperature, and the quality as well as quantity of light (Bals et al., 2008; Hawkins et al., 2008). Climate change and COVID-19 impact on indigenous fruits and vegetables has been huge and substantive (Fanzo & Miachon, 2023). The latter alters weather trends, such as wind patterns, and the intensity and frequency of weather extremes, such as droughts, heat waves, floods, or storms. All key components of food security are affected by climate change and, recently, COVID-19. Climate change is heavily linked to the high rise in malnutrition, poor health, hunger, and starvation, as well as food and water insecurity (Padulosi et al., 2011; Shumsky et al., 2014). However, the historical utilization of plants as health remedies for humans and animals is as old as civilization (Aziz et al., 2017; Maanda & Bhat, 2010; Mengistu & Hager, 2009). There is consensus in the literature that poverty remains one of Africa's most significant economic and social concerns (Omotayo & Aremu, 2020; Omotayo et al., 2020). Inadequate infrastructure, poor services, and structural constraints to economic development in marginalized villages hinder

DOI: 10.1201/9781003469766-2

progress and development in these areas (Stats SA, 2018). Ebert (2014) indicate that underutilized or traditional vegetables are best defined as locally important species for the sustainability of economies, human nutrition and health, and social systems. Local-based knowledge shows that traditional plants are indispensable in local people's lives (Mashela et al., 2013).

2.2 CURRENT CONSUMPTION TRENDS OF UNDERUTILIZED FRUITS AND VEGETABLES

Fruits and vegetables play an important role, especially during times of starvation, by providing an alternative source of nutrition (Akinnifesi et al., 2007; Bvenura & Sivakumar, 2017). According to FAO (2017), food security issues are especially severe in largely import-dependent countries of sub-Saharan Africa. Studies conducted in Southern Africa have shown that underutilized food and medicinal plants play a crucial role in the livelihoods of rural communities (Chivandi et al., 2015; Legwaila et al., 2011). Feyssa et al. (2011) articulated that many households consider wild edible plants significant to food security and the social-ecological systems that make gathering these natural resources possible and appropriately protected, managed, and valued to avoid overexploitation and degradation. Underutilized fruits and vegetables are essential for healthy ecosystems, providing the conditions and processes that sustain all life (De Vynck et al., 2016; Hawkins et al., 2008). Despite the fact that agricultural food production relies heavily on a few major crops and underutilized plants, it is frequently claimed that only 30 crops "feed the world." Of these, an estimated 30,000 species are edible, and about 7,000 have been cultivated or harvested for food by humans at some point (Vinceti et al., 2013).

Nevertheless, it is unclear how the current global food system will meet the growing demands of a population expected to reach 9 billion people by 2050 (Vinceti et al., 2013). National-level increases in food production alone cannot overcome food security deficits. Underutilized fruits and vegetables show substantive biodiversity, adapted to specific marginal soil and climatic conditions, which can often be grown with minimal external inputs (Hughes & Ebert, 2013; Shackleton et al., 2015). However, underutilized vegetables and fruits would be an additional advantage compared to the majority of other field crops (Table 2.1). The increasing dietary diversity and the intake of vegetables and fruits are widely recognized as a key strategy to address this problem (Boedecker et al., 2014; Frison et al., 2005). Yet

TABLE 2.1
Nutraceutical and Pharmacological Potential of Underutilized Fruits

Scientific Names Families	Nutraceutical Potential	Pharmacological Potential	References
Adansonia digitata L. **Malvaceae**	*A. digitata* fruits have high concentrations of lysine, thiamine, vitamin C, calcium, iron, and digestible carbs; it also contains crude protein and oil.	Baobab leaves to treat malaria and cure diarrhea or stomachache.	(De Caluwé et al., 2009; Van Wyk & Van Wyk, 1997)
Annona senegalensis Pers. **Annonaceae**	Vitamins and minerals.	Treats various infections, including pneumonia, yellow fever, tuberculosis, and diabetes.	(Konate et al., 2012).
Azadirachta indica **Meliaceae**	Protein, fat, carbohydrate, fiber.	Antidiabetic properties; anti-inflammatory properties; treatment of diabetes, eczema, and fever; anti-inflammatory, antiallergic, antioxidant, antidiabetic, antiviral, and anticancer activities; antileprosy activities, antimicrobial activity.	(Madaki et al., 2016; Sacarlal & Denning, 2018; Satyanarayana et al., 2015)

(Continued)

TABLE 2.1 (Continued)

Scientific Names Families	Nutraceutical Potential	Pharmacological Potential	References
Carissa macrocarpa (Eckl.) A.DC. **Apocynaceae**	The fruits of *C. macrocarpa* are high in important fatty acids, vitamin C, and several chemical elements, including calcium, potassium, magnesium, iron, and copper.	The leaves of *C. macrocarpa* cure diarrhea in cattle, as well as coughs and venereal illnesses.	(Omotayo et al., 2020; Ramasar et al., 2022; Souilem et al., 2019)
Diospyros lycioides Desf. **Ebenaceae**	The leaves and the fruit contain protein, carbohydrates, fiber, unsaturated fat–rich lipids, minerals, vitamins, phosphorus, magnesium, calcium, iron, potassium, sodium, zinc, copper, and manganese.	D. *lycioides* is a herb used for various diseases, chewing sticks, and sexually transmitted infections.	(Douglas, 2016; Fawole et al., 2009)
Diospyros simii (Kuntze) De Winter. **Ebenaceae**	–	–	(Grové et al., 2017; Magwede et al., 2019; Shai et al., 2020)
Dovyalis caffra (Hook. f. and Harv.) Sim **Salicaceae**	–	–	(Grové et al., 2017; Magwede et al., 2019; Shai et al., 2020)
Dovyalis zeyheri (Sond.) Warb. **Salicaceae**	Carbohydrate, crude fiber, protein, ash, moisture, and fat.	Bark and root are regarded as potent remedies against pain and rheumatism, chest pain, and in the treatment of amenorrhea.	(Aremu et al., 2019; Arnold & Gulumian, 1984; Grové et al., 2017; Magwede et al., 2019; Shai et al., 2020; Watt & Breyer-Brandwijk, 1962)
Gongronema latifolium Benth. **Asclepiadaceae**	The plant is a rich source of protein and minerals.	Maintaining weight, digestion, bone and muscle development, treating constipation, reducing cold symptoms, improving blood pressure, and promoting antidiarrheal, antibacterial, and antifungal activities.	(Mensah et al., 2008)
Gnetum africanum Welw. **Gnetaceae**	Rich source of protein, iron, calcium, and iodine.	Antiseptic, sore throat treatment, nausea, and pain management among women during childbirth; possesses anticarcinogenic, anti-inflammatory, and antioxidant properties.	(Ali et al., 2011; Isong et al., 1999; Onuminya et al., 2018)
Heinsia crinata (Afzel.) G. Taylor **Rubiaceae**	Vitamins and minerals.	Relief for arthritis and rheumatism pains.	(Onuminya et al., 2018)
Lasianthera africana Beauv. **Stemonuraceae**	Vitamins and minerals.	Antacid, analgesic, antiulcergenic, antidiabetic, antimalarial.	(Onuminya et al., 2018)
Lagenaria siceraria (Molina) Standl. **Cucurbitaceae**	The fruit is a rich source of ascorbic acid, beta-carotene, and a good source of vitamin B complex.	*L. siceraria* reduces depression, heart disease, liver disease, problems with urination, skin conditions, piles, diabetes, jaundice, and ulcers; furthermore, the juice helps adjust urea and uric acid concentrations in the kidney.	(Mehboob et al., 2022)

(Continued)

TABLE 2.1 (Continued)

Scientific Names Families	Nutraceutical Potential	Pharmacological Potential	References
Lactuca taraxacifolia Willd. **Asteraceae**	Rich source of vitamins, proteins, minerals, dietary fiber, essential fatty acids, and flavonoids.	Possesses antimicrobial, antiarthritic, antidiabetic, hypolipidemic, and antioxidant activity properties and is used in the treatment of epilepsy.	(Adejuwon et al., 2014; Adinortey et al., 2018; Ololade et al., 2017)
Lemeltonia cornuta (Mez and Sodiro) Barfuss and W.Till **Asteraceae**	Rich source of minerals and vitamins.	Cure for malaria and typhoid.	(Chacha & Laswai, 2020; Lyimo et al., 2003)
Ipomoea aquatica Forssk. **Convolvulaceae**	Protein, fat, carbohydrate, and fiber.	Anticancerous effects.	(Beheshti et al., 2021; Sarkar et al., 2022)
Manihot esculenta Crantz. **Euphorbiaceae**	Vitamins like B1, B2, C, and carotenoids, minerals, and protein.	Increases breast milk production in pregnant women and treats diarrhea.	(Fasuyi, 2005)
Mimusops zeyheri Sond. **Sapotaceae**	Minerals and vitamins (A and C); mineral elements potassium and calcium.	*M. zeyheri* is utilized for treating various sexually transmitted diseases, including wounds, sores, gonorrhea, candidiasis, ulcers, inflammation, bleeding gums, tuberculosis, and other related conditions.	(Matlala et al., 2024; Mngadi et al., 2019; Omotayo et al., 2020)
Momordica charantia L. **Cucurbitaceae**	Protein, fat, carbohydrate, and fiber.	Antimicrobial activities.	(Bakare et al., 2010; Leelaprakash et al., 2011)
Momordica foetida Schumach.	Rich source of vitamins (A, B1, B2, B3, and C) and minerals (calcium, iron, magnesium, and zinc).	Treatment of smallpox; relieves diarrhea and malaria, and removes toxins from the body.	(Chacha & Laswai, 2020)
Murraya koenigii Spreng. **Lamiaceae**	Carbohydrate, moisture content, crude fiber, ash, fats, and protein.	Dysentery, diabetes. and vomiting.	(Igara et al., 2016; Onuminya et al., 2018)
Myrianthus arboreus P.Beauv. **Urticaceae**	Source of protein.	Treat malaria, wounds, dysentery, skin infections, diarrhea, and muscle pain and relieve back pain; expel intestinal parasites.	(Agyare et al., 2014; Olonode et al., 2015)
Ocimum gratissimum L. **Lamiaceae**	*O. gratissimum* are excellent sources of minerals like calcium, phosphorus, sodium, potassium, magnesium, and trace elements like iron, zinc, and copper.	Fever, diarrhea, dysentery, pile, stomach upset.	(Onuminya et al., 2018; Pachkore & Dhale, 2012)
Operculina turpethum (L.) Silva Manso. **Convolvulaceae**	–	Anticancerous effects.	(Arora et al., 2017)
Parinari curatellifolia Planch. ex Benth. **Chrysobalanaceae**	–	Medicine for the treatment of pneumonia, wound infections, dressing of fractures, and dislocation.	(Magwede et al., 2019)
Pterocarpus mildbreadii Harms. **Fabaceae**	Vitamins and minerals	Skin infections.	(Onuminya et al., 2018)
Senecio biafrae Oliv. and Hiern. **Asteraceae**	Vitamins and minerals	Heart problems, cough, wound dressing, rheumatism, and tonic.	(Onuminya et al., 2018)

(Continued)

TABLE 2.1 (Continued)

Scientific Names Families	Nutraceutical Potential	Pharmacological Potential	References
Sclerocarya birrea Hochst. **Anacardiaceae**	The fruit contains considerable amino acids and fatty acids.	Treatment of heavy menstruation, bilharzias, coughs, weakness, sore eyes, and heart pains.	(Russo et al., 2018; Shackleton et al., 2002)
Sesbania sesban L. **Fabaceae**	Protein, fiber	Anti-inflammatory, anthelmintic, antidiarrheal, and antioxidant.	(Debela et al., 2011; Goswami et al., 2016)
Schinus molle L. Anacardiaceae	–	–	(Belhoussaine et al., 2022)
Solanum macrocarpon L. **Solanaceae**	Vitamins and minerals.	Ringworm, mouth ulcer, and flatulence treatment.	(Onuminya et al., 2018)
Stellaria media **Caryophyllaceae**	–	Anti-inflammatory activities.	(Bukola et al., 2012; Sarkar et al., 2022)
Strychnos spinosa L. **Loganiaceae**		–	(Madzimure et al., 2013)
Talinum triangulare (Jacq.) Willd. **Portulaceae**	Vitamins and minerals.	Potherbs, stomach upset, nutrient supplements.	(Onuminya et al., 2018)
Telfairia occidentalis Hook. **Cucurbitaceae**	Vitamins and minerals.	Blood tonic and nutrient supplement.	(Eseyin et al., 2014)
Vangueria infausta Burch. **Rubiaceae**		Gastrointestinal disorders, malaria, pneumonia, cough, menstrual problems, aphrodisiacs, parasitic worms, chest complaints, snake repellent, infertility, fever, candidiasis, and abdominal pains.	(Abosi et al., 2006; Chinsamy & Koitsiwe, 2016; Munodawafa et al., 2013; Nundkumar & Ojewole, 2002; Pollock, 2017)

Source: Scientific names for the plants were verified using World Flora Online (www.worldfloraonline.org/).

lack of market access, substantive production risks, and lack of knowledge constrain smallholders from better exploring the opportunities represented by these plants. Ethnobotanical studies are essential in recording information about the diversity of species used, the relative importance of each species, their different uses by Indigenous communities, seasonal availability, and the conservation needs of the identified species. Hence, this systematic review explores the contribution of underutilized fruits and vegetables to enhanced food and nutrition security in Southern Africa.

2.3 MATERIALS AND METHODS

This systematic review was developed using the guidelines from the Preferred Reporting Items for Systematic Reviews and Meta-Analyses (PRISMA) statements (Moher et al., 2009). In total, more than 2,746 articles were identified between 2023 and 2024 using a combination of keywords that had been specifically designed to explore underutilized fruits, vegetables, nutrition, and pharmacological uses. In this study, we used scholarly search engines and databases such as Scopus, Google Scholar, Web of Science, Google Search, JSTOR, and Sabinet to explore a wide variety of literature related to the review. Additional sources, such as books, dissertations, and theses, were retrieved from the Library and Information Centre of the University of Mpumalanga (UMP) in South Africa. Furthermore, the bibliographies from the retrieved articles were searched and saved on the Endnote reference manager. The vast majority of those articles publicly available appear to be type 1 articles (i.e., news media and personal or non-academic blogs). In line with the proposed methodology, these type 1 documents were discarded, leaving this

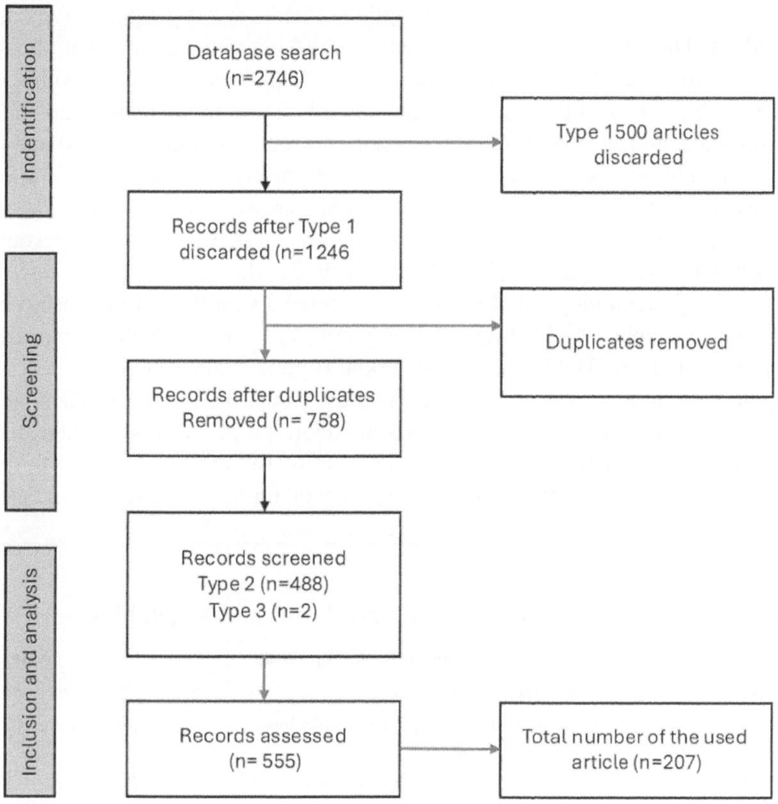

FIGURE 2.1 Preferred Reporting Items for Systematic Reviews and Meta-Analyses (PRISMA) flow diagram for the exclusion and inclusion of articles in the current review.

review with 1,246 articles. After the removal of duplicate articles (documents published simultaneously in several languages), 488 type 2 or higher articles were available, which were then entered into the database and reviewed. Finally, 207 studies were used to complete the study. A PRISMA-like flowchart is proposed in Figure 2.1 that summarizes the different steps of the protocol adopted for the review.

2.4 NUTRACEUTICAL CONSTITUENTS IN UNDERUTILIZED FRUITS

Globally, underutilized fruits are vital components of a healthy and balanced diet with various health benefits (Table 2.1). As recommended by the World Health Organization (WHO) board on nutrition, diet, and inhibition of chronic diseases, a minimum of 400 g, or five servings of 80 g, of fruits and vegetables (including a higher proportion of vegetables) is expected to be consumed per day, per person, to harness their full nutritional and health benefits (Hall et al., 2009; World Health Organization, 2003). Medicinal benefits harnessed from the consumption of fruits are innumerable, including prevention and easement of various nutrient deficiencies; improved immunity, lifespan, and mental health (Bellavia et al., 2013; Chowdhury et al., 2020; Conner et al., 2017); treatment and prevention of chronic diseases, such as heart disease, cancer, diabetes, and obesity (Collese et al., 2017; Fardet et al., 2019); improved digestive system (Lembo & Camilleri, 2003); maintenance of healthy eyesight and eye disease risk reduction (cataracts and macular degeneration) (Christen et al., 2005); improved cardiovascular health (Aune et al., 2017); prevention of asthma (Hosseini et al., 2017), hypertension (Li et al., 2016), and depression (Głąbska et al., 2020); health challenges; and reduction to risks of some cancers (Wang et al., 2017). Reports indicate that 14% of deaths from gastrointestinal cancer, 11% from ischemic heart disease, and about 9% from stroke or type 2 diabetes mellitus globally result from insufficient consumption of fruits and vegetables (Awuchi et al., 2020; Lock et al., 2005). The World Health Organization (2003) further reports that about 2.7 million deaths per year are associated with insufficient intake of fruits and vegetables worldwide.

Fruits are valuable sources of vitamins A, B, C, and E. They are also essential sources of fiber, calories, amino acids, and carotenoids (cherries, apricots, melons, and mangoes) and provide antioxidants for metabolic processing and detoxification of various carcinogenic diseases (Abobatta, 2021). Fruits and fruit products are very good sources of some rich micronutrients, such as calcium, iron, copper, phosphorus, magnesium, and potassium or zinc. They are excellent resources for soluble dietary fiber that reduces cholesterol and fats in the body while further easing bowel movements (Wang et al., 2022). Epidemiological studies have reported a significant relationship between consuming fruits and vegetables and healthy bones in young and older individuals, mainly due to their alkaline-forming properties (McGartland et al., 2004; Prynne et al., 2006).

According to Dhalaria et al. (2020), fruits possess antiproliferative, antidiabetic, anticancer, antimicrobial, anti-inflammatory, antiviral, and antioxidant properties due to their high content of polyphenolic compounds, such as catechins, anthocyanins, isoflavones, resveratrol, quercetin, and genistein (Table 2.2). Vegetables, on the other hand, are vital for improved health and the provision of

TABLE 2.2
Main Bioactive Components of the Underutilized Fruits and Vegetables

Fruit/Vegetable	Bioactive Compounds	Reference
Amaranthus cruentus L.	Trypsin, saponins	(Skwaryło-Bednarz et al., 2020)
Adansonia digitata L.	Epicatechin-(4β→8)-epicatechin (procyanidin B2),	(Silva et al., 2023)
Bidens pilosa Linn.	2,4-ditert-butylphenol, 4,6-ditert-Butyl-m-cresol and 2,2'-Methylenebis	(Ndiege et al., 2021)
Brassica rapa L.	Glucosinolates	(Shawon et al., 2020)
Carissa macrocarpa (Eckl.) A.DC.	Anthocyanins, cyanidin 3-O-β-sambubioside (Cy-3-Sa), and cyanidin 3-O-glucoside (Cy-3-G)	(Seke et al., 2022)
Ceratotheca trilobal L.	4-coumaric acid	(Masondo et al., 2019)
Cleome gynandra L.	Carvacrol, trans-phytol, and linalool	(Moyo & Aremu, 2022)
Citrillus lanatus (Thunb.) Matsum. and Nakai	Lycopene and citrulline	(Zamuz et al., 2021)
Cucurbita pepo L.	Rutin, kaempferol, isoqueratin, astragalin, myricetin, quercetin	(Kulczyński & Gramza-Michałowska, 2019)
Galinsoga parviflora Cav.	Caffeic acid	(Parzonko & Kiss, 2019)
Garcinia livingstonei T. Anderson	Morelloflavone	(Nossier et al., 2023)
Lasianthera africana P. Beauv.	Phytol and phytol acetate, squalene, caryophyllene, carmigrene	(Baba et al., 2022)
Moringa oleifera Lamm.	p-coumaric acid, p-coumaric acid ethyl ester	(Kashyap et al., 2022)
Parinari curatellifolia Planch. ex Benth	β-sitosterol	(Kulczyński & Gramza-Michałowska, 2019)
Sclerocarya birrea Hochst.	Condensed tannins	(Pfukwa et al., 2020)
Senna occidentalis (L.) Link	Quercetine	(Mishra et al., 2022)
Solanum nigrum L.	Withanolide A and withaferin A	(Preet et al., 2019)
Syzygium cordatum Hochst.	Myricetin rhamnoside, myrigalone-G pentoside, quercetin galloyl-pentoside, cryptostrobin, in which myrigalone-B and myrigalone-G	(Uddin et al., 2022)
Vernonia amygdalina Delile	Phytol	(Alara et al., 2019)
Vigna subterranea L. Verdc.	Quercetin	(Okafor et al., 2022)
Vigna unguiculata L. Walp.	β-carotene	(Wasswa & Muyonga, 2021)

essential macro- and micronutrients and vitamins (vitamin A, vitamin C, vitamin E, and vitamin K), minerals (iron, potassium, zinc, iodine, and calcium), antioxidants, phytosterols, polyphenolic compounds (phenolic compounds, flavonoids, bioactive peptides, etc.), and dietary fibers (Banwat et al., 2012; Poobalan et al., 2019). The bioactive medicinal natural compounds, such as lycopene, resveratrol, tannins, indoles, glucosinolates, polyphenols, phytoestrogens, carotenoids, and anthocyanins, present in vegetables make them essential components of the human diet for the prevention and treatment of various health issues, including prevention and treatment of debilitating diseases, such as cardiovascular diseases, diabetes, certain cancers, high blood pressure, diarrhea, and obesity (Poobalan et al., 2019).

2.5 INDIGENOUS AFRICAN GREEN LEAFY VEGETABLES

Indigenous green leafy vegetables (GLVs) are harvested from forests or other natural vegetation sites; collected from fields, meadows, roadside areas, and unused lands; or cultivated in personal home gardens. They are usually prepared by cooking and enjoyed in dishes like curry or soup or served fresh as salads (Easdown et al., 2014). The consumption habits of GLVs vary among households across different nations. In South Africa, these consumption patterns exhibit significant variability, influenced by factors such as poverty levels, urbanization rates, proximity to fresh produce markets, and seasonal variations. Lower-income households tend to rely more on these leafy GLVs compared to wealthier households (Uusiku et al., 2010). The phytochemicals present in indigenous GLV, particularly isothiocyanates (Pollock, 2017), help ameliorate inflammation and oxidative stress that leads to cancer, protect against carcinogenic substances, impede angiogenesis that contributes to tumor growth, and prevent the growth of cancer cells and gastrointestinal system carcinomas (Tao et al., 2018). The presence of phytochemicals in vegetables further helps reduce DNA damage, regulate hormones, ensure proper body metabolism, and boost the body's immune system. Most GLVs possess arrays of medicinal and health benefits, including antimicrobial, antibacterial, anti-inflammatory, anticancer, antidiabetic, cardiovascular-protective, and gastroprotective activities (Al Batran et al., 2013; Hasibuan et al., 2020; Johnson et al., 2017; Ojagbemi et al., 2021; Sarkar et al., 2022; Tosoc et al., 2021; Xu et al., 2019), among others

Hasibuan et al. (2020), in their studies on the anticancer activity of *V. amygdalina* leaves against 4T1 breast cancer cells, revealed the cytotoxicity effect of n-hexane, ethyl acetate, and ethanol fractions, with inhibition concentrations of 1,860.54 μg/ml, 25.04 μg/ml, and 1,940.84 μg/ml, respectively. Using the trypan blue test, cell morphology analysis, and MTT assay, Johnson et al. (2017) confirmed the cytotoxic effect of *V. amygdalina* methanolic extract on PC-3 human prostate carcinoma cells. In addition, the extract of *C. olitorius* has been found to reduce the size of melanoma and gastric cancer tumor cells in the human body (Tosoc et al., 2021).

Several studies have shown that various GLVs contain different anti-inflammatory and gastroprotective compounds (McGartland et al., 2004; Sarkar et al., 2023; Satheesh & Workneh Fanta, 2020). Extracts of *Amaranthus* spp. (Salvamani et al., 2016) and *C. olitorius* (Al Batran et al., 2013) have been reported to exhibit functional gastrointestinal and inhibitory activity against free radicals responsible for inflammation in the human body's defense system. The antidiabetic, hypoglycemic, and hypolipidemic properties of the extracts of varying vegetables, including *V. amygdalina* (Atangwho et al., 2014; Okoduwa et al., 2017), *C. speciosus* (Eliza et al., 2009), and *P. oleracea* (Bai et al., 2016), have also been reported in the management of diabetes.

In addition, several clinical trials have confirmed the health benefits of vegetable consumption in human physiology (Ojagbemi et al., 2021; Sarker et al., 2018a, 2018b). According to Ojagbemi et al. (2021), increasing GLV intake lowered the incidence of cardiovascular disease (CVD). These authors further recommended an increased advocacy for the consumption of GLVs for the management and prevention of cardiovascular diseases. In their study, Cheng et al. (2019) examined the possible relationship between plasma levels of lycopene in tomatoes and the risk of cardiovascular diseases. Their findings confirmed a significant reduction in systolic blood pressure after supplementation with lycopene. It was further affirmed that individuals with lycopene concentrations had lower risks of suffering a myocardial infarction and cardiovascular disease mortality. Bazzano et al. (2008), in their study on the nexus between the consumption of fruits and green leafy vegetables, found out that over 70,000 nurses between the age range of 38 and 63 and free of cardiovascular disease, cancer, and diabetes showed a lower risk of diabetes after consumption of various fruits and green leafy vegetables.

2.6 CULTIVATION AND CONSERVATION OF UNDERUTILIZED FRUITS AND VEGETABLES

The cultivation and conservation of underutilized fruits and vegetables are gaining popularity in the research community. This is because these crops are mostly consumed as food and used as medicines by rural communities at lower costs. Many of these neglected crops, including various grains, are highly suitable for cultivation and preservation. This is due to their extended shelf life, making them particularly favored by rural communities that rely on traditional storage methods (Chandra et al., 2020). Most rural communities use cheaper methods, such as ash made from plant products (wood, grasses, etc.), when storing and controlling pests on these crops. After harvesting the grains, they cover the dry grains with ash to prevent them from being infested by pests. This ash method also increases seed germination compared to the seeds treated with synthetic chemicals (Olorunmota et al., 2021). Apart from germination, climate change is another challenging factor in agriculture. The preserved or stored seeds in ash will also be conserved for trading purposes in the future or for use in times of need (Liverpool-Tasie & Parkhi, 2021).

Wild leafy vegetable cultivation is important because they are considered for their nutritive properties (Mudau et al., 2022). *T. divaricata* is one of the wild leafy vegetables which are underutilized in most rural and is very rich in nutrients, minerals, and vitamins; however, its cultivation information is very limited (Bulawa et al., 2022). According to Singh et al. (2020), these underutilized crops have a high-quality flavor that enhances their taste beyond their nutritional value and medicinal capabilities.

Cultivating and conserving these plants are vital because most of them are used as indigenous medicinal plants. *C. olitorius*, *S. nigrum*, and *C. gynandra*, among others, are some of the species cultivated. They are primarily used in rural communities by knowledgeable traditional healers or elders of specific communities.

2.7 MEDICINAL VALUES OF SOME WILD FRUITS AND VEGETABLES

Interestingly, the majority of the underutlized fruits and vegetables have gained more attention due to their pharmacological properties and benefits (Tables 2.1 and 2.3). For example, *Amaranthus* sp., one of the most widely used food and medicinal plant in Africa (Ballabio et al., 2011; Hussain et al., 2009; Muli et al., 2022; Onuminya et al., 2018; Sánchez-Urdaneta et al., 2020; Sarker et al., 2018a, 2018b; Welcome & Van Wyk, 2019). It is a high herb that treats various disorders, including anti-inflammatory, anticancerous, antidiabetic, cardioprotective, gastrointestinal properties; tapeworm expellant; and relief of pulmonary problems. Due to the long indigenous use and popularity of *Amaranthus* sp., it has been successfully commercialized and can currently be bought from various commercial stores (Emmanuel & Babalola, 2022). Since these underutilized fruits and vegetables are primarily found in rural communities, where there is poverty, food insecurity, unemployment, youth underdevelopment, and increasing malnutrition, these species can be used to improve livelihoods within rural communities (Mudau et al., 2022). Encouraging agricultural activities can potentially assist in alleviating poverty in poor and marginalized rural communities. Moreover, improved agricultural activities can also improve rural economic development through the creation of new value chains. Thus, more information-based research is needed to develop funding policies at the government level (Chandra et al., 2020).

TABLE 2.3
Nutraceutical and Pharmacological Potential of Underutilized Vegetables

Scientific Name and Family	Nutritional Compositions	Pharmacological Benefits	Reference
Abelmoschus esculentus **Malvaceae**	High fiber content.	The plant regulates the body's sugar level; has hypoglycemic, antimicrobial, anticancer, and antiulcer activities; used in the treatment of diabetes, ulcers, and hemorrhoids and recovery from cardiovascular disorders.	(Ansari et al., 2005; Durazzo et al., 2018; Mokganya & Tshisikhawe, 2019)
Asystasia gangetica L. **Acanthaceae**	Carbohydrates, protein, and lipid.	Roots possess diuretic activity and are also used in preparations prescribed for jaundice, hemoglobinuria, and gonorrhea.	(Ajuru et al., 2021; Janakiraman et al., 2021)
Amaranthus blitum L. Syn: *Amaranthus tricolor* L. **Amaranthaceae**	Contains ash, protein, fat, carbohydrate, and crude fiber.	The plant is used to combat allergies, inflammation, free radicals, and various diseases, with secondary metabolites like alkaloids controlling living system development and protecting animals.	(Naidoo et al., 2016; Nehal et al., 2016; Okwu, 2008; Schönfeldt & Pretorius, 2011; Welcome & Van Wyk, 2019)
Amaranthus cruentus L. Syn: *Amaranthus hybridus* L. **Amaranthaceae**	Rich in flavonols; less rich in tannins, protein, ash content, crude protein, crude lipid, crude fiber, and carbohydrate; sodium, potassium, calcium, magnesium, iron, zinc, phosphorus, and vitamins.	–	(Achigan-Dako et al., 2014; Akubugwo et al., 2007b; Welcome & Van Wyk, 2019)
Amaranthus dubius C. Mart. Ex Thell. **Amaranthaceae**	Protein, carbohydrates, dietary fiber, lipids rich in unsaturated fats, minerals, and vitamins.	–	(Odhav et al., 2007; Welcome & Van Wyk, 2019)
Amaranthus spp. **Amaranthaceae**	Protein, fat, carbohydrate, vitamins, and minerals.	Anti-inflammatory, anticancerous, antidiabetic, cardioprotective, gastrointestinal properties, tapeworm expellant, relief of pulmonary problems.	(Ballabio et al., 2011; Hussain et al., 2009; Muli et al., 2022; Onuminya et al., 2018; Sánchez-Urdaneta et al., 2020; Sarker et al., 2018a,2018b; Welcome & Van Wyk, 2019)
Basella alba L. **Basellaceae**	Vitamins and minerals.	Laxative; boosts blood supply and increases appetite; prevents constipation; relieves ulcerative pains; treats anemia, hypertension, malaria, coughs, and colds.	(Onuminya et al., 2018)
Bidens bipinnata L. **Asteraceae**	Calcium, magnesium, iron, potassium, zinc, vitamin C, and carotene.	*B. bipinnata* is a widely utilized medicinal plant for the treatment of diseases like malaria, sore throat, acute nephritis, and dysentery.	(Steyn et al., 2001; Zahara et al., 2022)

(Continued)

TABLE 2.3 (Continued)

Scientific Name and Family	Nutritional Compositions	Pharmacological Benefits	Reference
Bidens pilosa L. **Asteraceae**	Ash content, crude protein, crude lipid, crude fiber, and carbohydrate. Furthermore, leaves contain sodium, potassium, calcium, magnesium, iron, zinc, phosphorus, copper, manganese, and nitrogen.	Plants are used to treat respiratory infections, wounds, burns, diabetes, inflammation, enteritis, dysentery, pharyngitis, and other conditions, such as painkillers, diuretics, and antirheumatics.	(Adedapo et al., 2011; Arthur et al., 2012; Mokganya & Tshisikhawe, 2019; Welcome & Van Wyk, 2019)
Brassica rapa subsp. *chinensis* **Brassicaceae**	This includes primary metabolites, such as amino acids, plant sugars, and minerals, and secondary metabolites, such as vitamins A, B9, E, and K1, as well as glucosinolates.	–	(Welcome & Van Wyk, 2019; Zou et al., 2021)
Capsicum annuum L. **Solanaceae**	*C. annuum* contains high amounts of antioxidants (such as vitamins C and A), polyphenols, and capsaicinoids.	–	(Guevara et al., 2021)
Celosia argentea L. **Amaranthaceae**	Vitamins and minerals.	Cough and laxative.	(Onuminya et al., 2018)
Ceratotheca triloba (Bernh.) E. Mey. Ex Hook. f. **Pedaliacaeae**	Protein.	The plant is used to treat diarrhea, nausea, abdominal and menstrual cramps, and to induce abortion.	(Karim, 2012; Odhav et al., 2007)
Cleome gynandra L. **Cleomaceae**	Vitamins (A and C) and minerals (calcium and iron).	Stimulation of milk production in breastfeeding mothers.	(Mokganya & Tshisikhawe, 2019; Sowunmi & Afolayan, 2015)
Cleome gigantea L **Capparaceae**	The food contains protein, ash, fat, vitamin B2, ß-carotene, iron, zinc, magnesium, calcium, and phosphorus, along with other essential nutrients.	–	(Schönfeldt & Pretorius, 2011)
Cleome maxima L. **Capparaceae**	Zinc content, magnesium, and iron.	–	(Nesamvuni et al., 2001; Schönfeldt & Pretorius, 2011)
Citrullus lanatus (Thunb.) Matsum. and Nakai. **Cucurbitaceae**	Phytoconstituents in *C. lanatus* include flavonoids, anthocyanins, vitamins C and E, phenolic compounds, dietary fiber, and carotenoids.	*C. lanatus* is used as a purgative and emetic in various forms, like high dose, vermifuge, demulcent, diuretic, and tonic.	(Pennington & Fisher, 2010; Van der Hoeven et al., 2013)
Cocculus hirsutus (L.) Diels **Menispermaceae**	–	The plant has significant ethnomedicinal potential, potentially treating various ailments, such as premature ejaculation, taste issues, and eye burning sensations.	(Das & Bondya, 2016; Steyn et al., 2001)
Colocasia esculenta (L.) Schott **Araceae**	Tubers supply easily digestible starch and are known to contain substantial amounts of protein, vitamin C, thiamine, riboflavin, and niacin.	–	(Bassir, 1969; Lewu et al., 2010; Niba, 2003)

(Continued)

TABLE 2.3 (Continued)

Scientific Name and Family	Nutritional Compositions	Pharmacological Benefits	Reference
Crotalaria burkeana Benth. **Fabaceae**	–	*C. burkeana* is traditionally used to treat sexually transmitted diseases (STDs) and specific bacteria causing ear infections, specifically otitis media.	(Njoroge & Bussmann, 2006; Steyn et al., 2001)
Corchorus olitorius L. **Tiliaceae**	Source of carbohydrates, proteins, essential amino acids, vitamins (A, B, C, E); high in fiber and minerals (calcium and iron).	Gonorrhea, aches and pains, toothache, dysentery, fever, piles, tumors, chronic cystitis, purgative, restore appetite and strength, ease bowel movements, aid child delivery.	(Adebo et al., 2018; Dansi et al., 2009; Kamga et al., 2013; Onuminya et al., 2018; Schönfeldt & Pretorius, 2011)
Crassocephalum cerepidioides (Benth.) S.Moore **Asteraceae**	Vitamins and minerals.	Indigestion, stomachache, headache, stop nose bleeding.	(Onuminya et al., 2018)
Cucurbita maxima Dechesne **Cucurbitaceae**	The nutrient content includes vitamin B2, ß-carotene, iron, zinc, magnesium, calcium, and phosphorus, as well as moisture, protein, ash, and fat content.	*C. maxima* seeds are used as traditional medicine against irritable bladder and prostatic hyperplasia.	(Hamdi et al., 2017; Maanda & Bhat, 2010; Muchirah et al., 2018)
Cucurbita moschata Duchesne. **Cucurbitaceae**	*C. moschata* is rich in carotenoids, vitamins, dietary fiber, minerals, and phenolic compounds and is also rich in manganese, magnesium, and potassium.	*C. moschata* has numerous medicinal applications, including deworming, antidiabetes, antibacterial, anticancer, and antiobesity properties.	(Maanda & Bhat, 2010; Men et al., 2021)
Cucurbita pepo L. **Cucurbitaceae**	Potassium, calcium, and sodium are amongst the most abundant minerals, along with other essential vitamins and minerals.	Seeds are used to treat tapeworms, irritation of the skin, itches, and rashes.	(Men et al., 2021; Otang et al., 2014; Younis et al., 2000)
Dicerocaryum senecioides (Klotzsch) Abels **Pedaliacaeae**	–	The plant treats various ailments, such as inflammatory disorders, tuberculosis, and chronic cough.	(Maanda & Bhat, 2010; Semenya & Maroyi, 2019; Steyn et al., 2001)
Dicerocaryum zanguebaricum (Lour.) Merr. **Pedaliacaeae**	–	The leaves are used for cosmeceuticals.	(Maanda & Bhat, 2010; Ndhlovu et al., 2019; Setshego et al., 2020)
Emex australis Steinh. **Polygonaceae**	Carbohydrates and proteins.	Leaf decoction is utilized to treat stomach issues, like indigestion and colic, and to stimulate appetite and possesses antibacterial and anticancer properties.	(Gupta & Kaul, 2021; Makni et al., 2019; Nasr et al., 2018; Odhav et al., 2007; Saleem et al., 2021)
Ipomoea obscura (L.) Ker Gawl. **Convolvulaceae**	Fatty acid and lipids.	Treat diarrhea.	(Maanda & Bhat, 2010; Prabha & Gopalakrishnan, 2014)

(Continued)

TABLE 2.3 (Continued)

Scientific Name and Family	Nutritional Compositions	Pharmacological Benefits	Reference
Ipomoea batatas L. **Convolvulaceae**	This food is a rich source of β-carotene, minerals (magnesium, iron, copper), vitamins (B1, B6, C, E), and dietary fiber.	*I. batatas* has numerous health benefits, including antioxidant, cardioprotective, anti-inflammatory, anticancer, antidiabetic, antimicrobial, antiobesity, and prevention of vitamin A malnutrition.	(Alam, 2021; Bvenura & Afolayan, 2014)
Heliotropium steudneri Vatke **Boraginaceae**	–	Leaves are used as a paste for broken bones, swelling, and wounds.	(Moravec et al., 2014; Steyn et al., 2001)
Malva parviflora L. **Fabaceae**	Crude protein, crude fibers, total ash content, potassium content, calcium, and magnesium.	*M. parviflora* is known for its various health benefits, including its antimicrobial, antioxidant, antidiabetic, anticancer, anti-inflammatory, antihypertensive, and wound healing properties.	(Abdalla & Yousef, 2016; Maanda & Bhat, 2010; Magwede et al., 2019; Sharifi-Rad et al., 2020)
Malva sylvestris L. **Fabaceae**	Calcium, zinc, and iron.	Anti-inflammatory, antioxidant, anticomplementary, anticancer, and skin tissue integrity activity.	(Abdalla & Yousef, 2016; Gasparetto et al., 2011; Maanda & Bhat, 2010)
Laportea peduncularis (Wedd.) Chew. **Urticaceae**	Calcium, zinc, and iron; vitamins (β-carotene, vitamins C and E).	Rheumatoid arthritis, gout, eczema, benign prostatic hyperplasia, anemia, influenza, asthma, and diabetes.	(Mahlangeni et al., 2020; Steyn et al., 2001)
Momordica charantia Descourt **Cucurbitaceae**	*M. charantia* is rich in essential nutrients, such as total ash, crude fat, fiber, protein, carbohydrates, calories, calcium, magnesium, sodium, potassium, iron, zinc, manganese, copper, and phosphorus.	*M. charantia* has anticancer, antidiabetic, antihypertensive, anti-inflammation, and antioxidant activities.	(Jerald et al., 2012; Maanda & Bhat, 2010; Mokganya & Tshisikhawe, 2019)
Momordica foetida Schumach. **Cucurbitaceae**	Proteins, fibers, minerals (calcium, iron, magnesium, zinc), β-carotene, foliate, and ascorbic acid are essential components of a healthy diet.	The leaf is utilized in traditional medicine for treating various diseases, including stomach issues, smallpox, measles, body swellings, edema, gout, venomous stings, and pain-killing.	(Maanda & Bhat, 2010; Muronga et al., 2021; Nesamvuni et al., 2001; Soh et al., 2020; Steyn et al., 2001)
Nicandra physaloides (L.) Gaertn. **Solanaceae**	The fruits and the leaves of *N. physaloides* are high in carotenoids, proteins, and vitamins C and A. Additionally, it was discovered that the fruits were a good source of phenols, chlorophyll, anthocyanins, and lycopene.	–	(Kshirsagar & Bhogaonkar, 2015; Maanda & Bhat, 2010)

(Continued)

TABLE 2.3 (Continued)

Scientific Name and Family	Nutritional Compositions	Pharmacological Benefits	Reference
Obetia tenax (N.E.Br.) Friis **Urticaceae**	Proteins, vitamin C, and vitamin E.	Nettles are used to treat diabetes mellitus, gout, eczema, benign prostatic hyperplasia, anemia, influenza, and asthma, in addition to rheumatoid arthritis, and root pulp applied on snakebites.	(Maanda & Bhat, 2010; Mahlangeni et al., 2020; Mokganya & Tshisikhawe, 2019)
Oxygonum dregeanum Meisn. Syn: *Oxygonum sinuatum* (Hochst. and Steud ex Meisn.) Dammer **Polygonaceae**	–	Treatment for malaria and its symptoms.	(du Preez et al., 2020; Maanda & Bhat, 2010)
Pergularia daemia (Forssk.) Chiov. Pergularia **Apocynaceae**	–	*P. daemia* has traditionally been used as an expectorant, laxative, anthelmintic, and antipyretic. Additionally, it has been used to treat infantile diarrhea and intermittent malarial fevers.	(Karthishwaran & Mirunalini, 2010; Steyn et al., 2001)
Pentarrhinum inspidum E. Mey. **Asclepiadaceae**	It is rich in protein, ash content, crude protein, crude lipid, crude fiber, and carbohydrates; sodium, potassium, calcium, magnesium, iron, zinc, phosphorus, and vitamins.	Decoction is used in the treatment of anaplasmosis, and it has good antifungal activities.	(Maanda & Bhat, 2010; Mokganya & Tshisikhawe, 2019; Nanyingi et al., 2008)
Piliostigma thonningii (Schumach.) Milne-Redh. **Fabaceae**	It is rich in protein, ash content, crude protein, crude lipid, crude fiber, and carbohydrates; sodium, potassium, calcium, magnesium, iron, zinc, phosphorus, and vitamins.	The plant is used to treat a variety of infections, fevers, and inflammatory conditions.	(Hailemariam et al., 2021; Steyn et al., 2001)
Physalis viscosa L. **Solanaceae**	Vitamin C and mineral contents.	–	(Odhav et al., 2007)
Pouzolzia mixta Solms. **Urticaceae**	–	Pearl penile papules, infantile acropustolosis, eczema, boils, abscess, and contraception.	(Asong et al., 2019; Maanda & Bhat, 2010; Steyn et al., 2001; Van Wyk & Gericke, 2000)
Portulaca oleracea L. **Portulacaceae**	Alpha-linolenic acid, omega-3 fatty acids, pyridoxine, niacin, and riboflavin, as well as vitamins A, C, and B complex. Potassium is among the most abundant dietary minerals, followed by calcium, magnesium, phosphorus, and iron.	Topical treatment of inflammatory conditions.	(Gill & Valivety, 1997; Maanda & Bhat, 2010; Steyn et al., 2001)
Solanum nigrum L. **Solanaceae**	Crude fiber, protein, ash, and carbohydrates.	Antitumorigenic, antioxidant, anti-inflammatory, hepatoprotective, diuretic, and antipyretic agents.	(Akubugwo et al., 2007a; Bvenura & Afolayan, 2014; Jain et al., 2011)

(Continued)

TABLE 2.3 (Continued)

Scientific Name and Family	Nutritional Compositions	Pharmacological Benefits	Reference
Solanum nodiflorum Jacq. **Solanaceae**	Nutritional components found in *S. nodiflorum* include iron, manganese, copper, magnesium, calcium, phosphorus, potassium, and sodium.	Used medicinally.	(Mokganya & Tshisikhawe, 2019; Van Wijk et al., 2017)
Sida rhombifolia L. **Malvaceae**	–	Treating bronchial asthma, colds, the flu, chills, headaches, congestion in the nose, aching joints and bones, coughing and wheezing, and edema.	(Gawai & Tippat, 2024; Jenipher & Ayyanar, 2024; Steyn et al., 2001)
Sisymbrium thellingii O. Schulz **Brassicaceae**	–	–	(Bvenura & Afolayan, 2014)
Stellaria media L. **Caryophyllaceae**	–	–	(Bvenura & Afolayan, 2014)
Rumex crispus L. **Polygonaceae**	–	*R. crispus* is used for treating various health issues, such as hemostasis, infection, constipation, fever, diarrhea, stomach issues, edema, jaundice, and gastrointestinal, renal, antiviral, and inflammatory skin conditions.	(Bvenura & Afolayan, 2014; Jiang et al., 2012)
Taraxacum officinale Weber **Asteraceae**	Vitamin C, chlorophyll, and carotenoids.	The plant is used to treat stomach, urinary, menstrual, respiratory, diabetes, rheumatism, anemia, irregular menstruation, blood purification, regulation of digestion, appetite loss, and liver disorders.	(Bvenura & Afolayan, 2014; Mustafa et al., 2012; Pădureţ et al., 2016; Šarić-Kundalić et al., 2011)
Tropaeolum majus L. **Tropaeolaceae**	*T. majus* is a rich source of microelements, like potassium, phosphorus, calcium, and magnesium, and macroelements, like zinc, copper, and iron.	Anti-inflammatory treatments in animals.	(Junior et al., 2009; Maanda & Bhat, 2010; Silva et al., 2018)
Trema orientalis (L.) Blume **Ulmaceae**	Calcium and phosphorus.	*T. orientalis* has shown that it is used in the treatment of diabetes mellitus, respiratory diseases, oliguria, and malaria.	(Coates Palgrave, 1977; Maanda & Bhat, 2010; Narayan et al., 2013)
Tribulus terrestris L. **Zygophylaceae**	The plant has crude protein, fiber, carbohydrates, ash, calcium, and phosphorus.	*T. terrestris* has antimicrobial activity, antiurolithic activity, diuretic activity, aphrodisiac activity, antidiabetic activity, immunomodulatory activity, antioxidant activity, absorption enhancement, and central nervous system activity.	(Maanda & Bhat, 2010; Mashabela et al., 2024; Nath & Malik, 1970)

(Continued)

TABLE 2.3 (Continued)

Scientific Name and Family	Nutritional Compositions	Pharmacological Benefits	Reference
Tribulus zeyheri Sond. **Zygophylaceae**	–	–	(Maanda & Bhat, 2010; Mokganya & Tshisikhawe, 2019)
Urtica urens L. **Urticaceae**	Ashes, carbohydrates, proteins, lipids, sugars, anthocyanin, carotenoids, and fiber.	Treat diabetes mellitus, cancer, rheumatism, and toothache.	(Maanda & Bhat, 2010; Mzid et al., 2017; Nencu et al., 2015)
Waltheria indica L. **Malvaceae**	–	This includes pain, inflammation, conditions like diarrhea, dysentery, conjunctivitis, wounds, abscesses, epilepsy, convulsions, anemia, erectile dysfunctions, bladder ailments, and asthma.	(Mongalo, 2014; Steyn et al., 2001)
Withania somnifera (L.) Dunal **Solanaceae**	Ash, crude protein, carbohydrates, crude fiber, and crude fat.	*W. somnifera* is used to treat various health issues, including colds, coughs, ulcers, diabetes, conjunctivitis, epilepsy, insomnia, senile dementia, leprosy, nervous disorders, rheumatism, arthritis, intestinal infections, bronchitis, asthma, impotence, and HIV/AIDS suppression.	(Krishnamurthy & Sarala, 2010; Steyn et al., 2001; Umadevi et al., 2012)
Xanthosoma sagittifolium (L.) Schott. **Araceae**	This includes moisture, crude protein, crude fiber, crude fat, total ash, total carbohydrate, and gross energy, as well as minerals like iron, zinc, copper, magnesium, manganese, phosphorus, sodium, potassium, and calcium.	–	(Van der Hoeven et al., 2013; Wada et al., 2019)
Vigna unguiculata (L.) Walp. **Fabaceae**	V. *unguiculata* is rich in carbohydrates and proteins and has low-fat content, providing essential vitamins, minerals, folate, thiamin, and riboflavin.	Roots used in the treatment of epilepsy, chest pain, and dysmenorrhea and contains hypolipidemic activity.	(Mokganya & Tshisikhawe, 2019; Schönfeldt & Pretorius, 2011)

Source: Names for the plants were verified using World Flora Online (www.worldfloraonline.org/).

2.8 THE EFFECTS OF PROCESSING ON BOTH NUTRITIONAL AND SECONDARY METABOLITE PROFILES

The processing of foods can be non-thermal, using irradiation, ultraviolet treatment, high hydrostatic pressure treatment, and pulsed electric fields treatment, or thermal by roasting, blanching, and drying, or mechanical (Al-Juhaimi et al., 2018). Tylewicz et al. (2019) also highlight how these treatments allow for the processing of fruits and vegetables while increasing their quality and preserving their nutrients, taste, color, texture, and heat-sensitive nutrients. Furthermore, determining the most appropriate processing methods relies on the specific composition of the fruits or vegetables concerning their flavors, texture, color, nutritional content, and visual appeal (Ramjan & Ansari, 2018). Indigenous people also have their traditional method of processing, such as sun-drying, application of wood sand, burying under moistened soil, grinding with stone, frying, roasting, and fermentation (Is et al., 2017). Many traditional leafy vegetables also need to undergo processing due to their content of antinutritional elements, like tannins, saponins, phytates, oxalates, nitrates, cyanogenic glucosides, and glucosinolates. Indigenous communities also employ other traditional processing techniques, such as soaking, incorporating tamarind, and prolonged cooking, to detoxify vegetables, making them safe for human consumption (Easdown et al., 2014). The latter indicates that processing can significantly alter the composition and quantity of secondary metabolites (Thane & Reddy, 1997), resulting in either an increase or decrease in their levels or having no impact in some cases (Nicoli et al., 1999). For instance, cooking vegetables may lead to thermal degradation or loss of water-soluble phytochemicals, while drying fruits may affect the stability of heat-labile phytochemicals, such as polyphenols (Nakitto et al., 2024). However, Turkmen et al. (2005) found that cooking had no negative effect on the antioxidant activity and polyphenol content of vegetables, such as *Spinacia oleracea* L., *Phaseolus vulgaris* sensu Blanco, non L., *Capsicum* L., and *Brassica cretica* Lam.

2.9 COMMERCIALIZATION OF UNDERUTILIZED FRUITS AND VEGETABLES

The global demand for organic agricultural commodities has risen recently, partly because of concerns about food safety and growing environmental awareness in the midst of global warming (Akter et al., 2023). The demand for underutilized indigenous fruits and vegetables is rising among the upper- and middle-income groups in Southern Africa. Most people living in Southern Africa are increasingly becoming aware of underutilized fruits and vegetables, and their potential benefits, such as alleviating micronutrient deficiencies or hidden hunger (Mudau et al., 2022). These neglected crops are rich in bioactive compounds, with a low glycemic index, high dietary fiber, resistant starch, protein, and dense nutrients with greater micronutrients (Singh et al., 2022). With more than 200 edible native fruit tree species, Southern Africa has the potential to supply the world market with a wide range of novel food products. Normally, fruits from indigenous plants are typically eaten as a snack, except for times of famine, when they become the primary food source, and so play a much more significant role in the daily diet of rural people (Ramaano, 2021). The importance of commercialization is understanding that rural people are not able to eat farm produce right away, or in its unprocessed state; instead, it goes through the post-harvest processes, such as transportation, storage, marketing, and processing. Ineffective post-harvest management could result in post-harvest losses which can negatively impact on the farmers' welfare and food security.

The contribution of underutilized fruits and vegetables not only contributes to household consumption but can also generate income through selling surplus to the local markets. Evidence of a few businesses dealing in indigenous fruit products, such as fruit juice and seedlings, is already available (Ham et al., 2008). Smallholder farmers earn revenue from selling indigenous fruits and underutilized vegetables in urban and rural markets. Existing research demonstrates that selling fruits and vegetables gathered in the wild is a lucrative business (Ham et al., 2008). The returns to labor for collecting indigenous fruit trees are significantly higher than those for crop production, making it an effective method of labor allocation. For instance, *Uapaca kirkiana* collections in Zimbabwe have previously produced an annual average of \$50, whereas *S. birrea* collections in South Africa produced an average of \$78 (Ham et al., 2008). *Sclerocarya birrea* has also been gathered by communities in South Africa, earning \$180,000 per annum and accounting for more than 10% of the average household income (Ham, 2005). Each community harvested about 2,000 t of the plant. Underutilized plant species were historically employed for their positive traits, such as food, fiber, fodder, oil, or medicinal capabilities, but they typically lack national recognition and appreciation. These species

TABLE 2.4
Uses of Indigenous Fruits and Their By-Products

Fruit	By-Products of Indigenous Fruit	References
D. whyteana	Fruits are eaten as a snack.	(Magwede et al., 2019; Verschaeve et al., 2004)
S. birrea	Fruits are processed into juice, jam, jellies, sorbet, and alcoholic drinks.	(Kunene et al., 2020; Magwede et al., 2019; Ojewole et al., 2010)
C. macrocarpa	Jams, sauces, cordial, pies, and jellies.	(Singhurst & Holmes, 2010)
P. curatellifolia	Snack, cooked as porridge, or fermented for beverages, juice, or alcohol.	(Joulain et al., 2004; Kunene et al., 2020)
G. livingstonei	Pulps and juices.	(Kunene et al., 2020)
S. cordatum	Jam and/or jelly from a ripe fruit.	(Kunene et al., 2020; Maroyi, 2018)

can provide food security, nutrition, health, income production, and environmental services with proper mainstreaming. Some of the fruits, such as *S. birrea*, are presented in Table 2.4, stipulating the by-products that can be manufactured from different indigenous fruits. Despite various fruitful studies based on underutilized fruit and vegetable production and value-adding, most unique goods are not yet commercially viable due to ongoing constraints from production to marketing. Creating new livelihood strategies for the farming community and manufacturers can be facilitated by using underutilized fruits and vegetables to promote value-added goods for commercial use.

2.10 CONCLUSION AND PROSPECTS

Underutilized fruits and vegetables provide an alternative food source and income for local communities. These plant species are of excellent economic, medicinal, nutritional, organoleptic, and traditional value. In addition, several underutilized fruits and vegetables offer valuable remedies for ailments such as cancer, diabetes, jaundice, and asthma. As an essential component of a balanced diet, underutilized fruits and vegetables are a potential key to achieving household food and nutrition security. There is a great value in underutilized fruits and vegetables in local settings, but they are generally overlooked at higher levels of government in many countries of the world. The nutritional value of many underutilized vegetables is high and well-suited to low-input agriculture. For instance, the vitamin C and pro-vitamin A content of many underutilized vegetables is higher than that of commercially available exotic species. Cultivating underutilized vegetables and fruits is necessary since knowledge about their cultivation and management is not well-established. There is a need to prioritize conserving and sustainably utilizing underutilized fruits and vegetables and renaissance the usage of underutilized plants, which will aid disadvantaged populations and improve local livelihoods. Besides their characterization and evaluation, underutilized fruit and vegetable species have received limited recognition from farmers, researchers, plant breeders, and policymakers. There is a need to educate and commercialize these plants, not only to conserve plant biodiversity but also for human consumption.

2.11 ACKNOWLEDGMENTS

We are grateful to the University of Mpumalanga (UMP) for supporting this work.

REFERENCES

Abdalla, M., & Yousef, M. (2016). Effect of cooking on nutritive value of Jew's mallow (*Corchorus olitorius* L.) and mallow (*Malva parviflora* L.) leaves. *Alexandria Journal of Food Science and Technology, 13*, 1–10.

Abobatta, W. F. (2021). Nutritional and healthy benefits of fruits. *American Journal of Biomedical Science and Research, 40*, 31979–31983.

Abosi, A. O., Mbukwa, E., Majinda, R. R. T., Raseroka, B. H., Yenesew, A., Midiwo, J. O., Akala, H., Liyala, P., & Waters, N. C. (2006). *Vangueria infausta* root bark: In vivo and in vitro antiplasmodial activity. *British Journal of Biomedical Science, 63*, 129–133.

Achigan-Dako, E. G., Sogbohossou, O. E. D., & Maundu, P. (2014). Current knowledge on *Amaranthus* spp.: Research avenues for improved nutritional value and yield in leafy amaranths in Sub-Saharan Africa. *Euphytica, 197*, 303–317.

Adebo, H. O., Ahoton, L. E., Quenum, F. J., Adoukonou-Sagbadja, H., Bello, D. O., & Chrysostome, C. A. (2018). Ethnobotanical knowledge of jute (*Corchorus olitorius* L.) in Benin. *European Journal of Medicinal Plants, 26*, 1–11.

Adedapo, A., Jimoh, F., & Afolayan, A. (2011). Comparison of the nutritive value and biological activities of the acetone, methanol and water extracts of the leaves of Bidens pilosa and Chenopodium album. *Acta Poloniae Pharmaceutica – Drug Research, 68*, 83–92.

Adejuwon, A., Femi-Akinlosotu, O., Omirinde, J., Owolabi, O., & Afodun, A. (2014). *Launaea taraxacifolia* ameliorates cisplatin-induced hepato-renal injury. *European Journal of Medicinal Plants, 4*, 528.

Adinortey, M. B., Ansah, C., Weremfo, A., Adinortey, C. A., Adukpo, G. E., Ameyaw, E. O., & Nyarko, A. K. (2018). DNA damage protecting activity and antioxidant potential of *Launaea taraxacifolia* leaves extract. *Journal of Natural Science, Biology, and Medicine, 9*, 6.

Agyare, C., Ansah, A., Ossei, P. P. S., Apenteng, J. A., & Boakye, Y. D. (2014). Wound healing and anti-infective properties of *Myrianthus arboreus* and *Alchornea cordifolia. Medicinal Chemistry, 4*, 533–539.

Ajuru, M. G., Nmom, F., Kpekot, A., Morris, E., & Macfarlane, O. (2021). Phytochemical and proximate compositions of the leaf and root of *Asystasia gangetica* (L.) T. Anderson in Rivers state. *Journal of Biotechnological Sciences, 9*, 29–41.

Akinnifesi, F. K., Ajayi, O. C., Sileshi, G., Kadzere, I., & Akinnifesi, A. I. (2007). *Domesticating and commercializing indigenous fruit and nut tree crops for food security and income generation in Sub-Saharan Africa*. New Crops International Symposium.

Akter, S., Ali, S., Fekete-Farkas, M., Fogarassy, C., & Lakner, Z. (2023). Why organic food? Factors influence the organic food purchase intension in an emerging country (study from Northern part of Bangladesh). *Resources, 12*, 5.

Akubugwo, I., Obasi, A., & Ginika, S. (2007a). Nutritional potential of the leaves and seeds of Black nightshade-*Solanum nigrum* L. Var virginicum from Afikpo-Nigeria. *Pakistan Journal of Nutrition, 6*, 323–326.

Akubugwo, I., Obasi, N., Chinyere, G., & Ugbogu, A. (2007b). Nutritional and chemical value of Amaranthus hybridus L. leaves from Afikpo, Nigeria. *African Journal of Biotechnology, 24*.

Alam, M. K. (2021). A comprehensive review of sweet potato (*Ipomoea batatas* [L.] Lam): Revisiting the associated health benefits. *Trends in Food Science & Technology, 115*, 512–529.

Alara, O. R., Abdurahman, N. H., Ukaegbu, C. I., & Kabbashi, N. A. (2019). Extraction and characterization of bioactive compounds in Vernonia amygdalina leaf ethanolic extract comparing Soxhlet and microwave-assisted extraction techniques. *Journal of Taibah University for Science, 13*(1), 414–422.

Al Batran, R., Al-Bayaty, F., Ameen Abdulla, M., Jamil Al-Obaidi, M. M., Hajrezaei, M., Hassandarvish, P., Fouad, M., Golbabapour, S., & Talaee, S. (2013). Gastroprotective effects of C orchorus olitorius leaf extract against ethanol-induced gastric mucosal hemorrhagic lesions in rats. *Journal of Gastroenterology and Hepatology, 28*, 1321–1329.

Ali, F., Assanta, M. A., & Robert, C. (2011). *Gnetum africanum*: A wild food plant from the African forest with many nutritional and medicinal properties. *Journal of Medicinal Food, 14*, 1289–1297.

Al-Juhaimi, F., Ghafoor, K., Özcan, M. M., Jahurul, M., Babiker, E. E., Jinap, S., Sahena, F., Sharifudin, M., & Zaidul, I. (2018). Effect of various food processing and handling methods on preservation of natural antioxidants in fruits and vegetables. *Journal of Food Science and Technology, 55*, 3872–3880.

Ansari, N., Houlihan, L., Hussain, B., & Pieroni, A. (2005). Antioxidant activity of five vegetables traditionally consumed by South-Asian migrants in Bradford, Yorkshire, UK. *Phytotherapy Research: An International Journal Devoted to Pharmacological and Toxicological Evaluation of Natural Product Derivatives, 19*, 907–911.

Aremu, A. O., Ncama, K., & Omotayo, A. O. (2019). Ethnobotanical uses, biological activities and chemical properties of kei-apple [*Dovyalis caffra* (Hook. f. & Harv.) Sim]: An indigenous fruit tree of Southern Africa. *Journal of Ethnopharmacology, 241*, 111963.

Arnold, H.-J., & Gulumian, M. (1984). Pharmacopoeia of traditional medicine in Venda. *Journal of Ethnopharmacology, 12*, 35–74.

Arora, R., Bharti, V., Gaur, P., Aggarwal, S., Mittal, M., & Das, S. N. (2017). *Operculina turpethum* extract inhibits growth and proliferation by inhibiting NF-κB, COX-2 and cyclin D1 and induces apoptosis by up regulating P53 in oral cancer cells. *Archives of Oral Biology, 80*, 1–9.

Arthur, G., Naidoo, K., & Coopoosamy, R. (2012). *Bidens pilosa* L.: Agricultural and pharmaceutical importance. *Journal of Medicinal Plants Research*, *6*, 3282–3287.

Asong, J. A., Ndhlovu, P. T., Khosana, N. S., Aremu, A. O., & Otang-Mbeng, W. (2019). Medicinal plants used for skin-related diseases among the Batswanas in Ngaka Modiri Molema district municipality, South Africa. *South African Journal of Botany*, *126*, 11–20.

Atangwho, I. J., Yin, K. B., Umar, M. I., Ahmad, M., & Asmawi, M. Z. (2014). *Vernonia amygdalina* simultaneously suppresses gluconeogenesis and potentiates glucose oxidation via the pentose phosphate pathway in streptozotocin-induced diabetic rats. *BMC Complementary and Alternative Medicine*, *14*, 1–13.

Aune, D., Giovannucci, E., Boffetta, P., Fadnes, L. T., Keum, N., Norat, T., Greenwood, D. C., Riboli, E., Vatten, L. J., & Tonstad, S. (2017). Fruit and vegetable intake and the risk of cardiovascular disease, total cancer and all-cause mortality-a systematic review and dose-response meta-analysis of prospective studies. *International Journal of Epidemiology*, *46*, 1029–1056.

Awuchi, C. G., Igwe, V. S., & Amagwula, I. O. (2020). Nutritional diseases and nutrient toxicities: A systematic review of the diets and nutrition for prevention and treatment. *International Journal of Advanced Academic Research*, *6*, 1–46.

Aziz, M. A., Khan, A. H., Adnan, M., & Izatullah, I. (2017). Traditional uses of medicinal plants reported by the indigenous communities and local herbal practitioners of Bajaur agency, federally administrated tribal areas, Pakistan. *Journal of Ethnopharmacology*, *198*, 268–281.

Baba, H., Kashimawo, A. J., & Ibe, A. C. (2022). Phytochemical evaluation and GC-MS profiling of the dichloromethane and ethanol extracts of Ocimum gratissimum L. and Lasianthera Africana BEAUV. *Journal of Phytomedicine and Therapeutics*, *20*(2), 640–655.

Bai, Y., Zang, X., Ma, J., & Xu, G. (2016). Anti-diabetic effect of *Portulaca oleracea* L. Polysaccharideandits mechanism in diabetic rats. *International Journal of Molecular Sciences*, *17*, 1201.

Bakare, R., Magbagbeola, O., Akinwande, A., & Okunowo, O. (2010). Nutritional and chemical evaluation of Momordica charantia. *Journal of Medicinal Plants Research*, *4*, 2189–2193.

Ballabio, C., Uberti, F., Di Lorenzo, C., Brandolini, A., Penas, E., & Restani, P. (2011). Biochemical and immunochemical characterization of different varieties of amaranth (*Amaranthus* L. ssp.) as a safe ingredient for gluten-free products. *Journal of Agricultural and Food Chemistry*, *59*, 12969–12974.

Bals, C., Harmeling, S., Windfuhr, M., Anyango, S. O., Orindi, V., Rahman, F., Bijoy, M. R., Habib, N., & Kulsum, U. (2008). Climate change, food security and the right to adequate food. In T. Hirsch, H. Bent, & K. K. Seitz (Eds.), *Climate change and food security – the development of the debate* (pp. 1–242). Diakonisches Werk der Evangelischen Kirche in Deutschland.

Banwat, M. E., Lar, L. A., Daboer, J., Audu, S., & Lassa, S. (2012). Knowledge and intake of fruit and vegetables consumption among adults in an urban community in North Central Nigeria. *Nigerian Health Journal*, *12*, 12–15.

Bassir, O. (1969). Toxic substances in Nigerian foods. *West African Journal of Biological and Applied Chemistry*, *13*, 3–6.

Bazzano, L. A., Li, T. Y., Joshipura, K. J., & Hu, F. B. (2008). Intake of fruit, vegetables, and fruit juices and risk of *diabetes* in women. *Diabetes Care*, *31*, 1311–1317.

Beheshti, F., Shabani, A., Eidgahi, A. M., Kookhaei, P., Vazirian, M., & Safavi, M. (2021). Anticancer activity of *Ipomoea purpurea* leaves extracts in monolayer and three-dimensional cell culture. *Evidence-Based Complementary and Alternative Medicine*, *2021*, 1–14.

Belhoussaine, O., El Kourchi, C., Harhar, H., Bouyahya, A., El Yadini, A., Fozia, F., Alotaibi, A., Ullah, R., & Tabyaoui, M. (2022). Chemical composition, antioxidant, insecticidal activity, and comparative analysis of essential oils of leaves and fruits of *Schinus molle* and *Schinus terebinthifolius*. *Evidence-Based Complementary and Alternative Medicine*, *2022*, 4288890.

Bellavia, A., Larsson, S. C., Bottai, M., Wolk, A., & Orsini, N. (2013). Fruit and vegetable consumption and all-cause mortality: A dose-response analysis. *The American Journal of Clinical Nutrition*, *98*, 454–459.

Boedecker, J., Termote, C., Assogbadjo, A. E., van Damme, P., & Lachat, C. (2014). Dietary contribution of wild edible plants to women's diets in the buffer zone around the Lama forest, Benin–an underutilized potential. *Food Security*, *6*, 833–849.

Bukola, O., Adebowale, B., & Olayinka, A. (2012). Anti-inflammatory and analgesic effects of methanol extract of *Stellaria media* (L.) Vill leaf. *African Journal of Biomedical Research*, *15*, 181–186.

Bulawa, B., Sogoni, A., Jimoh, M. O., & Laubscher, C. P. (2022). Potassium application enhanced plant growth, mineral composition, proximate and phytochemical content in Trachyandra divaricata Kunth (sandkool). *Plants*, *11*(22), 3183.

Bvenura, C., & Afolayan, A. J. (2014). Ethnobotanical survey of wild vegetables in Mbashe and Nkonkobe municipalities, Eastern Cape Province, South Africa. *Acta Botanica Gallica*, *161*, 189–199.

Bvenura, C., & Sivakumar, D. (2017). The role of wild fruits and vegetables in delivering a balanced and healthy diet. *Food Research International, 99*, 15–30.

Chacha, J. S., & Laswai, H. S. (2020). Micronutrients potential of underutilized vegetables and their role in fighting hidden hunger. *International Journal of Food Science, 2020.*

Chandra, M. S., Naresh, R., Thenua, O., Singh, R., & Geethanjali, D. (2020). Improving resource conservation, productivity and profitability of neglected and underutilized crops in the breadbasket of India: A review. *Breeding Science, 70*(1), 19–31.

Cheng, H. M., Koutsidis, G., Lodge, J. K., Ashor, A. W., Siervo, M., & Lara, J. (2019). Lycopene and tomato and risk of cardiovascular diseases: A systematic review and meta-analysis of epidemiological evidence. *Critical Reviews in Food Science and Nutrition, 59*, 141–158.

Chinsamy, M., & Koitsiwe, M. (2016). Traditional knowledge of medicinal and food plant uses for sustainable community livelihoods: A case of Batswana communities in South Africa. *Journal of Social Sciences, 46*(2), 146–154.

Chivandi, E., Mukonowenzou, N., Nyakudya, T., & Erlwanger, K. H. (2015). Potential of indigenous fruit-bearing trees to curb malnutrition, improve household food security, income and community health in Sub-Saharan Africa: A review. *Food Research International, 76*, 980–985.

Chowdhury, M., Hossain, N., Kashem, M., Shahid, M. A., & Alame, A. (2020). Immune response in COVID-19: A review. *Journal of Infection and Public Health, 13*, 1619–1629.

Christen, W. G., Liu, S., Schaumberg, D. A., & Buring, J. E. (2005). Fruit and vegetable intake and the risk of cataracts in women. *The American Journal of Clinical Nutrition, 81*, 1417–1422.

Collese, T. S., Nascimento-Ferreira, M. V., de Moraes, A. C. F., Rendo-Urteaga, T., Bel-Serrat, S., Moreno, L. A., & Carvalho, H. B. (2017). Role of fruits and vegetables in adolescent cardiovascular health: A systematic review. *Nutrition Reviews, 75*, 339–349.

Conner, T. S., Brookie, K. L., Carr, A. C., Mainvil, L. A., & Vissers, M. C. (2017). Let them eat fruit! The effect of fruit and vegetable consumption on psychological well-being in young adults: A randomized controlled trial. *PLoS One, 12*, e0171206.

Dansi, A., Adjatin, A., Adoukonou-Sagbadja, H., Faladé, V., Adomou, A. C., Yedomonhan, H., Akpagana, K., & de Foucault, B. (2009). Traditional leafy vegetables in Benin: Folk nomenclature, species under threat and domestication. *Acta Botanica Gallica, 156*, 183–199.

Das, A., & Bondya, S. L. (2016). *Cocculus hirsutus* (L.) diels: Diverse uses by the tribal population of Dumka. *The Biobrio, 3*, 190–193.

Debela, E., Tolera, A., Eik, L. O., & Salte, R. (2011). Nutritive value of morphological fractions of *Sesbania sesban* and *Desmodium intortum. Tropical and Subtropical Agroecosystems, 14*, 793–805.

De Caluwé, E., De Smedt, S., Assogbadjo, A., Samson, R., Sinsin, B., & Damme, P. V. (2009). Ethnic differences in use value and use patterns of baobab (Adansonia digitata L.) in Northern Benin. *African Journal of Ecology, 47*, 433–440.

De Vynck, J. C., van Wyk, B. E., & Cowling, R. M. (2016). Indigenous edible plant use by contemporary Khoe-San descendants of South Africa's Cape South Coast. *South African Journal of Botany, 102*, 60–69.

Dhalaria, R., Verma, R., Kumar, D., Puri, S., Tapwal, A., Kumar, V., Nepovimova, E., & Kuca, K. (2020). Bioactive compounds of edible fruits with their anti-aging properties: A comprehensive review to prolong human life. *Antioxidants, 9*, 1123.

Douglas, K. (2016). Ethnobotanical medicinal plants used as chewing sticks among the Kenyan communities. *British Journal of Pharmaceutical Research, 13*, 1–8.

du Preez, I., Shingenge, W., & Mumbengegwi, D. R. (2020). Namibian plants used in the treatment of malaria and associated symptoms. In *African natural plant products: Vol. III. Discoveries and Innovations in Chemistry, Bioactivity, and Applications* (pp. 45–66). ACS Publications.

Durazzo, A., Lucarini, M., Novellino, E., Souto, E. B., Daliu, P., & Santini, A. (2018). *Abelmoschus esculentus* (L.): Bioactive components' beneficial properties-focused on antidiabetic role-for sustainable health applications. *Molecules, 24*, 38.

Easdown, W., Ravishankar, M., Kaur, D., & Bhushan, K. (2014). Traditional leafy vegetables of a tribal community in Jharkhand, India. *XXIX International Horticultural Congress on Horticulture: Sustaining Lives, Livelihoods and Landscapes (IHC2014), 1102*, 43–52.

Ebert, A. W. (2014). Potential of underutilized traditional vegetables and legume crops to contribute to food and nutritional security, income and more sustainable production systems. *Sustainability, 6*, 319–335.

Eliza, J., Daisy, P., Ignacimuthu, S., & Duraipandiyan, V. (2009). Antidiabetic and antilipidemic effect of eremanthin from *Costus speciosus* (Koen.) Sm., in STZ-induced diabetic rats. *Chemico-Biological Interactions, 182*, 67–72.

Emmanuel, O. C., & Babalola, O. O. (2022). Amaranth production and consumption in South Africa: The challenges of sustainability for food and nutrition security. *International Journal of Agricultural Sustainability*, *20*, 449–460.

Eseyin, O. A., Sattar, M. A., & Rathore, H. A. (2014). A review of the pharmacological and biological activities of the aerial parts of Telfairia occidentalis Hook. f.(Cucurbitaceae). *Tropical Journal of Pharmaceutical Research*, *13*(10), 1761–1769.

Fanzo, J., & Miachon, L. (2023). Harnessing the connectivity of climate change, food systems and diets: Taking action to improve human and planetary health. *Anthropocene*, *42*, 100381.

FAO. (2017). *The future of food and agriculture – Trends and challenges*. Food and Agriculture Organization.

Food and Agriculture Organization (FAO), International Fund for Agricultural Development (IFAD), United Nations Children's Fund (UNICEF), World Food Programme (WFP), World Health Organization (WHO). (2017). *The state of food security and nutrition in the world 2017: Building resilience for peace and food security* (pp. 1–132). Food and Agriculture Organization.

FAO & UNIDO. (2008, January). *Agricultural mechanization in Africa. Time for action, planning investment for enhanced agricultural productivity report of a expert group meeting* (pp. 1–25). Food and Agriculture Organization.

Fardet, A., Richonnet, C., & Mazur, A. (2019). Association between consumption of fruit or processed fruit and chronic diseases and their risk factors: A systematic review of meta-analyses. *Nutrition Reviews*, *77*, 376–387.

Fasuyi, A. O. (2005). Nutrient composition and processing effects on cassava leaf (Manihot esculenta, Crantz) antinutrients. *Pakistan Journal of Nutrition*, *4*, 37–42.

Fawole, O., Ndhlala, A., Amoo, S., Finnie, J., & Van Staden, J. (2009). Anti-inflammatory and phytochemical properties of twelve medicinal plants used for treating gastro-intestinal ailments in South Africa. *Journal of Ethnopharmacology*, *123*, 237–243.

Feyssa, D. H., Njoka, J. T., Asfaw, Z., & Nyangito, M. M. (2011). Seasonal availability and consumption of wild edible plants in semiarid Ethiopia: Implications to food security and climate change adaptation. *Journal of Horticulture and Forestry*, *3*, 138–149.

Frison, E., Smith, I., Cherfas, J., Eyzaguirre, P., & Johns, T. (2005). Using biodiversity for food, dietary diversity, better nutrition and health. *South African Journal of Clinical Nutrition*, *18*, 112–114.

Gasparetto, J. C., Martins, C. A. F., Hayashi, S. S., Otuky, M. F., & Pontarolo, R. (2011). Ethnobotanical and scientific aspects of *Malva sylvestris* L.: A millennial herbal medicine. *Journal of Pharmacy and Pharmacology*, *64*, 172–189.

Gawai, K. J., & Tippat, S. K. (2024). Ethnobotanical significance of several seasonal medicinal plants associated with Amravati suburban habitats. *Environment Conservation Journal*, *25*, 120–126.

Gill, I., & Valivety, R. (1997). Polyunsaturated fatty acids, part 1: Occurrence, biological activities and applications. *Trends in Biotechnology*, *15*, 401–409.

Głąbska, D., Guzek, D., Groele, B., & Gutkowska, K. (2020). Fruit and vegetable intake and mental health in adults: A systematic review. *Nutrients*, *12*, 115.

Goswami, S., Mishra, K., Singh, R. P., Singh, P., & Singh, P. (2016). Sesbania sesban, a plant with diverse therapeutic benefits: An overview. *Journal of Pharmaceutical Research & Education*, *1*, 111–121.

Grové, T., De Jager, K., & De Beer, M. (2017). Indigenous hosts of economically important fruit fly species (Diptera: Tephritidae) in South Africa. *Journal of Applied Entomology*, *141*, 817–824.

Guevara, L., Domínguez-Anaya, M. Á., Ortigosa, A., González-Gordo, S., Díaz, C., Vicente, F., Corpas, F. J., Pérez del Palacio, J., & Palma, J. M. (2021). Identification of compounds with potential therapeutic uses from sweet pepper (*Capsicum annuum* L.) fruits and their modulation by nitric oxide (NO). *International Journal of Molecular Sciences*, *22*, 4476.

Gupta, D., & Kaul, V. (2021). Phytochemical screening of the flower and leaf extracts of *Emex australis* from Jammu district (J&K, India): A preliminary analysis. *National Academy Science Letters*, *44*, 359–363.

Hailemariam, M. B., Woldu, Z., Asfaw, Z., & Lulekal, E. (2021). Ethnobotany of an indigenous tree *Piliostigma thonningii* (Schumach.) Milne-Redh.(Fabaceae) in the arid and semi-arid areas of South Omo Zone, Southern Ethiopia. *Journal of Ethnobiology and Ethnomedicine*, *17*, 1–18.

Hall, J. N., Moore, S., Harper, S. B., & Lynch, J. W. (2009). Global variability in fruit and vegetable consumption. *American Journal of Preventive Medicine*, *36*, 402–409 (e405).

Ham, C. (2005). Plant for food and drink. In M. Mander & M. McKenzie (Eds.), *Southern African trade directory of indigenous natural products* (pp. 17–22). Commercial Products from the Wild Group.

Ham, C., Akinnifesi, F. K., Franzel, S., Jordaan, D. D. P., Hansmann, C., Ajayi, O., & Kock, C. D. (2008). Opportunities for commercialization and enterprise development of indigenous fruits in Southern Africa. In *Indigenous fruit trees in the tropics: Domestication, utilization and commercialization* (pp. 254–272). CABI Wallingford.

Hamdi, K., Ben-Amor, J., Mokrani, K., Mezghanni, N., & Tarchoun, N. (2017). Assessment of the genetic diversity of some local squash *(Cucurbita maxima* Duchesne) populations revealed by agro-morphological and chemical traits. *Journal of New Sciences, Agriculture and Biotechnology, 42*, 2306–2317.

Hasibuan, P. A. Z., Harahap, U., Sitorus, P., & Satria, D. (2020). The anticancer activities of *Vernonia amygdalina* Delile. Leaves on 4T1 breast cancer cells through phosphoinositide 3-kinase (PI3K) pathway. *Heliyon, 6*, e04449.

Hawkins, B., Sharrock, S., & Havens, K. (2008). *Plants and climate change: Which future?* Botanic Gardens Conservation International.

Hosseini, B., Berthon, B. S., Wark, P., & Wood, L. G. (2017). Effects of fruit and vegetable consumption on risk of asthma, wheezing and immune responses: A systematic review and meta-analysis. *Nutrients, 9*, 341.

Hughes, J. D. A., & Ebert, A. W. (2013). Research and development of underutilized plant species: The role of vegetables in assuring food and nutritional security. *Acta Horticulture, 979*, 79–91.

Hussain, Z., Amresh, G., Singh, S., & Rao, C. V. (2009). Antidiarrheal and antiulcer activity of *Amaranthus spinosus* in experimental animals. *Pharmaceutical Biology, 47*, 932–939.

Igara, C., Omoboyowa, D., Ahuchaogu, A., Orji, N., & Ndukwe, M. (2016). Phytochemical and nutritional profile of *Murraya koenigii* (Linn) Spreng leaf. *Journal of Pharmacognosy and Phytochemistry, 5*, 07–09.

Is, A., Okoye, J., & Oni, K. (2017). Promotion of indigenous food preservation and processing knowledge and the challenge of food security in Africa. *Journal of Food Security, 5*, 75–87.

Isong, E., Adewusi, S., Nkanga, E., Umoh, E., & Offiong, E. (1999). Nutritional and phytogeriatological studies of three varieties of *Gnetum africanum* ('afang'). *Food Chemistry, 64*, 489–493.

Jain, R., Sharma, A., Gupta, S., Sarethy, I. P., & Gabrani, R. (2011). *Solanum nigrum*: Current perspectives on therapeutic properties. *Alternative Medicine Review, 16*, 78–85.

Janakiraman, N., Johnson, M., Almeida, R. S., & Melo, H. D. (2021). In vitro antioxidant potential of Abrus precatorius L. and *Asystasia gangetica* (L.) T. Anderson. *Letters in Applied NanoBioScience, 11*, 4071–4079.

Jenipher, C., & Ayyanar, M. (2024). Ethnobotanical analysis of medicinal plants used by Kani Tribals of Tirunelveli district (Tamil Nadu, India) in treating respiratory diseases. *Journal of Herbal Medicine, 43*, 100826.

Jerald, S. E., Pandey, A., Bigoniya, P., & Singh, S. (2012). Antifertility activity of *Momordica charantia* descourt pulp and seed hydroalcoholic extract. *Journal of Applied Pharmacy, 3*, 682–696.

Jiang, Y., Jiang, Z. B., Shao, G. J., Guo, D. C., Tian, Y., Song, J. J., & Lin, Y. Y. (2012). The extraction of main chemical compositions in *Rumex* root. *Advanced Materials Research, 382*, 372–374.

Johnson, W., Tchounwou, P. B., & Yedjou, C. G. (2017). Therapeutic mechanisms of *Vernonia amygdalina* delile in the treatment of prostate cancer. *Molecules, 22*, 1594.

Joulain, D., Casazza, A., Laurent, R., Portier, D., Guillamon, N., Pandya, R., Le, M., & Viljoen, A. (2004). Volatile flavor constituents of fruits from Southern Africa: Mobola plum (*Parinari curatellifolia*). *Journal of Agricultural and Food Chemistry, 52*, 2322–2325.

Junior, A. G., Boffo, M. A., Lourenço, E. L. B., Stefanello, M. E. A., Kassuya, C. A. L., & Marques, M. C. A. (2009). Natriuretic and diuretic effects of *Tropaeolum majus* (Tropaeolaceae) in rats. *Journal of Ethnopharmacology, 122*, 517–522.

Kamga, R. T., Kouamé, C., Atangana, A., Chagomoka, T., & Ndango, R. (2013). Nutritional evaluation of five African indigenous vegetables. *Journal of Horticultural Research, 21*, 99–106.

Karim, T. (2012). *The secretory apparatus of Ceratotheca triloba (Pedaliaceae): Morphology and chemical composition of the secretion* (pp. 1–90). University of Kwa-Zulu Natal.

Karthishwaran, K., & Mirunalini, S. (2010). Therapeutic potential of *Pergularia daemia* (Forsk.): The ayurvedic wonder. *International Journal of Pharmacology, 6*, 836–843.

Kashyap, P., Kumar, S., Riar, C. S., Jindal, N., Baniwal, P., Guiné, R. P., Correia, P. M., Mehra, R., & Kumar, H. (2022). Recent advances in Drumstick (Moringa oleifera) leaves bioactive compounds: Composition, health benefits, bioaccessibility, and dietary applications. *Antioxidants, 11*(2), 402.

Konate, A., Sawadogo, W., Dubruc, F., Caillard, O., Ouedraogo, M., & Guissou, I. (2012). Phytochemical and anticonvulsant properties of" *Annona senegalensis* "Pers. (Annonaceae), plant used in Burkina folk medicine to treat epilepsy and convulsions. *British Journal of Pharmacology and Toxicology, 3*, 245–250.

Krishnamurthy, S., & Sarala, P. (2010). Proximate nutritive values and mineral components of *Withania somnifera* (Linn.) Dunal. *Journal of Chemistry, 7*, 985–996.

Kshirsagar, P. P., & Bhogaonkar, P. Y. (2015). *Nicandra physaloides* (L.) Gaertn–A less known wild edible fruit. *International Journal of Current Pharmaceutical Research, 7*, 60–63.

Kulczyński, B., & Gramza-Michałowska, A. (2019). The profile of secondary metabolites and other bioactive compounds in Cucurbita pepo L. and Cucurbita moschata pumpkin cultivars. *Molecules*, *24*(16), 2945.

Kunene, E. N., Nxumalo, K. A., Ngwenya, M. P., & Masarirambi, M. T. (2020). Domesticating and commercialisation of indigenous fruit and nut tree crops for food security and income generation in the Kingdom of Eswatini. *Current Journal of Applied Science and Technology*, *39*, 37–52.

Leelaprakash, G., Rose, J. C., Gowtham, B., Javvaji, P. K., & Prasad, S. (2011). In vitro antimicrobial and antioxidant activity of *Momordica charantia* leaves. *Pharmacophore*, *2*, 244–252.

Legwaila, G., Mojeremane, W., Madisa, M., Mmolotsi, R., & Rampart, M. (2011). Potential of traditional food plants in rural household food security in Botswana. *Journal of Horticulture and Forestry*, *3*, 171–177.

Lembo, A., & Camilleri, M. (2003). Chronic constipation. *New England Journal of Medicine*, *349*, 1360–1368.

Lewu, M. N., Adebola, P. O., & Afolayan, A. J. (2010). Effect of cooking on the mineral contents and antinutritional factors in seven accessions of *Colocasia esculenta* (L.) Schott growing in South Africa. *Journal of Food Composition and Analysis*, *23*, 389–393.

Li, B., Li, F., Wang, L., & Zhang, D. (2016). Fruit and vegetables consumption and risk of hypertension: A meta-analysis. *The Journal of Clinical Hypertension*, *18*, 468–476.

Liverpool-Tasie, L. S. O., & Parkhi, C. M. (2021). Climate risk and technology adoption in the midstream of crop value chains: Evidence from Nigerian maize traders. *Journal of Agricultural Economics*, *72*(1), 158–179.

Lock, K., Pomerleau, J., Causer, L., Altmann, D. R., & McKee, M. (2005). The global burden of disease attributable to low consumption of fruit and vegetables: Implications for the global strategy on diet. *Bulletin of the World Health Organization*, *83*, 100–108.

Lyimo, M., Temu, R., & Mugula, J. (2003). Identification and nutrient composition of indigenous vegetables of Tanzania. *Plant Foods for Human Nutrition*, *58*, 85–92.

Maanda, M., & Bhat, R. (2010). Wild vegetable use by Vhavenda in the Venda region of Limpopo province, South Africa. *Phyton-Revista Internacional de Botanica Experimental*, *79*, 189.

Madaki, F., Kabiru, A., Bakare-Odunola, M., Mailafiya, S., Hamzah, R., & Edward, J. (2016). Phytochemical and proximate analyses of methanol leaf extract of neem Azadirachta indica. *Journal of Phytochemical and Proximate Analyses of Methanol Leaf Extract of Neem Azadirachta Indica*, *15*, 1–6.

Madzimure, J., Nyahangare, E. T., Hamudikuwanda, H., Hove, T., Belmain, S. R., Stevenson, P. C., & Mvumi, B. M. (2013). Efficacy of *Strychnos spinosa* (Lam.) and *Solanum incanum* L. aqueous fruit extracts against cattle ticks. *Tropical Animal Health and Production*, *45*, 1341–1347.

Magwede, K., Van Wyk, B. E., & Van Wyk, A. E. (2019). An inventory of Vhaveṋḓa useful plants. *South African Journal of Botany*, *122*, 57–89.

Mahlangeni, N. T., Moodley, R., & Jonnalagadda, S. B. (2020). Nutritional value, antioxidant and antidiabetic properties of nettles (*Laportea alatipes* and *Obetia tenax*). *Scientific Reports*, *10*, 9762.

Makni, S., Tounsi, S., Rezgui, F., Trigui, M., & Bouassida, K. Z. (2019). *Emex spinosa* (L.) Campd. ethyl acetate fractions effects on inflammation and oxidative stress markers in carrageenan induced paw oedema in mice. *Journal of Ethnopharmacology*, *234*, 216–224.

Maroyi, A. (2018). *Syzygium cordatum* hochst. Ex krauss: An overview of its ethnobotany, phytochemistry and pharmacological properties. *Molecules*, *23*, 1084.

Mashabela, N. M., Ndhlovu, T. P., & Otang-Mbeng, W. (2024). Metabolomic and chemometric profiles of *Tribulus terrestris* L. from three different locations in Mpumalanga province, South Africa. *Journal of Medicinal Plants for Economic Development*, *8*, 9.

Mashela, P. W., Tseke, P. E., Pofu, K. M., & Mafeo, T. P. (2013, October 14–17). Potential commercialisation of *Mimusops zeyheri*: An indigenous tree selected for research and development through indigenous knowledge systems. In *11th African Crop Science Proceedings, Sowing innovations for sustainable food and nutrition security in Africa* (Vol. 11, pp. 667–669). CABI Digital Library.

Masondo, N. A., Aremu, A. O., Kulkarni, M. G., Petřík, I., Plačková, L., Šubrtová, M., Novák, O., Grúz, J., Doležal, K., & Strnad, M. (2019). Elucidating the role of Kelpak® on the growth, phytohormone composition, and phenolic acids in macronutrient-stressed Ceratotheca triloba. *Journal of Applied Phycology*, *31*, 2687–2697.

Matlala, M. E., Ndhlovu, P. T., Mokgehle, S. N., & Otang-Mbeng, W. (2024). Ethnobotanical investigation of *Mimusops zeyheri*, an underutilized indigenous fruit tree in Gauteng Province, South Africa. *Sustainability*, *16*, 1410.

McGartland, C. P., Robson, P. J., Murray, L. J., Cran, G. W., Savage, M. J., Watkins, D. C., Rooney, M. M., & Boreham, C. A. (2004). Fruit and vegetable consumption and bone mineral density: The Northern Ireland young hearts project. *The American Journal of Clinical Nutrition*, *80*, 1019–1023.

Mehboob, M., Naureen, I., Saleem, A., & Amanat, A. (2022). Medicinal and nutritional importance of *Lagenaria siceraria* (Lauki). *Saudi Journal of Biomedical Research*, *7*, 67–73.

Men, X., Choi, S.-I., Han, X., Kwon, H.-Y., Jang, G.-W., Choi, Y.-E., Park, S.-M., & Lee, O.-H. (2021). Physicochemical, nutritional and functional properties of *Cucurbita moschata. Food Science and Biotechnology, 30*, 171–183.

Mengistu, F., & Hager, H. (2009). Wild edible fruit species cultural domain, informant species competence and preference in three districts of Amhara region, Ethiopia. *Ethnobotany Research and Applications, 6*, 487–502.

Mensah, J., Okoli, R., Ohaju-Obodo, J., & Eifediyi, K. (2008). Phytochemical, nutritional and medical properties of some leafy vegetables consumed by Edo people of Nigeria. *African Journal of Biotechnology, 7*, 2304–2309.

Mishra, A. K., Chahal, P., Singh, H., Kumar, A., & Mishra, A. (2022). Exploration of hepatoprotective and antidiabetic potential of Senna occidentalis L. and isolation of a bioactive compound by column chromatography. *South African Journal of Botany, 148*, 144–153.

Mngadi, S., Moodley, R., & Jonnalagadda, S. B. (2019). Elemental composition and nutritional value of the edible fruits of Transvaal red milkwood *(Mimusops zeyheri)* and impact of soil quality. *Environmental Monitoring and Assessment, 191*, 135.

Moher, D., Liberati, A., Tetzlaff, J., & Altman, D. G. (2009). Preferred reporting items for systematic reviews and meta-analyses: The PRISMA statement. *Annals of Internal Medicine, 151*, 264–266.

Mokganya, M. G., & Tshisikhawe, M. P. (2019). Medicinal uses of selected wild edible vegetables consumed by Vhavenda of the Vhembe District Municipality, South Africa. *South African Journal of Botany, 122*, 184–188.

Mongalo, N. (2014). Antibacterial activity of *Waltheria indica* Linn (Sterculiaceae), collected from Blouberg area, Limpopo province, South Africa. *African Journal of Biotechnology, 13*, 3199–3203.

Moravec, I., Fernandez, E., Vlkova, M., & Milella, L. (2014). Ethnobotany of medicinal plants of Northern Ethiopia. *Boletín Latinoamericano y del Caribe de Plantas Medicinales y Aromáticas, 13*, 126–134.

Moyo, M., & Aremu, A. O. (2022). Nutritional, phytochemical and diverse health-promoting qualities of Cleome gynandra. *Critical Reviews in Food Science and Nutrition, 62*(13), 3535–3552.

Muchirah, P. N., Waihenya, R., Muya, S., Abubakar, L., Ozwara, H., & Makokha, A. (2018). Characterization and anti-oxidant activity of *Cucurbita maxima* Duchesne pulp and seed extracts. *The Journal of Phytopharmacology, 7*, 134–140.

Mudau, F. N., Chimonyo, V. G. P., Modi, A. T., & Mabhaudhi, T. (2022). Neglected and underutilised crops: A systematic review of their potential as food and herbal medicinal crops in South Africa. *Frontiers in Pharmacology, 12*, 4054.

Muli, J. K., Neondo, J. O., Kamau, P. K., & Budambula, N. L. (2022). Genetic diversity and use of African indigenous vegetables especially slender leaf. *International Journal of Vegetable Science, 28*, 76–94.

Munodawafa, T., Chagonda, L. S., & Moyo, S. R. (2013). Antimicrobial and phytochemical screening of some Zimbabwean medicinal plants. *Journal of Biologically Active Products from Nature, 3*(5–6), 323–330.

Muronga, M., Quispe, C., Tshikhudo, P. P., Msagati, T. A., Mudau, F. N., Martorell, M., Salehi, B., Abdull Razis, A. F., Sunusi, U., & Kamal, R. M. (2021). Three selected edible crops of the genus *Momordica* as potential sources of phytochemicals: Biochemical, nutritional, and medicinal values. *Frontiers in Pharmacology, 12*, 625546.

Mustafa, B., Hajdari, A., Krasniqi, F., Hoxha, E., Ademi, H., Quave, C. L., & Pieroni, A. (2012). Medical ethnobotany of the Albanian Alps in Kosovo. *Journal of Ethnobiology and Ethnomedicine, 8*, 6.

Mzid, M., Ben Khedir, S., Bardaa, S., Sahnoun, Z., & Rebai, T. (2017). Chemical composition, phytochemical constituents, antioxidant and anti-inflammatory activities of *Urtica urens* L. leaves. *Archives of Physiology and Biochemistry, 123*, 93–104.

Naidoo, Y., Sadashiva, C., Naidoo, G., & Raghu, K. (2016). Antibacterial, antioxidant and phytochemical properties of the ethanolic extract of Ocimum obovatum E. Mey. ex Benth. *Indian Journal of Traditional Knowledge, 15*, 57–61.

Nakitto, A. M. S., Byaruhanga, Y. B., Wagner, A. E., & Muyonga, J. H. (2024). Effect of thermal processing on bioactive compounds contents and antioxidant capacities of unripe and ripe *Solanum anguivi* Lam. fruit accessions. *Journal of Applied Research on Medicinal and Aromatic Plants, 100535*.

Nanyingi, M. O., Mbaria, J. M., Lanyasunya, A. L., Wagate, C. G., Koros, K. B., Kaburia, H. F., Munenge, R. W., & Ogara, W. O. (2008). Ethnopharmacological survey of Samburu district, Kenya. *Journal of Ethnobiology and Ethnomedicine, 4*. https://doi.org/10.1186/1746-4269-1184-1114

Narayan, J., Kumar, K., Kumar, S., & Singh, P. (2013). Nutritional evaluation of jhunjhuna *(Trema orientalis)* tree leaves in black bengal kids. *Indian Journal of Animal Nutrition, 30*, 262–266.

Nasr, F. A., Abutaha, N., Al-Zahrani, M., Farooq, M., & Wadaan, M. A. (2018). Anticancer potential of plant extracts from Riyadh (Saudi Arabia) on MDA-MB-231 breast cancer cells. *African Journal of Traditional, Complementary and Alternative Medicines, 15*, 46–53.

Nath, K., & Malik, N. (1970). Chemical composition and nutritive value of *Tribulus terrestris* L. *Indian Journal of Animal Sciences, 40*, 434–437.

Ndhlovu, P. T., Mooki, O., Mbeng, W. O., & Aremu, A. O. (2019). Plant species used for cosmetic and cosmeceutical purposes by the Vhavenda women in Vhembe District Municipality, Limpopo, South Africa. *South African Journal of Botany, 122*, 422–431.

Ndiege, M. L., Kengara, F., & Maiyoh, G. K. (2021). Characterization of phenolic compounds from leaf extract of bidens pilosa Linn. Var. Radiata. *South Asian Research Journal of Natural Products, 4*(3), 44–58.

Nehal, N., Mann, S., & Gupta, R. K. (2016). Nutritional and phytochemical evaluation of A. lividus L. syn. *Amaranthus blitum* subsp. oleraceus (L.) Costea leaves. *Indian Journal of Traditional Knowledge, 15*, 669–674.

Nencu, I., Vlase, L., Istudor, V., & Mircea, T. (2015). Preliminary research regarding *Urtica urens* L. and *Urtica dioica* L. *Amino Acids, 63*, 710–715.

Nesamvuni, C., Steyn, N., & Potgieter, M. (2001). Nutritional value of wild, leafy plants consumed by the Vhavenda. *South African Journal of Science, 97*, 51–54.

Ngugi, G. (2000). Case study from Kenya on indigenous wild vegetables. In B. Grosskinsky & C. Gullick (Eds.), *Exploring the potential of indigenous wild food plants in Southern Sudan* (pp. 29–41). The Mitchell Group, Inc. (TMG).

Niba, L. L. (2003). Processing effects on susceptibility of starch to digestion in some dietary starch sources. *International Journal of Food Sciences and Nutrition, 54*, 97–109.

Nicoli, M. C., Anese, M., & Parpinel, M. (1999). Influence of processing on the antioxidant properties of fruit and vegetables. *Trends in Food Science and Technology, 10*, 94–100.

Njoroge, G. N., & Bussmann, R. W. (2006). Traditional management of ear, nose and throat (ENT) diseases in Central Kenya. *Journal of Ethnobiology and Ethnomedicine, 2*, 1–9.

Nossier, F. H., Ashour, R. M., Abdel-Monem, A. R., & Elsayed, A. M. (2023). *Garcinia livingstonei* T. Anderson: A potential source for bioactive constituents. *The Natural Products Journal, 17*, 1–17.

Nundkumar, N., & Ojewole, J. A. O. (2002). Studies on the antiplasmodial properties of some South African medicinal plants used as antimalarial remedies in Zulu folk medicine. *Methods & Findings in Experimental & Clinical Pharmacology, 24*, 397–340.

Odhav, B., Beekrum, S., Akula, U., & Baijnath, H. (2007). Preliminary assessment of nutritional value of traditional leafy vegetables in KwaZulu-Natal, South Africa. *Journal of Food Composition and Analysis, 20*(5), 430–435.

Ojagbemi, A., Okekunle, A. P., Olowoyo, P., Akpa, O. M., Akinyemi, R., Ovbiagele, B., & Owolabi, M. (2021). Dietary intakes of green leafy vegetables and incidence of cardiovascular diseases. *Cardiovascular Journal of Africa, 32*, 215–223.

Ojewole, J. A., Mawoza, T., Chiwororo, W. D., & Owira, P. M. (2010). *Sclerocarya birrea* (A. Rich) Hochst. ['Marula'](Anacardiaceae): A review of its phytochemistry, pharmacology and toxicology and its ethnomedicinal uses. *Phytotherapy Research: An International Journal Devoted to Pharmacological and Toxicological Evaluation of Natural Product Derivatives, 24*, 633–639.

Okafor, J. N., Jideani, V. A., Meyer, M., & Le Roes-Hill, M. (2022). Bioactive components in Bambara groundnut (*Vigna subterraenea* (L.) Verdc) as a potential source of nutraceutical ingredients. *Heliyon, 8*, e09024.

Okoduwa, S. I. R., Umar, I. A., James, D. B., & Inuwa, H. M. (2017). Validation of the antidiabetic effects of *Vernonia amygdalina* delile leaf fractions in fortified diet-fed streptozotocin-treated rat model of type-2 diabetes. *Journal of Diabetology, 8*, 74–85.

Okwu, D. E. (2008). Citrus fruits: A rich source of phytochemicals and their roles in human health. *International Journal of Chemical Science, 6*, 451–471.

Ololade, Z., Kuyooro, S., Ogunmola, O., & Abiona, O. (2017). Phytochemical, antioxidant, anti-arthritic, anti-inflammatory and bactericidal potentials of the leaf extract of Lactuca teraxacifolia. *Global Journal of Medical Research, 17*, 19–28.

Olonode, E. T., Aderibigbe, A. O., & Bakre, A. G. (2015). Anti-nociceptive activity of the crude extract of *Myrianthus arboreus* P. Beauv (Cecropiaceae) in mice. *Journal of Ethnopharmacology, 171*, 94–98.

Olorunmota, R., Ofuya, T., Idoko, J., & Ogundeji, B. (2021). Comparative effects of rice and melon wastes on Sitophillus zeamais, storage pest of maize grains in Nigeria. *Journal of Experimental Agriculture International, 43*(1), 114–123.

Omotayo, A. O., & Aremu, A. O. (2020). Underutilized African indigenous fruit trees and food–nutrition security: Opportunities, challenges, and prospects. *Food and Energy Security, e220*, 1–16.

Omotayo, A. O., Ndhlovu, P. T., Tshwene, S. C., & Aremu, A. O. (2020). Utilization pattern of indigenous and naturalized plants among some selected rural households of North West province, South Africa. *Plants, 9*, 953.

Onuminya, T., Shodiya, O., & Ehinju, P. (2018). Ethnobotanical survey and genetic conservation of leafy vegetables in Lagos, Nigeria. *International Journal of Biological and Chemical Sciences*, *12*, 689–702.

Otang, W. M., Grierson, D. S., & Afolayan, A. J. (2014). A survey of plants responsible for causing irritant contact dermatitis in the Amathole district, Eastern Cape, South Africa. *Journal of Ethnopharmacology*, *157*, 274–284.

Pachkore, G., & Dhale, D. (2012). Phytochemicals, vitamins and minerals content of three *Ocimum* species. *International Journal of Science Innovations and Discoveries*, *2*, 201–207.

Padulosi, S., Heywood, V., Hunter, D., & Jarvis, A. (2011). *Underutilized species and climate change: Current status and outlook*. Crop Adaptation to Climate Change; Wiley-Blackwell.

Păduret, S., Amariei, S., Gutt, G., & Piscuc, B. (2016). The evaluation of dandelion *(Taraxacum officinale)* properties as a valuable food ingredient. *Romanian Biotechnological Letters*, *21*, 11569–11575.

Palgrave, K. C. (1977). *Trees of Southern Africa*. C. Stroik Publishers.

Parzonko, A., & Kiss, A. K. (2019). Caffeic acid derivatives isolated from Galinsoga parviflora herb protected human dermal fibroblasts from UVA-radiation. *Phytomedicine*, *57*, 215–222.

Pennington, J. A. T., & Fisher, R. A. (2010). Food component profiles for fruit and vegetable subgroups. *Journal of Food Composition and Analysis*, *23*, 411–418.

Pfukwa, T. M., Chikwanha, O. C., Katiyatiya, C. L., Fawole, O. A., Manley, M., & Mapiye, C. (2020). Southern African indigenous fruits and their byproducts: Prospects as food antioxidants. *Journal of Functional Foods*, *75*, 104220.

Pollock, R. L. (2017). The effect of phytochemicals intake from green leafy vegetables on the incidence of gastrointestinal cancers: A meta-analysis. *Global Journal of Medical Research*, *17*, 1.

Poobalan, V., Praneetha, S., Arumugam, T., Kumaravadivel, N., & Jeyakumar, P. (2019). Medicinal properties of vegetable crops. *International Journal of Chemical Studies*, *7*, 1538–1542.

Prabha, P., & Gopalakrishnan, V. K. (2014). Phytochemical screening, functional groups and elemental analysis of leaf extract of *Ipomoea obscura* (L) ker-gawl. *International Journal of Pharmacy and Pharmaceutical Sciences*, *6*, 83–85.

Preet, R., Gupta, R. C., & Kaur, M. (2019). Simultaneous high-performance thin-layer chromatographic quantification of withaferin a and withanolide a in Solanum nigrum L. "Black nightshade". *JPC-Journal of Planar Chromatography-Modern TLC*, *32*(4), 339–342.

Prynne, C. J., Mishra, G. D., O'Connell, M. A., Muniz, G., Laskey, M. A., Yan, L., Prentice, A., & Ginty, F. (2006). Fruit and vegetable intakes and bone mineral status: A cross-sectional study in 5 age and sex cohorts. *The American Journal of Clinical Nutrition*, *83*, 1420–1428.

Ramaano, A. I. (2021). Potential for tourism to promote indigenous resources for community development in Musina Municipality, Vhembe District, Limpopo Province, South Africa. *Forestry Economics Review*, *3*, 53–78.

Ramasar, R., Naidoo, Y., Dewir, Y. H., & El-Banna, A. N. (2022). Seasonal change in phytochemical composition and biological activities of *Carissa macrocarpa* (Eckl.) A. DC. leaf extract. *Horticulturae*, *8*, 780.

Ramjan, M., & Ansari, M. T. (2018). Factors affecting of fruits, vegetables and its quality. *Journal of Medicinal Plants Studies*, *6*, 16–18.

Russo, D., Miglionico, R., Carmosino, M., Bisaccia, F., Andrade, P. B., Valentão, P., Milella, L., & Armentano, M. F. (2018). A comparative study on phytochemical profiles and biological activities of *Sclerocarya birrea* (A. Rich.) Hochst leaf and bark extracts. *International Journal of Molecular Sciences*, *19*, 186.

Sacarlal, J., & Denning, D. W. (2018). Estimated burden of serious fungal infections in Mozambique. *Journal of Fungi*, *4*, 75.

Saleem, H., Zengin, G., Sarfraz, M., Alafnan, A., Locatelli, M., Tartaglia, A., Ahmad, I., Mahomoodally, M. F., Khurshid, U., & Ahemad, N. (2021). Phytochemical composition and in-vitro pharmacological evaluation of Emex australis Steinh: A natural source of enzyme inhibitors. *South African Journal of Botany*, *143*, 374–381.

Salvamani, S., Gunasekaran, B., Shukor, M. Y., Shaharuddin, N. A., Sabullah, M. K., & Ahmad, S. A. (2016). Anti-HMG-CoA reductase, antioxidant, and anti-inflammatory activities of *Amaranthus viridis* leaf extract as a potential treatment for hypercholesterolemia. *Evidence-Based Complementary and Alternative Medicine*, *2016*. http://dx.doi.org/10.1155/2016/8090841

Sánchez-Urdaneta, A. B., Montero-Quintero, K. C., González-Redondo, P., Molina, E., Bracho-Bravo, B., & Moreno-Rojas, R. (2020). Hypolipidemic and hypoglycaemic effect of wholemeal bread with amaranth (*Amaranthus dubius* Mart. ex Thell.) on sprague dawley rats. *Foods*, *9*, 707.

Šarić-Kundalić, B., Dobeš, C., Klatte-Asselmeyer, V., & Saukel, J. (2011). Ethnobotanical survey of traditionally used plants in human therapy of east, north and north-east Bosnia and Herzegovina. *Journal of Ethnopharmacology*, *133*, 1051–1076.

Sarkar, T., Salauddin, M., Roy, S., Chakraborty, R., Rebezov, M., Shariati, M. A., Thiruvengadam, M., & Rengasamy, K. R. R. (2022). Underutilized green leafy vegetables: Frontier in fortified food development and nutrition. *Critical Reviews in Food Science and Nutrition*, 1–55.

Sarkar, T., Salauddin, M., Roy, S., Chakraborty, R., Rebezov, M., Shariati, M. A., Thiruvengadam, M., & Rengasamy, K. R. R. (2023). Underutilized green leafy vegetables: Frontier in fortified food development and nutrition. *Critical Reviews in Food Science and Nutrition*, *63*, 11679–11733.

Sarker, U., Islam, M. T., Rabbani, M. G., & Oba, S. (2018a). Phenotypic divergence in vegetable amaranth for total antioxidant capacity, antioxidant profile, dietary fiber, nutritional and agronomic traits. *Acta Agriculturae Scandinavica, Section B – Soil & Plant Science*, *68*, 67–76.

Sarker, U., Islam, M. T., Rabbani, M. G., & Oba, S. (2018b). Variability in total antioxidant capacity, antioxidant leaf pigments and foliage yield of vegetable *amaranth. Journal of Integrative Agriculture*, *17*, 1145–1153.

Satheesh, N., & Fanta, S. W. (2020). Kale: Review on nutritional composition, bio-active compounds, anti-nutritional factors, health beneficial properties and value-added products. *Cogent Food & Agriculture*, *6*, 1811048.

Satyanarayana, K., Sravanthi, K., Shaker, I. A., & Ponnulakshmi, R. (2015). Molecular approach to identify antidiabetic potential of Azadirachta indica. *Journal of Ayurveda and Integrative Medicine*, *6*, 165.

Schönfeldt, H. C., & Pretorius, B. (2011). The nutrient content of five traditional South African dark green leafy vegetables – A preliminary study. *Journal of Food Composition and Analysis*, *24*, 1141–1146.

Seke, F., Manhivi, V. E., Guerin, M., Shoko, T., Akinola, S. A., Garcia, C., Remize, F., Slabbert, R. M., & Sivakumar, D. (2022). Antioxidant activities of co-encapsulated natal plum (Carissa macrocarpa) juice inoculated with Ltp. plantarum 75 in different biopolymeric matrices after in vitro digestion. *Foods*, *11*(14), 2116.

Semenya, S. S., & Maroyi, A. (2019). Ethnobotanical survey of plants used by Bapedi traditional healers to treat tuberculosis and its opportunistic infections in the Limpopo province, South Africa. *South African Journal of Botany*, *122*, 401–421.

Setshego, M. V., Aremu, A. O., Mooki, O., & Otang-Mbeng, W. (2020). Natural resources used as folk cosmeceuticals among rural communities in Vhembe district municipality, Limpopo province, South Africa. *BMC Complementary Medicine and Therapies*, *20*, 81.

Shackleton, S. E., Shackleton, C., Netshiluvhi, T., Geach, B., Ballance, A., & Fairbanks, D. (2002). Use patterns and value of savanna resources in three rural villages in South Africa. *Economic Botany*, *56*, 130–146.

Shackleton, S. E., Ziervogel, G., Sallu, S., Gill, T., & Tschakert, P. (2015). Why is socially-just climate change adaptation in sub-Saharan Africa so challenging? A review of barriers identified from empirical cases. *Wiley Interdisciplinary Reviews: Climate Change*, *6*, 321–344.

Shai, K. N., Ncama, K., Ndhlovu, P. T., Struwig, M., & Aremu, A. O. (2020). An exploratory study on the diverse uses and benefits of locally-sourced fruit species in three villages of Mpumalanga province, South Africa. *Foods*, *9*, 1581.

Sharifi-Rad, J., Melgar-Lalanne, G., Hernández-Álvarez, A. J., Taheri, Y., Shaheen, S., Kregiel, D., Antolak, H., Pawlikowska, E., Brdar-Jokanović, M., & Rajkovic, J. (2020). Malva species: Insights on its chemical composition towards pharmacological applications. *Phytotherapy Research*, *34*, 546–567.

Shawon, R. A., Kang, B. S., Lee, S. G., Kim, S. K., Lee, H. J., Katrich, E., Gorinstein, S., & Ku, Y. G. (2020). Influence of drought stress on bioactive compounds, antioxidant enzymes and glucosinolate contents of Chinese cabbage (*Brassica* rapa). *Food Chemistry*, *308*, 125657.

Shumsky, S. A., Hickey, G. M., Pelletier, B., & Johns, T. (2014). Understanding the contribution of wild edible plants to rural social-ecological resilience in semi-arid Kenya. *Journal of Ecology Society*, *19*, 34.

Silva, L. F. L. E., Souza, D. C., Resende, L. V., Nassur, R. D. C. M., Samartini, C. Q., & Gonçalves, W. M. (2018). Nutritional evaluation of non-conventional vegetables in Brazil. *Anais da academia Brasileira de Ciencias*, *90*, 1775–1787.

Silva, M. L., Rita, K., Bernardo, M. A., Mesquita, M. F. D., Pintão, A. M., & Moncada, M. (2023). Adansonia digitata L.(Baobab) bioactive compounds, biological activities, and the potential effect on glycemia: A narrative review. *Nutrients*, *15*(9), 2170.

Singh, A., Singh, S., Saroj, P., Mishra, D., Yadav, V., & Kumar, R. (2020). Cultivation of underutilized fruit crops in hot semi-arid regions: Developments and challenges—A review. *Current Horticulture*, *8*(1), 12–23.

Singh, R. K., Sreenivasulu, N., & Prasad, M. (2022). Potential of underutilized crops to introduce the nutritional diversity and achieve zero hunger. *Functional & Integrative Genomics*, *22*, 1459–1465.

Singhurst, J. R., & Holmes, W. C. (2010). *Carissa macrocarpa* (Apocynaceae): New to the Texas flora. *Phytoneuron*, *19*, 1–3.

Skwaryło-Bednarz, B., Stępniak, P. M., Jamiołkowska, A., Kopacki, M., Krzepiłko, A., & Klikocka, H. (2020). Amaranth seeds as a source of nutrients and bioactive substances in human diet. *Acta Scientiarum Polonorum Hortorum Cultus*, *19*(6), 153–164.

Soh, D., Bakang, B. T., Tchouboun, E. N., Nganso, Y. O. D., Defokou, U. D., Sidjui, L. S., Ahmed, A., Teponno, R. B., Lateef, M., Ali, M. S., & Nyassé, B. (2020). New cucurbitane type triterpenes from *Momordica foetida* Schumach. (Cucurbitaceae). *Phytochemistry Letters*, *38*, 90–95.

Souilem, F., Dias, M. I., Barros, L., Calhelha, R. C., Alves, M. J., Harzallah-Skhiri, F., & Ferreira, I. C. F. R. (2019). Amantagula fruit (*Carissa macrocarpa* (Eckl.) A.DC.): Nutritional and phytochemical characterization. *Plant Foods for Human Nutrition*, *74*, 76–82.

Sowunmi, L. I., & Afolayan, A. J. (2015). Phytochemical constituents and antioxidant properties of acetone extract of *Cleome gynandra* (L.) growing in the Eastern Cape, South Africa. *African Journal of Traditional, Complementary and Alternative Medicines*, *12*(3), 1–8.

Stats SA. (2018). Provincial profile: North West community survey 2016. In R. Maluleke (Ed.), *Provincial profile: North West/statistics South Africa* (pp. 1–93). Statistics South Africa.

Steyn, N., Olivier, J., Winter, P., Burger, S., & Nesamvuni, C. (2001). A survey of wild, green, leafy vegetables and their potential in combating micronutrient deficiencies in rural populations: Research in action. *South African Journal of Science*, *97*, 276–278.

Tao, J., Li, Y., Li, S., & Li, H.-B. (2018). Plant foods for the prevention and management of colon cancer. *Journal of Functional Foods*, *42*, 95–110.

Thane, C., & Reddy, S. (1997). Processing of fruit and vegetables: Effect on carotenoids. *Nutrition and Food Science*, *97*, 58–65.

Tosoc, J. P. S., Nuñeza, O. M., Sudha, T., Darwish, N. H., & Mousa, S. A. (2021). Anticancer effects of the *Corchorus olitorius* aqueous extract and its bioactive compounds on human cancer cell lines. *Molecules*, *26*, 6033.

Turkmen, N., Sari, F., & Velioglu, Y. S. (2005). The effect of cooking methods on total phenolics and antioxidant activity of selected green vegetables. *Food Chemistry*, *93*, 713–718.

Tylewicz, U., Tappi, S., Nowacka, M., & Wiktor, A. (2019). Safety, quality, and processing of fruits and vegetables. *MDPI*, *569*.

Uddin, A. N., Hossain, F., Reza, A. A., Nasrin, M. S., & Alam, A. K. (2022). Traditional uses, pharmacological activities, and phytochemical constituents of the genus Syzygium: A review. *Food Science & Nutrition*, *10*(6), 1789–1819.

Umadevi, M., Rajeswari, R., Rahale, C. S., Selvavenkadesh, S., Pushpa, R., Kumar, K. S., & Bhowmik, D. (2012). Traditional and medicinal uses of *Withania somnifera*. *The Pharma Innovation*, *1*, 102.

Uusiku, N. P., Oelofse, A., Duodu, K. G., Bester, M. J., & Faber, M. (2010). Nutritional value of leafy vegetables of sub-Saharan Africa and their potential contribution to human health: A review. *Journal of Food Composition and Analysis*, *23*, 499–509.

Van der Hoeven, M., Osei, J., Greeff, M., Kruger, A., Faber, M., & Smuts, C. M. (2013). Indigenous and traditional plants: South African parents' knowledge, perceptions and uses and their children's sensory acceptance. *Journal of Ethnobiology and Ethnomedicine*, *9*, 1–12.

Van Wijk, Y., Tusenius, M. L., Rust, R., Cowling, R. M., & Wurz, S. (2017). Modern vegetation at the Klasies River Archaeological sites, Tsitsikamma Coast, South-Eastern Cape, South Africa: A reference collection. *Journal of Plant Ecology and Evolution*, *150*, 13–34.

Van Wyk, B., & Van Wyk, P. (1997). *Field guide to trees of Southern Africa*. Struik.

Van Wyk, B.-E., & Gericke, N. (2000). *People's plants. A guide to the useful plants of Southern Africa*. Briza Publications.

Verschaeve, L., Kestens, V., Taylor, J., Elgorashi, E., Maes, A., Van Puyvelde, L., De Kimpe, N., & Van Staden, J. (2004). Investigation of the antimutagenic effects of selected South African medicinal plant extracts. *Toxicology in Vitro*, *18*, 29–35.

Vinceti, B., Termote, C., Ickowitz, A., Powell, B., Kehlenbeck, K., & Hunter, D. (2013). The contribution of forests and trees to sustainable diets. *Journal of Sustainability*, *5*, 4797–4824.

Wada, E., Feyissa, T., & Tesfaye, K. (2019). Proximate, mineral and antinutrient contents of cocoyam (*Xanthosoma sagittifolium* (L.) Schott) from Ethiopia. *International Journal of Food Science*, *2019*, 8965476.

Wang, T., Cai, H., Sasazuki, S., Tsugane, S., Zheng, W., Cho, E. R., Jee, S. H., Michel, A., Pawlita, M., & Xiang, Y. B. (2017). Fruit and vegetable consumption, Helicobacter pylori antibodies, and gastric cancer risk: A pooled analysis of prospective studies in China, Japan, and Korea. *International Journal of Cancer*, *140*, 591–599.

Wang, X., Zhang, L., Qin, L., Wang, Y., Chen, F., Qu, C., & Miao, J. (2022). Physicochemical properties of the soluble dietary fiber from laminaria japonica and its role in the regulation of type 2 diabetes mice. *Nutrients*, *14*, 329.

Wasswa, M., & Muyonga, J. (2021). Influence of sun drying and a combination of boiling and sun drying on the retention of nutrients and bioactive compounds in cowpea (*Vigna unguiculata* (L). Walp) leaves. *African Journal of Biological Sciences*, *3*, 48–58.

Watt, J. M., & Breyer-Brandwijk, M. G. (1962). *The medicinal and poisonous plants of Southern and Eastern Africa*. Livingstone Ltd.

Welcome, A. K., & Van Wyk, B. E. (2019). An inventory and analysis of the food plants of Southern Africa. *South African Journal of Botany*, *122*, 136–179.

Willett, W., Rockström, J., Loken, B., Springmann, M., Lang, T., Vermeulen, S., Garnett, T., Tilman, D., DeClerck, F., Wood, A., & Jonell, M. (2019). Food in the anthropocene: The EAT–Lancet commission on healthy diets from sustainable food systems. *The Lancet*, *393*, 447–492.

World Health Organization. (2003). *Diet, nutrition, and the prevention of chronic diseases: Report of a joint WHO/FAO expert consultation*. WHO – World Health Organization.

Xu, Y.-B., Chen, G.-L., & Guo, M.-Q. (2019). Antioxidant and anti-inflammatory activities of the crude extracts of *Moringa oleifera* from Kenya and their correlations with flavonoids. *Antioxidants*, *8*, 296.

Younis, Y. M. H., Ghirmay, S., & Al-Shihry, S. S. (2000). African *Cucurbita pepo* L.: Properties of seed and variability in fatty acid composition of seed oil. *Phytochemistry*, *54*, 71–75.

Zahara, K., Bibi, Y., Arshad, M., Kaukab, G., Al Ayoubi, S., & Qayyum, A. (2022). In-vitro examination and isolation of antidiarrheal compounds using five bacterial strains from invasive species Bidens bipinnata L. *Saudi Journal of Biological Sciences*, *29*, 472–479.

Zamuz, S., Munekata, P. E., Gullón, B., Rocchetti, G., Montesano, D., & Lorenzo, J. M. (2021). Citrullus lanatus as source of bioactive components: An up-to-date review. *Trends in Food Science & Technology*, *111*, 208–222.

Zou, L., Tan, W. K., Du, Y., Lee, H. W., Liang, X., Lei, J., Striegel, L., Weber, N., Rychlik, M., & Ong, C. N. (2021). Nutritional metabolites in *Brassica rapa* subsp. chinensis var. parachinensis (choy sum) at three different growth stages: Microgreen, seedling and adult plant. *Food Chemistry*, *357*, 129535.

3 Exploring the Hidden Potential of Undervalued Food Plants for Nutrition and Health Security

[1*]Moji Taibat Bakare-Odunola, [2]Fausat Lola Kolawole, and [3]Rukayat Abiodun Oyegoke
[1]Department of Pharmaceutical and Medicinal Chemistry, Faculty of Pharmaceutical Sciences
[2]Department of Home Economics and Food Science, Faculty of Agriculture
[3]Department of Biochemistry, Faculty of Life Sciences
University of Ilorin, Nigeria
*Corresponding author: odunola.tm@unilorin.edu.ng

3.1 INTRODUCTION

Undervalued food plants are plants that can sustain life, generate energy, and enhance growth and health-sustaining characteristics (Table 3.1). These plants, often referred to as neglected or underutilized crops, play a crucial yet frequently overlooked role in the realms of food, nutrition, and public health (Sidiq et al., 2022). They encompass many vegetables, fruits, tubers, grains, and legumes with tremendous untapped potential (Sidiq et al., 2022).

The rapid growth of the world population is associated with a rise in malnutrition, a shortage of calories, and an escalation of health challenges (FAO, 2021). Out of the thousands of species available, approximately 200 have undergone extensive domestication, resulting in an overreliance on only a handful of them (Meyer et al., 2012). This means that the world's total food calories depends on this small fraction of species, making our food supply extremely vulnerable. This shows that inefficient calories, especially critical macronutrients required for normal body functioning, are depleting, causing various health issues, since food determines human health status. Therefore, variable plant biodiversity is crucial for food and nutrition security and health.

Consequently, the need for utilization of neglected and underutilized food plant species (Figure 3.1) for hunger, the deficiency of nutrients, and the health challenges cannot be over-emphasized (Knez et al., 2023). Food plants that are not fully utilized comprise wild, domesticated, or semi-domesticated varieties, holding promise to elevate the well-being of individuals by bolstering food and nutrition security as well as health. Even though these plants have the potential to enhance security, justice, and sovereignty, researchers, policymakers, and both local and global markets have yet to fully harness their benefits. This is largely due to their constrained competitiveness when compared to commodity crops within conventional agriculture, as highlighted by Ulian et al. (2020). Indeed, many of these underutilized food plants, also called "minor" or "orphan" crops, serve as multipurpose plants. Therefore, unveiling and increasing their inherent values have the potential to reveal their true role in food and nutrition security and health while protecting plant biodiversity (Diazgranados et al., 2020).

 DOI: 10.1201/9781003469766-3

TABLE 3.1
Phytochemical Constituents and Biological Activity of Some Undervalued Food Plants

Scientific Name	English Name	Distribution	Nutritional Constituents	Bioactive Constituents	Biological Activity
C. cajan	Pigeon pea	West Africa, East Africa, and Southern Africa	Protein, soluble and insoluble fibers, palmitic and linolenic acid, calcium, phosphorous, magnesium, vitamins B_1, B_2, B_3, C, and E	Phenolic and polyphenolic compounds	Antidiabetic, anticarcinogenic, anti-inflammatory properties (Reynolds et al., 2020; Locali-Pereira et al., 2023)
S. nigrum	Black/glossy nightshade	East, North, and Southern Africa	Calcium, magnesium, sodium, potassium, manganese, zinc, iron, dietary fiber, vitamins A, C, and B complex, β- carotene, lycopene, lutein, zeaxanthin, chlorophyll	Carotenoids, phenolics, tannins, flavonoids	Antioxidative properties (Agu & Agulanna, 2020); anticarcinogenic, hepatoprotective, antipyretic, anticataract, anti-inflammatory (Sangija & Wu, 2020), antigenotoxicant properties (Odongo et al., 2018)
D. guineense	Velvet tamarind	Sub-Saharan Africa, West Africa	B_2 (riboflavin), calcium, potassium, magnesium, iron, caloric contributor, fiber, essential oil	Phenolics, flavonoids, polyphenolics, tannins, alkaloids	Antiulcerogenic (Oyegoke & Oladiji, 2014), antibacterial, analgesic, and antimicrobial (Orji et al., 2012)
V. subterranea	Bambara nut	West Africa, Central African Republic, and Southern Africa	Calcium, iron, potassium, protein, energy and caloric contributor, presence of essential amino acids, fiber, fatty acids (i.e., palmitic, linolenic, and linolenic acids)	Phenolics, flavonoids, polyphenolics	Wound healing, anticataract (Khan et al., 2020), aphrodisiac potentials (Brink et al., 2006)
X. sagittifolium	Cocoyam	West Africa, East Africa	Soluble and insoluble fiber, starch, protein, iron, sodium, potassium, phosphorous, vitamin A, riboflavin, nicotinic acid	Flavonoids, polyphenols, antioxidants, polysaccharides	Antioxidant, anticancer, anti-inflammatory, and dyslipidemic properties (Mba & Agu, 2021)
D. cumminsii	Serendipity berry	West Africa and East Africa	Crude fiber, β-carotene, vitamin C, vitamin D, essential amino acid	Monellin, alkaloids, flavonoids, tannins	Antidiabetic and dieting properties, antioxidative and anti-inflammatory potentials (Oselebe & Nwankiti, 2007)
D. dumetorum	Bitter yam	Sub-Saharan and West Africa	Fiber, sugar, calcium, phosphorous, iron, vitamins	Phenolics, polyphenols, polysaccharides, steroids, saponins, allantoin, diosgenin	Anti-inflammatory, antiobesity, antidiabetic, antidyslipidemic properties, and prevention of non-alcoholic fatty liver disease (Wang et al., 2023)

FIGURE 3.1 Pictures of some undervalued food plants: (a) Cajanus *cajan*; (b) *Solanum nigrum*; (c) *Dialium guineense* fruit; (d) *Vigna subterranean*; (e) *Xanthosoma sagittifolium*; (f) *Dioscoreophyllum cumminsii*; (g) *Dioscorea dumetorum*.

Source: (a) Nkwonta et al. (2023); (b) Sangija and Wu (2020); (c) Ogbuewu et al. (2023); (d) Ajiloba et al. (2022); (e) Boakye et al. (2018); (f) Oselebe and Nwankiti (2007); (g) Wang et al. (2023).

In this chapter, underutilized food plants and their nutritional values, cultivation, and potential role as functional food plants to solve the current food and nutrition insecurity and health issues are reviewed.

3.2 UNDERUTILIZED FOOD PLANTS

Underutilized indigenous crops refer to locally grown species primarily cultivated in native communities. However, these crops have been experiencing a decline in popularity because they have not been integrated into the mainstream food system (Jahanshiri et al., 2020). The COVID-19 pandemic, along with other conflict and war crises worldwide, has intensified the overreliance on a small number of crops. This situation underscores the need for a shift toward underutilized crops, expanding our dietary variety and thereby diversifying our nutrient sources. Underutilized species, sometimes called "orphan crops," "forgotten foods," "minor crops," or "underresearched" food plants (Ulian et al., 2020), are

> wild, domesticated, or semi-domesticated plants that have a history of cultivation for human and animal food, oil, and medicinal properties, but whose potential to improve livelihoods, food security, justice and sovereignty has not been fully exploited by researchers, policy-makers and local and global markets due to their limited competitiveness with crops commonly traded in mainstream agriculture.
>
> (Ulian et al., 2020)

These species encompass a diverse group of food plants which include but are not limited to cereals, roots and tubers, pulses and legumes, oilseeds, fruits, and vegetables (Talabi et al., 2022). Although they are underutilized, they are very rich in micro- and macronutrients, as well as dietary fiber (Hussain et al., 2020). These nutrients are also well absorbed, digested, and utilized within the body's system to provide nutritional and health benefits (Hussain et al., 2020). Likewise, some of the plant parts of these species are of high economic value to various industries, which include construction, pharmaceutical, and livestock, among others (Ulian et al., 2020). Finally, concerning their cultivation, these species do not require intensive resources, can thrive on marginal land, and exhibit tolerance to various environmental stresses (AARINENA, 2021). These plants exhibit remarkable nutritional profiles, show resilience in diverse climates, and have a lower environmental footprint, making them a valuable resource in the face of challenges, such as climate change and population growth (AARINENA, 2021; Ayilara et al., 2022).

3.2.1 Categorization of Undervalued Food Plants

Undervalued food plants encompass an array of botanical categories, each offering unique benefits to the human diet and the environment. These categories include:

3.2.1.1 Vegetables

Vegetables are a rich source of essential vitamins, minerals, and dietary fiber. Undervalued vegetables like amaranth, malabar spinach, and moringa are nutrient powerhouses that can be cultivated with minimal resources, making them invaluable in combating malnutrition in resource-limited regions (Nkwonta et al., 2023). Their adaptability to various climates and soil conditions further underscores their importance.

3.2.1.2 Fruits

Undervalued fruits such as jackfruit, tamarind, and pawpaw often thrive in challenging conditions, providing essential vitamins and antioxidants (AARINENA, 2021). Their fruit-bearing capacity, even in arid or marginal areas, can contribute significantly to food and nutrition security (Ayilara et al., 2022).

3.2.1.3 Tubers

Tubers are sources of carbohydrates and are particularly well-suited for cultivation in harsh environments (Emesi et al., 2019). Their resistance to drought and adaptability to various soils make them crucial in addressing food shortages and ensuring dietary diversity (AARINENA, 2021).

3.2.1.4 Grains

Grains like teff, fonio, and amaranth are gluten-free, nutrient-dense alternatives to traditional staple grains, such as maize, wheat, and rice (Moreno et al., 2014). These crops are vital in regions where gluten intolerance is prevalent and can improve gluten-sensitive populations' nutrition (Nkwonta et al., 2023).

3.2.1.5 Legumes

Legumes such as Bambara groundnuts, winged bean, and lentils are a source of plant-based protein essential for vegetarians and areas where access to animal protein is limited (Bvenura & Sivakumar, 2017). Their ability to fix atmospheric nitrogen enhances soil fertility (Figure 3.2), making them

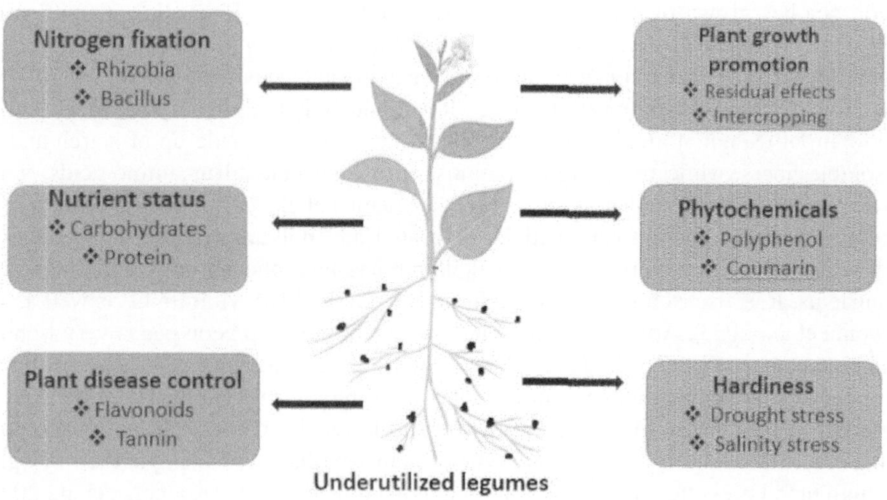

FIGURE 3.2 The mechanism of underutilized legumes in promoting food and nutrition security.

Source: Ayilara et al. (2022).

an environmentally friendly choice for sustainable agriculture (AARINENA, 2021; Ayilara et al., 2022). Each of these categories of undervalued food plants has a unique set of characteristics that can potentially address specific dietary and environmental challenges (Adeyanju et al., 2019).

3.3 SELECTED UNDERVALUED FOOD PLANTS

Due to their distinctive characteristics, the species in this chapter were selected based on the following criteria:

1. Belonging to one of the major groups of plant species
2. Possessing high nutritional values (macro- and micronutrients)
3. Possessing a nutrient–health maintenance interaction
4. High resilience to environmental stress, and ability to withstand the effect of climate change
5. Potential for increased local and global utilization
6. Local availability, for enhancement of local livelihoods

3.3.1 *Cajanus cajan* (L.) Millsp.

Cajanus cajan (pigeon pea), a legume belonging to the Fabaceae family, is a perennial diploid crop that is native to Africa (Nkwonta et al., 2023). In Nigeria, its local names are "otili" in Yoruba, "fio fio" in Igbo, and "aduwa" in the Hausa language. Pigeon pea seeds come in various colors, such as black, orange, and cream, and are bitter or sweet, depending on the variety (Upadhyaya et al., 2005). This crop is widely cultivated in tropical and semitropical regions of the world. They are commonly consumed in South Asia, Southeast Asia, Africa, Latin America, and the Caribbeans. Pigeon pea is the most prevalent fast-growing and adaptable pulse crop in tropical and subtropical climates (Kuraz, 2022). It is a drought-resistant pulse crop that thrives in locations with temperatures ranging from 20 to 40°C and less than 625 mm of annual rainfall, where other crops such as maize normally fail (Nong, 2022). Pigeon pea seeds contain about 20–22% protein and appreciable amounts of essential amino acids and minerals (Saxena et al., 2002). It is the world's fifth most well-known pulse crop, and India's second (Kuraz, 2022). It is rich in crude fiber, minerals (macro and micro), and vitamins (Sarkar et al., 2020). Being a staple food, pigeon pea is paradoxically underutilized (Fatokimi et al., 2021). Although the popularity of pigeon pea has grown over the years, it is still underutilized. Regarding its cultivation and growing characteristics, pigeon pea has a high resistance to drought and can thrive very well in soils with low fertility. Due to its deep taproots, it can tap nutrients from deeper layers of the soil (Abebe, 2022; Gichohi-Wainaina et al., 2022). It is moderately high in carbohydrates and protein but is low in fat (Singh et al., 2021). Its carbohydrate content is made up of starch and soluble and insoluble fibers, while the protein contains sulfur and non-sulfur amino acids, which are mostly the limiting amino acids in various legumes (Auma et al., 2019). The fat content of pigeon pea contains palmitic and linoleic acid for saturated and polyunsaturated fatty acids (Abebe, 2022). The pea is rich in minerals, such as calcium, potassium, phosphorous, and magnesium, for micronutrients. It is also rich in some B vitamins (B_1, B_2, and B_3), vitamin C, as well as vitamin E (Nkwonta et al., 2023). Apart from these primary metabolites, pigeon pea is very rich in polyphenolic compounds (Wei et al., 2022). These enriched nutrients in pigeon pea act via various biochemical mechanisms to confer health benefits to the biological system. The starch and dietary fiber present in pigeon pea have been discovered to lower the risk of various health concerns, including cardiovascular diseases, certain types of cancers, diabetes (owing to their low glycemic index), high blood pressure, and inflammation, among other benefits (Reynolds et al., 2020). The amino acid contents, especially of the sulfur-containing groups, are involved in various metabolic processes, including the formation of biologically important compounds, such as creatine, bile salts, and heme, among others (Caroline et al., 2023).

While the fat content of pigeon pea may be modest, the significance of its polyunsaturated fatty acids (PUFA), recognized as among the healthiest fatty acids for human physiology, should not be underestimated. The vitamin B content of the pea is absorbed and digested, thus playing a crucial role as a coenzyme in the oxidation–decarboxylation, oxidation–reduction, and transketolase reactions in human physiology (Caroline et al., 2023). The minerals in pigeon pea play critical biological roles, which include blood pressure control (potassium and magnesium), bone health (calcium and magnesium), and blood glucose control (magnesium). In addition, its phenolic contents have been implicated in its health-boosting properties, including antioxidant, anticarcinogenic, and cholesterol-lowering effects. Pigeon pea thus merits global utilization due to its nutritional value and potential health benefits.

3.3.2 *Solanum nigrum* L.

Solanum nigrum (black nightshade) is a short-lived perennial shrub, a vegetable belonging to the Solanaceae family (Alirea, 2021). This vegetable is distributed in the southern parts of Nigeria and is commonly found in forests or as weeds, although it is cultivated as a food crop in several continents, including Africa and North America (Alirea, 2021). This underutilized plant is utilized as a vegetable possibly because of its classification as a weed and the resulting lack of awareness regarding its nutritional properties (Bvenura & Afolayan, 2014). As a resilient weed, *S. nigrum* thrives in various soil conditions, including moist, dry, stony, shallow, or deep soil. It can be successfully cultivated in tropical and subtropical agro-climatic regions (Ghanari & Arianmehr, 2021). Black nightshade is rich in macro- and micronutrients, such as protein; carbohydrates; β-carotene; vitamins A, C, E, and B complex; minerals (iron, calcium, and zinc); and dietary fiber (Sangija & Wu, 2020). Furthermore, it is rich in bioactive compounds, particularly lutein, zeaxanthin, polyphenol, flavonoids, and chlorophylls, which possess antioxidant activity, antigenotoxicants, and anticancer properties (Sangija & Martin, 2021). β-carotene is a precursor of vitamin A, which is one of the functional carotenoids. Others include lycopene, among others. These carotenoids are found in this vegetable, and they have been implicated in various health benefits. High levels of carotenoids have been associated with a relatively low incidence of cardiovascular diseases, cataracts, cancer, and the prevention of the pathogenesis of Alzheimer's disease. In addition, high levels of lycopene have been associated with a decrease in cholesterol serum and have been shown to be effective against several common diseases, including atherosclerosis and heart disease (Yao et al., 2021). Carotenoids generally help reduce reactive oxygen species and prevent some types of cancer. Due to these factors, this underutilized vegetable is also known as an antioxidant vegetable. Carotenoids safeguard against age-related eye ailments and selectively filter light wavelengths, enhancing visual performance and shielding photoreceptors from light-induced harm (Lem et al., 2021). Likewise, lutein and zeaxanthin can inactivate free radicals and oxygen triplicates caused by light-induced cellular activity (Fuad et al., 2020). Chlorophyl, a natural pigment in this species, acts as a natural antioxidant, with the ability to scavenge free radicals and prevent several oxidative stress–related diseases, such as cancer, neurological disorders, inflammatory diseases, dermatitis, tissue damage, sepsis, cardiovascular disorders, decreased immune function, and aging (Sangija & Wu, 2020). In addition, chlorophyl can inhibit dehydration, which is the primary source of kidney stones (Jebir et al., 2023). A lot of biological activities are attributed to the vitamins present in the leaves. B vitamins function as coenzymes and help in energy production from carbohydrates and synthesis of neurotransmitters, fatty acids, and hormones (Tural & Tuzcu, 2023). While vitamins C and E act as natural antioxidants, vitamin C prevents scurvy, reduces gastric cancer, and stabilizes folate in food and plasma. Likewise, the minerals carry out various biological roles. For example, calcium contributes to bone development, while its deficiency leads to osteoporosis. Zinc contributes to the improvement of the immune system and the repairing of cell and organ integrity. Deficiency of zinc in humans results in growth retardation, bone maturation, delayed sexual maturity, diarrhea, skin lesions, reduced appetite, and increased vulnerability to infections (Prasad, 2020). These nutrients play a role in the biological function of the plant's leaves. Medicinally, the plant is utilized in managing various ailments, including pneumonia, toothache, stomachache, tonsillitis, ringworms, pain, inflammation, fever, and tumors (Ogundola et al., 2022). In addition, it serves as a tonic, antioxidant, anti-inflammatory, hepatoprotective, diuretic, and antipyretic. Hence, incorporating these leaves into one's diet provides an excellent nutritional supplement for maintaining a healthy body system.

3.3.3 *Dialium guineense* Willd.

Dialium guineense (velvet/black tamarind) belongs to the family Fabaceae. It is a woody plant that grows up to 15 m high in the rainforest region of West Africa (Ogbuewu et al., 2023). It has a dense leafy crown and smooth grayish bark. The leaves are hairy, and the flowers are usually whitish, while the fruits are less circular and flattened. The small black velvet fruits are very conspicuous and distinctive. The seed is embedded in a dry, brownish, sweet, and sour edible pulp (Ogbuewu et al., 2023). The fruit pulp, which is yellowish red, has a sweet-sour, astringent flavor, like baobab, but sweeter, and it is eaten raw by both humans and animals. *Dialium guineense* can be found in West African countries, such as Ghana, Sierra Leone, Senegal, Nigeria, Benin, Burkina Faso, and Cameroon (Orwa et al., 2009). It can also be found in the Central African Republic, Chad, Cote d'Ivoire, Equatorial Guinea, Guinea, Guinea-Bissau, Liberia, Mali, Niger, Sao Tome and Principe, Sudan, and Togo (Sileshi et al., 2023). The thirst-quenching, refreshing fruit pulp can also be soaked in water and taken as a beverage. Likewise, the fruit pulp is ground by the Hausa people of Nigeria and used as a condiment in drinking corn flour/pap. About 100 g of the dry fruit pulp is found to contain 4.0 g moisture, 8.3 g crude protein, 4.9 g crude lipids, 3.2 g ash, 0.6 g crude fiber, 79.6 g carbohydrates and yields 289.0 kcal of energy. The pulp has been found with high potassium (260 mg), sodium (390 mg), magnesium (300 mg), and iron (4.1 mg) content and low calcium (47.0 mg) and phosphorus (42.9 mg) per 100 g pulp, respectively. Since the leaves are bitter, they may be used to cook "Domoda," a Ghanaian dish with both a sweet and bitter taste (Sileshi et al., 2023). The tree is a source of firewood and charcoal (Sileshi et al., 2023). The bark and leaves of *D. guineense* possess medicinal properties and are used in treating several diseases. The fruits of the plant are traditionally chewed among some women in southeast Nigeria to improve lactation (Jiménez-González & Guerrero-Beltrán, 2021; Onah et al., 2022). *D. Guineans* is used as a chewing stick (indigenous toothbrush) among the Nigerian and Ghanaian populace (Onah et al., 2022). Also, the stem bark as well as the leaves are used as folkloric remedies for the treatment of infections, such as diarrhea, severe cough, bronchitis, wounds, stomachaches, malaria fever, jaundice, ulcers, and hemorrhoids (Onah et al., 2022). The fruit pulp of the plant is claimed to possess antiulcerogenic activity (Oyegoke et al., 2022). Some of the scientifically validated activities of the plant leaves, stem bark, and fruit pulp include their analgesic and antibacterial (Orji et al., 2012) and antioxidant and antimicrobial activities (Valentin et al., 2020). The acclaimed antiulcerogenic potential of the fruit pulp of *Dialium guineense* and its safety were validated by Oyegoke and Oladiji (2014). Likewise, the high concentration of some mineral micronutrients suggests that *D. guineense* fruits may potentially be used as a food source or fortifying ingredient for locally processed foods to reduce deficiencies in selected commonly deficient nutrients for at-risk populations.

3.3.4 *Vigna subterranea* (L.) Verdc

Vigna subterranea (Bambara groundnut) is a legume indigenous to Africa and is cultivated across the semi-arid sub-Saharan African region (Mateva et al., 2023). It is a hardy crop and has been recognized as an essential nutritious food source when food is scarce (Mbosso et al., 2020). This could be attributed to its climate-smart characteristics, including its ability to fix nitrogen and grow under adverse environmental conditions, such as poor soils and drought (Bitire et al., 2023). This nutrient-dense legume is sometimes termed a "complete food" due to its balanced macronutrient composition. Bambara groundnut contains about 64.4% carbohydrate, 23.6% protein, 6.5% fat, and

5.5% fiber and is rich in many minerals (Tan et al., 2020). It is relatively underutilized compared with major cash crops and has often been associated with small-scale, subsistence farming, with women being the major producers and processors (Mbosso et al., 2020).

3.3.5 *Xanthosoma sagittifolium* (L.) Schott

Xanthosoma sagittifolium (cocoyam) is among the world's six most important root and tuber crops (Tan et al., 2020). It is pantropical and has been domesticated in most communities in Oceania, Africa, and Asia (Okafor, 2023). Despite its high productivity levels and better storability compared to other tropical root and tuber crops (Adiele, 2023), it remains an underexploited food resource (Onyeka, 2014). Cocoyam is postulated to have superior nutritional value over other major root and tuber staples of West Africa, especially in terms of protein digestibility and mineral composition (calcium, phosphorous, and magnesium) (Ukwu et al., 2022).

3.3.6 *Dioscorea dumetorum* (Kunth) Pax

Dioscorea dumetorum (bitter yam/cluster yam/trifoliate yam/three-leaved yam) is a species of flowering plants in the yam family Dioscorea. It is native to sub-Saharan Africa and especially common in the tropical regions of West Africa, including Nigeria, Benin, and Ghana. Bitter yam is an underutilized yam species in many parts of Africa, including Nigeria. It may be cultivated or may grow in the wild (Abiodun & Akinoso, 2015) and comes in different colors, such as pale yellow, deep yellow, or white (Oyeyinka et al., 2018). The underutilization of bitter yam has been partly attributed to its bitter taste and hardening of the tubers that occurs during storage (Oyeyinka et al., 2018).

3.3.7 *Dioscoreophyllum cumminsii* (Stapf) Diels

Dioscoreophyllum cumminsii (serendipity berry) is a tropical rainforest vine. It belongs to the order Ranunculales and the family Menispermaceae. It is indigenous to a vast portion of tropical Africa, stretching from Sierra Leone in the west to Eritrea in the east, and from Angola and Mozambique in the south, and thrives at low elevations, ranging from sea level up to 400 m (Ojo et al., 2020).

3.4 PROCESSING AND NUTRITIONAL VALUE OF THE FOOD PLANTS

3.4.1 Vegetables

Vegetables like amaranth, Malabar spinach, and moringa are often cooked using various methods, such as boiling, steaming, sautéing, and stir-frying.

3.4.2 Fruits

Undervalued fruits like black sapole, jackfruit, tamarind, and pawpaw are often consumed fresh, providing a robust source of vitamins and antioxidants (Jiménez-González & Guerrero-Beltrán, 2021; Ayilara et al., 2022). It has been reported that some women chew the fruits of the tamarind to improve lactation and check genital infection in Nigeria (Jiménez-González & Guerrero-Beltrán, 2021). According to Sanusi et al. (2022), the pulp is edible and sweet, with low levels of ascorbic acid and tannins. The fruit has been ranked as a good source of plant protein and minerals (Jiménez-González & Guerrero-Beltrán, 2021; Sanusi et al., 2022). Snacks from velvet tamarind are majorly made with maize and sugar and cooked through steaming. Velvet tamarind pulp can be obtained and milled along with the maize to produce a snack. Maize is cleaned, cooked, and washed to remove the pericarp and then steeped and milled with the velvet tamarind fruit. The slurry is then cooked. Tamarind can be processed into various products, such as tamarind paste and juice, with culinary and health applications.

3.4.3 Tubers

Tubers like yacon, oca, and sunchokes are cooked through methods like boiling, roasting, and frying, and they can be preserved through drying or canning (Adeyanju et al., 2019). Cocoyam is postulated to possess superior nutritional value over other major root and tuber staples of West Africa, especially in terms of their protein digestibility and mineral compositions (calcium, phosphorous, and magnesium) (Ukwu et al., 2022). Snack from cocoyam is majorly consumed in the eastern parts of Nigeria. Ingredients needed include cocoyam, salt, seasoning, pepper, and vegetable oil. The cocoyam is peeled, washed, and cut into smaller pieces for easy crushing; the smaller pieces are then blended into a smooth paste, with the addition of water, salt, seasoning, and desired spices. The mixture is adequately mixed for even distribution of all ingredients, and then it is scooped into hot vegetable oil to fry.

3.4.3.1 Traditional Dishes

Traditional dishes incorporating tubers are often found in local culinary traditions and recipes specific to the regions where they are grown (Ebert et al., 2011). Bitter yam is an underutilized yam species in many parts of Africa. Fish roll is a snack from bitter yam for consumption. Ingredients include yam, fish, salt, bouillon cubes, pepper, flour for dusting the board, and oil for frying.

3.4.4 Grains

Grains like teff, fonio, and amaranth are often cooked by boiling or steaming or used in porridges and flatbreads (Nkwonta et al., 2023). These grains can be incorporated into a wide variety of dishes, including soups, stews, and grain-based dishes (Moreno et al., 2014).

3.4.5 Legumes

Legumes such as Bambara groundnut, winged bean, and lentil require soaking and cooking to make them edible and digestible. Legumes can be paired with complementary foods to improve overall nutritional value. Undervalued plants have significant nutritional and medicinal potential because they offer an opportunity to address malnutrition, contribute to health security, and preserve traditional knowledge. The synergy between these plants and traditional medicine practices can lead to more holistic and culturally sensitive health-care solutions, promoting the well-being of individuals and communities.

Pigeon pea can be used in the production of different foods and products, including snacks. Examples of snacks that can be produced from pigeon peas include cheula. *Cheula* is a crispy and crunchy snack produced from legumes because of their high fat content. It is produced by deep-frying and can be consumed by both children and adults, like "kulikuli," a similar snack produced mainly from grinding peanuts into a paste, spicing, and deep-frying. It is majorly consumed in southeastern Africa. Ingredients used include dried pigeon peas, maize flour, salt, curry, paprika, and desired spices.

3.5 SIGNIFICANCE OF UNDERVALUED FOOD PLANTS

Undervalued plants often possess diverse nutrient profiles containing essential vitamins, minerals, and macronutrients. Some are rich sources of protein, fiber, and healthy fats, while others provide valuable micronutrients, like vitamins A and C and iron. In addition, they can complement staple crops by providing nutrients lacking in conventional diets. For example, certain leafy greens and legumes can be vital sources of iron and other micronutrients. Some undervalued plants contain more bioavailable nutrients than those in common crops. This means the body can absorb and utilize these nutrients more effectively, enhancing their nutritional impact. These plants also possess the potential to combat malnutrition by diversifying diets and improving the overall nutritional status of communities, particularly in regions where access to a variety of foods is limited. Many undervalued plants are known to contain bioactive compounds with potential health benefits. These compounds may have anti-inflammatory, antioxidant, and antimicrobial properties. Also, undervalued plants have long been used in traditional medicine systems worldwide. Their medicinal properties have been employed to treat various ailments, from digestive disorders to skin conditions. Some undervalued plants have attracted the pharmaceutical industry's attention due to their bioactive compounds. They may serve as sources for developing new drugs and therapies. Undervalued plants can contribute to health security by providing alternative and natural remedies, reducing reliance on synthetic pharmaceuticals, and potentially addressing emerging health challenges. Additionally, undervalued plants play a crucial role in preserving traditional knowledge and practices related to health care. They are often deeply embedded in the cultural and healing traditions of Indigenous communities. Many traditional medicine systems emphasize a holistic approach to health, considering physical, mental, and spiritual well-being. Undervalued plants are integral to this approach, providing comprehensive health solutions. There is increasing recognition of the value of integrating traditional medicine and undervalued plants into modern health-care systems. This can lead to more accessible and culturally relevant health-care options. Promoting the use of undervalued plants in traditional medicine can empower local communities economically and socially. It can also help protect the rights and knowledge of Indigenous healers and practitioners.

3.6 CONCLUSION AND PROSPECTS

This chapter has provided insights into undervalued food plants, their classes, processing, nutritional status, and health importance. There is a need to expand the cultivation of undervalued food plants due to the plants' nutritional and climatic significance. It is hoped that this will create awareness among consumers and enhancement in production will be encouraged by several means to achieve food, nutrition, and health security in the future.

3.7 ACKNOWLEDGMENTS

We acknowledge the University of Ilorin, Nigeria, for providing an enabling environment and facilities to write this book chapter.

REFERENCES

Abebe, K. B. (2022). The dietary use of pigeon peas for human and animal diets. *The Scientific World Journal*, 1–12. https://doi.org/10.1155/2022/4873008

Abiodun, O. A., & Akinoso, R. (2015). Textural and sensory properties of trifoliate yam ioscorea dumetorum) flour and stiff dough 'amala'. *Journal of Food Science and Technology*, *52*(5), 2894–2901. https://doi.org/10.1007/s13197-014-1313-y

Adeyanju, J. A., Babarinde, G. O., Abioye, A. O., Olajire, A. S., & Bolarinwa, I. D. (2019). Cocoyam processing: Food uses and industrial benefits. *International Journal of Scientific & Engineering Research*, *10*(9), 156–161.

Adiele, J. G. (2023). Unveiling cocoyam potentials: Physiology and agronomy. *Nigerian Agricultural Journal*, *54*, 432–440.

Agu, C. M., & Agulanna, A. C. (2020). Kinetics and thermodynamics of extract from amaranth. In V. Y. Waisundara (Ed.), *Nutritional value of Amaranth* (pp. 147–165). http://creativecommon.org

Ajiloba, C. F., Olanrewaju, O. S., & Babalola O. O. (2022). Plant growth stages drive. *Journal of Fungi*, *6*, 126.

Alirea, T. (2021). Solanm nigrum and solanum physalifolium. In B. Chauhan (Ed.), *Biology and management of problematic crop weed species* (1st ed., pp. 357–373). Elsevier.

Association of Agricultural Research Institutions in the Near East and North Africa (AARINENA); The Asia-Pacific Association of Agricultural Research Institutions (APAARI); Crops for the Future (CFF); The Forum for Agricultural Research in Africa (FARA); The Global Forum for Agricultural Research and Innovation (GFAR); and The Alliance of Biodiversity International and CIAT. (2021). *Global manifesto on forgotten foods*. Retrieved November 1, 2022, from https://aarinena.org/wp-content/uploads/2021/08/Global-Manifesto-on-Forgotten-Foods-Final_0.pdf

Auma, I. C., Pradeilles, R., Blake, M., & Holdsworth, M. (2019). What can dietary patterns tell us about the nutrition transition and environmental sustainability of diets in Uganda? *Nutrients*, *11*, 342. https://doi.org/10.3390/nu11020342

Ayilara, M. S., Abberton, M., Oyatomi, O. A., Odeyemi, O., & Babalola, O. O. (2022). Potentials of underutilized legumes in food security. *Journal of Frontiers in Soil Science*, *2*, 1–12. https://doi.org/10.3389/fsoil.2022.1020193

Bitire, T. D., Abberton, M., Tella, E. O., Edemodu, A., Oyatomi, O., & Babalola, O. O. (2023). Impact of nitrogen-fixation bacteria on nitrogen-fixation efficiency of Bambara groundnut [Vigna subterranea (L) Verdc] genotypes. *Frontiers in Microbiology*, *1*, 15. https://doi.org/10.3389/fmicb.2023.1187250

Boakye, A. A., WireKo-Manu, F. D., Oduro, I., Ellis, W. O., Gudjonsdottir, M., & Chronakis, L. S. (2018). Utilizing Cocoyam (Xanthosoma sagittifolium) for food and nutrition security: A review. *Food Science & Nutrition*, *6*(4), 703–713.

Brink, M., Ramolemana, G. M., & Sibuga, K. P. (2006). *Vigna Subterrana* (L.) Verdc. record from protabase. In M. Brink & G. Belay (Eds.), *PROTA (Plant Resources of Tropical Africa/Resources vegetable de l' Afrique tropicale)*. PROTA. www.prota4u.org/database/protav8.asp?h=M11%2CM12%2CM25%2CM26%2CM27%2CM28

Bvenura, C., & Afolayan, A. J. (2014). Mineral uptake in Solanum nigrum L. cultivated on fertiliser amended soils of the Eastern Cape, South Africa. *Journal of Applied Botany and Food Quality*, *87*, 220–226.

Bvenura, C., & Sivakumar, D. (2017). The role of traditional food processing technologies in national food security. *Food Research International*, *104*, 174–194.

Caroline, R. H., Armaghan, S., Lois, B., Joshua, R. L., Jonathan, M. H., Harvey, M. A., & Lauren, C. B. (2023). Sulfur compounds: From plants to humans and their role in chronic disease prevention. *Critical Reviews in Food Science and Nutrition*, *63*(27), 8616–8638. https://doi.org/10.1080/10408398.2022.2057915

Diazgranados, M., Allkin, B., Black, N., Cámara-Leret, R., Canteiro, C., Carretero, J., & Ulian, T. (2020). *World checklist of useful plant species. Produced by the Royal Botanic Gardens, Kew.* https://doi.org/10.5063/F1CV4G34

Ebert, A. W., Wu, T., & Wang, S. (2011). *Vegetable amaranth (AmaranthusL.): AVRDC* (Publication no. 11–754). AVRDC – The World Vegetable Center.

Emesi, I. C., Zang, C. U., & Adeyanju, O. (2019). Analysis of Manihot esculentus Crantz tubers. A potential biomaterial in food processing. *Nigeria Journal of Current Research in Chemistry and Pharmaceutical Sciences*, *6*, 19–22.

FAO. (2021). Food security and nutrition. In *World food and agriculture organization of the United Nations-statistical yearbook* (pp. 248–293). Author.

Fatokimi, E. O., & Tanimonure, V. A. (2021). Analysis of the current situation and future outlooks for pigeon peas (*Cajanus Cajan*) production in Oyo State, Nigeria: A markov chain model approach. *Journal of Agriculture and Food Research*, *6*, 100218. (nth (*Amaranthus*.). AVRDC Publication no. 11–754. Tainan, Taiwan: AVRDC – The World Vegetable *Center Physalifolium J. Biology and Management of Problematic Crop Weed Species*, Elsevier 357–373).

Fuad, N. I. N., Sekar, M., Gan, S. H., Lum, P. T., Vaijanathappa, J., & Ravi, S. (2020). Lutein: A comprehensive review on its chemical, biological activities and therapeutic potentials. *Pharmacognosy Journal*, *12*, 1769–1778.

Ghanari, A., & Arianmehr, M. (2021). Spatial distribution of Amaranthus retroflexus and Solanum nigrum under different weed control methods. *Applied Field Crops Research*, *33*(4), 128–145.

Gichohi-Wainaina, W. N., Mremi, R., Chande, M., Msuya, J. M., Kumwenda, N. C., Muzanila, Y. C., . . . Okori, P. (2022). Drivers of pigeon peas consumption among school-aged children in Central Tanzania. *Frontiers in Sustainable Food Systems*, *5*, 726404. https://doi.org/10.3389/fsufs.2021.726404

Hussain, S., Jõudu, I., & Bhat, R. (2020). Dietary fibre from underutilized plant resources–A positive approach for valorization of fruit and vegetable wastes. *Sustainability*, *12*(13), 5401. https://doi.org/10.3390/su12135401

Jahanshiri, E., Mohd Nizar, N. M., Tengku, M., Suhairi, T. A. S., Gregory, P. J., Mohamed, A. S., Wimalasiri, E. M., & Azam-Ali, S. N. (2020). A land evaluation framework for agricultural diversification. *Sustainability*, *12*, 3110.

Jebir, R. M., & Mustafa, Y. F. (2023). Kidney stones: Natural remedies and lifestyle modifications to alleviate their burden. *International Urology and Nephrology*, *56*, 1025–1033. https://doi.org/10.1007/s11255-023-03764-1

Jiménez-González, O., & Guerrero-Beltrán, J. A. (2021). *Diospyros digyna* (black sapote), an undervalued fruit: A review. *ACS Food Science & Technology Journal*, *1*(1), 3–11.

Khan, M. M. H., Rafii, M. Y., Ramlee, S. I., Jushoh, M., & Mamun, A. (2020). Genetic variability, heritability clustering pattern exploration of Bambara groundnut (Vigna subterranean L. Verdc) accessions for the perfection of yield and yield-related traits. *BioMed Research International*, 2195797. https://doi.org/10.1155/2020/2195797

Knez, M., Ranić, M., & Gurinović, M. (2023). Underutilized plants increase biodiversity, improve food and nutrition security, reduce malnutrition, and enhance human health and well-being. Let's put them back on the plate. *Nutritional Reveiw*. https://doi.org/10.1093/nutrit/nuad103

Kuraz, B. (2022). The dietary use of pigeon peas for human and animal diets. *The Scientific World Journal*, 4873008. https://doi.org/10.1155/2022/4873008

Lem, D. W., Davey, P. G., Gierhart, D. L., & Rosen, R. B. (2021). A systematic review of carotenoids in the management of age-related macular degeneration. *Antioxidants*, *10*(8), 1255.

Locali-Pereira, A. R., Boire, A., Berton-Carabin, C., Taboga, S. R., & Solé-Jamault, V. R. (2023). Pigeon peas, an emerging source of plant-based proteins. *ACS Food Science & Technology*, *3*(11), 1777–1799.

Mateva, K. I., Tan, X. L., Halimi, R. A., Chai, H. H., Makonya, G. M., Gao, X., & Massawe, F. (2023). Bambara groundnut (Vigna subterranea (L.) Verdc.). In *Neglected and underutilized crops* (pp. 557–615). Academic Press.

Mba, C. J., & Agu, H. O. (2021). Developments on the bioactive components and food uses of the tubers *Colocasia esculenta* (L) Schott (Taro) and *Xanthosoma Sagittifolium* (L) Schott (Tanna). *Asian Journal of Food Science Journal*, *20*(1), 101–112.

Mbosso, C., Boulay, B., Padulosi, S., Meldrum, G., Mohamadou, Y., Niang, A., Coulibaly, H., Koreissi, Y., & Sidibé, A. (2020). Fonio and bambara groundnut value chains in mali: Issues, needs, and opportunities for their sustainable promotion. *Sustainability*, *2*(11), 4766.

Meyer, R. S., Duval, A. E., & Jensen, H. R. (2012). Patterns and processes in crop domestication: An historical review and quantitative analysis of 203 global food crops. *New Phytologist*, *196*(1), 29–48. https://doi.org/10.1111/j.1469-8137.2012.04253

Moreno, M. L., Comino, I., & Sousa, C. (2014). Alternative grains as potential raw material for gluten–free food development in the diet of celiac and gluten–sensitive patients. *Austin Journal of Nutrition and Food Sciences*, *2*(3), 1016. (ISSN: 2381–8980).

Nkwonta, C. G., Auma, C. I., & Gong, Y. (2023). Underutilised food crops for improving food security and nutrition health in Nigeria and Uganda. *Frontiers in Sustainable Food Systems*, 1–17. https://doi.org/10.3389/fsufs.2023.1126020

Nong, S. S. (2022). *The response of selected soil health variables to rainfed and irrigated maize-legume intercropping systems* [Theses and dissertations, Agriculture (Soil Science), University of Limpopo]. University of Limpopo. http://hdl.handle.net/10386/3914

Odongo, G., Schlotz, N., Baldermann, S., Neugart, S., Huyskens-Keil, S., Ngwene, B., & Lamy, E. (2018). African nightshade (*Solanum scabrum* Mill.): Impact of cultivation and plant processing on its health promoting potential as determined in a human liver cell model. *Nutrients*, *10*(10), 1532. https://doi.org/10.3390/nu10101532

Ogbuewu, I. P., Modisaojang-Mojanaga, M. M. C., Mokolopi, B. G., & Mbajiorgu, C. A. (2023). Nutritional and chemical composition of black velvet tamarind (Dialium guineense Wild) and its influence on animal production: A review. *Open Agriculture*, *8*(1), 20220174.

Ogundola, A. F., Bvenura, C., Ehigie, A. F., & Afolayan, A. J. (2022). Effects of soil types on phytochemical constituents and antioxidant properties of *Solanum nigrum*. *South African Journal of Botany*, *151*, 325–333.

Ojo, A., Enujiugha, V. N., Omogie, H. N. A., & Abiodun, O. A. (2020). Chemical composition of serendipity Berry (*Dioscoreophyllum cumminsii*) and miracle fruit (*Thaumatococcus daniellii*). *Asian Food Science Journal*, *19*(4), 10–15.

Okafor, V. N. (2023). Effective data and information management: The role of research librarians. *Centennal*, *133*.

Onah, G. T., Ajaegbu, E. E., & Enweani, I. B. (2022). Proximate, phytochemical and micronutrient compositions of Dialium guineense and Napoleona imperialis plant parts. *GSC Biological and Pharmaceutical Sciences*, *18*(3), 193–205.

Onyeka, J. (2014). *Status of Cocoyam (Colocasia esculenta and Xanthosoma spp.) in West and Central Africa: Production, household importance and the threat from leaf blight.* CGIAR.

Orji, J. O., Alo, M. N., Anyim, C., & Okonkwo, E. C. (2012). Antibacterial activities of crude leaf and bark extracts of "icheku" Dialium guineense on bacterial isolates from bronchitis patients. *Journal Pharmacy Biological Sciences*, *1*, 22–25.

Orwa, C., Mutua, A., Kindt, R., Jamnadass, R., & Anthony, S. (2009). Agroforestry database: A tree reference and selection guide version. *Scientific Research*, *4*, 1–5.

Oselebe, H. O., & Nwankiti, O. C. (2007). Cytology of root tips of Dioscoreophyllum cumminsii (STAPF) diel. *Agro-Science*, *4*(1). https://doi.org/10.4314/as.v4i1.1521

Oyegoke, R. A., & Oladiji, A. T. (2014). Anti-ulcerogenic activity of *Dialium guineense* Fruit Pulp Meal-based Diet in Aspirin-induced Ulcerogenic Rats. *Nigerian Journal of Biochemistry and Molecular Biology*, *29*(2), 77–93.

Oyegoke, R. A., & Oladiji, A. T. (2022). Safety evaluation of Dialium guineense fresh fruit pulp meal-based diet. *Biokemistri*, *33*(1).

Oyeyinka, S. A., Adeleke, O. F., Dauda, A. O., Abiodun, O. A., Kayode, R. M. O., & Adejuyitan, J. A. (2018). Flour composition and physicochemical properties of white and yellow bitter yam (*Dioscorea dumetorum*) starches. *Elsevier, 120*, 135–139.

Prasad, A. S. (2020). Lessons learned from experimental human model of zinc deficiency. *Journal of Immunology Research, 2020*, Article ID 9207279, 12 pages. https://doi.org/10.1155/2020/9207279

Reynolds, A. N., Akerman, A. P., & Mann, J. (2020). Dietary fibre and whole grains in diabetes management: Systematic review and meta-analyses. *PLoS Medicine, 17*, e1003053. https://doi.org/10.1371/journal.pmed.1003053

Sangija, F., & Martin, H. A. (2021). African nightshades (*Solanum nigrum* complex): The potential contribution to human nutrition and livelihoods in Sub-Saharan Africa. *Comprehensive Reviews in Food Science and Food Safety, 20*(4), 3284–3318. https://doi.org/10.1111/1541-4337.12756

Sangija, F., & Wu, W. (2020). Bamboo wine: Its production technology and potential as a sustainable health beverage. *Food Reviews International*. https://doi.org/10.1080/87559129.2020.1810699

Sanusi, B. A., Salaudeen, S., Olanite, W. A., Ariwoola, L. A., & Azeez, M. O. (2022). Determination of some selected engineering properties of tamarind (Tamarindus Indica) seed. *African Scholar Journal of Biotechnology and Agricultural Research (JBAR-1), 26*(1), 155–162.

Sarkar, S., Panda, S., Yadav, K. K., & Kandasamy, P. (2020). Pigeon peas (CajanusCajan)an important food legume in indian scenario – A review. *Legume Research, 43*(5), 601–610.

Saxena, K. B., Kumar, R. V., & Rao, P. V. (2002). Pigeon peas nutrition and its improvement *Journal of Crop Production, 5*, 227–260.

Sidiq, F. F., Coles, D., Hubbard, C., Clark, B., & Frewer, L. J. (2022). The role of traditional diets in promoting food security for indigenous peoples in low- and middle-income countries: A systematic review. (The 3rd international conference on agriculture and rural). *Development IOP Conf. Series: Earth and Environmental Science, 978*, 012001. (IOP Publishing). https://doi.org/10.1088/1755-1315/978/1/012001

Sileshi, G. W., Dagar, J. C., Akinnifesi, F. K., & Mng'omba, S. A. (2023). Potentials of indigenous fruit trees in enhancing nutrition, income and biodiversity conservation in African agroforestry. In *Agroforestry for sustainable intensification of agriculture in Asia and Africa* (pp. 321–361). Springer Nature.

Singh, M., Manickavasagan, A., Shobana, S., & Mohan, V. (2021). Glycaemic index of pulses and pulse-based products: A review. *Critical Reviews in Food Science and Nutrition, 61*, 1567–1588. https://doi.org/10.1080/10408398.2020.1762162

Talabi, A. O., Vikram, P., Thushar, S., Rahman, H., Ahmadzai, H., Nhamo, N., . . . Singh, R. K. (2022). Orphan crops: A best fit for dietary enrichment and diversification in highly deteriorated marginal environments. *Frontiers in Plant Science, 13*, 839704. https://doi.org/10.3389/fpls.2022.839704

Tan, X. L., Azam-Ali, S., Goh, E. V., Mustafa, M., Chai, H. H., Ho, W. K., & Massawe, F. (2020). Bambara groundnut: An underutilized leguminous crop for global food security and nutrition. *Frontiers in Nutrition, 7*, 601496.

Ukwu, C. P., Yahaya, A., Okere, P. C., Aladi, N. O., Odoemenam, V. U., Obikaonu, H. O., Uchegbu, M. C., & Okoli, I. C. (2022). The production, uses, nutritional and anti-nutritional characteristics of cocoyam as a potential feed ingredient in the tropics: A review. *Nigerian Journal of Animal Science, 24*(3), 91–111.

Ulian, T., Diazgranados, M., Pironon, S., Padulosi, S., Liu, U., Davies, L., . . . Mattana, E. (2020). Unlocking plant resources to support food security and promote sustainable agriculture. *Plants, People, Planet, 2*, 421–445. https://doi.org/10.1002/ppp3.10145

Upadhyaya, H. D., Pundir, R. P. S., Gowda, C. L. L., Reddy, K. N., & Singh, S. (2005). Geographical pattern of diversity for qualitative and quantitative traits in the pigeon peas germplasm collection. *Plant Genetic Resources, 3*, 332–352. https://doi.org/10.1079/PGR200587

Valentin, B. C., Amuri, B. S., Philippe, O. N., & Simbi, J. B. L. (2020). Antimicrobial, antioxidant activities and toxicity on Cavia porcellus of Dialium angolense Welw. Ex Oliv, a traditional medicinal plant from Bagira in Eastern of DR Congo. *GSC Biological and Pharmaceutical Sciences, 13*(2), 166–180.

Wang, Z., Zhao, S., Tao, S., Hou, G., Zhao, F., Tan, S., & Mang, Q. (2023). Dioscorea spp. Bioactive components and potential for the treatment of inflammation and metabolic diseases. *Molecules, 28*(6), 2878.

Wei, P., Zhao, F., Wang, Z., Wang, Q., Chai, X., Hou, G., & Meng Q. (2022). Sesame (*Sesamum indicum* L.): A comprehensive review of nutritional value, phytochemical composition, health benefits, development of food, and industrial applications. *Nutrients, 14*, 4079. https://doi.org/10.3390/nu14194079

Yao, Y., Goh, H. M., & Kim, J. E. (2021). The roles of Carotenoid consumption and bioavailaility in cardiovascular health. *Antioxidants, 10*(12), 1978. https://doi.org/10.3390/antiox1012198.

4 Optimizing Food and Nutrition Security

Exploring the Nutritional Potential of Finger Millet and Insect Protein for Enhanced Growth and Sensory Benefits in Broiler Chickens

*Peter Avwerosuo Ohwofasa, Callistus Bvenura, Oluwaseyi Moyode Aboyade, and *David Roy Katerere*
Department of Pharmaceutical Sciences, Tshwane University of Technology, 175 Nelson Mandela Drive, Private Bag X680, Pretoria 0001, South Africa
*Corresponding author: hwesa09@gmail.com

4.1 INTRODUCTION

Poultry meat is arguably the most important source of protein in Southern Africa (SAPA, 2013). The Food and Agricultural Organization projects that commercial poultry feed production will need to increase by 70% by 2050 to meet the growing world population and its demand for protein (FAO, 2006). Maize and soybean meal are the critical and commonly utilized ingredients in commercial poultry diets (FAO, 2006). However, their demand by both human and livestock has brought about an increase in their costs and ultimately impacted the cost of poultry feed (Bruinsma, 2003).

In South Africa, poultry feed costs have risen on average by 38%, with negative impacts on farmers, since between two-thirds and three-quarters of production costs are made up of feed (SAPA, 2013). Climate variability has also contributed to the rising costs of maize and soybean meal. The poultry industry is now confronted with viability challenges due to rising costs of conventional feed commodities which are imported and subject to unpredictable foreign exchange rates. These challenges have prompted the exploration of alternative protein and energy for poultry nutrition.

Several studies have shown that a few bowls of cereal are wealthy sources of protein and carbohydrate (vitality) due to their capacity to supply amino acids from a wide range of boundless and promptly accessible essential materials (D'Mello, 1995). A good example of such cereals is finger millet. This grain crop, also known as "rapoko" in Southern Africa, is an African native that originated in the highlands of Uganda and Ethiopia, where farmers have been growing it for thousands of years (Achaya, 2003). The annual world production of finger millet grain is 4.5 million tons, with Africa producing 2 million tons (Brink & Belay, 2006). The protein in millets is well balanced in essential amino acids that are appreciated for poultry diets. Lysine, methionine, and cystine levels in finger millet are about 2.86%, 1.75%, and 1.51% of the crude protein, respectively (Rachie & Peters, 1997). Compared to maize, finger millet has a higher carbohydrate (energy) content (Rachie &

DOI: 10.1201/9781003469766-4

Peters, 1997). Thus, the replacement of maize with finger millet in animal diets may reduce the dependence on maize and can also cut the costs related to poultry production.

Insects comprise the most extensive range of species worldwide. At all life stages, insects are rich sources of protein, essential amino acids, and fat (Makkar et al., 2014; Sánchez-Muros et al., 2014). Micro- and macronutrients present in insects can successfully be exploited for human consumption as well as constitute ingredients for animal feed (Zuidhof et al., 2003). In this study, we focus on emperor moth. The high protein content in insects has made them highly attractive for various livestock production systems and a possible alternative to the meat, fish, and soybean meal that currently comprise 60–70% of production costs (Henry et al., 2015; Veldkamp et al., 2012).

The poultry industry has expanded rapidly in developing countries in the last two decades. The use of insects in animal feed is one potential solution to improve the sustainability of animal diets. Grasshoppers, crickets, cockroaches, termites, lice, stink bugs, cicadas, aphids, scale insects, psyllids, beetles, caterpillars, flies, fleas, bees, wasps, and ants have all been used as complementary food sources for poultry (Ravindran & Blair, 1993). Several avian studies have demonstrated that termites serve as valuable sources of protein (Dao et al., 2020; Boafo et al., 2019; Abdou & Kyarimpa, 2023). Thus, insects are a low-cost, low-maintenance, high-protein poultry food source (Aganga et al., 2008).

We hypothesize that the complete or partial replacement of maize and soybean with finger millet, an undervalued African indigenous grain crop, and insect protein in chicken feed could help revitalize these underutilized indigenous resources in South Africa for human and animal food and nutrition security. Broiler chickens that are fed organic and healthy diets have the potential to provide organic, healthy, and nutritious meat and potentially positively impact sustainable food systems and better health outcomes.

Therefore, this study aimed to assess the effects on body weight, feed intake, and histomorphological features and taste of broiler chickens fed emperor moth larvae and finger millet.

4.2 MATERIALS AND METHODS

4.2.1 Ethics

All procedures used in this study were approved by the Animal Research Ethics Committee (AREC) of the Tshwane University of Technology and the Animal Ethics Committee (AEC) of the University of Pretoria, both in South Africa.

4.2.2 Chicken Housing and Environmental Conditions

In this study, 100 female day-old Cornish Plymouth Rocks were used. Birds were housed in 1.2 m × 3.7 m pens, with pine shavings as litter, in Pretoria East, South Africa. House temperature initially ranged between 27 and 32°C but was gradually reduced to 19°C over six weeks, as outlined by Daghir (2008). The chicks were given free access to water and diet of the different feed formulations. The chickens were monitored for their body weight gain (BWG) and feed intake (FI) for six weeks.

4.2.3 Experimental Diets

Three feeds were used in this study, as shown in Table 4.1. A commercial broiler diet based on maize meal and soybean meal was purchased from a local poultry feed supplier in Pretoria, South Africa, and this served as the control, or group 1. The second diet was a formulation composed of finger millet and insect protein (group 2), while the diet formulation with maize (YM) and insect protein was used in group 3. Formulation for feeds in groups 2 and 3 was done in three different phases, namely, starter, grower, and finisher phases.

TABLE 4.1
Nutrient Constituents of Ingredients

Nutrients %	Insect Protein	Finger Millet	Yellow Maize
Crude protein	55.33[a]	9.97[b]	8.73[b]
Carbohydrate	8.16[c]	75.2[a]	70.4[b]
Moisture	16.11[a]	12.2[b]	11.7[b]
Ash	9.77[a]	3.24[b]	1.24[c]
Crude fiber	12.6[a]	4.0	3.5
Mineral			
Calcium	15.1[b]	320.5[a]	1.0[c]
Iron	94.3[b]	114.9[a]	28.8[c]
Zinc	127.5[a]	33.9[b]	17.6[c]

Note: Different letters along the row represent significant differences at $p < 0.05$.

4.2.4 Proximate Analysis

Dried and ground samples of raw materials and finished formulations were subjected to proximate analysis to determine crude protein, crude fiber, moisture, and ash content, respectively, as described by Sundaram et al. (2001). Protein was analyzed by measuring nitrogen using the Kjeldahl method (BÜCHI, Auto-Kjeldahl K-370). Moisture was determined using a moisture analyzer. Fiber was determined using a fiber analyzer. The ash content was determined by incineration using a muffle furnace (Finetech, SA) at 600°C for 6 h.

4.2.5 Animal Performance

Health status and mortality were monitored daily throughout the experimental period. Live weight (LW) of the animals was recorded at an individual level at the beginning of the trial, on day 0 (baseline), and days 7, 14, 21, 28, 35, as well as day 42. Daily feed intake (DFI) was recorded daily. Feed conversion ratio (FCR) was determined for each growth period and for the overall experimental period. All measurements were made on a pen basis using a high-precision electronic scale (Sartorius Signum®). Dressed carcass weight was determined at the end of the study.

4.2.6 Histopathology

Six chickens from each group were subjected to histopathological examination at the Biomedical Research Centre, University of Pretoria. Samples of livers, kidneys, and hearts were collected and placed in a 10% buffered formalin solution for histopathological examination (Tursi et al., 2016). Organs were assessed for inflammatory and degenerative diseases.

4.2.7 Sensory Evaluation

Sensory profiling and evaluation on broiler breast meat were carried out according to ISO guidelines (ISO, 2007, 2012). Six experienced assessors from the Food Consulting Service Pty. participated in this study.

The samples of chicken breast meat (three each) from the three groups were refrigerated at 4°C between days 1 and 4 after slaughter until analyzed. Before cooking, samples were placed in color coded containers for identification and to evaluate the attributes of raw meat. The samples were then cooked and grilled at a preheated temperature of 170°C until an internal temperature of 80°C was reached. Samples were served individually to each assessor in hermetic plastic containers at 70 ± 5°C immediately after cooking.

4.2.8 Statistical Analysis

All quantitative data were recorded using Microsoft Excel. The performance data (dietary treatments) and feed intake of experimental diets and body weight gain were analyzed using one-way ANOVA and treated as significantly different at $p < 0.05$.

4.3 RESULTS

4.3.1 Proximate Analysis

The proximate compositions, including moisture, crude protein, crude fiber, fats, ash, and calcium content, of the raw materials are shown in Table 4.1. The percentage of crude protein recorded was higher in finger millet compared with yellow maize, although not statistically significant. Insect protein recorded the highest crude protein content. Crude fiber content was significantly higher in finger millet compared to yellow maize and insect protein. Similarly, carbohydrate and calcium contents of finger millet were higher compared to maize and insect protein. Moisture, crude fat, and ash were high in insect protein in comparison with finger millet and yellow maize, respectively.

The gross nutrient composition of the formulated diets used in the current study is presented in Table 4.2. Group 1 (MW/YM) recorded a higher crude protein content, followed by group 2 (FM/MW), and lastly, group 3 (control) in all the three phases (starter, grower, and finisher) of the feed formulation. Carbohydrate and energy value contents were higher in MW/YM in the starter phase compared to the control and YM/MW diets, and this was maintained up to the grower and finisher phases.

TABLE 4.2
Gross Nutrient Constituents of Formulated Diets

Nutrients %	Starter			Grower			Finisher		
	Control	MW/FM	MW/YM	Control	MW/FM	MW/YM	Control	MW/FM	MW/YM
Crude protein	18.99^{b}	24.07^{a}	26.48^{a}	17.02^{b}	20.89^{a}	26.03^{a}	17.0^{c}	19.03^{b}	23.69^{a}
Moisture	10.03^{a}	10.2^{a}	8.8^{b}	10.6	10.3	9.9	10.0	10.7	10.0
Ash	3.97^{b}	5.72^{a}	5.09^{a}	5.69	5.85	6.66	5.68	4.93	5.87
Crude fat	4.3^{b}	5.9^{b}	8.1^{a}	3.4^{b}	4.7^{b}	7.1^{a}	2.5^{c}	4.3^{b}	7.0^{a}
Crude fiber	4.2	5.4	4.2	6.2^{a}	4.8^{b}	4.0^{b}	7.0^{a}	4.1^{b}	3.8^{b}
Carbohydrate	58.2^{a}	48.7^{b}	47.3^{b}	57.1^{a}	53.5^{a}	46.3^{b}	54.5^{a}	57.0^{a}	49.5^{b}
Mineral									
Energy kcal/kg	1,414^{b}	1,406^{b}	1,508^{a}	1,330^{b}	1,387^{b}	1,445^{a}	1,350^{b}	1,393^{b}	1,456^{a}
Iron	279.54^{b}	415.76^{a}	452.66^{a}	395.16^{b}	502.50^{a}	362.04^{b}	387.2	384.40	376.44
Zinc	93.48^{c}	130.30^{b}	137.26^{a}	82.26^{c}	110.52^{b}	209.84^{a}	92.34^{c}	108.14^{b}	126.10^{a}
Sodium	141.57^{b}	203.24^{a}	215.38^{a}	172.54^{c}	182.00^{b}	868.60^{a}	173.83^{b}	106.96^{c}	637.68^{a}

Note: Different letters along the row per growth phase represent significant differences at $p < 0.05$.
Premix used in groups 1 and 2 feed contained vitamin A, 12,000,000 IU; vitamin D_3, 2,400,000 IU; vitamin E, 25,000,000 IU; vitamin K_3, 2.5 g; vitamin B1, 4.0 g; vitamin B_2, 6.5 g; niacin, 42.0 g; calcium, 13.0 g; vitamin B_{12}, 0.03 g; vitamin B_6, 5.0 g; choline, 303.029 g; folic acid, 1.2 g; ascorbic acid, 0.12 g; and zinc, 22.5 g.

4.3.2 Daily Feed Intake

There were significant differences ($p < 0.05$) across all the groups for daily feed intake and feed conversion ratios, as shown in Figure 4.1. Daily feed intake was higher ($p < 0.05$) for chickens fed with MW/YM diets (group 3) and chickens fed with maize diets (control and group 1) compared to those fed with finger millet–based diets (group 2). Feed conversion ratios of chickens that were fed the finger millet diets (group 2) were better ($p < 0.05$) than those of chickens that were fed the maize meal diets (control) and insect meal diet (group 3), respectively. This means that chickens that were fed the finger millet diet ate little but maintained a steady growth rate and body weight gain in comparison with chickens on the commercial diet (control) that had a higher feed intake.

4.3.3 Growth Performance and Body Weight Gain

Figure 4.2 shows the growth performance and body weight gain (BWG) of chickens that were fed the different formulated diets, including the commercial feed or control. There was a difference

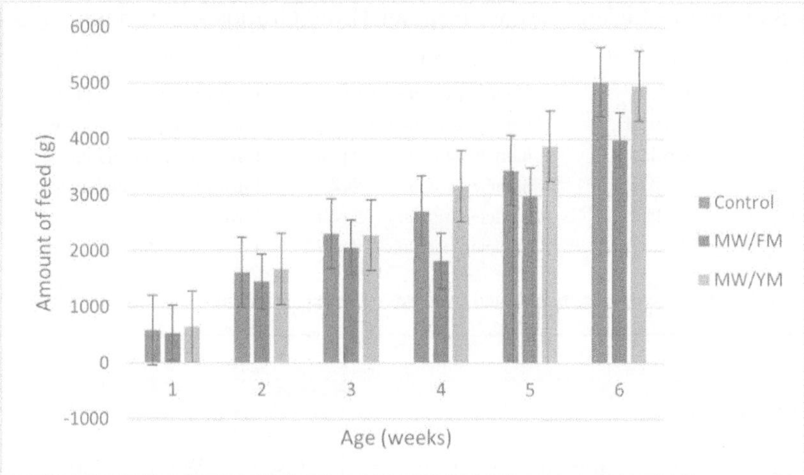

FIGURE 4.1 Weekly feed intake of the chickens.

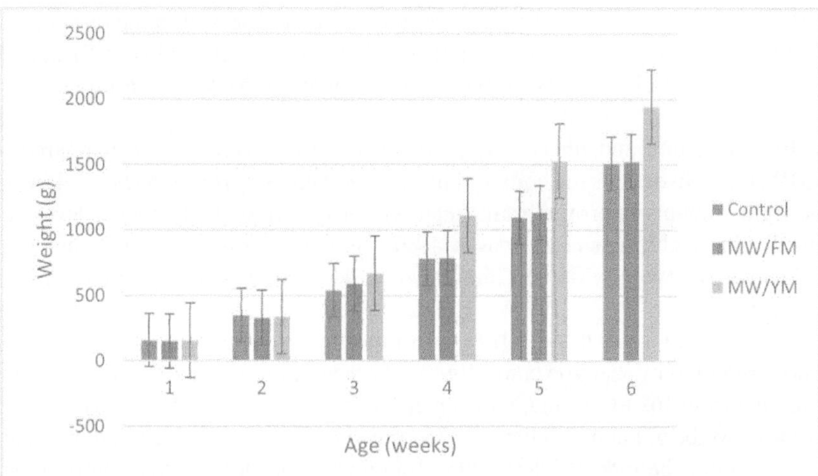

FIGURE 4.2 Growth rate and body weight gain of chickens that were fed the three different formulated diets.

($p < 0.05$) in body weight gain of chickens that were fed the MW/YM diet (group 3), which recorded the highest body weight compared to chickens that were fed the finger millet diet (group 2) and the control, which maintained a linear growth rate and body weight gain, respectively.

4.3.4 Histopathology

The livers from chickens in the three groups appeared normal, and a fine vacuolation of the hepatic cytoplasm was observed – this is considered normal. All the organs of the chickens in all the groups were analyzed for possible Newcastle disease and lesions. The results revealed that the organs from the three groups were pathologically normal and tested negative for the virulent Newcastle disease.

4.3.5 Sensory Evaluation

TABLE 4.3
Sensory Evaluation of Chicken Carcass from all Three Groups (Accredited Laboratory)

Sample/Groups	Description	Taste	Texture	Color
Group 1	Grilled	Characteristic, no off flavor	Firm yet tender and juicy	Typical color
	Cooked	Characteristic, no off flavor	Firm yet tender and juicy	Typical color
Group 2	Grilled	Characteristic, no off flavor	Firm yet tender and juicy	Typical color
	Cooked	Characteristic, no off flavor	Firm yet tender and juicy	Typical color
Group 3	Grilled	Characteristic, no off flavor	Firm yet tender and juicy	Typical color
	Cooked	Characteristic, no off flavor	Firm yet tender and juicy	Typical color

Note: No significant difference in taste, texture, and color was observed in chicken breast meat in all three groups.

4.4 DISCUSSION

The high protein content recorded in finger millet and the insect (MW) formulation was consistent with several studies that have reported similar results. Joshi and Katoch (1990) reported a high protein content in finger millet ranging from 5.6 to 12.7%. Ravindran (1991) and Rao et al. (1994) analyzed several varieties of finger millets and recorded similar results, which ranged from 4.9 to 15.58%, with a mean value of 9.7%. The high protein content observed in the insect protein (MW) was consistent with studies conducted by Madibela et al. (2009), Sekhwela (1989), and Ohiokpehai (2006), who reported similar results on the protein content of MW, which ranged between 44.5 and 58.9%.

Carbohydrate, calcium, and fiber contents in finger millet were higher compared to those of maize and MW. Carbohydrate is the body's main source of energy (Hargreaves, 1991). Carbohydrate taken in the right amounts improves brain function, muscle mass, body energy level, metabolism, and digestive health in chickens (Hargreaves, 1991). Studies conducted by Pore and Magar (1979), as well as Hulse et al. (1980), observed that finger millet is a good source of carbohydrate, with a range of 72–79.5%.

Calcium content (320.5 mg/100 g) observed in the finger millet used in the present study was similar to those reported by Vadivoo et al. (1998). These authors reported the level of calcium in finger millet ranging from 162 to 487 mg, with a mean value of 320.8 mg (Vadivoo et al., 1998). Foods high in calcium have been known to have a wide range of benefits, such as building and maintaining strong bones (Greenberg & Wrigley, 1995). In humans, besides positive contributions to dental health, studies have shown that calcium helps in the prevention of colon cancer and the reduction of obesity (Motro et al., 2001). Replacing yellow maize with finger millet could potentially result in positive effects on the healthy growth and development of chicks, particularly in terms of bone health. The high fiber content in finger millet (Table 4.1) correlated with those reported by Kamath and Belavady (1980). These authors reported about 12.6% Ca.

The crude protein analysis of the finger millet formulated diet was within the range reported by Barroso et al. (2014) and Bernard et al. (1997), who blended insect protein with various cereals. However, the present study observed that chickens that were fed the maize and insect protein diet (group 3) and control recorded a significantly higher feed intake (FI) ratio compared to chickens that were fed the finger millet diet (group 2). The high FI ratio observed in the present study correlates with the report of Nagra et al. (1990), who reported a high FI ratio in broiler chickens that were fed yellow maize. Chickens fed the finger millet diet recorded a lower FI ratio. Similar results were also observed by Rama Rao et al. (2002). These authors reported a significantly lower FI ratio in broiler chickens on the millet diet in comparison to those on the conventional maize-based diet.

Chickens fed the insect (MW) and maize (group 3) recorded significantly higher body weight gain (BWG) compared to the control and finger millet diets (group 2), both of which maintained a linear and equal growth rate and body weight gain. The higher BWG recorded in the broilers from group 3 can be attributed to the replacement of soybean with insect protein (MW). Similar observations were reported by Hwangbo et al. (2009), Loponte et al. (2017), and Madibela et al. (2009). These authors reported that broiler chickens that were fed diets where the soybean was replaced with insects such as mealworms (*Tenebrio molitor* and *Hermetia illucens*) grown on manure and housefly pupae (*Musca domestica*) significantly recorded higher growth rates and BWG.

Chickens fed with FM/MW diet (group 2) maintained a steady growth rate and BWG when compared with chickens that were fed the commercial feed or control diet. These findings potentially indicate that the feed conversion ratio of chickens fed the FM/MW diet (group 2) was better compared to chickens fed the insect protein and maize, as well as the commercial feed or control. Thus, chickens in group 2 ate little but maintained a stable BWG compared to chickens in the control that ate more. These findings agree with those of Davis et al. (2003), who observed that finger millet diet–fed chickens had a better feed conversion ratio and body weight gain compared to chickens fed the maize diet.

The high carcass weight observed in chicken fed MW/YM and FM/MW diets was potentially influenced by the replacement of dietary soybean with insect protein, such as MW. Bovera et al. (2016) and Biasato et al. (2016) previously reported a high carcass weight in broilers fed with mealworm larvae (*Tenebrio molitor*). Similar findings were obtained by Cullere et al. (2016) and Schiavone et al. (2017) in broiler quails that were fed diets with *Hermetia illucens* (black soldier fly) meal and fat, respectively.

There were no pathological findings in livers indicating that the diets were not safe. These findings can be attributed to good biosecurity, high-quality ingredients, and formulated diets that were free of microbial, chemical, and toxin contamination, as well as a balanced protein/energy mix. A balanced protein/energy mix and good biosecurity have a positive effect on chicken organs (Jacob et al., 2014).

There was no discernible difference in taste, texture, and color in broiler meat from the three different diets. The replacement of maize with finger millet has been shown not to affect the sensory characteristics of broiler meat (Davis et al., 2003; Medugu et al., 2010). Furthermore, organoleptic characteristics of broiler meat are not affected by supplementation of insect meal in their diet (Hwangbo et al., 2009; Awoniyi, 2007). Similarly, Aniebo et al. (2011) observed that the inclusion of insect meal (maggot meal) in fish diet resulted in no significant effect on texture, flavor, taste, and juiciness of treated and control group.

4.5 CONCLUSION AND PROSPECTS

The present study established the nutritional components of finger millets and insect protein broiler diets. It can also be concluded that maize and soybean components in poultry broiler feed can be totally or partially replaced with finger millet and insect protein (MW) without any negative effects on the performance and the organoleptic properties of the meat. If anything, feed intake was shown to decrease, while a significant increase in weight was observed. This is desirable, as it can impact on the overall cost of chicken production. Thus, an alternative broiler feed can be formulated from the product, thereby decreasing dependence on maize and soybean. Millets are considered resilient in the face of climate change, and growing them for the chicken industry would reduce imports. This would create a low carbon footprint for chicken farming while stimulating local farming systems.

4.6 ACKNOWLEDGMENTS

The authors would like to appreciate the National Research Foundation of South Africa (15052) for funding this project under the Indigenous Scholarships Grant.

REFERENCES

Abdou, K. D., & Kyarimpa, G. (2023). A review of non-conventional animal protein sources as poultry feeds. *International Journal of Veterinary Sciences and Animal Husbandry*, *8*(3), 243–252.

Achaya, K. T. (2003). *The story of our food*. Universities Press.

Aganga, A. A., Omphile, U. J., Malope, P., Chabanga, C. H., Motsamai, G. M., & Motsumi, L. G. (2008). Traditional poultry production and commercial broiler alternatives for small-holder farmers in Botswana. *Livestock Research for Rural Development*, *12*(4), 1–8.

Aniebo, A., Odukwe, C., Ebenebe, C., Ajuogu, P., Owen, O., & Onu, P. (2011). Effect of housefly larvae (Musca domestica) meal on the carcass and sensory qualities of the mud catfish, (*Clarias gariepinus*). *Advance Food Energy Security*, *1*(2).

Awoniyi, T. (2007). Health, nutritional and consumers' acceptability assurance of maggot meal inclusion in livestock diet: A review. *International Journal of Tropical Medicine*, *2*, 52–56.

Barroso, F. G., de Haro, C., Sánchez-Muros, M. J., Venegas, E., Martínez-Sánchez, A., & Pérez-Bañón, C. (2014). The potential of various insect species for use as food for fish. *Aquaculture*, *422*, 193–201.

Bernard, J. B., Allen, M. E., & Ullrey, D. E. (1997). Feeding captive insectivorous animals: Nutritional aspects of insects as food. *Nutrition Advisory Group Handbook, Fact Sheet*, *3*, 1–7.

Biasato, I., De Marco, M., Rotolo, L., Renna, M., Lussiana, C., Dabbou, S., Capucchio, M., Biasibetti, E., Costa, P., & Gai, F. (2016). Effects of dietary *Tenebrio molitor* meal inclusion in free-range chickens. *Journal of Animal Physiology and Animal Nutrition*, *100*, 1104–1112.

Boafo, H. A., Affedzie-Obresi, S., Gbemavo, D. S. J. C., Clottey, V. A., Nkegbe, E., Adu-Aboagye, G., & Kenis, M. (2019). Use of termites by farmers as poultry feed in Ghana. *Insects*, *10*(3), 69.

Bovera, F., Loponte, R., Marono, S., Piccolo, G., Parisi, G., Laconisi, V., Gasco, L., & Nizza, A. (2016). Use of larvae meal as protein source in broiler diet: Effect on growth performance, nutrient digestibility, and carcass and meat traits. *Journal of Animal Science*, *94*, 639–647.

Brink, M., & Belay, G. (2006). *Cereals and pulses* (Vol. 1). PROTA.

Bruinsma, J. (2003). *World agriculture: Towards 2015/2030: An FAO perspective*. Earthscan.

Cullere, M., Tasoniero, G., Giaccone, V., Miotti-Scapin, R., Claeys, E., De Smet, S., & Dallezotte, A. (2016). Black soldier fly as dietary protein source 75 for broiler quails: Apparent digestibility, excreta microbial load, feed choice, performance, carcass and meat traits. *Animal*, *10*(12), 1923–1930.

Daghir, N. J. (2008). *Poultry production in hot climates*. Cabi.

Dao, A. N. C., Sankara, F., Pousga, S., Coulibaly, K., Nacoulma, J. P., Ouedraogo, S., Kenis, M., & Somda, I. (2020). Traditional methods of harvesting termites used as poultry feed in Burkina Faso. *International Journal of Tropical Insect Science*, *40*, 109–118.

Davis, A. J., Dale, N. M., & Rerreira, F. J. (2003). Pearl millet as an alternative feed in broiler chickens diets. *Journal of Applied Poultry Research*, *12*, 137–144.

D'Mello, J. P. F. (1995). Antinutritional substances in legume seeds. In *Tropical legumes in animal nutrition*. CA International.

FAO. (2006). *Food security: Cereal and Maize* [Online]. Retrieved March 19, 2018, from www.fao.org/fileadmin/templates/wsfs/docs/Issues_papers/HLEF2050_Global_Agriculture.pdf

Greenberg, M. J., & Wrigley, W. M., Jr. (1995). *Chewing gum with dental health benefits employing calcium glycerophosphate* (U.S. Patent 5,378,131).

Hargreaves, M. (1991). Carbohydrates and exercise. *Journal of Sports Sciences*, *9*(1), 17–28.

Henry, M., Gasco, L., Piccolo, G., & Fountoulaki, E. (2015). Review on the use of insects in the diet of farmed fish: Past and future. *Animal Feed Science and Technology*, *203*, 1–22.

Hulse, J. H., Laing, E. M., & Pearson, O. E. (1980). *Sorghum and the millets: Their composition and nutritive value*. Academic Press.

Hwangbo, S., Choi, S. H., Kim, S. W., Son, D. S., Park, H. S., Lee, S. H., & Jo, I. H. (2009). Effects of crude protein levels in total mixed rations on growth performance and meat quality in growing Korean black goats. *Asian-Australasian Journal Animal Science*, *8*, 1133–1139.

International Organization for Standardization (ISO) 8586. (2012). *Sensory analysis: General guidance for the selection, training and monitoring of assessors*. Author.

International Organization for Standardization (ISO) 8589. (2007). *Sensory analysis: General guidance for the design of test rooms*. Author.

Jacob, J., Butcher, G., Mather, F., & Miles, R. (2014). *Avian influenza in poultry*. IFAS Extension University of Florida.

Joshi, H. C., & Katoch, K. K. (1990). Nutritive value of millets: A comparison with cereals and pseudocereals. *Himalayan Research Development*, *9*, 26–28.

Kamath, M. V., & Belavady, B. (1980). Unavailable carbohydrates of commonly consumed Indian foods. *Journal of the Science of Food and Agriculture*, *31*(2), 194–202.

Loponte, R., Nizza, S., Bovera, F., De Riu, N., Fliegerova, K., Lombardi, P., Vassalotti, G., Mastellone, V., Nizza, A., & Moniello, G. (2017). Growth performance, blood profiles and carcass traits of Barbary partridge (Alectoris barbara) fed two different insect larvae meals (*Tenebrio molitor* and *Hermetia illucens*). *Research in Veterinary Science*, *115*, 183–188.

Madibela, O. R., Mokwena, K. K., Nsoso, S. J., & Thema, T. F. (2009). Chemical composition of Mopane worm sampled at three sites in Botswana and subjected to different processing. *Tropical Animal Health and Production*, *41*(6), 935–942.

Makkar, H. P., Tran, G., Heuze, V., & Ankers, P. (2014). State-of-the-art on use of insects as animal feed. *Animal Feed Science and Technology*, *197*, 1–33.

Medugu, C. I., Kwari, I. D., Igwebuike, J., Nkama, I., Mohammed, I. D., & Hamaker, B. (2010). Performance and economics of production of broiler chickens fed sorghum or millet as replacement for maize in the semi-arid zone of Nigeria. *Agriculture and Biology Journal of North America*, *1*(3), 321–325.

Motro, M., Shemesh, J., & Grossman, E. (2001). Coronary benefits of calcium antagonist therapy for patients with hypertension. *Current Opinion in Cardiology*, *16*(6), 349–355.

Nagra, S. S., Chawla, J. S., & Phaul, P. S. (1990). Effect of feeding sorghum on growth performance of broilers. *Indian Journal of Poultry Science*, *25*, 138–144.

Ohiokpehai, O. (2006). Expanding the uses of phane, a nutritionally rich local food in Southern Africa. *International Journal of Food, Agriculture and Environment*, *4*(3&4), 26–32.

Pore, M. S., & Magar, N. G. (1979). Nutrient composition of hybrid varieties of finger millet [Eleusine coracana (Linn.) Gaertn., India]. *Indian Journal of Agricultural Sciences*, *49*(7), 526–531.

Rachie, K. O., & Peters, L. V. (1977). *The Eleusines-A review of the world literature*. International Crops Research Institute for Semi-arid Tropics.

Rao, A. S., Prabhu, U. H., Sampath, S. R., & Schiere, J. B. (1994). The effect of level of allowance on the intake and digestibility of finger millet (*Eleusine coracana*) straw in crossbred heifers. *Animal Feed Science and Technology*, *49*(1), 37–41.

Rao, S. V. R., Shyamsunder, A. K., Reddy, M. R., Raju, M. V., & Praharaj, N. K. (2002). Utilization of different millets replacing maize in coloured broiler chicken diet. *Indian Journal of Animal Nutrition*, *19*, 353–358.

Ravindran, G. (1991). Studies on millets: Proximate composition, mineral composition, and phytate and oxalate contents. *Food Chemistry*, *39*(1), 99–107.

Ravindran, V., & Blair, R. (1993). Feed resources for poultry production in Asia and the Pacific. III. Animal protein sources. *World's Poultry Science Journal*, *49*(3), 219–235.

Sánchez-Muros, M. J., Barroso, F. G., & Manzano-Agugliaro, F. (2014). Insect meal as renewable source of food for animal feeding: A review. *Journal of Cleaner Production*, *65*, 16–27.

SAPA. (2013). *Introduction to the South African poultry farming industry* [Online]. Retrieved January 20, 2017, from www.sapoultry.co.za/pdf-training/sa-farming-industry-introduction.pdf

Schiavone, A., Cullere, M., Demarco, M., Meneguz, M., Biasato, I., Bergagna, S., Dezzutto, D., Gai, F., Dabbou, S., & Gasco, L. (2017). Partial or total replacement of soybean oil by black soldier fly larvae (*Hermetia illucens L.*) fat in broiler diets: Effect on growth performances, feed-choice, blood traits, carcass characteristics and meat quality. *Italian Journal of Animal Science*, *16*(1), 93–100.

Sekhwela, M. B. M. (1989). The nutritive value of mophane bread–Mophane insect secretion (Maphote or Maboti). *Botswana Notes and Records*, *20*, 151–153.

Sundaram, T. K., Natrajan, A., & Chandrasekeran, D. (2001). Feed analytical techniques: Centre of advance studies in poultry science and animal feed analytical and quality control laboratory. *Veterinary College and Research Institute, Namakkal*, *4*, 9–65.

Tursi, A., Mastromarino, P., Capobianco, D., Elisei, W., Miccheli, A., Capuani, G., Tomassini, A., Campagna, G., Picchio, M., Giorgetti, G., & Fabiocchi, F. (2016). Assessment of fecal microbiota and fecal metabolome in symptomatic uncomplicated diverticular disease of the colon. *Journal of Clinical Gastroenterology*, *50*, S9–S12.

Vadivoo, A. S., Joseph, R., & Ganesan, N. M. (1998). Genetic variability and diversity for protein and calcium contents in finger millet (*Eleusine* coracana (L.) Gaertn) in relation to grain color. *Plant Foods for Human Nutrition*, *52*(4), 353–364.

Veldkamp, T., Van Duinkerken, G., Van Huis, A., Lakemond, C. M. M., Ottevanger, E., Bosch, G., & Van Boekel, T. (2012). *Insects as a sustainable feed ingredient in pig and poultry diets: A feasibility study.* Wageningen UR Livestock Research.

Zuidhof, M. J., Molnar, C. L., Morley, F. M., Wray, T. L., Robinson, F. E., Khan, B. A., Al-Ani, L., & Goonewaedene, L. A. (2003). Nutritive value of house fly (*Musca domestica*) larvae as a feed supplement for Turkey Poultry. *Animal Feed Science and Technology*, *105*, 225–230.

5 Characterization of Selected Nutritional Traits in Spider Plant (*Cleome gynandra* L.) Leaves

*[1]Fhatuwani Thovhogi, [2]Henry Okwudili Udeh, [3]Dorcus Maja, and *[1]Eastonce Tendayi Gwata*

[1]University of Venda, Faculty of Science, Engineering and Agriculture, Department of Plant and Soil Sciences, Private Bag X5050, Thohoyandou, 0950, South Africa

[2]State University of Medical and Applied Sciences, Faculty of Applied Natural Sciences, Department of Applied Microbiology and Brewing, Igbo Eno, Enugu State, Nigeria

[3]North West University, Faculty of Natural and Agricultural Sciences, Food Security and Safety Focus Area, Private Bag X2046, Mmabatho 2735, South Africa

*Corresponding author: ectgwata@gmail.com

5.1 INTRODUCTION

Spider plant (*Cleome gynandra* L.) is an indigenous African leafy vegetable that is valued for its high nutritional content. It is an annual herb which belongs to the kingdom Plantae, phylum Spermatophyta, division Magnoliophyta, class Magnoliopsida, order Brassicales, and family Cleomaceae (Sogbohossou et al., 2019). The genus *Cleome* is the largest group in Cleomaceae, with >200 annual or perennial herb and shrub species, from which approximately 50 occur in Africa (Sanchez-Acebo, 2005; Raju & Rani, 2016). Significant variations in morphological attributes of spider plant, such as leaf pubescence and stem color, exist (Figure 5.1) (Wu et al., 2018; Kwarteng et al., 2018a; Munene et al., 2018). In addition, a variation in the diploid chromosome number of the species was reported, ranging from $2n = 20$ to $2n = 34$ (Inda et al., 2008; Koshy & Mathew, 1985; van der Bergh et al., 2014; Omondi et al., 2017b).

Spider plant grows in a wide range of agro-ecologies in many parts of the world, particularly in Eastern and Southern Africa, West Africa, and South Asia (Thovhogi et al., 2021; Kwarteng et al., 2018b; Cernansky, 2015; Bala et al., 2010; Teklehaymanot & Giday, 2010). Recently, the medicinal utilization of spider plant in treating scorpion stings in the Middle East region (Saudi Arabia) was reported (Al-Asmari et al., 2017). Spider plant is also regarded as a weed of field and horticultural crops in other parts of the world, such as the Caribbean (Bahamas, Bermuda, Cuba), the Americas, Europe, China, Japan, Korea, New Zealand, and Australia (Saifullah et al., 2023; Cweya et al., 1997; Holm et al., 1991). The vegetable has a high geographical diversity.

In many countries in Southern Africa, the spider plant, like similar other leafy vegetables, is collected from the wild and is often regarded as a weed (van Rensburg et al., 2014; Tanor & Magama, 2020). It is adapted to numerous environmental stresses where most other leafy vegetables would

DOI: 10.1201/9781003469766-5

FIGURE 5.1 Variation in stem color among spider plant genotypes (left: purple stem; right: green stem).

fail (Slabbert et al., 2004; Hossain et al., 2022). However, currently, there are no reports of significant commercial production of this vegetable, particularly in Southern Africa; hence, it is available during the rainy season, which tends to limit its consumption. Nonetheless, dried leaves of the vegetable are often sold in informal markets (Thovhogi et al., 2021). In general, the popularity of indigenous leafy vegetables is influenced by availability, ease of preparation, taste, consistency, and appearance (van Rensburg et al., 2014). On the other hand, the utilization of indigenous vegetables such as the spider plant is often influenced negatively by the introduction of exotic leafy vegetables (Bvenura & Afolayan, 2015). Nonetheless, the spider plant has significant nutritional benefits over exotic vegetable species and provides dietary diversity, especially to poor communities (Thovhogi et al., 2023).

5.2 NUTRITIONAL VALUE

Apart from medicinal uses (Shah & Rahim, 2017; Sridhar et al., 2014; Ahouansinkpo et al., 2016; Ajaiyeoba, 2000) and biological importance (Ranjitha et al., 2014; Gaur et al., 2009; Runnie et al., 2004; Ajaiyeoba et al., 2001), the nutritional value of the spider plant stems mainly from the presence of a wide range of antioxidants, dietary fibers, flavonoids, minerals, and vitamins (Pilon-Smits et al., 2009; Jiménez-Aguilar & Grusak, 2015; Thovhogi et al., 2021; Houdegbe et al., 2022; Moyo & Aremu, 2022). The phytochemicals and nutrients occur in various organs of the plant, including the roots, stems, leaves, flowers, silique, and seeds (Thovhogi et al., 2021; Ravishankar et al., 2019; Neugart et al., 2017; Omondi et al., 2017b; Kasem & Fatahy, 2016; Nyalala et al., 2013; Songsak, & Lockwood. 2002). The mineral and vitamin content can vary, depending on the genotype of the plant (Jiménez-Aguilar & Grusak, 2015; Thovhogi et al., 2021). However, a significant geographic-independent variability in ascorbic acid content of spider plant was reported (Sogbohossou et al., 2019).

5.2.1 Genotypic Variation in Leaf Mineral Content

The variation in the mineral content of the spider plant is influenced partly by a wide range of factors, such as geographical location, season, genotype, and physiological state (Neugart et al., 2017; Greenfield & Southgate, 2003). A study conducted in South Africa under fertilizer-free conditions showed highly significant differences ($p \leq 0.001$) among the genotypes from different ecological regions for most of the mineral contents, including magnesium (Mg), phosphorous (P), aluminum (Al), copper (Cu), iron (Fe), manganese (Mn), and zinc (Zn) (Thovhogi et al., 2021). Zinc also occurs widely in Chinese cabbage (*Brassica pekinensis*) (12.40 mg/kg), in cabbage (*Brassica oleracea* var. *capitata*) (5.95 mg/kg) (Woldetsadik et al., 2017), as well as in lettuce (*Lactuca sativa* var. *crispa*) (56.9–94.4 mg/kg) and Ethiopian mustard (*Brassica carinata*) (66.3–109 mg/kg) (Li et al. (2014).

The Fe content was sixfold higher than the Zn leaf content, but the mean for the potentially toxic element Al was 202.72 mg/kg (Table 5.1) (Thovhogi et al., 2021). The leaf calcium (Ca) content (14.6 g/kg DW) was more than double that of both Mg and P. Among the exotic germplasms that were included in the study, the genotypes "GPS" contained the highest (6.5 g/kg DW) leaf Mg content, which was comparable to local genotypes "ML-2-DD" (6.4 g/kg DW) (Thovhogi et al., 2023) (Table 5.1).

TABLE 5.1
The Occurrence of Mineral Elements in *Cleome gynandra* Leaves

	g/kg DW				mg/kg DW				
Genotypes	N	Ca	Mg	P	Al	Cu	Fe	Mn	Zn
GPS (E)	67.3a	19.1a	6.5b	8.4ab	256.33bc	14.70bcd	409.33ab	123.33fg	71.33ab
UG-SF-15 (E)	66.7ab	15.6bcd	6.1bcd	6.8bcde	152.00ghij	15.37bc	275.67def	130.33fg	75.00a
PS (E)	65.6ab	15.5bcd	6.3bc	7.9abc	195.00defgh	14.47bcd	367.33abcd	109.67fg	65.33bc
ML-5-TGM (L)	64.4ab	14.8bcde	4.5efg	7.1abcd	146.67ghij	9.63gh	235.00f	199.33bcdefg	58.00def
ML-SF-29 (E)	63.7ab	16.3abc	5.5cbde	6.3cde	299.67ab	11.60defgh	431.33a	509.33a	50.33ghij
ML-3-KK (L)	62.6abc	13.7bcdef	4.9cdefg	7.3abcd	254.33bc	11.30defgh	373.33abc	169.33defg	52.00efghij
MP-B-3-CG (L)	61.4abcd	15.3bcd	4.7defg	6.8bcde	178.33efghi	11.57defgh	289.33cdef	134.00fg	47.00ijk
ML-13-SDM (L)	60.9abcd	15.3bcd	6.3bc	8.7a	201.33cdefg	15.70abc	346.67abcde	285.33bc	63.67cd
IP-3 (E)	60.5abcd	16.4abc	6.1bcd	6.2de	210.33cdef	10.20fgh	288.67cdef	169.33defg	52.00efghij
ML-14-MAG (L)	60.4abcd	16.6abc	5.2bcdef	7.1abcd	231.00cde	12.53cdefg	419.33ab	175.67cdefg	48.67hij
MP-B-1-CG (L)	60.1abcd	12.8defg	5.5bcde	7.3abcd	177.67efghi	12.30cdefgh	331.00bcde	217.00bcdef	59.00cde
MP-B-4-CG (L)	59.2abcd	14.6bcde	4.9cdefg	7.1abcd	246.33bcd	12.40cdefg	359.67abcd	157.33efg	55.33efgh
ML-12-TMP (L)	57.8abcd	15.7bcde	6.1bcd	7.8abcd	249.00bcd	13.83bcde	399.33ab	139.67fg	51.67fghij
Standard-2PL	55.8abcd	14.1bcde	6.1bcd	4.4f	135.67ij	16.27ab	342.33abcde	123.00fg	41.00kl
MP-B-2-CG (L)	55.8abcd	13.8bcdef	5.4bcde	6.8bcde	219.67cdef	13.40bcdef	352.00abcd	280.00bcd	54.00efghi
ML-2-DD (L)	54.8bcde	16.9ab	6.4bc	7.9abcd	316.33a	9.90gh	405.00ab	302.33b	50.33ghij

(Continued)

TABLE 5.1 (Continued)

	g/kg DW				mg/kg DW				
Genotypes	N	Ca	Mg	P	Al	Cu	Fe	Mn	Zn
ML-6-BTK (L)	54.7bcde	13.4cdef	4.7defg	5.3ef	167.67fghi	8.93h	254.33ef	137.00fg	48.67hij
Standard-1CP	51.2cde	15.2bcd	3.8fg	4.1f	124.33ij	10.67efgh	196.67f	264.33bcde	37.33l
Standard-3SC	50.0de	11.8efg	11.5a	3.9f	256.33bc	18.87a	405.67ab	436.33a	56.00efg
Standard-4JM	49.7de	10.8g	3.7fg	4.3f	97.00j	14.13bcd	236.00f	273.33bcd	37.00l
TZ-1 (E)	4.37e	9.8g	3.7g	5.3ef	142.33hij	9.17gh	229.00f	91.00g	45.67jk
Mean	58.4	14.5	5.6	6.5	202.73	12.71	330	200.5	53.3
R^2 (%)	52.99	62.88	82.72	75.44	82.44	70.61	70.82	79.89	88.29
C.V. (%)	12.56	13.89	16.23	15.63	16.65	16.33	17.13	32.7	8.3
LSD	12.77	3.5	1.6	1.8	46.39	3.54	56.27	85.85	5.17

Note: Means with the same letter in a column are not significantly different ($p < 0.05$). *E* = exotic; *L* = local; *S* = standard; *N* = nitrogen; *Mg* = magnesium; *Ca* = calcium; *P* = phosphorus; *Al* = aluminum; *Cu* = copper; *Fe* = iron; *Mn* = manganese; and *Zn* = zinc.

5.2.2 Relationships between Micro- and Macroelements in Spider Plant

A recent comparative study focusing on the presence of minerals, flavonoids, and total phenolic compounds in the leaves of common traditional vegetables demonstrated that the intensity of both micro- and macroelements in spider plant together with those of three other common leafy vegetables was comparable with the findings from similar studies (Thovhogi et al., 2023). The prevalence of differences in selected specific minerals was likely due to the environmental, ontogenic, and genetic factors. For instance, calcium ammonium nitrate fertilizer enhanced leaf Ca content, while farmyard manure decreased Fe content in spider plant (Omondi et al., 2017b). The inferiority of microelement content in spider plant in comparison with other common traditional leafy vegetables was consistent with findings from similar studies (El-Nakhel et al., 2021). However, where common leafy vegetables showed superiority macroelement content to spider plant potentially implied that consumers could benefit more from combining or alternating the various leafy vegetable types in their diets. It should be emphasized that the prescribed minimum daily intake amounts of individual nutrient elements are more important than the mere presence of the elements in the vegetable leaves. Moreover, several studies have elucidated the necessity of balancing the nutrient elements in diets to attain and maintain a healthy body (Mavengahama et al., 2014). In addition, the ultimate absorption and utilization of the nutrients by the human body depend on antinutrients and bioavailability of the specific phytochemicals. The microminerals are involved in the formation of erythrocyte cells, while macro-minerals, such as Ca and Mg, have a high potential to control blood pressure and the immune system (Sacks et al., 1998). In general, the findings showed significant variations among leafy vegetable species, suggesting that no single leafy vegetable type contains all the desirable minerals, flavonoids, and phenolic compounds. Therefore, human intervention, particularly through genetic enhancement, can develop novel genotypes of the leafy vegetables with enhanced preferred combinations of the nutrients. There may be some inverse relationships between desirable nutrients in some leafy vegetables that require the application of plant breeding techniques, such as mutation (Thangwana et al., 2021) and probably the clustered regularly interspaced short palindromic repeats (CRISPR) genome editing tools that can precisely and effectively modify plant variations using deletions, insertions, and substitutions in crop enhancement (Zhang et al., 2020).

The presence of quercetin-related compounds in spider plant in most studies is worth noting for several reasons. Quercetin is a potent antioxidant flavanol which occurs also in onion, broccoli, and citrus fruits and is protective against tissue injury caused by drug toxicities (Ramos et al., 2006; Brüll et al., 2015; David et al., 2016). It would be desirable to introgress and enhance this flavonol into other leafy vegetables without diminishing the already-existing beneficial nutrients in such vegetables. The significant positive relationships of pairs of mineral nutrients within a vegetable species (for instance, Zn and Al in the spider plant) potentially indicate that, likely, selection for increased content of either mineral may inadvertently result in the elevation of the content of the other one. Nonetheless, future studies could consider the ontogenic stages at which phytochemical analyses are carried out, partly because at the juvenile phase, the concentration of bio-compounds may increase rapidly partly due to a high uptake rate of the young roots as well as the high relative growth rate of the plants. The empirical evidence available regarding the distribution of phytonutrients in the spider plant supports the need to widen the range of leafy vegetables in the diets of consumers to optimize the spectrum of beneficial nutrients. Nonetheless, spider plants possess a superior spectrum of desirable phytonutrients, suggesting that it can be highly recommended to consumers.

5.3 OCCURRENCE OF BIOACTIVE MOLECULES

Several bioactive compounds have been reported in spider plant (Corradini et al., 2011; Thovhogi et al., 2021). For instance, total phenolics, which include non-flavonoids (phenolic acids and phenolic lipids) and flavonoids (anthocyanidins, flavonols, and flavones), were observed in various organs of the spider plant (Ravishankar et al., 2019; Neugart et al., 2017; Omondi et al., 2017a). The antioxidant activity in the spider plant is flavonoid-based, but the molecules can vary in concentration depending on genetic and environmental factors.

5.3.1 PHENOLIC COMPOUNDS

The phenolic compounds in the leaves of the spider plant possess redox properties responsible for antioxidant activity. Both phenolic and flavonoid compounds are secondary metabolites in the plant which contain an aromatic ring bearing at least one hydroxyl group, thus rendering them good electron donors in antioxidant activities (Frankel, 1997; Corradini et al., 2011). In addition, this vegetable possesses sulfur- and nitrogen-containing secondary metabolites (glucosinolates) that confer a bitter taste and vary in different plant organs (Cartea & Velasco, 2008). Moreover, some of the compounds can induce the synthesis of endogenous antioxidant molecules in cells of biological systems and exhibit free radical inhibition, oxygen scavenging, and peroxide decomposition (Muchuweti et al., 2007). In a study which examined spider plant leaf phytonutrient contents, a considerable variation in three phenolic compounds, namely, quercetin, quercetin-3-rutinoside, and quercetin-3-β-D-glucoside, was observed (Table 5.2) (Thovhogi et al., 2021). In the same study, a wide range of bioactive molecules was detected in the leaves of the vegetable, including quercetin (QC) (Table 5.3). Two spider plant local genotypes ("MP-B-2-CG" and "MP-B-1-CG") achieved significantly ($p < 0.05$) higher QC than all the exotic genotypes (Table 5.3). In comparison with the local genotypes, the highest QC content (1.76×10^{-2} mg/g) among the exotic genotypes was only 46.31% of the best local ("MP-B-2-CG") type. Moreover, the local genotypes were superior to the exotic ones and cowpea in terms of quercetin-3-rutinoside (Q3R), but quercetin-3-glucoside concentration differed significantly among the indigenous leafy vegetable types. In addition, the study identified crotonoside ($C_{10}H_{13}N_5O_5$) as well as kaempferol-3-O-rutinoside ($C_{27}H_{30}0_{15}$) in the spider plant (Table 5.3).

TABLE 5.2
Phenolic Compounds in *Cleome gynandra* Leaf Genotypes

Genotype	QC	Q3R	Q3βDG
	($\times 10^2$ mg/g)		
MP-B-2-CG (L)	3.80a	85.57ab	5.8b
MP-B-1-CG (L)	3.10ab	92.64a	4.45bc
ML-6-BTK (L)	2.17bc	75.49bc	3.69c
IP-3 (E)	1.76cd	65.34cd	4.53c
ML-2-DD (L)	1.71cde	66.00cd	1.93de
UG-SF-15 (E)	1.68cdef	61.45def	5.44b
ML-3-KK (L)	1.48cdef	60.25def	2.17d
MP-B-3-CG (L)	1.22cdefg	52.38ef	1.46def
ML-13-SDM (L)	0.92defg	63.57cde	0.76efg
PS (E)	0.90defg	63.85cde	1.80def
GPS (E)	0.85defg	63.88cde	1.65def
ML-12-TMP (L)	0.53defg	58.81def	0.00g
TZ-1 (E)	0.49efg	49.46f	0.61efg
MP-B-4-CG (L)	0.47efg	30.56g	0.45fg
ML-SF-29 (E)	0.45gf	61.03def	1.87de
Standard-1CP	0.14g	0.00i	0.00g
ML-14-MAG (L)	0.00g	18.52hg	1.09defg
Standard-2PL	0.00g	9.8hi	1.60def
Standard-3SC	0.00g	0.00i	0.00g
Standard-4JM	0.00g	0.00i	7.34a
Mean	1.08	48.93	2.3
R^2 (%)	91.25	98.78	97.19
C.V. (%)	36.92	7.97	19.15

Source: Adapted from Thovhogi et al. (2021).

Note: Means with the same letter in a column are not significantly different ($p < 0.05$). *E* = exotic; *L* = local; *QC* = quercetin; *Q3R* = quercetin-3-rutinoside; *Q3βDG* = quercetin-3-β-D-glucoside.

TABLE 5.3
Bioactive Molecules in *Cleome gynandra* and *Vigna unguiculata* Leaves

Rank	RT (Min)	Fragmentations [M-H]- (m/z)	Molecular Weight [M-H]- (m/z)	Molecular Formula	Molecular Name
			Spider plant (*Cleome gynandra*)		
1	3.87	111.0085	191.0184	$C_6H_8O_7$	Citric acid
2	5.68	110.0231	243.0605	$C_9H_{12}N_2O_6$	Uridine
3	6.55	150.0441	282.0850	$C_{10}H_{13}N_5O_5$	Crotonoside
4	15.95	301.0343	756.2035	$C_{33}H_{40}O_{20}$	Quercetin-3-robinobioside
5	15.97	301.0343	755.2135	$C_{33}H_{40}O_{20}$	Quercetin-3-robinobioside
6	16.11	301.0392	609.1432	$C_{27}H_{30}O_{17}$	Quercetin-3,4-di-O-glucoside
7	17.87	301.0392	609.1462	$C_{27}H_{30}O_{16}$	Quercetin-3-O-rutinoside
8	18.50	300.0275; 301.0359	463.0868	$C_{21}H_{20}O_{12}$	Quercetin-3-glocoside
9	19.52	285.044	593.1566	$C_{27}H_{30}0_{15}$	Kaempferol-3-O-rutinoside
10	24.19	151.0045	301.0371	$C^{15}H_{10}0_7$	Quercetin

(Continued)

TABLE 5.3 (Continued)

Rank	RT (Min)	Fragmentations [M-H]- (m/z)	Molecular Weight [M-H]- (m/z)	Molecular Formula	Molecular Name
			Cowpea (*Vigna unguiculata*) (Standard)		
1	1.78	203.0518; 185.0415	341.108	$C_{12}H_{22}O_{11}$	Sucrose
2	6.54	168.0508	282.081	$C_{10}H_{13}N_5O_5$	8-hydroxy-2-deoxy guanosine
3	8.78	301.0359	315.074	$C_{16}H_{12}O_{17}$	Quercetin-3-methyl ether
4	17.43	300.0316	609.137	$C_{27}H_{30}O_{16}$	Rutin/quercetin-3-O-rutinoside
5	18.18	329.1206	639.161	$C_{32}H_{32}O_{14}$	Chartresin
6	18.22	300.0291	463.083	$C_{21}H_{20}O_{12}$	Isoquercetin/quercetin-3-glucoside

Source: Adapted from Thovhogi et al. (2021).

Note: Cowpea is a common leafy vegetable, especially in sub-Saharan Africa, from where it originated (Coulibaly et al., 2002).

Quercetin is one of the main flavonoids in diets, and it is recognized as a versatile molecule with many pharmacological properties, such as cardiovascular, anticancer, antioxidant, neurological, antiviral (Jan et al., 2010; Maalik et al., 2014; David et al., 2016; Dabeek et al., 2019). However, the bioavailability of quercetin is relatively low (< 10%) (Kandemir et al., 2022), but the glucoside conjugate of quercetin, which is found mainly in onion, is the most available form (Dabeek et al., 2019). The low bioavailability is attributed to the hydrophobicity of the molecule, as indicated by a relatively low solubility in water (0.17–7 μg/ml), gastric fluids (5.5 μg/ml), and small intestine fluids (28.9 μg/ml) (Kaşıkcı & Bağdatlıoğlu, 2016). Consequently, quercetin tends to precipitate in these fluids, thus reducing its bioaccessibility. However, various kinds of encapsulation technologies have been developed to improve the bioavailability of quercetin. For instance, colloidal delivery systems which can be incorporated into aqueous-based foods, such as electrohydrodynamic atomization (Giannouli et al., 2018), electrospraying (Rodríguez-Félix et al., 2019), ultra-high-pressure homogenization (Chen et al., 2020), fluidized bed coating crystallization (Sheng et al., 2020), ionotropic gelation (Dey et al., 2020), and hot melt extrusion (Shi et al., 2020), were used successfully to enhance the bioavailability of quercetin.

The therapeutic potential of quercetin in the treatment of breast cancer has been reported (Oh et al., 2010). In a study involving male rats, quercetin, in conjunction with sulfasalazine, induced alterations in steroidogenic enzyme activity, which enhanced organ weight, sperm integrity, and plasma hormone management, among other beneficial activities (Osawe & Farombi, 2018). In addition, quercetin possesses antiviral activity during the early stage of infection by the influenza A virus (Wu et al., 2015). The abundance of phenolic compounds in the spider plant demonstrates the nutritional attributes associated with this leafy vegetable; hence, its inclusion in the diet can widen the spectrum of valuable nutrients available for human health. Nonetheless, some flavonoids, such as 3-hydroxypropyl glucosinolate and 4-methoxyglucobrassicin, that were reported in spider plant in other studies were not detected in the study of Thovhogi et al. (2021), and this is attributed to the principles of the assay. In contrast, the occurrence of crotonoside (glycoside) in spider plant leaves is of interest since it is a potent tyrosine inhibitor, promotes selective inhibition in acute myeloid leukemia cells (Li et al., 2017), and enhances antitumor activity (Lin et al., 2020; Yan et al., 2018).

The extracts from the spider plant leaves also show significant ($p \leq 0.001$) differences among the genotypes in both polyphenol content and antioxidant activity (Table 5.4) (Thovhogi et al., 2021). The highest (12.21 mg GAE/g) and lowest (9.86 mg GAE/g) total phenolic contents (TPC) were attained in "ML-13-SDM" and "ML-SF-29," respectively. In addition, the antioxidant activity determined by the ferric-reducing antioxidant power (FRAP) in the exotic spider plant genotypes was significantly ($p < 0.05$) lower than in the local genotype ("ML-6-BTK"). However, genotypes from

TABLE 5.4
Occurrence of Total Polyphenol Constituents and Antioxidant Activity in Some *Cleome gynandra* Genotypes

Genotype	TPC (mg GAE/g)	DPPH (%)
ML-13-SDM (L)	12.21a	19.13bcd
ML-12-TMP (L)	12.15a	18.93cd
GPS (E)	12.24a	11.90fgh
IP-3 (E)	12.14a	13.16efg
MP-B-4-CG (L)	12.04ab	11.34ghi
TZ-1 (E)	12.04ab	8.70hij
MP-B-1-CG (L)	11.87abc	15.14ef
ML-5-TGM (L)	11.63abcd	22.55ab
UG-SF-15 (E)	11.45abcde	15.17ef
ML-14-MAG (L)	11.27abcde	22.68a
Standard-3SC)	11.17abcde	5.06k
MP-B-3-CG (L)	10.76bcdef	11.59ghi
ML-3-KK (L)	10.76bcdef	5.37jk
Standard-4JM	10.62cdef	16.39de
ML-6-BTK (L)	10.38def	21.98abc
PS (E)	10.38def	8.07ijk
Standard-1CP	10.27ef	22.55ab
ML-SF-29 (E)	9.86f	9.83ghi
Standard-2PL	7.46g	4.87k
Mean	11.01	13.92
R^2 (%)	92.08	97.76
C.V. (%)	3.84	8.22
LSD	0.035	1.49

Source: Adapted from Thovhogi et al. (2021).

Note: Means with the same letter in a column are not significantly different ($p < 0.05$). *E* = exotic; *L* = local; *TPC* = total phenolic content; *DPPH* = 2,2-diphenyl-1-picrylhydrazyl) free radical scavenging activity

different agro-ecological zones show no significant differences in total phenolic content (Thovhogi et al., 2021). This is in disagreement with observations that were reported for other plant species, such as figwort (*Scrophularia striata*), in which both the antioxidant potential and the total phenol content were influenced by ecological factors, such elevation and temperature (Zargoosh et al., 2019). While several reasons can be attributed to these contrasting observations, it is important to point out that the synthesis and metabolism of secondary compounds in plants are complex and involve multiple enzymatic activities that are sensitive to temperature. For instance, in a study which examined total phenolic and flavonoid accumulation in the Siberian ginseng (*Eleutherococcus senticosus*), low temperatures significantly increased the activities of some enzymes, including superoxide dismutase, but high temperatures significantly decreased antioxidant enzyme activities (Shohael et al., 2006). In addition, the ontogenic factors which are integral components of the plant's internal developmental genetic circuits can contribute to the variation in metabolite accumulation (Li et al., 2020).

5.3.2 Relationship between Mineral, Flavonoid Content, and Antioxidant Activity

A positive relationship was detected between Cu and Zn and attributed partly to the role of Zn as a co-factor of the antioxidant enzyme Cu-Zn-superoxide dismutase (Wong et al., 2019; Skalny et al., 2015). The microelement Zn is critical in the human immune system, where it regulates intracellular signaling pathways (Wessels et al., 2017) and the activation of nicotinamide adenine dinucleotide phosphate (NADPH) oxidase, which is critical in phagocytosis (Hasegawa et al., 2000). Zinc is required in the normal synthesis, storage, and secretion of insulin in pancreatic β cells (Olechnowicz et al., 2018) and the glucose homeostasis of diabetic patients, since it plays a role in glycolysis stimulation, gluconeogenesis inhibition, as well as modulation of glucose in adipocytes (Jayawardena et al., 2012). Moreover, among several other biological functions in human physiology, the presence of chelation-free Zn prevents the formation of neutrophil extracellular traps, which are integral components of the DNA matrix, chromatin, and granule proteins that capture extracellular pathogenic protein (Brinkmann et al., 2004). However, while Zn is of immense biological importance, excessive amounts (Zn poisoning) can be detrimental to human health, manifesting in various symptoms, such as diarrhea, fever, lethargy, nausea, and vomiting (C Gupta & Gupta, 2011). Some studies reported that a daily intake of 150–450 mg of Zn led to diminished Cu utilization (hypocupremia), alteration in Fe function, as well as a reduction in the level of high-density lipoprotein (Mohammad et al., 2012; Nriagu, 2011; Chandra, 1984). Moreover, chronic enriched Zn intake is associated with various chronic effects in the gastrointestinal, hematological, and respiratory systems, along with alterations in cardiovascular and neurological systems, in human subjects (Nriagu, 2011). This is partly attributed to the presence of excess free radicals that interfere with the plasma membrane, serum ferritin, and hematocrit concentration (Nriagu, 2011).

Similarly, a positive relationship between quercetin and its derivatives (quercetin-3-rutinoside and quercetin-3-β-D-glucoside) has been observed (Table 5.4) (Thovhogi et al., 2021). However, there has been no strong correlation between the antioxidant activity (using FRAP and DPPH) with the phenolic content, suggesting that antioxidant activity is potentially influenced by other additional, non-phenolic

TABLE 5.5

Pearson's Correlation Coefficients for Mineral Content, Phenolic Compounds, and Antioxidant Activities in *Cleome gynandra*

	N	Ca	Mg	P	Al	Cu	Fe	Mn	Zn	QC	Q3R	Q3βDG	TPC	DPPH	FRAP	TFC
N	1.00															
Ca	0.72**	1.00														
Mg	0.58*	0.74**	1.00													
P	0.53*	0.57*	0.69*	1.00												
Al	0.29	0.56*	0.45	0.39	1.00											
Cu	0.46*	0.41	0.64*	0.68*	0.08	1.00										
Fe	0.41	0.60*	0.56*	0.58*	0.87**	0.43	1.00									
Mn	0.22	0.20	0.19	0.07	0.55*	−0.03	0.45*	1.00								
Zn	0.47*	0.38	0.59*	0.54*	−0.17	0.74**	0.05	−0.16	1.00							
QC	0.12	−0.26	0.06	−0.13	−0.22	−0.09	−0.29	0.10	0.08	1.00						
Q3R	0.12	−0.16	0.27	0.02	−0.14	0.01	−0.19	0.21	0.22	0.82**	1.00					
Q3βDG	0.23	−0.09	0.14	−0.30	−0.30	−0.01	−0.35	0.08	0.26	0.84**	0.64*	1.00				
TPC	−0.22	0.01	0.17	0.32	−0.15	0.22	−0.13	−0.36	0.25	−0.02	−0.06	−0.13	1.00			
DPPH	0.18	0.14	0.03	0.04	−0.29	−0.03	−0.17	0.00	−0.01	0.06	−0.06	0.07	0.21	1.00		
FRAP	0.11	−0.05	−0.15	−0.08	−0.34	−0.11	−0.28	−0.14	0.02	0.37	0.03	0.20	0.26	0.62*	1.00	
TFC	−0.13	−0.04	−0.14	−0.40	−0.25	−0.27	−0.39	−0.36	−0.03	0.33	−0.05	0.43*	0.48*	0.21	0.59*	1.00

Source: Adapted from Thovhogi et al. (2021).

Note: **, *Significant at the 1% and 5% probability level, respectively.

QC = quercetin; *Q3R* = quercetin-3-rutinoside; *Q3βDG* = quercetin-3-β-D-glucoside; *TPC* = total phenolic content; *TFC* = total flavonoid content; *DPPH* = 2,2-diphenyl-1-picrylhydrazyl free radical scavenging activity; *FRAP* = ferric-reducing antioxidant power assay.

compounds, such as ascorbates, reducing carbohydrates, tocopherols, carotenoids, and terpenes, probably acting synergistically to produce the total antioxidant activity (Babbar et al., 2011). But the significant positive relationship between total phenolic content, quercetin, quercetin-3-β-D-glucoside, and quercetin-3-rutinoside could be useful in spider plant selection programs that are aimed at the concomitant improvement of both total phenolic content and flavonoids in vegetables.

5.3.3 Variations in Leaf Polyphenol Content and Antioxidant Activity among Vegetable Species

In a study which compared the leaf minerals, total phenolic content, and antioxidant activity in spider plant with those of three common leafy vegetable species, the principal component (PC) analysis indicated that the first two PCs accounted for 79.74% of the total variation (Figure 5.2) (Thovhogi et al., 2023). A biplot analysis indicated that spider plant was strongly associated with Ca, N, P, and FRAP (Figure 5.2). The acute angles between each pair of the six mineral elements

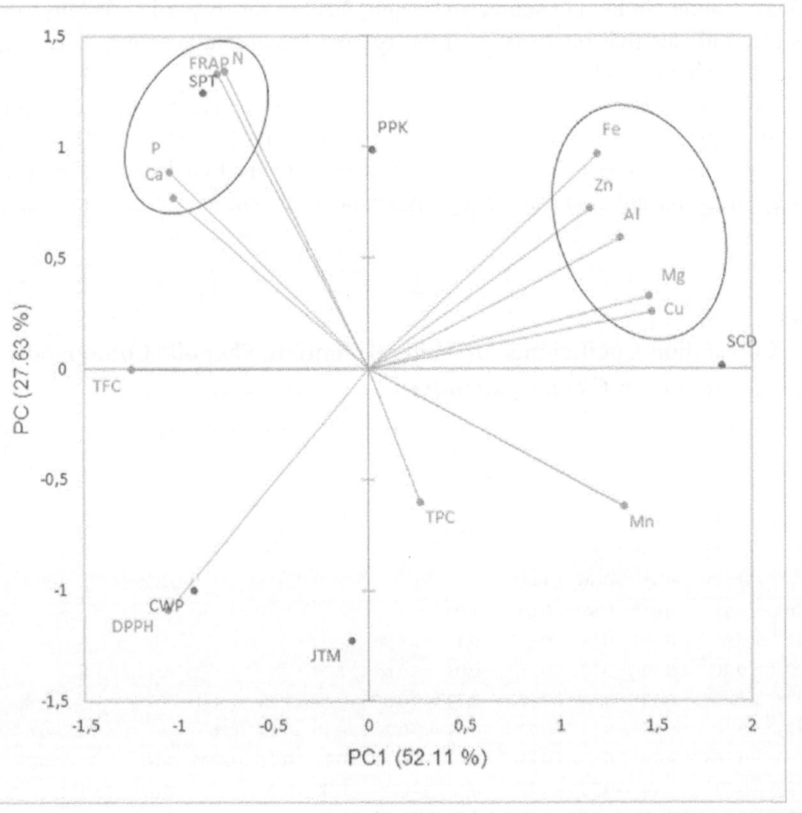

FIGURE 5.2 A biplot for principal component (PC) analysis of PC1 and PC2 indicating the variation between spider plant (SPT) and other common leafy vegetables, derived from the datasets of leaf minerals, total phenolic content, and antioxidant activity. The long lines suggest high correlation values, while the acute angles indicate positive correlation between each pair of variables. *CWP* = cowpea; *JTM* = jute mallow; *PPK* = pumpkin; *SCD* = Swiss chard (check); *Ca* = calcium; *Mg* = magnesium; *P* = phosphorus; *N* = nitrogen; *Al* = aluminum; *Cu* = copper; *Fe* = iron; *Mn* = manganese; *Zn* = zinc; *FC* = total flavonoid content; *TPC* = total phenolic content; *DPPH* = 2,2-diphenyl-1-picrylhydrazyl free radical scavenging activity; *FRAP* = ferric-reducing antioxidant power.

Source: Adapted from Thovhogi et al. (2023).

that were grouped in the top right quadrant suggested that there were positive correlations between them. However, both total phenolic content (TPC) and Mn showed no strong associations with any of the leafy vegetables.

5.4 PROSPECTS FOR IMPROVING THE NUTRITIONAL PROFILE OF SPIDER PLANT

The multipurpose and geographic diversity of spider plant demonstrated in the numerous studies documented in this chapter are sufficient to spur genetic improvement of the vegetable. Such genetic enhancement programs can selectively direct the specific focal areas of interest since it is difficult, if not impossible, to pyramid all the traits of interest in a single genotype. For instance, the development of improved genotypes high in nutritional therapeutic qualities will be desirable in folklore medicine in sub-Saharan Africa and other regions, where the vegetable is used widely for medicinal purposes partly due to limited access to modern medicine. The development of macro- and micronutrients as well as flavonoid-dense genotypes could also be useful in human nutrition. The reduction of leaf tannins through plant breeding methods could also improve the taste of the vegetable and arouse consumer interest (Thovhogi et al., 2021; Kutsukutsa et al., 2014). In expounding possible spider plant breeding approaches, a similar critical component, termed "definition of product targets," was proposed (Sogbohossou et al., 2018). The authors further elucidated the importance of integrating genomic tools as well as value chain development in the product development process. The genomic tools could explore the CRISPR technology (Zhang et al., 2020). Wide genetic distances that were observed among spider plant genotypes also indicated potential for genetic enhancement of this leafy vegetable. Perhaps the food processing technologies, particularly those which enhance the shelf life of the vegetable or preserve the nutrient profiles of the dried forms of the vegetable, could also be helpful for end users who require spider plant during the off-season period (Thovhogi et al., 2021).

5.5 CONCLUSIONS AND PROSPECTS

The spider plant is adapted to a wide ecological amplitude and often shows geographic diversity. In some regions, it is used mainly for human consumption and medicinal purposes, but in other areas, it is classified as a weed in field crops. The spider plant possesses numerous phytonutrients that are of immense biological importance. Therefore, this leafy vegetable is recommended for inclusion in human diets to contribute to health. There is merit in the genetic enhancement for specific nutrient profiles of this underutilized leafy vegetable. Such efforts can also spur the commercial production of the vegetable backed up by a viable seed system, especially in areas where consumers still collect it from the wild. Improvement of seed quality can also contribute to the adoption of the vegetable by farmers, probably on a commercial scale, which might ultimately contribute to shifting the mindset from regarding the spider plant as a weed species. In this regard, breeding programs that require seed testing and certification as pre-requisites for developing functional seed systems will be critical in the future. In the same breath, it is prudent to systematically collect and characterize wild accessions of the spider plant in various parts of the world for future genetic exploitation and to avert genetic erosion of the species, as well as concomitantly provide breeding programs with robust genetic resources for selection. We also believe that the spider plant will play a critical role in food security and nutrition for a variety of reasons. It is a typical climate-smart leafy vegetable and an affordable source of nutrition for resource-poor communities, especially those residing in harsh agro-ecologies where the vegetable thrives. Also, there will be some merit in conducting educational and awareness campaigns in communities about the nutritional value of the spider plant. Apart from community outreach programs, government policies that promote indigenous knowledge systems could also positively influence the future survival and utilization of this crop. Favorable policies can also positively impact the value chain of the spider plant, creating space for value addition (even simple sun-drying techniques), especially for ensuring off-season availability of the vegetable, as well as contributing to the improvement of rural economic growth and poverty alleviation. Moreover, such initiatives can reverse the current decline in the utilization of this vegetable. There are potential value chain players who can scale up production and marketing of the vegetable, thus circumventing the extinction of this edible species in the future.

5.6 ACKNOWLEDGMENTS

The authors would like to acknowledge the technical assistance rendered by N. Ntutshelo, at the Agricultural Research Council (Infruitec-Nietvoorbij), Stellenbosch, South Africa, in some of the original work cited in this chapter.

REFERENCES

Ahouansinkpo, E. J., Atanasso, A., Dansi, A., Adjatin, A., Azize, O., & Sanni, A. (2016). Ethnobotany, phytochemical screening and toxicity risk of *Cleome gynandra* and *Cleome viscosa*, two traditional leafy vegetables consumed in Benin. *International Journal of Current Microbiology and Applied Sciences, 5*, 813–829.

Ajaiyeoba, E. O. (2000). Phytochemical and antimicrobial studies of *Cleome gynandra* and *Buchholzia coriaceae* extracts. *African Journal of Biomedical Research, 3*, 161–165.

Ajaiyeoba, E. O., Onocha, O. A., & Olarenwaju, O. T. (2001). In vitro anthelmintic properties of Buchholzia coriaceae and Cleome gynandra extracts. *Pharmaceutical Biology, 39*, 217–220.

Al-Asmari, A., Manthiri, R. A., Abdo, N., Al-Duaiji, F. A., & Khan, H. A. (2017). Saudi medicinal plants for the treatment of scorpion sting envenomation. *Saudi Journal of Biological Sciences, 24*, 1204–1211.

Babbar, N., Oberoi, H. S., Uppal, D. S., & Patil, R. T. (2011). Total phenolic content and antioxidant capacity of extracts obtained from six important fruit residues. *Food Research International, 44*, 391–396.

Bala, A., Kar, B., Haldar, P. K., Mazumder, U. K., & Bera, S. (2010). Evaluation of anticancer activity of Cleome gynandra on Ehrlich's Ascites carcinoma treated mice. *Journal of Ethnopharmacology, 129*, 131–134.

Brinkmann, V., Reichard, U., Goosmann, C., Fauler, B., Uhlemann, Y., Weiss, D. S., Weinrauch, Y., & Zychlinsky, A. (2004). Neutrophil extracellular traps kill bacteria. *Science, 303*, 1532–1535.

Brüll, V., Burak, C., Stoffel-Wagner, B., Wolffram, S., Nickenig, G., Müller, C., Langguth, P., Alteheld, B., Fimmers, R., Naaf, S., Zimmermann, B., Stehle, P., & Egert, S. (2015). Effects of a quercetin-rich onion skin extract on 24 h ambulatory blood pressure and endothelial function in overweight-to-obese patients with (pre) hypertension: A randomised double-blinded placebo-controlled cross-over trial. *British Journal of Nutrition, 114*, 1263–1277.

Bvenura, C., & Afolayan, A. J. (2015). The role of wild vegetables in household food security in South Africa: A review. *Food Research International, 76*, 1001–1011.

Cartea, M. E., & Velasco, P. (2008). Glucosinolates in *Brassica* foods: Bioavailability in food and significance for human health. *Phytochemistry Reviews, 7*, 213–229.

Cernansky, R. (2015). The rise of Africa's super vegetables. *Nature, 522*(7555), 146–148.

Chandra, R. K. (1984). Excessive intake of zinc impairs immune responses. *JAMA, 252*, 1443–1446.

Chen, W., Ju, X., Aluko, R. E., Zou, Y., Wang, Z., Liu, M., & He, R. (2020). Rice bran protein-based nanoemulsion carrier for improving stability and bioavailability of quercetin. *Food Hydrocolloids, 108*, 106042.

Corradini, E., Foglia, P., Giansanti, P., Gubbiotti, R., Samperi, R., & Lagana, A. (2011). Flavonoids: Chemical properties and analytical methodologies of identification and quantitation in foods and plants. *Natural Product Research, 25*, 469–495.

Coulibaly, S., Pasquet, R. S., Papa, R., & Gepts, P. (2002). AFLP analysis of the phenetic organization and genetic diversity of cowpea [Vigna unguiculata (L.) Walp.] reveals extensive gene flow between wild and domesticated types. *Theoretical and Applied Genetics, 104*, 258–266.

Cweya, J. A., Heller, J., & Engels, J. (1997). *Cat's whiskers, Cleome gynandra L.* International Plant Genetic Resource Institute.

Dabeek, W. M., & Marra, M. V. (2019). Dietary quercetin and kaempferol: Bioavailability and potential cardiovascular-related bioactivity in humans. *Nutrients, 11*, 2288.

David, A. V. A., Arulmoli, R., & Parasuraman, S. (2016). Overviews of biological importance of quercetin: A bioactive flavonoid. *Pharmacognosy Reviews*, *10*, 84–89.

Dey, M., Ghosh, B., & Giri, T. K. (2020). Enhanced intestinal stability and pH sensitive release of quercetin in GIT through gellan gum hydrogels. *Colloids and Surfaces B: Biointerfaces*, *196*, 111341.

El-Nakhel, C., Pannico, A., Graziani, G., Giordano, M., Kyriacou, M. C., Ritieni, A., De Pascale, S., & Rouphael, Y. (2021). Mineral and antioxidant attributes of *Petroselinum crispum* at different stages of ontogeny: Microgreens vs. baby greens. *Agronomy*, *11*, 857.

Frankel, E. N. (1997). Nutritional benefits of flavonoids. In H. Ohigashi, T. Osawa, J. Terao, S. Watanabe, & T. Yoshikawa (Eds.), *Food factors for cancer prevention* (pp. 613–616). Springer.

Gaur, K., Kori, M. L., & Nema, R. K. (2009). Comparative screening of immunomodulatory activity of hydro-alcoholic extract of *Hibiscus rosa sinensis* Linn. and ethanolic extract of Cleome gynandra Linn. *Global Journal of Pharmacology*, *3*, 85–89.

Giannouli, M., Karagkiozaki, V., Pappa, F., Moutsios, I., Gravalidis, C., & Logothetidis, S. (2018). Fabrication of quercetin-loaded PLGA nanoparticles via electrohydrodynamic atomization for cardiovascular disease. *Materials Today: Proceedings*, *5*(8), 15998–16005.

Greenfield, H., & Southgate, D. A. T. (2003). *Food composition data. Production, management, and use* (2nd ed., pp. 78–79). Food and Agriculture Organization.

Gupta, C. U., & Gupta, C. S. (2011). Heavy metal toxicity in humans and its preventive and control measures. *Current Nutrition and Food Science*, *7*, 221–231.

Hasegawa, H., Suzuki, K., Suzuki, K., Nakaji, S., & Sugawara, K. (2000). Effects of zinc on the reactive oxygen species generating capacity of human neutrophils and on the serum opsonic activity in vitro. *Luminescence*, *15*, 321–327.

Holm, L. R. G., Plucknett, D. L., Pancho, J. V., & Herberger, J. P. A. (1991). *Geographical atlas of World weeds*. John Wiley and Sons.

Hossain, M. N., Sarker, U., Raihan, M. S., Al-Huqail, A. A., Siddiqui, M. H., & Oba, S. (2022). Influence of salinity stress on color parameters, leaf pigmentation, polyphenol and flavonoid contents, and antioxidant activity of Amaranthus lividus leafy vegetables. *Molecules*, *27*, 821.

Houdegbe, A. C., Achigan-Dako, E. G., Sogbohossou, E. O. D., Schranz, M. E., Odindo, A. O., & Sibiya, J. (2022). Leaf elemental composition analysis in spider plant [*Gynandropsis gynandra* L. (Briq.)] differentiates three nutritional groups. *Frontiers in Plant Science*, *13*, 841226.

Inda, L. A., Torrecilla, P., Catala, P., & Ruiz-Zapata, T. (2008). Phylogeny of Cleome L. and its close relatives *Podandrogyne* Ducke and *Polanisia* Raf (Cleomoideae, Cleomaceae) based on analysis of nuclear ITS sequences and morphology. *Plant Systematics and Evolution*, *274*, 111–126.

Jan, A. T., Kamli, M. R., Murtaza, I., Singh, J. B., Ali, A., & Haq, Q. M. (2010). Dietary flavonoid quercetin and associated health benefits – An overview. *Food Reviews International*, *26*, 302–317.

Jayawardena, R., Ranasinghe, P., Galappatthy, P., Malkanthi, R. L. D. K., Constantine, G. R., & Katulanda, P. (2012). Effects of zinc supplementation on diabetes mellitus: A systematic review and meta-analysis. *Diabetology and Metabolic Syndrome*, *4*, 1–12.

Jiménez-Aguilar, D. M., & Grusak, M. A. (2015). Evaluation of minerals, phytochemical compounds and antioxidant activity of Mexican, central American, and African green leafy vegetables. *Plant Foods Human Nutrition*, *70*, 357–364.

Kandemir, K., Tomas, M., McClements, D. J., & Capanoglu, E. (2022). Recent advances on the improvement of quercetin bioavailability. *Trends in Food Science and Technology*, *119*, 192–200.

Kasem, W. T., & Fatahy, S. (2016). Flavonoids and isoenzymes as chemotaxonomic markers in Cleome L. (Cleomaceae Bercht). *Current Botany*, *7*, 11.

Kaşıkcı, M. B., & Bağdatlıoğlu, N. (2016). Bioavailability of quercetin. *Current Research in Nutrition and Food Science Journal*, *4*, 146–151.

Koshy, J. K., & Mathew, P. M. (1985). Cytology of the genus Cleome. *Cytologia*, *50*, 283–288.

Kutsukutsa, R. T., Gasura, E., Mabasa, S., & Ngadze, E. (2014). Variability in condensed tannins and bitterness in spider plant genotypes. *African Crop Science Journal*, *22*, 275–280.

Kwarteng, A. O., Abogoom, J., Amoah, R. A., Nyadanu, D., Ghunney, T., Nyam, K. C., Ziyaaba, J. Z., Danso, E. O., Asiedu, D. D., & Whyte, T. (2018a). Current knowledge and breeding perspectives for the spider plant (*Cleome gynandra* L.): A potential for enhanced breeding of the plant in Africa. *Genetic Resources and Crop Evolution*, *65*, 1529–1550.

Kwarteng, A. O., Abogoom, J., Amoah, R. A., Nyadanu, D., Nyam, C. K., Ghunney, T., Awuah, E., Ziyaaba, J. Z., Ogunsanya, J. O., Orhin, E. E., & Asiedu, D. D. (2018b). Phenomic characterization of twenty-four accessions of spider plant (*Cleome gyandra* L.) the Upper East region of Ghana. *Scientia Horticulturae*, *235*, 124–131.

Li, P., Lu, Y. J., Chen, H., & Day, B. (2020). The lifecycle of the plant immune system. *Critical Reviews in Plant Sciences, 39*, 72–100.

Li, Y., Wang, H., Wang, H., Yin, F., Yang, X., & Hu, Y. (2014). Heavy metal pollution in vegetables grown in the vicinity of a multi-metal mining area in Gejiu, China: Total concentrations, speciation analysis, and health risk. *Environmental Science and Pollution Research, 21*, 12569–12582.

Li, Y. Z., Yu, S., Yan, P. A., Gong, D. Y., Wu, F. L., He, Z., Yuan, Y. Y., Zhao, A. Y., Tang, X., Zhang, R. Q., & Peng, C. (2017). Crotonoside exhibits selective post-inhibition effect in AML cells via inhibition of FLT3 and HDAC3/6. *Oncotarget, 8*, 103087–103099.

Lin, S. C., Lin, C. C., Li, S., Lin, W. Y., Lehman, C. W., Bracci, N. R., & Tsa, S. W. (2020). Alleviation of collagen-induced arthritis by crotonoside through modulation of dendritic cell differentiation and activation. *Plants, 9*, 1535.

Maalik, A., Khan, F. A., Mumtaz, A., Mehmood, A., Azhar, S., Atif, M., Karim, S., Altaf, Y., & Tariq, I. (2014). Pharmacological applications of quercetin and its derivatives: A short review. *Tropical Journal of Pharmaceutical Research, 13*, 1561–1566.

Mavengahama, S., de Clercq, W. P., & McLachlan, M. (2014). Trace element composition of two wild vegetables in response to soil-applied micronutrients. *South African Journal of Science, 110*, 1–5.

Mohammad, M. K., Zhou, Z., Cave, M., Barve, A., & McClain, C. J. (2012). Zinc and liver disease. *Nutrition in Clinical Practice, 27*, 8–20.

Moyo, M., & Aremu, A. O. (2022). Nutritional, phytochemical and diverse health-promoting qualities of Cleome gynandra. *Critical Reviews in Food Science and Nutrition, 62*, 3535–3552.

Muchuweti, M., Mupure, C., Ndhlala, A., Murenje, T., & Benhura, M. A. N. (2007). Screening of antioxidant and radical scavenging activity of *Vigna ungiculata*, *Bidens pilosa* and *Cleome gynandra*. *American Journal of Food Technology, 2*, 161–168.

Munene, K. A., Nzuve, F., Ambuko, J., & Odeny, D. (2018). Heritability analysis and phenotypic characterization of spider plant (*Cleome gynandra* L.) for yield. *Advances in Agriculture, 2018*, 8568424.

Neugart, S., Baldermann, S., Ngwene, B., Wasonga, J., & Schreiner, M. (2017). Indigenous leafy vegetables of Eastern Africa – a source of extraordinary secondary plant metabolites. *Food Research International, 100*, 411–422.

Nriagu, J. (2011). Zinc toxicity in humans. *Encyclopedia of Environmental Health, 2011*, 801–807.

Nyalala, S. O., Petersen, M. A., & Grout, B. (2013). Volatile compounds from leaves of the African spider plant (*Gynandropsis gynandra*) with bioactivity against spider mite (*Tetranychus urticae*). *Annals of Applied Biology, 162*, 290–298.

Oh, S. J., Kim, O., Lee, J. S., Kim, J. A., Kim, M. R., Choi, H. S., Shim, J. H., Kang, K. W., & Kim, Y. C. (2010). Inhibition of angiogenesis by quercetin in tamoxifen-resistant breast cancer cells. *Food and Chemical Toxicology, 48*, 3227–3234.

Olechnowicz, J., Tinkov, A., Skalny, A., & Suliburska, J. (2018). Zinc status is associated with inflammation, oxidative stress, lipid, and glucose metabolism. *Journal of Physiological Sciences, 68*, 19–31.

Omondi, E. O., Debener, T., Linde, M., Abukutsa-Onyango, M., Dinssa, F. F., & Winkelmann, T. (2017a). Mating biology, nuclear DNA content and genetic diversity in spider plant (*Cleome gynandra*) germplasm from various African countries. *Plant Breeding, 136*, 578–589.

Omondi, E. O., Engels, C., Nambafu, G., Schreiner, M., Neugart, S., Abukutsa-Onyango, M., & Winkelmann, T. (2017b). Nutritional compound analysis and morphological characterization of spider plant (*Cleome gynandra*) – An African indigenous leafy vegetable. *Food Research International, 100*, 284–295.

Osawe, S. O., & Farombi, E. O. (2018). Quercetin and rutin ameliorates sulphasalazine-induced spermiotoxicity, alterations in reproductive hormones and steroidogenic enzyme imbalance in rats. *Andrologia, 50*, e12981.

Pilon-Smits, E. A. H., Quinn, C. F., Tapken, W., Malagoli, M., & Schiavon, M. (2009). Physiological functions of beneficial elements. *Current Opinion in Plant Biology, 12*, 267–274.

Raju, A. J. S., & Rani, D. S. (2016). Reproductive ecology of Cleome gynandra and Cleome viscosa (Capparaceae). *Phytologia Balcanica, 22*, 15–28.

Ramos, F. A., Takaishi, Y., Shirotori, M., Kawaguchi, Y., Tsuchiya, K., & Shibata, H. (2006). Antibacterial and antioxidant activities of quercetin oxidation products from yellow onion (*Allium cepa*) skin. *The Journal of Agricultural and Food Chemistry, 54*, 3551–3557.

Ranjitha, J., Shalma, M., Donatus, M., & Vijayalakshmi, S. (2014). Isolation of novel phytoconstituents from the stem part of *Cleome gynandra* Linn. and their antimicrobial activity. *International Journal of Phytomedicine, 6*, 341–345.

Ravishankar, K. A., Prasanna, Y. V. V. M. A. L., & Kiranmayi, G. V. N. (2019). Comparative evaluation of in vitro antioxidant and in vivo hepatoprotective activities of ethanolic leaf and root extracts of *Cleome gynandra*. *Indian Drugs, 56*, 50.

Rodríguez-Félix, F., Del-Toro-Sánchez, C. L., Cinco-Moroyoqui, J. F., Juárez, J., Ruiz-Cruz, S., López-Ahumada, G. A., Carvajal-Millan, E., Castro-Enríquez, D. D., Barreras-Urbina, C. G., & Tapia-Hernández, J. A. (2019). Preparation and characterization of quercetin-loaded zein nanoparticles by electrospraying and study of in vitro bioavailability. *Journal of Food Science*, *84*, 2883–2897.

Runnie, I., Salleh, M. N., Mohamed, S., Head, R. J., & Abeywarden, M. Y. (2004). Vasorelaxation induced by common edible tropical plant extracts in isolated rat aorta and mesenteric vascular bed. *Journal of Ethnopharmacology*, *92*(2–3), 311–316.

Sacks, F. M., Willett, W. C., Smith, A., Brown, L. E., Rosner, B., & Moore, T. J. (1998). Effect on blood pressure of potassium, calcium, and magnesium in women with low habitual intake. *Hypertension*, *31*, 131–138.

Saifullah, K., Williams, A., & Adkins, S. (2023). Spider plant (*Cleome gynandra* L.): An emerging weed in the sweet corn–Brassica cropping system. *Agronomy*, *13*, 1430.

Sanchez-Acebo, L. (2005). A phylogenetic study of the new World Cleome L. (Brassicaceae-Cleomoideae). *Annals Missouri Botanical Garden*, *92*, 179–201.

Shah, A., & Rahim, S. (2017). Ethnomedicinal uses of plants for the treatment of malaria in Soon Valley, Khushab, Pakistan. *Journal of Ethnopharmacology*, *200*, 84–106.

Sheng, F., Chow, P. S., Hu, J., Cheng, S., Guo, L., & Dong, Y. (2020). Preparation of quercetin nanorod/microcrystalline cellulose formulation via fluid bed coating crystallization for dissolution enhancement. *International Journal of Pharmaceutics*, *576*, 118983.

Shi, X., Fan, N., Zhang, G., Sun, J., He, Z., & Li, J. (2020). Quercetin amorphous solid dispersions prepared by hot melt extrusion with enhanced solubility and intestinal absorption. *Pharmaceutical Development and Technology*, *25*, 472–481.

Shohael, A. M., Ali, M. B., Yu, K. W., Hahn, E. J., & Paek, K. Y. (2006). Effect of temperature on secondary metabolites production and antioxidant enzyme activities in *Eleutherococcus senticosus* somatic embryos. *Plant Cell, Tissue and Organ Culture*, *85*, 219–228.

Skalny, A. A., Tinkov, A. A., Medvedeva, Y. S., Alchinova, I. B., Karganov, M. Y., Skalny, A. V., & Nikonorov, A. A. (2015). Effect of short-term zinc supplementation on zinc and selenium tissue distribution and serum antioxidant enzymes. *Acta Scientiarum Polonorum, Technologia Alimentaria*, *14*, 269–276.

Slabbert, R., Spreeth, M., Krüger, G. H. J., & Bornman, C. H. (2004). Drought tolerance, traditional crops and biotechnology: Breeding towards sustainable development. *South African Journal of Botany*, *70*, 116–123.

Sogbohossou, E. O. D., Achigan-Dako, E. G., Maundu, P., Solberg, S., Deguenon, E. M. S, Mumm, R. H., Hale, I., Van Deynze, A., & Schranz, M. E. (2018). A roadmap for breeding orphan leafy vegetable species: A case study of *Gynandropsis gynandra* (Cleomaceae). *Horticulture Research*, *5*, 2.

Sogbohossou, E. O. D., Kortekaas, D., Achigan-Dako, E. G., Maundu, P., Stoilova, T., Deynze, V. A., de Vos, R. C. H., & Schranz, M. E. (2019). Association between vitamin content, plant morphology and geographical origin in a worldwide collection of the orphan crop *Cleome gynandra* (Cleomaceae). *Planta*, *250*, 933–947.

Songsak, T., & Lockwood, G. B. (2002). Glucosinolates of seven medicinal plants from Thailand. *Fitoterapi*, *73*, 209–216.

Sridhar, N., Kiran, B. V. V. S. S., Sasidhar, D. T., & Kanthal, L. K. (2014). In vitro antimicrobial screening of methanolic extracts of *Cleome chelidonii* and *Cleome gynandra*. *Bangladesh Journal of Pharmacology*, *9*, 161–166.

Tanor, E. B., & Magama, S. (2020). Assessment of the nutritional value of selected wild leafy vegetables growing in the Roma Valley, Lesotho. *European Journal of Nutrition and Food Safety*, *12*, 32–45.

Teklehaymanot, T., & Giday, M. (2010). Ethnobotanical study of wild edible plants of kara and kwego semi-pastoralist people in lower Omo River Valley, Debub Omo Zone, SNNPR, Ethiopia. *Journal of Ethnobiology and Ethnomedicine*, *6*, 23.

Thangwana, A., Gwata, E. T., & Zhou, M. M. (2021). Impact of chemical mutagenesis using ethyl methane sulphonate on tepary bean seedling vigour and adult plant performance. *Heliyon*, *7*, e06103.

Thovhogi, F., Gwata, E. T., Mchau, G. R. A., & Ntutshelo, N. (2021). Perceptions of end-users in Limpopo Province (South Africa) about the spider plant (*Cleome gynandra* L.). *Genetic Resources and Crop Evolution*, *68*, 605–614.

Thovhogi, F., Ntushelo, N., & Gwata, E. T. (2023). A comparative study of the presence of minerals, flavonoids and total phenolic compounds in the leaves of common traditional vegetables. *Applied Sciences*, *13*, 8503.

van der Bergh, E., Külahoglu, C., Bräutigam, A., Hibberd, J. M., Weber, A. P. M., Zhud, X., & Schranz, M. E. (2014). Gene and genome duplications and the origin of C4 photosynthesis: Birth of a trait in the leomaceae. *Current Plant Biology*, *1*, 2–9.

van Rensburg, W., Cloete, M., Gerrano, A., & Adebola, P. (2014). Have you considered eating your weeds? *American Journal of Plant Sciences*, *5*, 1110–1116.

Wessels, I., Maywald, M., & Rink, L. (2017). Zinc as a gatekeeper of immune function. *Nutrients*, *9*, 1286.

Woldetsadik, D., Drechsel, P., Keraita, B., Itanna, F., & Gebrekidan, H. (2017). Heavy metal accumulation and health risk assessment in wastewater-irrigated urban vegetable farming sites of Addis Ababa, Ethiopia. *International Journal of Food Contamination*, *4*, 9.

Wong, K. W., Yap, C. K., Nulit, R., Omar, H., Aris, A. Z., Cheng, W. H., Latif, M. T., & Leow, C. S. (2019). Zn in vegetables: A review and some insights. *Integrative Food, Nutrition and Metabolism*, *6*, 1–7.

Wu, T. H., Solberg, S. O., Yndgaard, F., & Chou, Y. Y. (2018). Morphological patterns in a world collection of Cleome gynandra. *Genetic Resources and Crop Evolution*, *65*, 271–283.

Wu, W., Li, R., Li, X., He, J., Jiang, S., Liu, S., & Yang, J. (2015). Quercetin as an antiviral agent inhibits influenza A virus (IAV) entry. *Viruses*, *8*, 1–18.

Yan, P., Zhang, L., Peng, C., & Zhang, R. (2018). Pharmacokinetics and tissue distribution of crotonoside. *Xenobiotica*, *48*, 28–36.

Zargoosh, Z., Ghavam, M., Bacchetta, G., & Tavili, A. (2019). Effects of ecological factors on the antioxidant potential and total phenol content of *Scrophularia striata* Boiss. *Scientific Reports*, *9*, 16021.

Zhang, S., Shen, J., Li, D., & Cheng, Y. (2020). Strategies in the delivery of Cas9 ribonucleoprotein for CRISPR/Cas9 genome editing. *Theranostics*, *11*, 614–648.

6 Are Wild Mushrooms a Panacea for Food and Nutrition Insecurity?

Nailoke Pauline Kadhila,[1] Isabella Shaningika Etuhole Ueitele,[2] Fimanekeni Ndaitavela Shivute,[2] Lydia Ndinelao Horn,[2] and Ahmad Cheikhyoussef[3]

[1]Department of Biochemistry, Microbiology, and Biotechnology, University of Namibia, Private Bag 13301, Windhoek, Namibia

[2]Zero Emission Research Initiative, Multidisciplinary Research Services, Centre for Research Services, University of Namibia, Private Bag 13301, Windhoek, Namibia

[3]Science and Technology Division, Multidisciplinary Research Services, Centre for Research Services, University of Namibia, Private Bag 13301, Windhoek, Namibia

*Corresponding author: npkadhila@unam.na

6.1 INTRODUCTION

Wild edible mushrooms have been used as food among different communities in many parts of the world due to their high-quality protein, rich fiber constituents, minerals, and vitamins, as well as low fat content and a high proportion of polyunsaturated fatty acids (Table 6.1) (Sitotaw et al., 2020). Their beneficial effects on human health and nutrition were recognized in early Greek, Egyptian, Roman, and Chinese civilizations (Das et al., 2021). Mushrooms contain components that can contribute to human wellness and mitigate threats that render the human body vulnerable to several life-threatening diseases (Živković et al., 2021). They have extensively been used as a food source for thousands of years due to their unique flavor, and recently, they have been used as functional food for the prevention of several human diseases. Mushrooms complement the human diet with various bioactive molecules (Table 6.2) that are deficient in foodstuffs derived from plants or animals (Bell et al., 2022). These bioactive molecules are responsible for mushrooms' pharmacological activities, contributing to their ability to prevent and treat various diseases. Increasing attention is being directed toward the diversity, ethnomycology, cultivation, and application of wild useful mushrooms worldwide. Folkloric medicine has long been used in traditional practice and a key cultural element of ethnic communities all over the world (Dulay et al., 2023). Basidiomycete mushrooms such as *Agaricus blazei* Murill, *Ganoderma lucidum*, *Hericium erinaceus*, and *Grifola frondosa* are consumed as immune response modifiers for the prevention of cancer and other conditions (Paloi et al., 2023).

Mushrooms are commonly used as nutritional support during chemotherapy and chronic inflammatory conditions, such as hepatitis, and other diseases (Hetland et al., 2020). Antiviral compounds isolated from mushrooms have potential activity against prominent viruses, such as human immunodeficiency virus, influenza, herpes simplex virus, and hepatitis B and C viruses, to name a few (Pradeep et al., 2019). Szú cko-Kociuba et al. (2023) reported that *H. erinaceus*, also known as *lion's mane mushroom*, is known for its strong bioactive properties. It shows promising potential as

DOI: 10.1201/9781003469766-6

TABLE 6.1
Nutritional Composition of Some Mushrooms

Mushroom Species	Nutritional Composition	References
Amanita	Ash (0.11–11.82%), carbohydrate (22.16–61.70%), protein (10.11–45.65%), fat (0.17–17.52%), fiber (1.18–30.30%), fatty acids (oleic acid: 58%), catechin (32.5 mg/g), ferulic acid (7 mg/g), p-coumaric acid (6 mg/g), and cinnamic acid (6.2 mg/g)	Kissanga et al., 2022 Kumla et al., 2023
Ganoderma	Protein (9–15%), carbohydrates (44%), minerals (Ca, Cu, Fe, Mg, Mn, K, Na, P, and Zn), dietary fiber (70%)	Singh et al., 2022; Yu et al., 2020; Garuba et al., 2020
Trametes	Protein (6–11%), carbohydrates (38–54%), minerals (Fe, Ca, Zn, Mn, Pb), fiber (4–20%), fiber (9–14%), fat (0.9–1.26%)	Upadhyaya et al., 2017; Mostafa et al., 2023
Oyster	Polysaccharides, lipopolysaccharides, proteins, peptides, glycoproteins, nucleosides, triterpenoids, lectins, lipids	Şöhretoğlu et al., 2023; Elkhateeb et al., 2022
Truffles	Proteins (5–27%), minerals (K, Fe, P, Ca), fats (2.15–9.33%), unsaturated fatty acids, carbohydrates (28–75%), fibers (7–13.2%), vitamin D, and metal-mineral contents	Khalifa et al., 2019; Elkhateeb et al., 2024
Termitomyces	Proteins (27–36%), ascorbic acid (10–18 mg/g), carbohydrates (37–51%), crude lipid (3–3.79%), crude fiber (3.08–9.46%), folic acid, niacin, thiamin, riboflavin, β-carotene and α-tocopherol, phosphorus, potassium, calcium, copper, and iron	Mattila et al., 2001; Nakalembe et al., 2015

TABLE 6.2
Biological Activities of Some Mushrooms

Mushroom Species	Biological Activities	References
Amanita	Antioxidant, neuroprotective effects	Doğan & Akbaş, 2013; Hu et al., 2021
Ganoderma	Antiallergic, anticancer, antidiabetic, anti-inflammatory, antibacterial, antifungal, antiviral, antioxidant, hepatoprotective, and immunomodulatory	Singh et al., 2020; Garuba et al., 2020
Trametes	Antiviral, antitumor, antiparasitic, antidiabetic, immunomodulation, antiobesity	Bains et al., 2021; Akgül et al., 2021
Oyster	Antioxidant, antidiabetic, protective osteoporosis, and cardiovascular	Kumla et al., 2023 Yadav et al., 2021
Truffles	Antioxidant, antimicrobial, anti-inflammatory, antidepressant, aphrodisiac	Doğan & Aydın, 2013; Elsayed et al., 2014; El Enshasy et al., 2013
Termitomyces	Antioxidant, anti-inflammatory, antimicrobial, anticancer, immunomodulation	Gebreyohannes et al., 2019; Tharavecharak et al., 2023 Araújo-Rodrigues et al., 2024; Ahmad et al., 2021; Caesario et al., 2023; Witasari et al., 2022

an excellent neuroprotective agent, capable of stimulating nerve growth factor release, regulating inflammatory processes, reducing oxidative stress, and safeguarding nerve cells from apoptosis. This chapter presents wild mushrooms as a panacea for food and nutrition security, with the purpose of highlighting the nutritional and biochemical potential of selected mushrooms that are abundant in Southern Africa. It was prepared by reviewing scholarly peer-reviewed articles from different databases, including, but not limited to, Google Scholar, PubMed, Scopus, and others.

6.2 *AMANITA* MUSHROOMS

Amanita is a genus of mushrooms under the family Amanitaceae. The classification of this genus is as follows: kingdom Fungi, division Basidiomycota, class Agaricomycetes, order Agaricales, and family Amanitaceae. They are mycorrhizal in nature and form symbiotic relationships with a wide range of deciduous and coniferous trees (de Mattos-Shipley et al., 2016). The *Amanita* genus is said to be comprised of about 500 described species, and there are fears that there is most likely a similar number of undescribed species (Zhang et al., 2015). *Amanita* mushrooms are important for both humans and ecosystems. The genus contains both poisonous, for example, *Amanita phalloides* (Vaill. ex Fr.), as well as edible, for example, *Amanita caesarea* (Scop.) Pers., species which are mostly found in the woodlands (de Mattos-Shipley et al., 2016; Luo et al., 2022). He et al. (2022) reported that the global annual number of fatalities due to mushroom poisoning is undetermined; however, Yilmaz et al. (2015), de Mattos-Shipley et al. (2016), and Falandysz et al. (2018) reported that the *Amanita* genus is responsible for 95% of mushroom poisoning incidents across the world. Ingestion of 50 g (three mushrooms) of *Amanita phalloides*, also known as the "death cap," accounts for most deaths due to mushroom poisoning (de Mattos-Shipley et al., 2016). Govorushko et al. (2019) outlined the incidence of mushroom poisoning and that incidence varies among different countries and depends on local traditions, lifestyle, nutritional factors, climate, and the occurrences of wild mushrooms, with the world highest mushroom poisoning incidences of 435 reported in China annually. According to Li et al. (2021), a total number of 10,036 mushroom poisoning was reported in China between 2010 and 2020, causing 38,676 illnesses, 21,967 hospitalizations, and 788 fatalities. The *Amanita* fungi species are known for causing ill effects on health and for their psychoactive qualities. For example, *Amanita muscaria* is an ectomycorrhizal mushroom that is native and common in the conifer and deciduous forests of the temperate and boreal zones of the Northern Hemisphere which is known for causing hallucinations due to a high bioconcentration of vanadium and cadmium in the fruiting bodies (Falandysz et al., 2018). In addition, *Amanita* can produce a wide range of mycotoxins which are classified into three groups of cyclic peptides, namely, amatoxins, phallotoxins, and virotoxins, which are known for inhibiting DNA-dependent RNA polymerase II, leading to impaired gene transcription, a process that reduces protein synthesis (de Mattos-Shipley et al., 2016). *Amanita* poisoning has also been reported in dogs after *A. exitialis* ingestion in Beijing, China (Sun et al., 2018). Hemorrhagic necrosis of hepatocytes is induced, leading to death 48–72 h after ingestion (Kaae et al., 2021). Because of its many negative effects, *A. muscaria* fruiting bodies are abundant in the areas where they grow and are available for research because people avoid touching them (Falandysz et al., 2018).

According to Falandysz et al. (2018), *A. muscaria* fruiting bodies can be consumed after slicing the stipes and caps into small pieces and blanching them in water for 15 or more minutes to lose their hallucinogens. In addition, *Amanita* species are used as an entheogen, a psychoactive, and known to cause a change in perception, mood, consciousness, cognition, or any behavior (Shapovalova et al., 2023; Winkelman, 2019). Hence, these fungi have been used for various purposes, such as religious, magical, shamanic, or spiritual, in many distinct parts of the world. According to Yilmaz et al. (2015), some of the deadly poisonous species that can also kill people and dogs include *Amanita abrupta*, *Amanita arocheae*, *Amanita exitialis* (Guangzhou destroying angel), *Amanita magnivelaris*, *Amanita ocreata* (western North American destroying angel), *Amanita subjunquillea* (East Asian death cap), European destroying angel (such as *Amanita virosa*), and others. The toxicity of most of the gunning in this genus is driven by the presence of compounds like amatoxins and phalloidins.

Amanita muscaria is a spectacular and well-recognizable ectomycorrhizal mushroom that is native and common in the conifer and deciduous forests of the temperate and boreal zones of the Northern Hemisphere. It has, in recent decades, been introduced by forestry into the Southern Hemisphere. This could be due to new species of trees or germplasms that are introduced to new areas carrying root diseases that are prone to exotic plantations caused predominantly by native pathogens (Wingfield et al., 2001). Most of the species in this genus form ectomycorrhizal relationships with vascular plants and play important roles in ecosystems (Zhang et al., 2015).

6.3 *GANODERMA* MUSHROOMS

The genus *Ganoderma* P. Karst is a popular and potent mushroom that consists of more than 250 species. *Ganoderma lucidum* is the most popular species in this genus, standing out from others, such as *G. sinense*, *G. applanatum*, *G. resinaceum*, *G. tuberculosum* Murill., *G. applanatum* Pat., *G. parvigibbosum* Welti and Courtec., *G. martinicense* Welti and Courtec., and *G. resinaceum* Boud. (Gong et al., 2019). The Asian people have benefited from *Ganoderma lucidum* (Lingzhi or Reishi) for over 2,000 years (Yuan et al., 2018). *Ganoderma* mushrooms are included in the Chinese and European pharmacopoeia, illustrating their importance. Therapeutic evidence of *Ganoderma* in Africa is documented too. In Ghana, *G. wiiroense*, *G. mbrekobenum*, and *G. enigmaticum* are used in ethnomedicine for immune regulation, treating hyperglycemia and liver disease (Peng et al., 2013). In the Democratic Republic of Congo, the safety of *G. applanatum* was demonstrated when up to 5,000 mg/kg methanol and ethanol extracts were administered to mice. Alanine aminotransferase and aspartate aminotransferase examinations afterward showed no resistance or alteration of major organs (Nyamangombe et al., 2023).

The medicinal properties of *Ganoderma* and various biological activities, such as immunomodulation, liver protection, antidiabetic, antitumor, and other neuroprotective effects, have been validated (Singh et al., 2022; Garuba et al., 2020). In vitro screening of methanol extracts derived from five *Ganoderma* species was conducted on human kidney (HK-2) cells to evaluate their potential protective properties against CisPt-induced tubulotoxicity. The findings revealed significant nephroprotective effects of *Ganoderma parvigibbosum* and *Ganoderma tuberculosum* against CisPt-induced tubulotoxicity in vitro (Sinaeve et al., 2022). Polysaccharides, nucleosides, triterpenoids, peptides, sterols, proteins, and alkaloids are responsible for these therapeutic activities, with polysaccharides and triterpenoids considered the most active (Liu et al., 2021).

Presently, the global market boasts a plethora of multimillion-dollar *Ganoderma* cultivation ventures alongside other mushroom varieties. Additionally, there is a burgeoning array of dietary supplements, functional foods, and cosmetics enriched with *Ganoderma* or its extracts, serving as the pivotal active ingredient in these products (Bishop et al., 2015; El Sheikha, 2022). However, commercialization and value addition of this mushroom are still scanty in many parts of Africa. Typically, the *Ganoderma* fruiting body is the preferred component for utilization. However, upon characterization of *Ganoderma* spores using LC-MS, triterpenoids emerged as the principal constituents of *Ganoderma lucidum* spores (Li et al., 2020).

6.4 *TRAMETES* MUSHROOMS

Like *Ganoderma*, *Trametes* has over 2,000 years of reputation as a complementary medicine or adjuvant in Asia. Polysaccharides are its main medicinal compound. With just over 50 species identified in this genus, *Trametes versicolor* is the most popular. *Trametes versicolor* (L.) Lloyd is a medicinal fungus also known as "yunzhi" in China and "turkey tail" in the West. *Trametes versicolor* stands as one of the most extensively researched medicinal mushrooms, spearheading the field by having its extracts, such as polysaccharide K (PSK) and polysaccharide P (PSP), approved and commercialized for cancer treatment (Lin et al., 2008; Orhan & Ustun, 2011; Harhaji et al., 2008; Lee et al., 2006; Lau et al., 2004).

The applications of *Trametes* are multiple, ranging from medicinal to industrial and textiles. These mushrooms produce enzymes such as laccases which are industrially used for bioremediation and biodegradation. The potential of *Trametes elegans* in food application was investigated by Eniolorunda (2016). According to this study, applying solvent extracts of *T. elegans* on pawpaw and banana fruit led to an extended shelf life of the fruit. Polypore mushrooms like *Ganoderma* and *Trametes* are tough and therefore not edible. These mushrooms and their products are therefore used as extracts and infusions of the whole basidiocarp or part thereof (Soares et al., 2013). Spores and mycelia are also considered to contain strong bioactive molecules.

Various studies have compared the composition of polar and non-polar solvent extracts, such as methanol, water, and ethyl acetate. There is a health and safety concern when using solvents such as methanol, although water has the disadvantage of getting contaminated. In their study, the hot water infusion of *T. gibbosa* contained a higher phenolic content than in the methanol extract (Bakhshi et al., 2022). According to Kachrimanidou et al. (2023) and Habtemariam (2020), the diversity of *Trametes* strains is unexplored and should be mined for novel polysaccharides from indigenous sources. Furthermore, they encourage research on the use of substrate from agricultural industrial waste to produce mycelial polysaccharides.

6.5 OYSTER MUSHROOM

Oyster mushrooms are macrofungi. They belong to the class Basidiomycota and the family Agaricaceae. They have a low fat content, high fiber, and protein, and they are natural sources of valuable food molecules, such as ergosterol, polyphenols, terpene and terpenoids, and mannitol and trehalose (Ayimbila et al., 2023; Devi et al., 2024). Except for button mushrooms, they are the second largest cultivated edible mushroom for various purposes, like food and medicine (Irfan et al., 2022). These edible fungi have been widely used as medicinal products in China, Japan, and Korea (Bell et al., 2022). Oyster mushrooms are made into dietary supplements and nutraceuticals that go beyond the usual health benefits of mushrooms (Lesa et al., 2022) and are becoming increasingly popular. Numerous mushroom extracts are used in folk medicine, like in cancer, which can be treated with the intake of mushrooms through the diet (Rizzo et al., 2021; Vetter, 2023). In the modern world, oyster mushrooms are known for their healing properties (Santhiya et al., 2022; Ayimbila et al., 2023). Oyster mushrooms are used as health enhancers more than any other mushroom (Kumar et al., 2022; Irfan et al., 2022). These mushrooms have a high potential to be used in functional meat for healthy and sustainable nutrition among consumers (Bulam et al., 2022) and as an ingredient in meat and starch-based food formulations, with varying degrees of success (Uukule, 2020).

Mushrooms and their by-products have been used to fortify various food products, as well as for use in animal feed, owing to their bioactive, therapeutic, and nutritional values (Kumar et al., 2022; Enyidi & Nwosu, 2023). Many biologically active compounds in this mushroom include polysaccharides, lipo-polysaccharides, proteins, peptides, glycoproteins, nucleosides, triterpenoids, lectins, lipids, and their derivatives (Şöhretoğlu et al., 2023; Elkhateeb et al., 2022). Oyster mushrooms are a good functional food for treating non-communicable diseases, such as osteoporosis and cardiovascular diseases (Yadav et al., 2021). The regular inclusion of mushrooms in one's diet can potentially help prevent degenerative diseases and maintain good health (Liuzzi et al., 2023). Therefore, edible mushrooms meet environmental and social requirements because their proteins are high in quality, low in cost, and widely available, making them suitable as sustainable alternative proteins (Niazi et al., 2021; Ayimbila et al., 2023).

6.6 TRUFFLES MUSHROOMS

Truffles (*Tuber* species) are a group of hypogeous ectomycorrhizal polyphyletic group of soil fungi belonging to Ascomycota, Basidiomycota, and Zygomycota with the ability to produce fruiting bodies that are appreciated for their high economic value and considered as one of the most expensive food types in the world (Zambonelli et al., 2016). There are more than 180 species under this group, with vast species diversity and soil biological characteristics (García-Montero et al., 2024). Jamali (2014) classified truffles into two broad types: tuber and desert. *Tuber melanosporum* Vittad., *Tuber aestivum* Vittad., *Tuber borchii* Vittad., and *Tuber magnatum* Pico are the common species for the tuber truffles, and *Terfezia metaxasi* Chatin, *Terfezia claveryi* Chatin, *Terfezia leonis* Tulasne, and *Terfezia boudieri* Chatin are the most common desert truffle species (Khalifa et al., 2019). Examples of the mycorrhizal fungi that are endemic to the semi-arid and arid geographical regions of the Mediterranean are *Tirmania* species and *Terfezia* species, while those from Southern Africa are known as *Kalaharituber pfeilii*; and this is why they are called "desert truffles" (Khalifa et al.,

2019; Haileka, 2015). Truffles contribute not only to food and nutrition security but also to public health security. Since time immemorial, they have been used in folkloric medicine to manage and treat various ailments and diseases, such as arthritis, skin and eye diseases, weakness, vomiting, wounds, colds, respiratory disorders, and rheumatism, among others (El Enshasy et al., 2013). The dried powder of truffles has been used for stomach ailments and open cut wounds, whereas the juice has been used for eye infection treatment (Lee et al., 2020). Therefore, truffles are relevant in the food and pharmaceutical industries (Veeraraghavan et al., 2022).

Several biological activities have been reported in truffles, including antioxidant, anti-inflammatory, antimicrobial, antimutagenic, and aphrodisiac activities (Dahham et al., 2018; Darwish et al., 2021). These biological activities can be linked to their bioactive compounds, such as phenolic, polysaccharide, triterpene, and steroids (Patel et al., 2017; Tejedor-Calvo et al., 2022). Truffles are rich in proteins (5–27%), minerals (K, Fe, P, Ca), fats (2.15–9.33%), unsaturated fatty acids (oleic and linoleic acid form more than 60% of total FA content), carbohydrates (28–75%), fibers (7–13.2%), vitamin D, and metal-mineral contents, which are important for human nutrition (Khalifa et al., 2019; Elkhateeb et al., 2024). Certain species of truffles, such as *Tuber maculatum*, have been reported as critical sources in the industrial production of important enzymes, such as amylase, xylanase, laccase, lipase, peroxidase, cellulase, and catalase, and *Tuber aestivum* as a producer of amylase, peroxidase, and catalase (Haileka, 2015). These truffles possess well-known advantages over other sources due to their higher yield, efficient production using inexpensive media, and catalytic activities (Thomas et al., 2021). Truffles contain a wide range of bioactive compounds, such as anthocyanins, amino acids, oligosaccharides, polysaccharides, flavonoids, carotenoids, pheromones, steroids, and volatile organic compounds (Doğan & Aydın, 2013; Khalifa et al., 2019; Yuliana et al., 2024). They possess unique volatile aromatic compounds with more than 200 volatile organic compounds (VOCs) that have been described from various truffle species, including butanedione, butyrate, dimethyl sulfide, dimethyl bisulfide, methylthiomethane, and hexenone (El Enshasy et al., 2013; Tejedor-Calvo et al., 2023).

6.7 *TERMITOMYCES* MUSHROOMS

The genus *Termitomyces* R. Heim belongs to the kingdom Fungi, phylum Basidiomycota, family Lyophyllaceae, and order Agaricales (Tharavecharak et al., 2023). *Termitomyces* mushrooms grow in association termites belonging to the subfamily Macrotermitinae, with many species found in Africa and Asia. They are uniquely important as food, and they hold promise in the development of nutritional supplements due to their ethnomedicinal values (Paloi et al., 2023). *Termitomyces* have historically served as valuable natural resources in various domains, including traditional medicine, human dietary practices, and livestock nutrition (Mediannikov et al., 2023). *Termitomyces* contain novel bioactive compounds with therapeutic antioxidant, immunomodulator, antirheumatic, antitumor, antimicrobial, and neuromodulator properties (Sharma et al., 2023).

6.7.1 Ethnomedicinal Significance of *Termitomyces* Species

Various ethnic groups prioritize differently when it comes to the utilization of natural resources; for example, some East Asian countries and Africa have well-documented traditional knowledge of mushrooms and have also found ways to use this knowledge in the present (Paloi et al., 2023). Several members of the genus *Termitomyces* are recognized for their ethnomedicinal importance to different ethnic groups and countries in Asia and Africa. For example, *T. microcarpus* is widely distributed across certain continents and can be employed in different ethnomedicinal applications. In Nigeria, especially among the Yoruba people, *T. microcarpus* is used to treat gonorrhea; in India, it treats fevers, colds, and fungal infections; while the native people of Tanzania and Nepal use the same species to boost the immune system and as an energy-stimulant tonic (Tibuhwa, 2012). In Cameroon, the same mushrooms are used to strengthen bones in children and for fever treatment (Paloi et al., 2023).

Latest studies indicate that *Termitomyces* are effective against rheumatism, high blood pressure, obesity, kwashiorkor, constipation, gastritis, and diarrhea, among others (Tharavecharak et al., 2023). The previous authors further added that the phenolic compounds and polysaccharides from the fruiting body of *Termitomyces* possess antioxidant, anti-inflammatory, antibacterial, and anticancer activities. In Namibia, Mukwata et al. (2023) and Kadhila-Muanding and Chimwamurombe (2012) reported that *Termitomyces schimperi* is used to heal diseases such as ringworms, kidney problems, sexually transmitted infections, heart diseases, severe diarrhea, cancer-related diseases, and in the prevention of miscarriages. Mukwata et al. (2023) reported that the pseudorhiza for the *Termitomyces* mushroom was the most important part used in healing various diseases reported in their study.

The extracts and derivatives of *Termitomyces* mushrooms have been reported to suppress the proliferation of malignant cells in vitro and to improve cancer response to cytotoxic therapy Tharavecharak et al. (2023). Anticancer activity of *Termitomyces* species derivatives has also been reported against several cancer cell lines from the liver, lungs, melanoma, colon, ovary, prostate, kidneys, central nervous system, breast, and leukemia, as reported by Tharavecharak et al. (2023). Caesario et al. (2023) added that *Termitomyces* species are used in Africa and Southeast Asia as blood pressure–lowering agents, as an anticholesterol, and for infections such as typhoid fever. They further reported that aqueous extract studies of fungus comb exerted antibiotic activities to some bacteria, such as *Salmonella enterica* and *Bacillus cereus*, and some highly resistant bacteria, such as methicillin-resistant *Staphylococcus aureus* and *Pseudomonas aeruginosa*.

6.8 CONCLUSION AND PROSPECTS

Owing to the value of mushrooms as food, it is very common to see people selling mushrooms alongside roads after the start of the rainy season in many African counties. In this chapter, it is evidenced that the consumption of mushrooms makes a valuable addition to unbalanced diets by providing essential nutrients. In addition, mushrooms provide raw materials for the development of functional food and nutraceuticals for health management. This is especially crucial in undernourished communities, providing a shift from calorie-dense foods with low nutrients to sustainably sourced mushrooms and mushroom products. Mushrooms as functional foods can help in the early intervention of sub-healthy states in humans to prevent the consequences of life-threatening diseases. Wild mushroom populations are declining due to diminishing land access and changing ecosystems by human activities, such as the removal of deadwood. Intense research on wild mushrooms, especially in Africa, is needed to expound the different roles of multiple active compounds and the pathways involved. Mushrooms have a short crop cycle; they can be cultivated in expediated quantities for nutrition and biomedical applications to address deadly infectious diseases worldwide. Sustainable utilization of wild mushrooms is encouraged, as well as domestication for cultivation. Documentation of wild medicinal mushrooms associated with indigenous knowledge can potentially contribute to their management and utilization to ensure sustainability. Another issue in the management and conservation of wild mushrooms is open-access harvesting. Therefore, policy guidelines are needed for the sustainable harvest of wild mushrooms in communities where such regulations do not currently exist. There is also a need for creating awareness on the diversity and importance of wild mushrooms in communities where basic mycological knowledge is limited. It is essential to note that some mushrooms are poisonous and potentially lethal, highlighting the need for proper education and extra caution in identifying species that can be consumed, as some edible mushrooms have look-alikes that can be fatal. The mushrooms market demand is growing regionally and internationally, which calls for intensive ethnomycological surveys and domestication for possible cultivation to meet the demand. Wild mushrooms can be processed into value-added nutritional food, either lonesome or in combination with local foods, to increase consumption for numerous health benefits and to benefit those who find fresh mushrooms not palatable. Lastly, the need for genetic conservation of wild mushrooms is acknowledged, highlighting the need to develop a mushroom gene bank not only in Southern Africa but also globally. Vigorous research on wild basidiomycetes for more insight and promising benefits to humankind is thus encouraged. Overall, wild mushrooms can enhance health and nutrition, but careful management and research are crucial for sustainable benefits. This chapter highlights that wild mushrooms are a potential panacea for food and nutrition insecurity not only in Namibia but also in the world.

6.9 ACKNOWLEDGMENTS

We are grateful to the Multidisciplinary Research and Biochemistry, Microbiology, and Biotechnology Departments of the University of Namibia for supporting this work.

REFERENCES

Ahmad, R., Sellathoroe, S., Ngadi, E., Saharuddin, T. T. S., Zakaria, I. I., Selvakumaran, S., & Wan-Mohtar, A. I. W. A. (2021). Isolation, identification, cultivation, and determination of antimicrobial β-glucan from a wild-termite mushroom Termitomyces heimii RFES 230662. *Biocatalysis and Agricultural Biotechnology*, *37*, 102187. https://doi.org/10.1016/j.bcab.2021.102187

Akgül, H., Aslan, A., Akata, I., Günal, S., Bal, C., & Baba, H. (2021). Phenolic content and biological activities of Trametes hirsuta. *Fresenius Environmental Bulletin*, *30*(4a), 4130–4135.

Araújo-Rodrigues, H., Sousa, A. S., Relvas, J. B., Tavaria, F. K., & Pintado, M. (2024). An overview on mushroom polysaccharides: Health-promoting properties, prebiotic and gut microbiota modulation effects and structure-function correlation. *Carbohydrate Polymers*, *333*, 121978. https://doi.org/10.1016/j.carbpol.2024.121978

Ayimbila, F., & Keawsompong, S. (2023). Nutritional quality and biological application of mushroom protein as a novel protein alternative. *Current Nutrition Reports*, *12*, 290–307. https://doi.org/10.1007/s13668-023-00468-x

Bains, A., Chawla, P., Kaur, S., Najda, A., Fogarasi, M., & Fogarasi, S. (2021). Bioactives from mushroom: Health attributes and food industry applications. *Materials*, *14*, 7640. https://doi.org/10.3390/ma14247640

Bell, V., Silva, C. R. P. G., Guina, J., & Fernandes, T. H. (2022). Mushrooms as future generation healthy foods. *Frontiers in Nutrition*, *9*, 1050099. https://doi.org/10.3389/fnut.2022.1050099

Bishop, K. S., Kao, C. H., Xu, Y., Glucina, M. P., Paterson, R. R., & Ferguson, L. R. (2015). From 2000 years of *Ganoderma lucidum* to recent developments in nutraceuticals. *Phytochemistry*, *114*, 56–65. https://doi.org/10.1016/j.phytochem.2015.02.015

Bulam, S., Üstün, N., & Pekşen, A. (2022). Oyster mushroom (Pleurotus ostreatus) as a healthy ingredient for sustainable functional food production. *Mantar Dergisi*, *13*(3), 131–143. https://doi.org/10.30708/mantar.1192063

Caesario, J., Hertanto, M. D., Susanto, H., Sudiana, K., Nandika, D., Karlinasari, L., Arinana, A., Batubara, I., Witasari, L. D., Rachmayanti, Y., Firmasyah, D., & Djoko Santoso, D. (2023). Hematological profile of mice after ethyl acetate extract of fungus comb of indo-malayan termite (Macrotermes gilvus Hagen) mound supplementation in regulating lipopolysaccharide-induced inflammatory response. *Malaysian Journal of Medicine and Health Sciences*, *19*(3), 57–63.

Dahham, S. S., Al-Rawi, S. S., Ibrahim, A. H., Abdul Majid, A. S., & Abdul Majid, A. M. S. (2018). Antioxidant, anticancer, apoptosis properties and chemical composition of black truffle Terfezia claveryi. *Saudi Journal of Biological Sciences*, *25*(8), 1524–1534. https://doi.org/10.1016/j.sjbs.2016.01.031

Darwish, R. S., Shawky, E., Nassar, K. M., Rashad ElSayed, R. M., Hussein, D. E., Ghareeb, D. A., & El Sohafy, S. M. (2021). Differential anti-inflammatory biomarkers of the desert truffles Terfezia claveryi and Tirmania nivea revealed via UPLC-QqQ-MS-based metabolomics combined to chemometrics. *LWT*, *150*, 111965. https://doi.org/10.1016/j.lwt.2021.111965

Das, A. K., Nanda, P. K., Dandapat, P., Bandyopadhyay, S., Gullón, P., Sivaraman, G. K., McClements, D. J., Gullón, B., & Lorenzo, J. M. (2021). Edible mushrooms as functional ingredients for development of healthier and more sustainable muscle foods: A flexitarian approach. *Molecules*, *26*, 2463. https://doi.org/10.3390/molecules26092463

de Mattos-Shipley, K. M. J., Ford, K. L., Alberti, F., Banks, A. M., Bailey, A. M., & Foster, G. D. (2016). The good, the bad and the tasty: The many roles of mushrooms. *Studies in Mycology*, *85*(1), 125–157. (33 Westerdijk Fungal Biodiversity Institute). https://doi.org/10.1016/j.simyco.2016.11.002

Devi, P. V., Islam, J., Narzary, P., Sharma, D., & Sultana, F. (2024). Bioactive compounds, nutraceutical values and its application in food product development of oyster mushroom. *Journal of Future Foods*, *4*(4), 335–342. https://doi.org/10.1016/j.jfutfo.2023.11.005

Doğan, H. H., & Akbaş, G. (2013). Biological activity and fatty acid composition of Caesar's mushroom. *Pharmaceutical Biology*, *51*(7), 863–871. https://doi.org/10.3109/13880209.2013.768272

Doğan, H. H., & Aydın, S. (2013). Determination of antimicrobial effect, antioxidant activity and phenolic contents of desert truffle in Turkey. *African Journal of Traditional, Complementary, and Alternative Medicines*, *10*(4), 52–58.

Dulay, R. M. R., Batangan, J. N., Kalaw, S. P., De Leon, A. M., Cabrera, E. C., Kimura, K., Eguchi, F., & Reyes, R. G. (2023). Records of wild mushrooms in the Philippines: A review. *Journal of Applied Biology and Biotechnology*, *11*(2), 11–32. https://doi.org/10.7324/JABB.2023.110202

El Enshasy, H., Elsayed, E. A., Aziz, R., & Wadaan, M. A. (2013). Mushrooms and truffles: Historical biofactories for complementary medicine in Africa and in the middle East. *Evidence-Based Complementary and Alternative Medicine*, 620451. https://doi.org/10.1155/2013/620451

Elkhateeb, W. A., & Daba, G. M. (2021). Medicinal mushroom what should we know. *International Journal of Pharmaceutical Chemistry and Analysis*, *9*(1), 1–19.

Elkhateeb, W. A., Soliman, A. A. F., Shaheen, M. N., Elmahdy, E. M., & Daba, G. M. (2024). Bioactive potentials of the truffle mushrooms *Tirmania nivea*, *Tirmania pinoyi* and *Tuber indicum*. *Egyptian Pharmaceutical Journal*, *23*(1), 94–102. https://doi.org/10.4103/epj.epj_204_23

Elsayed, E. A., El Enshasy, H., Wadaan, M. A., & Aziz, R. (2014). Mushrooms: A potential natural source of anti-inflammatory compounds for medical applications. *Mediators of Inflammation*. https://doi.org/10.1155/2014/805841

El Sheikha, A. F. (2022). Nutritional profile and health benefits of *Ganoderma lucidum* "lingzhi, reishi, or mannentake" as functional foods: Current scenario and future perspectives. *Foods (Basel, Switzerland)*, *11*(7), 1030. https://doi.org/10.3390/foods11071030

Eniolorunda, T. A. (2016). *Assessment of the preservative efficacy of extracts of Trametes elegans on pawpaw and banana fruits*. The Federal University of Technology.

Enyidi, U. D., & Nwosu, C. (2023). Effects of oyster mushroom, Lentinus sajor-caju on growth, gut histology and haematology of Clarias gariepinus. *Journal of Aquatic Sciences*, *38*(1).

Falandysz, J., Mędyk, M., & Treu, R. (2018). Bio-concentration potential and associations of heavy metals in *Amanita muscaria* (L.) Lam. from Northern regions of Poland. *Environmental Science and Pollution Research*, *25*, 25190–25206. https://doi.org/10.1007/s11356-018-2603-0

García-Montero, L. G., Monleón, V. J., Valverde-Asenjo, I., Menta, C., & Kuyper, T. W. (2024). Niche construction by two ectomycorrhizal truffle species (*Tuber aestivum* and *T. melanosporum*). *Soil Biology and Biochemistry*, *189*. https://doi.org/10.1016/j.soilbio.2023.109276

Garuba, T., Olahan, G. S., Lateef, A. A., Alaya, R. O., Awolowo, M., & Sulyman, A. (2020). Proximate composition and chemical profiles of reishi mushroom (*Ganoderma lucidum*). *Journal of Scientific Research*, *12*(1), 103–110. https://doi.org/10.3329/jsr.v12i1.42059

Gebreyohannes, G., Nyerere, A., Bii, C., & Sbhatu, D. B. (2019). Investigation of antioxidant and antimicrobial activities of different extracts of auricularia and termitomyces species of mushrooms. *Scientific World Journal*, *2019*, 7357048. https://doi.org/10.1155/2019/7357048

Gong, T., Yan, R., Kang, J., & Chen, R. (2019). Chemical components of ganoderma. *Advances in Experimental Medicine and Biology*, 59–106. https://doi.org/10.1007/978-981-13-9867-4_3

Govorushko, S., Rezaee, R., Dumanov, J., & Tsatsakis, A. (2019). Poisoning associated with the use of mushrooms: A review of the global pattern and main characteristics. *Food and Chemical Toxicology*, *128*, 267–279.

Habtemariam, S. (2020). *Trametes versicolor* (Synn. *Coriolus versicolor*) polysaccharides in cancer therapy: Targets and efficacy. *Biomedicines*, *8*(5), 135. https://doi.org/10.3390/biomedicines8050135

Haileka, V. L. (2015). *Screening, isolation and characterization of laccase enzymes from Namibian Termitomyces schimperi and Kalaharituber pfeilii* [Master's thesis, University of Namibia]. http://hdl.handle.net/11070/1451

Harhaji, L. J., Mijatović, S., Maksimović-Ivanić, D., Stojanović, I., Momcilović, M., Maksimović, V., Tufegdzić, S., Marjanović, Z., Mostarica-Stojković, M., Vucinić, Z., & Stosić-Grujicić, S. (2008). Anti-tumor effect of coriolus versicolor methanol extract against mouse B16 melanoma cells: In vitro and in vivo study. *Food and Chemical Toxicology*, *46*(5), 1825–1833. https://doi.org/10.1016/j.fct.2008.01.027

He, M. Q., Wang, M. Q., Chen, Z. H., Deng, W. Q., Li, T. H., Vizzini, A., Jeewon, R., Hyde, K. D., & Zhao, R. L. (2022). Potential benefits and harms: A review of poisonous mushrooms in the world. *Fungal Biology Reviews*, *42*, 56–68.

Hetland, G., Tangen, J. M., Mahmood, F., Mirlashari, M. R., Nissen-Meyer, L. S. H., Nentwich, I., Therkelsen, S. P., Tjønnfjord, G. E., & Johnson, E. (2020). Antitumor, anti-inflammatory and antiallergic effects of *Agaricus blazei* mushroom extract and the related medicinal basidiomycetes mushrooms, *Hericium erinaceus* and *Grifolafrondosa*: A review of preclinical and clinical studies. *Nutrients*, *12*(5), 1339. https://doi.org/10.3390/nu12051339

Hu, W., Li, Z., Wang, W., Song, M., Dong, R., Zhou, Y., Li, Y., & Wang, D. (2021). Structural characterization of polysaccharide purified from *Amanita caesarea* and its pharmacological basis for application in Alzheimer's disease: Endoplasmic reticulum stress. *Food & Function*, *12*(21), 11009–11023. https://doi.org/10.1039/d1fo01963e

Jamali, S. (2014). Taxonomy of the truffles. *Plant Science Today*, *1*(4), 219–222. https://horizonepublishing.com/journals/index.php/PST/article/view/55.

Jouybari, H. B., Bekhradnia, A., Mirzaee, F., Hosseinzadeh, M. H., & Habibi, E. (2022). Chemical composition of the lumpy bracket mushroom (Trametes gibbosa). *Research Journal of Pharmacognosy*, *9*(2), 19–27. https://doi.org/10.22127/rjp.2022.316499.1813

Kaae, J. A., Poppenga, R. H., & Hill, A. E. (2021). Physical examination, serum biochemical, and coagulation abnormalities, treatments, and outcomes for dogs with toxicosis from α-amanitin–containing mushrooms: 59 cases (2006–2019). *Journal of the American Veterinary Medical Association*, *258*(5), 502–509.

Kachrimanidou, V., Alexandri, M., Papapostolou, H., Papadaki, A., & Kopsahelis, N. (2023). Valorization of grape pomace for *Trametes versicolor* mycelial mass and polysaccharides production. *Sustainability*, *15*, 15080. https://doi.org/10.3390/su152015080

Kadhila-Muandingi, N. P., & Chimwamurombe, P. M. (2012). Uses of ganoderma and other mushrooms as medicine in Oshana and Ohangwena region of Northern Namibia. *Journal of Research in Agriculture*, *1*, 146–151.

Khalifa, S. A. M., Farag, M. A., Yosri, N., Sabir, J. S. M., Saeed, A., Al-Mousawi, S. M., . . . El-Seedi, H. R. (2019). Truffles: From Islamic culture to chemistry, pharmacology, and food trends in recent times. *Trends in Food Science and Technology*, *91*, 193–218. https://doi.org/10.1016/j.tifs.2019.07.008

Kissanga, R., Liberal, Â., Diniz, I., Rodrigues, A. S. B., Baptista-Ferreira, J. L., Batista, D., Ivanov, M., Soković, M., Ferreira, I. C. F. R., Fernandes, Â., Barros, L., & Catarino, L. (2022). Biochemical and molecular profiling of wild edible mushrooms from Huila, Angola. *Foods*, *11*(20), 3240. https://doi.org/10.3390/foods11203240

Kumar, H., Bhardwaj, K., Kuča, K., Sharifi-Rad, J., Verma, R., Machado, M., Kumar, D., & Cruz-Martins, N. (2022). Edible mushrooms' enrichment in food and feed: A mini review. *International Journal of Food Science & Technology*, *57*, 1386–1398. https://doi.org/10.1111/ijfs.15546

Kumla, J., Suwannarach, N., Liu, Y. S., Tanruean, K., & Lumyong, S. (2023). Survey of edible amanita in Northern Thailand and their nutritional value, total phenolic content, antioxidant and α-glucosidase inhibitory activities. *Journal of Fungi*, *9*(3). https://doi.org/10.3390/jof9030343

Lau, C. B., Ho, C. Y., Kim, C. F., Leung, K. N., Fung, K. P., Tse, T. F., Chan, H. H., & Chow, M. S. (2004). Cytotoxic activities of Coriolus versicolor (Yunzhi) extract on human leukemia and lymphoma cells by induction of apoptosis. *Life Science*, *75*(7), 797–808. https://doi.org/10.1016/j.lfs.2004.04.001

Lee, C., Yang, X., & Wan, J. (2006). The culture duration affects the immunomodulatory and anticancer effect of polysaccharopeptide derived from *Coriolus versicolor. Enzyme and Microbial Technology*, *38*, 14–21.

Lee, H., Nam, K., Zahra, Z., & Farooqi, M. Q. U. (2020). Potentials of truffles in nutritional and medicinal applications: A review. *Fungal Biology and Biotechnology*, *7*, 9. https://doi.org/10.1186/s40694-020-00097-x

Lesa, K. N., Khandaker, M. U., Faruque, M. R. I., Sharma, R., Islam, F., Mitra, S., & Emran, T. B. (2022). Nutritional value, medicinal importance, and health-promoting effects of dietary mushroom (Pleurotus ostreatus). *Journal of Food Quality*. https://doi.org/10.1155/2022/2454180

Li, W., Pires, S. M., Liu, Z., Liang, J., Wang, Y., Chen, W., Liu, C., Liu, J., Han, H., Fu, P., & Guo, Y. (2021). Mushroom poisoning outbreaks – China, 2010–2020. *China CDC Weekly*, *3*(24), 518.

Li, Z., Shi, Y., Zhang, X., Xu, J., Wang, H., Zhao, L., & Wang, Y. (2020). Screening immunoactive compounds of *Ganoderma lucidum* spores by mass spectrometry molecular networking combined with *in vivo* zebrafish assays. *Frontiers in Pharmacology*, *11*, 287. https://doi.org/10.3389/fphar.2020.00287

Lin, F. Y., Lai, Y. K., Yu, H. C., Chen, N. Y., Chang, C. Y., Lo, H. C., & Hsu, T. H. (2008). Effects of Lycium barbarum extract on production and immunomodulatory activity of the extracellular polysaccharopeptides from submerged fermentation culture of Coriolus versicolor. *Food Chemistry*, *110*(2), 446–453. https://doi.org/10.1016/j.foodchem.2008.02.023

Liu, Y. C., Tang, X. C., Hu, H. P., Chen, D. L., Xie, Y. Z., Liang, X. W., Li, X. M., Xiao, C., Huang, L. H., & Wu, Q. P. (2021). Genetic diversity and main functional composition of Lingzhi strains from main producing areas in China. *AMB Express*, *11*(1), Article 119. https://doi.org/10.1186/s13568-021-01280-y

Liuzzi, G. M., Petraglia, T., Latronico, T., Crescenzi, A., & Rossano, R. (2023). Antioxidant compounds from edible mushrooms as potential candidates for treating age-related neurodegenerative diseases. *Nutrients*, *15*(8), 1913.

Luo, H., Hallen-Adams, H. E., Lüli, Y., Sgambelluri, R. M., Li, X., Smith, M., Yang, Z. L., & Martin, F. M. (2022). Genes and evolutionary fates of the amanitin biosynthesis pathway in poisonous mushrooms. *Proceedings of the National Academy of Sciences*, *119*(20), https://doi.org/10.1073/pnas.2201113119

Mattila, P., Könkö, K., Eurola, M., Pihlava, J. M., Astola, J., Vahteristo, L., Hietaniemi, V., Kumpulainen, J., Valtonen, M., & Piironen, V. (2001). Contents of vitamins, mineral elements, and some phenolic compounds in cultivated mushrooms. *Journal of Agricultural and Food Chemistry*, *49*(5), 2343–2348. https://doi.org/10.1021/jf001525d

Mediannikov, O., Gaye, M., Armstrong, et al. (2023). Antibiotic activity of fungus comb extract from fungus-farming termites, PREPRINT (Version 1). *Research Square*. https://doi.org/10.21203/rs.3.rs-3467827/v1

Mostafa, Y. S., Širić, I., Alamri, S. A. M., Alrumman, S. A., Kumar, P., Abou Fayssal, S., Zjalić, S., Singh, R., & Eid, E. M. (2023). Assessment of metal elements and biochemical constituents of wild turkey tail (*Trametes versicolor*) mushrooms collected from the shivalik foothills of the Himalayas, India. *Forests*, *14*, 2247. https://doi.org/10.3390/f14112247

Mukwata, A., Kadhila, N. P., & Horn, L. N. (2023). Ethnomycological study of Termitomyces mushrooms at Judea-Lyabboroma and Katima rural in the Zambezi Region, Namibia. *International Science and Technology Journal of Namibia*, *16*, 54–64. https://journals.unam.edu.na/index.php/ISTJN/article/view/1858

Nakalembe, I., Kabasa, J. D., & Olila, D. (2015). Comparative nutrient composition of selected wild edible mushrooms from two agro-ecological zones, Uganda. *SpringerPlus*, *4*, 433. https://doi.org/10.1186/s40064-015-1188-z

Niazi, A. R., & Ghafoor, A. (2021). Different ways to exploit mushrooms: A review. *All Life*, *14*(1), 450–460. https://doi.org/10.1080/26895293.2021.1919570

Nyamangombe, G., Mayele, B., Nzembo, F., Nzunda, J.-P., & Amogu, J.-J. (2023). Evaluation of the toxicity of aqueous extract of the Ganoderma applanatum mushroom. *Moroccan Journal of Agricultural Sciences* (Moroccan Journal of Agricultural Sciences). https://doi.org/10.5281/zenodo.10426349

Orhan, I., & Ustun, O. (2011). Determination of total phenol content, antioxidant activity and acetylcholinesterase inhibition in selected mushrooms from Turkey. *Journal of Food Composition and Analysis*, *24*(3), 386–390.

Paloi, S., Kumla, J., Paloi, B. P., Srinuanpan, S., Hoijang, S., Karunarathna, S. C., Acharya, K., Suwannarach, N., & Lumyong, S. (2023). Termite mushrooms (Termitomyces), a potential source of nutrients and bioactive compounds exhibiting human health benefits: A review. *Journal of Fungi*, *9*, 112. https://doi.org/10.3390/jof9010112

Patel, S., Rauf, A., Khan, H., Khalid, S., & Mubarak, M. S. (2017). Potential health benefits of natural products derived from truffles: A review. *Trends in Food Science and Technology*, *70*, 1–8. https://doi.org/10.1016/j.tifs.2017.09.009

Peng, X. R., Liu, J. Q., Han, Z. H., Yuan, X. X., Luo, H. R., & Qiu, M. H. (2013). Protective effects of triterpenoids from Ganoderma resinaceum on H_2O_2-induced toxicity in HepG2 cells. *Food Chemistry*, *141*(2), 920–926. https://doi.org/10.1016/j.foodchem.2013.03.071

Pradeep, P., Manju, V., & Ahsan, M. F. (2019). Antiviral potency of mushroom preclinical and clinical studies. *Nutrients*, *12*(5), 1339. https://doi.org/10.3390/nu12051339

Rizzo, G., Goggi, S., Giampieri, F., & Baroni, L. (2021). A review of mushrooms in human nutrition and health. *Trends in Food Science and Technology*, *117*, 60–73. https://doi.org/10.1016/j.tifs.2020.12.025

Santhiya, K., Abinaya, S., Aswini, R., & Nitheshlee, M. (2022). Assessment of button and oyster mushroom nutritional quality using various drying methods. *Carpathian Journal of Food Science and Technology*, *14*(3), 171–181.

Shapovalova, V., Osyntseva, A., Shapovalov, V., & Nehretskii, S. (2023). Pharmacognostic, forensic and pharmaceutical, organizational and legal, clinical and pharmacological multidisciplinary study with an assessment of peculiarities of circulation (use) of smoking mixtures of spices and entheogens of amanita muscaria mushroom. *SSP Modern Pharmacy and Medicine*, *3*(3), 1–13.

Sharma, A., Sharma, V., Giri, A., Kumar, N., & Bhardwaj, P. (2023). Phytochemistry and nutritional composition of significant wild medicinal and edible mushrooms traditional uses and pharmacology. In Ajay Sharma, Garima Bhardwaj, Gulzar Ahmad Nayik (Eds.), *Royal society of chemistry* (Chapter 16, pp. 341–354). Royal Society of Chemistry. https://doi.org/10.1039/9781837672097.

Sinaeve, S., Husson, C., Antoine, M.-H., Welti, S., Stévigny, C., & Nortier, J. (2022). Nephroprotective effects of two ganoderma species methanolic extracts in an in vitro model of cisplatin induced tubulotoxicity. *Journal of Fungi*, *8*, 1002. https://doi.org/10.3390/jof8101002

Sitotaw, R., Lulekal, E., & Abate, D. (2020). Ethnomycological study of edible and medicinal mushrooms in Menge District, Asossa Zone, Benshangul Gumuz Region, Ethiopia. *Journal of Ethnobiology Ethnomedicine*, *16*, 11. https://doi.org/10.1186/s13002-020-00361-9

Soares, A. A., De Sá-Nakanishi, A. B., Bracht, A., Da Costa, S. M. G., Koehnlein, E. A., De Souza, C. G. M., & Peralta, R. M. (2013). Hepatoprotective effects of mushrooms. *Molecules*, *18*(7), 7609–7630. https://doi.org/10.3390/molecules18077609

Şöhretoğlu, D., & Kuruüzüm-Uz, A. (2023). Biological activities of some edible mushrooms. In D. C. Agrawal & M. Dhanasekaran (Eds.), *Mushrooms with therapeutic potentials*. Springer. https://doi.org/10.1007/978-981-19-9550-7_14

Sun, J., Niu, Y. M., Zhang, Y. T., Li, H. J., Yin, Y., Zhang, Y. Z., Ma, P. B., Zhou, J., Huang, L., Zhang, H. S., & Sun, C. Y. (2018). Toxicity and toxicokinetics of *Amanita exitialis* in beagle dogs. *Toxicon*, *143*, 59–67. https://doi.org/10.1016/j.toxicon.2018.01.008

Tejedor-Calvo, E., García-Barreda, S., Sánchez, S., Morte, A., Siles-Sánchez, M. L. N., Soler-Rivas, C., Santoyo, S., & Marco, P. (2022). Application of pressurized liquid extractions to obtain bioactive compounds from *Tuber aestivum* and *Terfezia claveryi*. *Foods*, *11*(3), 298. https://doi.org/10.3390/foods11030298

Tharavecharak, S. D., Alessandro-Gabazza, C. N., Toda, M., Yasuma, T., Tsuyama, T., Kamei, I., & Gabazza, E. C. (2023). Culture conditions for mycelial growth and anti-cancer properties of *Termitomyces*. *Mycobiology*, *51*(2), 94–108. https://doi.org/10.1080/12298093.2023.2187614

Thomas, P., Elkhateeb, W., & Daba, G. (2021). Industrial applications of truffles and truffle-like fungi. In *Advances in Macrofungi* (pp. 82–88). CRC Press.

Tibuhwa, D. D. (2012). Folk taxonomy and use of mushrooms in communities around Ngorongoro and Serengeti National Park, Tanzania. *Journal of Ethnobiology Ethnomedicine*, *8*, 36. https://doi.org/10.1186/1746-4269-8-36

Upadhyaya, J., Raut, J. K., & Koirala, N. (2017). *Analysis of nutritional and nutraceutical properties of wild grown mushrooms of Nepal*. EC Microbiology.

Uukule, E. N. (2020). *Nutritional composition and in vitro starch digestibility of crackers from pearl millet and wheat composted with mushrooms* [MSc thesis, University of Namibia].

Veeraraghavan, V. P., Hussain, S., Balakrishna, J. P., Dhawale, L., Kullappan, M., Ambrose, J. M., & Mohan, S. K. (2022). A comprehensive and critical review on ethnopharmacological importance of desert truffles: *Terfezia claveryi*, *Terfezia boudieri*, and *Tirmania nivea*. *Food Reviews International*, *38*, 846–865. https://doi.org/10.1080/87559129.2021.1889581

Vetter, J. (2023). The mushroom glucans: Molecules of high biological and medicinal importance. *Foods*, *12*(5), 1009. https://doi.org/10.3390/foods12051009

Wingfield, M. J., Slippers, B., Roux, J., & Wingfield, B. D. (2001). Worldwide movement of exotic forest fungi, especially in the tropics and the Southern hemisphere: This article examines the impact of fungal pathogens introduced in plantation forestry. *Bioscience*, *51*(2), 134–140.

Winkelman, M. (2019). Introduction: Evidence for entheogen use in prehistory and world religions. *Journal of Psychedelic Studies*, *3*(2), 43–62.

Witasari, L. D., Wahyu, K. W., Anugrahani, B. J., Kurniawan, D. C., Haryanto, A., Nandika, D., Karlinasari, L., Arinana, A., Batubara, I., Santoso, D., Rachmayanti, Y., Firmansyah, D., Sudiana, I. K., & Hertanto, D. M. (2022). Antimicrobial activities of fungus comb extracts isolated from Indomalayan termite (*Macrotermes gilvus* Hagen) mound. *AMB Express*, *12*(1), 14. https://doi.org/10.1186/s13568-022-01359-0

Yadav, D., & Negi, P. S. (2021). Bioactive components of mushrooms: Processing effects and health benefits. *Food Research International*, *148*, 110599. https://doi.org/10.1016/j.foodres.110599

Yilmaz, I., Ermis, F., Akata, I., & Kaya, E. (2015). A case study: What doses of Amanita phalloides and amatoxins are lethal to humans? *Wilderness and Environmental Medicine*, *26*(4), 491–496.

Yu, Q., Guo, M., Zhang, B., Wu, H., Zhang, Y., & Zhang, L. (2020). Analysis of nutritional composition in 23 kinds of edible fungi. *Journal of Food Quality*, *9*. https://doi.org/10.1155/2020/8821315

Yuliana, N. D., Saraswati, H., Regiyana, Y., Nora, A., Hutabarat, D. J. C., & Lioe, H. N. (2024). Advances in metabolomics to optimize quality parameters of culinary mushrooms. *Trends in Food Science & Technology*, *145*, 104378. https://doi.org/10.1016/j.tifs.2024.104378

Zambonelli, A., Iotti, M., & Murat, C. (2016). *True truffle (Tuber species) in the world* (Vol. 47). Springer International Publishing.

Zhang, P., Tang, L.-P., Cai, Q., & Xu, J.-P. (2015). A review on the diversity, phylogeography and population genetics of Amanita mushrooms. *Mycology*, *6*(2), 86–93. https://doi.org/10.1080/21501203.2015.1042536

Živković, J., Ivanov, M., Stojković, D., & Glamočlija, J. (2021). Ethnomycological investigation in Serbia: Astonishing realm of mycomedicines and mycofood. *Journal of Fungi*, *7*, 349. https://doi.org/10.3390/jof7050349

7 The Potential of Underutilized and Undervalued *Moringa Oleifera* Plant Accessions for Nutritional and Food Security

Nikita Patel,[1,2] *Stephen Oyedeji,*[1,2,3]
David Adedayo Animasaun,[3] *Kehinde Stephen Olorunmaiye,*[3] *and *Ramar Krishnamurthy*[1,2]
[1]C. G. Bhakta Institute of Biotechnology, Uka Tarsadia University, Pin 394350, Surat District, Gujarat State, India
[2] Kishorbhai Institute of Agriculture Sciences and Research Centre, Uka Tarsadia University Pin 394350, Surat District, Gujarat State, India
[3]Department of Plant Biology, Faculty of Life Sciences, University of Ilorin, 240003 Ilorin, Nigeria
*Corresponding author: krishnashanti@gmail.com

7.1 INTRODUCTION

Moringa (*Moringa oleifera* Lam.), sometimes referred to as the horseradish tree, drumstick tree, benzolive tree, or ben oil tree, is one of the most adaptable perennial plant species in the world. The plant's leaves, roots, pods, flowers, and seeds contain substantial amounts of nutrients with potential health benefits, including vitamins, minerals, and phytochemicals (Saini et al., 2016). As a result, this plant may be used to produce both extremely nutritious food and therapeutic goods for humans and animals (Zainab et al., 2020). The drumstick tree has a long history dating back to around 150 BC. Based on historical records, *M. oleifera* fruits and leaves were utilized by ancient queens and kings to keep their skin healthy and their minds sharp (Senthilkumar et al., 2018). While fighting in India, the ancient Mauritanian soldiers drank *M. oleifera* leaf extract, which was thought to be an elixir that would give them more energy and ease their pain and suffering (Trigo et al., 2020). These courageous warriors ultimately proved to be the ones to defeat Alexander the Great (Sujatha & Patel, 2017).

Malnutrition is presently an issue in many developing countries around the world, including Southern Hemisphere countries, such as China, India, and Brazil, the same geographical region which *M. oleifera* originated from. It is well recognized that increasing agricultural productivity helps to significantly lower hunger and malnutrition. Based on the projected total leaf biomass of a *M. oleifera* plantation (705 trees), 340 adults can potentially obtain their daily required caloric intake from a single harvest (Abuye et al., 2003). In this regard, *M. oleifera*'s high nutritional profile can help combat poverty, since cultivation of the plant can raise laborers' wages and provide jobs for those with low incomes (Němec et al., 2020; Bancessi et al., 2020).

In the current situation, when the amount of land that can be utilized to produce fodder is steadily declining because of the cultivation of cash crops, *Moringa oleifera* can be a significant source of high-quality fodder. This plant can be fed to animals as fodder or as a supplement to increase the feed's digestibility and dry matter intake (DMI). *Moringa oleifera* leaves may be added to other fodder crops

DOI: 10.1201/9781003469766-7

to increase the quality of the feed, which benefits animal performance. Because of its valuable nutritional content and efficient biomass production, *M. oleifera* can thus be utilized as a nutritional supplement to promote weight gain and milk production in cattle (Meena et al., 2010). From an industrial perspective, *M. oleifera* leaf powder is the most widely sold product since it is one of the most effective natural sources of calcium and iron and is regarded as a multivitamin supplement that contains high levels of amino acids, among other minerals. Various drying techniques have been researched for that purpose, including the use of a microwave in conjunction with hot air approach, convection or oven drying, sun drying, and shade drying (Dev et al., 2011). Therefore, the present chapter explores the post-harvest processing of *M. oleifera* accessions for nutraceutical and medicinal purposes to improve food and nutritional security while minimizing environmental degradation.

7.2 ORIGIN, DISTRIBUTION, AND GROWING CONDITIONS OF MORINGA

Moringa belongs to the family Moringaceae and has 13 species that originate from different parts of the world. The inflated water-storing trunks of four species – *Moringa drouhardii* (Madagascar), *M. ovalifolia* (Namibia and southwest Angola), *M. hilderandtii* (Madagascar), and *M. stenopetala* (Kenya and Ethiopia) – are their distinguishing feature (Popoola & Obembe, 2013). There are three species recognized for having thin trees that are tuberous in their juvenile stage: *M. peregrine*, *M. oleifera*, and *M. concanesis*. The remaining six species are tuberous in nature: *M. arborea*, *M. rivae*, *M. pygmaea*, *M. borziana*, *M. longituba*, and *M. peregrina* (Mallenakuppe et al., 2015). Moringa species are cultivated for commercial purposes in many parts of Africa and Asia, including India. The most widely cultivated species in the *Moringa* genus is *M. oleifera*. It is a deciduous softwood tree native to the foothills of the Himalayas. It is frequently cultivated in tropical parts of South Asia, including the Caribbean, sub-Saharan Africa, Latin America, northeastern Bangladesh, Nepal, Afghanistan, Sri Lanka, and northern Pakistan. Reports further indicate that this species is also cultivated in Florida and the Pacific Islands (Meena et al., 2010).

Moringa oleifera can be cultivated by stem cuttings, by transplanting, or by the direct seeding method. It is worthy of note that this crop has a minimal requirement for water and nutrients from the soil and may be propagated both sexually and asexually, making it reasonably easy to cultivate (Masih et al., 2019). The warm, semi-arid tropics provide the best growing conditions for *M. oleifera* since it is a crop that can withstand drought, thriving in regions with annual rainfall between 250 and 3,000 mm. It can survive a broad variety of environmental factors, such as infertile soils, extreme temperatures, and mild frosts (James & Zikankuba, 2017; Nouman et al., 2014).

7.3 AGRONOMIC PRACTICES

Depending on soil conditions, fertilizers may be applied to improve the growth of moringa. Both farmyard manure and chemical fertilizers have been used in the cultivation of the moringa plant. Typically, 1 ha of *Moringa oleifera* for fodder requires 150 kg of nitrogen, 60 kg of phosphorus, 40 kg of potash, 30 kg of sulfur, and 10 kg of zinc sulphate (Jattan et al., 2021). Split doses of nitrogen are also used, where 30 kg is applied during field preparation and, further, equal split doses are applied 45 days after seeding and after each cutting. To cultivate pods or seeds, 20 kg of farmyard manure must be added to each pit during seeding (Morton, 1991). Chemical fertilizers, such as 100 g of urea, superphosphate, and potash muriate per plant, must then be used to increase production.

Yield can be increased up to three times by applying 7.5 kg of farmyard manure and 0.37 kg of ammonium sulphate per tree. Weeding must be carried out at regular intervals for effective weed management. The amount of rain that moringa needs depends on its environment. It is recommended to supply the initial irrigation right after sowing. Irrigation must be carefully controlled in the early phases of growth for plants to establish themselves. Subsequently, it can be applied as needed. A biopesticide, such as neem seed kernel extract (5% solution), must be sprayed on the crop during infestation to control insect pests, like hairy caterpillars (Pérez-Rivera et al., 2021).

7.4 *MORINGA OLEIFERA* POST-HARVEST MANAGEMENT TECHNIQUES

Moringa is harvested when the plant reaches a height of 1 to 2 m if the major focus are the leaves. The pods are harvested once they reach a sufficient length (70 cm) and girth (6.3 cm), before developing fiber. For seed production, pods are harvested when they reach maturity, when the seeds turn brown. The post-harvest processing of moringa leaves, pods, or seeds depends on how they will be utilized for industrial purposes, food, fodder, or in cosmetics. In temperate or semi-temperate regions, where moringa is non-productive during the cold season, the leaves can be preserved at low temperature and consumed as green leafy vegetables. For leaf production, transportation plays a crucial role in ensuring high-quality leaves for consumption. Leaves can be transported by cutting the branches and transferring them directly to the processing center or by stripping the leaves from branches before transporting to the processing center.

The leaves of moringa are generally processed into a powder and replace coffee, tea, energy drinks, or juices. Moringa powder is also used in the preparation of noodles, bread, cookies, chips, or snacks. The drying of *Moringa oleifera* fruits using a solution of salt and turmeric increases shelf life without compromising nutritional content. A variety of processes, including freeze-drying, oven drying, microwave vacuum, spray drying, wind and sun drying, is also employed to preserve moringa fruits. Moringa pods and leaves can be dried through a variety of unit operations, such as washing, steeping in an antioxidant solution, blanching at 90–95°C for 3–4 min, cooling, sulfitation, and drying in a cabinet dryer at 60–70°C for 7–9 h with a moisture content of 5–8%, followed by milling (Neelavathi et al., 2022). Moringa seed oil, produced by cold-pressing or solvent extraction, has been utilized in skin treatments and ointments since ancient times. The oil is used to make cosmetics and fragrances. Oil cake, a by-product of oil production, is used as an organic fertilizer (Hekmat et al., 2015).

7.5 NUTRITIVE AND MEDICINAL POTENTIAL OF *MORINGA OLEIFERA*

Moringa oleifera is used to combat malnutrition among newborns and nursing mothers (Sokhela et al., 2023). It is particularly promising in tropical climates, where food is scarce (Gopalakrishnan et al., 2016). The plant's leaves may serve as a source of food and fodder even during dry periods. The benefits of moringa for human health, weight gain, and animal milk production make it a highly promising plant. Nearly all components of the moringa plant may be consumed or added to conventional herbal remedies. In some regions of Africa and India, the leaves and pods are often consumed (Stohs & Hartman, 2015). The leaves are a great source of a variety of minerals and vitamins (Table 7.1).

TABLE 7.1
Major Nutrients in *Moringa oleifera* Leaves and Seeds

Nutrients	Leaves	Seeds	References
Carbohydrates	41 mg/100 g	3.30 mg/100 g	Guzmán-Maldonado et al., 2020; Sodamade et al., 2017
Protein	20–29 mg/100 g	43.71 mg/100 g	Efiong et al., 2013
Calcium	2,016–2,620 mg/100 g	1.5 mg/100 g	Valdez-Solana et al., 2015; Sodamade et al., 2017
Iron	25–28 mg/100 g	0.9 mg/100 g	Guzmán-Maldonado et al., 2020; Sodamade et al., 2017
Vitamin E	15 mg/100 g	4.17 mg/100 g	Efiong et al., 2013; Gharsallah et al., 2023
Vitamin A	6.8 mg/100 g	24 mg/100 g	Gopalakrishnan et al., 2016
Vitamin C	10.8 mg/100 g	77 mg/100 g	Gopalakrishnan et al., 2016; Ahmed et al., 2016
Potassium	1.50 mg/100 g	5 mg/100 g	Moyo et al., 2011
Phosphorous	0.219 mg/100 g	0.11 mg/100 g	Sultana et al., 2015
Magnesium	320–340 mg/100 g	5.30 mg/100 g	Valdez-Solana et al., 2015; Sodamade et al., 2017

In addition to meeting the recommended daily requirements for vitamin B_6 (19%), iron (11%), riboflavin (B_2) (11%), and magnesium (8%), 21 g of chopped leaves contains 2 g of protein (Schmidhuber et al., 2018). Fresh leaves of the plant contain significant carotenoid constituents, including trans-lutein (about 30%), trans-b-carotene (about 18%), and trans-zeaxanthin (about 6%) (Saini et al., 2014). In addition to carotenoids, the leaves also contain a substantial amount of ascorbic acid (271 mg/100 g) and tocopherol (36.9 mg/100 g). *Moringa oleifera*'s nutritional composition also includes trace amounts of antinutrients, such tannins, phytates, saponins, and oxalates (Shih et al., 2011). These compounds are not hazardous or pernicious, but excessive use of these substances may cause problems with the absorption and digestion of many supplements, including calcium, magnesium, iron, zinc, and other nutrient elements. Compared to most legumes, such as soybeans, moringa seeds and leaves have lower levels of phytate and saponins and are thus nutritionally safer and healthier to consume (Gharsallah et al., 2023). The several health benefits (Table 7.2) associated with moringa, such as its anti-inflammatory, antimicrobial, antioxidant, and anticancerous properties, are highlighted in Figure 7.1.

TABLE 7.2
Important Medicinal Properties of *Moringa oleifera*

Medicinal Properties	Test Performed	Benefits/Mechanism	References
Antimicrobial property	Agar well diffusion assay	Inhibited *E. coli*, *S. aureus*, and other pathogens	Patel & Krishnamurthy, 2021; Abalaka et al., 2012
Antioxidant property	Oxygen radical absorbance capacity analysis, ABTS	319 µmol activity, 41%	Peñalver et al., 2022
Anticancer property	Lung cancer cells, cervical cancer cells	Inhibited tumor growth, reduced apoptosis	Jung, 2014; Cuellar-Nuñez et al., 2018
Antidiabetic property	Type 2 diabetes	Reduced blood and plasma glucose	Kumari, 2010
Hepatoprotective property	Wistar rats	Normal liver profile compared to control	Saalu et al., 2012; El-bakry et al., 2016
Anti-inflammatory and immunomodulatory property	Mice model	Reduced gene markers of inflammatory markers, enhanced cellular and humoral response	Vergara-Jimenez et al., 2017

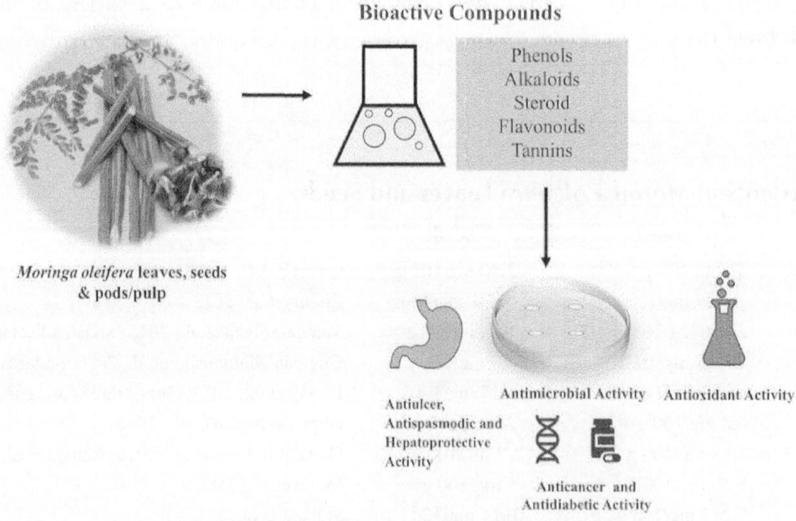

FIGURE 7.1 *Moringa oleifera* health benefits.

7.5.1 Anticancer and Antitumor Activity

Reactive oxygen species (ROS) are directly linked to cell death. Excessive ROS generation is a result of several environmental stresses, which eventually lead to increasing oxidative damage and cell death (Sharma et al., 2012). Three chemicals found in the leaves of moringa have been reported to possess anticancer properties: benzyl isothiocyanate, niazimicin, and glucosinolates. A bioactive substance from moringa leaves called "niazimicin" has been shown to possess some possible anticancer properties (Farooq et al., 2012). Niazimicin, 3-O-(6′-oleoyl-β-D-glucopyranoyl), 4-(α-L-rhamnosyloxy)-benzyl isothiocyanate, β-sitosterol, β-sitosterol-3-O-β-D-glucopyranoside, niazirin, and glycerol-1-(9-octadecnoate) are important bioactive chemicals that have been isolated from moringa seed extracts. The anticancer potential of benzoyl isothiocyanate has previously been reported (Farooq et al., 2012). Studies have demonstrated that benzoyl isothiocyanate produces intracellular ROS, which results in cell death (Lin et al., 2017). This could be one of the factors contributing to the effectiveness of moringa as a potential treatment for cancer (Patel & Krishnamurthy, 2023). The compounds D-allose and hexadecanoic acid (palmitic acid), also found in moringa leaves, inhibit the growth of cancer cells. D-allose produces a specific TXNIP protein which stabilizes the p27kip1 protein during the G1 phase (G1 cell cycle arrest), which slows the proliferation of cancer cells without damaging normal cells (Yamaguchi et al., 2008).

7.5.2 Antiulcer and Antispasmodic Activity

Moringa trees have long been used to treat gastrointestinal motility disorders due to their spasmolytic properties. Additionally, it has been shown that this plant's methanolic preparations prevent stomach ulcers (Abdu & Ashiru, 2021). Das et al. (2011) conducted research using two animal models to explore the potential antiulcer properties of water extracts derived from *M. oleifera*. The antiulcer effects of leaf extracts were evaluated in pyloric ligation and ibuprofen-induced stomach ulcer models at doses of 200 and 400 mg/kg. The ulcer index mean was used to evaluate the degree of stomach ulceration in both models. Both the models resulted in moderate to severe ulcers. The pylorus ligation procedure caused the greatest number of ulcers. In all ulcer models, famotidine and the *M. oleifera* extracts significantly ($p < 0.001$) decreased the ulcer index. The antiulcer effect of *M. oleifera* was comparable to that of conventional medications when tested using pylorus ligation and ibuprofen-induced ulcer techniques. The free and total acidities were dramatically ($p < 0.05$) decreased by famotidine and *M. oleifera* extract.

7.5.3 Antidiabetic Activity

Diabetes mellitus is a chronic condition caused by either a deficit of insulin or a lack of its activity. It eventually affects the body's metabolic functions by causing delayed hyperglycemia. It may cause significant tissue and vascular damage if left untreated, which can lead to major consequences, such as retinopathy, neuropathy, nephropathy, cardiovascular problems, and ulceration (Omonije et al., 2019). According to the World Health Organization (WHO, 2019), 150 million individuals worldwide have diabetes mellitus. By 2025, the figure might potentially exceed 300 million. There are two types of diabetes: types 1 and 2. *Type 1* diabetes is characterized by an entire lack of insulin production, necessitating the use of insulin replacements. The most prevalent kind of diabetes, known as type 2 diabetes, or non-insulin-dependent diabetes mellitus (NIDDM), is brought on by aberrant insulin production and resistance to the hormone.

Numerous investigations have demonstrated that moringa leaves can manage both forms of diabetes. A study of streptozotocin-induced rats suggested that the aqueous extract of leaves can potentially be utilized to treat insulin-resistant type 2 diabetes as well as type 1 diabetes (Vargas-Sánchez et al., 2019). With the aid of xanthine oxidase in beta cells, streptozotocin-induced ATP dephosphorylation results in the production of superoxides and free radicals. Type 2 diabetes and hyperglycemia are caused by these ROS because they destroy beta cells and decrease insulin production (Sissoko et al., 2020). The antioxidants in moringa leaves reduce these reactive oxygen species, shield beta cells from harm, and regulate hyperglycemia (Wang et al., 2022; Kamalakkannan & Prince, 2006).

7.5.4 Antimicrobial and Anthelmintic Activity

Moringa oleifera leaf, flower, root, bark, and stem bark extracts have been shown to exhibit anthelmintic and antibacterial properties (Tayo et al., 2014). Phytospermin from the leaves has been shown to possess potent antimicrobial and antifungal properties. Aqueous and ethanolic extracts of moringa leaves, flowers, and seeds have been reported to possess inhibitory activities against *Salmonella typhi A, Streptococcus, Klebsiella pneumoniae, Escherichia coli, Pseudomonas aeruginosa, Enterobacter* species, *Proteus mirabilis*, *Staphylococcus aureus*, and *Candida albicans* (Nepolean et al., 2009; Patel et al., 2023). Numerous studies have shown the anthelmintic action of *Moringa oleifera* flowers and leaves. For instance, alcoholic extracts from the leaves have been reported to suppress the Indian earthworm *Pheretima posthuma* (Rastogi et al., 2009). When 70% moringa leaf ethanol extract was tested for its dose-dependent antibacterial activity against *Staphylococcus epidermis*, a discernible change was observed as the extract concentration increased (Salihu et al., 2019). In another study, *E. coli*, *Shigella* spp., *Salmonella*, *Staphylococcus*, and *Enterococcus* isolated from the stool of patients were inhibited by the alcoholic extract of moringa at 100 mg/ml concentration (Mursyid et al., 2019; Salihu et al., 2019).

7.5.5 Hepatoprotective Activity

Detoxification and protein synthesis are two of the many critical processes the liver, a crucial organ, carries out (Apte & Krishnamurthy, 2011). If the liver's ability to carry out these tasks is hampered by various chemicals to which it is exposed, the results might be dire and may pose a hazard to life. Liver disease is a global issue, with 2 million deaths worldwide. Many of the medications that are prescribed in the conventional sense to treat liver problems are ineffective and have grave side effects. For example, Tylenol may cause uneasiness and various other problems (Katarey & Verma, 2016). Many times, the search for efficient hepatoprotective medications concentrates on medicinal plants that are used to treat a variety of liver conditions. Investigations were conducted on the hepatoprotective impact of *M. oleifera*'s ethanolic leaf extract on the histopathology of liver damage caused by paracetamol. Compared to the negative control group, which displayed noticeably deformed hepatic cords, necrotic cells, and obliterated sinusoids, Wistar rats treated with 500 mg/kg of the extracts displayed less damage (*Buraimoh* et al., 2011). Moringa leaf extracts have been shown to possess protective properties against liver injury and to help reduce liver fibrosis (Hamza, 2010; Aly et al., 2020; Khalid et al., 2023).

7.5.6 Anti-Inflammatory Activity

Inflammation is the body's natural defense against infection or damage. It is important as a defense mechanism, but if it persists for a long period, it might develop into a serious health problem. Chronic inflammation is thought to play a role in several chronic illnesses, such as cancer and heart disease (Majd et al., 2020). It has been demonstrated that moringa leaves, pods, and seeds possess anti-inflammatory qualities attributed to isothiocyanates (Patel & Krishnamurthy, 2023). However, only animal and test tube experiments have been conducted, and it is not clear whether moringa has similar anti-inflammatory properties in humans (Alhakmani et al., 2013). The *M. oleifera* tree is said to possess 36 naturally occurring anti-inflammatory chemicals. Aqueous root extracts of *M. oleifera* (750 mg/kg) inhibited carrageenan-induced edema in rats to a comparable level as that of standard drug indomethacin (Ndiaye et al., 2002; Saleem et al., 2020).

7.5.7 Antioxidant Activity

Antioxidants are substances that function in the human body to combat free radicals. Chronic disorders related to the heart and others such as diabetes may be exacerbated by oxidative stress, brought on by high amounts of free radicals (Peñalver et al., 2022). The leaves of the moringa plant contain several plant chemicals that are antioxidants. These include quercetin, a potent antioxidant that may also help decrease blood pressure, in addition to beta-carotene and vitamin C. According to a study, providing women 7 g of powdered moringa leaves daily for three months raised their blood antioxidant levels noticeably (Shah et al., 2015; Kushwaha et al., 2014). *Moringa oleifera*'s aqueous extract effectively scavenges DPPH, superoxide, and nitric oxide radicals and inhibits lipid peroxidation. Aqueous pod extract (750 mg/kg) and hydro-alcoholic leaf extracts (1,000 mg/kg) both have significant levels of flavonoids, phenolic compounds, and tannin. Their ethnomedical usage may be partially attributed to the polyphenolic components of this plant (Sharma & Singh, 2010).

7.6 FORTIFYING FOOD WITH *MORINGA OLEIFERA*

Food fortification or enrichment involves adding nutrients (such as trace elements and vitamins) to food. Numerous research has demonstrated the possible use of various sections of the moringa in cuisine, including soups, weaning meals, amala biscuits, bread, cakes, yoghurt, and cheese, among others (Karim et al., 2015; Oyeyinka & Oyeyinka, 2018). The inclusion of moringa leaves has been shown to considerably boost the nutritional value of several baked products. Sengev et al. (2013) incorporated 5% moringa leaves to wheat flour to fortify bread. These authors noticed that the fortified bread's protein and crude fiber content increased significantly, to 54 and 56%, respectively. However, other studies revealed that the fortified bread's crude fiber content increased significantly to 88%, while its protein content decreased to 17%. The protein and fat contents of tortilla chips enriched with 1, 3, and 5% moringa leaf flour increased by 50%. The two main fatty acids in the enriched tortilla chips were oleic and linoleic (Oyeyinka & Oyeyinka, 2018). The inclusion of moringa leaves significantly boosted total phenolic compounds and antioxidant activity in moringa bread (Chinma et al., 2014; Páramo-Calderón et al., 2019).

Moringa seeds or leaves have been used to fortify wheat biscuits and cookies. According to reports, wheat cookies prepared with 20% moringa seed flour had surface-breaking patterns and colors that resembled that of the control group (Riaz et al., 2022). Based on the weight of wheat flour, it can increase the nutritional content of baked goods like cookies by up to 9% without significantly affecting their sensory appeal. The technical characteristics of wheat flour cookies were enhanced by the inclusion of moringa leaf extract, which also had a favorable impact on their appeal (Ogunsina et al., 2011; Masih et al., 2019).

It has been reported that wheat-based cakes can be fortified with up to 20% moringa leaf protein. The use of moringa seeds and flowers for cake fortification shows greater potential, because they have a high nutritional content comparable to that of the leaves (Kolawole et al., 2013). According to Fombang and Saa (2016), when moringa-based tea samples were heated at 97°C for 35 min, they demonstrated a substantial number of phenolic compounds with 80 to 81% inhibition through DPPH assay.

7.7 AGRICULTURAL APPLICATIONS OF MORINGA

Moringa has various agricultural applications, which include the use of various parts as a honeybee forage crop, plant growth enhancer, natural pesticide, and animal feed (Figure 7.2).

Nectar found in moringa flowers is essential for honeybee survival. Because moringa blossoms at varying intensities throughout the year, honeybees can visit the flowers. During pollination, honeybees gather nectar from moringa flowers, which they then turn to honey (Masih et al., 2019). Moringa is used as an alternative forage crop. When the litchi flowering season in north India ends, moringa starts flowering. As a result, moringa may be utilized as a great fodder crop in addition to

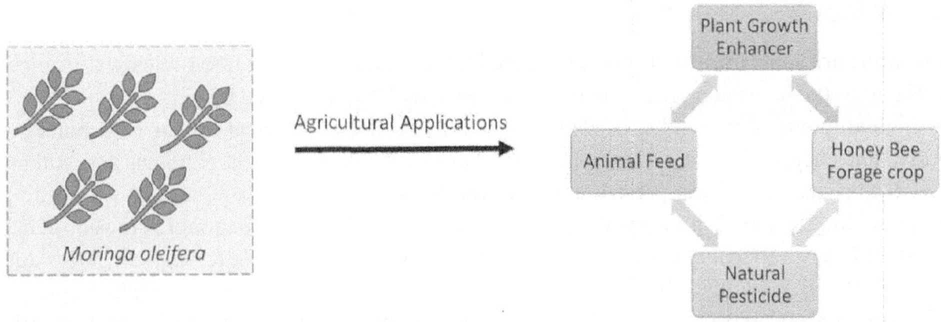

FIGURE 7.2 Agricultural applications of *Moringa oleifera*.

jamun, litchi, and other vegetables. Moringa leaves can also be used as a natural pesticide, where burying the leaves in the ground prevents seedlings from contracting a disease caused by *Pythium debaryanum* (Thurber and Fahey, 2009).

Moringa leaf extract (MLE) applied to the foliage of onions, bell peppers, maize, soy, coffee, tea, chili, and melon boosts yield by 25–30% because of the presence of K, Ca, carotenoids, phenols, and zeatin in the leaves. Foliar spraying should, however, be applied in conjunction with other agronomic techniques. According to Abd El-Hack et al. (2018), moringa extracts are inexpensive, natural plant growth boosters that make plants more resilient to harsh environmental circumstances. While using the MLE on different crops, several researchers have reported that spraying is advantageous (Batool et al., 2020). Additionally, it has been reported that moringa plants are more resistant and resilient to abiotic stress and pests (Abd El-Hack et al., 2018). The plants showed thicker roots, stems, and shoots and a longer lifespan. The green leaves and stems of moringa are easily consumed by animals, such as pigs, goats, sheep, and cattle. It raises the body weight and food and protein intake of Murrah buffalo calves by up to 60% (Aharwal et al., 2018). Bengal goats that were supplemented with moringa showed the highest body weight growth. Using the leaves as fodder also increased the animals' intake of nutrients and enhanced their digestibility. The highest levels of protein utilization, nutritional digestibility, and nitrogen utilization are found in a diet supplemented with moringa (Sultana et al., 2015). A mix of 10% *Leucaena* spp., 70% *Moringa* spp., plus 20% other leaves makes up the optimal diet for pigs. Moreover, moringa leaf meal has the potential to serve as an additive in the feeds for fish and poultry (Paguia et al., 2014; Sultana et al., 2015).

7.8 CONCLUSION AND PROSPECTS

Moringa oleifera offers several advantages along with potential applications. Moringa has a long history of use and benefits as a complete food, nutritional supplement, medicine, and animal feed. Although the moringa plant has immense potential, it faces significant challenges. There are still many questions regarding this tree, such as its most effective use in farming methods and the adverse impacts of antinutritional components on cattle and humans. This species' potential use as food and fodder may be limited by some antinutritional characteristics that have been documented. As a result, additional research and attention are needed to determine how it should be incorporated into diets. Apart from this, little investigation has been carried out on the agronomic techniques needed to produce large levels of fodder biomass. Consequently, to properly spotlight this undervalued tree, more attention should be directed to these areas of research on moringa. From modest subsistence farms to massive multinational holdings, the global moringa market has developed into an extremely different industry. Through complex and contemporary value chains, farmers' products are sold fresh not only in local marketplaces but also globally. Beyond its historical use in traditional meals and medicines, moringa is today recognized as a superfood and a popular supplement for health.

7.9 ACKNOWLEDGMENTS

We are thankful to the management of Uka Tarsadia University, Bardoli, Gujarat, India, for providing necessary research facilities to conduct this study.

REFERENCES

Abalaka, M. E., Daniyan, S. Y., Oyeleke, S. B., & Adeyemo, S. O. (2012). The antibacterial evaluation of Moringa oleifera leaf extracts on selected bacterial pathogens. *Journal of Microbiology Research*, *2*(2), 1–4.

Abd El-Hack, M. E., Alagawany, M., Elrys, A. S., Desoky, E. S. M., Tolba, H. M., Elnahal, A. S., Elnesr S. S., & Swelum, A. A. (2018). Effect of forage Moringa oleifera L. (moringa) on animal health and nutrition and its beneficial applications in soil, plants and water purification. *Agriculture*, *8*(9), 145.

Abdu, H., & Ashiru Garba, A. (2021). Proximate analysis and anti-ulcer activity of methanolic extract of *Moringa oleifera*. *African Scholar Journal of Agriculture and Agricultural Technology*, *20*(1), 188–201.

Abuye, C., Urga, K., Knapp, H., Selmar, D., Omwega, A. M., Imungi, J. K., & Winterhalter, P. (2003). A compositional study of Moringa stenopetala leaves. *East African Medical Journal*, *80*(5), 247–252.

Aharwal, B., Roy, B., Lakhani, G. P., Baghel, R. P. S., Saini, K. P. S., & Yadav, A. (2018). Effect of Moringa oleifera leaf meal on feed intake and growth performance of Murrah buffalo calves. *International Journal of Current Microbiology and Applied Sciences*, *7*(9), 1960–1973.

Ahmed, K. S., Banik, R., Hossain, M. H., & Jahan, I. A. (2016). Vitamin C (L-ascorbic acid) content in different parts of Moringa oleifera grown in Bangladesh. *American Chemical Science Journal*, *11*(1), 1–6.

Alhakmani, F., Kumar, S., & Khan, S. A. (2013). Estimation of total phenolic content, in–vitro antioxidant and anti–inflammatory activity of flowers of *Moringa oleifera*. *Asian Pacific Journal of Tropical Biomedicine*, *3*(8), 623–627.

Aly, O., Abouelfadl, D. M., Shaker, O. G., Hegazy, G. A., Fayez, A. M., & Zaki, H. H. (2020). Hepatoprotective effect of Moringa oleifera extract on TNF-α and TGF-β expression in acetaminophen-induced liver fibrosis in rats. *Egyptian Journal of Medical Human Genetics*, *21*, 1–9.

Apte, U., & Krishnamurthy, P. (2011). Detoxification functions of the liver. *Molecular Pathology of Liver Diseases*, 147–163.

Bancessi, A., Bancessi, Q., Baldé, A., & Catarino, L. (2020). Present and potential uses of *Moringa oleifera* as a multipurpose plant in Guinea-Bissau. *South African Journal of Botany*, *129*, 206–208.

Batool, S., Khan, S., & Basra, S. M. A. (2020). Foliar application of moringa leaf extract improves the growth of moringa seedlings in winter. *South African Journal of Botany*, *129*, 347–353.

Buraimoh, A. A., Bako, I. G., & Ibrahim, F. B. (2011). Hepatoprotective effect of ethanolic leave extract of Moringa oleifera on the histology of paracetamol induced liver damage in Wistar rats. *International Journal of Animal and Veterinary Advances*, *3*(1), 10–13.

Chinma, C. E., Abu, J. O., & Akoma, S. N. (2014). Effect of germinated tigernut and *Moringa* flour blends on the quality of wheat-based bread. *Journal of Food Processing and Preservation*, *38*(2), 721–727.

Cuellar-Nuñez, M. L., Luzardo-Ocampo, I., Campos-Vega, R., Gallegos-Corona, M. A., De Mejía, E. G., & Loarca-Piña, G. (2018). Physicochemical and nutraceutical properties of moringa (Moringa oleifera) leaves and their effects in an in vivo AOM/DSS-induced colorectal carcinogenesis model. *Food Research International*, *105*, 159–168.

Das Deepanjana, D. D., Devasrita Dash, D. D., Tatiyana Mandal, T. M., Anup Kishore, A. K., & Bairy, K. L. (2011). Protective effects of Moringa oleifera on experimentally induced gastric ulcers in rats. *Research Journal of Pharmaceutical, Biological and Chemical Sciences*, *2*(2), 50–55.

Dev, S. R. S., Geetha, P., Orsat, V., Gariépy, Y., & Raghavan, G. S. V. (2011). Effects of microwave-assisted hot air drying and conventional hot air drying on the drying kinetics, color, rehydration, and volatiles of Moringa oleifera. *Drying Technology*, *29*(12), 1452–1458.

Efiong, E. E., Igile, G. O., Mgbeje, B. I. A., Out, E. A., & Ebong, P. E. (2013). Hepatoprotective and antidiabetic effect of combined extracts of Moringa oleifera and Vernonia amygdalina in streptozotocin-induced diabetic albino Wistar rats. *Journal of Diabetes and Endocrinology*, *4*(4), 45–50.

El-bakry, K., Toson, E. S., Serag, M., & Aboser, M. (2016). Hepatoprotective effect of Moringa oleifera leaves extract against carbon tetrachloride-induced liver damage in rats. *World Journal of Pharmacy and Pharmaceutical Sciences*, *5*(5), 76–89.

Farooq, F., Rai, M., Tiwari, A., Khan, A. A., & Farooq, S. (2012). Medicinal properties of Moringa oleifera: An overview of promising healer. *Journal of Medicinal Plants Research*, *6*(27), 4368–4374.

Fombang, E. N., & Saa, W. R. (2016). Production of a functional tea from Moringa oleifera LAM leaf powder: Optimization of phenolic extraction using response surface methodology. *Nutrition and Food Science*, *6*, 556.

Gharsallah, K., Rezig, L., Rajoka, M. S. R., Mehwish, H. M., Ali, M. A., & Chew, S. C. (2023). Moringa oleifera: Processing, phytochemical composition, and industrial application. South African. *Journal of Botany, 160*, 180–193.

Gopalakrishnan, L., Doriya, K., & Kumar, D. S. (2016). Moringa oleifera: A review on nutritive importance and its medicinal application. *Food Science and Human Wellness, 5*(2), 49–56.

Guzmán-Maldonado, S. H., López-Manzano, M. J., Madera-Santana, T. J., Núñez-Colín, C. A., Grijalva-Verdugo, C. P., Villa-Lerma, A. G., & Rodríguez-Núñez, J. R. (2020). Nutritional characterization of Moringa oleifera leaves, seeds, husks and flowers from two regions of Mexico. *Agronomía Colombiana, 38*(2), 287–297.

Hamza, A. A. (2010). Ameliorative effects of *Moringa oleifera* Lam seed extract on liver fibrosis in rats. *Food and Chemical Toxicology, 48*(1), 345–355.

Hekmat, S., Morgan, K., Soltani, M., & Gough, R. (2015). Sensory evaluation of locally-grown fruit purees and inulin fibre on probiotic yogurt in Mwanza, Tanzania and the microbial analysis of probiotic yogurt fortified with Moringa oleifera. *Journal of Health, Population, and Nutrition, 33*(1), 60.

James, A., & Zikankuba, V. (2017). *Moringa oleifera* a potential tree for nutrition security in sub-Sahara Africa. *American Journal of Research Communication, 5*, 1–14.

Jattan, M., Kumari, N., Kumar, R., Kumar, A., Rani, B., Phogat, D. S., Kumar S., & Kumar, P. (2021). Moringa (Moringa oleifera L.): An underutilized and traditionally valued tree holding remarkable potential. *Journal of Horticultural Sciences, 16*(1), 1–13.

Jung, I. L. (2014). Soluble extract from Moringa oleifera leaves with a new anticancer activity. *PLoS One, 9*(4), e95492.

Kamalakkannan, N., & Prince, P. S. M. (2006). Antihyperglycaemic and antioxidant effect of rutin, a polyphenolic flavonoid, in streptozotocin-induced diabetic wistar rats. *Basic & Clinical Pharmacology & Toxicology, 98*(1), 97–103.

Karim, O., Kayode, R., Oyeyinka, S., & Oyeyinka, A. (2015). Physicochemical properties of stiff dough 'amala'prepared from plantain (*Musa Paradisca*) flour and *Moringa* (*Moringa oleifera*) leaf powder. *Hrana u Zdravlju i Bolesti: Znanstveno-Stručni Časopis za Nutricionizam i Dijetetiku, 4*(1), 48–58.

Katarey, D., & Verma, S. (2016). Drug-induced liver injury. *Clinical Medicine, 16*(Suppl. 6), s104.

Khalid, S., Arshad, M., Raza, K., Mahmood, S., Siddique, F., Aziz, N., Khalid, W., & Aqlan, F. (2023). Assessment of hepatoprotective, nephroprotective efficacy, and antioxidative potential of Moringa oleifera leaf powder and ethanolic extract against PCOS-induced female albino mice (Mus Musculus). *Food Science & Nutrition, 11*(11), 7206–7217.

Kolawole, F. L., Balogun, M. A., Opaleke, D. O., & Amali, H. E. (2013). An evaluation of nutritional and sensory qualities of wheat-moringa cake. *Agrosearch, 13*(1), 87–94.

Kumari, D. J. (2010). Hypoglycaemic effect of Moringa oleifera and Azadirachta indica in type 2 diabetes mellitus. *Bioscan, 5*(20), 211–214.

Kushwaha, S., Chawla, P., & Kochhar, A. (2014). Effect of supplementation of drumstick (*Moringa oleifera*) and amaranth (*Amaranthus tricolor*) leaves powder on antioxidant profile and oxidative status among postmenopausal women. *Journal of Food Science and Technology, 51*, 3464–3469.

Lin, J. F., Tsai, T. F., Yang, S. C., Lin, Y. C., Chen, H. E., Chou, K. Y., & Hwang, T. I. (2017). Benzyl isothiocyanate induces reactive oxygen species-initiated autophagy and apoptosis in human prostate cancer cells. *Oncotarget, 8*(12), 20220.

Majd, M., Saunders, E. F., & Engeland, C. G. (2020). Inflammation and the dimensions of depression: A review. *Frontiers in Neuroendocrinology, 56*, 100800.

Mallenakuppe, R., Homabalegowda, H., Gouri, M. D., Basavaraju, P. S., & Chandrashekharaiah, U. B. (2015). History, taxonomy and propagation of Moringa oleifera–A review. *Crops, 3*(3.28), 3–15.

Masih, L. P., Singh, S., Elamathi, S., Anandhi, P., & Abraham, T. (2019). *Moringa*: A multipurpose potential crop-A review. *Proceedings of the Indian National Science Academy, 85*(1), 589–601.

Meena, A. K., Ayushy, S., Ramanjeet, K., Bhavana, P., & Brijendra, S. (2010). *Moringa oleifera*: A review. *Journal of Pharmacy Research, 3*(4), 840–842.

Morton, J. F. (1991). The horseradish tree, *Moringa pterygosperma* (Moringaceae) – A boon to arid lands? *Economic Botany, 45*, 318–333.

Moyo, B., Masika, P. J., Hugo, A., & Muchenje, V. (2011). Nutritional characterization of Moringa (Moringa oleifera Lam.) leaves. *African Journal of Biotechnology, 10*(60), 12925–12933.

Mursyid, M., Annisa, R. N., Zahran, I., Langkong, J., & Kamaruddin, I. (2019). Antimicrobial activity of *Moringa leaf* (*Moringa oleifera* L.) extract against the growth of *Staphylococcus epidermidis*. *IOP Conference Series: Earth and Environmental Science, 343*(1), 012145.

Ndiaye, M., Dieye, A. M., Mariko, F., Tall, A., & Faye, B. (2002). Contribution to the study of the anti-inflammatory activity of Moringa oleifera (Moringaceae). *Dakar Medical, 47*(2), 210–212.

Neelavathi, R., Rani, C. I., Durgadevi, M., Ezhilmathi, S., Gnanasundari, K., Gokila, R., & Prabhu, M. (2022). Influence of blanching and drying methods on the retention of nutritional quality of dried moringa leaves. *Agriculture Association of Textile Chemical and Critical Reviews Journal, 2022*, 36–40.

Němec, P., Ungrová, M., Alem, S., Novák, J., & Habrová, H. (2020). Biomass production of a young plantation of *Moringa stenopetala* (Baker f.) Cufod. and *Moringa oleifera* Lam. in Southern Ethiopia. *South African Journal of Botany, 129*, 463–470.

Nepolean, P., Anitha, J., & Emilin, R. R. (2009). Isolation, analysis and identification of phytochemicals of antimicrobial activity of *Moringa oleifera* Lam. *Current Biotica, 3*(1), 33–37.

Nouman, W., Basra, S., Ahmed, M., Siddiqui, M. T., Yasmeen, A., Gull, T., & Alcayde, M. A. C. (2014). Potential of Moringa oleifera L. as livestock fodder crop: A review. *Turkish Journal of Agriculture and Forestry, 38*(1), 1–14.

Ogunsina, B. S., Radha, C., & Indrani, D. (2011). Quality characteristics of bread and cookies enriched with debittered *Moringa oleifera* seed flour. *International Journal of Food Sciences and Nutrition, 62*(2), 185–194.

Omonije, O. O., Saidu, A. N., & Muhammad, H. L. (2019). Anti-diabetic activities of *Chromolaena odorata* methanol root extract and its attenuation effect on diabetic induced hepatorenal impairments in rats. *Clinical Phytoscience, 5*(1), 1–10.

Oyeyinka, A. T., & Oyeyinka, S. A. (2018). Moringa oleifera as a food fortificant: Recent trends and prospects. *Journal of the Saudi Society of Agricultural Sciences, 17*(2), 127–136.

Paguia, H. M., Paguia, R. Q., Balba, C., & Flores, R. C. (2014). Utilization and evaluation of Moringa oleifera L. as poultry feeds. *APCBEE Procedia, 8*, 343–347.

Páramo-Calderón, D. E., Aparicio-Saguilán, A., Aguirre-Cruz, A., Carrillo-Ahumada, J., Hernández-Uribe, J. P., Acevedo-Tello, S., & Torruco-Uco, J. G. (2019). Tortilla added with *Moringa oleífera* flour: Physicochemical, texture properties and antioxidant capacity. *LWT, 100*, 409–415.

Patel, N., & Krishnamurthy, R. (2021). In-vitro phytochemical screening and bioactivity of moringa oleifera accessions. *Bioscience Biotechnology Research Communications, 14*(1), 1–5.

Patel, N., & Ramar, K. (2023). *Moringa oleifera* Accessions: Perspectives and application as nutraceuticals and phytomedicines. In *Bioprospecting of tropical medicinal plants* (pp. 463–479). Springer Nature.

Patel, N., Tarjani, C., & Ramar, K. (2023). Phytochemical constituents and antibacterial activity of different plant parts extracts of wild *Moringa oleifera* Lam. *Annals of Biology, 39*(2), 311–314.

Peñalver, R., Martínez-Zamora, L., Lorenzo, J. M., Ros, G., & Nieto, G. (2022). Nutritional and antioxidant properties of Moringa oleifera leaves in functional foods. *Foods, 11*(8), 1107.

Pérez-Rivera, E. P., Montes-Ávila, J., Castro-Tamayo, C. B., Portillo-Loera, J. J., & Castillo-López, R. I. (2021). Agronomical aspects of *Moringa oleifera* (*Moringa*). In *Biological and pharmacological properties of the genus moringa* (pp. 39–63). CRC Press.

Popoola, J. O., & Obembe, O. O. (2013). Local knowledge, use pattern and geographical distribution of Moringa oleifera Lam. (Moringaceae) in Nigeria. *Journal of Ethnopharmacology, 150*(2), 682–691.

Rastogi, T., Bhutda, V., Moon, K., Aswar, P. B., & Khadabadi, S. S. (2009). Comparative studies on anthelmintic activity of *Moringa oleifera* and *Vitex negundo*. *Asian Journal of Research in Chemistry, 2*(2), 181–182.

Riaz, A., Wahab, S., Ayub, M., Hashm, M. S., & Jabeen, F. (2022). Food to food fortification of whole wheat flour biscuit using moringa leaves powder. *Emirates Journal of Food and Agriculture, 34*(2), 128–134.

Saalu, L. C., Ogunlade, B., Ajayi, G. O., Oyewopo, A. O., Akunna, G. G., & Ogunmodede, O. S. (2012). The hepato-protective potentials of Moringa oleifera leaf extract on alcohol-induced hepato-toxicity in wistar rat. *American Journal of Biotechnology and Molecular Sciences, 2*(1), 6–14.

Saini, R. K., Shetty, N. P., & Giridhar, P. (2014). GC-FID/MS analysis of fatty acids in Indian cultivars of *Moringa oleifera*: Potential sources of PUFA. *Journal of the American Oil Chemists' Society, 91*(6), 1029–1034.

Saini, R. K., Sivanesan, I., & Keum, Y. S. (2016). Phytochemicals of *Moringa oleifera*: A review of their nutritional, therapeutic and industrial significance. *3 Biotech, 6*, 1–14.

Saleem, A., Saleem, M., & Akhtar, M. F. (2020). Antioxidant, anti-inflammatory and antiarthritic potential of Moringa oleifera Lam: An ethnomedicinal plant of Moringaceae family. *South African Journal of Botany, 128*, 246–256.

Salihu, A., Ali, M., & Abadallah, M. S. (2019). Antibacterial activity of *Moringa oleifera* leaf extracts against bacteria isolated from patients attending general Sani Abacha specialist hospital damaturu. *Allied Pharmaceutical Sciences, 1*(1), 61–66.

Schmidhuber, J., Sur, P., Fay, K., Huntley, B., Salama, J., Lee, A., & Afshin, A. (2018). The global nutrient database: Availability of macronutrients and micronutrients in 195 countries from 1980 to 2013. *The Lancet Planetary Health, 2*(8), e353–e368.

Sengev, A. I., Abu, J. O., & Gernah, D. I. (2013). Effect of *Moringa oleifera* leaf powder supplementation on some quality characteristics of wheat bread. *Food and Nutrition Sciences*, *4*(3), 270.

Senthilkumar, A., Karuvantevida, N., Rastrelli, L., Kurup, S. S., & Cheruth, A. J. (2018). Traditional uses, pharmacological efficacy, and phytochemistry of Moringa peregrina (Forssk.) Fiori – A review. *Frontiers in Pharmacology*, *9*, 327519.

Shah, M. A., Bosco, S. J. D., & Mir, S. A. (2015). Effect of *Moringa oleifera* leaf extract on the physicochemical properties of modified atmosphere packaged raw beef. *Food Packaging and Shelf Life*, *3*, 31–38.

Sharma, P., Jha, A. B., Dubey, R. S., & Pessarakli, M. (2012). Reactive oxygen species, oxidative damage, and antioxidative defense mechanism in plants under stressful conditions. *Journal of Botany*, 1–27.

Sharma, R., & Singh, V. J. (2010). In vivo antioxidant activity of Moringa oleifera leaf and pod extracts against carbon tetra chloride induced liver damage in albino mice. *Journal of Chemical and Pharmaceutical Research*, *2*(6), 275–283.

Shih, M. C., Chang, C. M., Kang, S. M., & Tsai, M. L. (2011). Effect of different parts (leaf, stem and stalk) and seasons (summer and winter) on the chemical compositions and antioxidant activity of *Moringa oleifera*. *International Journal of Molecular Sciences*, *12*(9), 6077–6088.

Sissoko, L., Diarra, N., Nientao, I., Stuart, B., Togola, A., Diallo, D., & Willcox, M. L. (2020). Moringa oleifera leaf powder for type 2 diabetes: A pilot clinical trial. *African Journal of Traditional, Complementary and Alternative Medicines*, *17*(2), 29–36.

Sodamade, A., Owonikoko, A., & Owoyemi, D. (2017). Nutrient contents and mineral composition of Moringa oleifera Seed. *International Journal of Communication Systems*, *5*(2), 205–207.

Sokhela, H., Govender, L., & Siwela, M. (2023). Complementary feeding practices and childhood malnutrition in South Africa: The potential of Moringa oleifera leaf powder as a fortificant: A narrative review. *Nutrients*, *15*(8), 2011.

Stohs, S. J., & Hartman, M. J. (2015). Review of the safety and efficacy of *Moringa oleifera*. *Phytotherapy Research*, *29*(6), 796–804.

Sujatha, B. K., & Patel, P. (2017). *Moringa oleifera*–nature's gold. *Imperial Journal of Interdisciplinary Research*, *3*(5), 1175–1179.

Sultana, N., Alimon, A. R., Huque, K. S., Baba, M., & Hossain, J. (2015). Evaluation of *Moringa* foliage (*Moringa oleifera*) as goat feed. *Iranian Journal of Applied Animal Science*, *5*(4), 865–871.

Tayo, G. M., Poné, J. W., Komtangi, M. C., Yondo, J., Ngangout, A. M., & Mbida, M. (2014). Anthelminthic activity of Moringa oleifera leaf extracts evaluated in vitro on four developmental stages of Haemonchus contortus from goats. *American Journal of Plant Sciences*, 2014.

Thurber, M. D., & Fahey, J. W. (2009). Adoption of Moringa oleifera to combat under-nutrition viewed through the lens of the "Diffusion of Innovations" theory. *Ecology of Food and Nutrition*, *48*(3), 212–225.

Trigo, C., Castello, M. L., Ortola, M. D., Garcia-Mares, F. J., & Desamparados Soriano, M. (2020). Moringa oleifera: An unknown crop in developed countries with great potential for industry and adapted to climate change. *Foods*, *10*(1), 31.

Valdez-Solana, M. A., Mejía-García, V. Y., Téllez-Valencia, A., García-Arenas, G., Salas-Pacheco, J., Alba-Romero, J. J., & Sierra-Campos, E. (2015). Nutritional content and elemental and phytochemical analyses of Moringa oleifera grown in Mexico. *Journal of Chemistry*, *2015*.

Vargas-Sánchez, K., Garay-Jaramillo, E., & González-Reyes, R. E. (2019). Effects of Moringa oleifera on glycaemia and insulin levels: A review of animal and human studies. *Nutrients*, *11*(12), 2907.

Vergara-Jimenez, M., Almatrafi, M. M., & Fernandez, M. L. (2017). Bioactive components in Moringa oleifera leaves protect against chronic disease. *Antioxidants*, *6*(4), 91.

Wang, F., Bao, Y., Zhang, C., Zhan, L., Khan, W., Siddiqua, S., Ahmad, S., Capanoglu, E., Skalicka-Woźniak, K., Zou, L., & Simal-Gandara, J. (2022). Bioactive components and anti-diabetic properties of *Moringa oleifera* Lam. *Critical Reviews in Food Science and Nutrition*, *62*(14), 3873–3897.

World Health Organization. (2019). *WHO global report on traditional and complementary medicine 2019*. World Health Organization.

Yamaguchi, F., Takata, M., Kamitori, K., Nonaka, M., Dong, Y., Sui, L., & Tokuda, M. (2008). Rare sugar D-allose induces specific up-regulation of TXNIP and subsequent G1 cell cycle arrest in hepatocellular carcinoma cells by stabilization of p27kip1. *International Journal of Oncology*, *32*(2), 377–385.

Zainab, B., Ayaz, Z., Alwahibi, M. S., Khan, S., Rizwana, H., Soliman, D. W., & Abbasi, A. M. (2020). In-silico elucidation of *Moringa oleifera* phytochemicals against diabetes mellitus. *Saudi Journal of Biological Sciences*, *27*(9), 2299–2307.

8 The Cultivation, Nutritional Polyphenolic Contents, Bioavailability, and Health Benefits of Finger Millet (*Eleusine coracana* (L.) Gaertn.) *An Updated Review*

**Linda Ifeanyichukwu Sowunmi, Yetunde Oyinkansade Babalola, and Paul Oluwatimilehin Ogunbamowo*
Biomedicinal Research Centre, Forestry Research Institute of Nigeria, Oyo State, Nigeria
*Corresponding author: ifysowunmi@gmail.com

8.1 INTRODUCTION

Finger millet (*Eleusine coracana* L.) is a perennial grass in the Poaceae family. It is a staple food crop cultivated by subsistence farmers in the semi-arid tropics and subtropics of the world under rain-fed conditions. Cultivated in several parts of Africa and Asia, including Nigeria, Ethiopia, Eritrea, Mozambique, Zimbabwe, Namibia, Senegal, Niger, Madagascar, India, Nepal, Malaysia, China, Japan, Iran, Afghanistan, and Sri Lanka, among others, finger millet is a nutritious cereal grain that possesses some health benefits (Puranik et al., 2017; Ceasar et al., 2018; Ramashia et al., 2019). *Bulo*, *ragi*, and *wimbi* are some of its common names in Uganda; *tamba*, *pwana*, and *sarga* in Nigeria; *dagussa* in Ethiopia; *ragi* and *mandua* in India; and *rapoko* in South Africa, and *African millet*, *birds' foot*, and *coracana* are some of its common names in English (Abioye et al., 2022). Regarding grain output in Africa and Asia, finger millet is ranked fourth, behind sorghum, pearl millet, and foxtail millet (Ramashia et al., 2018; Ceasar et al., 2018). There are 2.5 million and 1.2 million tons produced annually in Africa and India, respectively, of the 4.5 million tons produced annually (Ramashia et al., 2019).

Nigeria is the leading millet producer in Africa, with Niger and Mali following closely behind. Globally, it ranks as the third largest millet producer, trailing only China and India. The major millet-producing states in Nigeria include Kaduna, Yobe, Kano, and Borno. Due to its remarkable ability to thrive in challenging agro-climatic conditions, finger millet is highly valued by local farmers, particularly in semi-arid tropics and subtropics. In regions where conventional grain crops such as maize (*Zea mays*), wheat (*Triticum* spp.), and rice (*Oryza sativa*) struggle to grow in varied soil types, finger millet thrives very well. The general features of finger millet panicle and grain are shown in Figure 8.1.

DOI: 10.1201/9781003469766-8

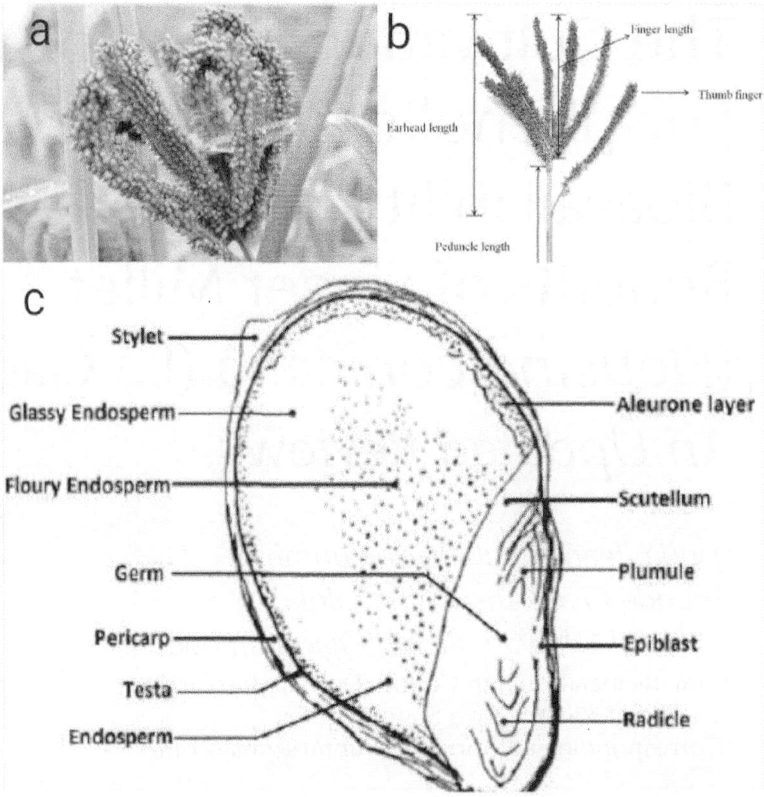

FIGURE 8.1 Finger millet: (a) matured panicle on field; (b) general description of panicle; (c) general features.

Source: (a) Dayakar Rao et al. (2017); (b) Patel et al. (2022); (c) Dayakar Rao et al. (2017).

8.1.1 Nutraceutical Potential of Finger Millet

Finger millet is predominantly grown for its grain, which serves as a staple food in different regions of the world. The wide range of applications of finger millet boasts its significant nutraceutical potential, making it a versatile ingredient for various food and beverage products. Its rich nutritional profile, coupled with bioactive compounds, contributes to its health-promoting properties. Several products have been developed from finger millet, ranging from bread and biscuits to fermented beverages, harnessing its unique qualities for enhanced nutrition and health benefits. Some of the products include:

1. *Bread and confectioneries.* Finger millet flour is commonly used in the production of bread, biscuits, and other baked goods due to its nutritional richness and gluten-free nature. These products offer a healthier alternative to traditional wheat-based items and cater to individuals with gluten intolerance or celiac disease (Gaikwad et al., 2023).
2. *Fermented foods.* Finger millet is a key ingredient in many fermented food products which not only enhances the flavor but also increases the bioavailability of nutrients and bioactive compounds present in finger millet, offering potential health benefits (Makokha et al., 2002).
3. *Snack foods and cereals.* Finger millet can be processed into various snack foods and cereals, including extruded snacks, crisps, and cereal flakes. These products provide convenient and nutritious snack options, contributing to dietary diversity and micronutrient intake (Shobana et al., 2013).

4. *Nutritional supplements.* Finger millet extracts and formulations are utilized to produce nutritional supplements and functional foods. These supplements may target specific health concerns, such as malnutrition, diabetes, and cardiovascular diseases. Finger millet–derived supplements offer a natural source of bioactive compounds with potential health benefits (Devi et al., 2014).

The outlook for finger millet seems promising, with several factors contributing to its potential growth and importance in agriculture, nutrition, and food security. These factors are bolstered by emerging trends, technological advancements, and evolving consumer preferences. Among the potential prospects are:

1. *Nutritional superiority.* Finger millet's exceptional nutritional profile, characterized by high protein, fiber, and micronutrient content, makes it an attractive option for addressing malnutrition and improving food security (Gaikwad et al., 2023). With growing awareness of the importance of nutrition-sensitive agriculture, finger millet is expected to gain prominence as a nutrient-dense food crop.
2. *Health and wellness trends.* The increasing focus on health and wellness among consumers is driving demand for nutritious and functional foods. Finger millet's nutraceutical properties, including its antioxidant, anti-inflammatory, and cardioprotective effects, align with consumer preferences for natural and functional ingredients (Shahidi & Chandrasekara, 2013). As consumers seek healthier dietary options, finger millet–based products are likely to witness greater acceptance and demand.
3. *Value-added products.* There is a growing trend toward the development of value-added products from finger millet, including baked goods, breakfast cereals, snacks, and beverages (Karki et al., 2020). Innovative processing techniques and product formulations are expected to expand the market for finger millet–based products, catering to diverse consumer preferences and dietary needs.
4. *Policy support and research investments.* Government initiatives and research investments aimed at promoting sustainable agriculture, improving food security, and enhancing nutritional outcomes are likely to bolster the cultivation and utilization of finger millet (Rawat et al., 2023). Supportive policies, subsidies, and research funding can incentivize farmers to adopt finger millet cultivation practices and facilitate the development of value chains and markets for finger millet products.

8.2 CULTIVATION

8.2.1 Agro-Climatic Requirements

The ability of finger millet to adapt to various agro-climatic conditions is one of its profound qualities. If rainfall is evenly distributed throughout the growing season, finger millet can be cultivated in areas with annual precipitation ranging from 500 to 1,000 mm (Gaikwad et al., 2023). The primary characteristic of finger millet lies in its remarkable tolerance capacity, enabling it to withstand diverse forms of stress, including heat, water stress, and nutritional deficiencies (Gupta, 2017). Although it is typically thought of as a drought-resilient crop, it requires a modest precipitation of less than or equal to 500 mm yearly when compared to other millets, like pearl millet, and other small grains, like sorghum (Prasad & Staggenborg, 2009). Studies have shown that finger millet favors wet circumstances, but it does poorly in locations that receive excessive rain (Satyavathi et al., 2021). Although yields can frequently be considerably increased under irrigation, it is typically produced as a rain-fed crop which can tolerate cooler and hotter circumstances. For finger millet to grow and mature, the ideal daytime and nighttime temperatures are 27–32°C and 22°C, respectively (Opole et al., 2018). These previous authors reported that the crucial temperatures are 18–32°C during the day and 18°C at night, respectively. Temperatures above 32°C prevent finger millet from blossoming.

8.2.2 Soil Requirements

Finger millet is grown in a range of soil types, spanning from fertile loam to poor shallow upland soils. It thrives best in porous, well-drained loam, light-red loam, and sandy loam soils with good fertility and water-retention capabilities. The soil should be enriched with organic matter. It can be grown in soils with a pH of 4.5–7.5. It can tolerate soil salinity better than other cereals (Soni et al., 2024).

8.3 NUTRITIONAL COMPOSITIONS

Finger millet is abundant in calcium, carbohydrates, fiber, and other minerals, as shown in Table 8.1. Sood et al. (2016) opined that this crop possesses better nutritional constituents than other grains, as it encompasses various essential nutrients crucial for human health.

8.3.1 Carbohydrates

Vagdevi et al. (2023) reported that finger millet has 1.04% free sugar, 65.5% starch, and 11.5% dietary fiber. The carbohydrate makeup of this crop is composed of 0.04–0.6% lignin, 1.4–1.8% cellulose, and 59.5–21.1% starch (Abioye et al., 2022). Additionally, according to the same authors, the starch makeup varies between 59.4 and 70.2% dry matter, with 80–85% amylopectin and 15% amylose. Singh and Raghuvanshi (2012) reported that 20–30% of the carbohydrates in finger millet are made up of non-starchy polysaccharides, and Abioye et al. (2022) reported that finger millet contains 0.03% non-reducing sugars and around 1.5% reducing sugars. Chowdary and Bisarya (2020) reported 65–83% carbohydrates in this crop. Sucrose is the primary sugar in finger millet, ranging from 0.20 to 0.24 g/100 g, according to reports (Himanshu et al., 2018; Hassan et al., 2021). The total sugar makeup of the grain ranges from 0.59 to 0.69 g/100 g.

8.3.2 Fiber

Due to its highly nutritious content, fiber helps speed up intestinal transit and preserve a healthy gastrointestinal system. This results in a decrease in blood serum lipoprotein cholesterol total value and blood postprandial glucose level. Finger millet's fiber content helps boost both immune system development and improved calcium bioavailability. In addition, this nutrient works effectively to

TABLE 8.1
Nutritional Composition of Finger Millet, Rice, and Wheat (per 100 g Edible Portion, 12% Moisture Content)

S/N	Particulars	Finger Millet	Maize	Wheat
1	Carbohydrates (g)	72.6	73	71
2	Protein (g)	7.7	9.2	11.6
3	Crude fiber (g)	3.6	2.8	2.0
4	Fat (g)	1.5	4.6	2.0
5	Ash (g)	2.6	1.2	1.6
6	Calcium (mg)	350	26	41
7	Riboflavin (mg)	0.19	0.20	0.10
8	Thiamine (mg)	0.42	0.38	0.41
9	Iron (mg)	3.9	2.7	3.5
10	Niacin (mg)	1.1	3.6	5.1

Source: www.fao.org.

combat colon cancer. Nonetheless, compared to many other grain varieties, finger millet has a larger percentage of nutritional fiber. The total dietary fiber content of finger millet is 22%, which is higher than that of other millets or cereal grains, which are 12.8%, 4.5%, 13.4%, and 12.6% for sorghum, rice, maize, and wheat, respectively (Rathore et al., 2016). Finger millets contain fiber, primarily in the pericarp and endosperm walls. In addition, this crop contains 18.6% fiber, mostly made up of lignin (7.9%), cellulose (4.7%), and non-cellulosic polysaccharides (6.1%), of which 4.7% and 1.5% are water-insoluble, respectively (Patel et al., 2016). Those who consume a single meal of this millet can work for extended periods because of its high fiber content, which slows down digestion. Furthermore, the high fiber content present in finger millet also plays a significant role in its low glycemic index (GI). Low-GI foods, characterized by gradual and steady increases in blood sugar levels, are advantageous for individuals with diabetes (Patel et al., 2016). These foods help manage blood sugar levels by preventing rapid spikes and crashes (Jenkins et al., 1981). By consuming finger millet, individuals can experience sustained energy release over extended periods due to its low-GI properties. This is attributed to the presence of fiber, particularly lignin, cellulose, and non-cellulosic polysaccharides, which contribute to its slow digestion rate, thus making it a valuable dietary option for individuals seeking to regulate blood sugar levels effectively, especially those with diabetes.

8.3.3 Fats

Studies indicate that the lipid content of finger millet typically hovers around 5.2%, comprising various fatty acids, such as oleic acid, palmitic acid, and linoleic acid (Jagati et al., 2021). Thapliyal and Singh (2015), as well as Rathore et al. (2019), have reported that approximately 1.3% to 1.8% of the grain's content consists of fat. Moreover, an in-depth analysis by Abioye et al. (2022) revealed a detailed breakdown of finger millet's lipid composition. According to their findings, finger millet grains encompass different lipid fractions, including free lipids (2.2%), bound lipids (2.4%), and structural lipids (0.6%). The relatively low fat composition of finger millet grains can be attributed to the small germ present in the grain structure. This characteristic not only contributes to the grain's nutritional profile but also enhances its suitability as a storage commodity, as it reduces the risk of fat rancidity (Mohamed et al., 2009; Muthamilarasan et al., 2016). Abioye et al. (2022) further elucidated that finger millet grains contain an average of 2.1% total crude fat, with a notable proportion of polyunsaturated fatty acids. Interestingly, the fat content of finger millet can vary between different variants, with brown and white variants exhibiting fat contents ranging from 1.2% to 1.4%. Unsaturated fatty acids constitute a significant portion of finger millet's fat content, accounting for approximately 74.4%, while saturated fatty acids make up the remaining 25.6% (Abioye et al., 2022). This composition of unsaturated fatty acids enhances the grain's potential health benefits and underscores its significance as a nutritious dietary option. Thus, finger millet's fat content, although relatively modest compared to that of other food sources, plays a crucial role in its nutritional profile and health implications.

8.3.4 Protein

Finger millet cultivars generally contain prolamins as the predominant protein. *Prolamins* are a group of storage proteins found in cereals, including millets, and they play a crucial role in providing essential amino acids and contributing to the nutritional quality of the grain (Patel et al., 2016). The specific composition of prolamins may vary slightly between different cultivars of finger millet, but they typically constitute much of the protein content in the grain. Additionally, finger millet cultivars may exhibit variations in protein content and amino acid profiles, which can influence their nutritional quality and suitability for different dietary needs. Some cultivars may have higher lysine and methionine contents, while others may excel in other essential amino acids. Overall, finger millet cultivars are valued for their protein-rich grains and their ability to provide essential amino acids, making them an important component of diets in regions where they are cultivated (Gaikwad et al., 2023).

Abioye et al. (2022) opined that the crude protein level of finger millet varied between 5.6 and 12.70%, and the protein content showed a mean value of 9.7% and ranged from 6.7 to 12.3%. Singh and Srivastava (2006) observed that protein concentrations ranged from 4.88 to 15.58% across 16 distinct cultivars that were examined. A range of 5 to 13% was reported by other authors (Chethan & Malleshi, 2007; Shobana et al., 2013; Devi et al., 2014; Chandra et al., 2016; Ramashia et al., 2019). Protein digestibility and the composition of essential amino acids are commonly used to evaluate the protein quality of various crops (Abioye et al., 2022). In finger millet, approximately 44.7% of the albuminoids consist of essential amino acids (Sood et al., 2017; Ramashia et al., 2018). Finger millet stands out for its high content of essential amino acids, like lysine and methionine, which are typically deficient in other crop varieties (Jagati et al., 2021; Vagdevi et al., 2023). Compared to other millet species, finger millet exhibits a higher concentration of methionine, with levels reaching 194 mg/g (Singh et al., 2012; Prashanth & Muralikrishna, 2014; Ramashia et al., 2019). According to Abioye et al. (2022), prolamin, which is the major finger millet protein fraction, constitutes between 24.6 and 36.2% of the whole protein. Finger millet has 99.1 mg of dissolvable proteins/100 g and a considerable measure of tryptophan and threonine, which are absent from other seeds (Jagati et al., 2021). Furthermore, finger millet protein is known for its digestibility and bioavailability. Devi et al. (2014) reported that the protein in finger millet is highly digestible and readily absorbed by the human body, making it an excellent source of plant-based protein for vegetarians and individuals with dietary restrictions. The range of concentrations of specific essential and non-essential amino acids present in finger millet, as determined by various researchers, is summarized in Table 8.2.

TABLE 8.2
Finger Millet Amino Acid Profiles

Amino Acid	Concentration (g/100 g)
Essential Amino Acid	
Isoleucine	4.3
Leucine	6.6–10.8
Lysine	2.2
Methionine	2.5–3.51
Phenylalanine	4.1–6.0
Threonine	3.4–4.3
Valine	6.3
Histidine	2.2–2.3
Tryptophan	NA
Non-Essential Amino Acid	
Alanine	6.0–6.2
Arginine	2.77–4.5
Aspartic acid	5.7–7.9
Cystine	1.7–2.5
Glutamic acid	20.3–27.1
Glycine	3.3
Serine	3.6–5.3
Tyrosine	2.79–3.6
Proline	7.0–9.9
***PER**	2.0

Source: Ramashia et al. (2019); Jagati et al. (2021).
*Protein efficiency ratio.

8.3.5 Mineral Constituents

The ash content of finger millet, a significant cereal grain, is noteworthy due to its nutritional significance and implications for food processing; also, the ash content of any food reflects its mineral content (Bakkali et al., 2009). Singh and Srivastava (2006) conducted a study examining the ash content of finger millet across 16 different cultivars, finding variability in ash levels ranging between 1.47% and 2.58%. Similarly, Thapliyal and Singh (2015) and Muthamilarasan et al. (2016) reported ash content within the range of 1.7 to 4.13% in their respective studies. Various studies have reported the mineral composition of finger millet, including phosphorus, potassium, magnesium, calcium, sodium, zinc, iron, manganese, and copper (Devi et al., 2014; Manjula et al., 2015; Ramashia et al., 2019; Vagdevi et al., 2023). These minerals play essential roles in metabolic processes and are crucial for overall health and well-being. Furthermore, finger millet has been recognized for its superior nutritional density compared to other staple grains, like rice and wheat (Ramashia et al., 2019). Research conducted by Shukla and Srivastava (2014), Manjula et al. (2015), Jayawardana et al. (2019), and Jagati et al. (2021) consistently demonstrate higher levels of calcium, phosphorus, iron, and magnesium in finger millet grains compared to other millet varieties. Finger millet is reported to contain phosphorus (130–283 mg/100 g), potassium (430–490 mg/100 g), magnesium (78–201 mg/100 g), calcium (162–350 mg/100 g), sodium (49 mg/100 g), zinc (2.3 mg/100 g), iron (3.3–14.39 mg/100 g), manganese (17.61–48.43 mg/100 g), and copper (0.47 mg/100 g), as reported by Devi et al. (2014), Manjula et al. (2015), Ramashia et al. (2019), and Vagdevi et al. (2023).

8.3.6 Vitamins

Millets are believed to contain potentially high levels of vitamin B. Jaybhaye et al. (2014) reported that 100 g of finger millet contained only 45 μg of vitamin B. Ray et al. (2016) found that the presence of β-carotene, a precursor to vitamin A, in finger millet is minimal, ranging from 0 to 1 μg. However, contrary findings have been noted by Thapliyal and Singh (2015) and Jagati et al. (2021), who observed the presence of vitamin A in finger millet equating to approximately six retinols. Additionally, finger millet contains both fat-soluble and water-soluble vitamins, including thiamine, niacin, riboflavin, tocopherol (recognized as vitamin E), and vitamin C. It is noteworthy that vitamin C is absent from other dried millet grains (Jagati et al., 2021). The germ and aleuronic layers of finger millet have been identified as having high concentrations of water-soluble vitamin B, while the germ specifically contains lipid-soluble vitamins, as highlighted by Jaybhaye et al. (2014) and Ray et al. (2016). However, Chappalwar et al. (2013) and Devi et al. (2014) reported that finger millet seeds are not a remarkably rich source of vitamins, despite containing small amounts of vitamins A and B. While the grains exhibit increased levels of riboflavin, niacin, thiamine, and folic acid, they are deficient in vitamin C (Abioye et al., 2022). The vitamin B_1 contents of different finger millet varieties have been documented, such as 0.27, 0.27, and 0.43 mg/100 g for the milky cream, brown, and black varieties, respectively (Ramashia et al., 2021).

8.3.7 Polyphenols in Finger Millet

In recent years, finger millet has gained considerable attention owing to its remarkable nutritional composition and potential health advantages. Known for its abundance of polyphenols, essential minerals, and other bioactive compounds, finger millet has emerged as a valuable functional food ingredient with numerous health-promoting properties (Balasubramaniam et al., 2019; Ramashia et al., 2019). Its high polyphenol content, particularly compared to other cereals like barley, rice, maize, and wheat, has been extensively studied (Oghbaei & Prakash, 2012). The polyphenols in finger millet are predominantly concentrated in the outer layers of the grain, including the testa, pericarp, and aleurone layers (Figure 8.1c), which form the bulk of the bran portion (Thapliyal & Singh, 2015).

Research indicates that phenolic compounds, which are more prevalent in the seed coat than in the flour fraction, are found in higher concentrations in the brown variety of finger millet compared to the white variety (Udeh et al., 2017). Xiang et al. (2019) studied the antioxidant characteristics of four Malawian varieties of finger millet, namely, white, brown, reddish, and red, and found that the brown variety exhibited the highest levels of total phenolic content, flavonoids, and catechin tannin. Results for the four types showed that the overall phenolic content was 292.29 and 302.42 mg ferric acid eq./100 g, the total flavonoids was 90.24–202.94 mg catechin acid eq./100 g, and the total catechin tannin was 31.76–83.59 mg catechin acid eq./100 g. Compared to the other types, the brown species exhibited the greatest levels of whole flavonoids, total phenolic makeup, and whole catechin tannin. Out of all the antioxidant qualities assessed, the white variant exhibited the threshold values (Xiang et al., 2019). This was anticipated since colored variants have higher levels of antioxidant-rich carotenoids and polyphenols (Thapliyal & Singh, 2015). Similarly, Lansakara et al. (2016) observed differences in flavonoid and antioxidant activities between Sri Lankan finger millet varieties named Ravana and Oshawa. The "Oshawa" variety exhibited overall levels of flavonoids and phenolics at 1.05 and 8.08 mg GAE/g, respectively (Jayawardana et al., 2018). In comparison, the phenolic content of Sri Lankan varieties "Bala" and "Wadimal" ranged between 160.20 and 181.39 mg GAE/100 g (Jayawardana et al., 2018). Notably, the "Wadimal" variety demonstrated higher radical scavenging activity (5.64) compared to "Bala" (4.67). However, in terms of antioxidant activity measured through the FRAP experiment, "Bala" exhibited superior performance, with a value of 58.18 mg/100 g GAE, while "Wadimal" scored 43.81 mg/100 g GAE (Jayawardana et al., 2018). Finger millet's phenolic compounds are categorized into free, conjugated, and bound forms, like those found in other cereal grains (Chethan et al., 2008; Chandrasekara & Shahidi, 2012). In furtherance, the total phenolic content of finger millet ranges from 265 to 373.15 mg/100 g, with varying levels of free, ester-substituted, etherified, and bound phenolic compounds (Sreeramulu et al., 2009; Shahidi & Chandrasekara, 2013). These phenolic compounds, including hydroxybenzoic and hydroxycinnamic acids, as well as flavonoids, contribute to the antioxidant properties of finger millet (Chandrasekara & Shahidi, 2011a). Recently, finger millet has garnered significant attention primarily due to its remarkable nutritional profile and potential health benefits. Studies have underscored its high content of polyphenols, essential minerals, and other bioactive compounds, positioning it as a valuable functional food ingredient with enormous health-promoting properties (Balasubramaniam et al., 2019; Ramashia et al., 2019). Furthermore, finger millet's suitability for individuals with specific dietary requirements, such as those seeking gluten-free alternatives or managing diabetes, has also contributed to its growing popularity (Kumar et al., 2016). In cereals, including finger millet, phenolic acids represent the primary polyphenols, contributing significantly to their phytochemical composition, while flavonoids, another subgroup of polyphenols, occur in small concentrations (Subba Rao & Muralikrishna, 2002). The proceeding sections discuss some classes of polyphenolic compound, including phenolic acids and flavonoids, found in finger millet, some of which are summarized in Table 8.3.

8.3.7.1 Phenolic Acids

Aromatic compounds with a carboxylic acid function and a single benzene ring are referred to as *phenolic acids*. Among grains, millets and sorghum exhibit the greatest diversity in these by-products of benzoic and cinnamic acids (Dykes & Rooney, 2007; Liu, 2007). Phenolic acids are classified into two types: hydroxycinnamic acids, derived from cinnamic acid; and hydroxybenzoic acids, originating from benzoic acid. In finger millet, the primary free phenolic fraction is composed of hydroxybenzoic acids, constituting approximately 27% of the phenolic acids in the seeds (Hithamani & Srinivasan, 2014; Himanshu et al., 2018). These include gallic, vanillic, gentisic, p-hydroxybenzoic, syringic, and protocatechuic acids, with protocatechuic acid identified as the predominant free phenolic acid (45 mg/100 g) (Hithamani & Srinivasan, 2014; Himanshu et al., 2018). Additionally, trace levels of hydroxycinnamic acids such as ferulic, caffeic, and p-coumaric acids have been detected in the free phenolic portion (Udeh et al., 2017; Himanshu et al., 2018). Common phenolic acids found in finger millet are summarized in Table 8.3.

TABLE 8.3
Phenolic Compounds in Finger Millet

Phenolic Compounds	Structural Class	Quantity (μg/g)	References
Hydroxybenzoic acids	Phenolic acids	–	Hithamani and Srinivasan (2014)
Gentisic acid	Phenolic acids	4.5	
Vanillic acid	Phenolic acids	20.0	Chethan et al. (2008)
Gallic acid	Phenolic acids	3.91–30.0	Chethan et al. (2008); Hithamani and Srinivasan (2014)
Syringic acid	Phenolic acids	10.0–60.0	Dykes and Rooney (2007); Hithamani and Srinivasan (2014)
Salicylic acid	Phenolic acids	5.12–413.0	Hithamani and Srinivasan (2014)
Protocatechuic acid	Phenolic acids	119.8–405.0	Chethan et al. (2008); Hithamani and Srinivasan (2014)
p-hydroxybenzoic acid	Phenolic acids	6.3–370.0	Dykes and Rooney (2007); Chethan et al. (2008); Hithamani and Srinivasan (2014)
Chlorogenic acid	Hydroxycinnamic acids	–	Shahidi and Chandrasekara (2013)
Caffeic acid	Hydroxycinnamic acids	5.9–10.4	Hithamani and Srinivasan (2014)
Sinapic acid	Hydroxycinnamic acids	11.0–24.8	Hithamani and Srinivasan (2014)
p-coumaric	Hydroxycinnamic acids	1.81–41.1	Chethan et al. (2008); Shahidi and Chandrasekara (2013)
Trans-cinnamic acid	Hydroxycinnamic acids	35–100.0	Chethan et al. (2008); Hithamani and Srinivasan (2014)
Trans-ferulic acid	Hydroxycinnamic acids	41–405.0	Shahidi and Chandrasekara (2013
Quercetin	Flavonoids	3	Chethan et al. (2008)
Catechin gallocatechin, **epicatechin, epigallocatechin, taxifolin,** vitexin, **tricin, luteolin, myricetin, apigenin, kaempferol, naringenin,** daidzein, **procyanidin B1, orientin,** isoorientin, **isovitexin, saponarin, violanthin, lucenin-1, saponarin, violanthin**	Flavonoids	–	Subba Rao and Muralikrishna (2002); Chethan et al. (2008); Shobana et al. (2009); Chandrasekara and Shahidi (2011b); Shahidi and Chandrasekara (2013)

8.3.8 Flavonoids

Flavonoids are a class of secondary metabolites found in plants which have a skeleton consisting of 15 carbons. One heterocyclic ring and two phenyl rings make up this structure. Thapliyal et al. (2015) and Himanshu et al. (2018) describe *flavonoids* as plant pigments with a basic C6-C3-C6 structure that are generated from phenylalanine. Studies have identified several flavonoids in finger millet, including quercetin, gallocatechin, epicatechin, and epigallocatechin (Udeh et al., 2017; Abioye et al., 2022). These flavonoids are known for their antioxidant properties and potential health benefits, making finger millet a valuable dietary component with various health-promoting effects. Compared to other millets, the concentration of flavonoids in finger millet has been demonstrated to be higher at 2,100 μg/g and to be mostly present in the soluble form (Chandrasekara et al., 2012; Shahidi & Chandrasekara, 2013). It has also been discovered that finger millet contains esterified forms of flavonoids, which differ significantly from the flavonoids found in other millets, such as kodo millet (Shahidi & Chandrasekara, 2013; Himanshu et al., 2018). These include a wide range of polyphenolic substances found in fruits and vegetables, such as isoflavones, flavanols, and anthocyanin pigments, all of which have a benzo-y-pyrone structure. Except for flavanols, which have a propensity to polymerize into condensed tannins, they are primarily found as glycosides. Das et al. (2020) categorized tannins as either hydrolyzable or condensed. While most hydrolyzable tannins are glucose or polyhydric alcohol esterified with gallic acid (gallotannins) or hexahydroxydiphenic acid (ellagitannins), most condensed tannins are polymers of flavan-3-ols (catechins) or flavan-3,4-diols (leucoanthocyanidins) (Himanshu et al., 2018). Flavonoids, which include anthocyanin, tannin, quercetin, and catechin, are components that are good for human health. Due to their pharmacological actions as radical scavengers, they are significant for human health (Himanshu et al., 2018; Abioye et al., 2022). The quantity and configuration of hydroxyl groups throughout the structure, as well as the existence of electron-donating and electron-withdrawing groups, determine the antioxidant capacity of flavonoids (Himanshu et al., 2018). Furthermore, a considerable number of oligomeric or polymeric flavonoids known as proanthocyanidins or condensed tannins are present in grain; procyanidin B1 and B2 are said to be the main dimers (Chandrasekara & Shahidi, 2012). The antioxidative, antiproliferative, antiallergic, and anti-inflammatory qualities of these phenolic compounds are well-known. Studies have shown that finger millet is the only millet that has been found to contain condensed tannins. Compared to white finger millets, which have tannins of 0.04–0.06% catechin equivalents, brown finger millets have 0.12–3.47% catechin equivalents (Himanshu et al., 2018).

8.3.9 Phytic Acid

Phytic acid, often known as *phytate*, is a naturally occurring antinutrient, and it is made up of myoinositol-1,2,3,4,5,6-hexakis dihydrogen phosphate. It is the main source of phosphorus storage, making about 1% to 5% of the weight of grains, legumes, nuts, oilseeds, and other foods (Himanshu et al., 2018). The phytate content in finger millet falls within the range of 0.679 to 0.693 g/100 g (Antony & Chandra, 1999).

Deosthale (2002) reported that finger millet contributes roughly 41% of the recommended daily intake of phytic phosphorus. According to Gemede and Ratta (2014), it is a secondary metabolic substance that serves as the seed's stored supply of nutrients during ripening. Because phytate is not absorbed by humans and cannot be hydrolyzed, it contributes to the reduced bioavailability of minerals. Positively divalent and trivalent mineral ions such as Zn^{2+}, $Fe^{2+/3+}$, Ca^{2+}, and Mg^{2+} have their bioavailability adversely impacted by phytate, a negatively charged ion that functions in pH-sensitive environments (Gemede & Ratta, 2014). Because they are influenced by growth circumstances, processing, harvesting techniques, and even testing methodologies, phytate levels in grains, legumes, nuts, and seeds vary greatly (Samtiya et al., 2021). Phytic acid is often considered an antinutrient due to its ability to chelate minerals and inhibit their absorption in the human body; it also possesses certain beneficial properties and health effects, including the following.

8.3.9.1 Presence in Finger Millets

Phytic acid is predominantly found in the outer layers of finger millet grains, particularly in the bran and germ. These parts of the grain contain higher concentrations of phytic acid compared to the starchy endosperm. Phytic acid serves as a phosphorus storage compound in plants, contributing to seed germination and growth (Shobana, 2013).

8.3.9.2 Antinutrient Properties

Phytic acid can bind to minerals such as calcium, iron, zinc, and magnesium, forming insoluble complexes called *phytates*. This reduces the bioavailability of these minerals, making them less absorbable in the gut. As a result, excessive consumption of phytic acid–rich foods may lead to mineral deficiencies, particularly in populations with poor diets or increased mineral requirements, such as children and pregnant women (Hurrell & Egli, 2010).

8.3.9.3 Implications on Health

Despite its antinutrient properties, phytic acid also exhibits several health benefits.

Antioxidant activity. Phytic acid possesses antioxidant properties and can scavenge harmful free radicals in the body. By neutralizing oxidative stress, phytic acid may help protect against various chronic diseases, including cancer, cardiovascular disorders, and neurodegenerative conditions (Adefegha, 2018).

Anticancer effects. Some studies suggest that phytic acid may exert anticancer effects by inhibiting the growth and proliferation of cancer cells, inducing apoptosis (programmed cell death), and suppressing tumor formation and metastasis (Nath et al., 2022).

Reduced risk of kidney stones. Phytic acid has been shown to inhibit the formation of calcium oxalate kidney stones by binding to calcium ions and preventing their crystallization. Regular consumption of phytic acid–rich foods may help reduce the risk of kidney stone formation (Pujol et al., 2023).

8.3.9.4 Mitigation Strategies

Several methods can help reduce the levels of phytic acid in finger millet and enhance mineral bioavailability.

Soaking. Soaking finger millet grains in water for a certain period can help activate phytase enzymes, which degrade phytic acid and reduce its levels. Discarding the soaking water before cooking can further decrease phytic acid content (Lestienne et al., 2005).

Sprouting. Sprouting finger millet seeds initiates enzymatic activity, including phytase activity, leading to the degradation of phytic acid. Sprouted finger millet grains are lower in phytic acid and higher in mineral bioavailability compared to unsprouted grains (Lestienne et al., 2005).

Fermentation. Fermentation of finger millet dough or batter by lactic acid bacteria can degrade phytic acid and enhance mineral release. Fermented finger millet products, such as idli and dosa, are popular traditional foods known for their improved nutritional profile (Shobana et al., 2009).

In a nutshell, while phytic acid in finger millet may hinder mineral absorption, it also offers various health benefits and can be mitigated through processing techniques like soaking, sprouting, and fermentation. Incorporating these methods into the preparation of finger millet–based foods can enhance their nutritional value and promote overall health.

8.4 BIOAVAILABILITY

The amount of nutrients that are taken and can be absorbed and used by typical metabolic processes is referred to as *bioavailability*. As such, the focus is no longer just on the nutrient content of the food when determining the adequacy of dietary intake but on the bioavailability of nutrients (Tharifkhan et al., 2021). It is well-known that finger millet is a good source of minerals. However, the presence of antinutrients, such as tannins, polyphenols, and phytates, limits the availability of nutrients for cellular absorption. Antinutritional factors are compounds found in food and animal feed that decrease the availability of nutrients (Soetan & Oyewole, 2009). Finger millet is thought to contain antinutritional compounds because they decrease the digestion of starch and protein, interfere with the availability of minerals, and inhibit amylolytic and proteolytic enzymes. Tannins, protease inhibitors, oxalates, and phytates are among the antinutritional elements in finger millet that are thought to have an adverse effect on how easily nutrients are absorbed (Kumar et al., 2016; Chauhan & Sarita, 2018). Furthermore, the presence of antinutritional chemicals in finger millet, such as phytates, phenols, and tannins, has been shown by Singh et al. (2012) to exert deleterious effects on nutrient consumption. It is thought that the antinutrient tannin reduces food and feed intake and efficiency (Kumar et al., 2016). Between 0.04 and 3.74% of the catechin equivalents, finger millet is thought to contain the most tannin of any millet (Abubakar et al., 2015; Kumar et al., 2016; Hassan et al., 2021). When finger millet is divided into brown and dark variations, the white seed types have reduced tannin concentrations (0.05%) than the brown and black forms (Abioye et al., 2022). Tannins, which are essential for the biological functions of living organisms and function as a defense against fungal onslaught, are present in the seed coat of finger millet (Devi et al., 2014). However, it has been asserted that it causes liver malfunction and compromises pancreas and thyroid function. When finger millet is used to make food items, the presence of tannin compounds can have an impact on the food products' color, flavor, nutritional digestibility, mineral absorption, and overall quality (Devi et al., 2014). Additionally, Shibairo et al. (2014) reported that the concentration of antinutrients present in finger millet may vary depending on the environment where it was cultivated. The bioavailability of the nutrients in finger millet can be affected by dietary and physiological variables. Various processing techniques can be used to reduce the levels of antinutritional components, making it more nutritious and improving its overall functional characteristics.

Abioye et al. (2022) reported that the phytate level of most of the finger millet ranged from 240 to 300 mg/100 g. Because of its potent binding ability, phytate has a significant impact on the nutritional and functional characteristics of food. While antinutritional components like tannins and others may hinder nutrient absorption, they also offer significant health benefits when consumed as part of a balanced diet. Quercetin, phytic acid, tannins, saponins, and phenolic acids present in finger millet contribute to its overall nutritional value and may help protect against chronic diseases, such as cardiovascular disorders, cancer, and inflammation.

Quercetin is a flavonol compound found abundantly in finger millet. It is well-known for its potent antioxidant properties, which help neutralize harmful free radicals in the body, thereby reducing oxidative stress and inflammation (Verma et al., 2023). Quercetin has been associated with numerous health benefits, including improved cardiovascular health, enhanced immune function, and potential anticancer effects (Deepika, 2022). Finger millet's richness in quercetin contributes to its overall antioxidant capacity and may play a role in preventing chronic diseases. Despite being classified as an antinutrient due to its ability to chelate minerals, phytic acid also offers several health benefits. Phytic acid acts as an antioxidant in the body, scavenging free radicals and reducing oxidative damage to cells and tissues (Graf & Eaton, 1990). Additionally, phytic acid has been shown to exhibit anticancer properties by inhibiting the growth and proliferation of cancer cells and inducing apoptosis (Nissar et al., 2017). In finger millet, phytic acid contributes to its overall antioxidant activity and may help protect against chronic diseases.

Tannins are polyphenolic compounds found in finger millet that possess antioxidant, anti-inflammatory, and antimicrobial properties (Scalbert, 1991). Tannins contribute to the astringent taste of finger millet and have been associated with various health benefits, including improved digestive health and reduced risk of cardiovascular disease (Vagdevi et al., 2023). Furthermore, tannins may exert protective effects against certain types of cancer by inhibiting the growth of cancer cells and preventing tumor formation (Rajasekar et al., 2021).

Saponins are bioactive compounds found in finger millet that have been reported to exhibit anti-inflammatory, hypocholesterolemic, and immunomodulatory effects (Rani et al., 2024). Saponins may help lower cholesterol levels, improve immune function, and reduce the risk of chronic diseases, such as cardiovascular disease and cancer (Marrelli et al., 2016). Additionally, saponins possess antimicrobial properties and may help protect against infections caused by bacteria, viruses, and fungi (Sparg et al., 2004).

Phenolic acids, including hydroxybenzoic and hydroxycinnamic acids, are present in finger millet and contribute to its antioxidant activity and health-promoting properties (Bello et al., 2022). These compounds have been associated with various health benefits, including anti-inflammatory, anticancer, and neuroprotective effects (Huang et al., 2009). Phenolic acids may also help improve cardiovascular health by reducing the risk of heart disease and hypertension (Panche et al., 2016).

8.5 FINGER MILLET HEALTH BENEFITS

Finger millet gains numerous advantageous qualities, including antibacterial, antioxidant, and antidiabetic effects, due to the presence of polyphenols (Patil et al., 2023), which have the potential to be used as a source of beneficial food supplements. According to Krishnan and Meera (2018), the brown finger millet type has higher levels of threonine, valine, and lysine, as well as 96% more phenolic acid than the white and other forms. Furthermore, Jagati et al. (2021) report that black finger millets contain 8.71 mg/g of dry-weight fatty acid and 8.47 mg/g of dry-weight protein. For diabetics, a diet including finger millet lowers blood sugar by 36%, and cholesterol by 36%. The potential health benefits of this plant are attributed to the presence of secondary metabolites and nutritional constituents. Some of the benefits of consuming finger millet, in addition to the ones mentioned earlier, are listed.

8.5.1 Antioxidant Properties

Finger millet is rich in antioxidants, including phenolic compounds, flavonoids, and tannins, which help neutralize harmful free radicals in the body. These antioxidants protect cells from oxidative damage, reducing the risk of chronic diseases, such as cancer, cardiovascular diseases, and neurodegenerative disorders (Balasubramaniam et al., 2019; Thapliyal & Singh, 2015).

8.5.2 Anti-Inflammatory Effects

Certain secondary metabolites found in finger millet, such as flavonoids and phenolic acids, possess anti-inflammatory properties. These compounds help reduce inflammation in the body which is associated with various chronic diseases, including arthritis, inflammatory bowel diseases, and metabolic syndrome (Chandrasekara & Shahidi, 2012).

8.5.3 Cardiovascular Health

Studies have shown that the amino acids found in finger millet, namely, methionine, threonine, and lecithin, help inhibit the production of new fat, lower cholesterol, and eliminate extra fat from the liver (Saleh et al., 2013; Prajapati et al., 2019). Once more, the serum triglyceride content of finger millet is low. By lowering plasma triglycerides, eating finger millet reduces the risk of cardiovascular disease (Chamoli et al., 2018; Jagati et al., 2021). The presence of antioxidants and dietary fiber in finger millet supports cardiovascular health. Antioxidants protect against oxidative damage to blood vessels, reducing the risk of atherosclerosis and hypertension. Additionally, dietary fiber helps lower cholesterol levels, promoting heart health and reducing the risk of heart disease (Soliman, 2019).

8.5.4 Control of Celiac Diseases

For individuals with a genetic predisposition, the consumption of gluten can trigger celiac disease, an autoimmune disorder. Given that finger millet is devoid of gluten (Chowdary & Bisarya, 2020), it serves as a viable option for those with celiac disease and other gluten-related sensitivities. Additionally, individuals who prefer to avoid gluten commonly found in wheat and other typical grains can find finger millet to be a suitable alternative (Chowdary & Bisarya, 2020; Jagati et al., 2021; Patil et al., 2023).

8.5.5 Source of Protein/Amino Acids

The body's ability to operate and heal itself depends heavily on amino acids. Threonine, methionine, valine, isoleucine, and tryptophan are abundant in finger millets (Thapliyal & Singh, 2015; Jagati et al., 2021). The following are the roles that amino acids play in the body, according to Jagati et al. (2021):

- Amino acids enhance the body's metabolism, support the equilibrium of nitrogen levels in the body, and aid in the coordination of muscle and tissue repair.
- Methionine is an essential amino acid that plays a critical role in the removal of excess fat, helps facilitate various bodily functions and processes, and acts as a sulfur provider in the body, which is required for the production of natural antioxidants like glutathione.
- Isoleucine helps repair muscle disorders and contributes to blood formation, which will help in bone formation and recovery of healthy skin.

8.5.6 Treatment of Anemia

Due to its high iron content, finger millet is often included in meals intended to aid in the recovery from anemia and malnourishment (Chamoli et al., 2018; Jagati et al., 2021).

8.5.7 Relaxation

Eating finger millet has been associated with inducing a natural relaxation response in the body. This ancient grain, rich in various nutrients and bioactive compounds, offers several health benefits, including its potential to promote relaxation and alleviate symptoms of anxiety, migraines, depression, and sleeplessness. Several studies have highlighted the calming effects of finger millet consumption. Chamoli et al. (2018) conducted research suggesting that finger millet possesses properties that contribute to relaxation and stress reduction. Additionally, Jagati et al. (2021) supported these findings, indicating that finger millet consumption can have a positive impact on mental well-being. The relaxation-inducing properties of finger millet can be attributed to its nutritional composition and bioactive constituents. Finger millet is rich in antioxidants, vitamins, minerals, and phytochemicals, all of which play essential roles in supporting overall health and well-being. These include magnesium, an essential mineral known for its calming effects on the nervous system, as well as B vitamins, which are important for neurotransmitter function and mood regulation. Furthermore, finger millet contains bioactive compounds, such as flavonoids and polyphenols, which have been studied for their potential to alleviate symptoms of anxiety and depression by modulating neurotransmitter activity and reducing oxidative stress in the brain. Incorporating finger millet into the diet in various forms, such as porridge, roti, or baked goods, may provide a natural and nutritious way to promote relaxation and support mental health. However, it is essential to consume finger millet as part of a balanced diet, alongside other nutrient-rich foods, to reap its full benefits for relaxation and overall well-being.

8.5.8 Supplement for Diabetes

The glycemic index (GI) of finger millet is an important consideration for individuals managing blood sugar levels or seeking foods with a lower glycemic impact. Research studies have provided insights into the GI of finger millet, offering valuable information for dietary planning and health management. Several studies have investigated the glycemic response of finger millet and have reported varying GI values. For instance, a study by Devi et al. (2014) found that finger millet had a GI value of approximately 65, indicating a moderate glycemic impact. Similarly, another study by Gaikwad et al. (2023) reported a GI value of around 68 for finger millet. The relatively low to moderate GI of finger millet can be attributed to its composition, including its complex carbohydrate structure, fiber content, and presence of certain bioactive compounds. These factors contribute to slower digestion and absorption of carbohydrates, resulting in a more gradual increase in blood sugar levels compared to high-GI foods. It is important to note that the GI of finger millet can vary depending on factors, such as processing methods, cooking techniques, and the specific variety of finger millet used. Additionally, individual differences in glycemic response may influence how different people react to consuming finger millet. Overall, finger millet can be considered a favorable dietary option for individuals seeking foods with a lower glycemic index. Incorporating finger millet in meals can help promote stable blood sugar levels and may offer various health benefits associated with better glycemic control. According to Chowdary and Bisarya (2020) and Jagati et al. (2021), the phytochemicals in finger millet slow down the digestive process, which helps manage blood sugar levels and boost antioxidant levels (Prajapati et al., 2019). The high fiber content of finger millet sets it apart from wheat and rice. Its low glycemic response also indicates that it has less capacity to raise blood sugar levels and increase starch absorptivity, thus regulating blood sugar levels and improving insulin sensitivity (Chamoli et al., 2018; Jagati et al., 2021; Patil et al., 2023).

According to Singh (2016), finger millet seed coats contain phenolic compounds that function as an inhibitor to limit the role of enzymes necessary for the digestion of complex carbohydrates, such as alpha glucosidase and amylase, therefore lowering postprandial hyperglycemia.

8.5.9 Digestive Health

The dietary fiber content of finger millet promotes digestive health by supporting regular bowel movements and preventing constipation. Additionally, fiber acts as a prebiotic, nourishing beneficial gut bacteria and maintaining a healthy gut microbiota, which is essential for overall digestive function (Balasubramaniam et al., 2019; Devi et al., 2014).

8.5.10 Bone Health

Finger millet is an excellent source of calcium, magnesium, and phosphorus, which are essential minerals for bone health. These nutrients help strengthen bones, prevent osteoporosis, and reduce the risk of fractures. Furthermore, the presence of bioactive compounds such as flavonoids and phenolic acids may also contribute to bone health by reducing bone resorption and promoting bone formation (Devi et al., 2014).

8.5.11 Weight Management

The high dietary fiber content and low fat content of finger millet make it an ideal food for weight management. Fiber helps promote satiety, preventing overeating and aiding in weight loss or weight maintenance. Additionally, finger millet is nutrient-dense, providing essential nutrients without excess calories, making it suitable for individuals looking to manage their weight (Balasubramaniam et al., 2019; Thapliyal & Singh, 2015). In summary, finger millet's health benefits are closely linked to its secondary metabolites and nutritional constituents, including antioxidants, dietary fiber, vitamins, and minerals. Regular consumption of finger millet as part of a balanced diet may help prevent chronic diseases, support overall health, and promote well-being.

8.5.12 Acts as Antiaging

Finger millet is rich in phenolic compounds and antioxidants, which play crucial roles in combating aging, metabolic syndrome, and overall health. Research by Chamoli et al. (2018), Jagati et al. (2021), and Patil et al. (2023) indicates that finger millet consumption helps prevent glycation and cross-linking of collagen, processes associated with aging. Regular intake of finger millet can mitigate the risk of malnutrition, manage degenerative diseases, and delay premature aging (Prajapati et al., 2019). Particularly, green finger millet offers advantages for individuals dealing with asthma, blood pressure, liver disorders, and heart problems (Chamoli et al., 2018). Finger millet is also beneficial for expectant mothers as it promotes better lactation. Moderate consumption of finger millet can help prevent the development of mucosal ulcers. Additionally, being a plant-based source of natural iron, finger millet contributes to improving children's hemoglobin levels (Chamoli et al., 2018). According to Chamoli et al. (2018) and Patil et al. (2023), the phenolic compounds in finger millet, particularly tannins, may provide resistance against fungal infections. These phenolic compounds, mainly present in the grain's outer layer, act as a structural barrier against fungal invasion. The high concentration of polyphenols in the seed coat gives acidic methanol extracts of the seed coat superior antifungal and antibacterial properties compared to extracts from whole wheat. Finger millet is known for its nutrient richness and its ability to support overall health. However, individuals with urinary calculi should limit their consumption of finger millet, as it may lead to increased oxalic acid production.

8.6 CONCLUSION AND PROSPECTS

This review has highlighted the profound significance of finger millet (*Eleusine coracana*) in the cultural, nutritional, and agronomic perspectives. It is a versatile cereal crop resilient to adverse conditions, making it an asset for farmers in semi-arid and tropical regions of Africa and Asia. The comprehensive analysis covered in this review explored the cultivation practices, nutritional composition, polyphenolic content, bioavailability, and associated health benefits of finger millet, showcasing its versatility and utility. Thus, nutritionally, finger millet emerges as a powerhouse of macronutrients, such as carbohydrates, proteins, and dietary fiber, complemented by essential micronutrients, like iron, calcium, and magnesium. The presence of bioactive compounds, particularly polyphenols and phytochemicals, further enhances its nutritional profile, promoting health and well-being. Previous studies underscore its potential in managing various health conditions, like diabetes, cardiovascular diseases, and obesity, attributed to its digestive aid properties, satiety promotion, and antioxidative capacity. Moreover, the agronomic advantages of finger millet, including its adaptability to challenging environmental conditions, short growth cycle, low water requirements, and minimal input needs, position it as a sustainable and climate-resilient crop with the capacity to bolster food security and livelihoods, especially in regions susceptible to climatic variability and resource limitations. Looking ahead, the increasing interest in advancing research and development initiatives aims to enhance the productivity, nutritional quality, and value addition of finger millet. Over the next decades, these endeavors should seek to unlock the full potential of finger millet as a nutritious staple, advocating for its broader cultivation and consumption. Future research prospects include genetic enhancement for increased yield and nutritional content, development of value-added products, exploration of novel culinary applications, and dissemination of agronomic best practices to empower smallholder farmers. In essence, this review sets the stage for ongoing exploration and utilization of finger millet as a pivotal component of global food security initiatives, offering solutions to nutritional deficiencies, health risks, and agricultural resilience amidst evolving environmental challenges.

8.7 ACKNOWLEDGMENTS

The authors appreciate the Biomedicinal Research Centre, Forestry Research Institute of Nigeria, for supporting this work.

REFERENCES

Abioye, V. F., Babarinde, G. O., Ogunlakin, G. O., Adejuyitan, J. A., Olatunde, S. J., & Abioye, A. O. (2022). Varietal and processing influence on nutritional and phytochemical properties of finger millet: A review. *Heliyon*, *8*, e12310.

Abubakar, A., Bala, S., Audu, E. A., Mohammad, S., Gero, M., & Lande, L. (2015). Characterization and the anti-nutritional composition of unprocessed finger millet (*Eleusine coracana*). *International Journal of Food and Nutrition Safety*, *6*, 117–124.

Adefegha, S. A. (2018). Functional foods and nutraceuticals as dietary intervention in chronic diseases; Novel perspectives for health promotion and disease prevention. *Journal of Dietary Supplements*, *15*(6), 977–1009.

Antony, U., & Chandra, T. S. (1999). Enzymatic treatment and use of starters for the nutrient enhancement in fermented flour of red and white varieties of finger millet (*Eleusine coracana*). *Journal of Agricultural and Food Chemistry*, *47*(5), 2016–2019.

Bakkali, K., Martos, N. R., Souhail, B., & Ballesteros, E. (2009). Characterization of trace metals in vegetables by graphite furnace atomic absorption spectrometry after closed vessel microwave digestion. *Food Chemistry*, *116*(2), 590–594. https://doi.org/10.1016/j.foodchem.2009.03.010

Balasubramaniam, V. G., Ayyappan, P., Sathvika, S., & Antony, U. (2019). Effect of enzyme pretreatment in the ultrasound-assisted extraction of finger millet polyphenols. *Journal of Food Science and Technology*, *52*, 3579–3589.

Bello, O. M., Ogbesejana, A. B., Balkisu, A., Osibemhe, M., Musa, B., & Oguntoye, S. O. (2022). Polyphenolic fractions from three millet types (Fonio, Finger millet, and Pearl millet): Their characterization and biological importance. *Clinical Complementary Medicine and Pharmacology*, *2*(1), 100020. https://doi.org/10.1016/j.ccmp.2022.100020

Ceasar, S. A., Maharajan, T., Krishna, T. P. A., Ramakrishnan, M., Roch, G. V., Satish, L., & Ignacimuthu, S. (2018). Finger millet (*Eleusine coracana* (L.) Gaertn.) Improvement: Current status future interventions of whole genome sequence. *Frontier Plant Science*, *9*, 1054. https://doi.org/10.3389/fpls.2018.01054

Chamoli, V., Badoni, A., Bahuguna, N., & Joshi, N. (2018). Finger millet (*Eluesine coracana*): Nutritional status, health benefits and processing status – A review. *Journal of Pharmacognosy and Phytochemistry*, *7*(5S), 80–83.

Chandra, D., Chandra, S., Pallavi, & Sharma, A. K. (2016). Review of finger millet (*Eleusine coracana* (L.) Gaertn): A powerhouse of health benefiting nutrients. *Food Science and Human Wellness*, *5*(3), 149–155. https://doi.org/10.1016/j.fshw.2016.05.004

Chandrasekara, A., & Shahidi, F. (2011a). Antiproliferative potential and DNA scission inhibitory activity of phenolics from whole millet grains. *Journal of Functional Foods*, *3*(3), 159–170. https://doi.org/10.1016/j.jff.2011.03.008

Chandrasekara, A., & Shahidi, F. (2011b). Bioactivities and antiradical properties of millet grains and hulls. *Journal of Agricultural and Food Chemistry*, *59*(17), 9563–9571. https://doi.org/10.1021/jf201849d

Chandrasekara, A., & Shahidi, F. (2012). Bioaccessibility and antioxidant potential of millet grain phenolics as affected by simulated in vitro digestion and microbial fermentation. *Journal of Functional Foods*, *4*(1), 226–237. https://doi.org/10.1016/j.jff.2011.11.001

Chappalwar, V. M., Peter, D., Bobde, H., & John, S. M. (2013). Quality characteristics of cookies prepared from oats and finger millet-based composite flour. *Engineering Science and Technology: An International Journal (Toronto, Ont.)*, *3*(4), 667–683.

Chauhan, E. S., & Sarita, S. (2018). Effects of processing (Germination and Popping) on the nutritional and anti-nutritional properties of finger millet (*Eleusine Coracana*). *Current Research in Nutrition and Food Science*, *6*(2), 566–572.

Chethan, S., & Malleshi, N. (2007). Finger millet polyphenols: Optimization of extraction and the effect of pH on their stability. *Food Chemistry*, *105*(2), 862–870. https://doi.org/10.1016/j.foodchem.2007.02.012

Chethan, S., Dharmesh, S. M., & Malleshi, N. G. (2008). Inhibition of aldose reductase from cataracted eye lenses by finger millet (*Eleusine coracana*) polyphenols. *Bioorganic & Medicinal Chemistry*, *16*(23), 10085–10090. https://doi.org/10.1016/j.bmc.2008.10.003

Chowdary, D. M., & Bisarya, D. (2020). Review of Finger millet (*Eleusine coracana* L.) on Nutrition and health benefits. *International Journal of All Research Education and Scientific Methods*, *8*(11), 1319–1323.

Das, A. K., Islam, M. N., Faruk, M. O., Ashaduzzaman, M., & Dungani, R. (2020). Review on tannins: Extraction proceses, applications and possibilities. *South African Journal of Botany*, *135*, 58–70.

Deepika, M. P. K. (2022). Health benefits of quercetin in age-related diseases. *Molecules, (Basel, Switzerland)*, *27*(8), 2498. https://doi.org/10.3390/molecules27082498

Deosthale, Y. G. (2002). *The nutritive value of foods and the significance of some household processes*, p. 6. www.unu.edu

Devi, P. B., Vijayabharathi, R., Sathyabama, S., Malleshi, N. G., & Priyadarisini, V. B. (2014). Health benefits of finger millet (*Eleusine coracana* L.) polyphenols and dietary fibre: A review. *Journal of Food Science and Technology*, *51*(6), 1021–1040. https://doi.org/10.1007/s13197-011-0584-9

Dykes, L., & Rooney, L. W. (2007). Phenolic compounds in cereal grains and their health benefits. *Cereal Foods World*, *52*, 105–111.

Gaikwad, V., Kaur, J., Rasane, P., Kaur, S., Singh, J., & Kumar, A. (2023). Nutritional significance of finger millet and its potential for using in functional products. *Foods and Raw Materials*, *12*(1), 110–123.

Gemede, H. F., & Ratta, N. (2014). Antinutritional factors in plant foods: Potential health benefits and adverse effects. *International Journal of Nutrition and Food Sciences*, *3*(4), 284–289.

Graf, E., & Eaton, J. W. (1990). Antioxidant functions of phytic acid. *Free Radical Biology and Medicine*, *8*(1), 61–69.

Gupta, S. M., Arora, S., Mirza, N., Pande, A., Lata, C., Puranik, S., Kumar, J., & Kumar, A. (2017). Finger millet: A "certain" crop for an "uncertain" future and a solution to food insecurity and hidden hunger under stressful environments. *Frontiers in Plant Science*, *8*, 643.

Hassan, Z. M., Sebola, N. A., & Mabelebele, M. (2021). The nutritional use of millet grain for food and feed: A review. *Agriculture and Food Security*, *10*, 16. https://doi.org/10.1186/s40066-020-00282-6

Himanshu, K., Chauhan, M., Sonawane, S. K., & Arya, S. S. (2018). Nutritional and nutraceutical properties of millets: A review. *Clinical Journal of Nutrition and Dietetics*, *1*(1), 1–10.

Hithamani, G., & Srinivasan, K. (2014): Effect of domestic processing on the polyphenol content and bioaccessibility in finger millet (Eleusine coracana) and pearl millet (Pennisetum glaucum). *Food Chemistry, 164*, 55–62.

Huang, W. Y., Cai, Y. Z., & Zhang, Y. (2009). Natural phenolic compounds from medicinal herbs and dietary plants: Potential use for cancer prevention. *Nutrition and Cancer, 62*(1), 1–20.

Hurrell, R., & Egli, I. (2010). Iron bioavailability and dietary reference values. *American Journal of Clinical Nutrition, 91*(5), 1461S–1467S.

Jagati, P., Mahapatra, I., & Dash, D. (2021). Finger millet (Ragi) as an essential dietary supplement with key health benefits: A review. *International Journal of Home Science, 7*(2), 94–100.

Jayawardana, L. S. A. S., Samarasekera, J. K. R. R. S., Gardhi, G. H. C. M., Gooneratnea, J., Mazumdarc, S. D., & Roopa, B. (2019). Dietary fibres, starch fractions and nutritional composition of finger millet varieties cultivated in Sri Lanka. *Journal of Food Composition and Analysis, 82*, 103249. https://doi.org/10.1016/j.jfca.2019.103249

Jayawardana, N., Wimalasiri, K. M. S., Samarasinghe, G., & Madhujith, T. (2018). Bound and total phenolic contents and antioxidant potential of selected Sri Lankan millet varieties. *Tropical Agricultural Research, 29*(3), 316–321.

Jaybhaye, R. V., Pardeshi, I. L., Vengaiah, P. C., & Srivastav, P. P. (2014). Processing and technology for millet-based food products: A review. *Journal of Ready to Eat Food, 1*(2), 32–48.

Jenkins, D. J., Wolever, T. M., Taylor, R. H., Barker, H., Fielden, H., Baldwin, J. M., Bowling, A. C., Newman, H. C., Jenkins, A. L., & Goff, D. V. (1981). Glycemic index of foods: A physiological basis for cssssarbohydrate exchange. *The American Journal of Clinical Nutrition, 34*(3), 362–366.

Karki, A., Chandra, A., Joshi, S., Rawat, P., & Sharma, S. (2020). An overview of finger millet (Eleusine coracana L.). *Journal of Pharmacognosy and Phytochemistry, 9*(4), 866–869.

Krishnan, R., & Meera, M. S. (2018). Pearl millet minerals: Effect of processing on bio-accessibility. *Journal of Food Science and Technology, 55*(9), 3362–3372.

Kumar, S. I., Babu, C. G., Reddy, V. C., & Swathi, B. (2016). Anti-nutritional factors in finger millet. *Journal of Nutrition and Food Science, 6*(3), 491.

Lansakara, L. H. M. P. R., Liyanage, R., Perera, K. A., Wijewardana, I., Jayawardena, B. C., & Vidanarachchi, J. K. (2016). Nutritional composition and health-related functional properties of *Eleusine coracana* (finger millet). *Procedia Food Science, 6*, 344–347. https://doi.org/10.1016/j.profoo.2016.02.069

Lestienne, I., Besançon, P., Caporiccio, B., Lullien-Péllerin, V., & Tréche, S. (2005). Iron and zinc in vitro availability in pearl millet flours (Pennisetum glaucum) with varying phytate, tannin, and fibre contents. *Journal of Agricultural and Food Chemistry, 53*(8), 3240–3247.

Liu, R. H. (2007). Whole grain phytochemicals and health. *Journal of Cereal Science, 46*, 207–219.

Makokha, A. O., Oniang'o, R. K., Njoroge, S. M., & Kamar, O. K. (2002). Effect of traditional fermentation and malting on phytic acid and mineral availability from sorghum (Sorghum bicolor) and finger millet (*Eleusine coracana*) grain varieties grown in Kenya. *Food and Nutrition Bulletin, 23*(3), 241–245.

Manjula, K., Bhagath, Y. B., & Nagalakshmi, K. (2015). Effect of radiation processing on bioactive components of finger millet flour (*Eleusine coracana* L.). *International Food Research Journal, 22*(2), 556–560.

Marrelli, M., Conforti, F., Araniti, F., & Statti, G. A. (2016). Effects of saponins on lipid metabolism: A review of potential health benefits in the treatment of obesity. *Molecules, 21*(10), 1404.

Mohamed, T. K., Zhu, K., Issoufou, A., Fatmata, T., & Zhou, H. (2009). Functionality, in vitro digestibility and physicochemical properties of two varieties of defatted foxtail millet protein concentrates. *International Journal of Molecular Sciences, 10*(12), 5224–5238.

Muthamilarasan, M., Dhaka, A., Yadav, R., & Prasad, M. (2016). Exploration of millet models for developing nutrient rich graminaceous crops. *Plant Science, 242*, 89–97.

Nath, H., Samtiya, M., & Dhewa, T. (2022). Beneficial attributes and adverse effects of major plant-based foods anti-nutrients on health: A review. *Human Nutrition & Metabolism, 28*, 200147. https://doi.org/10.1016/j.hnm.2022.200147

Nissar, J., Ahad, T., Naik, H. R., & Hussain, S. Z. (2017). A review phytic acid: As antinutrient or nutraceutical. *Journal of Pharmacognosy and Phytochemistry, 6*(6), 1554–1560.

Oghbaei, M., & Prakash, J. (2012). Bioaccessible nutrients and bioactive components from fortified products prepared using finger millet (*Eleusine coracana*). *Journal of the Science of Food and Agriculture, 92*(11), 2281–2290. https://doi.org/10.1002/jsfa.5622

Opole, R. A., Prasad, P. V. V., Djanaguiraman, M., Vimala, K., Kirkham, M. B., & Upadhyaya, H. D. (2018). Thresholds, sensitive stages and genetic variability of fingermillet to high temperature stress. *Journal of Agronomy and Crop Science, 204*(5), 477–492.

Panche, A. N., Diwan, A. D., & Chandra, S. R. (2016). Flavonoids: An overview. *Journal of Nutritional Science*, *5*, e47. https://doi.org/10.1017/jns.2016.41

Patel, I., Patel, K., Pinto, S., & Patel, S. (2016). Ragi: A power house of nutrients. *Research and Reviews: Journal of Diary Science and Technology*, *5*(3), 36–47.

Patel, K. K., Naik, R. K., Patel, G., & Pandey, S. (2022). Studies on some physical and engineering properties of finger millet for designing thresher. *Biological Forum – An International Journal*, *14*(4), 387–393.

Patil, P., Pratap, S. S., & Patel, P. (2023). Functional properties and health benefits of finger millet (*Eleusine coracana* L.): A review. *The Journal of Phytopharmacology*, *12*(3), 196–202.

Prajapati, M. R., Dave, P. H., & Sable, P. A. (2019). Nutritional and functional benefits of finger millet (*Eleusine coracana*). *International Journal of Pure and Applied Bioscience*, *7*(1), 342–345.

Prasad, P. V. V., & Staggenborg, S. A. (2009). Growth and production of sorghum and millets, in soils, plant growth and crop production. In W. H. Verheye (Ed.), *Encyclopedia of Life Support Systems (EOLSS), developed under the Auspices of the UNESCO*. Eolss Publishers.

Prashanth, M. R. S., & Muralikrishna, G. (2014). Arabinoxylan from finger millet (Eleusine coracana, v. Indaf 15) bran: Purification and characterization. *Carbohydrate Polymers*, *99*, 800–807. https://doi.org/10.1016/j.carbpol.2013.08.079

Pujol, A., Sanchis, P., Grases, F., & Masmiquel, L. (2023). Phytate intake, health and disease: "Let thy food be thy medicine and medicine be thy food". *Antioxidants*, *12*(1), 146. https://doi.org/10.3390/antiox12010146

Puranik, S., Kam, J., Sahu, P. P., Yadav, R., Srivastava, R. K., Ojulong, H., & Yadav, R. (2017). Harnessing finger millet to combat calcium deficiency in humans: Challenges and prospects. *Frontiers in Plant Science*, *8*, 1311.

Rajasekar, N., Sivanantham, A., Ravikumar, V., & Rajasekaran, S. (2021). An overview on the role of plant-derived tannins for the treatment of lung cancer. *Phytochemistry*, *188*, 112799. https://doi.org/10.1016/j.phytochem.2021.112799

Ramashia, S. E., Anyasi, T. A., Gwata, E. T., Meddows-Taylor, S., & Jideani, A. I. O. (2019). Processing, nutritional composition and health benefits of finger millet in Sub-Saharan Africa. *Food Science and Technology*, *39*(2), 253–266.

Ramashia, S. E., Gwata, E. T., Meddows-Taylor, S., Anyasi, T. A., & Jideani, A. I. O. (2018). Some physical and functional properties of finger millet (*Eleusine coracana*) obtained in Sub-Saharan Africa. *Food Research International*, *104*, 110–118. https://doi.org/10.1016/j.foodres.2017.09.065

Ramashia, S. E., Mashau, M. E., & Onipe, O. O. (2021). Millet cereal grains: Nutritional composition and utilisation in Sub-Saharan Africa. *Cereal Grains*, *1*.

Rani, R., Pandey, C., & Kapoor, N. (2024). Nurturing health through millet derived nutraceuticals. *Food and Humanity*, *14*, 100277.

Rao, B. D., Bhaskarachary, K., Christina, A. G. D., Devi, G. S., & Tonapi, V. A. (2017). *Nutritional and health benefits of millets* (p. 112). ICAR Indian Institute of Millets Research (IIMR).

Rathore, S., Singh, K., & Kumar, V. (2016). Millet grain processing, utilization and its role in health promotion: A review. *International Journal of Nutrition and Food Sciences*, *5*(5), 318–329.

Rathore, T., Singh, R., Kamble, D. B., Upadhyay, A., & Thangalakshmi, S. (2019). Review on finger millet: Processing and value addition. *The Pharma Innovation Journal*, *8*(4), 283–291.

Rawat, D. K., Prajapati, S. K., Kumar, P., Prajapati, B. K., Kumar, V., & Dayal, P. (2023). Policy and research recommendations for millets: Addressing challenges and production opportunities to ensure food and nutritional security. *Current Research in Agriculture and Farming*, *4*(3), 23–31.

Ray, M., Ghosh, K., Singh, S., & Mondal, K. C. (2016). Folk to functional: An explorative overview of rice-based fermented foods and beverages in India. *Journal of Ethnic Foods*, *3*(1), 5–18.

Saleh, S. M., Zhang, Q., Chen, J., & Shen, Q. (2013). Millet grains, nutritional quality, processing and potential health benefits. *Comprehensive Reviews in Food Science and Technology*, *12*(3), 281–295.

Samtiya, M., Soni, K., Chawla, S., Poonia, A., Sehgal, S., & Dhewa, T. (2021). Key Anti-nutrients of millet and their reduction strategies: An overview. *Acta Scientific Nutritional Health*, *5*(12), 68–80.

Satyavathi, C. T., Ambawat, S., Khandelwal, V., & Srivastava, R. K. (2021). Pearl millet: A climate-resilient nutricereal for mitigating hidden hunger and provide nutritional security. *Frontiers in Plant Science*, *12*, 659938. https://doi.org/10.3389/fpls.2021.659938

Scalbert, A. (1991). Antimicrobial properties of tannins. *Phytochemistry*, *30*, 3875–3883.

Shahidi, F., & Chandrasekara, A. (2013): Millet grain phenolics and their role in disease risk reduction and health promotion: A review. *Journal of Functional Foods*, *5*, 570–581.

Shibairo, S. I., Nyongesa, O., Onwonga, R., & Ambuko, J. (2014). Variation of nutritional and anti-nutritional contents in finger millet (*Eleusine coracana* (L.) Gaertn) genotypes. *IOSR Journal of Agriculture and Veterinary Science*, *7*, 6–12.

Shobana, S. (2009). *Investigations on the carbohydrate digestibility of finger millet (Eleusine coracana) with special reference to the influence of its seed coat constituents* [Doctoral dissertation, University of Mysore]. https://www.researchgate.net/publication/277218217_Investigations_on_the_Carbohydrate_Digestibility_of_Finger_Millet_Eleusine_Coracana_with_Special_Reference_to_the_Influence_of_its_Seed_Coat_Constituents

Shobana, S., Krishnaswamy, K., Sudha, V., Malleshi, N. G., Anjana, R. M., Palaniappan, L., & Mohan, V. (2013). Finger millet (Ragi, *Eleusine coracana* L.). Review of its nutritional properties, processing and plausible health benefits. *Advanced Food and Nutrition Research, 69*, 1–39.

Shukla, K., & Srivastava, S. (2014). Evaluation of finger millet incorporated noodles for nutritive value and glycemic index. *Journal of Food Science and Technology, 51*, 527–534.

Singh, E. (2016). Potential functional implications of finger millet (*Eleusine coracana*) in nutritional benefits, processing, health and diseases: A review. *International Journal of Home Science, 2*(21), 151–155.

Singh, K. P., Mishra, A., & Mishra, H. N. (2012). Fuzzy analysis of sensory attributes of bread prepared from millet-based composite flours. *Food Science and Technology (Campinas), 48*, 276–282.

Singh, P., & Raghuvanshi, R. S. (2012). Finger millet for food and nutritional security. *African Journal of Food Science, 6*(4), 77–84.

Singh, P., & Srivastava, S. (2006). Nutritional composition of sixteen new varieties of finger millet. *Journal of Community Mobilization Sustainable Development, 1*(2), 81–84.

Soetan, K. O., & Oyewole, O. E. (2009). The need for adequate processing to reduce the anti-nutritional factors in plants used as human foods and animal feeds. A review. *African Journal of Food Science, 3*(9), 223–232.

Soliman, G. A. (2019). Dietary fibre, atherosclerosis, and cardiovascular disease. *Nutrients, 11*(5), 1155.

Soni, R., Mondal, T., & Patel, H. (2024). *Package and practices of finger millet*. In *From field to harvest: A comprehensive crop production handbook* (Chapter 11, pp. 104–114, 1st ed.). Elite Publishing House.

Sood, S., Kant, L., & Pattnayak, A. (2017). Finger millet (*Eleusine coracana* (L.) Gaertn.): A minor crop for sustainable food and nutrition (Mini review). *Asian Journal of Chemistry, 29*(4), 707–710.

Sood, S., Komar, A., Babu, B. K., Gaur, V. S., Pandey, D., Kant, L., & Pattnayak, A. (2016). Gene, discovery and advances in Finger millet (*Eleusine coracana* (L.) Gaertn.). Genomics-An important nutricereal of future. *Frontiers in Plant Science, 7*, 1–17.

Sparg, S. G., Light, M. E., & Van Staden, J. (2004). Biological activities and distribution of plant saponins. *Journal of Ethnopharmacology, 94*, 219–243.

Sreeramulu, D., Reddy, C. V. K., & Raghunath, M. (2009). Antioxidant activity of commonly consumed cereals, millets, pulses and legumes in India. *Indian Journal of Biochemistry and Biophysics, 46*, 112–115.

Subba Rao, M. V., & Muralikrishna, G. (2002). Evaluation of the antioxidant properties of free and bound phenolic acids from native and malted finger millet (ragi, *Eleusine coracana* Indaf-15). *Journal of Agriculture and Food Chemistry, 50*(4), 889–892.

Thapliyal, V., & Singh, K. (2015). Finger millet: Potential millet for food security and power house of nutrients. *International or Research in Agriculture and Forestry, 2*(2), 22–33.

Tharifkhan, S. A., Perumal, A. B., Elumalai, A., Moses, J. A., & Anandharamakrishnan, C. (2021). Improvement of nutrient bioavailability in millets: Emphasis on the application of enzymes. *Journal of the Science of Food and Agriculture, 101*(12), 4869–4878.

Udeh, H. O., Duodu, K. G., & Jideani, A. I. O. (2017). Finger millet bioactive compounds, bioaccessibility and potential health effects – A review. *Czech Journal of Food Science, 35*, 7–17.

Vagdevi, A., Trupthi, S., Pandey, M., Reddy, N. B., & Singh, M. (2023). Nutritional qualities, processing and health benefits of finger millet (*Eleusine coracana* L.). *Pharma Innovation, 12*(1), 1291–1297.

Verma, R. C., Jitendrakumar, P. H., Ashoka, P., Shekhar, S., Pal, A., Mondal, K., & Singh, B. V. (2023). A review of millet grain phenolics, their health promotion and disease risk reduction. *International Journal of Plant and Soil Science, 35*(20), 863–874.

Xiang, J., Apea-Bah, F. B., Ndolo, V. U., Katundu, M. C., & Beta, T. (2019). Profile of phenolic compounds and antioxidant activity of finger millet varieties. *Food Chemistry, 275*, 361–368. https://doi.org/10.1016/j.foodchem.2018.09.120

9 Exploring the Potentials of Underutilized Leafy Vegetable *Corchorus olitorius* L.

**Alfred Maroyi,[1] Oluwagbenga Oluwasola Adeogun,[2] and Ruvimbo Jessy Mapaya[3]*
[1]Department of Botany, University of Fort Hare, Private Bag X1314, Alice 5700, South Africa
[2]Postharvest Plant Pathology Unit, Department of Botany, University of Lagos, Lagos, Nigeria
[3]Department of Applied Biosciences and Biotechnology, Faculty of Science and Technology, Midlands State University, Gweru, Zimbabwe
*Corresponding author: amaroyi@ufh.ac.za

9.1 INTRODUCTION

Corchorus olitorius L., commonly known as jute mallow, Jew's mallow, Nalita jute, Tossa jute, or bush okra, belongs to the subfamily Grewioideae of the Malvaceae family (Alverson et al., 1999; Bayer & Kubitzki, 2003). But in the past, *C. olitorius* had been assigned to the families Tiliaceae (Wild, 1963, 1984) and Sparrmanniaceae (Akoègninou et al., 2006; Heywood et al., 2007). Both Sparrmanniaceae and Tiliaceae families are not monophyletic, and therefore, these taxa were merged into the subfamily Grewioideae of the expanded Malvaceae family as proposed by the Angiosperm Phylogeny Group (APG, 2009). *Corchorus olitorius* is one of the leafy vegetables categorized as neglected and underutilized crop species in tropical Africa (Isuosuo et al., 2019). *Neglected* and *underutilized* crop species are often referred to as "indigenous," "local," "traditional," "minor," "orphan," or "promising" food plants. They are also defined as wild species and varieties, domesticated or semi-domesticated crops that have not been previously classified as major crops. These crops have previously been underresearched and ignored by breeders and policymakers, currently occupy low levels of utilization, and are mainly confined to smallholder farming communities (Padulosi et al., 2013; Isuosuo et al., 2019; Hunter et al., 2019). Recent research shows that some neglected and underutilized crop varieties could help transform agriculture and food systems, as some of these species possess superior nutrition content compared to major crops (Kobori & Rodriguez-Amaya, 2008; Hunter et al., 2019), are highly adapted to harsh and marginal environments (Rao et al., 2014), contribute to diversification and resilience of agro-ecological niches (Padulosi et al., 2013; Hunter et al., 2019), and have the potential to contribute to improved incomes and create opportunities for rural economic development (Mabhaudhi et al., 2017; Hunter et al., 2019). There are approximately 1,000 neglected and underutilized crop species in the world, and these plant species can serve as a buffer against drought in semi-arid regions of the world (Hossain et al., 2021). Therefore, it is within this context that the current study was undertaken on one of the neglected and underutilized crop varieties, *C. olitorius*, focusing on its ecological and physiological properties, nutrition, health benefits, diseases affecting the species, and post-harvest management operations.

DOI: 10.1201/9781003469766-9

9.2 LITERATURE SEARCH AND DATA COMPILATION

Literature search on taxonomy, ecology, physiology, uses, harvesting, phytochemistry, and pharmacological properties of *C. olitorius* was conducted using online databases, such as Google Scholar (https://scholar.google.com/), ScienceDirect® (www.sciencedirect.com/search), SpringerLink® (https://link.springer.com/), Web of Science™ (www.webofknowledge.com), PubMed® (https://pubmed.ncbi.nlm.nih.gov/), SciELO (https://search.scielo.org/), and Scopus® (www.scopus.com/). In addition to this, pre-electronic sources, such as books, journal articles, dissertations, book chapters, theses, and other scientific articles obtained from the university library were used. The keywords and descriptors used in the article search included "biological activities of *Corchorus olitorius*," "breeding of *Corchorus olitorius*," "pharmacological properties of *Corchorus olitorius*," "diseases and pests of *Corchorus olitorius*," "ethnobotany of *Corchorus olitorius*," "medicinal uses of *Corchorus olitorius*," "taxonomy of *Corchorus olitorius*," "phytochemistry of *Corchorus olitorius*," and "traditional uses of *Corchorus olitorius*."

9.3 TAXONOMICAL AND MORPHOLOGICAL CHARACTERISTICS OF *CORCHORUS OLITORIUS*

The *Corchorus* L. genus is fairly large, consisting of approximately 75 species (POWO, 2024) which are annual, perennial herbs or small shrubs with simple to stellate hairs and sometimes with woody rootstocks (Loumerem & Alercia, 2016). The genus *Corchorus* was originally described by Linnaeus in 1753, represented by four species, namely, *C. capsularis*, *C. hirsutus*, *C. olitorius*, and *C. siliquosus*, with *C. olitorius* as the type species of the genus (Loumerem & Alercia, 2016). *Corchorus* species occupy bushland, cultivated fields, forests, open savanna, riverine habitats, swamps, and woodlands in the tropics, subtropics, and warm, temperate regions of the world (Heywood, 1993; Bayer & Kubitzki, 2003; Lebrun & Stork, 2003; Moeaha, 2006). The greatest diversity of the genus across diverse habitats has been reported in tropical Africa, followed by Asia and Australia (Choudhary et al., 2017). Therefore, the African continent is considered the center of diversity of the genus, with the greatest number of wild *Corchorus* species, while the Indo-Burmese region is regarded as a secondary center of diversity of the genus (Fondio & Grubben, 2012). *Corchorus olitorius* is used as a major source of stem bast fibers (jute) in Asia, Eastern Europe, Latin America, and tropical African countries (Xiong, 2008; Zhang et al., 2019). The species ranks second to cotton (*Gossypium hirsutum* L.) as world vegetable fiber (Zhang et al., 2019). Cultivation of *C. olitorius* began about 5,000 years ago, providing 80% of total phloem and outer bark production in the world (Xiong, 2008; Zhang et al., 2019).

The genus name *Corchorus* is derived from the Greek word *korkhoros*, or *korkoros*, terms used for some herbaceous pot plants of uncertain origin or identity (Nethavhani, 2021). The species name *olitorius* is a Latin word which means "pertaining to vegetables" (Nethavhani, 2021). The synonyms of *C. olitorius* include *C. catharticus* Blanco, *C. decemangularis* Roxb. ex G.Don, *C. lanceolatus* G.Don, *C. longicarpus* G.Don, *C. malchairii* De Wild., *C. olitorius* L. var. *australiensis* Domin, *C. olitorius* L. var. *grandifolius* De Wild., *C. olitorius* L. var. *incisifolius* Asch. and Schweinf., *C. olitorius* L. var. *malchairii* (De Wild.) R.Wilczek, and *C. quinquelocularis* Moench (Wild, 1963, 1984; Robles & Nava, 2003; Alharbi et al., 2024). An evaluation of molecular characters, as well as macro- and micro-morphological traits, shows that *C. olitorius* is closely related to *C. aestuans* L. (Chakraborty & Palit, 2009), and therefore, these two species may be regarded as primary relatives.

Corchorus olitorius is a robust, erect annual herb, usually strongly branched and growing up to 5 m in height (Wild, 1963, 1984; Fondio & Grubben, 2012). Fondio and Grubben (2012) argued that there are two cultivar groups of *C. olitorius*, that is, plant specimens lower than 2 m in height, often heavily branched and used as vegetables, while the other cultivar is much taller, averaging 4–5 m in height, slightly branched at the top, and usually used as a source of stem fiber. The stems of *C. olitorius* are green to reddish in color, fibrous, and tough. The leaves are simple, alternate with stipules that are narrowly triangular with a terminal long point. The leaf blade is ovate to elliptical in shape, cuneate or obtuse, with setaceous appendages at the base, acuminate to acute at the apex. The leaf blade is thinly textured, mostly hairless on both surfaces, shiny, dark green in color, with crenate to serrate leaf margins. The inflorescence is axillary, solitary at the nodes, opposite the leaves, many-flowered, and bracteate. The flowers are bisexual, regular, and short-stalked. The sepals are free, narrowly obovate, while petals are obovate and yellow to orange in color. The fruit of *C. olitorius* is a cylindrical capsule, straight or slightly curved, ribbed with a short beak, usually dehiscing by many valves. The seeds are angular and pyramidal, dark gray in color, broadly elliptical to spherical in shape.

9.4 ECOLOGICAL AND PHYSIOLOGICAL CHARACTERISTICS OF *CORCHORUS OLITORIUS*

Corchorus olitorius is widely distributed in all tropical, subtropical, and temperate regions of the world. But the geographical origin of this plant is often disputed, as the species naturally occurs in several countries and has been cultivated for centuries in Asia, Eastern Europe, Latin America, and tropical Africa (Xiong, 2008; Loumerem & Alercia, 2016; Zhang et al., 2019). *Corchorus olitorius* occurs from sea level to 1,750 m above sea level in grassland, savannas, forests, coastal plains, swamp margins, riverbanks, irrigation channels, sandy riverbeds, and alluvial depressions (Wild, 1963, 1984; Robles & Nava, 2003; Alharbi et al., 2024). This underutilized vegetable is categorized as an arable land weed (Randall, 2017) and is common in disturbed locations, such as waste places, fallow land, and roadsides (Wild, 1984; Robles & Nava, 2003; Alharbi et al., 2024). The species prefers alluvial and loam soils rich in organic matter, slightly acidic or alkaline soils, with an optimum pH of 6.0 to 6.5, and grows poorly in clay, sandy, and stony soils (Loumerem & Alercia, 2016). *Corchorus olitorius* thrives best under hot and humid conditions and cannot tolerate shady environments (Loumerem & Alercia, 2016). Optimal growth conditions include annual rainfall range of 600 to 2,000 mm and temperatures ranging from 25 to 32°C (Fondio & Grubben, 2012).

This plant is propagated by seed, and growth of its seedlings is relatively fast (Fondio & Grubben, 2012; Loumerem & Alercia, 2016). The species achieves maximum physiological development under high humidity and temperatures (Weimer et al., 2023). The optimum temperatures for the species range from 25 to 32°C, and it cannot grow well below 15°C (Fondio & Grubben, 2012; Weimer et al., 2023). Research also shows that *C. olitorius* seedlings usually wilt and tend to flower prematurely when subjected to low humidity (≤19%) and soil moisture (<20%) (Weimer et al., 2023). Similarly, the plant is prone to stem and root rot when exposed to excessive moisture, especially when cultivated in water-logged and shady conditions (Weimer et al., 2023).

9.5 DISEASES ASSOCIATED WITH *CORCHORUS OLITORIUS*

Corchorus olitorius is susceptible to a spectrum of plant diseases, which can significantly impede its productivity and availability. Recent research demonstrated an escalating interest in exploring the diseases affecting *C. olitorius* to develop effective management strategies and ensure sustainable agricultural practices (Meena et al., 2015). The work of Nasim et al. (2017) identified powdery mildew caused by the fungus *Erysiphe corchori* which causes infections characterized by white powdery patches on the leaves, reducing photosynthetic efficiency, hindering growth, and ultimately, decreasing yield. Another study examined the prevalence of leaf blight induced by the bacterium *Xanthomonas campestris* in *C. olitorius* plants, demonstrating severe damage leading to withering, necrosis, and leaf loss (Biswas et al., 2013). Shoaib et al. (2022) highlighted the presence of charcoal rot caused by the fungus *Macrophomina phaseolina* in *C. olitorius* crops, noting its severity influenced by plant maturity. Additionally, minor diseases, such as anthracnose, damping off, and leaf spots, have been identified as affecting the species.

An overview of microorganisms causing diseases in *C. olitorius* is provided in Table 9.1. This is important particularly in tropical Africa, where such updated information focusing on infection status and management options for these diseases is limited or absent (Shoaib et al., 2022). To address this gap, there is a need to provide a comprehensive overview of key plant diseases affecting *C. olitorius* and the potential management strategies. Therefore, Table 9.1 provides a summary, emphasizing the various diseases, including powdery mildew, leaf blight, charcoal rot, anthracnose, damping off, and leaf spots, that can negatively impact *C. olitorius* growth and yield. The severity of these diseases depends on environmental factors and plant maturity, resulting in reduced photosynthetic capacity, stunted growth, wilting, necrosis, defoliation, and ultimately, diminished crop yield. The lack of up-to-date information on infection status and management strategies in specific regions underscores the need for further research and surveillance (Haque et al., 2015).

With the increasing demand for *C. olitorius* products, including higher-quality fiber, the management of diseases has become more crucial. Microorganisms capable of enzymatically degrading these components pose significant threats to food and fiber production. *Corchorus olitorius* naturally grows or is cultivated in seasons characterized by high rainfall, hot temperatures, and increased relative humidity, which not only promote extensive green biomass production but also render them vulnerable to various diseases. Efforts by entomologists and pathologists to select biological control measures have been hindered by the wide variation in meteorological conditions across different areas during normal seasonal cycles (Haque et al., 2015). Among the fungal diseases, stem rot caused by *Macrophomina phaseolina* is economically significant in *C. olitorius* cultivation. Management of fungal diseases in *C. olitorius* involves manipulating soil conditions

TABLE 9.1
Microorganisms Associated with Diseases in *Corchorus olitorius*

Name of Disease	Causal Organism	Class of Organism	Reference
Anthracnose	*Colletotrichum corchori* and *C. gloesporioides*	Fungal	Srivastava & Singh, 2020
Black band	*Botryodiplodia theobromae*	Fungal	Srivastava & Singh, 2020
Die back	*Diplodia corchori*	Fungal	Li et al., 2022
Golden jute mosaic	Whitefly vector *Begomovirus*	Viral	Ghosh et al., 2008
Hooghly wilt	*Macrophomina phaseolina* and *Ralstonia solanacearum*	Bacterial	Srivastava & Singh, 2020
Jute chlorosis	*Tobravirus*	Viral	Srivastava & Singh, 2020
Leaf blight	*Xanthomonas campestris*	Bacterial	Biswas et al., 2014
Leaf spot	*Phyllosticta*	Fungal	Rao, 1978
Powdery mildew	*Oidium* spp.	Fungal	Rao, 1978
Root gall nematode	*Meloidogyne incognita* and *M. javanica*	Nematodal	Srivastava & Singh, 2020
Root rot	*Macrophomina phaseolina*	Fungal	Bhattacharyya et al., 1985
Root rot	*Rhizoctonia* spp.	Fungal	Rao, 1978
Soft rot	*Agroathelia rolfsii*	Fungal	Ramasubramanian et al., 2010
Sooty mold	*Alternaria* spp., *Cercospora corchori*, and *Corynespora cassiicola*	Fungal	Srivastava & Singh, 2020
Stem gall	*Physoderma corchori*	Fungal	Rao, 1978
Stem rot	*Macrophomina phaseolina*	Fungal	Meena et al., 2015
Tip blight	*Curvularia subulata*	Fungal	Li et al., 2022
Wilt	*Ralstonia solanacearum*	Bacterial	Rao, 1978
Yellow vein mosaic disease	Monopartite *Begomovirus*	Viral	Mubeen et al., 2021

for soilborne diseases, treating seeds with fungicides for seed-borne diseases, foliar spraying of fungicides to reduce airborne spores, or a judicious combination of both. Removing and destroying infected plants and applying insecticides to control vectors can also reduce virus diseases.

9.6 POTENTIAL OF *CORCHORUS OLITORIUS* AS A LEAFY VEGETABLE

Corchorus olitorius is locally useful as a cheaply available leafy vegetable throughout its distributional range, and the species has potential for the development of new food and beverage products (Van Wyk, 2011). Leaves and young shoots of *C. olitorius* are used as a leafy vegetable in the African, Asian, European, and South American continents, as well as in countries such as Benin, Burkina Faso, Cameroon, Côte d'Ivoire, Ethiopia, Ghana, India, Japan, Kenya, Niger, Nigeria, Senegal, South Africa, South Sudan, Sudan, Tanzania, Uganda, and Zimbabwe (Table 9.2).

TABLE 9.2
***Corchorus* Species Used as Vegetables in Tropical and Subtropical Regions of the World**

Species	Part Used	Country/Region	Reference
C. aestuans L.	Leaves and roots	Benin, Cameroon, Ethiopia, India, Kenya, and Senegal	Westphal-Stevel, 1986; Mathieu & Meissa, 2007; Dansi et al., 2008; Benor et al., 2010; Kumari et al., 2019
C. asplenifolius Burch.	Leaves	Namibia, South Africa, and Zimbabwe	Wild et al., 1972; Maroyi, 2011a, 2011b, 2013; Van Wyk & Gericke, 2018; Magwede et al., 2019
C. capsularis L.	Leaves	India and Nepal	Manandhar, 2002; Choudhary et al., 2017; Kumari et al., 2019
C. confusus Wild	Leaves	Eswatini, South Africa, and Zimbabwe	Liengme, 1981; Gomez, 1988; Magwede et al., 2019
C. fascicularis Lam.	Leaves	Cameroon, Ethiopia, Kenya, and Tanzania	Westphal-Stevel, 1986; Maundu et al., 1999; Ruffo et al., 2002; Benor et al., 2010
C. olitorius L.	Leaves and young shoots	Africa, Asia, Benin, Burkina Faso, Cameroon, Côte d'Ivoire, Ethiopia, Europe, Ghana, India, Japan, Kenya, Niger, Nigeria, Senegal, South Africa, South America, South Sudan, Sudan, Tanzania, Uganda, and Zimbabwe	Wild et al., 1972; Westphal-Stevel, 1986; Furumoto et al., 2002; Mathieu & Meissa, 2007; Van Rensburg et al., 2007; Dansi et al., 2008; Benor et al., 2010; Choudhary et al., 2017; Nyadanu et al., 2017; Van Wyk & Gericke, 2018; Magwede et al., 2019; Maazou et al., 2023
C. pseudocapsularis Schweinf.	Leaves	Tanzania	Ruffo et al., 2002
C. schimperi Cufod.	Leaves	Ethiopia	Benor et al., 2010
C. tridens L.	Leaves	Benin, Burkina Faso, Cameroon, Ethiopia, Kenya, Namibia, Niger, Senegal, South Africa, Tanzania, tropical Africa, Yemen, and Zimbabwe	Wild et al., 1972; Liengme, 1981; Westphal-Stevel, 1986; Ruffo et al., 2002; Mathieu & Meissa, 2007; Van Rensburg et al., 2007; Dansi et al., 2008; Benor et al., 2010; Van Wyk & Gericke, 2018; Magwede et al., 2019; Maazou et al., 2023
C. trilocularis L.	Flowers, leaves, and stems	Burkina Faso, East Africa, Egypt, Ethiopia, India, Kenya, Mali, Middle East, South Africa, Tanzania, Uganda, Yemen, and Zimbabwe	Uphof, 1968; Wild et al., 1972; Liengme, 1981; Ruffo et al., 2002; Van Rensburg et al., 2007; Van der Maesen, 2007; Benor et al., 2010; Van Wyk & Gericke, 2018; Magwede et al., 2019
C. urticifolius Wight and Arn.	Leaves	Ethiopia, India, and Kenya	Benor et al., 2010; Choudhary et al., 2017

Corchorus olitorius is cultivated as a leafy vegetable in Bangladesh, Brazil, the Caribbean, China, Egypt, India, Japan, and the Middle East (Fondio & Grubben, 2012). The leaves and young shoots are cut into small pieces and boiled in water with salt and pepper and eaten as leafy vegetable or added to soups and stews (Ahmed, 2021). Young shoots and tender leaves are also eaten raw in green salads, while dried leaves are used as an herbal tea and soup thickener. A slimy, sticky sauce is prepared from the powdered and dried immature fruits, while the seeds are edible and used as a flavoring agent (Ahmed, 2021). The leaves and young shoots of *C. olitorius* may also be briefly cooked and then sun-dried and stored for later use or for trading on local informal markets. In some communities, *C. olitorius* is cooked with butter, milk, creeping foxglove (*Asystasia gangetica* (L.) T.Anderson), cat's whiskers (*Cleome gynandra* L.), slenderleaf (*Crotalaria brevidens* Benth.), rattlepod (*Crotalaria ochroleuca* G.Don), cocoyam leaves (*Colocasia esculenta* (L.) Schott) or tannia (*Xanthosoma sagittifolium* (L.) Schott), cowpeas (*Vigna unguiculata* (L.) Walp.), lemon (*Citrus limon* (L.) Osbeck), pepper (*Capsicum annuum* L. or *Capsicum frutescens* L.), pumpkin (*Cucurbita maxima* Duchesne, *Cucurbita moschata* Duchesne, or *Cucurbita pepo* L.), and sweet potatoes (*Ipomoea batatas* (L.) Lam.) (Maundu et al., 1999). Other *Corchorus* species also used as vegetables include *C. aestuans* L., *C. asplenifolius* Burch., *C. capsularis* L., *C. confusus* Wild, *C. fascicularis* Lam., *C. pseudocapsularis* Schweinf., *C. schimperi* Cufod., *C. tridens* L., *C. trilocularis* L., and *C. urticifolius* Wight and Arn. (Table 9.2).

9.7 NUTRITIONAL AND ETHNOPHARMACOLOGICAL PROPERTIES OF *CORCHORUS OLITORIUS*

Corchorus olitorius stands out as a valuable leafy vegetable, and analyses of the nutritional value of the species (Table 9.3) show that the leaves are a better source of energy; minerals, such as calcium, copper, iron, magnesium, manganese, potassium, and zinc; and classic nutrients, such as amino acids, carbohydrates, fat, fatty acids, fiber, lipids, niacin, protein, thiamine, and vitamin C than commercial vegetable crops, such as cabbage (*Brassica oleracea* L.). Several phytochemical compounds, such as alcohols, alkaloids, amino acids, cardiac glycosides, coumarins, esters, fatty acids, flavonoids, ionones, lactones, methylated xanthines, phenolic acids, steroids, and triterpenes (Table 9.4), have been isolated from the aerial parts, flowers, leaves, roots, seeds, and stems of

TABLE 9.3
Nutritional Composition of *Corchorus olitorius* and *Brassica oleracea* L.

Nutrients	*C. olitorius*	*B. oleracea*	References
Ash (%)	2.4–16.4	1.63–2.11	Idirs et al., 2009; Chaudhary & Verma, 2018
Calcium (mg/100 g)	1159.1	3.26–3.36	Acho et al., 2014; Chaudhary & Verma, 2018
Carbohydrate (%)	5.0–55.6	3.50–10.21	Idirs et al., 2009; Chaudhary & Verma, 2018
Copper (mg/100 g)	2.52	0.03–0.05	Idirs et al., 2009; Chaudhary & Verma, 2018
Energy (kcal/100 g)	200.8–277.4	38.53–61.12	Acho et al., 2014; Chaudhary & Verma, 2018
Fat (%)	0.3–1.0	0.53–0.68	Chaudhary & Verma, 2018; Ahmed, 2021
Fiber (%)	1.0–11.5	7.01–10.01	Acho et al., 2014; Chaudhary & Verma, 2018
Iron (mg/100 g)	27.4	1.37–1.44	Acho et al., 2014; Chaudhary & Verma, 2018
Lipids (%)	3.3–12.0	0.16–0.31	Idirs et al., 2009; Acho et al., 2014
Magnesium (mg/100 g)	572.0	3.05–3.80	Acho et al., 2014; Chaudhary & Verma, 2018
Manganese (mg/100 g)	5.95	0.04–0.09	Idirs et al., 2009; Chaudhary & Verma, 2018
Moisture (%)	82.0–88.3	87.85–89.40	Idirs et al., 2009; Chaudhary & Verma, 2018
Niacin (mg/100 g)	1.5	0.4–0.9	Hanif et al., 2006; Ahmed, 2021
Phosphorus (mg/100 g)	554.4	–	Acho et al., 2014
Potassium (mg/100 g)	1669.1–2814.2	1872.0–2398.50	Acho et al., 2014; Chaudhary & Verma, 2018

(Continued)

TABLE 9.3 (Continued)

Nutrients	*C. olitorius*	*B. oleracea*	References
Protein (%)	5.0–21.1	2.66–3.54	Idirs et al., 2009; Chaudhary & Verma, 2018
Riboflavin (mg/100 g)	0.3	0.05–0.08	Hanif et al., 2006; Ahmed, 2021
Sodium (mg/100 g)	54.56	1150.0	Idirs et al., 2009; Chaudhary & Verma, 2018
Thiamine (mg/100 g)	0.1	0.06–0.09	Hanif et al., 2006; Ahmed, 2021
Vitamin C (mg/100 g)	10.0–100.0	45.0–55.0	Hanif et al., 2006; Acho et al., 2014
Zinc (mg/100 g)	4.71	0.28–0.30	Idirs et al., 2009; Chaudhary & Verma, 2018
Fatty Acids			
Myristic acid (mg/100 g)	0.84–1.72	0.19–0.34	Zeghichi et al., 2003; Vega-Galvez et al., 2023
Palmitic acid (mg/100 g)	12.01–26.4	0.17–20.02	Zeghichi et al., 2003; Vega-Galvez et al., 2023
Palmitoleic acid (mg/100 g)	0.42–0.85	0.32–0.42	Zeghichi et al., 2003; Vega-Galvez et al., 2023
Margaric acid (mg/100 g)	0.18–0.27	0.52–0.99	Zeghichi et al., 2003; Vega-Galvez et al., 2023
Cis-10-Heptadecenoic acid (mg/100 g)	0.05–0.45	0.20–0.35	Zeghichi et al., 2003; Vega-Galvez et al., 2023
Stearic acid (mg/100 g)	1.70–3.37	3.14–5.95	Zeghichi et al., 2003; Vega-Galvez et al., 2023
Oleic acid (mg/100 g)	0.26–5.06	0.70–2.26	Zeghichi et al., 2003; Vega-Galvez et al., 2023
Arachidic acid (mg/100 g)	0.98–1.53	0.50–0.81	Zeghichi et al., 2003; Vega-Galvez et al., 2023
Erucic acid (mg/100 g)	0.53–1.11	0.06–0.11	Zeghichi et al., 2003; Vega-Galvez et al., 2023
Lignoceric acid (mg/100 g)	0.28–1.63	2.82	Zeghichi et al., 2003; Ayaz et al., 2006
Non-Essential Amino Acids			
Alanine (g/100 g)	3.06	0.25–1.31	Ndamitso et al., 2016; Vega-Galvez et al., 2023
Arginine (g/100 g)	4.87	0.23–3.90	Ndamitso et al., 2016; Vega-Galvez et al., 2023
Aspartic acid (g/100 g)	9.23	0.60–2.68	Ndamitso et al., 2016; Vega-Galvez et al., 2023
Glutamic acid (g/100 g)	7.34	1.25–6.42	Ndamitso et al., 2016; Vega-Galvez et al., 2023
Glycine (g/100 g)	3.34	0.35–0.94	Ndamitso et al., 2016; Vega-Galvez et al., 2023
Serine (g/100 g)	2.25	0.35–1.54	Ndamitso et al., 2016; Vega-Galvez et al., 2023
Tyrosine (g/100 g)	2.54	0.29–0.93	Ndamitso et al., 2016; Vega-Galvez et al., 2023
Essential Amino Acids			
Isoleucine (g/100 g)	3.55	0.39–2.05	Ndamitso et al., 2016; Vega-Galvez et al., 2023
Leucine (g/100 g)	4.97	0.05–0.45	Ndamitso et al., 2016; Vega-Galvez et al., 2023
Lysine (g/100 g)	3.93	0.29–1.32	Ndamitso et al., 2016; Vega-Galvez et al., 2023
Phenylalanine (g/100 g)	3.56	0.35–1.81	Ndamitso et al., 2016; Vega-Galvez et al., 2023
Threonine (g/100 g)	3.13	0.69–1.35	Ndamitso et al., 2016; Vega-Galvez et al., 2023
Valine (g/100 g)	3.74	0.25–1.06	Ndamitso et al., 2016; Vega-Galvez et al., 2023

TABLE 9.4
Summary of Phytochemical Compounds Isolated from *Corchorus olitorius*

Phytochemical Compound	Plant Part	Reference
Aliphatic Hydrocarbons		
1-tridecene	Whole plant	Alshabi et al., 2022
2-nonadecene	Whole plant	Alshabi et al., 2022
Tetradecane	Whole plant	Alshabi et al., 2022
Alkaloids		
1-(8-quinolinyl)-beta-carboline	Whole plant	Alshabi et al., 2022
Benzanilides		
3-O-caffeoyl-4-O-methylquinic acid	Leaves	Ncube et al., 2022
Benzoic Acid Allopyranosides		
Pseudolaroside A	Leaves	Ncube et al., 2022

(Continued)

TABLE 9.4 (Continued)

Phytochemical Compound	Plant Part	Reference
Cannabinoids		
Nabilone	Leaves	Abdallah et al., 2020
Cardiac Glycosides		
Andolitoriside	Leaves, roots, seeds, and stems	Hassan et al., 2019
Corchoroside A and B	Roots and seeds	Ramadevi, 2014
Coroloside	Leaves, roots, seeds, and stems	Hassan et al., 2019
Deglucocoroloside	Seeds	Khan et al., 2006
Digitoxin	Leaves	Abdallah et al., 2020
Digitoxigenin glycosides and triglycoside	Leaves and seeds	Ogawa et al., 2002
Erysimoside	Leaves, roots, seeds, and stems	Hassan et al., 2019
Evatromonoside	Seeds	Matsufuji et al., 2001
Gluco-olitoriside	Leaves, roots, seeds, and stems	Hassan et al., 2019
Glucoevatromonoside	Leaves, roots, seeds, and stems	Hassan et al., 2019
Oitoriside	Seeds	Matsufuji et al., 2001
Raffinose	Leaves, roots, seeds, and stems	Hassan et al., 2019
Strophanthidin glycosides and trioside	Leaves and seeds	Ogawa et al., 2002
Coumarins		
4,7-dihydroxycoumarin	Leaves and seeds	Khan et al., 2006
5-hydroxy-6-methoxycoumaric acid	Leaves	Ncube et al., 2022
7-hydroxy-4-methyl coumarin	Leaves	Abdallah et al., 2020
Aesculin	Leaves	Ncube et al., 2022
Cichoriin	Leaves	Yoshikawa et al., 1997
Scopolin	Leaves	Hasan & Kadhim, 2018
Cycloalkanes		
Quinic acid	Leaves	Ncube et al., 2022
Cyclohexanones		
2-(2-methyl-propenyl)-cyclohexanone	Leaves	Abdallah et al., 2020
Dicarboxylic Acids		
Camphoric acid	Leaves	Abdallah et al., 2020
Diterpene Alcohols		
Phytol acetate	Whole plant	Alshabi et al., 2022
Esters		
3-O-p-coumaroylquinic acid	Leaves	Ncube et al., 2022
Caffeoylmalic acid	Leaves	Ncube et al., 2022
Cyclopropane octanoic acid	Whole plant	Alshabi et al., 2022
Cynarine	Leaves, roots, and stems	Atalar et al., 2023
Dibromostigmastirol acetate	Whole plant	Alshabi et al., 2022
Linolenic acid methyl ester	Whole plant	Alshabi et al., 2022
Neochlorogenic acid	Leaves	Ncube et al., 2022
Fatty Alcohols		
1-tetradecanol	Whole plant	Alshabi et al., 2022
Flavonoids		
3-rhamnoside-7-(6"-succinylglucoside)	Leaves	Ncube et al., 2022
3,5-dicaffeoylquinic acid	Leaves and seeds	Yoshikawa et al., 1997,
Apigenin	Flowers and leaves	Yakoub et al., 2018
Apigenin 7-O-glucoside	Leaves	Yakoub et al., 2018
Astragalin	Aerial parts, flowers, leaves, and seeds	Ali et al., 2021
Biochanin A	Leaves, roots, and stems	Atalar et al., 2023
Catechins	Leaves and seeds	Yoshikawa et al., 1997

(Continued)

TABLE 9.4 (Continued)

Phytochemical Compound	Plant Part	Reference
Cirsiliol	Leaves	Yakoub et al., 2018
Cirsilineol	Leaves	Yakoub et al., 2018
Ellagic acid	Leaves	Ali et al., 2021
Fisetin	Leaves	Atalar et al., 2023
Glycitein	Leaves	Abdallah et al., 2020
Glyceryl monopalmitate	Leaves	Hassan et al., 2019
Hyperoside	Leaves, roots, and stems	Atalar et al., 2023
Isorhamnetin 3-glucosyl-(1–2)-galactoside	Leaves	Ncube et al., 2022
Juglanin	Aerial parts, flowers, and leaves	Hassan et al., 2019
Kaempferol	Flowers and leaves	Yakoub et al., 2018
Kaempferol-3-glucoside	Leaves and stems	Atalar et al., 2023
Kaempferol 3-(2”-rhamnosyl-6”-acetylgalactoside) 7-rhamnoside	Leaves	Ncube et al., 2022
Kaempferol-o-glycoside-o-sophoroside	Leaves	Ncube et al., 2022
Kaempferide	Leaves	Ncube et al., 2022
Luteolin	Flowers, leaves, and seeds	Yakoub et al., 2018
Luteolin-7-O-hexoside	Aerial parts	Handoussa et al., 2017
Myricetin	Flowers	Yakoub et al., 2018
Naringenin	Flowers and leaves	Yakoub et al., 2018
Naringin	Flowers and leaves	Yakoub et al., 2018
Quercetin	Aerial parts, flowers, leaves, and seeds	Ncube et al., 2022
Quercetin-3-galactoside	Flowers and leaves	Ncube et al., 2022
Quercetin-3-glucoside	Leaves, roots, and stems	Atalar et al., 2023
Quercetin-3-(6 malonyl galactose)	Leaves	Azuma et al., 1999;
Quercetin 3-(6 malonyl glucoside)	Aerial parts and leaves	Handoussa et al., 2017
Quercetin 3-β-glucoside	Aerial parts	Handoussa et al., 2017
Quercetin 3-glucosyl-(1–2)-galactoside	Leaves	Ncube et al., 2022
Quercetin 3-O-(6″-O-malonyl)-β-D-glucoside	Flowers	Yakoub et al., 2018
Quercetin-O-malonylglucoside-O-glycoside	Leaves	Ncube et al., 2022
Quercetin hydrate	Leaves	Ali et al., 2021
Quercimeritrin	Leaves	Atalar et al., 2023
Isoquercetin	Aerial parts, leaves, and seeds	Hassan et al., 2019
Rutin	Flowers	Yakoub et al., 2018
Rutin hydrate	Leaves	Ali et al., 2021
Tolifolin	Aerial parts and leaves	Hassan et al., 2019
Tricin	Leaves	Ncube et al., 2022
Ionones		
6S,9R-roseoside	Roots	Khan et al., 2006
Betulabuside A	Roots	Ramadevi & Ganapaty, 2012
Corchoiononside A, B, and C	Leaves and seeds	Khan et al., 2006
Keto Acids		
Oxalosuccinic acid	Leaves	Ncube et al., 2022
Ketone Diethers		
Levoglucosenone	Leaves	Abdallah et al., 2020
Lactones		
D-lyxo-D-manno-nononic-1,4-lactone	Leaves	Abdallah et al., 2020
D-saccharic acid 1,4-lactone	Leaves	Abdallah et al., 2020

(Continued)

TABLE 9.4 (Continued)

Phytochemical Compound	Plant Part	Reference
Methoxycinnamic Acids		
3,4-dimethoxycinnamic acid	Leaves	Abdallah et al., 2020
4-hydroxycinnamic acid	Leaves	Ncube et al., 2022
Trans-2,3-dimethoxycinnamic acid	Leaves	Abdallah et al., 2020
Methylated Xanthines		
Theophylline	Leaves	Abdallah et al., 2020
Phenolic Acids		
1,2-benzenedicarboxylic acid	Flowers	Driss et al., 2016
1,3-dicaffeoylquinic acid	Leaves	Ncube et al., 2022
1,3-di-O-caffeoylquinic acid	Leaves	Yakoub et al., 2018
1,4-dicaffeoylquinic acid	Leaves	Ncube et al., 2022
1,5-dicaffeoylquinic acid	Leaves and seeds	Handoussa et al., 2017
2-O-caffeoylhydroxycitric acid	Leaves	Ncube et al., 2022
3,4-Di-O-caffeoylquinic acid	Flowers and leaves	Yakoub et al., 2018
Trans-3-(4-hydroxy-3-methoxyphenyl) acrylic anhydride	Leaves	Taiwo et al., 2016
3,5-di-caffeoylquinic acid	Leaves and seeds	Ncube et al., 2022
4-hydroxybenzoic acid	Leaves and stems	Atalar et al., 2023
4-hydroxybenzaldehyde	Leaves, roots, and stems	Atalar et al., 2023
4-hydroxy-2-methoxybenzaldehyde	Leaves	Abdallah et al., 2020
4-O-caffeoylquinic acid	Leaves	Yakoub et al., 2018
4,5-dicaffeoylquinic acid	Leaves	Ncube et al., 2022
4,5-di-O-caffeoyquinic acid	Leaves	Yakoub et al., 2018
Caffeic acid	Flowers and leaves	Atalar et al., 2023
Catechin	Leaves	Ali et al., 2021
Chlorogenic acid	Flowers, leaves, roots, and seeds	Atalar et al., 2023
Cryptochlorogenic acid	Leaves	Ncube et al., 2022
Dicaffeoylquinic acids	Leaves	Guzzetti et al., 2021
Trans-ferulic acid	Flowers, leaves, roots, and stems	Atalar et al., 2023
Feruloylquinic acids	Leaves	Guzzetti et al., 2021
Gallic acid	Leaves	Yakoub et al., 2018
Gentisic acid 5-O-glucoside	Leaves	Ncube et al., 2022
Isochlorogenic acid	Flowers	Handoussa et al., 2017
Methyl-1,4,5-tri-O-caffeoyl quinate	Leaves	Taiwo et al., 2016
p-Coumaric acid	Leaves, roots, and stems	Atalar et al., 2023
Protocatechuic acid	Leaves	Yakoub et al., 2018
Quinic acid	Flowers, leaves, and seeds	Yakoub et al., 2018
Rosmarinic acid	Flowers and leaves	Yakoub et al., 2018
Syringic acid	Leaves	Ncube et al., 2022
Vanillic acid	Flowers	Ademiluyi et al., 2015
Phenylpropanoids		
Isoelemicin	Leaves	Abdallah et al., 2020
Sesquiterpenes		
3-oxo-7,8-dihydro-α-ionol	Leaves	Abdallah et al., 2020
β-cedrene	Flowers	Driss et al., 2016
Germacrene D	Flowers	Driss et al., 2016
Nerolidol	Flowers	Driss et al., 2016
Steroidal Derivatives		
Strophanthidin	Seeds	Ogawa et al., 2002
Ursolic acid	Roots and seeds	Furumoto et al., 2002

(Continued)

TABLE 9.4 (Continued)

Phytochemical Compound	Plant Part	Reference
Sterols		
β-sitosterol 3-O-β-D-glucopyranoside	Leaves	Hassan et al., 2019
β-sitosterol-3-glucoside	Leaves	Hassan et al., 2019
β-sitosterol-β-D-sitosterol-glucoside	Leaves	Hassan et al., 2019
β-sitosteryl fatty acid esters	Stems	Ragasa et al., 2016
β-sitosterol	Leaves, roots, seeds, and stems	Hassan et al., 2019
Stigmasterol	Seeds and stems	Ragasa et al., 2016
Terpenes		
Ascaridole epoxide	Leaves	Abdallah et al., 2020
Camphene	Leaves	Driss et al., 2016
Carvacrol methyl ether	Leaves	Yakoub et al., 2018
Citroside A	Leaves	Ncube et al., 2022
Dihydroterpineol	Leaves	Hassan et al., 2019
Cis-3-hexen-1-ol and cis-4-hexen-1-ol	Leaves	Zeid, 2002
Limonene	Leaves	Driss et al., 2016
Menthone	Leaves	Yakoub et al., 2018
β-myrcene	Leaves	Driss et al., 2016
α-phellandrene	Leaves	Driss et al., 2016
Phytol	Leaves and seeds	Ragasa et al., 2016
α-pinene	Leaves	Driss et al., 2016
Piperonal	Leaves	Yakoub et al., 2018
Sabinene	Leaves	Driss et al., 2016
α-Terpinene	Leaves	Driss et al., 2016
Terpinolene	Leaves	Zeid, 2002
Tetradecanal	Leaves	Zeid, 2002
Triterpenes		
β-Amyrin	Seeds	Hassan et al., 2019
Canarigenin	Seeds	Khan et al., 2006
Cannogenol	Seeds	Khan et al., 2006
Campesterol	Seeds	Hassan et al., 2019
Corchoroside A	Roots	Ramadevi, 2014
Coroloside	Seeds	Ogawa et al., 2002
Corosolic acid	Roots	Hassan et al., 2019
Crocin	Roots	Manzoor-i-Khuda & Habermehl, 1979
Corosolic acid	Roots and seeds	Khan et al., 2006
Diosgenin	Leaves, roots, and stems	Atalar et al., 2023
Erysimoside	Seeds	Ogawa et al., 2002
Glucoevatromonoside	Seeds	Ogawa et al., 2002
Helveticoside	Seeds	Ogawa et al., 2002
Oleanolic acid	Leaves and stems	Hassan et al., 2019
Olitoriside	Seeds	Ogawa et al., 2002
Oxo-corocin	Roots	Manzoor-i-Khuda & Habermehl, 1979

C. olitorius. These phytochemical compounds may act directly or indirectly to maintain health and prevent or treat disease or ailment. Moreover, the phytochemical compounds isolated from the species and its extract demonstrated promising effects in several biological assays using *in vitro* and *in vivo* systems (Table 9.5). These biological activities include α-amylase and α-glucosidase inhibitory, acetylcholinesterase, analgesic, antibacterial, antifungal, anticancer, antidiabetic, antihyperglycemic, anti-inflammatory, antioxidant, antiproliferative, cardioprotective, glutathione-s-transferase, hepatoprotective, immune-enhancing, and wound-healing effects (Table 9.5).

TABLE 9.5
Summary of Pharmacological Activities of the Extracts and Phytochemical Compounds Isolated from *Corchorus olitorius*

Activity Tested	Extract	Plant Part	Model	Effect	Reference
α-amylase inhibitory	Aqueous	Leaves	α-amylase inhibition assay	Extract exhibited dose-dependent activities	Ademiluyi et al., 2015
α-glucosidase inhibitory	Aqueous	Leaves	α-glucosidase inhibition assay	Extract exhibited dose-dependent activities	Ademiluyi et al., 2015
Acetylcholinesterase	Methanol	Leaves, roots, and stems	Acetylcholinesterase enzyme inhibition	Exhibited activities with half-maximal inhibitory concentration (IC_{50}) values ranging from 0.16 to 0.20 mg/ml	Atalar et al., 2023
Analgesic	Methanol	Aerial parts	Writhing test	Dose-dependent activities exhibited by the extract	Parvin et al., 2015
Antibacterial	Methanol	Seeds	Agar diffusion	Extracts exhibited activities against *Staphylococcus aureus*, *Bacillus punilus*, *Salmonella typhi*, *S. thyphimurium*, *Shigella boydii*, *S. sonnei*, *S. dysenteriae*, *Klebsiella pneumoniae*, *Escherichia coli*, and *Vibrio cholerae*, with zone of inhibition of 4.0–25.0 mm	Pal et al., 2006
	Aqueous, petroleum ether, methanol, and ethyl acetate	Leaves	Agar diffusion	Extracts exhibited activities against *S. aureus*, *Bacillus cereus*, *B. subtilis*, *Micrococcus luteus*, *Enterococcus faecium*, *E. faecalis*, *Proteus vulgaris*, *Pseudomonas gingeri*, *E. coli*, *Enterobacter aerogenes*, *S. typhimurium*, and *Yersinia enterocolitica*, with zone of inhibition of 12.0–20.0 mm	Ilhan et al., 2007
	Aqueous, diethyl ether, and methanol	Flowers and leaves	Agar diffusion	Extracts exhibited activities against *E. coli*, *S. typhimurium*, *B. subtilis*, *S. aureus*, *E. faecalis*, *Bacillus thuringiensis*, *P. aeruginosa*, *M. luteus*, *K. pneumoniae*, *Enterobacter* spp., and *Actinomyces* spp., with zone of inhibition of 12.5–16.7 mm	Driss et al., 2016
	Chloroform, ethanol, and methanol	Leaves	Agar diffusion	Exhibited activities against *E. fecealis*, *B. subtilis*, *E. coli*, *P. vulgaris*, *Klebsiella* spp., and *Serratia marcescens*, with zone of inhibition of 12.0–22.0 mm	Mohammed, 2016
	Hexane	Leaves	Agar diffusion	Extracts exhibited activities against *B. subtilis* and *S.* aureus, with zone of inhibition of 8.0–12.0 mm	Özdenefe et al., 2018
	Ethanol, water, and ethanol–water (1:1)	Leaves	Agar diffusion	Extracts exhibited activities against *S. aureus*, *M. luteus*, *B. cereus*, *E. coli*, *K. pneumoniae*, *Salmonella enterica*, *S. typhi*, and *Enterobacter* spp., with zone of inhibition of 2.75–34.5 mm	Yakoub et al., 2018
	Ethanol	Leaves	Agar diffusion	Extracts exhibited activities against *B. subtilis*, *S. aureus*, and *E. coli*, with zone of inhibition of 16.0–22.0 mm	Mouas et al., 2021

(Continued)

TABLE 9.5 (Continued)

Activity Tested	Extract	Plant Part	Model	Effect	Reference
	Aqueous and ethanol	Leaves	Agar diffusion	Extracts exhibited activities against *E. coli*, with zone of inhibition of 6.23–16.30 mm	Khan et al., 2022
	Aqueous, diethyl ether, and methanol	Leaves	Agar diffusion	Extracts exhibited activities against *S. aureus*, *Streptococcus pneumoniae*, and *E. coli*, with zone of inhibition of 11.0–16.0 mm	Nakaziba et al., 2022
	Methanol	Leaves	Microdilution assay	Exhibited activities against *S. aureus*, *B. cereus*, *B. subtilis*, *E. coli*, and *P. aeruginosa*, with IC_{50} values of 2.2–8.1 µg/ml	Atalar et al., 2023
Antifungal	Aqueous, petroleum ether, methanol, and ethyl acetate	Leaves	Agar diffusion	Extracts exhibited activities against *Candida albicans*, *Aspergillus niger*, *Botrytis cinerea*, *Geotrichum candidum*, and *Fusarium graminarum*, with zone of inhibition of 11.0–20.0 mm	Ilhan et al., 2007
	Methanol	Leaves	Microdilution assay	Exhibited activities with IC_{50} values of 4.0 µg/ml against *C. albicans*	Atalar et al., 2023
Anticancer	Methyl-1,4,5-tri-O-caffeoyl quinate and trans-3-(4-hydroxy-3-methoxyphenyl) acrylic anhydride	Whole plant	Alamar Blue assay	The compounds exhibited moderate activities against the cancer cell lines HeLa, HL460, and PC3	Taiwo et al., 2016
	Aqueous, chlorogenic acid, and isoquercetin	Leaves	MTT (3-(4,5-dimethylthiazol-2-yl)-2,5-diphenyltetrazolium bromide) cytotoxicity assay	Extract and phytochemical compounds exhibited activities with IC_{50} values of 2.54–373.53 mg/ml against human melanoma (A-375), gastric cancer (AGS), and pancreatic cancer (SUIT-2)	Tosoc et al., 2021
	Aqueous, ethyl acetate, hexane, and methanol	Whole plant	MTT assay	Extracts exhibited activities against non-cancerous L929 cells and cancerous MCF-7 and A549 cell lines with IC_{50} values of 12.45–294.03 µg/ml	Alshabi et al., 2022
	Methanol	Leaves	MTT assay	Exhibited activities against human ovarian (skov-3) cell lines, with values of 37.1% and 42.1% at 250 µg/ml and 500 µg/ml concentrations, respectively	Atalar et al., 2023
Antidiabetic	Ethanol	Leaves	*In vivo* through inhibition of α-amylase enzyme	Extract decreased α-amylase activity	Abdallah et al., 2020
	Methanol	Leaves	α-amylase and β-glucosidase inhibition assays	Extract exhibited activities with IC_{50} value of 27.95–41.64 µg/ml	Chigurupati et al., 2020
Antihyperglycemic	Methanol	Aerial parts	Oral glucose tolerance tests	Extract exhibited activities	Parvin et al., 2015
	Aqueous	Leaves	Oral glucose tolerance tests	Extract decreased serum glucose level	Abdallah et al., 2020

(Continued)

TABLE 9.5 (Continued)

Activity Tested	Extract	Plant Part	Model	Effect	Reference
Anti-inflammatory	Aqueous	Leaves and whole plant	Cyclooxygenase (COX-2) inhibitory	Extracts suppressed the expression of COX-2 in a dose-dependent manner	Yan et al., 2013
Antioxidant	Aqueous, ethanol, and petroleum ether	Leaves	2-diphenyl-1-picrylhydrazyl (DPPH) assay	Exhibited activities with IC_{50} values of 0.0054–0.0086 μg/ml	Hussien et al., 2017
	Aqueous, ethanol, and ethanol–water (1:1)	Leaves	DPPH and ferric reducing antioxidant potential (FRAP) assays	Extract exhibited activities against DPPH and FRAP	Yakoub et al., 2018
	Ethanol	Leaves	DPPH, FRAP, and 2,2-azino-bis (3-ethylbenzothiazoline-6-sulphonic acid (ABTS) assays	Extract exhibited activities against DPPH, FRAP, and ABTS	Biswas et al., 2020
	Methanol	Leaves	ABTS and DPPH assays	Exhibited activities with IC_{50} values of 33.19–37.65 μg/ml	Chigurupati et al., 2020
	Ethanol	Leaves	DPPH assay	Extract exhibited activities with IC_{50} values of 86.4 μg/ml	Mouas et al., 2021
	Aqueous	Leaves	ABTS assay	Extract exhibited antioxidant activities of 81.89–122.23 μM TE/g	Ncube et al., 2022
	Aqueous	Leaves	ABTS, DPPH, and FRAP assays	Extracts exhibited activities against ABTS, DPPH, and FRAP	Ncube et al., 2022
	Ethanol	Leaves, roots, and stems	ABTS, DPPH, and FRAP assays	Extracts exhibited activities against ABTS, DPPH, and FRAP	Biswas et al., 2023
Antiproliferative	Essential oils	Leaves	MTT assay	Essential oils exhibited activities against human breast cancer cells (MCF7), mouse melanoma cells (B16), cervical cancer line (HeLa), and human lung cancer cells (LoVo), with IC_{50} value of 4.0 μg/ml	Driss et al., 2016
Cardioprotective	Aqueous	Leaves	Cardioprotective assays	Extracts exhibited cardioprotective activities against sodium arsenite–induced oxidative damages in cardiac tissues	Das et al., 2010
Glutathione-S-transferase	Methanol	Leaves, roots, and stems	Glutathione-S-transferase enzyme inhibition	Exhibited activities with IC_{50} values ranging from 0.17 to 0.22 mg/ml	Atalar et al., 2023
Hepatoprotective	Ethanol	Leaves	Serum enzyme assay	Extract exhibited hepatoprotective activities	Ujah et al., 2014
Immune enhancing	Crude polysaccharide	Leaves	*In vitro* immune-enhancing and bone marrow cell proliferation and production assays	Polysaccharides enhanced immunological function via increase in bone marrow cell proliferation activity as well as immunoglobulin A and cytokines production	Lee et al., 2021
Wound healing	Aqueous and powder	Leaves	Excision wound model	Both extracts produced 100.0% wound contraction	Barku et al., 2013

Besides its nutritional composition, *C. olitorius* is widely valued for its diverse medicinal uses against several diseases and conditions (Table 9.6). Most parts of the species, namely, flowers, leaves, roots, seeds, shoots, stems, and twigs, are used as antidote, blood purifier, colic, lactogogue, post-natal, purgative, and tonic. The decoctions of *C. olitorius* are commonly used as herbal medicine against anemia, ascites, asthma, diabetes, female infertility, fever, gastrointestinal problems, heart problems, loss of appetite, malaria, piles, sexual problems, skin infections, toothache, ulcers, visceral obstruction, waist pain, and wounds (Table 9.6). The slippery texture of the cooked leaves

TABLE 9.6
Medicinal Uses of Different Plant Parts of *Corchorus olitorius* in Different Countries

Medicinal Uses	Plant Parts	Country	Reference
Anemia	Leaves	Ghana	Nyadanu et al., 2017
Antidote	Whole plant	India	Trivedi, 2009
Ascites	Whole plant	India	Trivedi, 2009
Asthma	Leaves	Ghana	Nyadanu et al., 2017
Blood purifier	Leaves	Nigeria	Aiyeloja & Bello, 2006
Child malnutrition and food for sick people, pregnant women, and nursing mothers	Leaves	Benin, Botswana, Nigeria, and South Africa	Velempini et al., 2003; Oyedele et al., 2006; Bvenura & Afolayan, 2014; Adebo et al., 2018
Colic	Leaves, seeds, or whole plant	Benin	Adebo et al., 2018
Diabetes	Leaves or seeds	Benin and Nigeria	Mustafa et al., 2014; Lawin et al., 2015
Female infertility	Leaf, seed, or whole plant	Benin, Ghana, and Nigeria	Soladoye et al., 2014; Nyadanu et al., 2017; Adebo et al., 2018
Fever	Flowers, leaves, roots, seeds, stems, or whole plant	Benin, Ghana, India, and Nigeria	Trivedi, 2009; Fondio & Grubben, 2012; Nyadanu et al., 2017; Adebo et al., 2018
Gastrointestinal problems (constipation, diarrhea, and stomach problems)	Leaves	Benin, Ethiopia, Ghana, and Tanzania	Benor et al., 2010; Fondio & Grubben, 2012; Nyadanu et al., 2017; Adebo et al., 2018
Heart problems	Twigs or whole plant	Benin and Kenya	Fondio & Grubben, 2012; Adebo et al., 2018
Lactogogue	Leaves	Ghana	Nyadanu et al., 2017
Loss of appetite	Leaves	Ghana	Nyadanu et al., 2017
Malaria	Leaves, roots, stems, or whole plant	Benin	Adebo et al., 2018
Piles	Leaves or whole plant	Ghana and India	Trivedi, 2009; Nyadanu et al., 2017
Post-natal	Shoots	Nigeria	Ogwu et al., 2017
Purgative	Seeds or whole plant	India and Nigeria	Trivedi, 2009; Fondio & Grubben, 2012
Sexual problems	Not specified	Benin	Adebo et al., 2018
Skin infections (measles and rash)	Leaves	Nigeria	Sonibare et al., 2009; Mowobi et al., 2016
Toothache	Roots	Kenya	Fondio & Grubben, 2012
Tonic	Roots	Kenya	Fondio & Grubben, 2012
Ulcers	Leaves, seeds, or whole plant	Benin	Adebo et al., 2018
Visceral obstruction	Whole plant	India	Trivedi, 2009
Waist pain	Leaves	Ghana	Nyadanu et al., 2017
Wounds	Leaves	Ghana	Barku et al., 2013, 2014

of *C. olitorius* is particularly good for children and sick people who usually have difficulties in swallowing. Therefore, the leaves of *C. olitorius* are used against child malnutrition in Benin (Adebo et al., 2018), food for pregnant women and nursing mothers in Nigeria (Oyedele et al., 2006), and food for sick people in Botswana (Velempini et al., 2003) and South Africa (Bvenura & Afolayan, 2014). Clearly, diversified diets that include a regular intake of leafy vegetables such as *C. olitorius* have great potential in ensuring food and nutrition security due to the potential health benefits offered by the species beyond its nutritional value.

9.8 POST-HARVEST MANAGEMENT OF *CORCHORUS OLITORIUS*

Post-harvest management and processing play vital roles in mitigating the impact of pathogens and environmental factors, thus ensuring the quality maintenance of *C. olitorius*. A myriad of techniques exists for enhancing and preserving *C. olitorius*, encompassing sun drying, solar drying, canning, vacuum packaging, minimal processing, chilling, freezing, temperature control, utilization of resistant genotypes, biocontrol agents, and irradiation (Van't Hag et al., 2020; Tripathi et al., 2022; Cefola & Pace, 2023). In a study by Shitanda and Wanjala (2006), the influence of various drying methods, including shade, sun, hoover, and freeze-drying, on the quality of *C. olitorius* was investigated. It was observed that freeze-drying resulted in superior color retention, increased ascorbic acid content, and enhanced reconstitution properties compared to other drying techniques. This underscores the significance of drying in extending the shelf life of jute by reducing moisture content, thus impeding microbial decay and enzymatic reactions.

Mutuli and Mbuge (2015) demonstrated that employing gentle drying conditions at 40°C with superabsorbent polymers effectively preserved *C. olitorius* leaf structures and vitamins. This method maintained a microbial-safe moisture content (≤14% dry basis) crucial for addressing micronutrient deficiencies during droughts. Dynamic vapor sorption analysis and various moisture sorption models were utilized to ascertain equilibrium moisture content based on water activity. Masarirambi et al. (2010) noted that the peak nutritional value of *C. olitorius* is attained when consumed fresh, surpassing the benefits provided by vegetables like cabbage (*B. oleracea*). However, the availability of fresh vegetables is limited to the rainy season due to physical and biochemical alterations post-harvest, resulting in diminished nutritional content, altered taste properties, and the onset of decay (Masarirambi et al., 2010).

Ohmic heat processing emerges as an innovative approach to food preservation, effectively controlling microorganisms while minimizing surface drying and overcooking, thus offering promising advantages for food producers (Tian et al., 2018). Tulio et al. (2002) conducted an investigation into the impact of various storage temperatures on the quality of *C. olitorius* leaves. The study utilized a range of temperatures (1 and 8°C, 15 and 20°C, as well as 30°C) to assess the freshness and stability of *C. olitorius* leaves. It was found that higher temperatures maintained the freshness and stability of the leaves, while lower temperatures resulted in chilling injury characterized by yellowing and detachment from stems. Optimal storage temperatures between 8 and 15°C were observed to minimize chilling injury and senescence. The study by Tulio et al. (2002) highlights the critical role of temperature control in preserving the quality of *C. olitorius* leaves. In contrast, the study of Dey et al. (2021) demonstrated that drying temperature significantly influences the stability of nutrients in *C. olitorius*, and particularly, temperature significantly influences the stability of nutrients in *C. olitorius* and particularly vitamin C. Elevated temperatures during the drying process accelerate reactions, leading to increased rates of nutrient degradation. Even at shorter drying times, higher temperatures can result in substantial nutrient losses. Furthermore, the stability of different nutrients varies during drying, with β-carotene exhibiting the highest stability and vitamin C being the least stable.

Das et al. (2021) established that *C. olitorius* is susceptible to various phytopathogens, notably *Macrophomina phaseolina*, which induces root and stem rot diseases. Nonetheless, a repertoire of antagonistic microorganisms has demonstrated efficacy in managing agricultural diseases, encompassing *Bradyrhizobium* spp., *Rhizobium meliloti* species, and diverse species of *Trichoderma*, including *T. harzianum* and *T. viride*. Their documented antagonistic activity against *Macrophomina phaseolina* underscores their pivotal role in post-harvest disease control strategies. These beneficial microorganisms employ intricate mechanisms to elicit their effects. They engage in resource competition, impeding the proliferation of deleterious organisms by limiting nutrient availability and space. Furthermore, they synthesize antimicrobial metabolites, selectively targeting and diminishing pathogen populations. Additionally, these microorganisms have the capacity to induce systemic resistance in jute plants, bolstering their post-harvest defense mechanisms against pathogenic incursions (Ghosh et al., 2018; Majumder et al., 2018).

Hosen et al. (2016) also alluded to the effect of antagonistic organisms, such as *Aspergillus niger* and *Trichoderma viride*, on *C. olitorius. Colletotrichum gloeosporioides* and *Agroathelia rolfsii* associated with jute plants are responsible for anthracnose and soft rot diseases in jute plants. They employed dual culture techniques to assess the inhibitory effects of *A. niger* and *T. viride* on the growth of *C. gloeosporioides* and *A. rolfsii*. Their results showed promising inhibitory effects, especially by *T. viride*, suggesting its potential as a biocontrol agent against these pathogens. Additionally, their study examined the impact of volatile and non-volatile metabolites released by these fungi, demonstrating significant growth inhibition of the pathogens. *Trichoderma viride* particularly stands out for its strong antagonistic effect, attributed to its rapid growth, sporulation, and toxin-producing capacity.

In recent years, significant research by Sivakumar et al. (2021) and de Oliveira et al. (2023) has underscored the impact of biocontrol agents, including nanoemulsions derived from edible coatings integrated with essential oils, as well as essential oils and plant extracts individually, on diseases such as anthracnose affecting fruits and vegetables. Nanoemulsions have garnered significant attention for their potential to combat pathogens responsible for soft rot, anthracnose, and stem rot, thus prolonging the shelf life of produce. These colloidal dispersions, characterized by their stability and ability to enhance the bioavailability of active agents like essential oils and plant extracts, offer controlled release mechanisms and protection against degradation. Moreover, encapsulating essential oils and plant extracts within nanoemulsions, facilitated by materials such as edible coatings, emerges as a reliable method to preserve their volatility and enhance their bioactivity without compromising the sensory qualities of food (Matche & Adeogun, 2022).

Essential oils and plant extracts have individually demonstrated efficacy against phytopathogens causing plant diseases. Essential oils, complex mixtures primarily comprising low-molecular-weight terpenic compounds, are soluble in nonpolar solvents and possess distinct fragrances owing to their volatility (Peralta-Ruiz et al., 2023). Generally considered safe for consumption, they must not compromise sensory or organoleptic properties or pose toxicity risks (Matche & Adeogun, 2022). Literature supports the effectiveness of essential oils from aromatic plants in combating anthracnose caused by *Colletotrichum* spp. in vegetables such as *C. olitorius*. However, further exploration of their antifungal mechanisms, individual component effects, and potential phytotoxicity is needed (Kumar & Kudachikar, 2018; Purkait et al., 2020; Peralta-Ruiz et al., 2023). Notably, studies by Matche and Adeogun (2022) and Peralta-Ruiz et al. (2023) have highlighted the inhibitory effects of essential oils from thyme, savory, mint, cinnamon, eucalyptus, and lavender, with savory and thyme exhibiting significant control over *C. gloeosporioides* growth. Plant extracts, recognized for their antimicrobial properties, offer a promising alternative to synthetic fungicides for managing *C. gloeosporioides*. Numerous studies have validated their antifungal activity *in vitro* against *C. gloeosporioides* and *in situ* for post-harvest anthracnose management in fruits and vegetables (Peralta-Ruiz et al., 2023; Rahman et al., 2022). Various extraction methods utilizing solvents like methanol, chloroform, and ethanol have significantly inhibited the growth of *C. gloeosporioides*. Additionally, when applied as edible coatings, certain plant extracts have demonstrated the ability to alleviate anthracnose symptoms in fruits, highlighting their potential as innovative post-harvest treatments (Rahman et al., 2022). Kaur et al. (2022) and Majumder et al. (2018) posited the impact of transgenic *C. olitorius* on post-harvest management of diseases in the species, such as stem rot and root rot. This approach strategically integrates genes linked to plant defense mechanisms, particularly those crucial in phenylpropanoid metabolism and phytohormone signaling pathways, like jasmonic acid and ethylene. Additionally, investigating RNA interference (RNAi) technology offers a promising avenue to suppress genes responsible for lignin production, potentially enhancing jute's resistance to necrotrophic diseases.

Plants such as *C. olitorius* have been established to exhibit sophisticated stress response mechanisms, which include the participation of short RNAs (sRNAs), such as microRNAs (miRNAs), in post-transcriptional gene regulation. Utilizing host plant resistance, possibly mediated by sRNAs, emerges as a viable strategy in managing *Macrophomina* infections in jute crops. Furthermore, the *Agrobacterium*-mediated genetic transfer is a notable method for addressing post-harvest diseases in jute plants. This process involves the introduction of desired genes into plants, including *C. olitorius*, to confer disease resistance (Meena et al., 2015; Majumder et al., 2018; Kaur et al., 2022).

9.9 CONCLUSION AND PROSPECTS

There is increasing consensus that wild foods such as *C. olitorius* have potential to significantly contribute to alleviating malnutrition and hunger. Therefore, promotion of such species through domestication and integration into diets is the most practical and sustainable way to achieve this goal. This review adds valuable literature on the importance of *C. olitorius* for local livelihood needs and the contribution of the species to environmental sustainability in rural, peri-urban, and marginalized areas, where the species is collected from the wild, domesticated, and managed in home gardens. Since *C. olitorius* has been the mainstay of human diets for centuries, the species could make an important contribution to combating micro- and macronutrient malnutrition as well as providing food security for many low-income households in developing countries. Unfortunately, *C. olitorius* is currently underutilized and has been neglected by researchers, agricultural extension officers, and policymakers. Cultivation of *C. olitorius*, and its promotion and integration into the human diet, could assist in addressing dietary deficiencies in impoverished households in developing countries. Additionally, there is a need for in-depth research on the chemical, nutritional, and toxicological properties of the species, its micro- and macronutrients, as well as the secondary metabolites of the species, and its products after modification by various processing techniques. The dual nutritional and medicinal role of *C. olitorius* has significance in the maintenance and promotion of good health and protection against diseases and ailments. Research on *C. olitorius* over the past decades on health-promoting properties of the species contributed to the increased consumption of the species as both leafy vegetable and herbal medicine. Significant research has been made in the past 50 years into the phytochemistry and pharmacological properties of *C. olitorius*. Based on available literature, *C. olitorius* has great potential as a high-value nutraceutical and an important source of several bioactive compounds which can be used as dietary supplements which characterize functional foods. These studies have shown *C. olitorius* to display various chemical and biological activities, some of which justify its ethnopharmacological utilization in tropical Africa and Asia. In light of this evidence, it is therefore essential to conduct detailed phytochemical studies, especially the mechanisms of action of its bioactive constituents, to illustrate the correlation between ethnomedicinal applications of the species and its biological activities. There is a need for extensive *in vivo* experiments and clinical studies to validate the existing pharmacological activities of the species. Future research should assess the nutraceutical properties of *C. olitorius* based on traditional uses of the species. Considerable research into the genetics and breeding of *C. olitorius* has been done, including other related *Corchorus* species widely used as leafy vegetables, such as *C. aestuans*, *C. asplenifolius*, *C. capsularis*, *C. confusus*, *C. fascicularis*, *C. pseudocapsularis*, *C. schimperi*, *C. tridens*, *C. trilocularis*, and *C. urticifolius*. Evaluation of *C. olitorius* germplasm showed large variability in agronomically important characters, such as seed yield per plant, crop duration, days to maturity, resistance to pests, yield of the species under rain-fed cultivation, and handling of the species after harvest. Future research should also focus on improving planting material availability, crop yield levels, yield stability, and improved pest control. In countries where research has linked health and diet benefits associated with intake of *C. olitorius*, demand for this species has increased. It is therefore likely that *C. olitorius* production will expand in many tropical African and Asian countries in the future. *Corchorus olitorius* does not seem to be overcollected, threatened with extinction, or in danger of genetic erosion. The large genetic diversity of *C. olitorius* gives ample scope for improved crop yields through rapid multiplication techniques. Demand for *C. olitorius* is likely to remain strong, and if efforts of domestication are successful, this species is likely to become a promising functional crop in tropical Africa and Asia.

9.10 ACKNOWLEDGMENTS

The study was funded in part by the University of Fort Hare, grant R188.

REFERENCES

Abdallah, H. M. I., Jaleel, G. A. A., Mohammed, H. S., Mahmoud, S. S., Yassin, N. A., el Din, A. A. G., Shaffie, N. M., & Bassyouni, F. A. (2020). Phytochemical screening, gas chromatography-mass spectrometry analysis, and antidiabetic effects of *Corchorus olitorius* leaves in rats. *Macedonian Journal of Medical Sciences*, *8*, 385–394.

Acho, C. F., Zoue, L. T., Akpa, E. E., Yapo, A. G., & Niamke, S. L. (2014). Leafy vegetables consumed in Southern Côte d'Ivoire: A source of high value nutrients. *Journal of Animal and Plant Sciences*, *20*, 3159–3170.

Adebo, H. O., Ahoton, L. E., Quenum, F. J., Adoukonou-Sagbadja, H., Bello, D. O., & Chrysostome, C. A. (2018). Ethnobotanical knowledge of jute (*Corchorus olitorius* L.) in Benin. *European Journal of Medicinal Plants*, *26*, 1–11.

Ademiluyi, A. O., Oboh, G., Aragbaiye, F. P., Oyeleye, S. I., & Ogunsuyi, O. B. (2015). Antioxidant properties and in vitro a-amylase and a-glucosidase inhibitory properties of phenolics constituents from different varieties of *Corchorus* spp. *Journal of Taibah University Medical Sciences*, *10*, 278–287.

Ahmed, F. (2021). Nutraceutical potential of molokhia (*Corchorus olitorius* L.): A versatile green leafy vegetable. *Pharmacognosy Research*, *13*, 1–12.

Aiyeloja, A. A., & Bello, O. A. (2006). Ethnobotanical potentials of common herbs in Nigeria: A case study of Enugu state. *Educational Research and Reviews*, *1*, 16-22.

Akoègninou, A., Van der Burg, W. J., & Van der Maesen, L. J. G. (2006). *Flore analytique du Bénin*. Backhuys Publishers.

Alharbi, S. A., Al-Hawshabi, O. S. S., & Badry, M. O. (2024). Flora and phytochorolgy of Lahij Governorate of Yemen: 2-taxonomic revision of *Corchorus* L. (Grewioideae: Malvaceae *sensu lato*) in Toor Al-Baha District. *Egyptian Journal of Botany*, *64*(1), 375–385.

Ali, M., Ahmed, K., Hossain, H., Roy, B., Rokeya, B., Rahman, M., Jahan, I., & Rahman, M. (2021). Total antioxidant capacity and profiling of polyphenolic compounds in jute leaves by HPLC-DAD. *Food Research*, *5*, 343–348.

Alshabi, A. M., Alkahtani, S. A., Shaikh, I. A., Orabi, M. A. A., Abdel-Wahab, B. A., Walbi, I. A., Habeeb, M. S., Khateeb, M. M., Hoskeri, J. H., Shettar, A. K., & Asdaq, S. M. B. (2022). Phytochemicals from *Corchorus olitorius* methanolic extract induce apoptotic cell death via activation of caspase-3, anti-Bcl-2 activity, and DNA degradation in breast and lung cancer cell lines. *Journal of King Saud University Science*, *34*, 102238.

Alverson, W. S., Whitlock, B. A., Nyppeler, R., Bayer, C., & Baum, D. A. (1999). Phylogeny of the core Malvales: Evidence from ndh F sequence data. *American Journal of Botany*, *86*(10), 1474–1486.

Angiosperm Phylogeny Group (APG). (2009). An update of the Angiosperm Phylogeny Group classification for the orders and families of flowering plants: APG III. *Botanical Journal of the Linnean Society*, *16*, 105–121.

Atalar, M. N., Erenler, R., Turkan, F., Alma, M. H., Demirtas, I., Baran, A., Kandemir, S. I., Kekec, A. I., & Saltan, F. Z. (2023). Phytochemical analysis and biological activity of *Corchorus olitorius* L.: Quantitative analysis of bioactive compounds by LC–MS/MS, antibacterial, enzyme inhibition, and cytotoxic activities. *European Journal of Integrative Medicine*, *62*, 102290.

Ayaz, F. A., Glew, R. H., Millson, M., Huang, H. S., Chuang, L. T., Sanz, C., & Hayırlıoglu-Ayaz, S. (2006). Nutrient contents of kale (*Brassica oleraceae* L. var. *acephala* DC.). *Food Chemistry, 96*, 572–579.

Azuma, K., Nakayama, M., Koshioka, M., Ippoushi, K., Yamaguchi, Y., Kohata, K., Yamauchi, Y., Ito, H., & Higashio, H. (1999). Phenolic antioxidants from the leaves of *Corchorus olitorius* L. *Journal of Agriculture and Food Chemistry, 47*, 3963–3966.

Barku, V. Y. A., Boye, A., & Quansah, N. (2013). Antioxidant and wound healing studies on the extracts of *Corchorus olitorius* leaf. *World Essays Journal, 1*(3), 67–73.

Barku, V. Y. A., Opoku-Boahen, Y., & Dali, G. A. (2014). Ehnobotanical study of wound healing plants in Kpando traditional area, Ghana. *International Journal of Phytomedicine, 6*(4), 564–572.

Bayer, C., & Kubitzki, K. (2003). Malvaceae. In K. Kubitzki & C. Bayer (Eds.), *The families and genera of vascular plant, flowering plant dicotyledon, Malvales, Capparales and non betalain Caryophyllales, Malvaceae* (Vol. 5, pp. 225–311). Springer-Verlag.

Benor, S., Blattner, F. R., Demissew, S., & Hammer, K. (2010). Collection and Ethnobotanical investigation of *Corchorus* species in Ethiopia: Potential leafy vegetables for dry regions. *Genetic Resources and Crop Evolution, 57*(2), 293–306.

Bhattacharyya, D., Basu, S., Chattapadhyay, J. P., & Bose, S. K. (1985). Biocontrol of *Macrophomina* root-rot disease of jute by an antagonistic organism, *Aspergillus versicolor. Plant Soil, 87*, 435–438.

Biswas, A., Dey, S., Li, D., Liu, Y., Zhang, J., Huang, S., Pan, G., & Deng, Y. (2020). Comparison of phytochemical profile, mineral content, and *in vitro* antioxidant activities of *Corchorus capsularis* and *Corchorus olitorius* leaf extracts from different populations. *Journal of Food Quality, 2020*, Article ID 2931097s.

Biswas, A., Dey, S., Xiao, A., Huang, S., Birhanie, Z. M., Deng, Y., Liu, L., & Li, D. (2023). Phytochemical content and antioxidant activity of different anatomical parts of *Corchorus olitorius* and *C. capsularis* during different phenological stages. *Heliyon, 9*, e16494.

Biswas, C., Dey, P., Bera, A., Satpathy, S., & Mahapatra, B. S. (2013). First report of bacterial leaf spot caused by *Xanthomonas campestris* pv. *olitorii* on jute grown for seed in India. *Plant Disease, 97*(8), 1109.

Biswas, C., Dey, P., & Satpathy, S. (2014). First report of bacterial leaf blight of jute (Corchorus capsularis L.) caused by Xanthomonas campestris pv. capsularii in India. *Archives of Phytopathology and Plant Protection, 47*(13), 1600–1602.

Bvenura, C., & Afolayan, A. J. (2014). Ethnobotanical survey of wild vegetables in Mbashe and Nkonkobe municipalities, Eastern Cape Province, South Africa. *Acta Botanica Gallica, 161*(2), 189–199.

Cefola, M., & Pace, B. (2023). Advances postharvest preservation technology. *Foods, 12*(8), 1664.

Chakraborty, A., & Palit, P. (2009). Morphological and molecular characterization of nine species of *Corchorus* (L.) grown in India with special reference to their interrelationship. *Journal of the Botanical Society of Bengal, 63*, 89–99.

Chaudhary, P., & Verma, R. (2018). Nutritional evaluation of organically and conventionally grown kale (Brassica oleraceae L. var. acephala). *Applied Biological Research, 20*(3), 250–255.

Chigurupati, S., Aladhadh, H. S., Alhowail, A., Selvarajan, K. K., & Bhatia, S. (2020). Phytochemical composition, antioxidant and antidiabetic potential of methanolic extract from *Corchorus olitorius* Linn. grown in Saudi Arabia. *Medicinal Plants International Journal of Phytomedicines and Related Industries, 12*(1), 71–76.

Choudhary, S. B., Sharma, H. K., Kumar, A. A., Maruthi, R. T., & Karmakar, P. G. (2017). The genus *Corchorus* L. (Malvaceae) in India: Species distribution and ethnobotany. *Genetic Resources and Crop Evolution, 64*(7), 1675–1686.

Dansi, A., Adjatin, A., Adoukonou-Sagbadja, H., Faladé, V., Yedomonhan, H., Odou, D., & Dossou, B. (2008). Traditional leafy vegetables and their use in the Benin Republic. *Genetic Resources and Crop Evolution, 55*, 1239–1256.

Das, A. K., Sahu, R., Dua, T. K., Bag, S., Gangopadhyay, M., Sinha, M. K., & Dewanjee, S. (2010). Arsenic-induced myocardial injury: Protective role of *Corchorus olitorius* leaves. *Food and Chemical Toxicology, 48*, 1210–1217.

Das, S., Ghosh, A., & Mukherjee, A. (2021). Nanoencapsulation-based edible coating of essential oils as a novel green strategy against fungal spoilage, mycotoxin contamination, and quality deterioration of stored fruits: An overview. *Frontiers in Microbiology, 12*, 768414.

de Oliveira, T. S., Costa, A. M. M., Cabral, L. M. C., Freitas-Silva, O., Rosenthal, A., & Tonon, R. V. (2023). Anthracnose controlled by essential oils: Are nanoemulsion-based films and coatings a viable and efficient technology for tropical fruit preservation? *Foods, 12*(2), 279.

Dey, S., Biswas, A., Huang, S., Li, D., Liu, L., Deng, Y., Xiao, A., Birhanie, Z. M., Zhang, J., Li, J., & Gong, Y. (2021). Low temperature effect on different varieties of *Corchorus capsularis* and *Corchorus olitorius* at seedling stage. *Agronomy, 11*(12), 2547.

Driss, D., Kaoubaa, M., Mansour, R. B., Kallel, F., Abdelmalek, B. E., & Chaabouni, S. E. (2016). Antioxidant, antimutagenic and cytotoxic properties of essential oil from *Corchorus olitorius* L. flowers and leaf. *Free Radicals and Antioxidants*, *6*, 34–43.

Fondio, L., & Grubben, G. J. H. (2012). *Corchorus olitorius* L. In G. J. H. Grubben & O. A. Denton (Eds.), *Plant resources of tropical Africa 2. Vegetables* (pp. 217–221). Prota Foundation.

Furumoto, T., Wang, R., Okazaki, K., Afm, F., Ali, M. I., Kondo, A., & Fukui, H. (2002). Antitumor promoters in leaves of jute (*Corchorus capsularis* and *Corchorus olitorius*). *Food Science and Technology Research*, *8*(3), 239–243.

Ghosh, R., Paul, S., Das, S., Palit, P., Acharyya, S., Das, A., Mir, J. I., Ghosh, S. K., & Roy, A. (2008). Molecular evidence for existence of a New World begomovirus associated with yellow mosaic disease of *Corchorus capsularis* in India. *Australasian Plant Disease Notes*, *3*(1), 59–62.

Ghosh, T., Bıswas, M., Dhara, S., & Ghosh, C. (2018). Mechanisms involve in jute resistance to *Macrophomina phaseolina*: Strategies for developing resistant jute varieties. *Plant Cell Biotechnology and Molecular Biology*, *19*, 91–106.

Gomez, M. I. (1988). A resource inventory of indigenous and traditional foods in Zimbabwe. *Zambezia*, *15*(1), 379–622.

Guzzetti, L., Panzeri, D., Ulaszewska, M., Sacco, G., Forcella, M., Fusi, P., Tommasi, N., Fiorini, A., Campone, L., & Labra, M. (2021). Assessment of dietary bioactive phenolic compounds and agricultural sustainability of an African leafy vegetable *Corchorus olitorius* L. *Frontiers in Nutrition*, *8*, 667812.

Handoussa, H., Hanafi, R. S., El-Khatib, A. H., Linscheid, M. W., Mahran, L. G., & Ayoub, N. A. (2017). Computer-assisted HPLC method development using DryLab for determination of major phenolic components in *Corchorus olitorius* and *Vitis vinifera* by using HPLC-PDA-ESI-TOF-MSn. *Research and Reviews Journal of Botanical Sciences*, *6*, 9–16.

Hanif, R., Iqbal, Z., Iqbal, M., Hanif, S., & Rasheed, M. (2006). Use of vegetables as nutritional food: Role in human health. *Journal of Agricultural and Biological Science*, *1*, 22–26.

Haque, S. M. A., Ali, S., Ali, M. M., Rahman, M. A., & Hossain, I. (2015). Effect of storage materials on seed quality and health in Jute variety O-9897. *African Journal of Agricultural Research*, *10*(21), 2190–2196.

Hasan, H. T., & Kadhim, E. J. (2018). Phytochemical investigation of leaves and seeds of *Corchorus olitorius* L. cultivated in Iraq. *Asian Journal of Pharmaceutical and Clinical Research*, *11*, 408–417.

Hassan, A. Z., Mekhael, M. K., Hanna, A. G., Simon, A., Tóth, G., & Duddeck, H. (2019). Phytochemical investigation of *Corchorus olitorius* and *Corchorus capsularis* (family Tiliaceae) that grow in Egypt. *Egyptian Pharmaceutical Journal*, *18*, 123–134.

Heywood, V. H. (1993). *Flowering plants of the world*. Mayflower.

Heywood, V. H., Brummitt, R. K., Culham, A., & Seberg, O. (2007). *Flowering plant families of the world*. Royal Botanical Gardens.

Hosen, S., Shamsi, S., & Bashar, M. A. (2016). *In vitro* biological control of *Colletotrichum gloeosporioides* (Penz.) Sacc. and *Sclerotium rolfsii* Sacc., causal agent of Anthracnose and soft rot of *Corchorus capsularis* L. *Bangladesh Journal of Botany*, *45*(1), 175–180.

Hossain, A., Islam, M. T., Maitra, S., Majumder, D., Garai, S., Mondal, M., Ahmed, A., Roy, A., Skalicky, M., Brestic, M., & Islam, T. (2021). Neglected and underutilized crop species: Are they future smart crops in fighting poverty, hunger and malnutrition under changing climate? In S. M. Zargar, A. Masi, & R. K. Salgotra (Eds.), *Neglected and underutilized crops: Towards nutritional security and sustainability* (pp. 1–50). Springer.

Hunter, D., Borelli, T., Beltrame, D. M. O., Oliveira, C. N. S., Coradin, L., Wasike, V. W., Wasilwa, L., Mwai, J., Manjella, A., Samarasinghe, G. W. L., Madhujith, T., Nadeeshani, H. V. H., Tan, A., Ay, S. T., Güzelsoy, N., Lauridsen, N., Gee, E., & Tartanac, F. (2019). The potential of neglected and underutilized species for improving diets and nutrition. *Planta*, *250*, 709–729.

Hussien, N. M., Labib, S. E., El-Massry, R. A., & Hefnawy, H. T. M. (2017). Phytochemical studies and antioxidant activity of leaves extracts of *Corchorus olitorius* L. (molokhia). *Zagazig Journal of Agricultural Research*, *44*, 2231–2239.

Idirs, S., Yisa, J., & Ndamitso, M. (2009). Nutritional composition of *Corchorus olitorius* leaves. *Animal Production Research Advances*, *5*(2), 83–86.

Ilhan, S., Savarõglu, F., & Çolak, F. (2007). Antibacterial and antifungal activity of *Corchorus olitorius* L. (molokhia) extracts. *International Journal of Natural and Engineering and Science*, *1*, 59–61.

Isuosuo, C. C., Akaneme, F. I., & Abu, N. E. (2019). Nutritional evaluation of the seeds of *Corchorus olitorius*: A neglected and underutilized species in Nigeria. *Pakistan Journal of Nutrition*, *18*, 692–703.

Kaur, S., Samota, M. K., Choudhary, M., Choudhary, M., Pandey, A. K., Sharma, A., & Thakur, J. (2022). How do plants defend themselves against pathogens: Biochemical mechanisms and genetic interventions. *Physiology and Molecular Biology of Plants*, *28*(2), 485–504.

Khan, M., Bano, S., Javed, K., & Mueed, M. A. (2006). A comprehensive review on the chemistry and pharmacology of *Corchorus* species: A source of cardiac glycosides, triterpenoids, ionones, flavonoids, coumarins, steroids and some other compounds. *Journal of Scientific and Industrial Research*, *65*, 283–298.

Khan, M. M. I., Akhtar, B. J., & Ahmad, M. (2022). Pharmacognostic studies and antibacterial activity of *Corchorus olitorius* L. leaf. *Pharmacognosy Research*, *14*(4), 474–482.

Kobori, N. C., & Rodriguez-Amaya, D. B. (2008). Uncultivated Brazilian green leaves are richer sources of carotenoids than are commercially produced leafy vegetables. *Food Nutrition Bulletin*, *29*(4), 320–328.

Kumar, A., & Kudachikar, V. B. (2018). Antifungal properties of essential oils against anthracnose disease: A critical appraisal. *Journal of Plant Diseases and Protection*, *125*(2), 133–144.

Kumari, N., Choudhary, S. B., Sharma, H. K., Singh, B. K., & Kumar, A. A. (2019). Health-promoting properties of *Corchorus* leaves: A review. *Journal of Herbal Medicine*, *15*, 100240.

Lawin, I. F., Laleye, F. O. A., Agbani, O. P., & Assogbadjo, A. E. (2015). Ethnobotanical assessment of the plant species used in the treatment of diabetes in the Sudano-Guinean zone of Benin. *Journal of Animal and Plant Science*, *26*(3), 4108–4123.

Lebrun, J. P., & Stork, A. L. (2003). Tropical african flowering plants. In *Ecology and distribution 1*. Conservatoire et Jardin Botaniques.

Lee, H.-B., Son, S.-U., Lee, J.-E., Lee, S.-H., Kang, C.-H., Kim, Y.-S., Shin, K.-S., & Park, H.-Y. (2021). Characterization, prebiotic and immune-enhancing activities of rhamnogalacturonan-I-rich polysaccharide fraction from molokhia leaves. *International Journal of Biological Macromolecule*, *175*, 443–450.

Li, D., Ahmed, F., Wu, N., & Sethi, A. I. (2022). Yolo-JD: A deep learning network for jute diseases and pests detection from images. *Plants*, *11*(7), 937.

Liengme, C. A. (1981). Plants used by Tsonga people of Gazankulu. *Bothalia*, *13*, 501–518.

Loumerem, M., & Alercia, A. (2016). Descriptors for jute (*Corchorus olitorius* L.). *Genetic Resources and Crop Evolution*, *63*(7), 1103–1111.

Maazou, R., Rabiou, H., Mahamane, M., Issaharou-Matchi, I., Mahamane, A., & Lykke, A. M. (2023). Diversity and socio-cultural importance of wild food herbs and cyanobacteria in the Lake Chad Basin (Niger). *Ethnobotany Research and Applications*, *25*, 60.

Mabhaudhi, T., Chimonyo, V. G. P., Chibarabada, T. P., & Modi, A. T. (2017). Developing a roadmap for improving neglected and underutilized crops: A case study of South Africa. *Frontiers in Plant Science*, *8*, 2143.

Magwede, K., Van Wyk, B.-E., & Van Wyk, A. E. (2019). An inventory of Vhaveṋda useful plants. *South African Journal of Botany*, *122*, 57–89.

Majumder, S., Datta, K., Sarkar, C., Saha, S. C., & Datta, S. K. (2018). The development of *Macrophomina phaseolina* (fungus) resistant and glufosinate (herbicide) tolerant transgenic jute. *Frontiers in Plant Science*, *9*, 920.

Manandhar, N. P. (2002). *Plants and people of Nepal*. Timber Press.

Manzoor-i-Khuda, M., & Habermehl, G. (1979). Chemical constituents of *Corchorus capsularis* and *C. olitorius* (jute plant), part iv isolation of corosolic acid, ursolic acid and oxo-corosin and correlation of corosin with tormentic acid. *Zeitschrift Für Naturforschung B*, *34*, 1320–1325.

Maroyi, A. (2011a). Potential role of traditional vegetables in household food security: A case study from Zimbabwe. *African Journal of Agricultural Research*, *6*, 5720–5728.

Maroyi, A. (2011b). The gathering and consumption of wild edible plants in Nhema communal area, Midlands province, Zimbabwe. *Ecology of Food and Nutrition*, *50*, 506–525.

Maroyi, A. (2013). Use of weeds as traditional vegetables in Shurugwi district, Zimbabwe. *Journal of Ethnobiology and Ethnomedicine*, *9*, 60.

Masarirambi, M. T., Mavuso, V., Songwe, V. D., Nkambule, T. P., & Mhazo, N. (2010). Indigenous post-harvest handling and processing of traditional vegetables in Swaziland: A review. *African Journal of Agricultural Research*, *5*(24), 3333–3341.

Matche, R. S., & Adeogun, O. O. (2022). Physicochemical characterisations of nanoencapsulated *Eucalyptus globulus* oil with gum Arabic and gum Arabic nanocapsule and their biocontrol effect on anthracnose disease of *Syzygium malaccense* fruits. *Scientific African*, e01421.

Mathieu, G., & Meissa, D. (2007). Traditional leafy vegetables in Senegal: Diversity and medicinal uses. *African Journal of Traditional Complimentary and Alternative Medicine*, *4*(4), 469–475.

Matsufuji, H., Sakai, S., Chino, M., Goda, Y., Toyoda, M., & Takeda, M. (2001). Relationship between cardiac glycoside contents and color of *Corchorus olitorius* seeds. *Journal of Health Science*, *47*, 89–93.

Maundu, P., Ngugi, G., & Kabuye, C. H. (1999). *Traditional food plants of Kenya*. Kenya Resource Centre for Indigenous Knowledge (KENRIK), National Museum of Kenya, Nairobi.

Meena, P. N., De, R. K., Roy, A., Gotyal, B. S., Satpathy, S., & Mitra, S. (2015). Evaluation of stem rot disease in jute (*Corchorus olitorius*) germplasm caused by *Macrophomina phaseolina* (tassi) goid. *Journal of Applied and Natural Science*, *7*(2), 857–859.

Moeaha, M. J. (2006). Malvaceae: A new species of *Corchorus* in Southern Africa. *Bothalia*, *36*(2), 168–171.

Mohammed, R. M. O. (2016). Phytochemical investigation of antimicrobial and antioxidant activity leaves extracts of *Corchorus olitorius*. *Open Access Library Journal*, *3*, e2225.

Mouas, T. N., Kabouche, Z., Benabid, N., & Bendal, M. (2021, March 15–31). Investigations on bioactive compounds and *in vitro* biological potent of *Corchorus olitorius*. L from Algerian cultivar. In *Proceedings of the 1st international electronic conference on biological diversity, ecology and evolution*. MDPI. https://doi.org/10.3390/BDEE2021-09430

Mowobi, G. G., Abubakar, S., Osuji, C., Etim, V. N., Ogechi, N., & Egya, J. J. (2016). Ethnobotanical survey of medicinal plants used for the treatment of skin disease in Keffi, Nigeria. *American Journal of Phytomedicine and Clinical Therapeutics*, *4*(2), 73–90.

Mubeen, M., Iftikhar, Y., Abbas, A., Abbas, M., Zafar-ul-Hye, M., Sajid, A., & Bakhtawar, F. (2021). Yellow vein mosaic disease in okra (*Abelmoschus esculentus* L.): An overview on causal agent, vector and management. *Phyton*, *90*(6), 1573.

Mustafa, A. A., Fawibe, O. O., Ajiboye, A. A., & Agboola, D. A. (2014). Ethnobotanical survey of medicinal plants used in the treatment of diabetes in Irepodun local government area of Osun State, Nigeria. *Greener Journal of Biological Sciences*, *4*(2), 59–68.

Mutuli, G. P., & Mbuge, D. O. (2015). Drying characteristics and energy requirement of drying cowpea leaves and jute mallow vegetables. *Agricultural Engineering International CIGR Journal*, *17*(4), 265–272.

Nakaziba, R., Lubega, A., Ogwal-Okeng, J., & Alele, P. E. (2022). Phytochemical analysis, acute toxicity, as well as antihyperglycemic and antidiabetic activities of *Corchorus olitorius* L. leaf extracts. *The Scientific World Journal*, *2022*, Article 1376817.

Nasim, A., Hosen, S., & Bashar, M. (2017). Incidence of diseases in germplasms *Corchorus olitorius* L. and control of fungal pathogens. *Dhaka University Journal of Biological Sciences*, *26*(2), 189–198.

Ncube, B., Dlamini, B., & Beswa, D. (2022). Effect of common cooking and drying methods on phytochemical and antioxidant properties of *Corchorus olitorius* identified using liquid chromatography-mass spectrometry (LC-MS). *Molecules*, *27*, 90.

Ndamitso, M. M., Jagaba, M., Musah, J. I., Mann, A., Mohammed, E. B., & Umar, A. (2016). Chemical composition, antinutrient contents and functional properties of *Corchorus olitorius* and *Melochia corchorifolia*. *FUW Trends in Science and Technology Journal*, *1*, 96–100.

Nethavhani, M. (2021). *Corchorus olitorius*. Retrieved November 17, 2023, from https://pza.sanbi.org/corchorus-olitorius#

Nyadanu, D., Amoah, R. A., Kwarteng, A. O., Akromah, R., Aboagye, L. M., Adu-Dapaah, H., Dansi, A., Lotsu, F., & Tsama, A. (2017). Domestication of jute mallow (*Corchorus olitorius* L.): Ethnobotany, production constraints and phenomics of local cultivars in Ghana. *Genetic Resources and Crop Evolution*, *64*(6), 1313–1329.

Ogawa, M., Hayasi, K., Tomimori, S., Konishi, N., & Nakayama, O. (2002). Contents of strophanthidin glycosides and digitoxigenin glycosides in "moroheiya" (*Corchorus olitorius* L.) and its products, *Nippon Shokuhin Kagaku Kogaku Kaishi*, *49*, 282–287.

Ogwu, M. C., Osawaru, M. E., & Obahiagbon, G. E. (2017). Ethnobotanical survey of medicinal plants used for traditional reproductive care by Usen people of Edo State, Nigeria. *Malaya Journal of Biosciences*, *4*, 17–29.

Oyedele, D. J., Asonugho, C., & Awotoye, O. O. (2006). Heavy metals in soil and accumulation by edible vegetables after phosphate fertilizer application. *Electronic Journal of Environmental Agricultural and Food Chemistry*, *5*, 1446–1453.

Özdenefe, M. S., Muhammed, A., Süer, K., Güler, E., Aysun, H., & Takcı, M. (2018). Determination of antimicrobial activity of *Corchorus olitorius* leaf extracts. *Cyprus Journal of Medical Sciences*, *3*, 159–164.

Padulosi, S., Thompson, J., & Rudebjer, P. (2013). *Fighting poverty, hunger and malnutrition with neglected and underutilized species (NUS): Needs, challenges and the way forward*. Bioversity International.

Pal, D. K., Mandal, M., Senthilkumar, G. P., & Padhiari, A. (2006). Antibacterial activity of *Cuscuta reflexa* stem and *Corchorus olitorius* seed. *Fitoterapia*, *77*, 589–591.

Parvin, S., Marzan, M., Rahman, S., Das, A. K., Haque, S., & Rahmatullah, M. (2015). Preliminary phytochemical screening, antihyperglycemic, analgesic and toxicity studies on methanolic extract of aerial parts of *Corchorus olitorius* L. *Journal of Applied Pharmaceutical Science*, *5*(9), 68–71.

Peralta-Ruiz, Y., Rossi, C., Grande-Tovar, C. D., & Chaves-López, C. (2023). Green management of postharvest anthracnose caused by colletotrichum gloeosporioides. *Journal of Fungi*, *9*(6), 623.

POWO (Plants of the World Online). (2024). *Facilitated by the Royal Botanic Gardens, Kew, Richmond.* Retrieved December 23, 2023, from https://powo.science.kew.org/taxon/urn:lsid:ipni.org:names:212914-1
Purkait, A., Worede, R. E., Baral, D., Hazra, D. K., Panja, B. N., Biswas, P. K., & Kole, R. K. (2020). Development of nanoemulsion formulation of mustard oil, its chemical characterization and evaluation against post harvest anthracnose pathogens. *Indian Phytopathology, 73*(3), 449–460.
Ragasa, C. Y., Vivar, J. L. A., Tan, M. C. S., & Shen, C. C. (2016). Chemical constituents of *Corchorus olitorius* L. *International Journal of Pharmacognosy and Phytochemical Research, 8*, 2085–2089.
Rahman, S., Talukder, F. U., & Islam, N. (2022). Pesticidal effect of selected plant extracts on *Polyphagotarsonemus latus* (banks) infestation in *Corchorus olitorius* L. jute. *Malaysian Journal of Sustainable Agriculture, 6*, 22–28.
Ramadevi, D. (2014). Phytochemical examination of *Corchorus capsularis* leaves. *International Journal of Current Pharmaceutical Research, 4*, 42–46.
Ramadevi, D., & Ganapaty, S. (2012). Phytochemical examination of *Corchorus aestuans* (Tiliaceae) capsule. *International Journal of Pharmacognosy and Phytochemical Research, 4*, 54–58.
Ramasubramanian, T., Sarkar, S. K., Satpathy, S., Laha, S. K., De, R. K., Gotyal, B. S., & Biswas, C. (2010). Integrated pest and disease management for jute and allied fibres. *Indian Farming, 60*(3), 33–47.
Randall, R. P. (2017). *A global compendium of weeds* (3rd ed.). R.P. Randall.
Rao, N. K., Shahid, M., Al Shankiti, A., & Elouafi, I. (2014). Neglected and underutilized species for food and income security in marginal environments. *Acta Horticulturae, 1051*, 91–103.
Rao, V. G. (1978). Diseases of fibre crops in India. *Sydowia, 30*, 164–185.
Robles, J. A. C., & Nava, R. F. (2003). New records of *Corchorus* (Tiliaceae) for Mexico. *Sida Contributions to Botany, 20*(3), 1299–1309.
Ruffo, C. K., Birnie, A., & Tengnas, B. (2002). *Edible wild plants of Tanzania*. Relma Technical Handbook Series 27.
Shitanda, D., & Wanjala, N. V. (2006). Effect of different drying methods on the quality of jute (*Corchorus olitorius* L.). *Drying Technology, 24*(1), 95–98.
Shoaib, A., Khan, K. A., Awan, Z. A., Jan, B. L., & Kaushik, P. (2022). Integrated management of charcoal rot disease in susceptible genotypes of mungbean with soil application of micronutrient zinc and green manure (prickly sesban). *Frontiers in Microbiology, 13*, 899224.
Sivakumar, D., Gunes, T. N., & Romanazzi, G. (2021). A comprehensive review on the impact of edible coatings, essential oils, and their nano formulations on postharvest decay anthracnose of avocados, mangoes, and papayas. *Frontiers in Microbiology, 12*, 711092.
Soladoye, M. O., Chukwuma, E. C., Sulaiman, O. M., & Feyisola, R. T. (2014). Ethnobotanical survey of plants used in the traditional treatment of female infertility in Southwestern Nigeria. *Ethnobotany Research and Applications, 12*, 81–90.
Sonibare, M. A., Moody, J. O., Adesanya, E. O., Momordica, L., Seem, P. B., & Ocimum, L. (2009). Use of medicinal plants for the treatment of measles in Nigeria. *Journal of Ethnopharmacology, 122*, 268–272.
Srivastava, J. N., & Singh, A. K. (2020). *Diseases of field crops diagnosis and management: Volume 1: Cereals, small millets, and fiber crops*. Apple Academic Press.
Taiwo, B. J., Taiwo, G. O., Olubiyi, O. O., & Fatokun, A. A. (2016). Polyphenolic compounds with anti-tumour potential from *Corchorus olitorius* (L.) Tiliaceae, a Nigerian leaf vegetable. *Bioorganic and Medicinal Chemistry Letters, 26*, 3404–3410.
Tian, X., Yu, Q., Wu, W., & Dai, R. (2018). Inactivation of microorganisms in foods by ohmic heating: A review. *Journal of Food Protection, 81*(7), 1093–1107.
Tosoc, J. P. S., Nuñeza, O. M., Sudha, T., Darwish, N. H. E., & Mousa, S. A. (2021). Anticancer effects of the *Corchorus olitorius* aqueous extract and its bioactive compounds on human cancer cell lines. *Molecules, 26*, 6033.
Tripathi, A. N., Tiwari, S. K., & Behera, T. K. (2022). Postharvest diseases of vegetable crops and their management. In M. Ahiduzzaman (Ed.), *Postharvest technology: Recent advances, new perspectives and applications* (pp. 13–51). Intech Open. https://doi.org/10.5772/intechopen.101852
Trivedi, P. C. (2009). *Medicinal plants utilisation and conservation*. Aavishkar Publishers.
Tulio, A. Z., Ose, K., Chachin, K., & Ueda, Y. (2002). Effects of storage temperatures on the postharvest quality of jute leaves (*Corchorus olitorius* L.). *Postharvest Biology and Technology, 26*, 329–338.
Ujah, O. F., Ipav, S. S., Ayaebene, C. S., & Ujah, I. R. (2014). Phytochemistry and hepatoprotective effect of ethanolic leaf extract of *Corchorus olitorius* on carbon tetrachloride induced toxicity. *European Journal of Medicinal Plants, 4*(8), 882–892.
Uphof, J. C. F. (1968). *Dictionary of economic plants*. Verlag J. Cramer.

Van der Maesen, L. J. G. (2007). Opportunities for indigenous African vegetables. *Acta Horticulturae*, *752*, 391–396.

Van Rensburg, W. S. J., Van Averbeke, W., Slabbert, R., Faber, M., Van Jaarsveld, P. J., Van Heerden, I., Wenhold, F., & Oelofse, A. (2007). African leafy vegetables in South Africa. *Water SA*, *33*, 317–326.

Van Wyk, B.-E. (2011). The potential of South African plants in the development of new food and beverage products. *South African Journal of Botany*, *77*, 857–868.

Van Wyk, B.-E., & Gericke, N. (2018). *Peoples's plants: A guide to useful plants of Southern Africa*. Briza Publications.

Van't Hag, L., Danthe, J., Handschin, S., Mutuli, G. P., Mbuge, D., & Mezzenga, R. (2020). Drying of African leafy vegetables for their effective preservation: The difference in moisture sorption isotherms explained by their microstructure. *Food and Function*, *11*(1), 955–964.

Vega-Galvez, A., Gomez-Perez, L. S. Zepeda, F., Vidal, R. L., Grunenwald, F., Mejías, N., Pasten, A., Araya, M., & Ah-Hen, K. S. (2023). Assessment of bio-compounds content, antioxidant activity, and neuroprotective effect of red cabbage (*Brassica oleracea* var. *capitata rubra*) processed by convective drying at different temperatures. *Antioxidants*, *12*, 1789.

Velempini, P., Riddoch, I., & Batisani, N. (2003). Seed treatments for enhancing germination of wild okra (*Corchorus olitorius*). *Experimental Agriculture*, *39*, 441–447.

Weimer, N., Kumaga, F. K., & Nkansah, G. O. (2023). Effect of seed rate, planting method and location on the yield of *Corchorus olitorius* in the Greater Accra Region. *African Journal of Hortricultural Science*, *22*, 19–34.

Westphal-Stevel, J. M. C. (1986). Local vegetables in Cameroon: *Corchorus* species used as a vegetable. *Acta Horticulturae*, *182*, 423–425.

Wild, H. (1963). Tiliaceae. In A. W. Exell, A. Fernandes, & H. Wild (Eds.), *Flora Zambesiaca 2 (Part 1)* (pp. 33–91). Crown Agents for Oversea Governments and Administrations.

Wild, H. (1984). Tiliaceae. In O. A. Leistner (Ed.), *Flora of Southern Africa 21 part 1* (pp. 1–44). Botanical Research Institue.

Wild, H., Biegel, H. M., & Mavi, S. (1972). *A Rhodesian botanical dictionary of African and English plant names*. Government Printer.

Xiong, H. P. (2008). *Breeding sciences of bast and leaf fiber crops*. Chinese Agricultural Science and Technology Press.

Yakoub, A. R. B., Abdehedi, O., Jridi, M., Elfalleh, W., Nasri, M., & Ferchichi, A. (2018). Flavonoids, phenols, antioxidant, and antimicrobial activities in various extracts from Tossa jute leave (*Corchorus olitorus* L.). *Industrial Crops and Products*, *118*, 206–213.

Yan, Y.-Y., Wang, Y.-W., Chen, S.-L., Zhuang, S.-R., & Wang, C.-K. (2013). Anti-inflammatory effects of phenolic crude extracts from five fractions of *Corchorus olitorius* L. *Food Chemistry*, *138*, 1008–1014.

Yoshikawa, M., Shimada, H., Saka, M., & Yoshizumi, S. (1997). Medicinal food stuffs. V. 1 Moroheiya. (1): Absolute stereostructures of corchoionosides A, B, and C, histamine release inhibitors from the leaves of Vietnamese *Corchorus olitorius* L. (Tiliaceae). *Chemical and Pharmaceutical Bulletin*, *45*, 464–469.

Zeghichi, S., Kallithraka, S., & Simopoulos, A. P. (2003). Nutritional composition of molokhia (*Corchorus olitorius*) and stamnagathi (*Cichorium spinosum*). *World Review of Nutrition and Dietetics*, *91*, 1–21.

Zeid, A. H. S. A. (2002). Stress metabolites from *Corchorus olitorius* L. leaves in response to certain stress agents. *Food Chemistry*, *76*, 187–195.

Zhang, L., Ibrahim, A. K., Niyitanga, S., Zhang, L., & Qi, J. (2019). Jute (*Corchorus* spp.) breeding. In J. M. Al-Khayri, S. M. Jain, & D. V. Johnson (Eds.), *Advances in plant breeding strategies: Industrial and food crops* (pp. 85–113). Springer.

10 Exploring the Nutrient Profiles in Overlooked Food Plants for Enhanced Food Security and Nutrition

Callistus Bvenura and Learnmore Kambizi
Department of Horticulture, Faculty of Applied Sciences, Cape Peninsula University of Technology, Bellville 7535, South Africa
*Corresponding author: callistusb@gmail.com

10.1 INTRODUCTION

Wild vegetables have been a part of the human diet since the beginning of time. Human consumption of these vegetables and the ability of the species to meet nutrient needs have been documented for a long time. It is estimated that, at least, 1 billion people globally still include wild foods in their diets (Aberoumand & Deokule, 2010). In South Africa, Wehmeyer and Rose (1983) identified more than 100 plant species that are being used as wild vegetables. In many places of the developing world, there has been a low trend in the consumption of wild vegetables. This reduction in the consumption of wild vegetables has been attributed to seasonal availability and culture in Nigeria, while in South Africa, culture, taste, and affordability have been cited as the major causes (WHO, 2003; Hart et al., 2005; Shackleton et al., 2009; Faber et al., 2010). Wild vegetables are usually consumed by the rural populace as supplements to their diet since the vegetables usually naturally grow on cultivated or fallow fields, thereby making them easily accessible. However, the urban populace pays less attention to wild vegetables in favor of the conventional ones because of easy accessibility of the latter. The use of herbicides and pesticides, as well as the excessive cultivation of the fields, has led to the decline in the availability of wild vegetables and the subsequent decline in their knowledge (Odhav et al., 2007). Also, the perception, especially among young people, that such vegetables are foods for the poor causes a lack of interest in the cultivation and nutritional importance of these plants (Vorster et al., 2007). Yet studies have revealed that wild vegetables have numerous beneficial nutritional values often better than those of the domesticated exotic breeds like spinach and cabbage (Odhav et al., 2007; Lewu & Mavengahama, 2010). The use of wild vegetables could be lost with time if their knowledge, especially on the identification, nutrient value, and methods of preparation and preservation, is not passed down to the younger generations and properly documented. Therefore, this study was conducted to investigate the nutritional compositions of some of the wild vegetables growing in the Eastern Cape Province of South Africa, to create an awareness of some important aspects of these neglected food plants. The inclusion of these vegetables in the diet could help alleviate the food and nutrition security concerns being faced by the inhabitants of the province.

 DOI: 10.1201/9781003469766-10

10.2 MATERIALS AND METHODS

10.2.1 Plant Collection and Preparation

Fourteen wild vegetables were collected fresh from the Mbashe and Nkonkobe municipalities of the Eastern Cape during the rainy season, when the wild vegetables were vegetatively growing in home gardens and the fields. The plants were initially identified by their vernacular names (Xhosa) and later validated at the University of Fort Hare Herbarium (Bvenura & Afolayan, 2014). Young and tender shoots were plucked from the mother plant, as practiced locally, stored in khaki paper sampling bags, and transported to the laboratory, where they were washed thoroughly with distilled water. Leaves of the same plant from the two municipalities were combined to make a single composite sample, oven-dried at 40°C to a constant weight, and homogenized using a 2 mm sieve Polymix (PX-MFC 90 D) electric grinder, after which they were sealed in polyethylene bags and stored in the refrigerator at 4°C until needed for the various analysis.

10.2.2 Proximate Analysis

10.2.2.1 Determination of Ash Content

About 5 g of the powdered plant leaf sample was weighed into a previously weighed crucible. This was incinerated in an E-Range muffle furnace with TOHO P4 program at 550°C for 12 h. The final weight of the sample was used to calculate the ash content as follows:

ash content (%) = [final weight of sample after incineration (g) /(5 g)] × 100%

(Antia et al., 2006)

10.2.2.2 Determination of Crude Lipid

Crude lipid was determined as described by Antia et al. (2006). About 5 g of the powdered sample was measured into a 250 ml beaker; 100 ml of diethyl ether was added, covered with aluminum foil, and shaken in an orbital shaker for 24 h. Filtration followed this process, and the supernatant was decanted. Another 100 ml of diethyl ether was added to the residue and shaken for another 24 h. The residue obtained after filtration was the lipid-free sample and was calculated as:

crude lipid = (weight of sample after diethyl ether extraction/initial sample weight) × 100%

10.2.2.3 Determination of Crude Fiber

The AOAC (1984) method was used in estimating the crude fiber. About 5 g of the powdered sample was weighed into a beaker and digested in 100 ml of 1.25% sulfuric acid for 30 min. The acid-digested sample was allowed to cool, and then filtered. The residue was collected into a beaker and further digested in 100 ml of 1.25% sodium hydroxide. The sample was filtered, and the residue dried in an oven at 100°C to a constant weight. The dried residue was then incinerated in a muffle furnace for 24 h at 550°C. The crude fiber was obtained from the loss in weight on ignition of dried residue remaining after the digestion of fat-free samples (AOAC, 1984):

% fiber = (loss of weight on ignition/initial sample weight) × 100%

10.2.2.4 Determination of Vitamin C

10.2.2.4.1 Preparation of Iodine Solution

The iodometric titration method was used to determine the vitamin C content. About 5 g of potassium iodide and 0.268 g of potassium iodate were dissolved in 200 ml distilled water in a 400 ml beaker, followed by the addition of 30 ml of 3 M sulfuric acid. The mixture was poured into a 500 ml graduated cylinder and then diluted to a final volume of 500 ml with distilled water. Vitamin C standard solution was prepared by dissolving 0.250 g vitamin C in 100 ml water. This was made up to 250 ml with distilled water.

10.2.2.4.2 Standardization of Iodine with Vitamin C Standard Solution

About 25 ml of vitamin C standard solution was measured into a 125 ml Erlenmeyer flask, following which ten drops of 1% starch solution were added as the indicator. This was titrated against the acidified potassium iodide iodine solution until the end point (the first blue color that showed after at least 20 s of swirling) was reached.

10.2.2.4.3 Vitamin C Determination in the Samples

About 5 g of fresh leaf samples were macerated in 20 ml of distilled water. The mixture was filtered and the filtrate collected in a 50 ml volumetric flask and made to the mark with distilled water. About 10 ml of the sample solution was transferred into an Erlenmeyer flask, and ten drops of 1% starch were added and titrated against the acidified potassium iodide solution.

10.2.2.5 Determination of Phytate Components

Phytate was determined according to the method of Wheeler and Ferrel (1971). A sample of finely ground plant material measuring 4 g was soaked in 100 ml of 2% hydrochloric acid for 3 h and then filtered through Whatman No. 43 filter paper. About 25 ml of the filtrate was measured into a conical flask, and 5 ml of 0.3% ammonium thiocyanate solution was added as indicator, followed by the addition of 53.5 ml distilled water. This was titrated against a 1,000 ppm standard iron (III) chloride solution until a brownish yellow color persisted for 5 min. The phytate content was calculated from the iron determinations, assuming a 4:6 iron to phytate molecular ratios, and multiplied by a constant of 3.55 (Vijayakumari et al., 1996).

10.2.2.6 Determination of Protein

About 0.5 g of finely ground vegetable samples was placed in dry, clean digestion tubes, and 5 ml of the digestion mixture comprising one part $HClO_4$ + two parts HNO_3 was added. This mixture was digested at 230°C on a digestion block for 70 min, allowed to cool down, and made up to 100 ml volume with distilled water. The concentration of nitrogen was then determined using the Inductively Coupled Plasma Optical Emission Spectrometer (ICP OES). Percentage crude protein was obtained by multiplying the nitrogen value by a factor of 6.25 (AGRILASA, 2008).

10.2.3 Mineral Analysis

Macro-minerals (phosphorus, potassium, calcium, sodium, and magnesium) and micro-minerals (copper, iron, zinc, and manganese) were determined using the method described for the determination of nitrogen.

10.3 STATISTICAL ANALYSIS

Data of the nutrient concentrations of various wild vegetables were subjected to statistical analysis using MNITAB Release 12. A one-way analysis of variance was used to compare the means of various nutrient concentrations among the wild vegetables. Means were segregated using Duncan's multiple range test. The means were treated as significantly different at $p < 0.05$.

10.4 RESULTS

10.4.1 Proximate Composition

The proximate compositions of the wild vegetables gathered from Mbashe and Nkonkobe municipalities in the Eastern Cape Province of South Africa varied significantly ($p < 0.05$) among the vegetables (Table 10.1). Crude fiber ranged between 0.91 and 6.79 g/100 g, crude protein between 0.47 and

1.53 g/100 g, crude lipid between 0.14 and 3.83 g/100 g, phytate between 1.09 and 8.92 g/100 g, and ash content between 0.38 and 1.79 g/100 g. *Tulbaghia violacea* had the highest protein, and *Sonchus oleraceus* the highest lipid content. The lowest protein content was recorded in *Chenopodium murale*, while the highest fiber and ash contents were recorded in *Solanum nigrum* and *Chenopodium murale*, respectively. The lowest ash values were observed in *Solanum nigrum*.

10.4.2 Mineral Composition

The mineral compositions of the 14 vegetables significantly differed among the vegetables, as shown in Table 10.2. The mean concentration of nutrients in all the vegetables decreased in the order vitamin C > K > Ca > Mg > Na > P > Fe > Zn > Mn > Cu. *Chenopodium murale* contained the highest K, Mg, and P, while *Physalis peruviana* contained the highest concentration of Fe and vitamin C. Cu was remarkably low in all the wild vegetables.

TABLE 10.1
Proximate Compositions (g/100 g) of 14 Vegetables in Mbashe and Nkonkobe Municipalities in the Eastern Cape Province of South Africa

Vegetable Species	†Phytate	Crude Protein	Crude Lipid	Crude Fiber	Ash
Bidens pilosa L.	2.4±6.51^{c}	1.19±0.03^{ab}	1.60±0.22^{d}	2.14±0.32^{c}	0.38±0.03^{b}
Centella coriacea Nannfd.	2.31±1.31^{a}	1.1±0.05^{ab}	0.52±0.02^{b}	1.09±0.03^{a}	1.74±0.05^{c}
Chenopodium album L.	8.92±0.38^{a}	0.48±0.01^{a}	1.15±0.19^{b}	1.68±0.02^{ac}	1.38±0.04^{d}
Chenopodium murale L.	3.07±4.30^{b}	0.47±0.01^{a}	3.50±0.03^{a}	1.35±0.01^{a}	1.79±0.02^{c}
Cotula heterocarpa DC.	2.23±0.57^{ab}	1.02±0.14^{ab}	0.84±0.05^{b}	1.01±0.02^{a}	1.44±0.01^{d}
Galinsoga parviflora Cav.	2.98±0.81^{c}	0.94±0.08^{ab}	2.51±0.46^{e}	1.66±0.29^{c}	0.45±0.03^{b}
Hypochaeris radicata L.	1.09±0.81^{a}	0.82±0.07^{ab}	0.14±0.03^{f}	1.00±0.01^{a}	0.80±0.16^{a}
Physalis peruviana L.	4.19±0.85^{ab}	1.02±0.14^{ab}	1.93±0.51^{de}	1.53±0.03^{ac}	1.39±0.24^{d}
Rumex obtusifolius L.	4.86±1.05^{a}	0.48±0.01^{a}	0.35±0.11^{c}	1.08±0.01^{a}	1.24±0.01^{d}
Solunum nigrum L.	2.34±0.50^{a}	0.56±0.02^{a}	0.58±0.02^{b}	6.79±0.01^{b}	0.39±0.02^{b}
Sonchus oleraceus L.	1.50±5.37^{d}	1.09±0.06^{ab}	3.83±0.04^{a}	1.11±0.02^{a}	0.77±0.05^{a}
Stellaria media L.	4.64±0.61^{b}	0.51±0.00^{a}	2.56±0.06^{e}	1.26±0.01^{a}	1.55±0.01^{d}
Tulbaghia violacea Harv.	1.55±2.14^{c}	1.53±0.13^{ab}	0.64±0.04^{b}	0.91±0.04^{a}	1.29±0.0.02^{d}
Urtica urens L.	2.51±0.90^{a}	0.87±0.13^{ab}	0.64±0.02^{b}	2.35±0.02^{c}	1.67±0.0.01^{cd}

Note: Different letters down the same column represent significant differences at $p < 0.05$.

TABLE 10.2
Mineral Compositions (g/100 g) of 14 Vegetables in Mbashe and Nkonkobe Municipalities in the Eastern Cape Province of South Africa

WVS	Mineral Element								
	Ca	Mg	K	Na	P	Fe	Zn	Mn	Vit C
B. pilosa	14.70 ± 3.48^{e}	3.46 ± 1.51^{cb}	71.95 ± 74.00^{cd}	0.28 ± 1.99^{a}	2.57 ± 0.67^{f}	0.87 ± 0.53^{c}	0.28 ± 2.73^{b}	0.13 ± 0.06^{ab}	0.042 ± 0.06^{a}
C. coriacea	14.39 ± 1.33^{e}	16.69 ± 6.93^{f}	110.88 ± 54.30^{fg}	2.54 ± 0.98^{c}	4.29 ± 0.11^{g}	0.26 ± 0.19^{a}	0.15 ± 3.76^{a}	0.15 ± 0.04^{a}	0.125 ± 0.06^{a}
C. album	10.40±2.70^{cd}	2.69±0.20^{cb}	16.80±21.50^{a}	22.01 ± 1.86^{g}	2.10±0.19^{a}	0.68±0.19^{b}	0.12±0.18^{a}	0.21±1.80^{b}	0.050 ± 0.03^{a}
C. murale	7.27±4.66^{ab}	2.72±0.66^{b}	61.65±29.40^{c}	12.44± 2.78^{e}	3.56±0.26^{e}	0.76±0.40^{bc}	0.16±0.31^{a}	0.07±0.03^{a}	0.058 ± 0.06^{a}

(Continued)

TABLE 10.2 (Continued)

WVS	Mineral Element								
C. heterocarpa	11.34±4.26^{d}	22.79±19.64^{h}	125.97±44.20^{g}	5.37±2.14^{cd}	4.76±0.56^{i}	0.59±0.73^{a}	0.14±0.23^{a}	0.05±0.51^{a}	0.117 ± 0.00^{b}
G. parviflora	24.77±3.41^{g}	4.04±1.51^{c}	53.31±114.00^{b}	0.38±1.13^{a}	4.51±0.74^{h}	0.53±0.07^{b}	0.14±0.30^{a}	0.13±0.08^{ab}	0.047 ± 0.15^{d}
H. radicata	9.14±6.80^{c}	4.10±0.17^{c}	52.00±11.10^{b}	19.54± 6.00^{f}	2.57±0.55^{b}	0.42±0.15^{b}	0.28±0.22^{b}	0.07±0.06^{a}	0.048 ± 0.02^{a}
P. peruviana	13.11±7.50^{e}	9.70±7.21^{e}	84.63±85.20^{cd}	0.25±2.82^{a}	2.89±0.91^{c}	2.60±2.29^{e}	0.34±0.22^{c}	0.10±0.49^{a}	0.225 ± 0.03^{a}
R. obtusifolius	8.03±3.59^{b}	4.93±1.56^{c}	73.28±63.10^{c}	0.86±0.65^{a}	4.59±0.71^{h}	0.44±0.80^{ab}	0.14±0.29^{a}	0.06±0.85^{a}	0.108 ± 0.10^{c}
S. nigrum	6.70±6.98^{a}	6.05±1.56^{d}	121.52±73.80^{g}	3.83±1.79^{c}	4.43±1.18^{g}	0.77±0.48^{cb}	0.30±3.91^{bc}	0.08±0.29^{a}	0.042 ± 0.10^{a}
S. oleraceus	16.98±4.59^{f}	3.72±1.46^{cb}	76.91±80.40^{d}	0.65±1.09^{ab}	3.88±0.70^{f}	0.43±0.66^{b}	0.12±0.50^{a}	0.07±0.02^{a}	0.014 ± 0.04^{a}
S. media	12.25±4.78^{de}	4.15±0.54^{c}	50.6±5.20^{b}	18.73±2.37^{f}	3.17±0.74^{d}	0.56±0.59^{b}	0.14±0.39^{a}	0.04±0.41^{a}	0.058 ± 0.15^{b}
T. violacea	7.36±3.98^{ab}	1.54±0.80^{a}	80.44±38.70^{cd}	3.72±2.39^{c}	3.65±1.54^{e}	0.21±0.19^{a}	0.19±0.48^{a}	0.02±0.02^{a}	0.162 ± 0.06^{b}
U. urens	34.84±3.65^{i}	4.25±1.58^{c}	66.99±12.90^{c}	0.41±1.20^{a}	4.68±0.56^{ih}	0.35±0.17^{ab}	0.16±0.47^{a}	0.10±0.04^{a}	0.045 ± 0.02^{a}
Range	6.70–34.84	1.54–22.79	50.6–125.97	0.25–18.73	2.10–4.76	0.21–2.60	0.12–0.34	0.04–0.60	0.01–0.23

Note: Different letters down the same column represent significant differences at $p < 0.05$.
*Cu ranged between 0.01 and 0.02 g/100 g in all the samples.

10.5 DISCUSSION

10.5.1 Proximate Composition

The proximate analysis revealed that all the 14 wild vegetables are rich in the various nutrients investigated, but in varying proportions. Ash, protein, lipid, and fiber contents in *Sonchus oleraceus* were lower than earlier reported by Jimoh et al. (2011). Odhav et al. (2007) investigated the proximate composition of 20 wild plants and reported higher values for protein (3–7 g/100 g) and ash (1.74–4.91 g/100 g), but lipid (0.2–2.7 g/100 g) and fiber (1.21–2.92 g/100 g) were within the range of the present study. *Sonchus oleraceus* and *Chenopodium murale* showed appreciable lipid contents in this study. Lyimo et al. (2003) reported a lipid range of between 0.1 and 1.0 g/100 g in 30 wild vegetables, but lower than observed in the current study. According to Antia et al. (2006), lipids increase the palatability of food by absorbing and retaining flavors. A diet providing 1–2% of its calorific energy as fat is said to be sufficient for humans, as excess fat consumption is implicated in cardiovascular disorders, such as atherosclerosis, cancer, and aging (Sharma et al., 2012). The considerable amount of lipids in some of these vegetables in relation to the NHMRC (2005) values would therefore improve the palatability of the vegetables and reduce the risk of some diseases. Lyimo et al. (2003) reported a protein range between 0.6 and 5.0 g/100 g in 30 wild vegetables; again, this was higher than the current study, with a range of between 0.47 and 1.53 g/100 g. Protein is essential for growth and repair of muscles, bones, skin, tendons, ligaments, hair, and eyes, amongst other tissues in humans. The protein values from this study indicate that the wild vegetables can supply a fraction of the recommended daily intake (RDI) values (NHMRC, 2005). Children need to consume at least 150 g of vegetables per day to harness the required nutrients (NHMRC, 2003). For example, *Sonchus oleraceus*, *Tulbaghia violacea*, and *Bidens pilosa* can supply 5.0, 5.5, and 7.1%, respectively, of children's RDI values. A high ash content in plants reflects high mineral content (Aberoumand & Deokule, 2010). The high ash content of *Chenopodium murale* compared to other vegetables in this study is an indication of high mineral value of the vegetable. High-fiber diets have been linked with lower serum cholesterol concentrations, lower risk of coronary heart disease, reduced blood pressure, enhanced weight control, reduced risk of certain forms of cancer, and improved gastrointestinal function (Anderson et al., 1994). *Solanum nigrum*, *Urtica urens*, and *Bidens pilosa* can supply about 67.9, 23.5, and 21.4% of the required daily amount in men if a quantity of about 300 g is consumed. The concentration of fiber would also presumably not lead to high-fiber-related conditions and diseases. Phytate is an antinutrient which has a strong ability to chelate multivalent metal ions, especially Zn, Ca, and Fe, leading to their poor bioavailability (Gupta et al., 2006). Although the presence of phytate could decrease mineral absorption in humans, this antinutrient is said to be heat-labile (Akwaowa et al., 2000). It is therefore conceivable that the high phytate content in *Chenopodium album* (8.92 g/100 g) and *Physalis peruviana* (4.19 g/100 g) will be significantly reduced during cooking. Additionally, leafy vegetables are generally considered to be superior sources of mineral supplements and therefore would ideally lower the effect antinutrients would have on the availability of some nutrients (Odhav et al., 2007).

10.5.2 Mineral and Vitamin C Composition

Vitamin C was remarkably high in *Physalis peruviana* (0.225 g/100 g), which makes the vegetable a better source of vitamin C compared to guava (0.188 g) and orange (0.07 g) fruits, as well as some leafy vegetables, such as broccoli (0.039 g) (Brand et al., 1982). The results of the current study are comparable with what was reported by Lyimo et al. (2003) in a study of 30 wild vegetables in Tanzania. These authors found that *Galinsoga parviflora*, *Bidens pilosa*, and *Solunum nigrum*, respectively, contained 0.054, 0.059, and 0.234 g/100 g vitamin C. *Sonchus oleraceus*, which had the lowest vitamin C concentration, has the potential to supply about 31% of vitamin C RDI, while *Physalis peruviana*, which had the highest concentration, can supply about 500% of the required vitamin C RDI in adults. Vitamin C prevents tissue damage and aids in the recovery from several ailments and diseases, including colds, cough, influenza, sores, wounds, and skin diseases, among others (Ogunlesi et al., 2010). According to Lopez and Martos (2004), vitamin C improves Fe availability, therefore reducing the risk of iron-deficiency anemia. Potassium was also remarkably high in the wild vegetables, while Cu was low in the present study. In studies previously conducted by Bvenura and Afolayan (2012) in the Nkonkobe municipality, Cu was low in cabbage, spinach, and carrot. The previous and current results possibly indicate the low levels of the mineral in the soil of the study area. Jimoh et al. (2011) and Kawada et al. (2002) reported slightly lower mineral values compared to the present study, except for Mn, Cu, and Zn, which were slightly higher. In Iran, lower levels of mineral nutrients were found in some wild vegetables as compared to the present study (Aberoumand & Deokule, 2010). Research done in Akure, Nigeria, indicated a slightly higher concentration of mineral nutrients compared to this study (Aletor et al., 2002). In KwaZulu Natal, South Africa, Odhav et al. (2007) found high levels of Ca, P, and Mg with many wild vegetables exceeding 1,000 mg/100 g. Variations in the nutrient compositions of edible plants are influenced by various factors, including farming practices, prevailing environmental conditions, including soil manipulation using organic and inorganic fertilizers, and the age of the plants at harvest (Nordeide et al., 1996). In addition, some minerals, such as Zn, decrease with advancing plant age, while other minerals, such as Fe and Mn, reportedly increase with increasing plant age (Tiffin, 1971). In this study, vegetables were collected from fields where the soil has been manipulated by the addition of organic and mineral fertilizers, and only young fresh plants were collected. These factors may have contributed to the amount of minerals in the vegetables. The findings of this study also indicated that all vegetable samples had Na–K ratios of less than 1. Yang et al. (2011) linked a high Na–K ratio with increased risk of cardiovascular diseases, leading to mortality. These authors further reported that a high Na intake and a low K intake are linked to high blood pressure. The consumption of these vegetables would therefore not only help lower blood pressure, especially among the elderly, but also boost their immune system.

In sub-Saharan Africa, South Africa has the fourth highest number of people living with HIV/AIDS, with an estimated 5.6 million (17.8%) people infected. The Eastern Cape Province is the sixth most affected, with 9% of the population living with the virus (UN/AIDS, 2010). Researchers around the world have strongly linked HIV to nutrition; they view nutrition as a fundamental intervention in boosting the immune system of the infected in the early stages and ongoing treatment of the disease (Elbein, 1995; Tinnerello, 1998; Charles, 2009). Among other nutrients, people living with HIV are more susceptible to low levels of Zn and Fe in their blood (Barnett, 2006). As shown by the appreciable amounts of these minerals in this study, a wild vegetable–inclusive diet, for example, comprising *T. violacea*, *S. oleraceus*, and *S. nigrum* would presumably improve the nutrition of HIV/AIDS patients.

The high levels of P observed in vegetables from this study may also increase the mineral's availability in human nutrition. Mn is an activator and constituent of several enzymes and occurs in very low quantities in humans, though this mineral's importance cannot be overlooked (Medeiros & Wildman, 2000). The Mn content observed in this study, though low, may supplement its presence in the diet. In humans, Mg is a critical co-factor in more than 300 enzymatic reactions in the body, while in plants, the most recognized role of the element is in photosynthesis, where the element must be incorporated into the chlorophyll molecule before chlorophyll is effective at gathering light for photosynthetic carbon reduction reactions (Schachter, 2012; Wilkinson et al., 1990). Therefore, the abundance of Mg in wild vegetables is essential in ensuring a healthy plant and the subsequent availability of other minerals and their supply in the human diet. In relation to the RDI values and the NMHRC (2003) recommended quantities for children, the mineral elements can be sufficiently supplied per 150 g cooked portion. However, one of the drawbacks with wild vegetables is their seasonal availability. These nutritionally rich foods are usually available during rainy season, but this shortfall can be overcome by gathering in large quantities when available, drying, and storing for off-season consumption. Furthermore, a more sustainable solution that may ensure a continuous fresh supply of these wild foods is to cultivate them in home gardens.

10.6 CONCLUSION AND PROSPECTS

The results of this study reveal that the leaves of the 14 wild vegetables are rich in minerals. However, the vegetables differ in nutrient contents. Regardless, *Sonchus oleraceus*, which had the lowest vitamin C concentration, has the potential to supply about 31% of vitamin C RDI, while *Physalis peruviana*, which had the highest concentration, can supply about 500% of the required vitamin C RDI in adults per 300 g of serving. *Solanum nigrum*, *Urtica urens*, and *Bidens pilosa* can supply about 67.9, 23.5, and 21.4%, respectively, of the required daily amount of fiber in men per 300 g serving. *Sonchus oleraceus*, *Tulbaghia violacea*, and *Bidens pilosa* can supply 5.0, 5.5, and 7.1%, respectively, of children's RDI values for protein per 150 g of serving. When compared to other vegetables, *Chenopodium murale* is a good source of Mg, K, and P, while *Physalis peruviana* is a good source of Cu and Fe. Furthermore, *Physalis peruviana* is a good source of vitamin C as well as protein, and *Solanum nigrum* is a good source of fiber. Mixing these vegetables when cooking them, as is practiced locally, has the potential to meet the human requirements of nutrients for growth and development daily, which, in turn, helps overcome nutritional deficiency problems which are prevalent especially in poor rural areas. These vegetables are therefore recommended for consumption alone or in combination with other vegetables.

10.7 ACKNOWLEDGMENTS

The authors are grateful to the National Research Foundation of South Africa for funding this work.

REFERENCES

Aberoumand, A., & Deokule, S. S. (2010). Preliminary studies on proximate and mineral composition of Marchubeh stem (*Asparagus officinalis*) vegetable consumed in the Behbahan of Iran. *World Journal of Dairy & Food Sciences*, 9(2), 127–130.

AGRILASA. (2008). *Agri laboratory association of Southern Africa. Method 6.1.2: Wet ashing.* Plant and Feed Analysis Handbook.

Akwaowa, E. U., Ndon, B. A., & Etuk, E. U. (2000). Minerals and antinutrients in fluted pumpkin (*Telfairia occidentalis* Hook f.). *Food Chemistry, 70*, 235–240.

Aletor, O., Oshodi, A. A., & Ipinmoroti, K. (2002). Chemical composition of common leafy vegetables and functional properties of their leaf protein concentrates. *Food Chemistry, 78*, 63–68.

Anderson, J. W., Smith, B. M., & Gustafson, N. J. (1994). Health benefits and practical aspects of high-fibre diets: Review article. *American Journal of Clinical Nutrition, 59*, 1242S–1247S.

Antia, B. S., Akpan, E. J., Okon, P. A., & Umoren, I. U. (2006). Nutritive and anti-nutritive evaluation of sweet potatoes (*Ipomoea batatas*) leaves. *Pakistan Journal of Nutrition, 5*(2), 166–168.

Association of Official Analytical Chemists (AOAC). (1984). *Official methods of analysis* (14th ed.). Author.

Barnett, T. (2006). HIV/AIDS, nutrition and food security: Looking to future challenges. In S. Gillespie (Ed.), *AIDS, poverty and hunger: Challenges and responses* (pp. 341–348). (Proceedings of the International Conference on HIV/AIDS and food and nutrition Security, Durban, South Africa, April 14–16 2005). International Food Policy Research Institute.

Brand, J. C., Cherikoff, V., Lee, A., & Truswell, A. S. (1982). An outstanding food source of vitamin C. *Lancet, 2*(8303), 873.

Bvenura, C., & Afolayan, A. J. (2012). Analysis of heavy metal contamination in home gardens in Alice, Nkonkobe Municipality, Eastern Cape, South Africa. *South African Journal of Science, 108*(9/10), 1–6.

Bvenura, C., & Afolayan, A. J. (2014). Ethnobotanical survey of wild vegetables in Mbashe and Nkonkobe municipalities, Eastern Cape Province, South Africa. *Acta Botanica Gallica: Botany Letters, 161*(2), 189–199.

Charles, P. C. (2009). Calcium absorption and calcium availability. *Journal of Internal Medicine, 231*(2), 161–168.

Elbein, R. C. (1995). Nutrition and HIV infection. A continuum of care. *Journal of the American Podiatric Medical Association, 85*(2), 434–438.

Faber, M., Oelofse, A., van Jaarsveld, P. J., Wenhold, F. A. M., & van Rensburg, J. W. S. (2010). African leafy vegetables consumed by households in the Limpopo and KwaZulu-Natal provinces in South Africa. *South African Journal of Clinical Nutrition, 23*(1), 30–38.

Gupta, S., Lakshimi, J. A., & Prakash, J. (2006). In vitro bioavailability of calcium and iron from selected green leafy vegetables. *Journal of the Science of Food and Agriculture, 86*(13), 2147–2152.

Hart, R. E. C., Azubuike, C. U., Barmalaa, I. S., & Achinewhu, S. C. (2005). Vegetable consumption pattern of households in selected areas of the Old Rivers State in Nigeria. *African Journal of Food, Agriculture and Nutrition Development, 5*(1), 1–19.

Jimoh, F. O., Adedapo, A. A., & Afolayan, A. J. (2011). Comparison of nutritive value, antioxidant and antibacterial activities of *Sonchus asper* and *oleracea*. *Records of Natural Products, 5*(1), 29–42.

Kawada, T., Lee, Y., Suzuki, S., & Rivai, I. F. (2002). Copper in carrots by soil type and area in Japan: A baseline study. *Journal of Trace Elements in Medicine and Biology, 16*(3), 179–182.

Lewu, F. B., & Mavengahama, S. (2010). Wild vegetables in Northern Kwa-Zulu Natal, South Africa: Current status of production and research needs. *Scientific Research Essays, 5*(20), 3044–3048.

Lopez, M. A. A., & Martos, F. C. (2004). Iron availability: An updated review. *International Journal of Food Sciences and Nutrition, 55*(8), 597–606.

Lyimo, M., Temu, R. P. C., & Mugula, J. K. (2003). Identification and nutrient composition of indigenous vegetables of Tanzania. *Plant Foods for Human Nutrition, 58*, 85–92.

Medeiros, D. M., & Wildman, R. E. C. (2000). *Advanced human nutrition* (2nd ed.). Jones and Bartlet Learning, LLC.

NHMRC. (2003). *Dietary guidelines for children and adolescents in Australia: Incorporating the infant feeding guidelines for health workers.* Author.

NHMRC. (2005). *Nutrient Reference values for Australia and New Zealand: Including recommended dietary intakes.* Author.

Nordeide, M. B., Hatloy, A., Folling, M., Lied, E., & Oshaug, A. (1996). Nutrient composition and nutritional importance of green leaves and wild foods in an agricultural district, Koitiala, in Southern Mali. *International Journal of Food Science and Nutrition, 47*, 455–478.

Odhav, B., Beekrum, S., Akula, U., & Baijnath, H. (2007). Preliminary assessment of nutritional value of traditional leafy vegetables in KwaZulu-Natal, South Africa. *Journal of Food Composition and Analysis, 20*, 430–435.

Ogunlesi, M., Okiei, W., Azez, H., Obakachi, V., Osunsanmi, M., & Nkenchor, G. (2010). Vitamin C contents of tropical vegetables and foods determined by voltammetric and titrimetric methods and their relevance to the medicinal uses of the plants. *International Journal of Electrochemical Science, 5*, 105–115.

Schachter, S. M. (2012). *The importance of magnesium to human nutrition*. Retrieved January 26, 2012, from www.healthy.net/scr/article.aspx?Id=541

Shackleton, C. M., Pasquini, M., & Drescher, A. W. (2009). *African indigenous vegetables in urban agriculture*. Earthscan Publishers.

Sharma, N., Gupta, P. C., & Rao, C. V. (2012). Nutrient content, mineral content and antioxidant activity of *Amaranthus viridis* and *Moringa oleifera* leaves. *Research Journal of Medicinal Plants*, *6*(3), 253–259.

Tiffin, L. O. (1971, April 20–22). Translocation of micronutrients in plants. In J. J. Mortveldvedt, P. M. Giordano, & W. L. Lindsay (Eds.), *Micronutrients in agriculture, Proceedings of a symposium held at Muscle Shoals, Alabama USA*. Soil Science Society of America, Inc.

Tinnerello, D. (1998). HIV and nutrition. *Body Positive*, *11*(9), 24–27, 30–31.

UN/AIDS. (2010). *Report on the global AIDS epidemic*. Retrieved January 18, 2012, from www.unaids.org/globalreport/documents/20101123_GlobalReport_full_en.pdf

Vijayakumari, K., Siddhuraju, P., & Janardhanan, K. (1996). Effect of soaking, cooking and autoclaving on phytic acid and oligosaccharide of the tribal pulse, *Mucunna monosperma* DC. Ex. Wight. *Food Chemistry*, *55*(2), 173–177.

Vorster, I. H. J., Venter, L. S., & van Rensburg, J. W. S. (2007). The importance of traditional leafy vegetables in South Africa. *African Journal of Food, Agriculture and Nutrition Development*, *7*(4), 1–13.

Wehmeyer, A. S., & Rose, E. F. (1983). Important indigenous plants used in the Transkei as food supplements. *Bothalia*, *14*(3/4), 613–615.

Wheeler, V. E., & Ferrel, F. E. (1971). A method of phytic acid determination in wheat fraction. *Cereal Chemistry*, *48*, 312–316.

WHO. (2003). *Technical report series: Diet, nutrition and the prevention of chronic diseases* (No. 916). Joint FAO/WHO Expert Consultation.

Wilkinson, S. R., Welch, R. M., Maryland, H. F., & Grunes, D. L. (1990). Magnesium in plants: Uptake, distribution, function and utilisation by man and animals. In H. Sigel & A. Sigel (Eds.), *Metal ions in biological systems: Compendium on magnesium and its role in biology, nutrition and physiology* (Vol. 26). Marcel Dekker, INC New York and Base.

Yang, Q., Liu, T., Kuklina, E. V., Flanders, W. D., Hong, Y., Gillespie, C., Chang, M. H., Gwinn, M., Dowling, N., Khoury, M. J., & Hu, F. B. (2011). Sodium and potassium intake and mortality among US adults: Prospective data from the third national health and nutrition examination survey. *Archives of Internal Medicine*, *171*(13), 1183–1191.

11 Mineral Constituents and Bioactive Compounds of Finger Millet Grain

[1]Dorcus Maja, [2]Henry Okwudili Udeh, [3]Tsakani Maria Maphosa, [4]Evison Bhebhe, and [3*]Eastonce Tendayi Gwata

[1]North West University, Faculty of Natural and Agricultural Sciences, Food Security and Safety Focus Area, Private Bag X2046, Mmabatho 2735, South Africa

[2]State University of Medical and Applied Sciences, Faculty of Applied Natural Sciences, Department of Applied Microbiology and Brewing, Igbo Eno, Enugu State, Nigeria

[3]University of Venda, Faculty of Science, Engineering and Agriculture, Department of Plant and Soil Sciences, Private Bag X5050, Thohoyandou, 0950, South Africa

[4]University of Venda, Faculty of Science, Engineering and Agriculture, Department of Animal Science, Private Bag X5050, Thohoyandou, 0950, South Africa

*Corresponding author: ectgwata@gmail.com

11.1 INTRODUCTION

Finger millet (FM) (*Eleusine coracana* [L.] Gaertn.) (2n = 4x = 36) is one of the most important millets in the world that are a staple food to millions of people in resource-poor economies of Africa and Asia and the semi-arid tropical regions. It originated in Eastern Africa, in the region between western Uganda and the Ethiopian highlands, and subsequently spread to the Indian subcontinent about 3,000 years ago (Hilu et al., 1979). FM is an annual herbaceous, climate-resilient, and self-fertilized cereal crop.

The inflorescence of the FM plant consists of fingerlike-shaped panicles. It has a small-seeded naked caryopsis (measuring about 1.2–1.8 mm in diameter), mostly spherical in shape. A thin membranous pericarp which is loosely attached covering the seed usually detaches during harvesting or simple abrasion (Chethan & Malleshi, 2007). Taxonomically, FM belongs to the family *Poaceae* with five races: *coracana*, *vulgaris*, *elongata*, *plana*, and *compacta* (Seetharam et al., 1986; Tatham et al., 1996). The race *coracana* is particularly adapted to the arid and semi-arid agro-ecosystem of the eastern highlands of Africa and Ghats of India, and the race *vulgaris* is cultivated mainly in Uganda, Ethiopia, and Southern Africa (Seetharam et al., 1986). Most FM cultivars possess either one of two distinct grain colors, namely, brown or white (cream), or shades of brown.

11.2 POPULARITY OF FINGER MILLET

In global production, FM ranks fourth in importance, after sorghum, pearl millet, and foxtail millet (Anthony Ceasar et al., 2018). FM accounts for approximately 10% of the world's total 30 million tons of millet produced (Kalsi et al., 2023). Production of the crop is popular due to its broad

DOI: 10.1201/9781003469766-11

agro-ecological amplitude, which includes the mountainous regions of Nepal (3,420 m ASL) to the lowland communal areas in Zimbabwe (1,100 m ASL) and Madagascar (<750 m ASL) (Luitel et al., 2017; Rurinda et al., 2014; Orr et al., 2016). Unlike other cereal crops, FM is well adapted to poor and infertile soils (Gupta et al., 2017). It can withstand harsh climatic conditions, including soil moisture stress conditions that are prevalent in semi-arid tropical areas. FM can be cultivated in almost all types of soils and climatic conditions, including alkaline soils with high pH (Dass et al., 2013). Smallholder FM farmers on light, sandy soils often apply merely cattle manure instead of chemical fertilizers. Therefore, FM is a low-input crop that can be cultivated by resource-poor farmers in harsh agro-ecological areas.

From a nutritional perspective, FM is popular partly due to its unique nutritional composition, abundance of beneficial bioactive compounds, and their functional use in a variety of food and pharmaceutical products. Recently, the Indian government declared FM, along with other millets, as "nutri-cereals" for production, consumption, and trade (Chaudhary & Mudgal, 2020). The nutrients and bioactive components of FM are among the highest of all cereals, including rice and wheat. The proximate nutritional composition of FM contains about 65–75% carbohydrates, 5–8% protein, 1.3–1.8% fat, 15–20% dietary fiber, 2.5–3.5% minerals, and 1–2% ether extractives (a portion of dry matter extracted with ether) (Devi et al., 2014; Kalsi et al., 2023). The major portion of FM carbohydrate includes starch (60%), pentosans (6.2–7.2%), cellulose (1.4–1.8%), and lignin (0.04–0.6%) (Kalsi et al., 2023). Although FM is low in protein content compared to other millets, it has a considerable balanced amino acid content with prolamin, which constitutes up to 24.6–36.2% of the total protein (Lupien, 1990). It is estimated that 44.7% of the amino acids in FM are essential amino acids (Kalsi et al., 2023). Moreover, the bioactive components of FM are ascribed mainly to its total dietary fiber content, minerals, vitamins, and polyphenolic compounds. The bioactive compounds of FM, including ferulic acid–rich arabinoxylans or feraxans, ferulic acid, and quercetin, exhibit significant therapeutic and nutraceutical effects (Udeh et al., 2017). Several *in vitro* and *in vivo* studies have demonstrated the antioxidative, anti-inflammatory, antimicrobial, antidiabetic, wound healing, antiosteoporosis, anticataractogenic, and antilithiatic or nephroprotective properties of FM (Chethan & Malleshi, 2007; Shobana & Malleshi, 2007; Banerjee et al., 2012; Chaudhary & Mudgal, 2020). Data on the nutritional composition, processing methods, phenolic profile, and use of FM in food and pharmaceutical products have increased tremendously over the years.

11.3 TRADITIONAL PROCESSING OF FINGER MILLET GRAIN

Depending on the cultivar and its intrinsic properties, FM can be processed using various traditional and novel technologies, such as soaking, germination, malting, fermentation, popping or puffing, and extrusion, to produce weaning and geriatric foods, beverages, ready-to-eat foods, and some pharmaceutical formulations (Subba Rao & Muralikrishna, 2001; Shiihii et al., 2011). In Southern Africa, traditional opaque beer is brewed from FM flour. In some cases, the flour is also used for preparing porridge and an assortment of traditional beverages and confections (Gabaza et al., 2018; Sood et al., 2017). For instance, in many parts of Zimbabwe, a thick porridge (popularly known as *sadza reZviyo*) is prepared from FM flour and consumed with a side dish (Gabaza et al., 2018). Sometimes, the flour is fermented to improve the aroma, nutritional value, and digestibility as well as to lower the content of antinutrients (Tsafrakidou et al., 2020). In general, cereal grains such as FM contain low levels of organoleptically active compounds, conferring a rather flat flavor, which is often unpleasant to consumers. The fermented cereal products benefit consumers, particularly those who follow low-lipid or vegan diets, since they present probiotic delivery vehicles, enhancing their palatability through flavor changes (Wuyts et al., 2020). The enzymatic activity during fermentation generates volatile (carboxylic acids, alcohols, aldehydes, ketones, and esters) or non-volatile compounds (sugars and carboxylic acids), which contribute to the aroma or sweet and sour taste of the product, especially in beverages (Peyer et al., 2016). The fermentation increases the dry matter content of cereal-based porridges, for instance, thus optimizing macronutrient balance and availability of minerals for absorption (Songré-Ouattara et al., 2010).

11.4 MINERAL CONTENT

The germplasm we evaluated in South Africa showed significant differences ($p < 0.01$) in a wide range of minerals and their interrelationships. In the study, the variation in manganese (Mn) grain content ranged between 66.0 and 142.87 μg/g (Figure 11.1). In addition, there was more iron (Fe) than zinc (Zn) in most (>90.0%) of the genotypes, while among the macroelements, calcium (Ca) showed superior abundance than magnesium (Mg) (Figure 11.2). Genotypic variations in grain mineral content were observed in previous studies, suggesting the potential for developing elite cultivars of FM with improved grain mineral content. High Ca in FM is desirable for reducing the deficiency of the mineral, particularly in developing countries in Africa and Asia (Maharajan et al., 2021; Sharma et al., 2017; Kumssa et al., 2015). It is critical in regulatory functions, including the activation of enzymes, coagulation of blood, contraction and relaxation of muscles, transmission of nerve impulses, as well as maintenance of hormonal secretion in the human body (Pravina et al., 2013). Ca is also associated with protection against cancers, such as breast, ovarian, and prostate (Gao et al., 2005; Lin et al., 2007). Moreover, Ca is important in bone (skeletal) development (Chan et al., 2006; Puranik et al., 2017). Diets that are low in Ca can lead to widespread osteoporosis and osteopenia, especially in elderly people (Bhatia, 2008; Pettifor, 2008).

In another study which evaluated both minerals and phytic acid constituents in Indian FM, both Ca and Mg content ranged from 207.65 to 368.00 mg/100 g and from 143.24 to 192.94 mg/100 g, respectively (Hiremath et al., 2018). However, in the same study, the mean values for Fe and Zn content were 3.00 mg/100 g and 2.79 mg/100 g, respectively. In the germplasm from South Africa, there was a negative but significant ($p < 0.05$) correlation between Mg and each of Zn and Mn, but a positive correlation was observed between Zn and sulfur (S) (Figure 11.3).

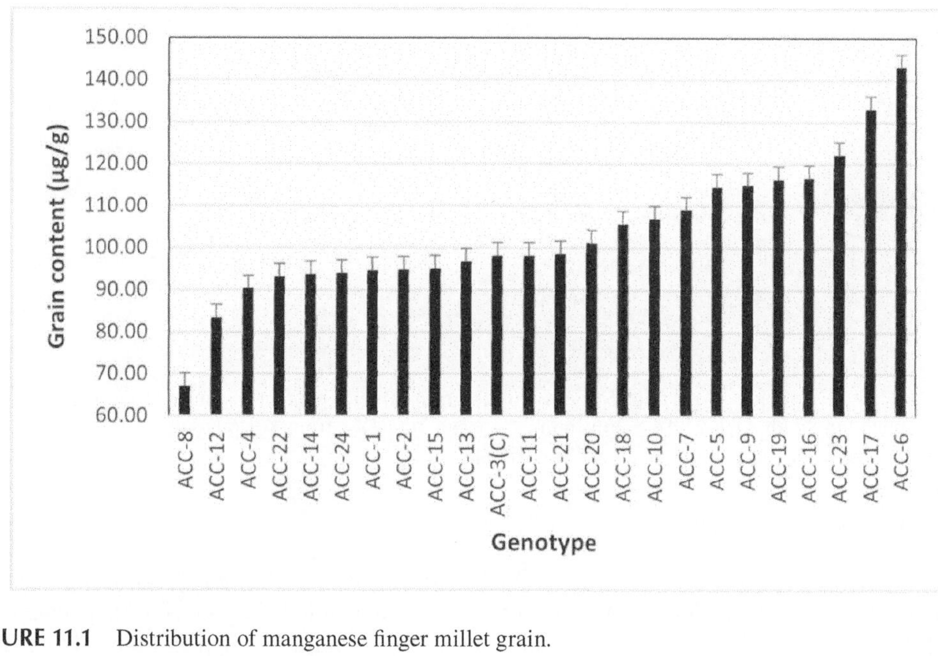

FIGURE 11.1 Distribution of manganese finger millet grain.

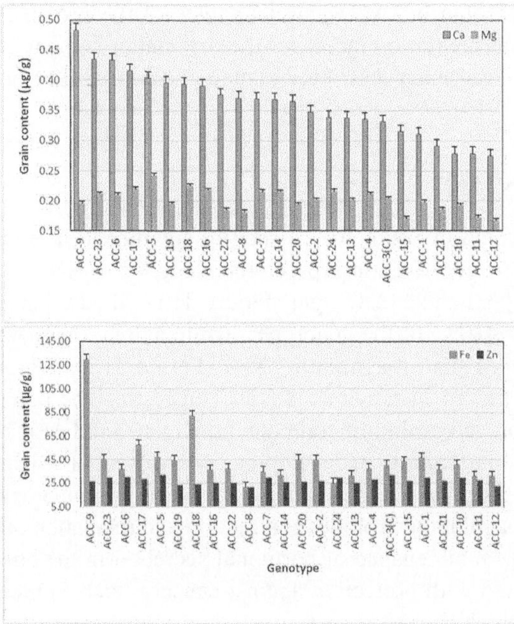

FIGURE 11.2 Distribution of grain mineral content in finger millet genotypes. (*Ca* = calcium; *Fe* = iron; *Mg* = magnesium; *Zn* = zinc.)

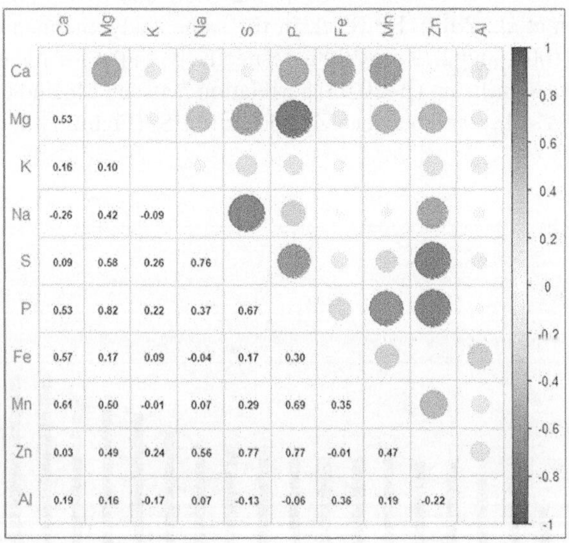

FIGURE 11.3 Coefficient of correlation between the grain mineral content in finger millet genotypes that were evaluated in South Africa. (*Al* = aluminum; *Ca* = calcium; *Fe* = iron; *K* = potassium; *Mg* = magnesium; *Mn* = manganese; *Na* = sodium; *S* = sulfur; *Zn* = zinc.)

11.5 VITAMIN CONTENT

Vitamins are grouped into fat- and water-soluble forms, and their deficiency can result in health problems (Biebinger et al., 2008). FM has both fat- and water-soluble vitamins (Chappalwar et al., 2013). FM grain contains a range of vitamins, including vitamin B complex (Table 11.1) (Devi

TABLE 11.1
Vitamin Content in Finger Millet Grain

Vitamin	(mg/100 g)	Reference
Vitamin A (retinol)	6.0	Siwela, 2009
Vitamin B_1 (thiamine)	0.2–0.48	Saleh et al., 2013; Ramashia et al., 2019
Vitamin B_2 (riboflavin)	0.12	Devi et al., 2014
Vitamin C (ascorbic acid)	0.0–1.0	Siwela, 2009
Niacin	1.0–1.30	Saleh et al., 2013

et al., 2014; Saleh et al., 2013; Siwela et al., 2009). Reports from various studies show that FM is a good source of vitamin B complex, particularly niacin, thiamine, and riboflavin, in the amounts of 1.11 mg/100 g, 0.48 mg/100 g, and 0.12 mg/100 g, respectively (Ibrahem & Sayed, 2023). Another study showed that the "Indaf" cultivar contained 0.229 mg/100 g thiamine, 0.093 mg/100 g riboflavin, 2.22 mg/100 g niacin, and 0.11 mg/100 g folic or ascorbic acid (Malleshi & Klopfenstein, 1998). FM also contains vitamin A (Vidyavati et al., 2004). This ancient grain contains six retinol equivalents of vitamin A and 45 µg/100 g of carotene, and tocopherol at a concentration of 4.1 mg/100 g (Thapliyal & Singh, 2015). FM grain also contains high levels of provitamin A (Bachar et al., 2013). However, it lacks vitamin C in particular (Vidyavati et al., 2004; Siwela, 2009; Chappalwar et al., 2013). In addition, changes in both thiamine and riboflavin content in malting of cereals such as FM were reported (Chavan & Kadam, 1989). Similar observations indicating a significant increase in niacin and ascorbic acid content on malting in millets have been reported (Malleshi & Klopfenstein, 1998).

11.6 PROTEIN

FM contains nearly 7.0% protein (Muthamilarasan et al., 2016). However, wide variations in protein content ranging between 5.6 and 12.7% have been reported in several studies (Bhatt et al., 2003). Protein quality is largely a function of its essential amino acids. Prolamin (the storage protein component with high proline amino acid content) constitutes the largest fraction of FM protein, of which 44.7% are essential amino acids (Sood et al., 2017). This quantity is higher than the 33.9% essential amino acids recommended dietary protein (FAO, 1991). In comparison with other millets, FM is better balanced in essential amino acids, partly because it possesses more lysine, threonine, and valine (Ravindran, 1992). Both lysine and methionine are abundant in FM (Figure 11.4) (McDonough et al., 2000). The prolamin fraction also contains a higher proportion of glutamic acid, leucine, isoleucine, phenylalanine, proline, and valine, but it is low in arginine and glycine (Lupien, 1990). Studies based on *in vitro* protein digestibility report values ranging from 50.0 to 88.1% (Antony & Chandra, 1999). However, the availability of protein from FM is generally limited by the occurrence of enzyme inhibitors, phenols, phytates, and tannins which bind to it (Bandyopadhyay et al., 2012; Antony & Chandra, 1999). Nonetheless, soaking, roasting, boiling, germination, and fermentation, as well as malting, significantly reduce tannin content (Onyango et al., 2013). Phytic acid binds to the minerals derived from the grain or encountered in the digestive tract (Raboy, 2000). However, partial retention of phytates is desirable for health benefits, such as antidiabetic, antioxidant, and anticancer effects (Nissar et al., 2017).

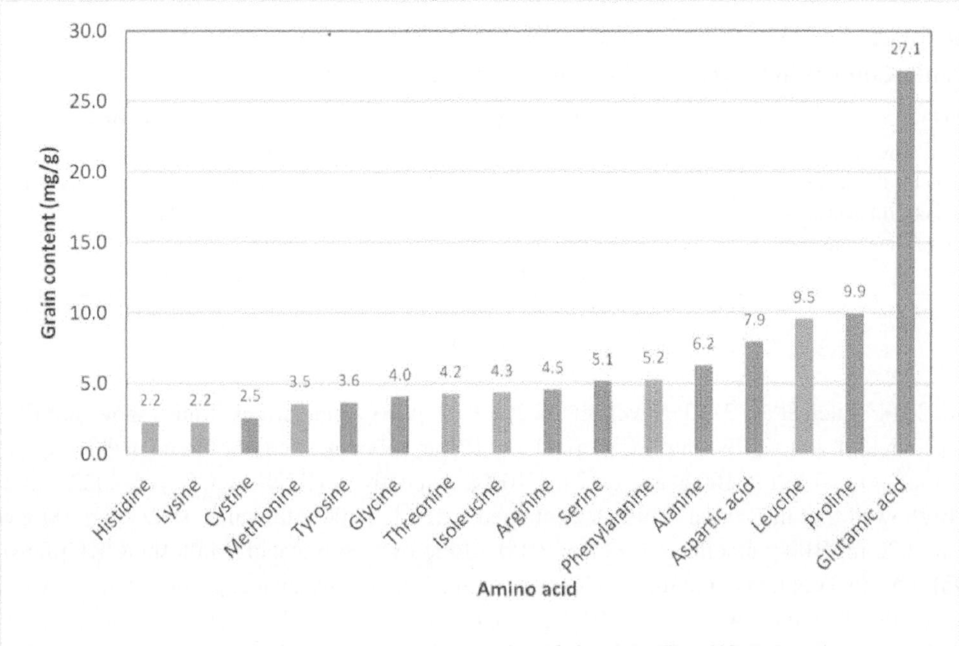

FIGURE 11.4 Amino acid content in finger millet grain. Green and red bars represent essential and non-essential amino acids, respectively.

Source: Adapted from Amadou et al. (2013); Fernandez et al. (2003); McDonough et al. (2000); Serna-Saldivar (2010); Siwela (2009); Thapliyal and Singh (2015).

11.7 FINGER MILLET BIOACTIVE COMPOUNDS

FM contains biologically active compounds that are also nutritionally important. These compounds, which include dietary fiber, minerals, and phenolic compounds, among others, have been found to have positive health benefits in several FM trial studies. The extraction of starch and polyphenolic components from FM is already making waves in the food and pharmaceutical industries.

11.7.1 Dietary Fiber Fractions of Finger Millet

Interest in dietary fiber as a bioactive compound is based on the belief that it has a positive effect on health and quality of life; hence, considering its physiological potencies, it has become a food component of interest. Dietary fiber is a multi-component mixture of plant origin. It forms the main food constituent that influences the rate and extent to which blood glucose increases after ingestion of a carbohydrate or analogous carbohydrate food (Fardet, 2010; Udeh et al., 2017). Primary plant dietary fiber fractions include non-starch polysaccharides, non-α-glucan oligosaccharide, resistance starch, some polyols, and modified starches (Lafiandra et al., 2014). Total dietary fibers constitute about 18.6–22% of FM grain, and they include the non-starch polysaccharides – water-soluble and water-insoluble polysaccharides, hemicellulose A and B, cellulose, pectin, and lignin (Table 11.2) (Subba Rao & Muralikrishna, 2007; Shobana & Malleshi, 2007). The non-starch polysaccharides are largely made up of arabinoxylans together with β-d-glucans, and both make up the main fraction of water-soluble fiber of the grain (Subba Rao & Muralikrishna, 2006). The sugars identified in the polysaccharides are arabinose, xylose, galactose, and glucose, with mannose and rhamnose being present in minute amounts.

TABLE 11.2
Dietary Fiber: A Multi-Component Mixture of Plant Origin

Dietary Fiber

Component	Amount (%)	Reference
Total dietary fiber	22.0	Shobana and Malleshi (2007)
Soluble dietary fiber	2.50	Shobana and Malleshi (2007)
Insoluble dietary fiber	19.70	Shobana and Malleshi (2007)
Water-soluble	0.13	Subba Rao and Muralikrishna (2001)
Hemicellulose A	1.40	Subba Rao and Muralikrishna (2001)
Hemicellulose B	1.90	Subba Rao and Muralikrishna (2001)

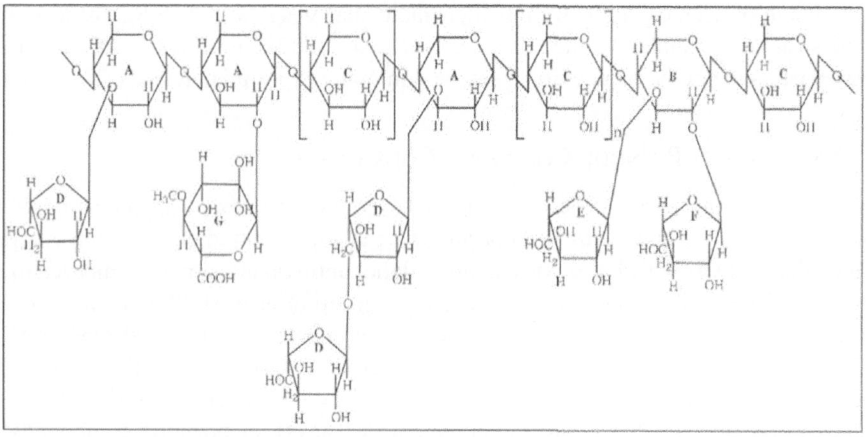

FIGURE 11.5 Structure of finger millet arabinoxylan.

Source: Adapted from Subba Rao and MUralikrishna (2004).

11.7.2 Finger Millet Arabinoxylans

Arabinoxylan (AX) is the principal non-starch polysaccharide present in the cell wall of cereal grains. The AXs are highly branched in nature, owing to the side chains of the polymer containing not only high amounts of terminally linked arabinose but also high amounts of uronic and ferulic acids, galactose, and minute amounts of xylose (Subba Rao & Muralikrishna, 2004; Izydorczyk & Dexter, 2008). Generally, the structure of cereal arabinoxylans is made up of a linear backbone of β-d-xylopyranosyl residues linked through a 1-to-4 glycosidic bond, with attached residues of α-l-arabinopyranosyl at carbon position 3 or 2 or both (Bartlomiej et al., 2011; Lafiandra et al., 2014).

FM is rich in AX and contains both the water-soluble and water-insoluble forms, with a low molecular weight of about 139.9–140 kDa (Subba Rao & Muralikrishna, 2006). FM AX is similar to that of sorghum, rice, and maize AX and is more complex or branched in structure compared to wheat and barley AX. The purified AX isolated from FM is structurally similar to that of other cereals (Fischer et al., 2004). The backbone of the polysaccharide constitutes of 1,4-β-d-linked xylan units, with majority of arabinose and xylose residue substitutions at carbon position 3 (Figure 11.5) (Subba Rao & Muralikrishna, 2004). The previous study showed that the polysaccharide is not only highly branched in structure but is also composed of high-arabinose substitutions at C-3 (disubstituted) or C-2 (monosubstituted) or both, together with xylose residues present in the mono- and unsubstituted forms. A similar structural orientation of the residues has been observed for the oligosaccharide obtained from the AX. The arabinose substitutions were at C-3 (monosubstituted) and at both C-3 and C-2 in the disubstituted forms. Similar xylose substitutions (mono- and unsubstituted) have also been reported along with a very minute number of disubstitutions (Subba Rao & Muralikrishna, 2004). It has also been found that the polysaccharide contains up to 10% of uronic acid in the form of glucuronic acid, which is linked to the xylose residues at C-2 (Subba Rao & Muralikrishna, 2004, 2006).

FM AXs exhibit high antioxidant activities because of their bound phenolic acids. The antioxidative property has been associated with the presence of bound ferulic acids which form a cross-linkage by oxidation to an adjacent arabinoxylan chain. Subba Rao and Muralikrishna (2006) showed that the bound ferulic acids were covalently attached to terminal arabinose residues at C-3 of the xylan backbone. These substitutions often result in the formation of ferulate dimers or, rarely, triferulate residues, which are very stable and possess high antioxidant properties (Chandrasekara & Shahidi, 2011; Lafiandra et al., 2014). These residues are collectively termed "feraxans" due to the abundance of bound ferulic acids. Their high antioxidant activity is also influenced by the presence of a high amount of uronic acid as well as their molecular weights (Subba Rao & Muralikrishna, 2007). The structural complexity, the bound ferulic acids, and the uronic acid of arabinoxylan are believed to participate in the bioactivities observed in many experimental trials (Udeh et al., 2017).

11.7.3 Finger Millet Phenolic Compound Constituents

Phenolic compounds are among the most highly diversified groups of phytochemicals found in plant foods and, as such, constitute an important part of the human diet. They are compounds of significant value that have displayed considerably higher antioxidant activity compared to vitamins (Hatcher et al., 2008). These polyphenols are a wide group of at least 10,000 complex molecules containing one or more aromatic rings with one or even more compound hydroxyl units attached to them (Bello et al., 2022). They represent several groups of compounds that include hydroxybenzoic acids, hydroxycinnamic acids, flavonoids, stilbenes, and lignans (Naczk & Shahidi, 2004; Chandrasekara & Shahidi, 2012). FM contains a wide variety of phenolic compounds, and the widely studied phenolic compounds of the grain are phenolic acids and flavonoids (Table 11.3).

TABLE 11.3
Phenolic Compounds in Finger Millet

	Phenolic Compound	Quantity (μg/g)	Reference
Phenolic Acids	**Hydroxybenzoic Acids**		
	Gentisic	4.50	HithamAni and Srinivasan (2014)
	Vanillic	20.0	Chethan et al. (2008)
	Gallic	3.91–30.0	Hithamani and Srinivasan (2014)
	Syringic	10.0–60.0	Dykes and Rooney (2007);
	Salicylic	5.12–413.0	HithamAni and Srinivasan (2014)
	Protocatechuic	119.8–405.0	Hithamani and Srinivasa (2014)
	p-hydroxybenzoic	6.3–370.0	Dykes and Rooney (2007); Hithamani and Srinivasan (2014)
	Hydroxycinnamic Acids		
	Chlorogenic	–	Shahidi and Chandrasekara (2013)
	Caffeic	5.9–10.4	HithamAni and Srinivasan (2014)
	Sinapic	11.0–24.8	HithamAni and Srinivasan (2014)
	p-coumaric	1.81–41.1	Chethan et al. (2008); Shahidi and Chandrasekara (2013)
	Trans-cinnamic	35–100.0	HithamAni and Srinivasan (2014)
	Trans-ferulic	41–405.0	Shahidi and Chandrasekara (2013)

(Continued)

TABLE 11.3 (Continued)

	Phenolic Compound	Quantity (μg/g)	Reference
Flavonoids	Quercetin	3.0	Chethan et al. (2008)
	Hesperetin	96.3–96.7	Udeh et al. (2018)
	Catechin gallocatechin, epicatechin, epigallocatechin, taxifolin, vitexin, tricin, luteolin, myricetin, apigenin, kaempferol, naringenin, daidzein, procyanidin B1, orientin, isoorientin, isovitexin, saponarin, violanthin, lucenin-1, saponarin, violanthin		Subba Rao and Muralikrishna (2002); Chethan et al. (2008); Chandrasekara and Shahidi (2011); Banerjee et al. (2012); Shahidi and Chandrasekara (2013)

These values might vary based on experimental conditions, assay methods, and grain type used. FM brown cultivars contain a high number of phenolic compounds compared to the white cultivars, which are abundant in the seed coat (Chethan & Malleshi, 2007). An amount of 0.1% was confirmed in the brown FM cultivar compared to the white counterpart with 0.003% (Sripriya et al., 1996; Chethan & Malleshi, 2007).

Ramachandra et al. (1977) also showed that the brown FM cultivar contains about 0.12–3.47% of proanthocyanins, in contrast to the white cultivar, which has 0.04–0.06%. FM phenolic compounds occur in free, conjugated, and bound forms (Chethan & Malleshi, 2007; Chethan et al., 2008; Chandrasekara & Shahidi, 2012). The amounts of the free, esterified, etherified, and bound phenolic compounds in FM are 1,970, 216, 81, and 536 μg/g, respectively, with the total phenolic content ranging from 2.65 to 7.09 μg/g (Sreeramulu et al., 2009; Shahidi & Chandrasekara, 2013; Udeh et al., 2018). The soluble and insoluble fractions have been derived from the sum of the hydroxybenzoic and hydroxycinnamic acids and flavonoids as determined in the report of Chandrasekara and Shahidi (2011).

11.7.4 Phenolic Acids

Phenolic acids are derivatives of benzoic and cinnamic acids. They are found in all cereals, with sorghum and millets exhibiting the widest variety (Dykes & Rooney, 2007; Lin, 2007; Udeh et al., 2017). There are two classes of phenolic acids: *hydroxybenzoic acids*, which are derived from benzoic acid, and *hydroxycinnamic acids*, which are derived from cinnamic acid. The hydroxybenzoic acids form the main free phenolic fraction of FM and constitute about 70–71% of the phenolic acids of the millet. The phenolic acids of FM include gallic acid, protocatechuic, *p*-hydroxybenzoic, vanillic, gentisic, and syringic acids, with protocatechuic acid identified as the major free phenolic acid (4.5 mg/g) (Subba Rao & Muralikrishna, 2002; Chandrasekara et al., 2012; Hithamani & Srinivasan, 2014; Udeh et al., 2018). However, minute amounts of hydroxycinnamic acids – ferulic, caffeic, and *p*-coumaric acids – are often present as free phenolic fractions.

Conversely, hydroxycinnamic acids form the major bound phenolic acids of FM, and they include caffeic, chlorogenic, sinapic, cinnamic (*trans*-cinnamic), coumaric (*p*-coumaric), and ferulic (*trans*-ferulic) acids (Subba Rao & Muralikrishna, 2002; Chethan & Malleshi, 2007; Dykes & Rooney, 2007; Chandrasekara et al., 2012; Shahidi & Chandrasekara, 2013; Hithamani & Srivivasan, 2014). They constitute about 30% of the phenolic acids of the grain. Ferulic acid makes up the most abundant bound phenolic acid in FM, followed by caffeic and coumaric acids, with their occurrence in the order of 18.6, 1.64, and 0.012–0.0125 mg/g flour, respectively (Subba Rao & Muralikrishna, 2001, 2002; Dykes & Rooney, 2006). Hydroxybenzoic, protocatechuic, and syringic acids are also found in trace amounts in the bound fraction.

11.7.5 Flavonoids

FM is rich in flavonoid compounds. A variety of flavonoids has been identified from the grain and leaves of this millet. FM flavonoids are mainly present in the soluble form and are higher by about 2,100 μg/g compared to other millets (Chandrasekara et al., 2012; Shahidi & Chandrasekara, 2013). Esterified forms of flavonoids are also present in FM, which are different from flavonoids of other millets, such as kodo millet (Shahidi & Chandrasekara, 2015). Some of the major flavonoids that occur in FM are quercetin, catechin, gallocatechin, epicatechin, and epigallocatechin. In a recent study, hesperetin, which is an important dietary flavonoid present in fruits, was found in significant amounts of 96.3 μg/g and 9.2 μg/g in the raw and malted FM extracts, respectively (Udeh et al., 2018). Proanthocyanidins or condensed tannins, which are oligomeric or polymeric flavonoids, are found in significant amounts in FM, with procyanidin B1 and B2 being the major dimers (Chandrasekara & Shahidi, 2012; Shahidi & Chandrasekara, 2015). These phenolic compounds are recognized for their antioxidative, antiproliferative, antiallergic, and anti-inflammatory properties. However, there are no structural representations of these phenolic compounds peculiar to finger millet; instead, they are presumed to adopt a similar structural identity and activity of the compounds in themselves. Therefore, there is merit in characterizing the structural forms of these compounds in FM, which could be relevant in predicting their bioactivities.

11.8 IMPACT OF PROCESSING ON BIOACCESSIBLE FINGER MILLET BIOACTIVE COMPOUNDS

The bioaccessibility of cereal grain nutrients is a novel research area owing to its arbitral description of the importance of food nutrients and health. Although controversies still linger on the working definition of the term, recent studies have somewhat adopted the definition to be the fraction or amount of food substance of the food matrix that is soluble and available for absorption in the gastrointestinal environment (Cardoso et al., 2015). The release and subsequent activity of bioactive compounds of foods are dependent on their accessibility to enzymatic attack either from the food or the gastrointestinal tract. The gastrointestinal release of cereal bioactive compounds is very low due to their association with cell wall matrix, such as dietary fiber, polyphenols, and other antinutritional factors, like phytate and tannin (Udeh et al., 2017). Multiple factors such as food source and subsequent chemical interactions with other phytochemicals and bioactive molecules present in the food tend to interfere with the bioavailability of nutrients. Often, some nutrients are mixed with different macromolecules, such as carbohydrates, lipids, and proteins, resulting in the formation of food matrices. On the other hand, the bioaccessibility of the nutrients depends largely on the compounds released from the food matrix or absorbed in the small intestine, thus making them potentially bioavailable for beneficial effects. In addition, processing the food source using the extrusion technology, for instance, can also alter heat-labile bioactive ingredients (Conti-Silva et al., 2012; Seth & Rajamanickam, 2012).

Recent studies have demonstrated the bioaccessibility potential of FM bioactives, such as minerals and phenolic compounds (Tatala et al., 2007; Platel et al., 2010; Chandrasekara & Shahidi, 2012; Hithamani & Srinivasan, 2014). A study that was conducted to determine the impact of malting on bioaccessibility of minerals in FM observed a decline in mineral content after malting, with a subsequent increase in the bioaccessibility of Fe and Ca (Platel et al., 2010). The malting process reduces antinutritional factors like phytate and tannin by activating endogenous enzymes such as phytase and tannase, which break down the antinutrients and subsequently reduce their amounts. The decrease in mineral content was attributed to the effect of malting on phytate and other antinutritional factors that form complexes with the minerals (Platel et al., 2010). However, the increase in bioaccessible Fe and Ca may be due to the ionization reaction of the minerals during hydrolysis of the phytate–mineral complex. A similar increase in bioaccessible Fe from 0.75 to 1.25 mg/100 g after germination of FM has been reported (Tatala et al., 2007). The increase in bioaccessible Fe was due to the activation of endogenous enzymes, such as esterase and phytase, which consequently break down polyphenols and phytate–mineral complexes, thus releasing the mineral.

Besides the food structure and antinutritional factors, the processing method is also an important determinant of the bioaccessibility of cereal bioactive compounds. Some of the processing methods, such as milling, sprouting, roasting, enzymatic digestion, malting, and fermentation, are effective in the release of these compounds by increasing their surface area ratio, inducing the activity of endogenous enzymes and the bioconversion of the bioactive compounds to more active forms (Udeh et al., 2017). Except for roasting, the effects of sprouting, roasting, pressure cooking, open-pan boiling, and microwave heating on the FM phenolic profiles and their bioaccessibility were shown to reduce the total phenolic acid and flavonoid content (Hithamani & Srinivasan, 2014). The reduction of the phenolic content, especially for the sprouted grains, has been attributed to the activation of endogenous enzymes, the hydrolysis of phenolic complexes, and the subsequent loss of the released phenolic compounds during soaking. On the other hand, sprouting and roasting have been shown to increase the bioaccessibility of phenolic compounds in FM (Hithamani & Srinivasan, 2014).

In a study investigating the effect of simulated gastrointestinal pH conditions, gastric and gastrointestinal digestion, and colonic fermentation on the bioaccessibility of FM phenolic compounds, high bioaccessible total phenolics were observed for the gastric and gastrointestinal digesta compared to the pH treatment and colonic fermentation (Chandrasekara & Shahidi, 2012). The observed increase was attributed partly to digestion of proteins and the subsequent release of their bound phenolics present in the grain. However, the activity of endogenous enzymes, such as proteases and esterases, could also play a role in the release of the phenolic compounds from the cell wall matrix during digestion (Chandrasekara & Shahidi, 2012).

11.9 BIOACTIVITY OF FINGER MILLET GRAIN

Studies involving FM have demonstrated its several therapeutic effects that are relevant to food- and lifestyle-dependent diseases, such as diabetes, obesity, hypertension, and cardiovascular and gastrointestinal tract disorders (Geetha & Parvathi, 1990; Tovey, 1994; Shobana et al., 2010; Chaudhary & Mudgal, 2020). Due to its gluten-free nature and low-glycemic-index status, together with the high dietary fiber, Ca, Fe, Mg, polyphenolic, contents and natural microflora, FM is utilized as a healthy food alternative for gluten-intolerant or celiac and diabetic patients, as well as in probiotic and prebiotic food for health promotion (Udeh et al., 2018; Chaudhary & Mudgal, 2020; Kalsi et al., 2023). The high fiber content of FM can also contribute to weight management in obese patients by promoting satiety and reducing overall calorie intake. For instance, diabetes mellitus is a complex metabolic disorder characterized by high systemic glucose levels resulting from impaired insulin secretion and consequent alteration in carbohydrate, protein, and lipid metabolism (Kumari & Sumathi, 2002; Devi et al., 2014; Chaudhary & Mudgal, 2020). According to the International Diabetes Federation estimates, the number of people suffering from diabetes mellitus is expected to rise to 642 million in 2024 (Chaudhary & Mudgal, 2020). In the treatment of the condition, one preventive approach is to decrease postprandial hyperglycemia, or glucose surge, by blocking or reducing the action of carbohydrate-hydrolyzing enzymes. The enzymes α-amylase and α-glucosidase are important in the conversion of carbohydrates into glucose for absorption in the human gut, and their modulation can be relevant in managing diabetic and other metabolic health conditions. Several epidemiological studies have demonstrated that regular consumption of FM alleviates diabetes-related complications. These claims are based on the role of the bioactive compounds in the metabolism of food materials, especially carbohydrates. Phenolic compounds of FM have a potential modulatory effect on the breakdown of carbohydrates either by reacting with proteins and enzymes or altering the properties of biopolymers. The high dietary fiber lowers postprandial glucose and insulin responses as it forms unabsorbable complexes with available carbohydrates (Lafiandra et al., 2014). These phenomena often affect carbohydrate digestibility and result in the delay of the systemic absorption of glucose, which ultimately controls postprandial blood glucose surge. The consumption of FM-based diets has been shown to significantly lower the blood plasma glucose levels in individuals with non-insulin-dependent diabetes mellitus (Kumari & Sumathi, 2002). Moreover, the supplementation of diets with FM consistently has been shown to lead to a reduction in fasting and postprandial glucose levels compared with supplementation with other millets (Geetha & Parvathi, 1990). Regular consumption of FM food products decreases fasting glucose by 32% and mitigates insulin resistance by 43% (Chaudhary & Mudgal, 2020). In addition, improved glycemic response and carbohydrate digestibility of processed FM have been reported (Chandra et al., 2016). This previous study also demonstrated that malting and fermentation were critical in the conversion of starch into less-absorbable forms of dextrin and maltose.

Aldose reductase is also one of the key enzymes involved in diabetes-induced cataractogenesis (Devi et al., 2014). The enzyme was reversibly inhibited by the phenolic content of FM seed coat in a non-competitive mode (Chethan & Malleshi, 2007; Chethan et al., 2008). It was suggested that the mechanism of inhibition of FM polyphenols on aldose reductase was via the enzymatic conversion of glyceraldehyde to glycerol, glucose, and sorbitol. The structural function study showed that the hydroxyl group at position 4 and the *O*-methyl group of polyphenols were responsible for the inhibition of the enzyme. In another similar study, FM phenolic extracts were responsible for exerting a mixed non-competitive mode of inhibition on malt amylases, whereas the individual phenolic compounds showed an uncompetitive inhibition (Chethan et al., 2008). *Trans*-cinnamic acid showed the highest degree of inhibition (79.2%), while a weak inhibition of ~56% was reported for syringic acid. The mode of inhibition was suggested to be dependent on the concentration of phenolic compounds as well as the number and position of hydroxyl groups (Rohn et al., 2002; Chethan et al., 2008). The competition of phenolic compounds with substrate for enzyme active site or enzyme substrate specificity could also account for the observation.

The oxidation of glycated collagen or glycoxidation presents a complication risk in a diabetic condition. The findings from a study focusing on the effect of FM methanolic extract on glycation or collagen cross-linking showed that the glycation process was inhibited for collagen that was incubated with glucose (50 mM) and 3 mg of the FM extract. The inhibitory effect of FM extract was attributed to the antioxidative activities of phenolic compounds and other phytochemicals, as well as the high mineral content of the FM extract, since minerals such as Ca and Mg also show antioxidative properties. In another study on the effect of FM methanolic extract on collagen cross-linking, the extract strongly inhibited the glycation reaction compared to the synthetic antioxidants (aminoguanidine, butylated hydroxyanisole) (Hegde et al., 2005). Among other compounds, the inhibitory effect was specifically attributed to ferulic acid, which offers renal protection through improved glycemic control and renal structural changes involved in the inhibition of oxidative stress (Hegde et al., 2005; Choi et al., 2011).

The evidence on FM antihyperglycemic, hypocholesterolemic, nephroprotective, and anticataractogenic properties has been clearly documented (Shobana et al., 2010). In the study, reduced fasting hyperglycemia and partial reversal of abnormalities in the serum albumin, urea, and creatinine status were observed in streptozotocin-induced diabetic rats. Hypercholesterolemia, hypertriacylglycerolemia, nephropathy, and neuropathy were significantly reversed in a diabetic group of rats fed with diet containing FM seed coat matter (Shobana et al., 2010; Devi et al., 2014). Furthermore, FM-based diets enhanced the antioxidant status and control of blood glucose levels in diabetes-induced rats (Rajasekaran et al., 2004). These observations were associated with the high dietary fiber content of the grain and the presence of antinutritional factors known to reduce starch digestibility and glucose diffusion. The phenolic compounds and minerals present in the grain could also account for the observed effect, as they are potential inhibitors of carbohydrate-degrading enzymes. In addition, plausible antihypertensive activities of FM diet were established (Jeong et al., 2024). Hypertension is also one of the most significant risk factors for cardiovascular diseases. As the blood pressure remains high in hypertensive subjects, the vascular rhythm increases and vasoconstriction dysfunction occurs, leading to the need for revascularization. Cardiac revascularization is characterized by prolonged high blood pressure that increases the heart workload, leading to cardiac remodeling through structural and functional changes, including cardiac hypertrophy and myocardial fibrosis, in which the heart muscles become large and thick. It has been postulated that antihypertensive effects can be promoted through a mechanism by which natural antioxidants restrain free radicals and reactive oxygen species production.

Angiotensin-converting enzyme is a major target for hypertension in the renin-angiotensin-aldosterone system. The enzyme converts angiotensin I into angiotensin II, which causes vasoconstriction and blood pressure elevation. The use of synthetic drugs in the treatment and management of hypertension has slowly been redirected to natural nutraceuticals with angiotensin-converting enzyme inhibitory properties due to their side effects. FM is a potent antihypertensive nutraceutical for regulating the renin–angiotensin system and simultaneously inhibiting oxidative stress (Park et al., 2021). In the previous study, FM supplementation at 250 mg/kg bw and 500 mg/kg bw exhibited a significant control over the renin–angiotensin system by reducing the angiotensin-converting enzyme levels, ameliorating vascular remodeling, and drastically reducing the systolic blood pressure. The underlying mechanism of the FM extract in the regulation of angiotensin-converting enzyme level in the renin–angiotensin system has been attributed to increased systemic antioxidant capacity. Similarly, a reduction in serum angiotensin-converting enzyme activity and angiotensin II levels as well as a decrease in systolic and diastolic blood pressure after feeding hypertensive rats with FM composite meal have been reported (Han et al., 2023). The effect of the meal in maintaining blood pressure was associated with its elevated levels of polyphenols, such as ferulic acid and proanthocyanidins, as well as its higher calcium content resulting from the blended diet.

A study on the antimicrobial activity of FM polyphenol extract as well as the individual phenolic compounds found that millet extract exhibited proliferative inhibitory activities against *Escherichia coli*, *Bacillus cereus*, *Staphylococcus aureus*, *Listeria monocytogenes*, *Streptococcus pyogenes*, *Proteus mirabilis*, *Pseudomonas aeruginosa*, *Serratia marsescens*, *Klebsiella pneumonia*, and *Yersinia enterocolitica*. It was suggested that the high FM phenolic content may have caused the inhibition of microbial enzymes and the oxidation of the microbial membranes, thus resulting in the proliferation of bacterial cells. In this study, the phenolic compound quercetin was found to exert the highest antiproliferative effect among the phenolic compounds evaluated. This compound is widely known for its pharmaceutical properties, including antioxidant, antiapoptotic, and antiproliferative effects.

11.10 CONCLUSIONS AND PROSPECTS

To date, voluminous useful information about FM has been generated by researchers from various perspectives, including production, nutrition and health, genetic improvement, and marketing, among others. Our perspective on the future of FM is that the production and utilization are set to increase significantly, depending on key drivers. For instance, the current major shifts in climatic conditions will inevitably drive the adoption of drought-tolerant crops such as FM to ensure food and nutrition security in many rural households, particularly in sub-Saharan Africa. Moreover, the increase in our knowledge about the nutritional value of FM is spurring home-level utilization of the crop for simple products, such as homemade bread, non-alcoholic beverages, and porridge, among others. Recently, improved FM varieties have been released for commercial production in several African countries. This development augers well for the adoption of the crop by farmers; hence, we anticipate a steady upward production trajectory of FM. As conducive policies for promoting production and consumption of underutilized crops such as FM, trade volumes of the grain or value-added products on the international and domestic markets are likely to increase significantly. National policies that incentivize farmers, traders, and other end users in the value chain are critical in shaping the future of FM production. The abundance of bioaccessible nutrients in FM suggests that it can be used as a functional food ingredient to support animal and human health. However, to firmly confirm the health-related claims, more research is necessary, particularly on human subjects, as well as the underlying mechanisms of action. It is imperative to underscore the significance of conducting more research on the multifarious uses of FM for food and feed production, as this will promote the grain's consumption, particularly in areas with limited resources. From a technical perspective, modern value-addition technologies, particularly in the brewing and confectionary industries, are markedly efficient, offering better packaging for long-distance transport and extended shelf life of products, thus presenting good opportunities for reaching huge global markets. Moreover, the research support galvanized by second-generation biotechnology tools can expedite cultivar development and access by growers. The improved future FM varieties will likely boost both crop productivity and nutritional attributes of this valuable grain cereal.

11.11 ACKNOWLEDGMENTS

We wish to acknowledge the National Research Foundation (South Africa) and the Water Research Commission (South Africa) for partially funding some of the research work on finger millet that was conducted by the first author (Dorcus Maja).

REFERENCES

Amadou, I., Gounga, M. E., & Le, G. W. (2013). Millets: Nutritional composition, some health benefits and processing – A review. *Food Science and Nutrition, 25*, 501–508.

Antony, U., & Chandra, T. S. (1999). Enzymatic treatment and use of starters for the nutrient enhancement in fermented flour of red and white varieties of finger millet (*Eleusine coracana*). *Journal of Agricultural and Food Chemistry, 47*, 2016–2019.

Bachar, K., Mansour, E., Khaled, A. B., Abid, M., Haddad, M., Yahya, L. B., El Jarray, N., & Ferchichi, A. (2013). Fiber content and mineral composition of the finger millet of the Oasis of Gabes Tunisia. *Journal of Agricultural Science, 5*, 219.

Bandyopadhyay, P., Ghosh, A. K., & Ghosh, C. (2012). Recent developments on polyphenol-protein interactions: Effects on tea and coffee taste, antioxidant properties and the digestive system. *Journal of Functional Foods, 3*, 592–605.

Banerjee, S., Sanjay, K. R., Chethan, S., & Malleshi, N. G. (2012). Finger millet (*Eleusine coracana*) polyphenols: Investigation of their antioxidant capacity and antimicrobial activity. *African Journal of Food Science, 6*, 362–374.

Bartlomiej, S., Justyna, R., & Ewa, N. (2011). Bioactive compounds in cereal grains – occurrence, structure, technological significance and nutritional benefits – A review. *Food Science and Technology International, 18*, 559–568.

Bello, O. M., Ogbesejana, A. B., Balkisu, A., Osibemhe, M., Musa, B., & Oguntoye, S. O. (2022). Polyphenolic fractions from three millet types (fonio, finger millet, and pearl millet): their characterization and biological importance. *Clinical Complementary Medicine and Pharmacology, 2*, 100020.

Bhatia, V. (2008). Dietary calcium intake-a critical reappraisal. *Indian Journal of Medical Research, 127*, 269–273.

Bhatt, A., Singh, V., Shrotria, P. K. & Baskheti, D. C. (2003). Coarse grains of Uttaranchal: Ensuring sustainable food and nutritional security. *Indian Farmer's Digest, 7*(1), 34–38.

Biebinger, R., Hurrell, R. F. & Ottaway, P. (2008). Vitamin and mineral fortification of foods. In P. B. Ottaway (Ed.), *Food fortification and supplementation* (pp. 27–40). Woodhead Publishing.

Cardoso, C., Afonso, C., Lourenco, H., Costa, S., & Nunes, M. L. (2015). Bioaccessibility assessment methodologies and their consequences for the risk–benefit evaluation of food. *Trends in Food Science and Technology, 41*, 5–23.

Ceasar, A. S., Maharaja, T., Krishna, A. T. P., Ramakhrishnan, M., Roch, V. G., Satish, L., & Ignacimuthu, S. (2018). Finger millet (*Eleusine coracana* (L.) Gaertn.) improvement: Current status and future interventions of whole genome sequence. *Frontiers in Plant Science, 9*, 1054.

Chan, G. M., McElligott, K., McNaught, T., & Gill, G. (2006). Effects of dietary calcium intervention on adolescent mothers and newborns: A randomized controlled trial. *Obstetrics and Gynecology, 108*, 565–571.

Chandra, D., Chandra, S., & Pallavi Sharma, A. K. (2016). Review of finger millet (*Eleusine coracana* (L.) Gaertn): A powerhouse of health benefiting nutrients. *Food Science and Human Wellness, 5*, 149–155.

Chandrasekara, A., Naczk, M., & Shahidi, F. (2012). Effect of processing on the antioxidant activity of millet grains. *Food Chemistry, 133*, 1–9.

Chandrasekara, A., & Shahidi, F. (2011). Antiproliferative potential and DNA scission inhibitory activity of phenolics from whole millet grains. *Journal of Functional Foods, 3*, 159–170.

Chandrasekara, A., & Shahidi, F. (2012). Bioaccessibility and antioxidant potential of millet grain phenolics as affected by simulated in vitro digestion and microbial fermentation. *Journal of Functional Foods, 4*, 226–237.

Chappalwar, V. M., Peter, D., Bobde, H., & John, S. M. (2013). Quality characteristics of cookies prepared from oats and finger millet based composite flour. *Engineering Science and Technology: An International Journal, 3*, 677–683.

Chaudhary, J. K., & Mudgal, S. (2020). Antidiabetic and hypolipidemic action of finger millet (*Eleusine coracana*) enriched probiotic fermented milk: An in vivo rat study. *Food Technology and Biotechnology, 58*, 192–202.

Chavan, J. K., & Kadam, S. S. (1989). Nutritional improvement of cereals by sprouting. *Critical Reviews in Food Science and Nutrition, 28*, 401–437.

Chethan, S., Dharmesh, S. M., & Malleshi, N. G. (2008). Inhibition of aldose reductase from cataracted eye lenses by finger millet (*Eleusine coracana*) polyphenols. *Bioorganic and Medicinal Chemistry, 16*, 10085–10090.

Chethan, S., & Malleshi, N. G. (2007). Finger millet polyphenols: Characterization and their nutraceutical potential. *American Journal of Food Technology, 2*, 582–592.

Choi, R., Kim, B. H., Naowaboot, J., Lee, M. Y., Hyun, M. R., Cho, E. J., Lee, E. S., Lee, E. Y., Yang, Y. C., & Chung, C. H. (2011). Effects of ferulic acid on diabetic nepropathy in a rat model of type 2 diabetes. *Experimental and Molecular Medicine, 43*, 676–683.

Conti-Silva, A. C., Bastos, D. H. M., & Arêas, A. G. (2012). The effects of extrusion conditions and the addition of volatile compounds and flavour enhancers to corn grits on the retention of the volatile compounds and texture of extrudates. *International Journal of Food Science and Technology, 47*, 1896–1902.

Dass, A., Sudhishri, S., & Lenka, N. K. (2013). Integrated nutrient management to improve finger millet productivity and soil conditions in hilly region of eastern India. *Journal of Crop Improvement, 27*, 528–546.

Devi, P. B., Vijayabharathi, R., Sathyabama, S., Malleshi, N. G., & Priyadarisini, V. B. (2014). Health benefits of finger millet (*Eleusine coracana* L.) polyphenols and dietary fiber: A review. *Journal of Food Science and Technology, 51*, 1021–1040.

Dykes, L., & Rooney, L. W. (2006). Sorghum and millet phenols and antioxidants. *Journal of Cereal Science, 44*, 236–251.

Dykes, L., & Rooney, L. W. (2007). Phenolic compounds in cereal grains and their health benefits. *Cereal Foods World, 52*, 105–111.

FAO. (1991). *Food, nutrition and agriculture-2/3-nutrient requirements, recent developments in protein quality evaluation*, Rome, Italy.

Fardet, A. (2010). New hypotheses for the health-protective mechanisms of whole-grain cereals: What is beyond fibre? *Nutrition Research Review, 23*, 65–134.

Fernandez, D. R., Vanderjagt, D. J., Millson, M., Huang, Y., Chuang, L., Pastuszyn, A., & Glew, R. H. (2003). Fatty acid, amino acid and trace mineral composition of *Eleusine coracana* (pwana) seeds from Northern Nigeria. *Plant Foods for Human Nutrition, 58*, 1–10.

Fischer, M. H., Yu, N., Gray, G. R., Ralph, J., Anderson, L., & Marlett, J. A. (2004). The gel-forming polysaccharide of psyllium husk (*Plantago ovata* Forsk). *Carbohydrate Research, 339*(11), 2009–2017.

Gabaza, M., Shumoy, H., Louwagie, L., Muchuweti, M., Vandamme, P., Du Laing, G., & Raes, K. (2018). Traditional fermentation and cooking of finger millet: Implications on mineral binders and subsequent bioaccessibility. *Journal of Food Composition and Analysis, 68*, 87–94.

Gao, X., LaValley, M. P., & Tucker, K. L. (2005). Prospective studies of dairy product and calcium intakes and prostate cancer risk: A meta-analysis. *Journal of the National Cancer Institute, 97*, 1768–1777.

Geetha, C., & Parvathi, E. P. (1990). Hypoglycaemic effect of millet incorporated breakfast on selected non-insulin dependent diabetic patients. *The Indian Journal of Nutrition and Dietetics, 27*, 316–320.

Gupta, S. M., Arora, S., Mirza, N., Pande, A., Lata, C., Puranik, S., Kumar, J., & Kumar, A. (2017). Finger Millet: A "certain" crop for an "uncertain" future and a solution to food insecurity and hidden hunger under stressful environments. *Frontiers in Plant Science, 8*, 643.

Han, B., Jeong, E. W., Han, A., Baek, Y., Kim, H., & Lee, H. G. (2023). Effects of roasting and extrusion puffing on the antihypertensive activity of blended grains in spontaneously hypertensive rats. *Journal of Food Biochemistry, 23*, 1–10.

Hatcher, H., Planalp, R., Cho, J., Torti, F. M., & Torti, S. V. (2008). Curcumin: From ancient medicine to current clinical trials. *Cellular and Molecular Life Sciences, 65*, 1631–1652.

Hegde, P. S., Anitha, B., & Chandra, T. S. (2005). *In vivo* effect of whole grain flour of finger millet (*Eleusine coracana*) and kodo millet (*Paspalum scrobiculatum*) on rat dermal wound healing. *Indian Journal of Experimental Biology, 43*, 254–258.

Hilu, K. W., De Wet, J. M. J., & Harlan, J. R. (1979). Archaeobotanical studies of *Eleusine coracana* ssp. Coracana (finger millet). *American Journal of Botany, 66*, 330–333.

Hiremath, N., Geetha, K., Vikram, S. R., Nanja, Y. A., Joshi, N., & Shivaleela, H. B. (2018). Minerals content in finger millet [*Eleusine coracana* (L.) Gaertn]: A future grain for nutritional security. *International Journal of Current Microbiology and Applied Sciences, 7*, 3448–3455.

Hithamani, G., & Srinivasan, K. (2014). Effect of domestic processing on the polyphenol content and bioaccessibility in finger millet (*Eleusine coracana*) and pearl millet (*Pennisetum glaucum*). *Food Chemistry, 164*, 55–62.

Ibrahem, E. S., & Sayed, R. S. (2023). Characteristics of finger millet (Eleusine coracana L.) flour and its effect on obese rats: Biological and applied study. *Egyptian Journal of Food Science, 51*, 113–125.

Izydorczyk, M. S., & Dexter, J. E. (2008). Barley β-glucans and arabinoxylans: Molecular structure, physicochemical properties and uses in food products – A review. *Food Research International, 41*, 850–868.

Jeong, E., Yun, D., Baek, Y., Kim, H., & Lee, H. G. (2024). Antihypertensive effects of the combined extract of sorghum bicolor, *Vignia angularis*, and *Eleusine coracana* in a spontaneously hypertensive rats. *Scientific Reports, 14*, 803–806.

Kalsi, R., Bhasin, J., Goksen, G., & Kashyap, P. (2023). Exploration of nutritional, pharmacological, and the processing trends for volarization of finger millet (*Eleusine coracana*): A review. *Food Science and Nutrition, 11*, 1–18.

Kumari, L., & Sumathi, S. (2002). Effect of consumption of finger millet on hyperglycemia in non-insulin dependent diabetes mellitus (NIDDM) subjects. *Plant Foods for Human Nutrition, 57*, 205–213.

Kumssa, D. B., Joy, E. J., Ander, E. L., Watts, M. J., Young, S. D., Walker, S., & Broadley, M. R. (2015). Dietary calcium and zinc deficiency risks are decreasing but remain prevalent. *Scientific Reports, 5*, 1–11.

Lafiandra, D., Riccardi, G., & Shewry, P. R. (2014). Review – Improving cereal grain carbohydrates for diet and health. *Journal of Cereal Science, 59*, 312–326.

Lin, J., Manson, J. E., Lee, I. M., Cook, N. R., Buring, J. E., & Zhang, S. M. (2007). Intakes of calcium and vitamin D and breast cancer risk in women. *Archives of Internal Medicine 167*, 1050–1059.

Luitel, D. R., Siwakoti, M., Jha, P. K., Jha, A. K., & Krakauer, N. (2017). Climate, distribution and cropping pattern of finger millet in Nepal: A review. *International Journal of Agriculture and Environmental Research, 3*, 3854–3868.

Lupien, J. R. (1990). *Sorghum and millets in human nutrition*. FAO, ICRISAT. (At: ao.org., 86: 17).

Maharajan, T., Antony Ceasar, S., Ajeesh Krishna, T. P., & Ignacimuthu, S. (2021). Finger millet [*Eleusine coracana* (L.) Gaertn]: An orphan crop with a potential to alleviate the calcium deficiency in the semi-arid tropics of Asia and Africa. *Frontiers in Sustainable Food Systems, 5*, 684447.

Malleshi, N. G., & Klopfenstein, C. F. (1998). Nutrient composition, amino acid and vitamin contents of malted sorghum, pearl millet, finger millet and their rootlets. *International Journal of Food Sciences and Nutrition, 49*, 415–422.

McDonough, C. M., Rooney, L. W., & Serna-Saldivar, S. O. (2000). The millets. In K. Kulp & J.G. Ponte Jr. (Eds.), *Handbook of cereal science and technology, revised and expanded* (pp. 177–201). CRC Press.

Muthamilarasan, M., Dhaka, A., Yadav, R., & Prasad, M. (2016). Exploration of millet models for developing nutrient rich graminaceous crops. *Plant Science, 242*, 89–97.

Naczk, M., & Shahidi, F. (2004). Extraction and analysis of phenolics in food. *Journal of Chromatography, 1054*, 95–111.

Nissar, J., Ahad, T., Naik, H. R., & Hussain, S. Z. (2017). A review phytic acid: As antinutrient or nutraceutical. *Journal of Pharmacognosy and Phytochemistry, 6*, 1554–1560.

Onyango, C. A., Ochanda, S. O., Mwasaru, M. A., Ochieng, J. K., Mathooko, F. M., & Kinyuru, J. N. (2013). Effects of malting and fermentation on anti-nutrient reduction and protein digestibility of red sorghum, white sorghum and pearl millet. *Journal of Food Research, 2*, 41–49.

Orr, A., Mwema, C., Gierend, A., & Nedumaran, S. (2016). *Sorghum and millets in Eastern and Southern Africa: Facts, trends and outlook* (Working Paper Series No. 62). ICRISAT.

Park, S. Y., Jeong, E. W., Yang, Y. S., Kim, H., Go, G., & Lee, H. G. (2021). Finger millet ethanol extracts prevent hypertension by inhibiting the angiotensin-converting enzyme level and enhancing the antioxidant capacity in spontaneously hypertensive rats. *Antioxidant (MDPI), 10*, 1766–1777.

Pettifor, J. M. (2008). Vitamin D and/or calcium deficiency rickets in infants and children: A global perspective. *Indian Journal of Medical Research, 127*, 245–249.

Peyer, L. C., Zannini, E., & Arendt, E. K. (2016). Lactic acid bacteria as sensory biomodulators for fermented cereal-based beverages. *Trends Food Science and Technology, 54*, 17–25.

Platel, K., Eipeson, S. W., & Srinivasan, K. (2010). Bioaccessible mineral content of malted finger millet (*Eleusine coracana*), wheat (*Triticum aestivum*), and barley (*Hordeum vulgare*). *Journal of Agricultural and Food Chemistry, 58*, 8100–8103.

Pravina, P., Sayaji, D., & Avinash, M. (2013). Calcium and its role in human body. *International Journal of Pharmaceutical and Bio-Medical Science, 4*, 659–668.

Puranik, S., Kam, J., Sahu, P. P., Yadav, R., Srivastava, R. K., Ojulong, H. F., & Yadav, R. S. (2017). Harnessing finger millet to combat calcium deficiency in humans: Challenges and prospects. *Frontiers in Plant Science, 8*, 1–16.

Raboy, V. (2000). Low-phytic-acid grains. *Food and Nutrition Bulletin, 21*, 423–427.

Rajasekaran, N. S., Nithya, M., Rose, C., & Chandra, T. S. (2004). The effect of finger millet feeding on the early responses during the process of wound healing in diabetic rats. *Biochimica Biophysica Acta, 1689*, 190–201.

Ramachandra, G., Virupaksha, T. K., & Shadaksharaswamy, M. (1977). Relation between tannin levels and in vitro protein digestibility in finger millet (*Eleusine coracana* Gaertn.). *Journal of Agricultural and Food Chemistry, 25*, 1101–1104.

Ramashia, S. E., Anyasi, T. A., Gwata, E. T., Meddows-Taylor, S., & Jideani, A. I. O. (2019). Processing, nutritional composition and health benefits of finger millet in subSaharan Africa. *Food Science Technology, 39*, 253–266.

Ravindran, G. (1992). Seed protein of millets: Amino acid composition, proteinase inhibitors and in-vitro protein digestibility. *Food Chemistry, 44*, 13–17.

Rohn, S., Rawel, H. M., & Kroll, J. (2002). Inhibitory effects of plant phenols on the activity of selected enzymes. *Journal of Agricultural and Food Chemistry, 50*, 3566–3571.

Rurinda, J., Mapfumo, P., van Wijk, M. T., Mtambanengwe, F., Rufino, M. C., Chikowo, R., & Giller, K. E. (2014). Comparative assessment of maize, finger millet and sorghum for household food security in the face of increasing climatic risk. *European Journal of Agronomy, 55*, 29–41.

Saleh, A. S., Zhang, Q., Chen, J., & Shen, Q. (2013). Millet grains: Nutritional quality, processing, and potential health benefits. *Comprehensive Reviews in Food Science and Food Safety, 12*, 281–295.

Seetharam, A., Riley, K. W., & Harinarayana, G. (1986). Small millets in global agriculture. In *Proceedings of the first international small millets workshop, Oct 29–Nov 2* (pp. 21–23). Oxford & IBH Publishing Co. Pvt. Ltd.

Serna-Saldivar, S. O. (2010). *Cereal grains properties, processing and nutritional attributes* (pp. 538–562). CRC Press.

Seth, D., & Rajamanickam, G. (2012). Development of extruded snacks using soy, sorghum, millet and rice blend – A response surface methodology approach. *International Journal of Food Science and Technology, 47*, 1526–1531.

Shahidi, F., & Chandrasekara, A. (2013). Millet grain phenolics and their role in disease risk reduction and health promotion: A review. *Journal of Functional Foods, 5*, 570–581.

Shahidi, F., & Chandrasekara, A. (2015). Processing of millet grains and effects on non-nutrient antioxidant compounds. In V. Preedy (Ed.), *Processing and impact on active components in food* (pp. 345–352). Academic Press.

Sharma, D., Jamra, G., Singh, U. M., Sood, S., & Kumar, A. (2017). Calcium biofortification: Three pronged molecular approaches for dissecting complex trait of calcium nutrition in finger millet (*Eleusine coracana*) for devising strategies of enrichment of food crops. *Frontiers in Plant Science, 7*, 1–18.

Shiihii, S. U., Musa, H., Bhati, P. G., & Martins, E. (2011). Evaluation of physicochemical properties of *Eleusine coracana* starch. *Nigerian Journal of Pharmaceutical Sciences, 10*, 91–102.

Shobana, S., Harsha, M. R., Platel, K., Srinivasan, K., & Malleshi, N. G. (2010). Amelioration of hyperglycaemia and its associated complications by finger millet (*Eleusine coracana* L.) seed coat matter in streptozotocin-induced diabetic rats. *British Journal of Nutrition, 104*, 1787–1795.

Shobana, S., & Malleshi, N. G. (2007). Preparation and functional properties of decorticated finger millet (*Eleusine coracana*). *Journal of Food Engineering, 79*, 529–538.

Siwela, M., Taylor, J. R. N., & Duodu, K. G. (2009). Phenolic content, antioxidant activity and sensory acceptability of wheat-finger millet composite cookies. *Journal of Food Science and Technology, 2*, 52–55.

Songré-Ouattara, L. T., Mouquet-Rivier, C., Humblot, C., Rochette, I., Diawara, B., & Guyot, J. P. (2010). Ability of selected lactic acid bacteria to ferment a pearl millet–soybean slurry to produce gruels for complementary foods for young children. *Journal of Food Science, 75*, 261–269.

Sood, S., Kant, L., & Pattnayak, A. (2017). Finger millet [*Eleusine coracana* (L.) Gaertn.]: A minor crop for sustainable food and nutrition. (Mini Review). *Asian Journal of Chemistry, 29*, 707–710.

Sreeramulu, D., Reddy, C. V. K., & Raghunath, M. (2009). Antioxidant activity of commonly consumed cereals, millets, pulses and legumes in India. *Indian Journal of Biochemistry and Biophysics*, *46*, 112–115.

Sripriya, G., Chandrasekharan, K., Murty, V. S., & Chandra, T. S. (1996). ESR spectroscopic studies on free radical quenching action of finger millet (*Eleusine coracana*). *Food Chemistry*, *7*, 537–540.

Subba Rao, M. V. S. S. T. S., & Muralikrishna, G. (2001). Non-starch polysaccharides and bound phenolic acids from native and malted finger millet (Ragi, *Eleusine coracana*, Indaf – 15). *Food Chemistry*, *72*, 187–192.

Subba Rao, M. V. S. S. T. S., & Muralikrishna, G. (2002). Evaluation of the antioxidant properties of free and bound phenolic acids from native and malted finger millet (Ragi, *Eleucine coracana* Indaf-15). *Journal of Agricultural and Food Chemistry*, *50*, 889–892.

Subba Rao, M. V. S. S. T. S., & Muralikrishna, G. (2004). Structural analysis of arabinoxylans isolated from native and malted finger millet (*Eleusine coracana*, ragi). *Carbohydrate Research*, *339*, 2457–2463.

Subba Rao, R. S. P., & Muralikrishna, G. (2006). Water soluble feruloyl arabinoxylans from rice and ragi: Changes upon malting and their consequence on antioxidant activity. *Phytochemistry*, *67*, 91–99.

Subba Rao, R. S. P., & Muralikrishna, G. (2007). Structural characteristics of water-soluble feruloyl arabinoxylans from rice (*Oryza sativa*) and ragi (finger millet, *Eleusine coracana*): Variations upon malting. *Food Chemistry*, *104*, 1160–1170.

Tatala, S., Ndossi, G., Ash, D., & Mamiro, P. (2007). Effect of germination of finger millet on nutritional value of foods and effect of food supplement on nutrition and anaemia status in Tanzanian children. *Tanzanian Health Research Bulletin*, *9*, 77–86.

Tatham, A. S., Fido, R. J., Moore, C. M., Kasarda, D. D., Kuzmicky, D. D., Keen, J. N., & Shewry, P. R. (1996). Characterization of the major prolamins of tef (*Eragrostis tef*) and finger Millet (*Eleusine coracana*). *Journal of Cereal Science*, *24*, 65–71.

Thapliyal, V., & Singh, K. (2015). Finger millet: Potential millet for food security and power house of nutrients. *International Research in Agriculture and Forestry*, *2*, 22–33.

Tovey, F. I. (1994). Diet and duodenal ulcer. *Journal of Gastroenterology and Hepatology*, *9*, 177–185.

Tsafrakidou, P., Michaelidou, A. M. G., & Biliaderis, C. (2020). Fermented cereal-based products: Nutritional aspects, possible impact on gut microbiota and health implications *Foods*, *9*, 734–758.

Udeh, H. O., Duodu, K. G., & Jideani, A. I. O. (2017). Finger millet bioactive compounds, bioaccessibility and potential health effects – A review. *Czech Journal of Food Science*, *35*, 7–17.

Udeh, H. O., Duodu, K. G., & Jideani, A. I. O. (2018). Malting period effect on the phenolic composition and antioxidant activity of finger millet (*Eleusine Coracana* L. Gaertn) Flour. *Molecules*, *23*, 2091–2103.

Vidyavati, H. G., Begum, J., Vijayakumari, J., Gokavi, S., & Begum, S. (2004). Utilization of finger millet in preparation of Papad. *Journal of Food Science and Technology*, *41*, 379–382.

Wuyts, S., Van Beeck, W., Allonsius, C. N., van den Broek, M. F., & Lebeer, S. (2020). Applications of plant-based fermented foods and their microbes. *Current Opinion in Biotechnology*, *61*, 45–52.

12 Underutilized Plants and Hidden Hunger

Suitability of Hydroponics for Achieving Diet Diversity

Oluwadamilola Mariam Baruwa,[1]
Samuel Oloruntoba Bamigboye,[1] *Avela Sogoni,*[2]
Abdulazeez Opeyemi Lawal,[1] *and Muhali Olaide Jimoh*[1*]
[1]Department of Plant Science, Olabisi Onabanjo University, PMB 2002, Ago-Iwoye 120107, Nigeria
[2]Department of Horticultural Sciences, Cape Peninsula University of Technology, Symphony Way, Bellville 7535, South Africa
*Corresponding author: moajay006@gmail.com

12.1 INTRODUCTION

Reduction in dietary diversity is one of the major causes of hidden hunger in the world, and more than 2 billion people worldwide suffer from this phenomenon (Li & Zhang, 2024; Weffort & Lamounier, 2024). *Dietary diversity* is a measure of food range belonging to different food groups consumed by an individual over a defined time (Amao et al., 2023; Samosir et al., 2023). On the other hand, *hidden hunger* describes the non-explicit need for one or more of the 26 micronutrients that are essential for adequate body function (Teklu et al., 2023). Globally, more than 2 billion people are faced with hidden hunger arising from micronutrient deficiencies, and its effects can be devastating, leading to mental impairment, poor health, low productivity, and even death (Harding et al., 2018; Lopes et al., 2023; Tulchinsky, 2010). Poor diet is a common source of hidden hunger. Diets that mostly consist of staple crops, such as maize, wheat, rice, and cassava, provide a large share of energy but relatively low amounts of essential vitamins and minerals, frequently resulting in hidden hunger (Ali & Bhattacharjee, 2023). Poverty also plays a major role as it limits access to diverse nutritious foods (Headey et al., 2023; Zezza & Tasciotti, 2010). Energy-dense food such as wheat, rice, and potatoes are consumed due to their affordable price range in low-income households, leading to a reduction in dietary diversity and micronutrient deficiencies (Pingali et al., 2019). Increasing dietary diversity is one of the most effective ways to sustainably prevent hidden hunger normally caused by the deficiency in iron, zinc, and vitamin A (Li et al., 2020). Hydroponics can serve as a simplified method of promoting dietary diversity, often through home gardening, consequently decreasing hidden hunger by some percentage (Jomova et al., 2022). A minimal hydroponic setup can grow up to 60 plants ranging from tomatoes to vegetables, fruits, and herbs within a short period of time.

The first noted use of hydroponics was conducted by W. F. Gericke of the University of California, who first coined the term *hydroponics* (Gericke, 1938). The word is derived from two Greek terms: *hydro* (meaning "water") and *ponos* (meaning "labor"). Thus, *hydroponics* literally means "water working." Urbanization reduces available land that would have been devoted to farming (Masters

DOI: 10.1201/9781003469766-12

et al., 2013; Zhou et al., 2021). People living in the cities also have less access to fresh produce. Instead of waiting for seasonal farmers markets and community gardens, however, hydroponics seems to promise a year-round source of diverse vegetables (Rufí-Salís et al., 2020). According to the United Nations (2019), adaptation is necessary to meet the demands of a growing population, especially in areas where there is lack of farmable land and resources. When hydroponics is adopted, it helps solve a lot of agricultural problems, including hidden hunger. Thus, this chapter is aimed at exploring the potential use of vegetables and the suitability of hydroponics in achieving diet diversity.

12.2 LITERATURE SEARCH AND DATA COMPILATION

A comprehensive literature search was conducted electronically using Google Scholar and Web of Science, as described by Jimoh et al. (2022). The search was limited to articles published between 2005 and 2024, with a focus on studies written in English, although a few references from older literature were considered. Search terms included "diet diversity," "hydroponics," and "hidden hunger." Boolean operators were used to combine search terms for optimal retrieval of relevant literature. Initial screening of search results was performed based on titles and abstracts to identify potentially relevant articles. Articles that did not meet the inclusion criteria were excluded.

12.3 DIET DIVERSITY AND ITS RELATIONSHIP WITH HIDDEN HUNGER

Diet diversity plays a crucial role in addressing hidden hunger (Li & Zhang, 2024). Large portions of sub-Saharan Africa and the South Asian subcontinent are recognized as hot spots with a high prevalence of hidden hunger, primarily due to limited knowledge of diverse nutrient indices in dietary intake, leading to an increasing reliance in processed foods (Harding et al., 2018). In Latin America and the Caribbeans, rates are comparatively low, as diets are less reliant on singular staples and are influenced more by widespread deployment of micronutrient interventions, nutrition education, and basic health services, although increasing income causes a decline in the relative importance of food consumption in these regions (Keats et al., 2019).

Despite the growing complexity of the world's malnutrition challenges, developing nations are transitioning from traditional diets, centered on minimally processed foods, to consuming highly processed, energy-dense, and nutrient-poor foods and beverages (Chin et al., 2024). This shift contributes to obesity and the rise of diet-related chronic diseases. With this nutrition transition, many developing countries face a phenomenon known as the "triple burden" of malnutrition–undernourishment, micronutrient deficiencies, and obesity (Barrett, 2021). In higher-income and more urbanized countries, hidden hunger can coexist with overweight or obesity when a person consumes too much dietary energy from macronutrients such as fats and carbohydrates (Eggersdorfer et al., 2018). While it may seem paradoxical, an obese child can suffer from hidden hunger. A person might have access to enough calories but lack adequate micronutrients. This phenomenon has been defined as "hidden hunger" because its symptoms are not always obvious, and people may not even be aware of it.

Increasing dietary diversity is one of the most effective ways to sustainably prevent hidden hunger. Diversifying one's diet is crucial for ensuring a healthy intake of a balanced combination of macronutrients (carbohydrates, fats, and protein), essential micronutrients, and dietary fiber (Li et al., 2021). These essential nutrients may be derived from a variety of cereals, legumes, fruits, vegetables, and animal-source foods, which can provide sufficient nutrition for most individuals.

12.4 HIDDEN HUNGER AND ITS GLOBAL IMPACT

The most recent estimates from FAO indicate that 840 million people do not receive enough energy from their diets to meet their nutritional needs, and most of these people live in developing countries (Akano et al., 2023). Adequate energy intake from the food supply doesn't guarantee the absence of micronutrient deficiencies in populations (Hameed et al., 2023). In this situation, people are not considered "hungry" in the classical sense, but their diets may be lacking in one or more essential micronutrients. Indeed, the consequences of subclinical nutrient deficiency are increasingly being recognized for their impact on health, particularly in terms of immune system functioning, growth, and cognitive development (Miller et al., 2020). Despite their significant effects, these deficiencies often go unnoticed within communities, earning them the moniker "hidden hunger." This term highlights the subtle yet profound impact of insufficient micronutrient intake on individuals, particularly in developing countries, where dietary diversity is limited.

Micronutrient deficiencies are particularly prevalent in areas where diets lack variety (Panzeri et al., 2024). When individuals cannot afford or access a diverse array of foods, they are more likely to experience deficiencies in essential vitamins and minerals (Beal et al., 2017). Fruits, vegetables, and animal-source foods are rich sources of micronutrients, and their absence in the diet increases the risk of deficiencies (Miller & Welch, 2013). Furthermore, it is important to note that the absorption of certain vitamins, particularly vitamins A, D, E, and K, requires fat intake (Dowling, 2020; Sajovic et al., 2022). Without an adequate intake of fats, even if the diet contains enough of these vitamins, their absorption may be compromised, leading to deficiencies despite apparent consumption (Eilander et al., 2015).

Vitamin A is crucial for various bodily functions, including normal growth, tissue repair, vision, and immune system functioning (Dowling, 2020). It exists in the form of retinol and carotene. Retinol is found in foods like eggs, milk, and fish, while carotene is a precursor to vitamin A found in yellow fruits and vegetables, green leafy vegetables, and red palm oil (Sajovic et al., 2022). Retinol is more easily absorbed by the body than carotene, although consuming dietary fat alongside carotene-rich foods can enhance its bioavailability. Vitamin A deficiency is still the leading cause of preventable blindness in children (Yisak et al., 2020). However, clinical forms of the deficiency are now becoming less frequent, and detection of subclinical deficiency is gaining more importance. It is estimated that improving the vitamin A status in children can potentially decrease the overall child mortality rates by 25%, measles death rates by 50%, and death caused by diarrhea by 40% (Baye et al., 2022).

Recently, global attention to zinc deficiency has also increased significantly (Chasapis et al., 2020). Despite this, there remains a lack of comprehensive information regarding the prevalence of zinc deficiency, although it is believed to be widespread in areas where dietary diversity is lacking (Gupta et al., 2020). Zinc plays a crucial role in the body, serving as an essential component in over 300 enzymes necessary for metabolism (Chasapis et al., 2020). The National Institutes of Health (NIH) recommends that males 19 years and older aim for 11 mg of zinc daily, and that females in this age group consume 8 mg (Barreca et al., 2023). They recommend pregnant females to consume 11 mg daily, while those who are breastfeeding should consume 12 mg. Fish, meat, and eggs are well-known sources of zinc, although it is important to note that certain plant-based foods, such as cereals and legumes, also contain zinc (Adelowo, 2022). However, the zinc content in these foods tends to be lower, and some may contain phytates, which can inhibit zinc absorption. Therefore, relying solely on these foods for zinc intake may increase the risk of deficiency, making it crucial for individuals to adopt diversification.

Iron deficiency remains the primary cause of anemia, although deficiencies in vitamin B_{12}, congenital defects in red cells, blood losses during reproduction, and infections by parasites like malaria, or infestations such as hookworm, can also lead to anemia (Makhmonov et al., 2021). Hemoglobin levels serve as the most widely used indicator for screening iron-deficiency anemia, providing extensive data globally (De Franceschi et al., 2017). The true prevalence of iron deficiency in a population will be larger than the level of clinically detectable iron-deficiency anemia, because most individuals are likely to be iron deficient long before there is a detectable drop in blood hemoglobin levels (Makhmonov et al., 2021). Nutritional iron deficiency, or habitual iron intake that is insufficient to cover requirements, is the most common cause of iron deficiency (FAO et al., 2023). Dietary iron exists in two forms: heme iron, abundant in animal-source foods like meat, poultry, and fish, and non-heme iron, found in cereals, pulses, fruits, and vegetables. Ascorbic acid, present in many fruits and vegetables, enhances iron absorption (Hurrell & Egli, 2010; Vogt et al., 2021).

When individuals cannot afford a diverse diet rich in fruits, vegetables, and animal-source foods containing essential micronutrients, deficiencies become inevitable. Addressing micronutrient deficiencies is best tackled through food-based approaches that enhance overall dietary quality, rather than targeting single nutrients (Henson & Agnew, 2021; Lopes et al., 2023). This holistic approach is supported by several factors. First, there are complex nutrient–nutrient interactions that increase bioavailability when nutrients are consumed simultaneously. For example, iron absorption is increased when it is combined with vitamin C (Basrowi & Dilantika, 2021). However, ongoing research reveals the protective benefits of phytochemicals and antioxidants readily available through consuming a diverse array of fruits and vegetables. As scientific understanding of nutrition and disease deepens, an expanding range of nutrients with diverse roles in maintaining health is being implicated, highlighting the importance of a varied and balanced diet.

12.5 UNDERUTILIZED VEGETABLES AND THEIR POTENTIAL IN COMBATING HIDDEN HUNGER

Since the commencement of the green revolution in the 1960s, agriculture has primarily prioritized the advancement and growth of essential food crops, such as rice, wheat, and maize (Tian et al., 2021). While the green revolution led to a significant boost in food output in developed nations, its impact in Africa has often been seen as limited (Frankema, 2014). The primary emphasis of the green revolution focused on improving staple crops like grains and traditional horticulture crops, while wild vegetables were neglected (Ebert, 2014; Sahoo et al., 2021). Throughout history, wild plants have served as important sources of nutrition during periods of drought and as dietary reinforcements, particularly in communities that rely on hunting and gathering (Borelli et al., 2020). Wild vegetables, along with other wild plant foods, are often known as the "hidden harvest" since they are gathered directly from the wild, such as agricultural fields and marshy regions, without any intentional cultivation for food purposes (Ghosh-Jerath et al., 2021).

Wild vegetables, such as African kale (*Brassica carinata*), jute mallow (*Corchorus carinata*), slender leaf (*Crotalaria brevidens*), leafy amaranth (*Amaranthus* spp.), dune spinach (*Tetragonia decumbens*), wild cabbage (*Trachyandra divaricata*), cowpea (*Vigna unguiculata*), pumpkin (*Cucurbita moschata*), African nightshade (*Solanum scabrum*), and spider plant (*Cleome gynandra*), are known to be a primary source of essential nutrients and have the potential to combat widespread nutritional shortages (Afolayan & Jimoh, 2009; Bulawa et al., 2022; Bvenura & Sivakumar, 2017; Cebani et al., 2024; Ntuli, 2019; Salami et al., 2022; Sogoni, 2020). These species contain essential elements, including macronutrients and micronutrients, which are necessary for important cellular functions and the periodic repair of damaged tissues (Punchay et al., 2020). Leafy vegetables aid in the restoration of energy depletion caused by oxidative stress by capturing free radicals and their corresponding biomolecules (Medina-Lozano et al., 2021). Nevertheless, the failure to fully use green vegetables has led to conclusive proof of metabolic abnormalities, specifically classified as malnutrition, undernutrition, and stunting. Thus, the inclusion of these neglected leafy greens in human diets is necessary in reducing hidden hunger resulting from micronutrient deficiencies.

The worldwide deficit of iron, vitamin A, zinc, and iodine is very frequent in African nations, notably among young children (Friesen et al., 2020; Harika et al., 2017). Occurrences of stunted development, low birth weight, and underweight in children have been the principal focus of government efforts with an aim to dramatically decrease such occurrences (Lemoine & Tounian, 2020). Wild vegetables have been reported to possess high amount of minerals, dietary fiber, and a range of vitamins, including vitamins A, C, E, B_1, B_6, and B_9 (Chyne et al., 2019; Lata et al., 2023). They are also considered for their nutraceutical value, as they aid in avoiding or slowing down the development of certain gastrointestinal illnesses and chronic diseases (Salami et al., 2022). Moreover, these vegetables require little effort in cultivation and are adapted to changing climatic conditions. Thus, their cultivation in hydroponic systems is crucial in bridging the gap of diet diversity. However, up to date, there are still few studies conducted on these wild vegetables in a hydroponic setup. and the following table shows the few species that have been successfully grown in these systems.

12.6 HYDROPONICS: BASICS AND PRINCIPLES

Hydroponics is the cultivation of plants by placing the roots in liquid nutrient solutions rather than in soil (Sathyanarayan et al., 2023). Depending on one's knowledge regarding plant-growing techniques, it may be startling to learn that plants do not need dirt to grow. After all, most would consider dirt, sunlight, and water as the trifecta required for plant growth. The technique of hydroponics turns this basic principle that has been around since our ancestors transitioned from hunter-gatherers to farmers (Sambo et al., 2019). Indeed, while hydroponics might seem like a product of modern technology, its origins trace back much further (Kumar & Singh, 2023). The first known use of hydroponics dates to Babylon's Hanging Gardens, although the veracity of this account is subject to individual interpretation (Bomford, 2023).

Today, botanists use many different types of hydroponics: nutrient film techniques (NFT), aeroponics, aeration, flood and drain, trickle feed method, and tube culture methods. NFT involves pumping a nutrient solution over the roots, which then drains back into the reservoir (Ngxabi et al., 2021). Aeroponics involves misting exposed roots with a nutrient solution every few minutes; therefore, to prevent the roots from drying out, misting must be continual (Tunio et al., 2021). The aeration method utilizes an aquarium air pump to deliver oxygen to the roots; the roots themselves are suspended in a nutrient solution (Butcher et al., 2017). The flood-and-drain system involves intermittently immersing the roots in a nutrient solution and then draining the solution back through a valve (Serio et al., 2022).

The trickle feed method has a nutrient solution pumped through tubes to deliver the solution to each plant container (Maya Waiba et al., 2020). The tube culture technique is a variant of trickle feed where a vertical bag is filled with aggregate media, holes are made for each plant, and an irrigation tube is placed at the top. To many, the world of hydroponics with its various systems and complexities may seem confusing and foreign (Sharma et al., 2018). One may question what is wrong with the traditional system of raising plants in soil. While traditional soil-based agriculture is effective, hydroponic systems offer several advantages, including year-round growing seasons, water conservation, and reduced pesticide usage (Majid et al., 2021). These benefits can lead to increased plant yields, aiding in feeding a growing global population, which currently stands at 7.3 billion. Adaptation to methods like hydroponics is crucial, particularly in regions with limited arable land and resources, to meet increasing food demands (Goodman & Minner, 2019). This is emphasized by the United Nations as a necessity in addressing future food security challenges.

Urbanization reduces available land that would otherwise be devoted to farming. People living in the cities also have less access to fresh produce. Instead of waiting for seasonal farmers markets and community gardens, hydroponics seems to promise a year-round source of vegetables (Goodman & Minner, 2019). In addition, hydroponics is advantageous because it helps conserve water. This may seem counterintuitive at first – after all, hydroponics grows plants in water – however, hydroponics conserves water because it allows for the precise water and nutrient application directly to the roots of each plant (Kumar & Singh, 2023). Another benefit of hydroponics is the lessened need for pesticides (Rajaseger et al., 2023). The absence of soil eliminates the risk of soil-based diseases, as well as exposure to insects and animals, which are common concerns in traditional agriculture (Pascual et al., 2018). This reduced reliance on pesticides, coupled with the year-round growing capability and water conservation features of hydroponics, makes it an appealing solution for cultivating plants in the face of population growth and agricultural challenges.

12.7 COMPARING THE ROLE OF HYDROPONICS WITH TRADITIONAL FARMING IN SUSTAINABLE AGRICULTURE

When growing plants in soil, the nutrition of the soil is highly variable and can be unavailable to plants due to poor soil structure (Pomoni et al., 2023) – also, the soil can be difficult to sample, test, and adjust – while plants grown in soilless conditions benefit from controlled and stable nutrition, with nutrient availability easily adjusted and monitored (Song et al., 2022). Unlike soil-based cultivation, where plant spacing is restricted by soil nutrition and available light, soilless systems allow for closer and vertical planting, maximizing space usage (Viljoen et al., 2021). Furthermore, weed control and cultivation tasks common in soil-based farming are unnecessary in soilless systems, as there are no weeds to contend with and no soil to till.

Water usage efficiency is significantly enhanced in soilless conditions compared to traditional soil-based farming (Kannan et al., 2022; Song et al., 2022). Hydroponic greenhouses use about ten times less water than in field cultivation (Kannan et al., 2022). In traditional farming, the entire field is sprayed with water, with irrigation eventually seeping into the roots. However, in hydroponics, less water can be used because it goes directly to the roots.

In soil-based cultivation, fruit quality can suffer from deficiencies in essential nutrients like calcium and potassium, resulting in poorer-quality fruit with shorter shelf life. However, in soilless conditions such as hydroponics, fruits tend to be firmer and have a longer shelf life (El-Essawy et al., 2019). This enables growers to harvest vine-ripened fruit and transport it over longer distances. Moreover, studies have indicated that the nutritional value of fruits and vegetables grown hydroponically is equal to or even higher than those of grown in soil, further highlighting the benefits of soilless cultivation methods (Song et al., 2022).

12.8 ROLE OF HYDROPONICS IN DIETARY DIVERSIFICATION

The consumption of fruits and vegetables significantly reduces the risk of many types of chronic disease in humans (Wallace et al., 2020). Various bioactive compounds and nutrients, including beta-carotene and antioxidants found in vegetables, offer advantageous effects on health (Kawashima & Valente Soares, 2003; Punchay et al., 2020; Sahoo et al., 2021). Therefore, employing eco-friendly cultivation methods like hydroponics presents the potential to enhance the concentration of health-promoting compounds and elevate the overall quality of fruits and vegetables. Dietary diversity indirectly addresses hidden hunger by providing essential nutrients crucial for overall health (Gödecke et al., 2018). Dark-green leafy vegetables, fruit, and vegetables, such as tomatoes, and herbs like oregano, thyme, and fennel serve as significant sources of provitamin A and zinc, while peas and spinach contribute to iron intake (Yahia et al., 2019). Notably, all these plants can be cultivated hydroponically. Contrary to popular belief, hidden hunger is not confined to underdeveloped regions; it significantly affects both developing and developed regions of the world, often unnoticed. A primary contributor to hidden hunger is poor dietary choices, particularly diets heavily reliant on staple crops like maize, wheat, rice, and cassava. While these crops provide ample energy, they often lack essential vitamins and minerals, consequently exacerbating the prevalence of hidden hunger (Lenaerts & Demont, 2021).

A recent FAO report, "The State of Food Insecurity in the World," states that "[a] stock-taking of where we stand on reducing hunger and malnutrition shows that progress in hunger reduction at the global level has continued but that food insecurity is still a challenge to be conquered" (FAO et al., 2023). With increasing concerns about the poor levels of food security, subsistence farming should be encouraged using hydroponics. In South Africa, an estimated 20.7% of the households engage in agriculture, and 65% of these households use agriculture purely as a subsistence strategy to meet household food demand (Tibesigwa & Visser, 2016). FAO et al. (2023) further states:

> For decades, poverty and food insecurity were primarily associated with rural areas. Nevertheless, recent analyses reveal a significant increase in urban poverty, which has often been underestimated. Unlike rural settings, urban environments witness a more direct translation of income insufficiency into food scarcity.

Access to land is one of the main obstacles of urban agriculture (Tornaghi, 2017). Employing hydroponics for household farming addresses the issue of limited land availability while simultaneously boosting yield per plant (Tornaghi, 2017). By cultivating various crop varieties hydroponically, families can enhance their dietary diversity and use the additional income earned by selling the surplus crops to access animal-based food sources (Lal, 2020).

12.9 ECONOMIC VIABILITY OF HYDROPONIC FARMING

According to the United Nations Food and Agriculture Organization (FAO), the world population should reach 9 billion by 2050, and 70% increase in food production would be required to feed such a growing population in the coming years (da Cunha Dias et al., 2021). Faced with these threats, new food production techniques are constantly being created to feed an expanding population with declining land resources (Giller et al., 2021). Despite criticism by proponents of agro-ecology approaches, hydroponics is promising as a soil protection technique because of its minimal use of soil (Viljoen et al., 2021).

Two aspects help increase crop yield, *viz.*, maximizing the number of plants per area and enhancing production per plant. In hydroponics, vertical layering allows for increased plant density per cultivated unit area compared to traditional soil-based cultivation, which typically utilizes only one layer (Touliatos et al., 2016). Additionally, production per plant is optimized in hydroponics through precise control of nutrients, light, and CO_2 levels (Manos & Xydis, 2019). Scientific studies, such as the one by Barbosa et al. (2015) in Table 12.1, have demonstrated significantly higher yields for hydroponically cultivated lettuce compared to soil-grown lettuce. These findings suggest that hydroponic systems not only increase yield but also have the potential to enhance the nutritional quality of plants through meticulous control throughout the cultivation process.

Policymakers have prioritized reforming the agricultural sector to address its negative externalities while simultaneously seeking to mitigate inconsistencies and improve food security for the poor and malnourished (Barel-Shaked & Buda, 2024; Kennedy et al., 2021). This emphasis has led to the exploration of alternative production systems and opportunities aimed at generating new employment, fostering economic growth, and increasing incomes (Barel-Shaked & Buda, 2024). Agricultural diversification indicates the changes in crop-mix, enterprises-mix, and activity-mix at the household level (Sonawane et al., 2022). It is considered a shift of resources from monoculture to a large mix of crops and livestock. The need for agricultural diversification arises due to various factors. For example, Chand (1999) found that in Punjab, labor absorption was 307 hours per acre in potato and 225 hours in other vegetable crops, whereas the corresponding figure for traditional crops like wheat, barley, and mustard cultivation was only 127 hours. Now imagine using hydroponics to cultivate these crops.

TABLE 12.1
Wild Vegetable Species That Have Been Cultivated Successfully in Hydroponics

Wild Species	Hydroponic System	Yield (Grams/Fresh Leaves/Plant)	Refences
Tetragonia decumbens	Nutrient film technique and drip system	210 and 245	(Nkcukankcuka et al., 2022, 2021; Sogoni et al., 2021)
Trachyandra divaricata	Nutrient film technique	183	(Ngxabi et al., 2021)
Rumex acetosa	Floating system	209.5	(Ceccanti et al., 2018)
Sanguisorba minor	Floating system	202.7	(Ceccanti et al., 2018)
Diplotaxis tenuifolia	Floating system	133.3	(Ceccanti et al., 2018)
Amaranthus cruentus	Nutrient film and ebb-and-flow system	89.9	(Croft et al., 2017)

TABLE 12.2
Comparing Yield of Hydroponically Grown and Soil-Grown Crops

Edible Crops	Hydroponic Yield	Soil Yield
Lettuce	41 $kg/m^2/y$	3.9 $kg/m^2/y$
Lettuce lamb	$1{,}585 \pm 304$ g/m^2	$1{,}203 \pm 304$ g/m^2
Chad	246.78 g/plan	228.22 g/plant
Eggplant	7.36 kg/m^2	6.47 kg/m^2

(Continued)

TABLE 12.2 (Continued)

Edible Crops	Hydroponic Yield	Soil Yield
Wheat	1,828 kg/ac	1,762 kg/ac in 7 days
Tomato var. Miramar	7,844 g/plant	5,118 g/plant
Cherry tomato	4,741.83 g/plant	3,513.58 g/plant
Basil	388.14 g/plan	326.64 g/plant
Parsley	414.64 g/plant	342.04 g/plant
Cayenne pepper	10.8 fruits/plant	8.6 fruits/plant
Sage	2,286 g/m^2	863 g/m^2
German chamomile	3,015 g/m^2	962 g/m^2
Jordanian chamomile	906 g/m^2	411 g/m^2
Thyme	1,283 g/m^2	518 g/m^2
Mint	3,208 g/m^2	1,320 g/m2
Strawberry	85 fruits	70 fruits
Sweet pepper	10.95 kg yield/m^2	9.40 kg yield/m^2
Bell pepper	1,277.88 g/plant	834.54 g/plant
Cucumber	4,727.38 g/plant	4,427.38 g/plant

Source: Barbosa et al. (2015).

12.10 HYDROPONICS CHALLENGES

12.10.1 Financial Implications

While hydroponics offers numerous advantages, there are significant barriers hindering its widespread adoption. One major obstacle is the cost associated with implementing hydroponic systems. There is a considerable initial capital investment required, as highlighted by Kannan et al. (2022). Additionally, the costs vary depending on the type and scale of the hydroponic operation, whether it is for home or commercial use. While hydroponics may be more financially feasible for individual home growers, the expense escalates significantly for commercial-scale operations. This cost factor poses a challenge for the broader adoption of hydroponic agriculture (Gumisiriza et al., 2022).

12.10.2 Abiotic Stresses

Abiotic stresses such as temperature, lack of oxygen, and salinity buildup can have a significant impact on both shoot and root tissues in hydroponic cultivation due to their rapid accessibility (Geary et al., 2015). These stresses primarily influence the composition of plants. High temperatures, for example, affect the nutrient uptake and metabolism in hydroponically cultivated lettuce by decreasing the leaf area and nitrogen content, while cool temperatures increase the sugar, malic acid, and potassium content (Gent, 2016).

Salinity, among the abiotic stresses, is also a major limiting factor for plant growth in hydroponics. Studies have shown that salinity reduces root water uptake due to osmotic effects, leading to water stress (Alfosea-Simón et al., 2020). However, salinity can positively affect tomato fruit quality while negatively impacting plant growth and fruit production. Furthermore, increases in light intensity and continuous light duration can enhance oxidative stress, as evidenced by increased production of reactive oxygen species (ROS) and lipid peroxidation (Zha et al., 2019). These findings highlight the importance of understanding and managing abiotic stresses in hydroponic systems to optimize plant growth and yield.

12.10.3 Biotic Stresses

In hydroponic systems, the likelihood of pest infestation is higher due to the conducive environment for various pests (Li et al., 2023). Managing different types of pests requires varying control methods, underscoring the importance of understanding pests and their control measures. Pest control in hydroponics follows similar procedures as in soil-based cultivation, emphasizing the necessity of daily monitoring to identify and address damaging pest incidences economically (Niu & Masabni, 2022). Common pests in hydroponic crop production include spider mites, aphids, thrips, whiteflies, and fungus gnats (Smagghe et al., 2023). Effective pest management programs prioritize prevention over control measures.

12.11 HYDROPONICS PROSPECTS

According to the World Food Programme, an estimated 795 million people worldwide lack sufficient food to maintain a healthy and active lifestyle, equating to approximately one in nine individuals globally (Mc Carthy et al., 2018). While hunger is a multifaceted problem, hydroponics has emerged as a potential component of the solution (Sousa et al., 2024). In a trial overseen by the Food and Agriculture Organization (FAO) of the United Nations, hydroponic micro-gardens were established in Senegal as part of efforts to explore innovative approaches to food production (de Marinis et al., 2023). Senegal is one of the world's most food-insecure countries, with 17.2% of its population being underweight from the years 2008 to 2012 (Ilboudo Nébié et al., 2021). To help combat hunger in Senegal, the FAO taught families how to set up rooftop hydroponic gardens. The plants were supported by an aggregate of gravel and groundnut shells or through holes punched in sheets of Styrofoam which floated on the water in the tray. Despite the simple construction, each 1 m^2 box provided up to 100 pounds of tomatoes each year. The micro-gardens helped supplement the families' regular diets while also providing income, as the extra produce could be sold at the market. These micro-gardens were also relatively affordable. Although this test run was conducted on a small scale, its results look promising in reducing hunger in developing nations.

Nonetheless, the adoption of hydroponics in developing countries still faces many challenges, such as lack of knowledge transfer and financial support from the government. Moreover, holistic research focusing on hydroponic cultivation of vegetables in African countries has been ignored by policymakers and researchers, although this cultivation system is currently attracting interest in other parts of the world due to its high production returns in confined city spaces. Therefore, amending such policies and enhancing research funding and collaboration between universities and government research institutions could promote intensive research on cost-effective hydroponic structures. This will then unlock the resource value of these production systems through validated findings to meet the needs of potential consumers and further promote their small-scale commercialization in developing nations. Likewise, appropriate awareness and education on the benefits of hydroponic cultivation can potentially boost their use. This can be achieved through the inclusion of agricultural education curricula in universities that deal with the construction and assembling of hydroponic structures in both commercial and communal areas. Consequently, this will upgrade the knowledge base of rural dwellers and low-income groups on the importance of hydroponics and its benefits in improving nutrients diversity.

12.12 CONCLUSION AND PROSPECTS

In this chapter, literature on the potential use of vegetables and the suitability of hydroponics in achieving diet diversity was reviewed with a view to reveal its potential in cultivating various vegetable species to enhance dietary diversity within households. Micronutrient deficiencies are particularly prevalent in areas where diets lack variety. This is due to the lack of access to a variety of diverse foods, which results in deficiencies in essential vitamins and minerals. The outcome of the literature search shows that incorporating various leafy vegetables in diets can alleviate micronutrients deficiencies. Moreover, the use of hydroponics in crop cultivation has immense benefits when compared to traditional farming practices. Thus, hydroponics is suitable for cultivating various vegetables, with the aim of tackling micronutrients deficiencies.

12.13 ACKNOWLEDGMENTS

The authors wish to acknowledge Olabisi Onabanjo University, Ago-Iwoye, Nigeria, and Cape Peninsula University of Technology, Bellville, South Africa, for supporting this study with library resources.

REFERENCES

Adelowo, A. (2022). A critical review of diet supplementation: The rationale, concerns and guiding principles in modern healthcare. *Nigerian Journal of Basic and Clinical Sciences*. https://doi.org/10.4103/njbcs.njbcs_31_21

Afolayan, A. J., & Jimoh, F. O. (2009). Nutritional quality of some wild leafy vegetables in South Africa. *International Journal of Food Sciences and Nutrition*, *60*, 424–431. https://doi.org/10.1080/09637480701777928

Akano, O. I., Oderinde, F. O., & Omotayo, A. O. (2023). Agricultural yield, food nutrition and dietary energy supply in Nigeria: Evidence from nationally representative data. *Journal of Agriculture and Food Research*, *11*, 100525. https://doi.org/10.1016/J.JAFR.2023.100525

Alfosea-Simón, M., Zavala-Gonzalez, E. A., Camara-Zapata, J. M., Martínez-Nicolás, J. J., Simón, I., Simón-Grao, S., & García-Sánchez, F. (2020). Effect of foliar application of amino acids on the salinity tolerance of tomato plants cultivated under hydroponic system. *Scientia Horticulturae*, *272*, 109509. https://doi.org/10.1016/J.SCIENTA.2020.109509

Ali, A., & Bhattacharjee, B. (2023). Nutrition security, constraints, and agro-diversification strategies of neglected and underutilized crops to fight global hidden hunger. *Frontiers in Nutrition*, *10*, 1144439. https://doi.org/10.3389/FNUT.2023.1144439/BIBTEX

Amao, I. O., Ogunniyi, A. I., Mavrotas, G., & Omotayo, A. O. (2023). Factors affecting food security among households in Nigeria: The role of crop diversity. *Sustainability*, *15*, 8534. https://doi.org/10.3390/SU15118534

Barbosa, G. L., Almeida Gadelha, F. D., Kublik, N., Proctor, A., Reichelm, L., Weissinger, E., Wohlleb, G. M., & Halden, R. U. (2015). Comparison of land, water, and energy requirements of lettuce grown using hydroponic vs. conventional agricultural methods. *International Journal of Environmental Research and Public Health*, *12*, 6879–6891. https://doi.org/10.3390/IJERPH120606879

Barel-Shaked, S., & Buda, E. (2024). Exploring the nexus of agricultural policy reforms and food security. *Frontiers in Sustainable Food Systems*, *8*, 1366807. https://doi.org/10.3389/FSUFS.2024.1366807

Barreca, S., Orecchio, S., Orecchio, S., Abbate, I., & Pellerito, C. (2023). Macro and micro elements in traditional meals of Mediterranean diet: Determination, estimated intake by population, risk assessment and chemometric analysis. *Journal of Food Composition and Analysis*, *123*, 105541. https://doi.org/10.1016/J.JFCA.2023.105541

Barrett, C. B. (2021). Overcoming global food security challenges through science and solidarity. *American Journal of Agricultural Economics*, *103*, 422–447. https://doi.org/10.1111/AJAE.12160

Basrowi, R. W., & Dilantika, C. (2021). Optimizing iron adequacy and absorption to prevent iron deficiency anemia: The role of combination of fortified iron and vitamin C. *World Nutrition Journal*, *5*, 33–39. https://doi.org/10.25220/WNJ.V05.S1.0005

Baye, K., Laillou, A., Seyoum, Y., Zvandaziva, C., Chimanya, K., & Nyawo, M. (2022). Estimates of child mortality reductions attributed to vitamin A supplementation in Sub-Saharan Africa: Scale up, scale back, or refocus? *The American Journal of Clinical Nutrition*, *116*, 426–434. https://doi.org/10.1093/AJCN/NQAC082

Beal, T., Massiot, E., Arsenault, J. E., Smith, M. R., & Hijmans, R. J. (2017). Global trends in dietary micronutrient supplies and estimated prevalence of inadequate intakes. *PLoS One*, *12*, e0175554. https://doi.org/10.1371/JOURNAL.PONE.0175554

Bomford, M. (2023). More bytes per acre: Do vertical farming's land sparing promises stand on solid ground? *Agric Human Values*, *40*, 879–895. https://doi.org/10.1007/S10460-023-10472-0/FIGURES/1

Borelli, T., Hunter, D., Powell, B., Ulian, T., Mattana, E., Termote, C., Pawera, L., Beltrame, D., Penafiel, D., Tan, A., Taylor, M., & Engels, J. (2020). Born to eat wild: An integrated conservation approach to secure wild food plants for food security and nutrition. *Plants*, *9*, 1299. https://doi.org/10.3390/PLANTS9101299

Bulawa, B., Sogoni, A., Jimoh, M. O., & Laubscher, C. P. (2022). Potassium application enhanced plant growth, mineral composition, proximate and phytochemical content in trachyandra divaricata kunth (Sandkool). *Plants*, *11*, 3183. https://doi.org/10.3390/PLANTS11223183

Butcher, J. D., Laubscher, C. P., & Coetzee, J. C. (2017). A study of oxygenation techniques and the chlorophyll responses of Pelargonium tomentosum grown in deep water culture hydroponics. *HortScience*, *52*, 952–957. https://doi.org/10.21273/HORTSCI11707-16

Bvenura, C., & Sivakumar, D. (2017). The role of wild fruits and vegetables in delivering a balanced and healthy diet. *Food Research International*, *99*, 15–30. https://doi.org/10.1016/J.FOODRES.2017.06.046

Cebani, S., Jimoh, M. O., Sogoni, A., Wilmot, C. M., & Laubscher, C. P. (2024). Nutrients and phytochemical density in Mesembryanthemum crystallinum L. cultivated in growing media supplemented with dosages of nitrogen fertilizer. *Saudi Journal of Biological Sciences*, *31*, 103876. https://doi.org/10.1016/J.SJBS.2023.103876

Ceccanti, C., Landi, M., Benvenuti, S., Pardossi, A., & Guidi, L. (2018). Mediterranean wild edible plants: Weeds or "new functional crops"? *Molecules*, *23*, 2299. https://doi.org/10.3390/MOLECULES23092299

Chand, R. (1999). *Effects of trade liberalization on agriculture in India: Commodity aspects*. CGPRT Centre.

Chasapis, C. T., Ntoupa, P. S. A., Spiliopoulou, C. A., & Stefanidou, M. E. (2020). Recent aspects of the effects of zinc on human health. *Archives of Toxicology*, *94*(5), 1443–1460. https://doi.org/10.1007/S00204-020-02702-9

Chin, C. F., Goh, E. V., Clarke, A. C., Ang, M. L. E., Show, P. L., Supramaniam, C. V., Ali, A., Azam-Ali, S., Beyer, R., Culaba, A., Munawaroh, H. S. H., Kairo, K., Khoo, K. S., Khoo, G. C., Matanjun, P., Massawe, F., Lim, M. H., Sankaran, R., Sriariyanun, M., Menbrebe, B. N. Q., Noi, P., & Lechner, A. M. (2024). Transforming food systems in Maritime Southeast Asia and Pacific small Island developing states to support food security and sustainable healthy diets. *Frontiers in Sustainable Food Systems*, *8*, 1304317. https://doi.org/10.3389/FSUFS.2024.1304317/BIBTEX

Chyne, D. A. L., Ananthan, R., & Longvah, T. (2019). Food compositional analysis of indigenous foods consumed by the Khasi of Meghalaya, North-East India. *Journal of Food Composition and Analysis*, *77*, 91–100. https://doi.org/10.1016/J.JFCA.2019.01.008

Croft, M. M., Hallett, S. G., & Marshall, M. I. (2017). Hydroponic production of vegetable Amaranth (Amaranthus cruentus) for improving nutritional security and economic viability in Kenya. *Renewable Agriculture and Food Systems*, *32*, 552–561. https://doi.org/10.1017/S1742170516000478

da Cunha Dias, T. A., Lora, E. E. S., Maya, D. M. Y., & del Olmo, O. A. (2021). Global potential assessment of available land for bioenergy projects in 2050 within food security limits. *Land Use Policy*, *105*, 105346. https://doi.org/10.1016/J.LANDUSEPOL.2021.105346

De Franceschi, L., Iolascon, A., Taher, A., & Cappellini, M. D. (2017). Clinical management of iron deficiency anemia in adults: Systemic review on advances in diagnosis and treatment. *European Journal of Internal Medicine*, *42*, 16–23. https://doi.org/10.1016/j.ejim.2017.04.018

de Marinis, P., Mazzocchi, C., & Sali, G. (2023). Perception of microgardens in Dakar, Senegal. *Journal of Agriculture and Environment for International Development*, *114*, 77–98. https://doi.org/10.12895/JAEID.20201.1202

de Sousa, R., Bragança, L., da Silva, M. V., & Oliveira, R. S. (2024). Challenges and solutions for sustainable food systems: The potential of home hydroponics. *Sustainability*, *16*, 817. https://doi.org/10.3390/SU16020817

Dowling, J. E. (2020). Vitamin A: Its many roles – from vision and synaptic plasticity to infant mortality. *Journal of Comparative Physiology A: Neuroethology, Sensory, Neural, and Behavioral Physiology*, *206*, 389–399. https://doi.org/10.1007/S00359-020-01403-Z/FIGURES/8

Ebert, A. (2014). Potential of underutilized traditional vegetables and legume crops to contribute to food and nutritional security, income and more sustainable production systems. *Sustainability*, *6*, 319–335. https://doi.org/10.3390/su6010319

Eggersdorfer, M., Akobundu, U., Bailey, R. L., Shlisky, J., Beaudreault, A. R., Bergeron, G., Blancato, R. B., Blumberg, J. B., Bourassa, M. W., Gomes, F., Jensen, G., Johnson, M. A., Mackay, D., Marshall, K., Meydani, S. N., & Tucker, K. L. (2018). Hidden hunger: Solutions for America's aging populations. *Nutrients*, *10*, 1210. https://doi.org/10.3390/NU10091210

Eilander, A., Harika, R. K., & Zock, P. L. (2015). Intake and sources of dietary fatty acids in Europe: Are current population intakes of fats aligned with dietary recommendations? *European Journal of Lipid Science and Technology, 117*, 1370–1377. https://doi.org/10.1002/ejlt.201400513

El-Essawy, H., Nasr, P., & Sewilam, H. (2019). Aquaponics: A sustainable alternative to conventional agriculture in Egypt – A pilot scale investigation. *Environmental Science and Pollution Research, 26*, 15872–15883. https://doi.org/10.1007/S11356-019-04970-0/TABLES/5

FAO, IFAD, UNICEF, WFP, & WHO. (2023). *The state of food security and nutrition in the World 2023*. Author. https://doi.org/10.4060/cc3017en

Frankema, E. (2014). Africa and the green revolution a global historical perspective. *NJAS: Wageningen Journal of Life Sciences, 70*, 17–24. https://doi.org/10.1016/J.NJAS.2014.01.003

Friesen, V. M., Mbuya, M. N. N., Aaron, G. J., Pachón, H., Adegoke, O., Noor, R. A., Swart, R., Kaaya, A., Wieringa, F. T., & Neufeld, L. M. (2020). Fortified foods are major contributors to apparent intakes of vitamin A and iodine, but not iron, in diets of women of reproductive age in 4 African countries. *Journal of Nutrition, 150*, 2183–2190. https://doi.org/10.1093/JN/NXAA167

Geary, B., Clark, J., Hopkins, B. G., & Jolley, V. D. (2015). Deficient, adequate and excess nitrogen levels established in hydroponics for biotic and abiotic stress-interaction studies in potato. *Journal of Plant Nutrition, 38*, 41–50. https://doi.org/10.1080/01904167.2014.912323

Gent, M. P. N. (2016). Effect of temperature on composition of hydroponic lettuce. *Acta Horticulturae, 1123*, 95–100. https://doi.org/10.17660/ACTAHORTIC.2016.1123.13

Gericke, W. F. (1938). Crop production without soil. *Nature, 141*, 536–540. https://doi.org/10.1038/141536A0

Ghosh-Jerath, S., Kapoor, R., Barman, S., Singh, G., Singh, A., Downs, S., & Fanzo, J. (2021). Traditional food environment and factors affecting indigenous food consumption in munda tribal community of Jharkhand, India. *Frontiers in Nutrition, 7*, 600470. https://doi.org/10.3389/FNUT.2020.600470/BIBTEX

Giller, K. E., Delaune, T., Silva, J. V., Descheemaeker, K., van de Ven, G., Schut, A. G. T., van Wijk, M., Hammond, J., Hochman, Z., Taulya, G., Chikowo, R., Narayanan, S., Kishore, A., Bresciani, F., Teixeira, H. M., Andersson, J. A., & van Ittersum, M. K. (2021). The future of farming: Who will produce our food? *Food Security, 13*, 1073–1099. https://doi.org/10.1007/S12571-021-01184-6/FIGURES/8

Gödecke, T., Stein, A. J., & Qaim, M. (2018). The global burden of chronic and hidden hunger: Trends and determinants. *Global Food Security Index, 17*, 21–29. https://doi.org/10.1016/J.GFS.2018.03.004

Goodman, W., & Minner, J. (2019). Will the urban agricultural revolution be vertical and soilless? A case study of controlled environment agriculture in New York City. *Land Use Policy, 83*, 160–173. https://doi.org/10.1016/J.LANDUSEPOL.2018.12.038

Gumisiriza, M. S., Kabirizi, J. M. L., Mugerwa, M., Ndakidemi, P. A., & Mbega, E. R. (2022). Can soilless farming feed urban East Africa? An assessment of the benefits and challenges of hydroponics in Uganda and Tanzania. *Environmental Challenges, 6*, 100413. https://doi.org/10.1016/J.ENVC.2021.100413

Gupta, S., Brazier, A. K. M., & Lowe, N. M. (2020). Zinc deficiency in low- and middle-income countries: Prevalence and approaches for mitigation. *Journal of Human Nutrition and Dietetics, 33*, 624–643. https://doi.org/10.1111/JHN.12791

Hameed, A., Padda, I. U. H., & Karim, S. (2023). Spatial analysis of food and nutrition security in Pakistan: A holistic pathway towards zero hunger policies. *GeoJournal, 88*, 2563–2585. https://doi.org/10.1007/S10708-022-10762-W/TABLES/7

Harding, K. L., Aguayo, V. M., & Webb, P. (2018). Hidden hunger in South Asia: A review of recent trends and persistent challenges. *Public Health Nutrition, 21*, 785–795. https://doi.org/10.1017/S1368980017003202

Harika, R., Faber, M., Samuel, F., Kimiywe, J., Mulugeta, A., & Eilander, A. (2017). Micronutrient status and dietary intake of iron, vitamin a, iodine, folate and zinc in women of reproductive age and pregnant women in Ethiopia, Kenya, Nigeria and South Africa: A systematic review of data from 2005 to 2015. *Nutrients, 9*, 1096. https://doi.org/10.3390/NU9101096

Headey, D. D., Ecker, O., Comstock, A. R., & Ruel, M. T. (2023). Poverty, price and preference barriers to improving diets in Sub-Saharan Africa. *Global Food Security, 36*, 100664. https://doi.org/10.1016/J.GFS.2022.100664

Henson, S., & Agnew, J. (2021). Are market-based solutions a viable strategy for addressing micronutrient deficiency? Lessons from case studies in Sub-Saharan Africa and South Asia. *Development Policy Review, 39*, 233–249. https://doi.org/10.1111/DPR.12492

Hurrell, R., & Egli, I. (2010). Iron bioavailability and dietary reference values. *The American Journal of Clinical Nutrition, 91*, 1461S–1467S. https://doi.org/10.3945/AJCN.2010.28674F

Ilboudo Nébié, E. K., Ba, D., & Giannini, A. (2021). Food security and climate shocks in Senegal: Who and where are the most vulnerable households? *Global Food Security, 29*, 100513. https://doi.org/10.1016/J.GFS.2021.100513

Jimoh, M. O., Okaiyeto, K., Oguntibeju, O. O., & Laubscher, C. P. (2022). A systematic review on Amaranthus-related research. *Horticulturae*, *8*(3), 239.

Jomova, K., Makova, M., Alomar, S. Y., Alwasel, S. H., Nepovimova, E., Kuca, K., Rhodes, C. J., & Valko, M. (2022). Essential metals in health and disease. *Chemico-Biological Interactions*, *367*, 110173. https://doi.org/10.1016/J.CBI.2022.110173

Kannan, M., Elavarasan, G., Balamurugan, A., Dhanusiya, B., & Freedon, D. (2022). Hydroponic farming – A state of art for the future agriculture. *Materials Today: Proceedings*, *68*, 2163–2166. https://doi.org/10.1016/J.MATPR.2022.08.416

Kawashima, L. M., & Valente Soares, L. M. (2003). Mineral profile of raw and cooked leafy vegetables consumed in Southern Brazil. *Journal of Food Composition and Analysis*, *16*, 605–611. https://doi.org/10.1016/S0889-1575(03)00057-7

Keats, E. C., Neufeld, L. M., Garrett, G. S., Mbuya, M. N. N., & Bhutta, Z. A. (2019). Improved micronutrient status and health outcomes in low- and middle-income countries following large-scale fortification: Evidence from a systematic review and meta-analysis. *The American Journal of Clinical Nutrition*, *109*, 1696–1708. https://doi.org/10.1093/AJCN/NQZ023

Kennedy, E., Webb, P., Block, S., Griffin, T., Mozaffarian, D., & Kyte, R. (2021). Transforming food systems: The missing pieces needed to make them work. *Current Developments in Nutrition*, *5*, nzaa177. https://doi.org/10.1093/CDN/NZAA177

Kumar, V. A., & Singh, J. (2023). Trends in hydroponics practice/technology in horticultural crops: A review. *International Journal of Plant & Soil Science*, *35*, 57–65. https://doi.org/10.9734/IJPSS/2023/V35I22759

Lal, R. (2020). Home gardening and urban agriculture for advancing food and nutritional security in response to the COVID-19 pandemic. *Food Security*, *12*, 871–876. https://doi.org/10.1007/S12571-020-01058-3/FIGURES/2

Lata, S., Yadav, A., Kumar, P., & Yadav, M. (2023). Biodiversity of underutilized fruits of the Himalayas. *Genetic Resources and Crop Evolution*, *70*, 71–94. https://doi.org/10.1007/S10722-022-01476-6/TABLES/4

Lemoine, A., & Tounian, P. (2020). Childhood anemia and iron deficiency in Sub-Saharan Africa – risk factors and prevention: A review. *Archives de Pédiatrie*, *27*, 490–496. https://doi.org/10.1016/J.ARCPED.2020.08.004

Lenaerts, B., & Demont, M. (2021). The global burden of chronic and hidden hunger revisited: New panel data evidence spanning 1990–2017. *Global Food Security*, *28*, 100480. https://doi.org/10.1016/J.GFS.2020.100480

Li, H., Mo, J., Wang, X., Pan, B., Xu, S., Li, S., Zheng, X., & Lu, W. (2023). IPS (In-Plant System) delivery of double-stranded vitellogenin and vitellogenin receptor via hydroponics for pest control in diaphorina citri kuwayama (Hemiptera: Psyllidae). *International Journal of Molecular Sciences*, *24*, 9497. https://doi.org/10.3390/IJMS24119497/S1

Li, S., & Zhang, Y. (2024). More unequal food distribution in low-income countries exacerbates global hunger risk. *Sustainable Production and Consumption*, *46*, 108–118. https://doi.org/10.1016/J.SPC.2024.02.009

Li, X., Yadav, R., & Siddique, K. H. M. (2020). Neglected and underutilized crop species: The key to improving dietary diversity and fighting hunger and malnutrition in Asia and the Pacific. *Frontiers in Nutrition*, *7*, 593711. https://doi.org/10.3389/FNUT.2020.593711/BIBTEX

Li, Y., Bahadur, R., Ahuja, J., Pehrsson, P., & Harnly, J. (2021). Macro-and micronutrients in raw plant foods: The similarities of foods and implication for dietary diversification. *Journal of Food Composition and Analysis*, *102*, 103993. https://doi.org/10.1016/J.JFCA.2021.103993

Lopes, S. O., Abrantes, L. C. S., Azevedo, F. M., Morais, N. S., Morais, D. C., Gonçalves, V. S. S., Fontes, E. A. F., Franceschini, S. C. C., & Priore, S. E. (2023). Food insecurity and micronutrient deficiency in adults: A systematic review and meta-analysis. *Nutrients*, *15*, 1074. https://doi.org/10.3390/NU15051074/S1

Majid, M., Khan, J. N., Ahmad Shah, Q. M., Masoodi, K. Z., Afroza, B., & Parvaze, S. (2021). Evaluation of hydroponic systems for the cultivation of Lettuce (Lactuca sativa L., var. Longifolia) and comparison with protected soil-based cultivation. *Agricultural Water Treatment Systems*, *245*, 106572. https://doi.org/10.1016/J.AGWAT.2020.106572

Makhmonov, L. S., Kh Mamatkulova, F., Berdiyarova, M. B., & Shomurodov, K. E. (2021). The main causes of anemia in iron and Vitamin b 12 deficiency associated with helicobacter pylori. *NVEO – Natural Volatiles & Essential Oils*, *8*, 10167–10174.

Manos, D. P., & Xydis, G. (2019). Hydroponics: Are we moving towards that direction only because of the environment? A discussion on forecasting and a systems review. *Environmental Science and Pollution Research*, *26*, 12662–12672. https://doi.org/10.1007/S11356-019-04933-5/FIGURES/3

Masters, W. A., Djurfeldt, A. A., De Haan, C., Hazell, P., Jayne, T., Jirström, M., & Reardon, T. (2013). Urbanization and farm size in Asia and Africa: Implications for food security and agricultural research. *Global Food Security*, *2*, 156–165. https://doi.org/10.1016/J.GFS.2013.07.002

Maya Waiba, K., Sharma, P., Sharma, A., Chadha, S., Kaur, M., & Author, C. (2020). Soil-less vegetable cultivation: A review. *Journal of Pharmacognosy and Phytochemistry*, *9*, 631–636.

Mc Carthy, U., Uysal, I., Badia-Melis, R., Mercier, S., O'Donnell, C., & Ktenioudaki, A. (2018). Global food security – Issues, challenges and technological solutions. *Trends in Food Science & Technology*, *77*, 11–20. https://doi.org/10.1016/J.TIFS.2018.05.002

Medina-Lozano, I., Bertolín, J. R., & Díaz, A. (2021). Nutritional value of commercial and traditional lettuce (Lactuca sativa L.) and wild relatives: Vitamin C and anthocyanin content. *Food Chemistry*, *359*, 129864. https://doi.org/10.1016/J.FOODCHEM.2021.129864

Miller, B. D. D., & Welch, R. M. (2013). Food system strategies for preventing micronutrient malnutrition. *Food Policy*, *42*, 115–128. https://doi.org/10.1016/J.FOODPOL.2013.06.008

Miller, R., Wentzel, A. R., & Richards, G. A. (2020). COVID-19: NAD+ deficiency may predispose the aged, obese and type2 diabetics to mortality through its effect on SIRT1 activity. *Medical Hypotheses*, *144*. https://doi.org/10.1016/j.mehy.2020.110044

Ngxabi, S., Jimoh, M. O., Kambizi, L., & Laubscher, C. P. (2021). Growth characteristics, phytochemical contents, and antioxidant capacity of Trachyandra ciliata (L.f) kunth grown in hydroponics under varying degrees of salinity. *Horticulturae*, *7*, 244. https://doi.org/10.3390/HORTICULTURAE7080244

Niu, G., & Masabni, J. (2022). Hydroponics. *Plant Factory Basics, Applications and Advances*, 153–166. https://doi.org/10.1016/B978-0-323-85152-7.00023-9

Nkcukankcuka, M., Jimoh, M. O., Griesel, G., & Laubscher, C. P. (2021). Growth characteristics, chlorophyll content and nutrients uptake in Tetragonia decumbens Mill. cultivated under different fertigation regimes in hydroponics. *Crop & Pasture Science*, *73*, 67–76. https://doi.org/10.1071/CP20511

Nkcukankcuka, M., Laubschera, C. P., & Wilmot, C. M. (2022). Potential of hydroponically cultivated Tetragonia decumbens Mill. as a new urban food crop: An overview. *Acta Horticulturae*, *1356*, 295–302. https://doi.org/10.17660/ACTAHORTIC.2022.1356.35

Ntuli, N. R. (2019). Nutrient content of scarcely known wild leafy vegetables from Northern KwaZulu-Natal, South Africa. *South African Journal of Botany*, *127*, 19–24. https://doi.org/10.1016/J.SAJB.2019.08.033

Panzeri, C., Pecoraro, L., Dianin, A., Sboarina, A., Arnone, O. C., Piacentini, G., & Pietrobelli, A. (2024). Potential micronutrient deficiencies in the first 1000 days of life: The pediatrician on the side of the weakest. *Current Obesity Reports*, 1–14. https://doi.org/10.1007/S13679-024-00554-3/TABLES/3

Pascual, M. P., Lorenzo, G. A., Gabriel, A. G., Pascual, M. P., Lorenzo, G. A., & Gabriel, A. G. (2018). Vertical farming using hydroponic system: Toward a sustainable onion production in Nueva Ecija, Philippines. *Open Journal of Ecology*, *8*, 25–41. https://doi.org/10.4236/OJE.2018.81003

Pingali, P., Aiyar, A., Abraham, M., & Rahman, A. (2019). *Diet diversity and the declining importance of staple grains* (pp. 73–91). Springer Nature. https://doi.org/10.1007/978-3-030-14409-8_4

Pomoni, D. I., Koukou, M. K., Vrachopoulos, M. G., & Vasiliadis, L. (2023). A review of hydroponics and conventional agriculture based on energy and water consumption, environmental impact, and land use. *Energies*, *16*, 1690. https://doi.org/10.3390/EN16041690

Punchay, K., Inta, A., Tiansawat, P., Balslev, H., & Wangpakapattanawong, P. (2020). Nutrient and mineral compositions of wild leafy vegetables of the karen and lawa communities in Thailand. *Foods*, *9*, 1748. https://doi.org/10.3390/FOODS9121748

Rajaseger, G., Chan, K. L., Tan, K. Y., Ramasamy, S., Khin, M. C., Amaladoss, A., & Haribhai, P. K. (2023). Hydroponics: Current trends in sustainable crop production. *Bioinformation*, *19*, 925. https://doi.org/10.6026/97320630019925

Rufí-Salís, M., Calvo, M. J., Petit-Boix, A., Villalba, G., & Gabarrell, X. (2020). Exploring nutrient recovery from hydroponics in urban agriculture: An environmental assessment. *Resources, Conservation and Recycling*, *155*, 104683. https://doi.org/10.1016/J.RESCONREC.2020.104683

Sahoo, G., Swamy, S., Rout, S., Wani, A., & Mishra, A. (2021). Exploitation of wild leafy vegetables and under-utilized fruits: Consequences for food and nutritional security. *Annals of the Romanian Society for Cell Biology*, *25*, 5656–5668.

Sajovic, J., Meglič, A., Glavač, D., Markelj, Š., Hawlina, M., & Fakin, A. (2022). The role of Vitamin A in retinal diseases. *International Journal of Molecular Sciences*, *23*, 1014. https://doi.org/10.3390/IJMS23031014

Salami, S. O., Adegbaju, O. D., Idris, O. A., Jimoh, M. O., Olatunji, T. L., Omonona, S., Orimoloye, I. R., Adetunji, A. E., Olusola, A., Maboeta, M. S., & Laubscher, C. P. (2022). South African wild fruits and vegetables under a changing climate: The implications on health and economy. *South African Journal of Botany*, *145*, 13–27. https://doi.org/10.1016/J.SAJB.2021.08.038

Sambo, P., Nicoletto, C., Giro, A., Pii, Y., Valentinuzzi, F., Mimmo, T., Lugli, P., Orzes, G., Mazzetto, F., Astolfi, S., Terzano, R., & Cesco, S. (2019). Hydroponic solutions for soilless production systems: Issues and opportunities in a smart agriculture perspective. *Frontiers in Plant Science, 10*, 465257. https://doi.org/10.3389/FPLS.2019.00923/BIBTEX

Samosir, O. B., Radjiman, D. S., & Aninditya, F. (2023). Food consumption diversity and nutritional status among children aged 6–23 months in Indonesia: The analysis of the results of the 2018 basic health research. *PLoS One, 18*, e0281426. https://doi.org/10.1371/JOURNAL.PONE.0281426

Sathyanarayan, S. R., Warke, V. G., Mahajan, G. B., & Annapure, U. S. (2023). Soil free nutrient availability to plants. *Journal of Plant Nutrition, 46*, 801–814. https://doi.org/10.1080/01904167.2022.2071736

Serio, F., Manuel, R., Machado, A., Fussy, A., & Papenbrock, J. (2022). An overview of soil and soilless cultivation techniques – Chances, challenges and the neglected question of sustainability. *Plants, 11*, 1153. https://doi.org/10.3390/PLANTS11091153

Sharma, N., Acharya, S., Kumar, K., Singh, N., & Chaurasia, O. P. (2018). Hydroponics as an advanced technique for vegetable production: An overview. *Journal of Soil and Water Conservation, 17*, 364. https://doi.org/10.5958/2455-7145.2018.00056.5

Smagghe, F., Spooner-Hart, R., Chen, Z. H., & Donovan-Mak, M. (2023). Biological control of arthropod pests in protected cropping by employing entomopathogens: Efficiency, production and safety. *Biological Control, 186*, 105337. https://doi.org/10.1016/J.BIOCONTROL.2023.105337

Sogoni, A. (2020). *The effect of salinity and substrates on the growth parameters and antioxidant potential of Tetragonia decumbens (Dune spinach) for horticultural applications (Msc).* Cape Peninsula University of Technology.

Sogoni, A., Jimoh, M. O., Kambizi, L., & Laubscher, C. P. (2021). The impact of salt stress on plant growth, mineral composition, and antioxidant activity in Tetragonia decumbens mill.: An underutilized edible halophyte in South Africa. *Horticulturae, 7*, 140. https://doi.org/10.3390/horticulturae7060140

Sonawane, K. G., More, S. S., Perke, D. S., & Chavan, R. V. (2022). Techniques and status of crop diversification: A review. *Journal of Pharmacognosy and Phytochemistry, 11*, 258–262.

Song, S., Hou, Y., Lim, R. B. H., Gaw, L. Y. F., Richards, D. R., & Tan, H. T. W. (2022). Comparison of vegetable production, resource-use efficiency and environmental performance of high-technology and conventional farming systems for urban agriculture in the tropical city of Singapore. *Science of the Total Environment, 807*, 150621. https://doi.org/10.1016/J.SCITOTENV.2021.150621

Teklu, D., Gashu, D., Joy, E. J. M., Amede, T., & Broadley, M. R. (2023). Effectiveness of agronomic biofortification strategy in fighting against hidden hunger. *Agronomy, 13*, 2173. https://doi.org/10.3390/AGRONOMY13082173

Tian, Z., Wang, J. W., Li, J., & Han, B. (2021). Designing future crops: Challenges and strategies for sustainable agriculture. *The Plant Journal, 105*, 1165–1178. https://doi.org/10.1111/TPJ.15107

Tibesigwa, B., & Visser, M. (2016). Small-holder farming, food security and climate change in South Africa: Male-female and urban-rural differences. *Environment for Development*, 1–12.

Tornaghi, C. (2017). Urban agriculture in the food-disabling city: (Re)defining urban food justice, reimagining a politics of empowerment. *Antipode, 49*, 781–801. https://doi.org/10.1111/ANTI.12291

Touliatos, D., Dodd, I. C., & Mcainsh, M. (2016). Vertical farming increases lettuce yield per unit area compared to conventional horizontal hydroponics. *Food Energy Security, 5*, 184–191. https://doi.org/10.1002/FES3.83

Tulchinsky, T. H. (2010). Micronutrient deficiency conditions: Global health issues. *Public Health Reviews, 32*, 243–255. https://doi.org/10.1007/BF03391600/METRICS

Tunio, M. H., Gao, J., Lakhiar, I. A., Solangi, K. A., Qureshi, W. A., Shaikh, S. A., & Chen, J. (2021). Influence of atomization nozzles and spraying intervals on growth, biomass yield, and nutrient uptake of butter-head lettuce under aeroponics system. *Agronomy, 11*, 97. https://doi.org/10.3390/AGRONOMY11010097

Viljoen, C. C., Jimoh, M. O., & Laubscher, C. P. (2021). Studies of vegetative growth, inflorescence development and eco-dormancy formation of abscission layers in Streptocarpus formosus (Gesneriaceae). *Horticulturae, 7*, 120. https://doi.org/10.3390/HORTICULTURAE7060120

Vogt, A. C. S., Arsiwala, T., Mohsen, M., Vogel, M., Manolova, V., & Bachmann, M. F. (2021). On iron metabolism and its regulation. *International Journal of Molecular Sciences, 22*, 4591. https://doi.org/10.3390/IJMS22094591

Wallace, T. C., Bailey, R. L., Blumberg, J. B., Burton-Freeman, B., Chen, C. y. O., Crowe-White, K. M., Drewnowski, A., Hooshmand, S., Johnson, E., Lewis, R., Murray, R., Shapses, S. A., & Wang, D. D. (2020). Fruits, vegetables, and health: A comprehensive narrative, umbrella review of the science and recommendations for enhanced public policy to improve intake. *Critical Reviews in Food Science and Nutrition, 60*, 2174–2211. https://doi.org/10.1080/10408398.2019.1632258

Weffort, V. R. S., & Lamounier, J. A. (2024). Hidden hunger – A narrative review. *The Journal of Pediatrics (Rio J)*, *100*, S10–S17. https://doi.org/10.1016/J.JPED.2023.08.009

Yahia, E. M., García-Solís, P., & MaldonadoCelis, M. E. (2019). *Contribution offruits and vegetables to human nutrition and health. postharvest physiology and biochemistry of fruits and vegetables* (pp. 19–45). Woodhead Publishing. https://doi.org/10.1016/B978-0-12-813278-4.00002-6

Yisak, H., Elmneh, R., Taklual, W., Ewunetei, A., & Kefale, B. (2020). Prevalence and associated factors of clinical vitamin a deficiency among pre-school children 1–5 years of age in rural kebeles in farta district, south gondar zone, Ethiopia: A mixed methods study. *Journal of Multidisciplinary Healthcare*, *13*, 1191–1201. https://doi.org/10.2147/JMDH.S279571

Zezza, A., & Tasciotti, L. (2010). Urban agriculture, poverty, and food security: Empirical evidence from a sample of developing countries. *Food Policy*, *35*, 265–273. https://doi.org/10.1016/J.FOODPOL.2010.04.007

Zha, L., Liu, W., Zhang, Y., Zhou, C., & Shao, M. (2019). Morphological and physiological stress responses of lettuce to different intensities of continuous light. *Frontiers in Plant Sciences*, *10*, 472308. https://doi.org/10.3389/FPLS.2019.01440/BIBTEX

Zhou, B. B., Aggarwal, R., Wu, J., & Lv, L. (2021). Urbanization-associated farmland loss: A macro-micro comparative study in China. *Land Use Policy*, *101*, 105228. https://doi.org/10.1016/J.LANDUSEPOL.2020.105228

13 Optimizing Nitrogen Fertilizer for Enhanced Growth, Mineral Enrichment, and Yield of *Phaseolus vulgaris* in Hydroponics for Improved Food and Nutrition Security

**Sihle Ngxabi, Avela Sogoni, and Felix Nchu*
Department of Horticultural Sciences, Faculty of Applied Sciences, Cape Peninsula University of Technology, South Africa
*Corresponding author: sogoniavis@gmail.com

13.1 INTRODUCTION

Generally, environmental conditions such as light, humidity, and temperature are the main determinants of the rate of transpiration in plants (Daniels et al., 2011; Sogoni et al., 2023). However, studies indicate that nutrient availability, especially nitrogen, in plants controls transpiration rates and, therefore, mass flow acquisition of nutrients (Matimati et al., 2014). Nitrogen is one of the major regulating nutrients for most crops and is an essential element vital for successful plant growth (Anas et al., 2020). Studies have shown that accurate nitrogen concentration in the root zone effectively contributes to the overall growth, yield, and quality of plants (Naku et al., 2019). The effect of nitrogen on plants has been extensively studied in the past years, but little attention has been given to exact nitrogen concentrations which promote growth for different edible plant species (Näsholm et al., 2009). Nutrient regulation of transpiration may function through intensive regulation of root hydraulic conductance through control of aquaporins by nitrates and shoot stomatal conductance through the production of nitric oxide (Naku et al., 2019).

Liu et al. (2014) and Matimati et al. (2014) found that although nitrogen is an essential element for successful plant growth, excessive accumulation in the foliage tends to be poisonous on the plant and is not ideal for human health. This then stimulates the need to determine the exact nitrogen concentrations required to cultivate edible plants. Likewise, low or absence of nitrogen results in nitrogen deficiency, which may be easily noticed by chlorosis and stunted growth (Paponov et al., 2021).

The hydroponic cultivation of plants has many advantages. Besides using less water and being more precise, it is possible to manipulate nutrient contents in plant growth media and ambient environmental conditions to optimize crop yield (Ventura & Sagi, 2013). Furthermore, it is possible to grow some crops that are out of season, as well as minimize contamination of the environment with organic fertilizers and pesticides (Ngxabi et al., 2021).

Phaseolus vulgaris belongs to the Fabaceae family and is an herbaceous annual plant grown worldwide for its edible dry seed, its unripe fruit, and occasionally, its leaves, which are used as a

DOI: 10.1201/9781003469766-13

vegetable in some countries. In South Africa, it is a very important food crop because of its high protein content and dietary benefits (Muedi et al., 2015). Yet South Africa produces only 75% of the dry beans consumed in the country, and the rest are imported from other countries. This is due to its watering requirements of about 400 to 500 mm of annual rainfall when grown in fields, which most provinces cannot attain (Muedi et al., 2015). Moreover, rapid population growth, weakening agricultural lands, unmanageable practices, global climate change, and water scarcity suggest a need to explore technological solutions to achieve sustainable and efficient ways to produce important crops, such as *P. vulgaris*. It has been found that plants grown in hydroponics use ten times less water than field-grown plants (Gaikwad et al., 2020). This is because, in soil-based cultivation, water leaches out, while water is collected and recycled in hydroponics. Gaikwad et al. (2020) also reported that hydroponic plants grow faster than conventionally grown plants because all required nutrients are readily available in the solution. It is therefore paramount to dilute accurate ratios of nutrients in the hydroponics aqueous solutions for maximum production.

Therefore, the present study was conducted to determine the optimum nitrogen concentration for cultivating *P. vulgaris* in hydroponics to develop the plant's cultivation protocols. This research is of high importance to growers because the results may be used as a manual to increase yield by regulating nitrogen ratios in the medium and decreasing water loss by balancing the nutrients.

13.2 MATERIALS AND METHODS

13.2.1 Experimental Location

The experiment was conducted in the research greenhouse of the Department of Horticultural Sciences at the Cape Peninsula University of Technology (CPUT), Bellville Campus, Cape Town, South Africa, located at 33°55'56" S, 18°38'25" E. The research greenhouse is equipped with environmental controls, with temperatures set to range from 21 to 26°C during the day and from 12 to 18°C at night. The average humidity was maintained at 60–70% during the experimental period.

13.2.2 Plant Material and Experimental Design

Seeds of *P. vulgaris* cv. Star 2000 were obtained from the Starke Ayres commercial garden center, Rondebosch, Cape Town, South Africa. Viable seeds were sown in two small seed trays. The medium used for seed sowing was river sand, coco coir, and vermiculite (1:1:1). A layer of course bark was laid in the seed tray before the medium was added, to prevent leaching of the medium. Approximately 2 kg of the medium was added to each tray. The seeds were evenly broadcast on the trays, followed by the application of a thin layer of vermiculite on top of the seeds. The trays were then watered with Captab (4 g/L) to prevent the development of fungal diseases; thereafter, they were placed on a heating bed under mist irrigation sprayers for germination. After the emergence of the first set of true leaves, 20 germinated seedlings were washed using tap water to remove soil and other debris. They were then potted in black plastic pots containing river sand, coco coir, and vermiculite (1:1:1) and, thereafter, were arranged into four treatments each with five replicates. The pots were all hand-irrigated daily with 100 ml of the nutrient solution per pot.

13.2.2.1 Treatments

Nutrifeed fertilizer (10 g/5 L) sourced from Starke Ayres, Cape Town, was used as the standard fertilizer and as a source of base nutrients for all treatments since it is mostly used by South African vegetable growers. The Nutrifeed fertilizer used for making the nutrient solution contained all the necessary nutrients needed for plant growth, as shown in Table 13.1. Nitrogen was supplemented in the form of ammonium nitrate and injected in the nutrient solution at varying ratios in different treatments (64 ppm, 128 ppm, and 256 ppm). The control was sustained and irrigated with Nutrifeed only.

TABLE 13.1
Mineral Composition of the Nutrifeed Fertilizer

Nutrifeed Mineral Composition	
Content	**Mg/kg**
N	65
K	130
P	27
Ca	70
Cu	20
Fe	1,500
Mo	10
Mg	22
Mn	240
S	75
B	240
Zn	240

13.2.3 Determination of Plant Growth

13.2.3.1 Shoot Length and Leaf Number

To determine new growth, shoot length and leaf number were employed as variables. The length of each shoot was measured on a weekly basis with a 30 cm ruler from the substrate level to the tip of the tallest shoot, and the number of leaves was tallied.

13.2.3.2 Plant Weight

The weight of the pods, shoots, and roots were measured using a standard laboratory scale (RADWAG® Model PS 750.R2) after four weeks of growth. Pods, shoots, and roots were separated, and the fresh/wet weights of the individual samples were recorded (Sogoni et al., 2021).

13.2.3.3 Photosynthetic Pigment

Chlorophyll *a* and chlorophyll *b* were estimated using spectrophotometric procedures, as stated by Wang et al. (2023). Concisely, 100 mg of finely ground fresh leaves was extracted with 5 ml of 99.5% dimethyl sulfoxide (DMSO; Sigma). The determination of the photosynthetic pigment was conducted using the absorbance of the supernatants at 649 and 665 nm, as described in the work of Lichtenthaler and Wellburn (1983).

13.2.4 Determination of Pod Mineral Composition

The mineral composition was determined using an inductively coupled plasma optical emission spectrometer (ICP-OES; Varian 710–ES series, SMM Instruments, Cape Town, South Africa) following the methods previously described by Bulawa et al. (2022).

13.3 STATISTICAL ANALYSIS

The experimental data were analyzed using a one-way analysis of variance (ANOVA), and Fisher's least significant difference was used to compare means, at $P \leq 0.05$ level of significance, between treatments. All analyses were computed on the STATISTICA version 13.5.0.17 statistical software.

13.4 RESULTS

13.4.1 Effect of Ammonium Nitrate on *P. vulgaris* Plant Growth

13.4.1.1 Number of Leaves and Height

The results showed that *P. vulgaris* number of leaves and height were not significantly influenced by ammonium nitrate application from week 1 to week 4. Both the number of leaves and the height were not significantly different from one another, as seen in Figure 13.1a and b.

13.4.1.2 Pod Yield

Ammonium nitrate had a positive effect on *P. vulgaris* pod yield. The pod yield gradually increased with an increase in ammonium nitrate application. The control had the lowest pod yield (10.6) when compared to the other treatments. However, the highest pod yield (14.31) was reported in plants supplied with the highest concentration of ammonium nitrate (Figure 13.2).

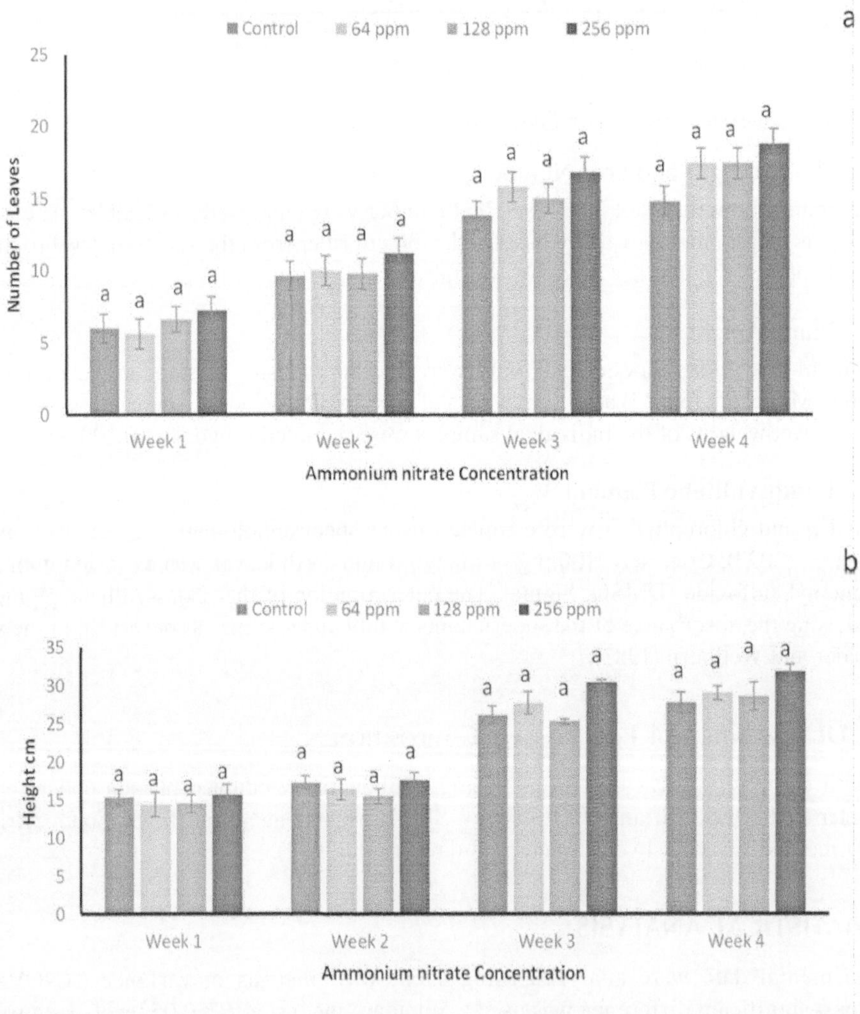

FIGURE 13.1 Effect of ammonium nitrate application on the (a) number of leaves and (b) height of *P. vulgaris* from week 1 to week 4. Bars with the same letters indicate that means are not significantly different ($p \leq 0.05$).

13.4.1.3 Shoot and Root Fresh Weight

The responses of *P. vulgaris* shoots and roots to ammonium nitrate were variable. However, the shoots were positively affected by the increase of ammonium nitrate application, and as such, shoot fresh weight increased with an increase in ammonium nitrate application, with the highest shoot weight of 13.7 obtained from the 256 ppm treatment (Figure 13.3a). Contrary to this, the root's wet weight was negatively affected by ammonium nitrate application. The highest root weight was obtained from the control plants (Figure 13.3b).

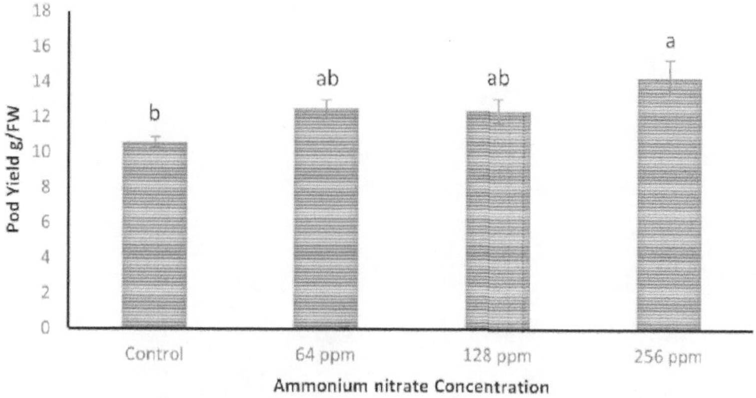

FIGURE 13.2 Effect of ammonium nitrate application on pod yield of *P. vulgaris* at four weeks post-treatment. Bars with different letters indicate that means are significantly different ($p \leq 0.05$).

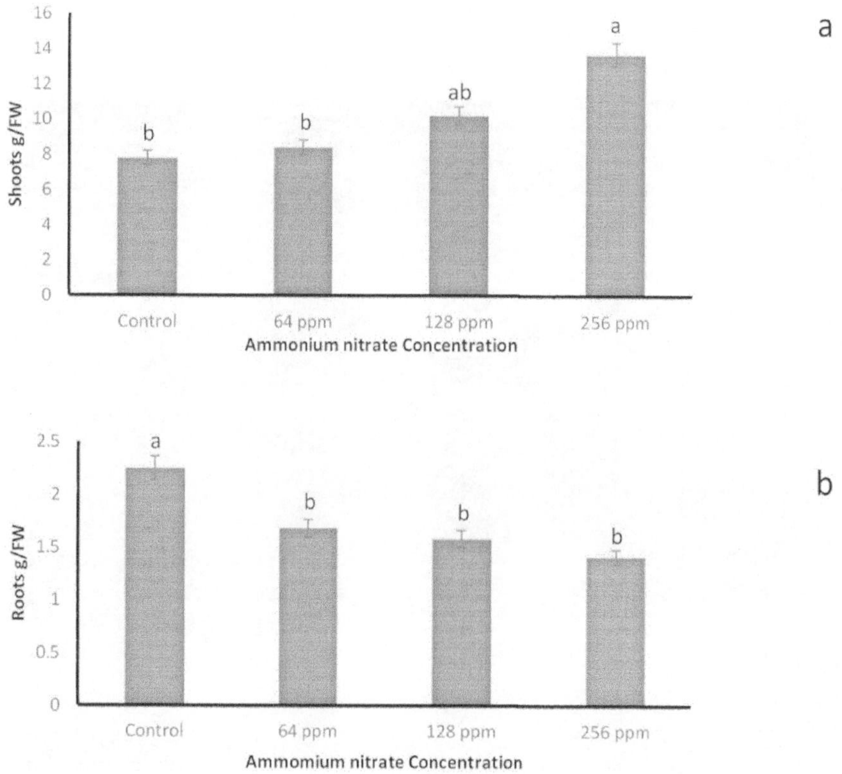

FIGURE 13.3 Effect of ammonium nitrate on (a) shoots and (b) roots of *P. vulgaris* at four weeks post-treatment. Bars with the same letters indicate that means are not significantly different ($p \leq 0.05$).

13.4.1.4 Photosynthetic Pigment

The application of ammonium nitrate had a significant effect on the photosynthetic pigments of *P. vulgaris* leaves, as shown in Figure 13.4a. An increase of ammonium nitrate in the nutrient solution enhanced the chlorophyll content up to 128 ppm, then a substantial decline was recorded at the highest concentration of 256 ppm. Nevertheless, it must be noted that even though a decrease in chlorophyll content was observed at the highest concentration, these values were substantially higher than in the control. This substantial increase in photosynthetic pigments in plants treated with ammonium nitrate is visually depicted in Figure 13.4b.

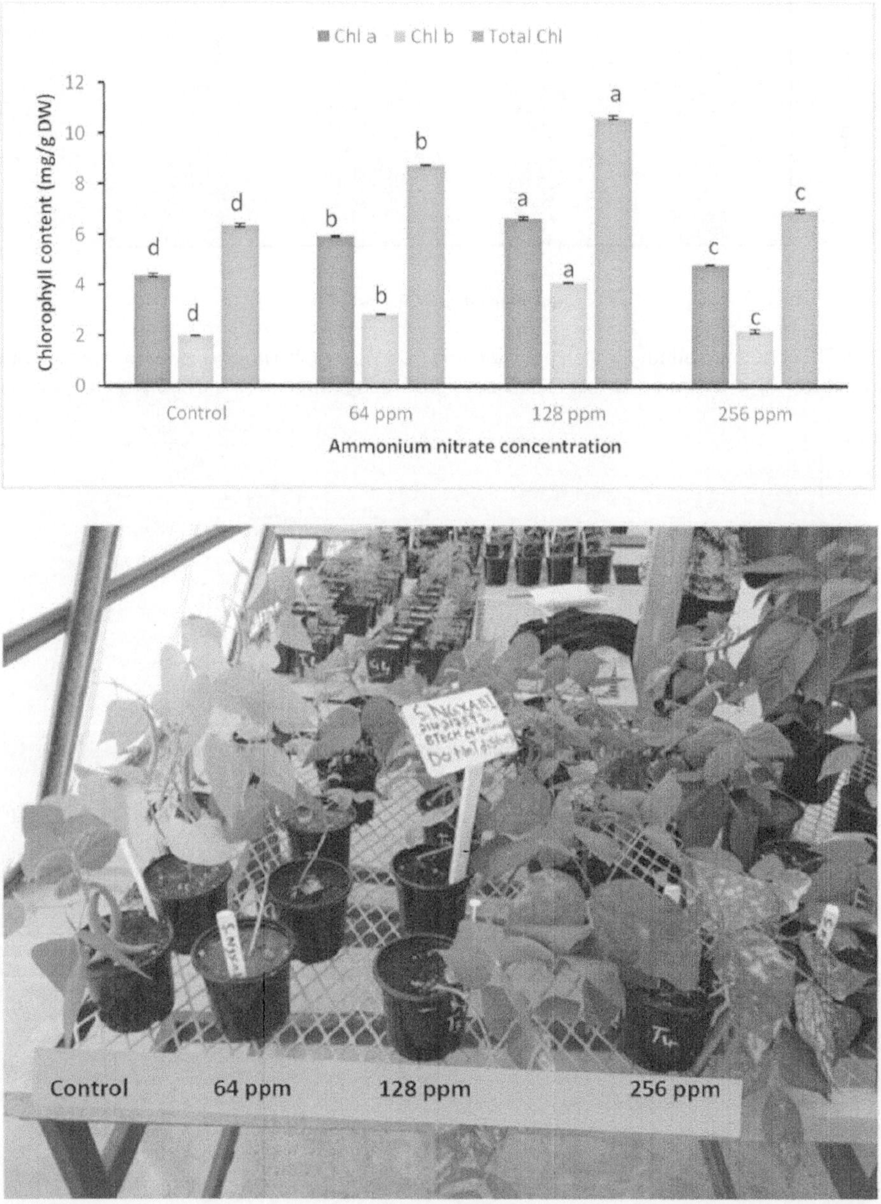

FIGURE 13.4 Changes in leaf chlorophyll pigments as influenced by ammonium nitrate application (a and b). Bars with different letters indicate that means are significantly different ($p \leq 0.05$).

13.4.1.5 Effect of Ammonium Nitrate on *P. vulgaris* Mineral Compositions

The additional application of nitrogen in the form of ammonium nitrate had a significant effect on the regulation of both the micro- and macronutrient compositions in *P. vulgaris* pods, as shown in Table 13.2. Plants irrigated with various doses of ammonium nitrate had a substantial increase in calcium, nitrogen, phosphorus, potassium, zinc, and iron compared to control plants. In contrast, manganese and copper were significantly reduced by the application of ammonium nitrate. The highest yields in Ca, N, P, Zn, and Fe were all recorded in plants irrigated with 128 ppm of ammonium nitrate, while the least were recorded in control plants. This was not the case for Mn and Cu, however, where the highest yield was recorded in control plants, and these values were substantially higher than in all plants treated with ammonium nitrate.

13.5 DISCUSSION

Nitrogen (N) supplementation has been proven in previous studies to exert a considerable impact on shoot morphology and nutritional contents of nursery plants (Razaq et al., 2017). The growth-promoting effect of N increases cytokinin production, which influences cell wall flexibility and cell growth (Carranca et al., 2018; Wang et al., 2021). The present study demonstrated that N supplementation in the form of ammonium nitrate can increase growth, can enhance minerals and yield to a certain extent, but has a negative effect at higher levels. These results agree with many studies arguing that adequate N supplementation in the soil is important for balanced growth and productivity (Giannoulis et al., 2021; Matimati et al., 2014; Wang et al., 2021). However, the form/type of N application has also been reported to influence root development and growth (Midolo et al., 2019). For instance, nitrate is said to enhance root growth by influencing root elongation and branching through signaling molecules that direct genome-wide gene expression changes for root growth, whilst ammonium inhibits root elongation (Sun et al., 2017; Vega et al., 2019). In this study, it can be assumed that the presence of ammonium in the supplemented ammonium nitrate fertilizer might be the cause of the gradual root inhibition.

Moreover, excessive ammonium application has been reported to cause ammoniacal syndrome, characterized by decreased photosynthesis, which is normally shown by necrotic patches on the leaves and, eventually, stunted plant growth (Ghafoor et al., 2021). This was the case in this study, as leaves of the plants irrigated with the highest ammonium nitrate concentration (256 ppm) had a reduced

TABLE 13.2
Effect of Ammonium Nitrate Application on Mineral Composition of *P. vulgaris* Pods, Expressed in Mg/100 g DW

Treatments	Ca	N	P	K
Control	109.5 ± 0.5^c	558 ± 2^d	327.50 ± 1.6^d	400.5 ± 2.5^d
NH_4NO_3 64 ppm	119 ± 1^b	589 ± 1.2^c	331 ± 1.9^c	452.6 ± 2.5^c
NH_4NO_3 128 ppm	129 ± 1^a	742 ± 2.1^a	346.50 ± 1.5^a	528.6 ± 1.6^a
NH_4NO_3 256 ppm	122 ± 0.5^b	661 ± 2.6^b	340.6 ± 2.4^b	502 ± 1.8^b

Treatments	Zn	Mn	Cu	Fe
Control	4.15 ± 0.15^c	4.27 ± 0.04^a	0.49 ± 0.005^a	7.15 ± 0.03^c
NH_4NO_3 64 ppm	6.35 ± 0.08^b	3.21 ± 0.015^b	0.37 ± 0.006^b	8.85 ± 0.05^a
NH_4NO_3 128 ppm	6.78 ± 0.01^a	2.95 ± 0.02^c	0.4 ± 0.01^b	7.3 ± 0.02^b
NH_4NO_3 256 ppm	6.5 ± 0.03^{ab}	2.42 ± 0.02^d	0.36 ± 0.01^b	5.61 ± 0.01^d

Note: Means with the different letters along the columns are significantly different at $p \leq 0.05$.

chlorophyll content and revealed nitrogen toxicity through necrotic patches on the leaves. This decrease in chlorophyll content could be attributed to disruptions in photosynthetic machinery, pigment malfunction, pigment–protein complex instability, and structural damage in the light-harvesting complex. Furthermore, excessive ammonium nitrate content can cause reactive oxygen species formation in chloroplasts, which breaks the double bonds of unsaturated fatty acids, causing chloroplast membrane damage and chlorophyll leakage from thylakoids (Sachdev et al., 2021). Conversely, a low application of N potentially caused leaf chlorosis, a symptom that is associated with nitrogen deficiency, which was visible in the control plants – proving that low availability of nitrogen results in inadequate mass flow acquisition of nutrients (Naku et al., 2019). Thus, it is crucial to determine the optimum N doses for increased plant growth, minerals, and vitamins in common beans (*Phaseolus vulgaris* L.)

Furthermore, the common bean is a significant legume crop mostly consumed by humans (Mallor et al., 2018; Nadeem et al., 2021) due to its significant source of protein, fiber, carbohydrates, vitamins, and minerals and accounts for about half of the grain legumes eaten globally (Chaturvedi et al., 2024; Ganesan & Xu, 2017; Uebersax et al., 2023). Thus, it is critical to find alternative ways to increase its biomass and nutrient compositions. In the present study, the mineral compositions of the common bean, such as calcium, nitrogen, phosphorus, zinc, and iron, increased with additional application of N in the form of ammonium nitrate. A similar study was also conducted by Chaturvedi et al. (2024) on common bean supplemented with organic nutrients rich in nitrogen, which enhanced the yield and economic production returns. Thus, emphasizing the critical role of nutrient management in improving mass flow acquisition of nutrients in plants for optimum yield, attributing parameters under hydroponic cultivation.

13.6 CONCLUSION AND PROSPECTS

The results of this study showed that there is a significant advantage of using a nutrient solution incorporated with ammonium nitrate to increase the yield of *P. vulgaris*. Plants irrigated with a nutrient solution incorporated with 128 ppm of ammonium nitrate revealed a significant increase in growth, minerals, and yield. Therefore, it is evident that there is potential of increasing the yield of *P. vulgaris* by manipulating the application of nitrogen in the nutrient solution in hydroponics. Thus, these findings are crucial due to the growing awareness worldwide of the importance of food nutrition and security in the context of the current decline in freshwater supplies, the increasing global population, and the rising occurrence of drought. Hydroponic cultivation has been reported to enhance crop production in confined spaces whilst using less water. Such advantages will contribute in establishing food security in developing countries like South Africa, where malnutrition still prevails. Hence, further research on the manipulation of nutrients in hydroponic cultivation of other leafy vegetables in South Africa is needed for food security. More innovative technologies like hydroponics are recommended in the future to mitigate the detrimental effects of drought and wasteful usage of fertilizer.

13.7 ACKNOWLEDGMENTS

This work was financially supported by the Department of Horticultural Sciences of the Cape Peninsula University of Technology through Prof. Nchu's NRF-linked research grant.

REFERENCES

Anas, M., Liao, F., Verma, K. K., Sarwar, M. A., Mahmood, A., Chen, Z. L., Li, Q., Zeng, X. P., Liu, Y., & Li, Y. R. (2020). Fate of nitrogen in agriculture and environment: Agronomic, eco-physiological and molecular approaches to improve nitrogen use efficiency. *Biological Research*, *53*(1), 1–20. https://doi.org/10.1186/S40659-020-00312-4

Bulawa, B., Sogoni, A., Jimoh, M. O., & Laubscher, C. P. (2022). Potassium application enhanced plant growth, mineral composition, proximate and phytochemical content in Trachyandra divaricata Kunth (Sandkool). *Plants*, *11*, 3183. https://doi.org/10.3390/PLANTS11223183

Carranca, C., Brunetto, G., & Tagliavini, M. (2018). Nitrogen nutrition of fruit trees to reconcile productivity and environmental concerns. *Plants (Basel), 7.* https://doi.org/10.3390/PLANTS7010004

Chaturvedi, A., Chaturvedi, S., Mishra, T., Pandey, V., & Sharma, N. (2024). Enhancing productivity of Rajma (Phaseolus vulgaris L.) by managing soil acidity and organic nutrient management under arunachal hill conditions. *International Journal of Plant & Soil Science, 36*, 378–382. https://doi.org/10.9734/ijpss/2024/v36i34436

Daniels, C. W., Rautenbach, F., Mabusela, W. T., Valentine, A. J., & Marnewick, J. L. (2011). Comparative antioxidant-capacity and -content of leaves, bulbs, roots, flowers and fruit of Gethyllis multifolia L. Bolus and G. villosa Thunb. species. *South African Journal of Botany, 77*, 711–717. https://doi.org/10.1016/J.SAJB.2011.03.005

Gaikwad, D., Gaikwad, D. J., Priyadarsini, S., & Mallick, B. (2020). Effects of different hydroponics systems and growing media on physiological parameters of spinach. *Article in International Journal of Current Microbiology and Applied Sciences, 9*, 1409–1414. https://doi.org/10.20546/ijcmas.2020.905.160

Ganesan, K., & Xu, B. (2017). Polyphenol-rich dry common beans (Phaseolus vulgaris L.) and their health benefits. *International Journal of Molecular Sciences, 18*, 2331. https://doi.org/10.3390/IJMS18112331

Ghafoor, I., Habib-ur-Rahman, M., Ali, M., Afzal, M., Ahmed, W., Gaiser, T., & Ghaffar, A. (2021). Slow-release nitrogen fertilizers enhance growth, yield, NUE in wheat crop and reduce nitrogen losses under an arid environment. *Environmental Science and Pollution Research, 28*, 43528–43543. https://doi.org/10.1007/S11356-021-13700-4/TABLES/3

Giannoulis, K. D., Skoufogianni, E., Bartzialis, D., Solomou, A. D., & Danalatos, N. G. (2021). Growth and productivity of Salvia officinalis L. under Mediterranean climatic conditions depends on biofertilizer, nitrogen fertilization, and sowing density. *Industrial Crops and Products, 160*, 113136. https://doi.org/10.1016/J.INDCROP.2020.113136

Lichtenthaler, H. K., & Wellburn, A. R. (1983). Determinations of total carotenoids and chlorophylls a and b of leaf extracts in different solvents. *Biochemical Society Transactions, 11*, 591–592. https://doi.org/10.1042/BST0110591

Liu, C. W., Sung, Y., Chen, B. C., & Lai, H. Y. (2014). Effects of nitrogen fertilizers on the growth and nitrate content of Lettuce (Lactuca sativa L.). *International Journal of Environmental Research and Public Health, 11*, 4427–4440. https://doi.org/10.3390/IJERPH110404427

Mallor, C., Barberán, M., & Aibar, J. (2018). Recovery of a common bean landrace (phaseolus vulgaris L.) for commercial purposes. *Frontiers in Plant Science, 871*, 402112. https://doi.org/10.3389/FPLS.2018.01440/BIBTEX

Matimati, I., Verboom, G. A., & Cramer, M. D. (2014). Nitrogen regulation of transpiration controls mass-flow acquisition of nutrients. *Journal of Experimental Botany, 65*, 159–168. https://doi.org/10.1093/JXB/ERT367

Midolo, G., Alkemade, R., Schipper, A. M., Benítez-López, A., Perring, M. P., & De Vries, W. (2019). Impacts of nitrogen addition on plant species richness and abundance: A global meta-analysis. *Global Ecology and Biogeography, 28*, 398–413. https://doi.org/10.1111/GEB.12856

Muedi, H. T. H., Fourie, D., & McLaren, N. W. (2015). Distribution and severity of bacterial brown spot on dry beans in South Africa: An update. *South African Journal of Science, 111*, 1–6. https://doi.org/10.17159/SAJS.2015/20140015

Nadeem, M. A., Yeken, M. Z., Shahid, M. Q., Habyarimana, E., Yılmaz, H., Alsaleh, A., Hatipoğlu, R., Çilesiz, Y., Khawar, K. M., Ludidi, N., Ercişli, S., Aasim, M., Karaköy, T., & Baloch, F. S. (2021). Common bean as a potential crop for future food security: An overview of past, current and future contributions in genomics, transcriptomics, transgenics and proteomics. *Biotechnology & Biotechnological Equipment, 35*, 758–786. https://doi.org/10.1080/13102818.2021.1920462

Naku, M., Kambizi, L., Matimati, I., Naku, M., Kambizi, L., & Matimati, I. (2019). Functional roles of ammonium (NH4+) and nitrate (NO3–) in regulation of day- and night-time transpiration in Phaseolus vulgaris. *Functional Plant Biology, 46*, 806–815. https://doi.org/10.1071/FP17179

Näsholm, T., Kielland, K., & Ganeteg, U. (2009). Uptake of organic nitrogen by plants. *New Phytologist, 182*, 31–48. https://doi.org/10.1111/J.1469-8137.2008.02751.X

Ngxabi, S., Jimoh, M. O., Kambizi, L., & Laubscher, C. P. (2021). Edibility of Trachyandra ciliata (l.f.) kunth-an underutilized vegetable from south african Western Cape Coast: A review. *African Journal of Food, Agriculture, Nutrition and Development, 21*, 18644–18661. https://doi.org/10.18697/AJFAND.104.20810

Paponov, M., Arakelyan, A., Dobrev, P. I., Verheul, M. J., & Paponov, I. A. (2021). Nitrogen deficiency and synergism between continuous light and root ammonium supply modulate distinct but overlapping patterns of phytohormone composition in xylem sap of tomato plants. *Plants, 10*, 573. https://doi.org/10.3390/PLANTS10030573

Razaq, M., Zhang, P., Shen, H. L., & Salahuddin. (2017). Influence of nitrogen and phosphorous on the growth and root morphology of Acer mono. *PLoS One*, *12*, e0171321. https://doi.org/10.1371/JOURNAL.PONE.0171321

Sachdev, S., Ansari, S. A., Ansari, M. I., Fujita, M., & Hasanuzzaman, M. (2021). Abiotic stress and reactive oxygen species: Generation, signaling, and defense mechanisms. *Antioxidants*, *10*, 277. https://doi.org/10.3390/ANTIOX10020277

Sogoni, A., Jimoh, M. O., Kambizi, L., & Laubscher, C. (2021). The impact of salt stress on plant growth, mineral composition, and antioxidant activity in *Tetragonia decumbens* Mill.: An underutilized edible halophyte in South Africa. *Horticulturae*, *7*, 140. https://doi.org/10.3390/horticulturae7060140

Sogoni, A., Jimoh, M. O., Keyster, M., Kambizi, L., & Laubscher, C. P. (2023). Salinity induced leaf anatomical responses and chemical composition in Tetragonia decumbens Mill.: An underutilized Edible Halophyte in South Africa. *Russian Journal of Plant Physiology*, *70*, 1–9. https://doi.org/10.1134/S1021443723601775

Sun, C. H., Yu, J. Q., & Hu, D. G. (2017). Nitrate: A crucial signal during lateral roots development. *Frontiers in Plant Science*, *8*, 254579. https://doi.org/10.3389/FPLS.2017.00485/BIBTEX

Uebersax, M. A., Cichy, K. A., Gomez, F. E., Porch, T. G., Heitholt, J., Osorno, J. M., Kamfwa, K., Snapp, S. S., & Bales, S. (2023). Dry beans (Phaseolus vulgaris L.) as a vital component of sustainable agriculture and food security – A review. *Legume Science*, *5*, e155. https://doi.org/10.1002/LEG3.155

Vega, A., O'Brien, J. A., & Gutiérrez, R. A. (2019). Nitrate and hormonal signaling crosstalk for plant growth and development. *Current Opinion in Plant Biology*, *52*, 155–163. https://doi.org/10.1016/J.PBI.2019.10.001

Ventura, Y., & Sagi, M. (2013). Halophyte crop cultivation: The case for salicornia and sarcocornia. *Environmental and Experimental Botany*, *92*, 144–153. https://doi.org/10.1016/j.envexpbot.2012.07.010

Wang, G. Y., Hu, Y. X., Liu, Y. X., Ahmad, S., & Zhou, X. B. (2021). Effects of supplement irrigation and nitrogen application levels on soil carbon–Nitrogen content and yield of one-year double cropping maize in subtropical region. *Water*, *13*, 1180. https://doi.org/10.3390/W13091180

Wang, Y., Ma, W., Fu, H., Li, L., Ruan, X., & Zhang, X. (2023). Effects of Salinity stress on growth and physiological parameters and related gene expression in different ecotypes of Sesuvium portulacastrum on Hainan Island. *Genes (Basel)*, *14*, 1336. https://doi.org/10.3390/GENES14071336/S1

14 Novel Applications of Undervalued Herbal Teas for Food and Nutrition Security and the Prevention of Malaria

*Whitney Shingenge, *Iwanette du Preez-Bruwer, and Davis Mumbengegwi*
Multidisciplinary Research Services, Centre for Research Services, University of Namibia, 340 Mandume Ndemufayo Avenue, Pioneers Park, Windhoek, Namibia
*Corresponding author email: ibruwer@unam.na

14.1 INTRODUCTION

Tea is a hot beverage thought to have originated in China and prepared by pouring hot water over the leaves of the *Camellia sinensis* plant (Shafiq et al., 2024). *Herbal tea*, on the other hand, is prepared from the infusion of various plant parts, including roots, stems, leaves, flowers, fruits, and whole plants, and may consist of various plants that are not members of the *Camellia* genus (Long et al., 2024). The demand for herbal teas is growing as more people adopt healthier lifestyles associated with the daily consumption of the beverage, which is affordable and can contribute to overall well-being. Herbal tea is known as *tisane* in French, a term that means "tea without tea" (Shikha et al., 2019) that is consumed for its health benefits (Joshi et al., 2023). Health benefits of herbal tea include boosting the immune system; reducing the risk of cancer, diabetes, heart diseases, and strokes; weight management; enhancing mental capacity; and stress reduction (Hosen & Madhu, 2023). The consumption of herbal teas also aides digestion, easing stomach ailments, while cleansing the body, promoting energy and wellness, preventing colds as well as activating internal organs (Hailu, 2016) and regulating blood glucose (Joshi et al., 2023). Infectious diseases like malaria can also be treated using herbal teas. *Artemisia annua* tea has been proven effective in treating malaria in clinical trials in Uganda and the Democratic Republic of Congo (Ogwang et al., 2012). The health benefits of tea vary depending on the type of tea, the amount consumed, and individual factors. Tea should be used as part of a healthy, balanced diet and lifestyle, but even though it can possess certain health benefits, it is not a cure (Hosen & Madhu, 2023).

Malaria is an infectious disease transmitted by anopheline mosquitoes. It is a global public health problem (Tajuddeen & Van Heerden, 2019), particularly in sub-Saharan Africa, where 93.6% of global malaria cases and 95.4% of deaths occur (WHO, 2023). In 2022, there were an estimated 249 million malaria cases and 608,000 deaths globally (WHO, 2023). Namibia, situated within sub-Saharan Africa, is moving toward the elimination of malaria (Jacobson et al., 2019). But despite advancements, the task of eliminating malaria still faces obstacles, including the emergence of diseases like COVID-19 (Rogerson et al., 2020) and the increasing spread of drug resistance to existing antimalarials, alongside challenges in eliminating insecticide resistance (WHO, 2023). Several studies conducted in Namibia have identified populations that are at high risk of malaria

DOI: 10.1201/9781003469766-14

infection. These include school-aged children and adults with outdoor occupational exposure, for example, agricultural workers, cow herders, and unemployed people, among others (Smith et al., 2017; Jacobson et al., 2019; Mwema, 2021). These populations do not usually have access to the main malaria interventions, such as insecticides or mosquito nets (Smith et al., 2021). These findings emphasize the need for tailored approaches to improve the overall reach and efficacy of public health initiatives and address transmission within specific subpopulations.

The regular administration of antimalarial medications at below-therapeutic doses to prevent the disease is known as *chemoprophylaxis*. It is recommended for individuals who reside in malaria-endemic areas and are at risk of contracting the disease to use preventative measures such as chemoprophylaxis (WHO, 2021). Atovaquone-proguanil, doxycycline, mefloquine, and tafenoquine are the first-line drugs that are used as chemoprophylaxis for travelers to malaria-endemic regions. However, concerns regarding their safety and high costs, as well as limited availability, remain major issues prohibiting their use (DeVos & Dunn, 2024). Alternatively, the use of medicinal plants for chemoprophylaxis has been documented (Nagendrappa et al., 2017; Argemi et al., 2019), including the use of *Artemisia annua* (Cohen et al., 2023). A study conducted in Bo, Sierra Leone, revealed that reputable local herbalist communities actively sought traditional medicines and were also encouraged to utilize herbal blends (Ranasinghe et al., 2015). Considering that traditional medicines are occasionally used in combination, this chapter focuses on three medicinal plants used in Namibia that were blended to create a polyherbal tea that may help ameliorate the strain of providing care on the country's health-care system while also lowering the risk of malaria transmission.

The first part of this book chapter addresses the importance of medicinal plants, including the use of the mentioned three plants, for malaria treatment/prevention in Namibia, thus contributing to the use of *undervalued plants* as herbal teas for nutrition and in the management and treatment of malaria. The second part focuses on the formulation of these three plants as a polyherbal tea that can contribute to reducing the transmission of the disease and ease the burden of service provision on the Namibian health system and overall health, food, and nutrition security. The third and final part focuses on the findings of the study – the efficacy of the herbal tea formulations (biological activity against a clinical strain of *Plasmodium falciparum*), the safety of five lead herbal tea formulations (cytotoxicity), and the organoleptic properties of five herbal tea formulations (sensory evaluation) to determine the appeal of the tea formulations.

14.2 PLANTS USED FOR THE TREATMENT OF MALARIA

Plants have a long history of use (Najmi et al., 2022). They have been used anecdotally to cure various ailments and treat symptoms of many diseases (Mahomoodally, 2013), although some are still undervalued or disregarded (Karunamoorthi et al., 2013). However, because of cultural acceptance and beliefs, accessibility, and affordability of herbal medicines, plants are still being relied on in rural settings, especially in Africa (Okaiyeto & Oguntibeju, 2021). Medicinal plants are currently generating a lot of interest because of their therapeutic properties and because they are rich sources of phytochemicals that could lead to the development of new medicines (Okaiyeto & Oguntibeju, 2021). Some plants have been linked to antimalarial drug discovery as early as 1820, when quinine was first isolated from the bark of the *Cinchona* species (Tse et al., 2019). *Artemisia annua* is another plant with a long history of use for the treatment of malaria. It was used to treat fever by the Chinese as early as AD 317–420 and was discovered during the period of 1967–1980 by the Chinese military research program 523 (Czechowski et al., 2020). This was discovered while searching for medicine to treat soldiers infected by malaria during the Vietnam War (Lee et al., 2023). An extensive review of literature in 2004 revealed that over 1,277 plants from 160 families were traditionally used to treat malaria (Willcox et al., 2004); however, this number has increased with a widespread interest in antimalarial plants.

Researchers are exploring the use of plant-based traditional medicine to reduce malaria transmission and aid treatment despite these medicines not being incorporated into the national health-care systems of many countries. Some herbalists in western Uganda use several medicinal herbs, such *Azadirachta indica*, in tea or as a prophylactic against malaria on a regular basis (Gumisiriza et al., 2023). In Crocker Range, Sabah, Malaysia, 50 medicinal plants are used as tea by the local people to treat various ailments, including malaria and its symptoms (Ahmad & Ismail, 2003). These plants include *Scoparia dulcis*, taken to reduce fever and for the treatment of malaise, and *Asclepias curassavica* L., whose root decoction is taken as a tea to treat fever and bronchitis. *Blumea balsamifera* leaf tea is taken for fevers, while the roots of *Hyptis capitata* are boiled and taken as tea for treating fevers and colds. A plant decoction of *Physalis minima* is taken as a tea for treating malaria. A decoction of the roots of *Polygala paniculata* is boiled in water and taken as a tea to cure cough and fever. In northern Ghana, herbal teas and baths made with neem, pawpaw, guava, and eucalyptus leaves are commonly used as a first-line treatment at home against malaria (Adongo et al., 2005). In Ethiopia, 51 plant species (28 families) are relied on for malaria treatment (Nigussie & Wale, 2022); in Uganda, 45 plants from 26 families are utilized for herbal remedies that treat malaria and its symptoms (Tabuti et al., 2023); and in Cameroon, 49 species of medicinal plants from 27 families have been identified (Pierre et al., 2011).

The antimalarial efficacy of various herbal teas has been assessed against the malaria parasite, revealing their antiplasmodial properties. Audomkasok et al. (2014) reported that crude extracts of green tea, safflower, and mulberry possess antihemolysis effect induced by malaria infection. In a separate study, the clinical effectiveness and safety of a tea bag formulation containing the root of *Cryptolepis sanguinolenta* were demonstrated (Bugyei et al., 2010). The findings of this study showed that the formulation was non-toxic and highly effective in the treatment of uncomplicated malaria. Approximately 33% of plants used in Uganda to treat malaria were tested for efficacy or safety (Angupale et al., 2023). *M. pyrifolia* showed extremely strong antiplasmodial activities but was not investigated *in vivo* (Angupale et al., 2023). Numerous reports indicate that *Artemisia annua* (Qinghao) is used in the form of herbal tea or infusions to treat malaria and its symptoms, occasionally serving as a prophylactic measure (Willcox et al., 2006). Argemi et al. (2019) reported that 8 out of 14 (57.1%) patients who received *Artemisia* dry leaf infusions in varied regimens were cured. In Nigeria, a study showed that extracts of "Agbo iba," a polyherbal remedy, possessed prophylactic activity against a murine model of malaria with no apparent side effects (Nwabuisi, 2002). In malaria-endemic Odisha, India, the use of herbal medicinal plants, particularly those employed in traditional medicine for the prevention of malaria, is widespread. Amongst them is the traditional plant-based malaria prophylactic 74 (TPMP74) (Nagendrappa et al., 2017), which is a coded polyherbal traditional treatment that is frequently administered as an aqueous extract (decoction). All the plants in TPMP74 have long been valued for their therapeutic benefits in Ayurvedic medicine. The *Plasmodium yoelii* hepatic stages have been shown to be inhibited by TPMP74 in vitro (50% inhibitory concentration = 0.74 mg/ml) (Nagendrappa et al., 2017).

Currently, the use of traditional medicines or medicinal plants is not recognized in the national health-care system in Namibia as a treatment or a preventative measure for many ailments, including malaria (Christians, 2020). Nonetheless, more than 100 plants have been identified as being used for the treatment of malaria-related symptoms, such as fever and headache (Du Preez et al., 2020). These include *Guibourtia coleosperma* and *Diospyros chamaethamnus*, which were used to make herbal formulations for the study presented in this chapter. *Hibiscus cannabinus*, although not previously reported for its antimalarial properties in Namibia, has been extensively studied and used in the treatment of many ailments, including malaria (Salihu et al., 2020; Vasavi et al., 2019). *Hibiscus cannabinus* is included for its widely accepted organoleptic properties.

14.2.1 *Guibourtia coleosperma*

Guibourtia coleosperma (Figure 14.1) belongs to the Fabaceae family. It is known as *omusii* in Oshiwambo, *usivi* in Kwangali, *bastermopanie* in Afrikaans, *muzauli* in SiLozi, and *false mopane* in English. It is an indigenous forest species indispensable to rural communities of Namibia, as it provides many benefits, such as food (fruits and oils), traditional medicines, fuelwood, and construction materials (Heita et al., 2019). The leaves of *G. coleosperma* are frequently used as cough cure (Cheikhyoussef & Embashu, 2013), while the root extract is used to treat wounds (Von Koenen, 2007). The fruits are also used to increase the nutritional value of infant porridge (Cheikhyoussef & Embashu, 2013). The Kalahari Desert Bushmen use the bark to prepare a decoction to treat diarrhea. Blood clotting, stomach problems, and constipation are additional applications, among others (Cheikhyoussef & Embashu, 2013).

G. coleosperma is also used to cure headaches, fever, coughing, and malaria in the Ohangwena region of Namibia (Nafuka & Mumbengegwi, 2013). In the Zambezi region of Namibia, the roots are used to treat febrile disorders in young children. Du Preez (2016) studied the antimalarial properties, both *in vivo* and *in vitro*, of this plant. In this study, the antimalarial activities of the plant extracts were determined by quantifying parasitemia and percentage growth inhibition of the chloroquine-sensitive *P. falciparum* D10 strain. *G. coleosperma* extracts exhibited antiplasmodial

FIGURE 14.1 *G. coleosperma* growing in the wild in the Kavango region of Namibia.
Source: © Iwanette du Preez-Bruwer.

activities with IC_{50} values of 31.61 and 28.17 μg/ml for aqueous and organic, respectively (du Preez, 2016). *In vivo* antimalarial evaluation of the plant extracts using Peter's four-day suppressive test and a repository test for prophylactic activities tests showed that *G. coleosperma* extracts reduced the plasmodium load in infected mice.

14.2.2 *Diospyros chamaethamnus*

Diospyros chamaethamnus (Figure 14.2) belongs to the Ebenaceae family. In other languages, it is also known as *mukushuwa* in SiLozi, *sikwewo* in Kwangali, and *tsi'xa-i* in Khoekhoegowab, or *sand apple* in English. This plant was chosen for this study because in traditional medicine, *D. chamaethamnus* roots are applied topically to skin incisions to treat headaches, and in the Zambezi region, they can also be taken orally or as a vapor bath. An extensive study by Du Preez (2016) revealed that extracts of *D. chamaethamnus* have antimalarial properties, both in vivo and in vitro. In this study, the antimalarial activities of the plant extracts were determined by quantifying parasitemia and percentage growth inhibition of the chloroquine-sensitive *P. falciparum* D10 strain. The *D. chamaethamnus* aqueous and organic extracts had antiplasmodial activity with IC_{50} values of 18.30 and 19.51 g/ml, respectively (du Preez, 2016). In vivo antimalarial evaluation of *D. chamaethamnus* extracts using Peter's four-day suppressive test showed a reduction in mean parasite count in some of the treatment groups. In addition, this study also tested the prophylactic activities of the crude extracts of *D. chamaethamnus* using the repository test and showed that *D. chamaethamnus* extracts reduced the parasite load in the mice.

FIGURE 14.2 *D. chamaethamnus* in the wild in the Zambezi region of Namibia.

Source: © Iwanette du Preez-Bruwer.

14.2.3 *Hibiscus cannabinus*

Hibiscus cannabinus (Figure 14.3), popularly known as *kenaf* globally and *mutete* in Namibia, is an herbaceous plant that belongs to the Malvaceae family. The plant is native to East, West, and Southern African countries, including Namibia (Vasavi et al., 2019), where it can be found in the upper Zambezi region (Fanshawe, 2010). This plant has a wide range of uses, including medicinal applications (Salihu et al., 2020). In Africa and India, kenaf has long been used as traditional folk medicine to treat a variety of illnesses, such as fatigue and anemia (Sim & Nyam, 2021). The calyces, or flowers, are used as an emollient, and an infusion for palliative treatment, and the juice for nausea (Vasavi et al., 2019). Furthermore, the flower extracts are effective in reducing oxidative stress, free radical–induced DNA damage, and bone ailments, such as osteoarthritis (Navale et al., 2021).

Moreover, kenaf produces a wide range of bioactive chemicals with known antioxidant, cardioprotective, anti-inflammatory, antihypertensive, antiproliferative, and antitumor activity (Rekaya et al., 2016). In addition, pharmacological studies have shown that kenaf also has antibacterial, hepatoprotective, antiulcer, antityrosinase, and antihyperlipidemic properties (Sim & Nyam, 2021). Research has also been conducted on tea derived from the leaves of *H. cannabinus*, revealing significant antioxidant activity (Kho et al., 2019). This plant in particular was selected for the study to enhance the flavor of the herbal formulation, given its historical usage as a culinary additive in meals (Sim & Nyam, 2021) and due to the pleasant taste associated with its flowers.

FIGURE 14.3 *Hibiscus cannabinus* in its natural habitat.

Source: © Whitney Shingenge.

14.3 THE ROLE OF HERBAL TEAS IN FOOD SECURITY AND NUTRITION

Medicinal plants are undervalued food plants, yet they contribute to a healthy lifestyle and the overall well-being of communities in rural and resource-poor settings. Lifestyle changes relating to urbanization have led to unhealthy diets and sedentary lifestyles, mainly attributed to the influence of the Western culture. This has led to a rise in non-communicable diseases, such as diabetes, cancer, heart diseases, high blood pressure, and nutrient deficiency (Clemente-Suárez et al., 2023). Functional foods play a significant role in enhancing well-being and health, as they can mitigate the risk of diseases. They typically contain essential macronutrients or micronutrients, along with non-essential nutrients or non-nutritive components (Shaikh, 2022). Teas are considered functional foods (Y. Liu et al., 2023) with health-enhancing claims becoming increasingly important in the growth of this market. Herbal teas are important for improving food security and nutrition because of their nutritional content and possible health advantages. Drinking herbal teas has been linked to several health benefits, such as glycemic management, antioxidant activity and a decrease in the incidence of chronic illnesses (Wang et al., 2022), regulation of gut microbiota and possible therapeutic effects (Vamanu et al., 2021), and blood glucose regulation, making them beneficial for people interested in controlling their blood sugar levels (Liu et al., 2023). Furthermore, research has shown that herbal teas are nutritionally valuable, offering vital vitamins and minerals that support a balanced diet and help address dietary deficiencies (Chukwuma et al., 2019).

In Africa, herbal beverages are essential for improving nutrition and health, especially when consumed as herbal teas. These drinks, which are frequently made from indigenous plants, have several health advantages and nutritional value. For example, rooibos and honeybush are two South African tisanes that have achieved commercial success and are acknowledged for their promise in the treatment of diseases such as type 2 diabetes mellitus (Ajuwon et al., 2018). Furthermore, Africans have long utilized herbal teas made from plants like *Lippia javanica* for their ethnomedicinal uses, which include treating ailments including fever, diarrhea, and colds (Maroyi, 2017). Additionally, because of its pleasant flavor and ability to prevent oxidative hepatic cell damage, *Myrothamnus flabellifolius* herbal tea has demonstrated commercialization potential (Chukwuma et al., 2019). The wide variety of herbal teas available in Africa gives consumers several ways to include these nutrient-dense drinks in their regular diets. These herbal beverages are an essential part of a balanced diet in Africa since they not only increase nutritional intake but also may possess positive health effects.

14.4 MATERIAL AND METHODS

14.4.1 Plant Collection and Formulation of Herbal Teas

It is important to know the safety and efficacy of herbal tea formulations used in the treatment and prevention of malaria, as well as their acceptability to potential end users. The results outlined in this chapter delineate the characteristics of nine herbal formulations created from dried and ground plant material sourced from *D. chamaethamnus*, *G. coleosperma*, and *H. cannabinus*. The rationale for using *G. coleosperma* and *D. chamaethamnus* in the herbal tea formulation is that they have known antimalarial prophylactic activities (C. I. Du Preez, 2016). *H. cannabinus* flowers (calyces) were used as flavoring agents for *D. chamaethamnus* and *G. coleosperma* plants. Before commencement of the research activities, a research collection permit was obtained from the Ministry of Environment, Forestry, and Tourism (MEFT) of Namibia, and ethical approval was sought and obtained from the University of Namibia Research Ethics Committee.

Voucher specimens of the plants were prepared using a plant press and deposited at the National Herbarium at the National Botanical Research Institute (NBRI) of Namibia for the confirmation of the scientific identities of the plants. The plants were confirmed to be *G. coleosperma* (voucher specimen number BRL 104), *H. cannabinus* (voucher specimen number BRL 128), and

TABLE 14.1
Quantities of Dried Ground Material from *H. cannabinus, G. coleosperma*, and *D. chamaethamnus* in the Nine Herbal Tea Formulations

Formulation	*Hibiscus cannabinus* (Calyces) (%)	*Guibourtia coleosperma* (Roots) (%)	*Diospyros chamaethamnus* (Leaves) (%)
WS1	30	35	35
WS2	20	40	40
WS3	10	45	45
WS4	30	70	–
WS5	30	–	70
WS6	100	–	–
WS7	–	100	–
WS8	–	–	100
WS9	–	50	50

D. chamaethamnus (voucher specimen number BRL 81). The mature leaves of *D. chamaethamnus*, the roots of *G. coleosperma*, and the calyces of *H. cannabinus* were collected. The plant material was dried at room temperature for up to four weeks in a dark room to preserve the semi-volatile compounds and protect the light-sensitive compounds (Krakowska-Sieprawska et al., 2022). Once completely dry, the plant material was ground into a powder, from which nine herbal formulations were prepared using varying amounts of plant material (in percentages) of the three plants together, as indicated in Table 14.1. The percentages were determined using Design–Expert, a statistical method used for preparing polyherbal formulations (Balekundri et al., 2020). The herbal tea formulations were mixed by shaking the content of each formulation in a labelled Ziplock bag until a homogenous mixture was obtained.

14.4.2 Preparation of Herbal Tea Aqueous Extracts

Extracts of the nine herbal formulations were prepared as tea to determine various activities and phytochemicals. Specifically, aqueous extracts were prepared from each formulation by adding 2 g plant material to a tea plunger with 250 ml boiling distilled water to simulate one serving of tea. Each formulation was left to infuse for 5 min, after which it was filtered using Whatman filter paper (110 mm). Once cooled, the concentrated dried tea extract was obtained by rotary evaporation (Heidolph GI, Germany) at reduced pressure and 60°C, followed by freeze-drying (Christ Alpha 1–2 LD Plus, Germany). The dried plant extracts were stored at −20°C (freezer) in an airtight container until assays were conducted.

14.4.3 Phytochemical Analysis of the Herbal Tea Formulations

The presence or absence of selected classes of antiplasmodial phytochemicals was determined in the extracts of the nine herbal tea formulations using thin-layer chromatography (TLC) (Nafuka & Mumbengegwi, 2013; du Preez, 2016). The aqueous plant extracts for analysis were prepared by reconstituting 8 mg/ml of plant extract in 4 ml of distilled water and vortexing until completely dissolved. The extracts were spotted onto aluminum silica–backed TLC plates (20 × 20 cm, Merck, silica gel 60 F254 TM) using thin capillary tubes. The spots were made 1 cm apart and 1 cm from the bottom of the TLC plates. Solvent systems were prepared for each of the phytochemicals to be tested, and the TLC plates were then placed carefully inside the TLC tank with the respective solvent, ensuring that the solvent system is not above the 1 cm line. The plates were then developed, dried, and sprayed with various spraying agents and observed for changes in color as the reagent underwent a chemical reaction with the compounds on the TLC plate to convert it into a colored compound, enabling the spot to be seen with the naked eye. The plates were also viewed under UV light at 360 nm.

14.4.4 Antimalarial Activity of the Herbal Tea Formulations

Parasitemia and the SYBR Green assay were used to screen the herbal formulations for antimalarial activity using chloroquine-resistant *Plasmodium falciparum* D10 (American Type Culture Collection USA). The strain was maintained in continuous culture in sealed flasks using the method of Trager and Jensen (1976), as modified by du Preez et al. (2020). The assays were carried out using cultures with parasitemia of 1% and hematocrit of 2%. Briefly, the aqueous extracts of the nine herbal tea formulations were resuspended in distilled water at 1 mg/ml and filter-sterilized using a 0.2 μm Millipore nylon syringe filter. The tea extracts were added to each flask at a concentration of 100 μg/ml. The flasks were incubated for 48 and 72 h at 37°C. A negative control (infected red blood suspension only) and a positive control (10 μg/ml chloroquine) were also used. All assays were in triplicates.

14.4.5 Cytotoxicity of the Herbal Tea Formulations

The WST-1 assay was used to measure cytotoxicity using NIH/3T3 fibroblast cells (CRL-1658) obtained from the ATCC. The formulations with the highest antiplasmodial activity were screened, and these were WS1, WS2, WS3, WS4, and WS8. The concentration of the cells was adjusted to a seeding density of 5×10^2. One hundred microliters of cell suspension was seeded into each well in a 96-well plate and incubated for 24 h, whereafter the extracts of the herbal tea formulations were incubated with the cells for 72 h at 25, 50, and 100 μg/ml. With an initial concentration of 100 μg/ml, herbal formulation extracts were serially diluted in the 96-well plate. Herbal tea formulations were incubated with the fibroblast cells in the 96-well plate for 72 h. After the incubation period, the plates were removed from the incubator, and the cells were washed with PBS. Once the PBS was aspirated, 100 μL of freshly prepared WST-1 buffer solution (with 10% WST-1 stain) was added to each well. At this point, 10 μL of 10% Triton-X was added to the positive control wells. The plate was incubated for 1 h at 37°C, after which absorbance was recorded at 440 nm. The percentage cell viability was calculated in relation to the viability of cells without treatment (control wells) using the following formula: % percentage viability = OD sample − OD control × 100%.

14.4.6 Sensory Evaluation of the Herbal Tea Formulations

Sensory evaluation measures how the senses perceive food, including smell, taste, touch, and sound (Dubey et al., 2020). There are two types of sensory tests, namely, analytical tests, which are based on visible differences, and affective tests, which are based on individual acceptability and preference (Sinesio, 2005; Drake, 2022). Regardless of the health or nutritional benefits, consumers are often unwilling to buy food that lacks sensory appeal. It is important to pay attention to the sensory qualities of functional foods when creating new products (Baker et al., 2022). Herbal tea can be made with a wide variety of plant materials that have distinct sensory properties. Affective tests were carried out to determine the acceptance and preference of five herbal formulations (WS1, WS2, WS3, WS8, and WS4).

Fifty untrained panelists of age ranges 18–24, 25–35, 36–49, 50–65, and >65 years were recruited from the University of Namibia's main campus. The panel members were selected based on their willingness to participate in the research. All panelists were provided with information about the project and the intended objectives of the sensory evaluation. They were then asked for their consent to participate in the study. The panelists were instructed to evaluate the herbal tea formulations by scoring them based on five attributes of the infusions – color, aroma, flavor, aftertaste, and overall acceptability – and to record their acceptance and preferences accordingly. The panelists who did not clearly understand the meaning of an attribute were provided with an explanation. The panelists then scored their acceptance of the attributes using a five-point hedonic scale, with 1 meaning "dislike very much," 2 meaning "dislike slightly," 3 meaning "neither like nor dislike," 4 meaning "like slightly," and 5 meaning "like very much." The tea formulations were each given a three-digit code, and 20 ml of each formulation (Figure 14.7) was randomly served to the panelists. The panelists were required to rinse their mouths with water in between each tasting and wait 90 s before each tasting. They were not required to swallow the 20 ml of each sample but were asked to hold about 10 ml of the sample in their mouth for 5 s and swallow small quantities to appreciate the full sensory characteristics of the beverage. The panelists were given the opportunity for repeat tasting if needed.

14.5 RESULTS AND DISCUSSION

14.5.1 Phytochemical Analysis

Thin-layer chromatography showed that all compounds tested for were present in five of the nine herbal tea formulations, as shown in Table 14.2. The formulations WS4, WS6, WS7, and WS9, which contained only one or two of the three plants under investigation, did not show the presence of the tested compounds. WS4 also showed fewer phytochemicals than WS5 even though they both contained two of the three plants. WS5 contained 30% *H. cannabinus* calyces and 70% *D. chamaethamnus* leaves, whereas WS4 also contained 30% *H. cannabinus* calyces and 70% *G. coleosperma* root (Table 14.1). In the formulations containing *G. coleosperma*, coumarins and steroids were detected, and this is consistent with the literature (Dushimemaria et al., 2017). Another study by Nafuka and Mumbengegwi (2013) also showed that anthraquinones were detected in the roots of *G. coleosperma*. The aqueous extracts of *D. chamaethamnus* contain alkaloids, anthraquinones, flavonoids, and steroids, which is also consistent with previously reported studies (Dushimemaria et al., 2017). *H. cannabinus* calyces were shown to contain alkaloids, anthraquinones, coumarins, and terpenoids in this study. Similarly, phytochemical screening of *H. cannabinus* has revealed the presence of phenolics, tannins, saponins, alkaloids, and steroids (Agbor et al., 2005; Obouayeba et al., 2015). Despite being used as a flavor component, *H. cannabinus* added to the phytochemical composition of the formulations, depending on amounts used.

TABLE 14.2
Phytochemical Profiles of the Nine Herbal Tea Formulations Showing the Presence or Absence of Different Classes of Antimalarial Compounds

Formulation	Al	An	C	F	Sa	St	T
WS 1	+	+	+	+	+	+	+
WS 2	+	+	+	+	+	+	+
WS 3	+	+	+	+	+	+	+
WS 4	+	+	+	–	–	–	+
WS 5	+	+	+	+	+	+	+
WS 6	+	+	+	–	–	–	+
WS 7	–	–	+	–	+	+	+
WS 8	+	+	+	+	+	+	+
WS9	+	–	+	+	+	+	+

Note: The profile of compounds is denoted as present (+) and not present (–). *Al* = alkaloids; *An* = anthraquinones; *C* = coumarins; *F* = flavonoids; *S* = saponins; *St* = steroids; *T* = terpenoids.

14.5.1.1 Antimalarial Activity, Percentage Parasitemia

The antimalarial activity of the nine formulations was measured by determining the parasitemia after each time interval by preparing thin smears. After 48 h, WS2 and WS4 recorded the highest percentage inhibition (42.9%), while WS6 recorded the lowest percentage inhibition (4%) (Figure 14.4). The percentage inhibition increased after 72 h, for which most of the formulations recorded the highest percentage inhibition. These findings were consistent with the results obtained from the 72 h fluorochromatic (SYBR Green) assay. WS4 recorded the highest percentage inhibition of 78%, while WS1 recorded the lowest percentage of inhibition (28.5%).

14.5.1.2 Antimalarial Activity, Percentage Inhibition

Inhibition of parasite growth was observed for all formulations post-48 h of incubation with the extracts and controls. WS3 showed the highest percentage inhibition of 24% (Figure 14.4), while WS6 recorded the lowest percentage of inhibition of all the formulations (3.1%). A high percentage of inhibition indicates high activity, and this was observed for the formulations WS1, WS2, WS3, WS4, and WS8. Over time, the formulations showed an increase in percentage inhibition, with WS1 recording the highest percentage inhibition (35%), after 72 h, while WS6 had the lowest percentage inhibition (−1.4%) at this time point (Figure 14.5).

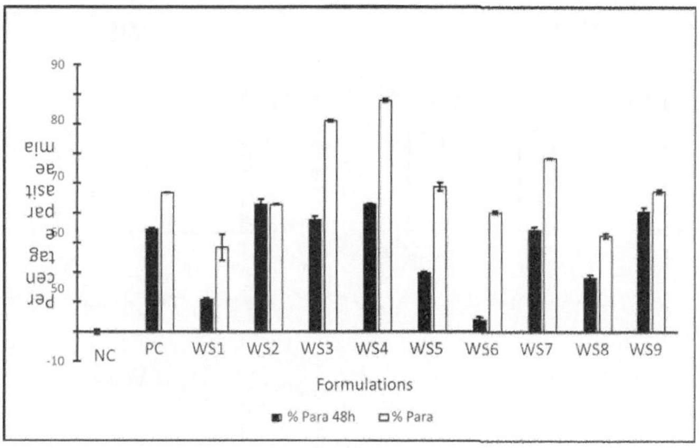

FIGURE 14.4 Parasitemia (%) of the nine herbal formulations at 48 h and 72 h. The control used was chloroquine at 10 µg/ml. *NC* = non-treated parasites; *PC* = chloroquine (10 µg/ml); *Para* = parasitemia.

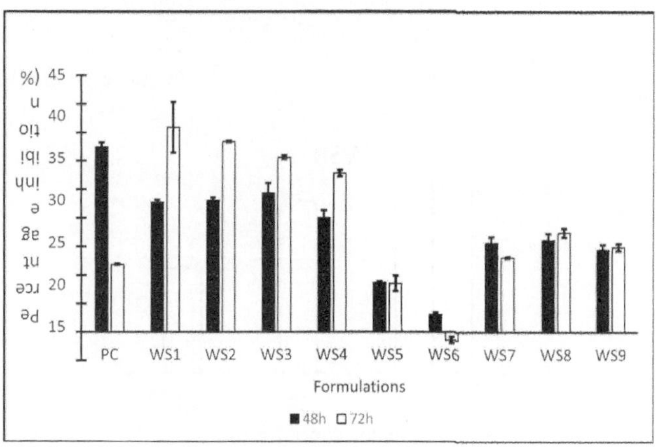

FIGURE 14.5 Parasite growth inhibition (%) for the nine herbal tea formulations using the SYBR Green assay at 100 µg/ml at 48 and 72 h. The control used was chloroquine at 10 µg/ml.

The results of the present study show that combinations of three or more plants in the formulations have higher activity than when each plant is used alone. The tea formulations with a mixture of three plants showed antiplasmodial activity based on the presence of selected classes of antimalarial compounds (WS1, WS2, and WS3), while tea formulations containing one or two plants had fewer classes of phytochemical compounds detected, as a proxy for antimalarial activity. These findings support the long use of polyherbal remedies in cultural systems. Combining plants can reduce the required dose of each plant and prevent the negative effects of high doses, and combining various phytochemicals provides greater protection (Malongane et al., 2017). Several studies have also reported the use of a combination of plants in the treatment of malaria, for example, in Ghana. According to a qualitative survey, Ghanaian herbalists reported that using a combination of herbs was more effective in treating malaria than relying on a single species (Ranasinghe et al., 2015).

14.5.2 Cytotoxicity Assay

There were no cytotoxic effects observed for the herbal formulations (Figure 14.6); however, cell proliferation was observed for all the formulations. WS1 showed the highest percentage growth of 21.6% at 25 μg/ml, and this was shown to decrease with increasing concentration. WS4 and WS8

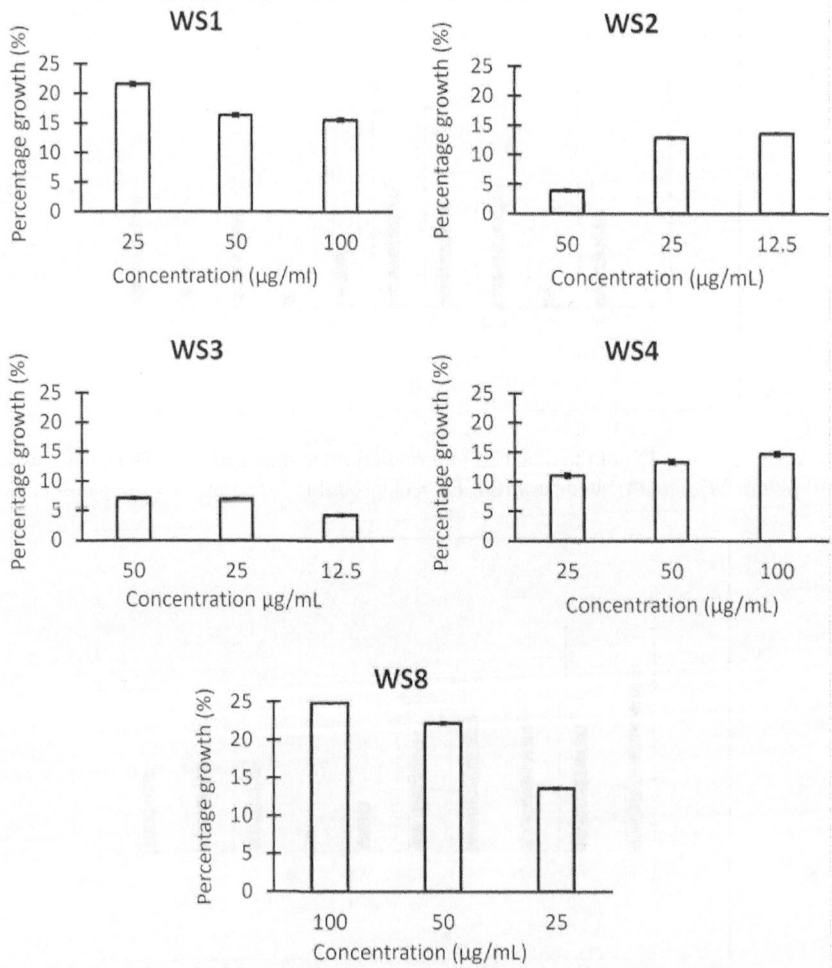

FIGURE 14.6 Percentage growth of the NIH/3T3 fibroblast cells induced by the formulations WS1, WS2, WS3, WS4, and WS8 using the WST-1 cytotoxicity assay.

showed the highest percentage growth at 14.71 and 24.92%, respectively, at 100 μg/ml, and this was shown to decrease with decreasing concentration. WS2 showed the highest percentage growth of 13.52% at 12.5 μg/ml and decreased activity with increasing concentration. WS3 showed the highest percentage growth of 7.15% at 50 μg/ml. For WS2 and WS3, proliferation was recorded at lower concentrations (≤50 μg/ml). Previous studies have indicated that plant extracts, as well as polyherbal formulations, can promote cell growth (Ramamoorthy et al., 2019).

14.5.3 Sensory Evaluation

14.5.3.1 Color

In terms of color, WS4 was the most preferred herbal tea from the formulations. There was no significant difference between the different age groups ($p = 0.000$). For the age groups 36–49, WS3 and WS4 recorded the highest, with a mean score of 4.33 (Figure 14.8). A similar result was also

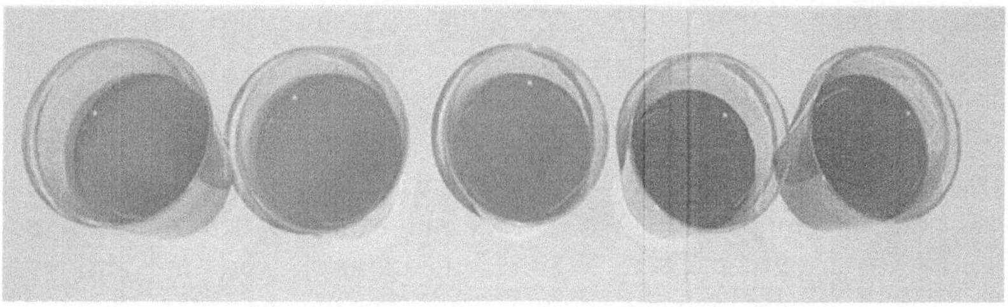

FIGURE 14.7 Herbal formulations as served to the panelist. In order from left to right: WS1, WS2, WS3, WS4, and WS8.

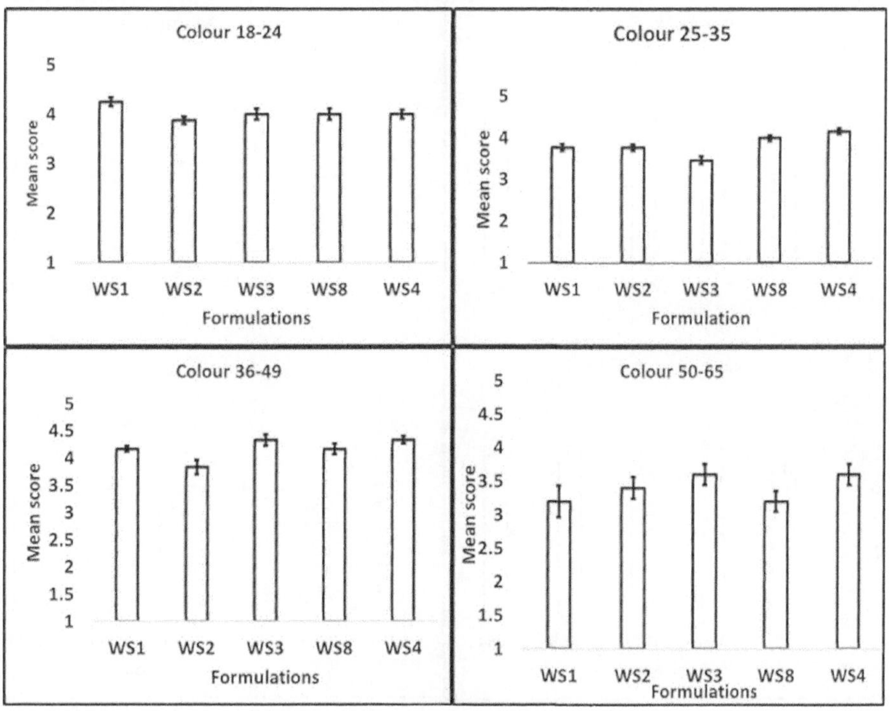

FIGURE 14.8 The mean score for the color of the formulations WS1, WS2, WS3, WS4, and WS8 that was preferred amongst age groups 18–24, 25–35, 36–49, and 50–65.

observed for the age group 50–65, with the highest mean score being recorded for WS3 and WS4 (4.60). For the age group 18–24, however, WS1 recorded the highest mean score of 4.25. On the other hand, WS4 had the highest mean score of 4.15 in the age group 25–35.

14.5.3.2 Aroma

For the evaluation of the aroma of the five herbal formulations, there was no significant difference between the different age groups ($p = 0.163$). In the age group 18–24, WS4 obtained a mean score of 3.875 (Figure 14.9). In the age group 25–35, WS1 obtained a mean score of 3.77. In the age group 36–49, WS8 obtained a mean score of 4.167. In the age group 50–65, both WS2 and WS4 obtained a mean score of 3.4.

14.5.3.3 Flavor

For the evaluation of the flavor of the herbal teas, there was no significant difference between the different age groups ($p = 0.560$). For the age group 18–24, WS1 had a mean score of 4.125 (Figure 14.10). For the age group 25–35, WS4 had a mean score of 3.667. For the age group 36–49, WS2 and WS8 had a mean score of 3.667 each. For the age group 50–65, WS1 and WS2 had a score of 3.8 each.

14.5.3.4 Aftertaste

For the aftertaste evaluation, there was no significant difference between the different age groups ($p = 0.132$). For the age group 18–25, WS2, WS3, and WS4 recorded a mean score of 3.5 each (Figure 14.11). For the age group 25–35, WS2 recorded a mean score of 3.54. For the age group 36–49, WS1 and WS3 recorded a mean score of 3.833. For the age group 50–65, WS1 and WS4 recorded a mean score of 3.6.

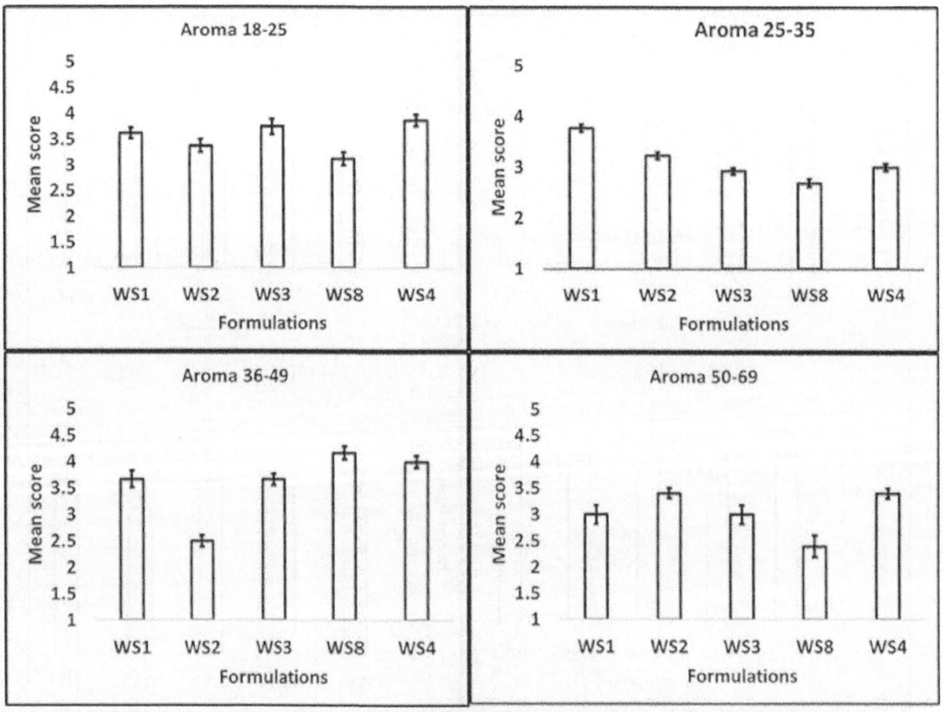

FIGURE 14.9 Mean score for aroma for the formulations WS1, WS2, WS3, and WS4 amongst age groups 18–24, 25–35, 36–49, and 50–65.

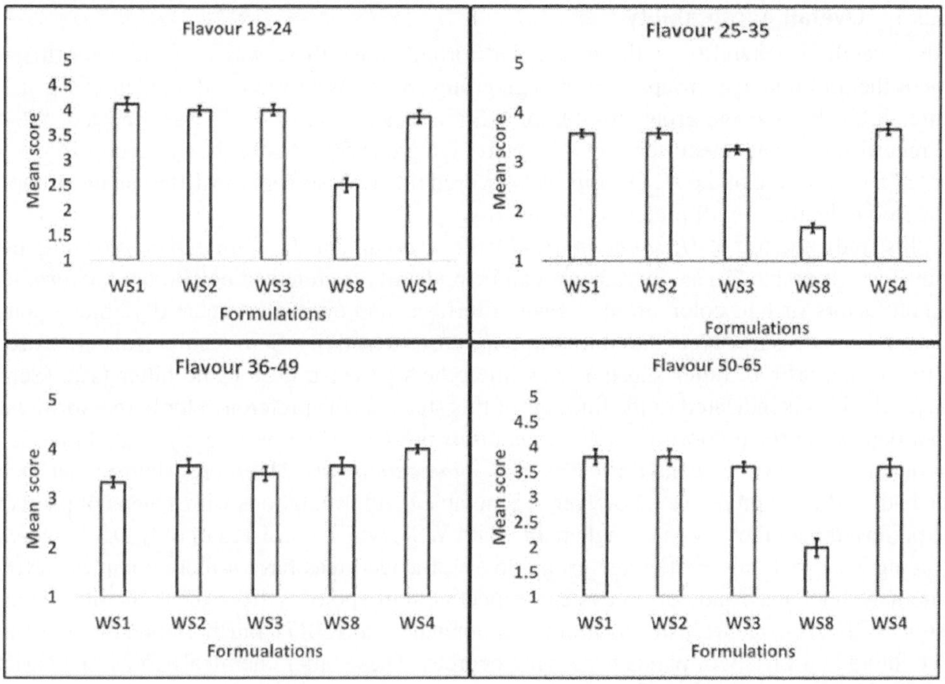

FIGURE 14.10 Mean score for flavor for the formulations WS1, WS2, WS3, WS4, and WS8 for the age groups 18–24, 25–35, 36–49, and 50–65.

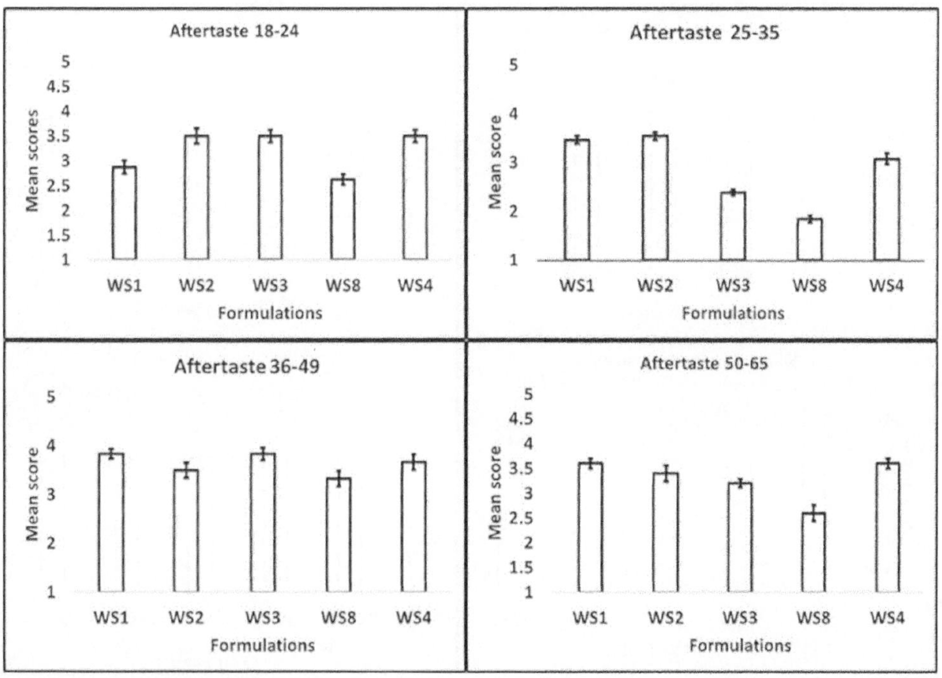

FIGURE 14.11 Mean score for aftertaste for the formulations WS1, WS2, WS3, WS4, and WS8 for the age groups 18–24, 25–35, 36–49, and 50–65.

14.5.3.5 Overall Acceptability

For the overall acceptability of the five herbal formulations, there was no significant difference between the different age groups. For the age group 18–25, WS4 recorded a mean score of 3.969 (Figure 14.12). For the age group 36–49, for WS4, a mean score of 4.333 was reported. WS4 and WS2 recorded the same mean score of 4.2 for the age group 50–65. For the age group 26–35, WS1 recorded a mean score of 4.096. Overall, WS4 scored the highest from the different age groups and can be said to be the overall preferred formulation.

In this study, the use of *H. cannabinus*, *G. coleosperma*, and *D. chamaethamnus* in herbal tea formulas is acceptable. These ingredients can help identify a preferred polyherbal tea formulation based on factors such as color, aroma, flavor, aftertaste, and overall acceptability among potential end users. Humans show great variability in bitter taste sensitivity. Some individuals are extremely sensitive to the taste of bitter substances, while others perceive little or no bitter taste (Schienle et al., 2021). This is indicated in the findings of this study, as the preference for herbal formulations did not depend on the proportion of *H. cannabinus* present, which was to mask the bitter taste of the roots of *G. coleosperma* and the leaves of *D. chamaethamnus*. The scores showed that each age group had a different preference; however, it is notable that formulations with a blend of plants were more preferred than those with a single plant, with WS8, which consisted of only *D. chamaethamnus*, being preferred only by the age group 36–49. Blended teas have a more complex flavor and aroma than single teas, and consumers can expect not only potency from the components but also synergistic effects from the taste and aroma (Malongane et al., 2017). Furthermore, bitter chemicals can be found in a variety of plants to varying degrees. These bitter chemicals come in a variety of chemical forms and evoke bitter sensations of varying intensity (Schienle et al., 2015). This is consistent with literature, as the study of Drewnowski (2001) showed that the complexity and enjoyment of tea and other beverages are enhanced by interactions between sweet, bitter, and sour tastes, as well as between taste and volatile flavor components.

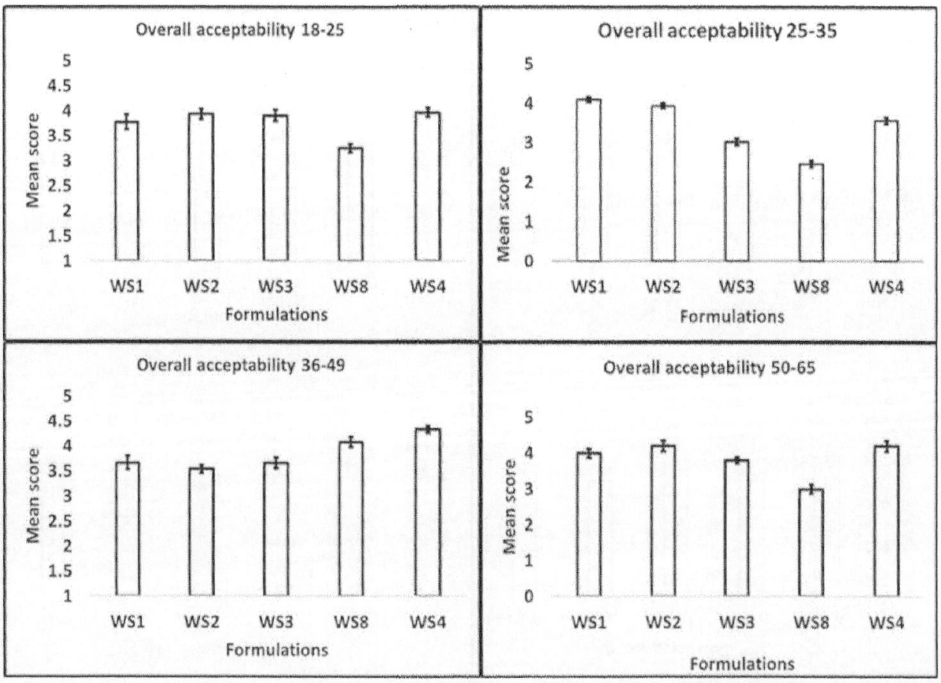

FIGURE 14.12 Mean score for overall acceptability for the formulations WS1, WS2, WS3, WS4, and WS8 for the age groups 18–25, 25–35, 36–49, and 50–65.

14.6 CONCLUSION AND PROSPECTS

The present study highlights the importance of adding value to medicinal plants in the development of a marketable product. The formulation of herbal teas possessing antiplasmodial properties that people can enjoy can be of great benefit, especially to high-risk groups living in malaria-endemic regions. This report details nine herbal formulations containing active ingredients that exhibit activity against the malaria-causing parasite. The organoleptic properties of five formulations were found to be widely acceptable during a sensory evaluation based on color, flavor, aroma, and aftertaste. Furthermore, the formulations were found to possess health benefits, as they enhanced cell proliferation, meaning, that they are potentially beneficial for gastric ailments, such as ulcers. This approach is the first of its kind in Namibia and should be strengthened by further studies to develop these prototypes for mainstreaming including safety and efficacy in vivo, their shelf life in terms of efficacy, safety, and stability. More work can also be done to enhance the aroma and flavor of the teas, to make them more competitive with established commercial herbal tea brands. Namibia has a rich history of IKS, with medicinal plants contributing to primary health care, general well-being, food security, and recreation. Changing lifestyles have eroded the value of many of these plants, with knowledge being lost as older generations pass away. This study was conducted as part of efforts to reverse this trend of undervaluing medicinal IK – by adding value to traditionally used plants to address current problems, such as the scourge of malaria. The changing values in communities as well as the increased consumer awareness require validation of any potential products, more so those with origins in IK. This approach where medicinal plants used to treat malaria were formulated as polyherbal teas and evaluated for their antimalarial, sensory/organoleptic properties, and acceptability has led to a prototype product that fits these requirements. Local communities can benefit socio-economically not only from the use of the tea but also as part of the value addition chain, which includes plant harvesting and preliminary processing. This not only will potentially result in the preservation of IK but will also increase the use of underutilized herbs as tea. The formulation of this tea also falls in line with SDG3, which aims to ensure healthy lives and promote well-being for all ages, as well as Namibia's vision 2030 of health and development. With indigenous knowledge systems being promoted, this study can also be used as a template for the formulation of other medicinal teas using local plants.

14.7 ACKNOWLEDGMENTS

The authors would like to the Multi-Disciplinary Research Services at UNAM, for providing financial support to conduct this study; the Biomedical Research Laboratory team of phlebotomists, volunteer donors, and sensory evaluation panelists, for their time and support in implementing this study; the Biochemistry, Microbiology, and Biotechnology Department at UNAM; and the indigenous knowledge holders in the Zambezi and Kavango East regions of Namibia, for sharing their knowledge and for their continued active participation in the preservation and advancement of IK.

REFERENCES

Adongo, P. B., Kirkwood, B., & Kendall, C. (2005). How local community knowledge about malaria affects insecticide-treated net use in Northern Ghana. *Tropical Medicine & International Health: TM & IH*, *10*(4), 366–378. https://doi.org/10.1111/j.1365-3156.2005.01361.x

Agbor, G. A., Oben, J. E., & Ngogang, J. Y. (2005). Haematinic activity of Hibiscus cannabinus. *African Journal of Biotechnology*, *4*(8), 833–837.

Ahmad, F., & Ismail, G. (2003). Medicinal plants used by kadazandusun communities around crocker range'. In *1–10 ASEAN Review of Biodiversity and Environmetal Conversation (ARBEC)*. www.semanticscholar.org/paper/MEDICINAL-PLANTS-USED-BY-KADAZANDUSUN-COMMUNITIES-Ahmad-Ismail/dfbce8ce99c319e54200e38f01c76c559a3783e3

Ajuwon, O. R., Ademola, O. A., & Gbenga, A. A. (2018). The potential of South African herbal tisanes, rooibos and honeybush in the management of type 2 diabetes mellitus. *Molecules (Basel, Switzerland)*, *23*(12), 3207. https://doi.org/10.3390/molecules23123207

Angupale, J. R., Tusiimire, J., & Ngwuluka, N. C. (2023). A review of efficacy and safety of Ugandan antimalarial plants with application of RITAM score. *Malar Journal*, *22*(1), 97. https://doi.org/10.1186/s12936-023-04486-6. PMID: 36932389; PMCID: PMC10021060

Argemi, X., Sandrine, H., Harold, N., Ophélie, B., Christian, C., & Christophe, R. (2019). Imported Plasmodium falciparum malaria following non-pharmaceutical forms of artemisia annua prophylaxis. *Journal of Travel Medicine*, *26*(8), taz073. https://doi.org/10.1093/jtm/taz073

Audomkasok, S., Singpha, W., Chachiyo, S., & Somsak, V. (2014). Antihemolytic activities of green tea, safflower, and mulberry extracts during Plasmodium berghei infection in mice. *Journal of Pathogens*, *2014*, 203154. https://doi.org/10.1155/2014/203154

Baker, M. T., Lu, P., Parrella, J. A., & Leggette, H. R. (2022). Consumer acceptance toward functional foods: A scoping review. *International Journal of Environmental Research and Public Health*, *19*(3), 1217. https://doi.org/10.3390/ijerph19031217

Balekundri, A., Shahapuri, A., & Patil, M. (2020). Poly-herbal tablet formulation by design expert tool and in vitro anti-lipase activity. *Future Journal of Pharmaceutical Sciences*, *6*(1), 125. https://doi.org/10.1186/s43094-020-00131-0

Bugyei, K. A., Boye, G. L., & Addy, M. E. (2010). Clinical efficacy of a tea-bag formulation of Cryptolepis Sanguinolenta root in the treatment of acute uncomplicated falciparum malaria. *Ghana Medical Journal*, *44*(1), 3–9.

Cheikhyoussef, A., & Embashu, W. (2013). Ethnobotanical knowledge on indigenous fruits in Ohangwena and Oshikoto regions in Northern Namibia. *Journal of Ethnobiology and Ethnomedicine*, *9*(1), 34. https://doi.org/10.1186/1746-4269-9-34

Christians, F. (2020). Country profile – Primary healthcare and family medicine in Namibia'. *African Journal of Primary Health Care & Family Medicine*, *12*(1), 2242. https://doi.org/10.4102/phcfm.v12i1.2242

Chukwuma, C. I., Matsabisa, M. G., Rautenbach, F., Rademan, S., Oyedemi, S. O., Chaudhary, S. K., & Javu, M. (2019). Evaluation of the nutritional composition of Myrothamnus Flabellifolius (Welw.) herbal tea and its protective effect against oxidative hepatic cell injury. *Journal of Food Biochemistry*, *43*(11). https://doi.org/10.1111/jfbc.13026

Clemente-Suárez, V. J., Beltrán-Velasco, A. I., Redondo-Flórez, L., Martín-Rodríguez, A., & Tornero-Aguilera, J. F. (2023). Global impacts of Western diet and its effects on metabolism and health: A narrative review. *Nutrients*, *15*(12), 2749. https://doi.org/10.3390/nu15122749

Cohen, O., Boutrou, M., Nacher, M., Caumes, E., Djossou, F., & Epelboin, L. (2023). A severe case of plasmodium a imported by a french traveler from Cameroon to French Guiana despite regular intake of Artemisia Annua herbal tea. *Revista Do Instituto de Medicina Tropical de São Paulo*, *65*(January). (Instituto de Medicina Tropical de São Paulo: e3). https://doi.org/10.1590/S1678-9946202365003

Czechowski, T., Weathers, P. J., Brodelius, P. E., Brown, G. D., & Graham, I. A. (2020). 'Editorial: Artemisinin – from traditional chinese medicine to artemisinin combination therapies; Four decades of research on the biochemistry, physiology, and breeding of Artemisia Annua. *Frontiers in Plant Science*, *11*. www.frontiersin.org/journals/plant-science/articles/10.3389/fpls.2020.594565

De Vos, E., & Dunn, N. (2024). Malaria prophylaxis. In *StatPearls*. StatPearls Publishing. www.ncbi.nlm.nih.gov/books/NBK551639/

Drake, M. A. (2022). Sensory evaluation. In P. L. H. McSweeney & J. P. McNamara (Eds.), *Encyclopedia of dairy sciences* (3rd ed., pp. 572–576). Academic Press. https://doi.org/10.1016/B978-0-12-818766-1.00151-3

Drewnowski, A. (2001). The science and complexity of bitter taste. *Nutrition Reviews*, *59*(6), 163–169. https://doi.org/10.1111/j.1753-4887.2001.tb07007.x

Dubey, K. K., Janve, M., Ray, A., & Singhal, R. S. (2020). Chapter 4 – Ready-to-drink tea. In C. M. Galanakis (Ed.), *Trends in non-alcoholic beverages* (pp. 101–140). Academic Press. https://doi.org/10.1016/B978-0-12-816938-4.00004-5

Du Preez, C. I. (2016). *Ethnopharmacological assessment of Guibourtia Coleosperma and Diospyros Chamaethamnus extracts as alternative treatment options for malaria* [PhD, University of Namibia].

Du Preez, I., Stefan, L., & Mumbengegwi, D. R. (2020). Chemical composition and inhibitory effects of Guibourtia Coleosperma against Plasmodium parasites in vitro. In *African Natural Plant Products: Vol. III. Discoveries and innovations in chemistry, bioactivity, and applications* (1361, pp. 153–170). ACS Symposium Series 1361; American Chemical Society. https://doi.org/10.1021/bk-2020-1361.ch007

Dushimemaria, F., Iwanette, D. P., & Mumbengegwi, D. (2017). Randomized anticancer and cytotoxicity activities of Guibourtia coleosperma and Diospyros chamaethamnus. *African Journal of Traditional, Complementary, and Alternative Medicines*, *14*(4), 1–7. https://doi.org/10.21010/ajtcam.v14i4.1

Fanshawe, D. (2010). *Vegetation descriptions of the upper Zambezi districts of Zambia* (22). Occasional Publications in Biodiversity Biodiversity Foundation for Africa.

Fawzi, M. M. (2013). Traditional medicines in Africa: An appraisal of ten potent african medicinal plants (Research article). In *Evidence-based complementary and alternative medicine*. https://doi.org/10.1155/2013/617459

Gumisiriza, H., Olet, E. A., Mukasa, P., Lejju, J. B., & Omara, T. (2023). Ethnomedicinal plants used for malaria treatment in Rukungiri district, Western Uganda. *Tropical Medicine and Health*, *51*(August), 49. https://doi.org/10.1186/s41182-023-00541-9

Hailu, H. (2016). *Developing of herbal tea with wild abyssinian thyme (Thyme Schimperi L.) grown in high lands of Southern Tigray, North Ethiopia*. Harmaya University.

Heita, H. T. N., Ham, H., & Mensah, S. (2019). Effects of seed viability and pre-treatments on seed germination of two indigenous species – Strychnos Cocculoides baker and Guibourtia Coleosperma (Benth.) leonard in Namibia. *Agroforestry Systems*, *93*(6), 2173–2180. https://doi.org/10.1007/s10457-018-0333-4

Hosen, M., & Madhu, B. (2023). *Health benefits of herbal tea: A review*. Daffodil International University. https://doi.org/10.13140/RG.2.2.34928.43526

Jacobson, J. O., Smith, J. L., Cueto, C., Chisenga, M., Roberts, K., Hsiang, M. S., Gosling, R., Mumbengegwi, D., & Bennett, A. (2019). Assessing malaria risk at night-time venues in a low-transmission setting: A time-location sampling study in Zambezi, Namibia. *Malaria Journal*, *18*(1). (BioMed Central). https://doi.org/10.1186/s12936-019-2807-x

Joshi, D. D., Deb, L., Somkuwar, B. G., & Rana, V. S. (2023). Relevance of indian traditional tisanes in the management of type 2 diabetes mellitus: A review. *Saudi Pharmaceutical Journal*, *31*(5), 626–638. https://doi.org/10.1016/j.jsps.2023.03.003

Karunamoorthi, K., Jegajeevanram, K., Vijayalakshmi, J., & Mengistie, E. (2013). Traditional medicinal plants: A source of phytotherapeutic modality in resource-constrained health care settings. *Journal of Evidence-Based Complementary & Alternative Medicine*, *18*(1), 67–74. https://doi.org/10.1177/2156587212460241

Kho, K., Sim, Y. Y., & Nyam, K. L. (2019). Antioxidant activities of tea prepared from kenaf (Hibiscus Cannabinus L. KR9) leaves at different maturity stages. *Journal of Food Measurement and Characterization*, *13*(3), 2009–2016. https://doi.org/10.1007/s11694-019-00121-8

Krakowska-Sieprawska, A., Kiełbasa, A., Rafińska, K., Ligor, M., & Buszewski, B. (2022). Modern methods of pre-treatment of plant material for the extraction of bioactive compounds. *Molecules*, *27*(3), 730. https://doi.org/10.3390/molecules27030730. PMID: 35163995; PMCID: PMC8840492

Lee, B. J., Morne, W., Richard, K. H., & van der Kooy, F. (2023). Discovery of Artemisinin in *Artemisia Annua*, its current production, and relevance to Sub-Saharan Africa. *South African Journal of Botany*, *153*(February), 21–27. https://doi.org/10.1016/j.sajb.2022.12.017

Liu, H.-T., Yun-Xuan, C., Tzu-Yun, C., Ya-Peng, W., Tsung-Han, W., Yen-Jung, L., Pi-Hsin, C., et al. (2023). Effects of herbal tea (Platostoma Palustre) on blood glucose regulation in vivo. *GSC Advanced Research and Reviews*, *15*(1), 7–14. https://doi.org/10.30574/gscarr.2023.15.1.0063

Liu, Y., Chunyan, G., Erhuan, Z., Shi, R., Liu, Q., Zhang, M., Zhang, K., & Minhui, Li. (2023). Review on herbal tea as a functional food: Classification, active compounds, biological activity, and industrial status. *Journal of Future Foods*, *3*(3), 206–219. https://doi.org/10.1016/j.jfutfo.2023.02.002

Long, X., Ranjitkar, S., Waldstein, A., Wu, H., Li, Q., & Geng, Y. (2024). Preliminary Exploration of herbal tea products based on traditional knowledge and hypotheses concerning herbal tea selection: A case study in Southwest Guizhou, China. *Journal of Ethnobiology and Ethnomedicine*, *20*(1), 1. https://doi.org/10.1186/s13002-023-00645-w

Malongane, F., McGaw, L. J., & Mudau, F. N. (2017). The synergistic potential of various teas, herbs and therapeutic drugs in health improvement: A review. *Journal of the Science of Food and Agriculture*, *97*(14), 4679–4689. https://doi.org/10.1002/jsfa.8472

Maroyi, A. (2017). Lippia Javanica (Burm.f.) Spreng.: Traditional and commercial uses and phytochemical and pharmacological significance in the African and Indian subcontinent. *Evidence-Based Complementary and Alternative Medicine: eCAM*, *2017*, 6746071. https://doi.org/10.1155/2017/6746071

Mwema, T. (2021). *Anopheles vectors species composition, their biting cycle and role of human behaviour in malaria transmission in an endemic region from Namibia* [Thesis, University of Namibia]. https://repository.unam.edu.na/handle/11070/3000

Nafuka, S. N., & Mumbengegwi, D. R. (2013). Phytochemical analysis and in vitro anti-plasmodial activity of selected ethnomedicinal plants used to treat malaria associated symptoms in Northern Namibia. *International Science and Technology Journal of Namibia*, *2*(1), 078–093.

Nagendrappa, P. B., Annamalai, P., Naik, M., Mahajan, V., Mathur, A., Susanta, G., Gay, F., & Venkatasubramanian, P. (2017). A prospective comparative field study to evaluate the efficacy of a traditional plant-based malaria prophylaxis. *Journal of Intercultural Ethnopharmacology*, *6*(1), 36–41. https://doi.org/10.5455/jice.20161112021406

Najmi, A., Javed, S. A., Al Bratty, M., & Alhazmi, H. A. (2022). Modern approaches in the discovery and development of plant-based natural products and their analogues as potential therapeutic agents. *Molecules, 27*(2), 349. https://doi.org/10.3390/molecules27020349

Navale, A., Jadhav, R., & Vikhe, P. (2021). Hibiscus cannabinus a pharmacological: A review. *World Journal of Pharmaceutical Research, 10*(10), 36–41. https://doi.org/10.20959/wjpr202110-17900

Nigussie, G., & Wale, M. (2022). Medicinal plants used in traditional treatment of malaria in Ethiopia: A review of ethnomedicine, anti-malarial and toxicity studies. *Malaria Journal, 21*(1), 262. https://doi.org/10.1186/s12936-022-04264-w

Nwabuisi, C. (2002). Prophylactic effect of multi-herbal extract 'Agbo-Iba' on malaria induced in mice. *East African Medical Journal, 79*(August), 343–346. https://doi.org/10.4314/eamj.v79i7.8836

Obouayeba, A. P., Diarrassouba, M., Soumahin, E. F., & Kouakou, T. (2015). Phytochemical analysis, purification and identification of Hibiscus Anthocyanins. *Journal of Parmaceutical, Chemical and Biological Sciences, 3*(2), 156–168.

Ogwang, P., Ogwal, J. O., Kasasa, S., Olila, D., Ejobi, F., Kabasa, J., & Obua, C. (2012). Artemisia Annua L. infusion consumed once a week reduces risk of multiple episodes of malaria: A randomised trial in a Ugandan community. *Tropical Journal of Pharmaceutical Research, 11*(June), 445–453.

Okaiyeto, K., & Oguntibeju, O. O. (2021). African herbal medicines: Adverse effects and cytotoxic potentials with different therapeutic applications. *International Journal of Environmental Research and Public Health, 18*(11), 5988.https://doi.org/10.3390/ijerph18115988

Pierre, S., Tchobsala, V., Fohouo, T., Nloga, A., Michael, A., & Jean, M. (2011). Medicinal plants used in traditional treatment of malaria in Cameroon. *Journal of Ecology and the Natural Environment, 3*, 104–117.

Ramamoorthy, R., Muthalagu, M., Andra, S., Ravichandran, B., & Narayanasamy, M. (2019). Investigation on antimicrobial, antioxidant and cytotoxicity properties of triple bark extract formulated using traditional medicinal plants. *SN Applied Sciences, 1*(7), 772. https://doi.org/10.1007/s42452-019-0791-y

Ranasinghe, S., Ansumana, R., Lamin, J. M., Bockarie, A. S., Bangura, U., Buanie, J. A. G., Stenger, D. A., & Jacobsen, K. H. (2015). Herbs and herbal combinations used to treat suspected malaria in Bo, Sierra Leone. *Journal of Ethnopharmacology, 166*(May), 200–204. https://doi.org/10.1016/j.jep.2015.03.028

Rekaya, A., Hanana, M., Mzid, R., Hamrouni, L., Khouja, M. L., & Hanachi, A. (2016). Hibiscus Cannabinus L. – " Kenaf ": A review paper. *Journal of Natural Fibers, 14*(December), 1–19. https://doi.org/10.1080/15440478.2016.1240639

Rogerson, S. J., Beeson, J. G., Laman, M., Poespoprodjo, J. R., William, T., Simpson, J. A., Price, R. N., et al. (2020). Identifying and combating the impacts of COVID-19 on malaria. *BMC Medicine, 18*(1), 239. https://doi.org/10.1186/s12916-020-01710-x

Salihu, S., Otitolaiye, C. A., & Hizbullah, M. U. (2020). In vitro antimalarial and antibacterial activities of methanol stem bark extract of frankincense tree (Boswellia Dalzielii) and leaves extract of Kenaf (Hibiscus Cannabinus). *Asian Journal of Biochemistry, Genetics and Molecular Biology, 5*(2), 19–23. https://doi.org/10.9734/ajbgmb/2020/v5i230126

Schienle, A., Arendasy, M., & Schwab, D. (2015). Disgust responses to bitter compounds: The role of disgust sensitivity. *Chemosensory Perception, 8*(4), 167–173. https://doi.org/10.1007/s12078-015-9186-7

Schienle, A., Osmani, F., & Schlintl, C. (2021). Disgust propensity and the bitter aftertaste response. *Chemosensory Perception, 14*(2), 57–63. https://doi.org/10.1007/s12078-020-09283-y

Shafiq, M., Maheen, N., Sadiq, S., Rizvi, S. S. B., & Abbas, M. T. (2024). Tea. In L. P. Awasthi (Ed.), *Viral diseases of field and horticultural crops* (pp. 279–283). Academic Press. https://doi.org/10.1016/B978-0-323-90899-3.00003-3

Shaikh, S. (2022). Sources and health benefits of functional food components. In *Current topics in functional food*. IntechOpen. https://doi.org/10.5772/intechopen.104091

Shikha, R., Pawan, J., & Manju, K. (2019). Review article on tisanes: Recreational use of local plants. *International Journal for Science and Advance Research in Technology, 5*(2), 1–19.

Sim, Y. Y., & Nyam, K. L. (2021). '*Hibiscus Cannabinus* L. (Kenaf) studies: Nutritional composition, phytochemistry, pharmacology, and potential applications. *Food Chemistry, 344*(May), 128582. https://doi.org/10.1016/j.foodchem.2020.128582

Sinesio, F. (2005). Sensory evaluation. In P. Worsfold, A. Townshend, & C. Poole (Eds.), *Encyclopedia of analytical science* (2nd ed., pp. 283–290). Elsevier. https://doi.org/10.1016/B0-12-369397-7/00561-6

Smith, J. L., Auala, J., Haindongo, E., Uusiku, P., Gosling, R., Kleinschmidt, I., Mumbengegwi, D., & Sturrock, H. J. W. (2017). Malaria risk in young male travellers but local transmission persists: A case–control study in low transmission Namibia. *Malaria Journal, 16*(1), 70. https://doi.org/10.1186/s12936-017-1719-x

Smith, J. L., Mumbengegwi, D., Haindongo, E., Cueto, C., Roberts, K. W., Gosling, R., Uusiku, P., Kleinschmidt, I., Bennett, A., & Sturrock, H. J. (2021). Malaria risk factors in Northern Namibia: The importance of occupation, age and mobility in characterizing high-risk populations. *PLoS One*, *16*(6), e0252690. https://doi.org/10.1371/journal.pone.0252690

Tabuti, J. R. S., Obakiro, S. B., Nabatanzi, A., Anywar, G., Nambejja, C., Mutyaba, M. R., Omara, T., & Waako, P. (2023). Medicinal plants used for treatment of malaria by indigenous communities of Tororo district, Eastern Uganda. *Tropical Medicine and Health*, *51*(1), 34. https://doi.org/10.1186/s41182-023-00526-8

Tajuddeen, N., & Heerden, F. R. V. (2019). Antiplasmodial natural products: An update. *Malaria Journal*, *18*(1), 404. https://doi.org/10.1186/s12936-019-3026-1

Trager, W., & Jensen, J. B. (1976). Human malaria parasites in continuous culture. *Science*, *193*, 673–675.http://dx.doi.org/10.1126/science.781840

Tse, E. G., Korsik, M., & Todd, M. H. (2019). The past, present and future of anti-malarial medicines. *Malaria Journal*, *18*(1), 93. https://doi.org/10.1186/s12936-019-2724-z

Vamanu, E., Dinu, L. D., Pelinescu, D. R., & Gatea, F. (2021). Therapeutic properties of edible mushrooms and herbal teas in gut microbiota modulation. *Microorganisms*, *9*(6), 1262. https://doi.org/10.3390/microorganisms9061262

Vasavi, C. L., Jyothi, A. S., Sravani, P. T., Chand, P., Adil, S. K., Raja, R. R., Dr., & Baba, K. H., Dr. (2019). Hibiscus Cannabinus and Hibiscus Sabdariffa phyto phamacognostical review. *Journal of Pharmacognosy and Phytochemistry*, *8*(1) (AkiNik Publications) 313–318.

Von Koenen, E. (2007). *Medicinal, poisonous, and edible plants in Namibia* (Vol. 4, Edition Namibia). Klaus Hess Publishers.

Wang, P., Ma, X. M., Geng, K., Jiang, Z. Z., Yan, P. Y., & Xu, Y. (2022). Effects of camellia tea and herbal tea on cardiometabolic risk in patients with type 2 diabetes mellitus: A systematic review and meta-analysis of randomized controlled trials. *Phytotherapy Research: PTR*, *36*(11), 4051–4062. https://doi.org/10.1002/ptr.7572

Willcox, M., Bodeker, G., & Burford, G. (2004). An overview of ethnobotanical studies on plants used for the treatment of malaria. In *Traditional medicinal plant and malaria* (pp. 187–197). CRC Press.

Willcox, M., Falquet, J., Ferreira, J. F. S., Gilbert, B., Hsu, E., Magalhães, P., Plaizier-Vercammen, J., Sharma, V., Wright, C., & Yaode, W. (2006). Artemisia annua as a herbal tea for malaria. *African Journal of Traditional, Complementary, and Alternative Medicines*, *4*, 121–123.

World Health Organization (WHO). (2021). *World malaria report 2021*. World Health Organization. https://iris.who.int/handle/10665/350147. License: CC BY-NC-SA 3.0 IGO.

World Health Organization (WHO). (2023). *World malaria report 2023*. Author. www.who.int/publications-detail-redirect/9789240086173

15 Cultural Perspectives, Nutritional and Health Values of Wild Edible Plants among the Batswana of Southern Africa

Keamogela Innocentia Nko,[1, 2] *James Lwambi Mwinga,*[1] *John Awungnjia Asong,*[3] *Sonja Louise Venter,*[2] *Stephen Oluwaseun Amoo,*[2, 3] *and *Adeyemi Oladapo Aremu*[1, 4]
[1]Indigenous Knowledge Systems Centre, Faculty of Natural and Agricultural Sciences, North-West University, Private Bag X2046, Mmabatho 2790, South Africa
[2]Agricultural Research Council – Vegetables, Industrial, and Medicinal Plants, Private Bag X293, Pretoria, 0001, South Africa
[3]Unit for Environmental Sciences and Management, Faculty of Natural and Agricultural Sciences, North-West University, Private Bag X1290, Potchefstroom 2531, South Africa
[4]School of Life Sciences, College of Agriculture, Engineering, and Science, University of KwaZulu-Natal, Private Bag X54001, Durban 4000, South Africa
*Corresponding author: Oladapo.Aremu@nwu.ac.za

15.1 INTRODUCTION

Wild edible plants are known to grow naturally in the environment and are collected for diverse uses, including food and medicine (Ju et al., 2013; Liu et al., 2023). These plants have remained important for the health and survival of humans for centuries (Fentahun & Hager, 2009). Wild edible plants are important when food is scarce, especially during famine, drought, and war (Liu et al., 2023; Oduor et al., 2023). Recently, people have recognized that wild edible plants can help make farming and food systems more sustainable. They can provide more food, supply better nutrition, and help local communities make a living (Kanter et al., 2023).

Wild edible plants, particularly vegetables, are an important source of food in the rural parts of South Africa (Jansen van Rensburg et al., 2004, 2009). In South Africa, the use of wild edible plants for nutrition has been part of the culture for a long time, especially among the Khoi and the San, who are the well-known Indigenous inhabitant in southern Africa (Bvenura & Afolayan, 2014).

Generally, wild edible plants have received limited scientific research when compared to cultivated plants. However, increasing interests on wild edible plants have been evident in recent times. Particularly, there is an increasing attention on the documentation of wild edible plants across

 DOI: 10.1201/9781003469766-15

the provinces in South Africa. For example, in the Eastern Cape Province, Bhat et al. (2002) and Bvenura and Afolayan (2015) listed 36 and 22 wild edible plants (fruits and vegetables), respectively. Potgieter (2001) reported 24 wild leafy plants in Limpopo Province. Another study identified 32 wild edible species in North West Province and Limpopo Province (Steyn et al., 2001). In the northern KwaZulu-Natal, 72 traditional vegetables consisting of 34 naturalized aliens and 38 native plants were recorded (Ntuli et al., 2012).

Welcome and Van Wyk (2019) listed 1,740 plants used as food in southern Africa. This illustrates the rich biodiversity used as food among different ethnic groups in southern Africa. As indicated by Mander et al. (1995), the Batswana have known used their local plants for a long time. The Batswana people originated in Central and Eastern Africa and then migrated to southern Africa in the 1300s, while they also descended mainly from the Bantu-speaking tribes, along with the Khoi-San (Cornwell, 1988). This chapter entails a comprehensive overview on the cultural perspectives and nutritional and health potential of locally sourced wild edible plants among the Batswana in southern Africa.

15.2 MATERIALS AND METHODS

To identify and generate comprehensive baseline information on the cultural perspectives and nutritional and health values of wild edible plants among the Batswana of southern Africa, a web-based systematic literature search was conducted to obtain literature published from January 2000 to January 2024. The systematic review was conducted following the PRISMA guidelines for reporting systematic reviews and meta-analysis (Moher et al., 2009). Electronic databases such as Google Scholar (n = 397), ScienceDirect (n = 40), and Scopus (n = 4) were used to search for literature, including journal articles, books, theses, and dissertations. These databases were searched using keywords/phrases, such as *wild edible plants*, *underutilized plants*, and *neglected plants*, coupled with *Batswana*. The scientific names of wild edible plants were verified using the World Flora Online (https://www.worldfloraonline.org, WFO, 2024). The resultant hits were screened by carefully reviewing the titles and abstracts of articles to determine their relevance and eligibility for inclusion in this chapter. Selected research articles were downloaded for further assessment. The critical assessment of these articles involved evaluating how they met the predetermined inclusion criteria. Review and non-English articles were excluded from this study. To generate evidence on the nutritional and health values, empirical data for each of the recorded wild edible plants were individually searched online.

15.3 RESULTS AND DISCUSSION

15.3.1 Literature Search Output

A total of 438 studies was recorded from various scientific databases (Figure 15.1), which included journal articles, thesis, books, and dissertations. After screening, 425 studies were excluded based on the inclusion and exclusion criteria, while 13 studies remained eligible (Figure 15.1). In this chapter, the eligible studies covered two countries, namely, Botswana and South Africa (Table 15.1). Botswana had a higher number of studies relative to South Africa. Understanding the different uses of wild edible plants is important, as it ensures that the existing knowledge is preserved, identifying their localities and taxonomy (Cruz et al., 2013). The generated inventory from the eligible studies potentially serves as a valuable reference for future research.

15.3.2 Botanical Overview of the Documented Wild Edible Plants

The eligible studies resulted in an inventory of 70 wild edible plants (Table 15.2). *Sclerocarya birrea* (6), *Grewia flavescens* (5), *Cleome gynandra* (4), *Gasteria bicolor* (4), *Ziziphus mucronata* (4), *Allium thunbergii* (3), *Ximenia caffra* (3), *Vangueria infausta* (3), *Ximenia americana* (3), *Strychnos*

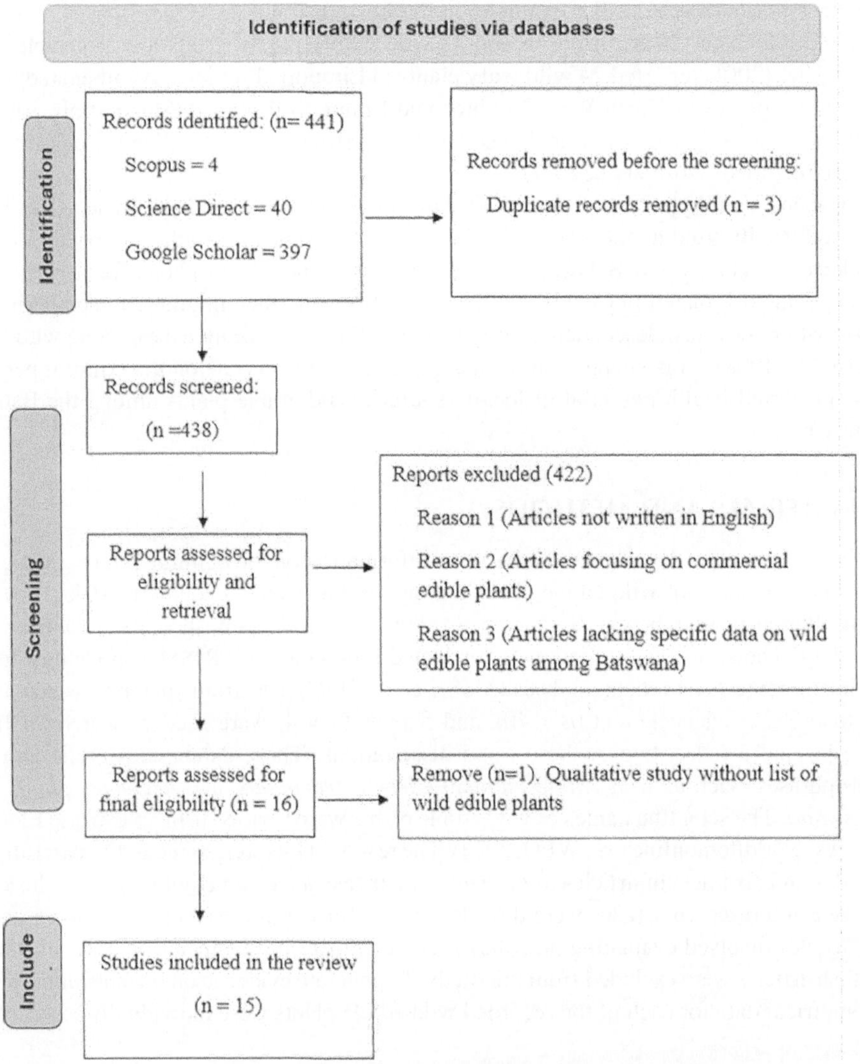

FIGURE 15.1 Approach used for the screening and selection of articles on wild edible plants among the Batswana of southern Africa (January 2000 to January 2024).

TABLE 15.1
Overview of Literature Documenting Wild Edible Plants among the Batswana of southern Africa from January 2000 to January 2024

Author (Year)	Title of the Study	Country	No. of Plants	No. of Families	Selection Criteria	Participant Type	Methodological Framework
Badimo et al. (2015)	"Utilization of edible wild plants and their contribution to household income in Gweta Village, central Botswana"	Botswana	19	12	Food and medicinal uses	Village chief, elders, women, village development committee members, and some government agents	Interviews with key informants and household survey

(Continued)

TABLE 15.1 (Continued)

Author (Year)	Title of the Study	Country	No. of Plants	No. of Families	Selection Criteria	Participant Type	Methodological Framework
Bultosa et al. (2020)	"Plant-based traditional foods and beverages of Ramotswa Village, Botswana"	Botswana	21	13	Foods, beverages, and medicinal uses	Residents in the village with knowledge on traditional foods/dishes and beverages	Questionnaire interviews and focus group discussions
Danley (2006)	"Letters of the Bush: A case study of traditional Setswana herbal medicine"	Botswana	1	1	Food and medicine	Knowledge holders	Interviews
Matenge (2011)	"Utilisation of traditional and indigenous foods in the North West Province of South Africa"	South Africa	8	6	Food	Elderly individuals	Questionnaires and focus group discussions
Mojeremane and Tshwenyane (2004)	"*Azanza garckeana*: A valuable edible indigenous fruit tree of Botswana"	Botswana	1	1	Food and income	Unspecified	Unspecified
Motlhanka and Makhabu (2011)	"Medicinal and edible wild fruit plants of Botswana as emerging new crop opportunities"	Botswana	13	11	Food and medicine	Traditional healers, village community trusts, and old people	Interviewing
Motlhanka et al. (2008)	"Edible indigenous wild fruit plants of Eastern Botswana"	Botswana	14	11	Food and medicinal uses	Unspecified	Interviews and field surveys
Neudeck et al. (2012)	"The contribution of edible wild plants to food security, dietary diversity and income of households in Shorobe Village, Northern Botswana"	Botswana	21	14	Food and medicinal uses	Participants knowledgeable about wild edible plants identified with the assistance of local people and authorities	Interviews and focus group discussions
Olarewaju et al. (2021)	"Food and medicinal relevance of Cucurbitaceae species in Eastern and Southern Africa"	South Africa	6	1	Food and medicine	Unspecified	Unspecified
Omotayo et al. (2020)	"Utilization pattern of indigenous and naturalized plants among some selected rural households of North-West Province, South Africa"	South Africa	5	5	Food and medicine	Youth and elders	Face-to-face interviews and questionnaires
Pitso and Lebese (2014)	"Traditional uses of wild edible plants in arid areas of South Africa"	South Africa	21	15	Food and medicine	Knowledge holders	Focus group discussions and in-depth interviews

(Continued)

TABLE 15.1 (Continued)

Author (Year)	Title of the Study	Country	No. of Plants	No. of Families	Selection Criteria	Participant Type	Methodological Framework
Tselaesele et al. (2023)	"Plant-based traditional foods and beverages of Gumare village, Botswana"	Botswana	28	20	Foods, beverages	Residents in the village with knowledge on traditional foods/ dishes and beverages	Semi-structured questionnaire for a survey and focus group discussions
Van der Hoeven et al. (2013)	"Indigenous and traditional plants: South African parents' knowledge, perceptions and uses and their children's sensory acceptance"	South Africa	15	13	Food and medicine	Parents	Structured questionnaire prior to the focus group discussions

TABLE 15.2
Inventory of Wild Edible Plants Used by the Batswana of southern Africa

Scientific Name	#Family	Local Name(s)	Edible Part(s)	References
Adansonia digitata L.	Bombacaceae	Mmowana, mowana	Fruit, seeds	Motlhanka et al. (2008); Badimo et al. (2015); Tselaesele et al. (2023)
Agaricus spp.	Agaricaceae	Mabowa	Fruit	Bultosa et al. (2020)
Amaranthus hybridus L.	Amaranthaceae	Theepe	Leaves	Matenge (2011)
Amaranthus thunbergii Moq.	Amaranthaceae	Theepe	Leaves, unspecified	Badimo et al. (2015); Tselaesele et al. (2023); Neudeck et al. (2012)
Azanza garckeana (F. Hoffm.) Exell and Hillc.	Malvaceae	Moroja, morojwa	Unspecified, fruit	Neudeck et al. (2012); Mojeremane and Tshwenyane (2004); Badimo et al. (2015)
Berchemia discolor (Klotzsch) Hemsl.	Rhamnaceae	Motsintsila, motsentsela	Fruit	Badimo et al. (2015); Tselaesele et al. (2023)
Berchemia zeyheri (Sond.) Grubov	Rhamnaceae	Moe	Fruit	Bultosa et al. (2020)
Boscia albitrunca (Burch.) Gilg and Gilg-Ben.	Capparaceae	Motopi, motlopi	Roots, fruit	Motlhanka et al. (2008); Bultosa et al. (2020)
Bridelia mollis Hutch.	Phyllanthaceae	Mokopakopa, moka-manawa	Fruit	Motlhanka et al. (2008); Tselaesele et al. (2023)
Ceropegia rendallii N.E. Br	Apocynaceae	Serowa	Tubers, roots	Tselaesele et al. (2023); Badimo et al. (2015)
Ceropegia spp.	Apocynaceae	Serowa	Unspecified	Neudeck et al. (2012)
Citrullus lanatus (Thunb.) Matsum. and Nakai	Cucurbitaceae	Mokate, lekatane	Fruit	Bultosa et al. (2020); Matenge (2011)
Cleome gynandra L	Capparaceae	Leketa, rothwe, lerotho	Leaves, unspecified	Tselaesele et al. (2023); Neudeck et al. (2012); Badimo et al. (2015); Matenge (2011)
Coccinia sessilifolia (Sond.) Cogn.	Cucurbitaceae	Legabala	Fruit	Bultosa et al. (2020)

(Continued)

TABLE 15.2 (Continued)

Scientific Name	#Family	Local Name(s)	Edible Part(s)	References
Corchorus olitorius L.	Malvaceae	Delele	Leaves, unspecified	Badimo et al. (2015); Neudeck et al. (2012)
Cucurbita spp.	Cucurbitaceae	Morogo wa lephutshe	Leaves	Matenge (2011)
Cyperus fulgens C.B. Clark.	Cyperaceae	Monakaladi	Unspecified	Neudeck et al. (2012)
Cyperus papyrus L	Cyperaceae	Koma	Roots, stem	Tselaesele et al. (2023)
Diospyros lycioides Desf.	Ebenaceae	Lethanyu	Fruit	Omotayo et al. (2020)
Diospyros mespiliformis Hochst. Ex A. D	Ebenaceae	Mokhutsomu	Fruit, unspecified	Tselaesele et al. (2023); Neudeck et al. (2012)
Echinocystis lobata (Michx.) Torr. and Gray	Cucurbitaceae	Mogaba	Fruit	Matenge (2011)
Ehretia rigida (thumb.) Druce	Boraginaceae	Morobe	Fruit	Bultosa et al. (2020)
Englerophytum magalismontanum (Sond.) T.D.Penn.	Sapotaceae	Motlhatswa	Fruit	Bultosa et al. (2020)
Ficus spp.	Moraceae	Mochabachabana	Fruit	Tselaesele et al. (2023)
Ficus sycomorus L.	Moraceae	Mochaba/ motshaba	Fruit, stem exudates, unspecified	Tselaesele et al. (2023); Neudeck et al. (2012)
Fockea angustifolia K. Schum.	Apocynaceae	Leruswa	Tubers, roots, unspecified	Badimo et al. (2015); Neudeck et al. (2012)
Garcinia livingstonei T. Anderson	Clusiaceae	Motsaodi	Unspecified, fruit	Neudeck et al. (2012)
Grewia avellana Heirn	Malvaceae (Tiliaceae)	Marago-a-bahumagadi	Fruit	Badimo et al. (2015)
Grewia bicolor Juss	Malvaceae (Tiliaceae)	Mogwana	Fruit, tubers, unspecified	Tselaesele et al. (2023); Neudeck et al. (2012); Bultosa et al. (2020); Badimo et al. (2015)
Grewia flava DC	Malvaceae (Tiliaceae)	Moretlwa/ moretwa	Fruit, unspecified	Tselaesele et al. (2023); Neudeck et al. (2012); Motlhanka et al. (2008); Badimo et al. (2015); Bultosa et al. (2020)
Grewia flavescens Juss. var. *flavescens*	Malvaceae (Tiliaceae)	Mokgomphatha/ motsotsojane	Fruit, seeds, unspecified	Tselaesele et al. (2023); Neudeck et al. (2012); Motlhanka et al. (2008)
Grewia retinervis Burret	Malvaceae (Tiliaceae)	Motsotsojane	Fruit, unspecified	Badimo et al. (2015); Bultosa et al. (2020); Neudeck et al. (2012)
Grewia schinzii K. Schum	Malvaceae (Tiliaceae)	Mokgomphate	Fruit	Badimo et al. (2015)
Hyphaene petersiana Klotzsch ex Mart.	Arecaceae	Mokolwane	Fruit, unspecified	Badimo et al. (2015); Neudeck et al. (2012)
Ipomea batatas (L.) Lam	Convolvulaceae	Kwele	Roots, tubers	Matenge (2011)
Kalanchoe rotundifolia (Haw.) Haw.	Crassulaceae	Sereledi	Fruit	Bultosa et al. (2020)
Kigelia africana (Lam.) Benth.	Bignoniaceae	Moporota	Flower, fruit	Tselaesele et al. (2023)
Mentha aquatica L	Lamiaceae	Kwena	Leaves	Tselaesele et al. (2023)
Mimusops zeyheri Sond.	Sapotaceae	Mmupudu	Fruit	Badimo et al. (2015)
Momordica balsamina L	Cucurbitaceae	Mampimpinyane	Fruit, roots	Tselaesele et al. (2023)
Nymphaea nouchali Burm.f.	Nymphaeaceae	Tswii	Unspecified	Neudeck et al. (2012)
Nyphaeae Burm.f. var. *caerulea* (Savigny) Verdc	Nymphaeaceae	Tswii	Tubers	Tselaesele et al. (2023)
Opuntia spp.	Cactaceae	Ntoroko	Fruit	Bultosa et al. (2020)
Pappea capensis Eckl. and Zeyh.	Sapindaceae	Molalakgaka, mopenoeng	Seeds, fruit	Motlhanka et al. (2008); Bultosa et al. (2020)

(Continued)

TABLE 15.2 (Continued)

Scientific Name	#Family	Local Name(s)	Edible Part(s)	References
Parinari curatellifolia Planch. ex Benth.	Chrysobalanaceae	Mobola	Fruit	Omotayo et al. (2020)
Pentarrhinum insipidum E. Mey	Apocynaceae	Lefe	Leaves	Matenge (2011)
Phoenix reclinata Jacq	Arecaceae	Tsaro/nkadi	Fruit, roots, tubers, buds	Tselaesele et al. (2023)
Rhus pyroides Burch	Anacardiaceae	Mogodiri	Fruit	Bultosa et al. (2020)
Rhus quartiniana A. Rich.	Anacardiaceae	Moropaphiri	Unspecified	Neudeck et al. (2012)
Rhus tenuinervis Engl. (synonym *Searsia tenuinervis* (Engl.) Mofet)	Anacardiaceae	Morupaphiri	Fruit	Tselaesele et al. (2023)
Schinziophyton rautanenii (Schinz) Radcl.-Sm.	Euphorbiaceae	Mokongwa	Fruit	Badimo et al. (2015)
Sclerocarya birrea (A. Rich.) Hochst.	Anacardiaceae	Morula	Fruit, unspecified	Badimo et al. (2015); Tselaesele et al. (2023); Neudeck et al. (2012); Omotayo et al. (2020); Motlhanka et al. (2008); Bultosa et al. (2020)
Stachys aethiopica L	Lamiaceae	Dikgobe tsa badisana	Fruit, leaves	Tselaesele et al. (2023)
Stomatostemma monteiroae (Oliv.) N.E.Br.	Apocynaceae	Mosata	Fruit	Motlhanka et al. (2008)
Strychnos cocculoides Baker	Loganiaceae	Moruta, mogorogorwane	Fruit pulp	Motlhanka et al. (2008); Tselaesele et al. (2023); Bultosa et al. (2020)
Strychnos madagascariensis	Loganiaceae	Mogwagwa	Fruit pulp	Motlhanka et al. (2008)
Strychnos spinosa Lam	Loganiaceae	Nsala	Fruit	Omotayo et al. (2020)
Strychnos spp.	Loganiaceae	Mogorogoro	Fruit, seeds	Tselaesele et al. (2023)
Trochomeria debilis	Cucurbitaceae	Mogabaditswane, mogakangwaga	Fruit, tubers	Bultosa et al. (2020)
Tylosema esculentum (Burch.) A. Schreiber	Fabaceae	Morama	Fruit	Badimo et al. (2015)
Typha capensis (Rohrb.) N.E. Br	Typhaceae	Tsita	Root stalks (rhizomes)	Tselaesele et al. (2023)
Urochloa mosambicensis (Hack.) Dandy	Poaceae	Phoka	Unspecified	Neudeck et al. (2012)
Vangueria infausta Burch	Rubiaceae	Mmilo	Fruit	Tselaesele et al. (2023); Motlhanka et al. (2008); Omotayo et al. (2020)
Vangueriopsis lanciflora (Hiern.) Robyns ex R.D. Good	Rubiaceae	Mmupudu	Unspecified	Neudeck et al. (2012)
Vigna unguiculata L	Fabaceae	Morogo wa dinawa	Leaves	Matenge (2011)
Withania somnifera (L.) Dunal	Solanaceae	Khusubele	Fruit	Bultosa et al. (2020)
Ximenia americana L.	Olacaceae	Morotologapudi, moretologa	Fruit, unspecified	Motlhanka et al. (2008); Neudeck et al. (2012); Badimo et al. (2015)
Ximenia caffra Sonder	Olacaceae	Morotologakgomo	Fruit	Motlhanka et al. (2008); Badimo et al. (2015)
Ximenia spp.	Olacaceae	Moretologa	Fruit	Tselaesele et al. (2023)
Ziziphus mucronata Willd.	Rhamnaceae	Mokgalo	Fruit, leaves, roots, bark, seeds	Motlhanka et al. (2008); Tselaesele et al. (2023); Bultosa et al. (2020); Badimo et al. (2015)

Note: The botanical names for the wild edible plants were verified using the World Flora Online (https://www.worldfloraonline.org) (WFO, 2024).

#Family: name provided in bracket were used in the original article which we updated using the World Flora Online.

cocculoides (3), *Grewia retinervis* (3), *Azanza garckeana* (3), and *Adansonia digitata* (3) were the most frequently mentioned plants. These 13 wild edible plants are considered of high potential for food and nutrition as well as good health and well-being among the Batswana of Botswana and South Africa. The generated 70 plant species were from 34 plant families (Table 15.2). The most dominant plant families were Malvaceae (8), Cucurbitaceae (6), Apocynaceae (5), Loganiaceae (4), and Anacardiaceae (4). Malvaceae is recognized as one of the largest genera of flowering plants. Many plants belonging to the family Malvaceae are known for their high economic and medicinal properties (Erarslan & Koçyiğit, 2019).

In terms of plant parts, wild edible plants with "fruit" as the edible part dominated (58.1%), followed by leaves (9.3%), tubers (9.3%), and roots (9.3%). Other plant parts consumed were stem (2.3%), seeds (1.1%), rhizomes (1.1%), flowers (1.1%), and buds (1.1%). For the wild edible plants, the fruits were popularly consumed as snacks (83%), while a few are prepared as a whole meal (e.g., *A. digitata* fruits mixed with fresh milk). Leafy vegetables are usually eaten with meals, while tubers are often consumed to quench thirst (Badimo et al., 2015).

15.3.3 Diverse Uses of Wild Edible Plants among the Batswana

Different ethnic groups use wild edible plants in many different ways (Kanter et al., 2023). Researchers have found that paying attention to these plants and the knowledge about them, especially from local communities, can contribute to food security, mitigate climate change, and enhance the public health systems. In this section, the overarching benefits related to the uses of wild edible plants among the Batswana are discussed.

15.3.3.1 Cultural Significance and Miscellaneous Uses of Wild Edible Plants

The relationship between wild edible plants and the culture of local community is well recognized (Liu et al., 2023). Indigenous knowledge about wild edible plants exhibits a strong pattern of linguistic uniqueness in all regions (Cámara-Leret & Bascompte, 2021). The availability of local names for the recorded wild edible plants is an indication of their importance among the Batswana. Some wild edible plants have different local names in Botswana and South Africa. For example, *Citrullus lanatus* is known as mokate and lekatane in Botswana and South Africa, respectively.

Wild edible plants are also used for spiritual and ritual purposes. For instance, the leaves of *Celtis africana* (modutu) were used to purify blood, particularly for women who had miscarriages. It is believed that consuming its roots or leaves can enhance sexual desire, and the plant was used to improve the fertility of livestock (Pitso & Lebese, 2014).

Another important edible plant is *Rhus pyroides* (mogodiri), which has many different uses. Traditionally, it is used to "cleanse" people who are sick from sleeping with a widow. In Botswana, *Zaluzianskya marlothii* (phekolola) is used to get rid of evil spirits and to "cleanse" widows and widowers ritually (Danley, 2006). In Botswana, there is a belief that cutting down *V. infausta* (mmilo) for firewood could bring bad luck to the community, cause thunderstorms, and make cows only give birth to male calves (Bultosa et al., 2020). *Phoenix reclinata* is used for magical purposes (Tselaesele et al., 2023).

15.3.3.2 Traditional Uses of Wild Edible Plants as a Source of Food and Nutrition

The leaves of *Physalis peruviana* are used to add flavor to starchy foods. According to Pitso and Lebese (2014), people use *Z. mucronata* (mokgalo) in different ways. The ripe fruit is consumed fresh or dried, while the ground form is included into a meal to make porridge. During the Anglo–Boer War, soldiers used the ground and roasted seeds as a coffee substitute. Young leaves and shoots were added to salads or cooked as a leafy vegetable (Pitso & Lebese, 2014). People also made strong alcohol from the fruits. The fruit of *V. infausta* (mmilo or mothwanyê) was eaten raw or fermented to make beer. The fruits of *G. flava* and *G. bicolor* were used to make jam, juice, wine, or dried for processed products (Welcome & Van Wyk, 2019). They can be eaten fresh, dried for later, used as a snack with raisins, or to make traditional alcoholic drinks called khadi (Tselaesele et al., 2023). *Adansonia digitata* (baobab) is a highly valued tree because its fruit, seeds, leaves, bark, flowers, and taproots are used for nutrition. In Botswana, particularly in Gumare Village, people eat the fruit and seeds of *A. digitata* (baobab) for nourishment (Tselaesele et al., 2023). The fruit of *B. discolor* (motsentsela, motsintsila) is eaten fresh, dried, or processed into a fermented beverage. Likewise, the ripe fruit of *F. sycomorus* is eaten fresh, dried, or fermented into beverages (Bultosa et al., 2020).

The fruit of *S. cocculoides* (monkey orange) is widely eaten in southern Africa. People consume the fruit fresh after picking, or they dry it to make leather, jams, juices, or ferment it into wine and beer. They also use it to make fritters, muffins, or cook it with maize meal porridge (Ngadze et al., 2017). The fruit of *Garcinia livingstonei* (motsaodi) is eaten as a local food and can be turned into a tasty alcoholic drink by fermenting it. Batswana women use the fruit of *P. capensis* (mothata) to make jam, jelly, alcoholic drinks, and vinegar (Tselaesele et al., 2023). *Momordica balsamina* (African cucumber, balsam apple) is used for food, whereby the leaves, fruit pulp, and roots are commonly consumed among households (Danley, 2006). The young leaves are cooked and eaten with porridge. The fruit of *B. mollis* (mokamanawa) is eaten fresh or dried or used to make jam or juice (Tselaesele et al., 2023). Additionally, the bark can be eaten as a vegetable.

15.3.3.3 Medicinal Uses of Wild Edible Plants for Health and Well-Being among the Batswana

Existing records indicate that the Batswana people use wild plants for medicine. Particularly, two plants, *M. balsamina* and *P. peruviana*, are well-known for their medicinal properties (Van der Hoeven et al., 2013). Even though the Batswana mainly eat the fruit of *M. balsamina* as a snack, the leaves are used to treat eye problems in children and in burns, itching, and dysentery. In addition, the roots are used to treat coughs, diarrhea, fever, malaria, worms, and stomach pains (Van der Hoeven et al., 2013).

Traditional remedies made from the leaves, bark, and roots of *Z. mucronata* (mokgalo) are used to treat dysentery, respiratory issues, and lumbago (a type of back pain). The roots and leaves of *V. infausta* (mothwanyê) can be boiled into a tea and used to treat ringworm and pneumonia and to enhance bowel movement (Mongalo et al., 2020). The Batswana women use oil from *P. capensis* seeds to treat baldness and ringworms and to lubricate gums (Tselaesele et al., 2023). Tea made from the leaves of *A. digitata* (baobab) is used to treat diarrhea, fever, inflammation, kidney and bladder problems; to clean the blood; and in asthma and insect bites in Botswana (Gray, 2004). The fruits are used to treat infections and high levels of lipids in the blood (Ismail et al., 2021). Furthermore, the seeds are used to treat fever, diarrhea, coughs, and wounds (Rahul et al., 2015). *Phoenix reclinata* is used to treat diverse health conditions, including infections, injuries, inflammation, and digestive and central nervous system issues in humans (Tselaesele et al., 2023). Furthermore, the plant is popular in ethnoveterinary medicine for managing livestock.

In traditional medicine, the Batswana use the fruit of *B. discolor* to treat bleeding gums, while the bark is used for infertility, malaria, nosebleeds, itchy skin, and toothache (Cheikhyoussef & Maroyi, 2017). For *Ficus* species, different parts, such as the fruit, leaves, bark, and roots, are used to fight against bacterial infections, protect against damage from oxidation, protect the brain and liver, lower blood pressure, and help control diabetes (Earnest et al., 2018). *Stachys aethiopica* leaves are used as tea to treat internal hemorrhoids, flu, women's health problems, and liver issues.

15.3.4 Nutritional Attributes of Wild Edible Plants

Ensuring access to the nutritional constituents of wild edible plants is crucial to enhance their significance in addressing food and nutritional security (Oduor et al., 2023). Wild edible plants are a rich source of energy, protein, beta-carotene, vitamin C, fiber, calcium, iron, and zinc. Information about the nutritional value of about 57% (40 out of 70) of the wild edible plants used by Batswana has been reported (Table 15.3).

TABLE 15.3
Overview of the Nutritional Constituents of Some Wild Edible Plants Consumed by the Batswana of southern Africa

Scientific Name	Energy (kcal/ 100 g)	Protein (g/100 g)	B-carotene (mg/100 g)	Vitamin C (mg/100 g)	Fiber (g/100 g)	Calcium (mg/100 g)	Iron (mg/100 g)	Zinc (mg/100 g)	References
Adansonia digitata	21	0.04	NF	38.4	14.94	862	11.13	8.41	Monteiro et al. (2022)
Agaricus spp.	NF	NF	NF	NF	NF	NF	NF	NF	NF
Amaranthus hybridus	268.92	17.92	3.29	25.40	8.61	44.15	13.58	3.80	(Akubugwo et al., 2007)
Amaranthus thunbergii	NF	NF	7.26	69.11	NF	NF	NF	NF	Mokganya et al. (2021)
Azanza garckeana	NF	NF	NF	NF	NF	NF	NF	NF	NF
Berchemia discolor	NF	NF	NF	NF	NF	NF	NF	NF	NF
Berchemia zeyheri	370	1.1	0.1	6.5	0.7	75.8	1.0	0.2	Wehmeyer (1986)
Boscia albitrunca	449.0	6.5	NF	6.6	3.6	63.6	4.2	0.8	Maroyi (2019)
Bridelia mollis	41	21.7	NF	2.4	55.1	0.31	81.0	0.17	Maroyi (2019)
Ceropegia rendallii	NF	NF	NF	NF	NF	NF	NF	NF	NF
Citrullus lanatus	2,340	28.3	NF	NF	4.68	100	7.3	5.2	Kaul (2011)
Citrullus vulgaris	NF	NF	NF	NF	NF	NF	NF	NF	NF
Cleome gynandra	220	5.68	4,117	37.0	4.48	393	14.3	1.0	Mashamaite et al. (2022)
Coccinia sessilifolia	NF	NF	NF	NF	NF	NF	NF	NF	NF
Corchorus olitorius	1,197	17.5	NF	6.6	12.9	46.5	5.01	1.24	(Adesina et al., 2022)
Cucurbita spp.	250	5.7	NF	NF	18	27.9	28.2	0.85	Karanja et al. (2013)
Cyperus fulgens	NF	NF	NF	NF	NF	NF	NF	NF	NF
Cyperus papyrus	NF	NF	NF	NF	NF	NF	NF	NF	NF
Diospyros lycioides	70.70	9.10	0.10	45.20	3.50	1.10	1.00	0.30	Maroyi (2018a)
Diospyros mespiliformis	440	2.6	NF	NF	NF	6.0	0.14	0.17	Nyambe et al. (2019)
Echinocystis lobata	NF	NF	NF	NF	NF	NF	NF	NF	NF
Ehretia rigida	215	1.7	NF	6.9	0.7	30.5	0.9	0.2	Maroyi (2023)
Englerophytum magalismontanum	NF	NF	NF	NF	NF	NF	NF	NF	NF
Ficus spp.	92	1.03	NF	5.17	1.37	316.5	25.9	12.6	Chiteva and Wairagu (2013)
Ficus sycomorus	114	17.24	NF	NF	31.54	90.78	11.65	9.58	Nkafamiya et al. (2010)
Fockea angustifolia	NF	NF	NF	NF	NF	NF	NF	NF	NF
Garcinia livingstonei	NF	NF	NF	NF	NF	NF	NF	NF	NF

(Continued)

TABLE 15.3 (Continued)

Scientific Name	Energy (kcal/ 100 g)	Protein (g/100 g)	B-carotene (mg/100 g)	Vitamin C (mg/100 g)	Fiber (g/100 g)	Calcium (mg/100 g)	Iron (mg/100 g)	Zinc (mg/100 g)	References
Grewia avellana	790	7.7	NF	NF	20.5	790	20.8	1.9	Elhassan and Yagi (2010)
Grewia bicolor	20.73	7.00	NF	NF	33.15	582.0	20.73	1.59	Tairo (2021)
Grewia flava	91.30	1.57	1.48	4.385	5.53	136	140.8	144	Sinha et al. (2015)
Grewia flavescens	269	8.7	NF	NF	42.8	269	26.9	1.1	Elhassan and Yagi (2010)
Grewia retinervis	817	6.7	NF	NF	25.5	536	29.6	2.5	Elhassan and Yagi (2010)
Grewia schinzii	NF	NF	NF	NF	NF	NF	NF	NF	NF
Hyphaene petersiana	114	1.6	NF	NF	NF	5.01	0.14	0.18	Nyambe et al. (2019)
Ipomea batatas	86	1.57	8,509	2.4	3.0	30	0.61	0.3	Cartabiano-Leite et al. (2020)
Kalanchoe rotundifolia	NF	NF	NF	NF	NF	NF	NF	NF	NF
Kigelia africana	62	16.31	NF	NF	21.09	50.2	3.57	2.09	Oseni and Williams (2018)
Mentha aquatica	5	0.38	NF	1.5	0.8	23	1.35	0.12	Selmi et al. (2022)
Mimusops zeyheri	510	NF	NF	NF	NF	464	15.3	2.87	Mngadi et al. (2019)
Momordica balsamina	189.22	0.07	NF	NF	1.23	7.21	60.3	3.18	Hassan and Umar (2006)
Nymphaea nouchali	76.5	10.76	115.22	3.12	0.64	148.55	1.98	1.33	Anand et al. (2019)
Nyphaeae Burm.f. var. *caerulea*	NF	NF	NF	NF	NF	NF	NF	NF	NF
Opuntia spp.	94	4.35	12.6	5.17	20	0.69	0.20	0.07	Alshaikhi et al. (2024)
Pappea capensis	42	NF	650	19,310	NF	NF	2.6	27.8	Karau et al. (2012)
Parinari curatellifolia	NF	NF	NF	NF	NF	NF	NF	NF	NF
Pentarrhinum insipidum	NF	NF	NF	NF	NF	NF	NF	NF	NF
Phoenix reclinata	60.44	10.39	NF	NF	16.52	1,547.38	1,119.98	491.28	Adenekan et al. (2018)
Rhus pyroides	NF	NF	NF	NF	NF	NF	NF	NF	NF
Rhus quartiniana	NF	NF	NF	NF	NF	NF	NF	NF	NF
Rhus tenuinervis synonym *Searsia tenuinervis*	NF	NF	NF	NF	NF	NF	NF	NF	NF
Schinziophyton rautanenii	NF	NF	NF	NF	NF	NF	NF	NF	NF
Sclerocarya birrea	700	20	NF	62	NF	480	2.5	NF	Hiwilepo-van Hal (2013)

(Continued)

TABLE 15.3 (Continued)

Scientific Name	Energy (kcal/100 g)	Protein (g/100 g)	B-carotene (mg/100 g)	Vitamin C (mg/100 g)	Fiber (g/100 g)	Calcium (mg/100 g)	Iron (mg/100 g)	Zinc (mg/100 g)	References
Stachys aethiopica	NF	NF	NF	NF	NF	NF	NF	NF	NF
Stomatostemma monteiroae	NF	NF	NF	NF	NF	NF	NF	NF	NF
Strychnos cocculoides	NF	NF	NF	NF	NF	NF	NF	NF	NF
Strychnos madagascariensis	791	8.3	NF	NF	53.0	148.03	15.78	1.36	Van Rayne et al. (2023)
Strychnos spinosa	1,923	3.3–11.7	NF	880	16.7	0.15	0.14	0.01	Van Rayne et al. (2023)
Strychnos spp.	791	8.3	NF	NF	53.0	148.03	15.78	1.36	Van Rayne et al. (2023)
Trochomeria debilis	NF	NF	NF	NF	NF	NF	NF	NF	NF
Tylosema esculentum	2.66	31.8	NF	10.31	16	4.2	40	100	Bower et al. (1988)
Typha capensis	NF	NF	NF	NF	NF	NF	NF	NF	NF
Urochloa mosambicensis	NF	NF	NF	NF	NF	NF	NF	NF	NF
Vangueria infausta	1,445	3.01	NF	11.50	10.20	0.20	0.09	0.02	Maroyi (2018b)
Vangueriopsis lanciflora	NF	NF	NF	NF	NF	NF	NF	NF	NF
Vigna unguiculata	55.98	57.84	NF	9	3.40	651	54	35	Alfa et al. (2020)
Withania somnifera	NF	5.312	NF	NF	11.38	NF	740.0	52.9	Krishnamurthy and Sarala (2010)
Ximenia americana	74.8	7.26	NF	21.12	3.0	0.65	NF	NF	Muhammad et al. (2019)
Ximenia caffra	NF	NF	NF	49.2	1	23.2	20.4	1	Maluleke et al. (2024)
Ximenia spp.	NF	NF	NF	NF	NF	NF	NF	NF	NF
Ziziphus mucronata	2,008.58	18.9	NF	NF	6.1	38.056	5.019	NF	Yerima and Markus (2020)

Note: NF = not found.

Wild edible plants such as *C. gynandra*, *Diospyros lycioides*, and *P. capensis* have a high quantity of vitamin C (ascorbic acid), ranging from 46 to 80 mg/100 g. The consumption of these plants can help meet the daily recommended intake of 45 mg/day (WHO, 2004). However, some species, such as *Bridelia mollis* and *Corchorus olitorius*, have less amount of vitamin C. Wild edible plants are also rich in other vitamins, such as A, D, E, and K (Duguma, 2020). They contain important mineral elements (iron [Fe], calcium [Ca], and zinc [Zn]) and proteins, which play a vital role in growth and development. For example, *V. unguiculata* is a valuable high-protein wild edible plant containing about 57.84 g/100 g. Proteins are crucial for many functions in the body, and their quality depends on having the right balance of essential amino acids, which our bodies cannot make on their own and need to get from food (Khan et al., 2016).

15.3.5 Biological Effects and Phytochemicals of Wild Edible Plants

Diverse biological activities for wild edible plants and associated phytochemicals have been reported (Table 15.4). Examples of the phytochemicals are polyphenols, flavonoids, and carotenoids. Flavonoids, for instance, can act as antioxidants by scavenging free radicals and chelating metal ions, thus preventing oxidative stress and cellular damage (Kumar & Pandey, 2013). The fact that these wild edible plants possess flavonoids suggests that they could be natural sources of substances that fight inflammation and act as antioxidants. Wild edible plants such as *B. discolor*, *Ceropegia* spp., *F. sycomorus*, and *Grewia* species are known to exert antifungal, antimicrobial, and antiulcer effects. Antimicrobial effects were associated with wild edible fruits, such as those of *C. lanatus*, *B. mollis*, and *D. lycioides* plants. Antidiabetic, anti-inflammatory, and anticancer effects have been linked with wild edible fruits from *S. madagascariensis*, *R. pyroides*, and *R. quartiniana*. Some of the bioactive compounds in wild edible plants confer diverse biological activities, such as anticancer (Potočnjak et al., 2018), antiallergic (Kobayashi et al., 2016), antibacterial (Guimarães et al., 2019), and antioxidant (Wang et al., 2019). Furthermore, some of these wild edible plants are high in carotenoids, which are strong antioxidants and are important for maintaining good eyesight and protecting the eyes from damage caused by harmful molecules (Fiedor & Burda, 2014). The consumption of these wild edible plants provides a rich natural source of carotenoids, which can lower the chances of developing age-related eye problems, especially age-related macular degeneration, and other eye issues (Fiedor & Burda, 2014).

TABLE 15.4
Phytochemical and Biological Activities in Wild Edible Plants Consumed by the Batswana of southern Africa

Scientific Name	Phytochemicals	Biological Activities
Adansonia digitata	Phenolics acids, flavonoids, and lipids (Ibrahim et al., 2021)	Antioxidant, anti-inflammatory, analgesic, and antimicrobial activities (Braca et al., 2018)
Agaricus spp.	Alkaloid, carotenoids, steroid, triterpenoids, fatty acid, emodins, flavonoid, coumarin, anthracene glycoside, tannins, saponins (Pavithra et al., 2016)	Antioxidant and antimicrobial activities (Pavithra et al., 2016)
Amaranthus hybridus	Flavonoids, steroids, terpenoids, and cardiac glycoside (Maiyo, 2009)	Antimicrobial activity (Maiyo, 2009)
Amaranthus thunbergii	Lysine-rich protein, β-carotene (Chen et al., 2022)	Antioxidant activity (Chen et al., 2022)
Azanza garckeana	Phenol, flavonoids, and alkaloids (Yusuf et al., 2020)	Anti-inflammatory, analgesic, antiarthritic, and antioxidant activities (Yusuf et al., 2020)
Berchemia discolor	Tannins, alkaloids, flavonoids, saponins, and phenols (Hettiarachchi et al., 2006)	Antimicrobial activities (Gundidza & Sibanda, 1991)
Berchemia zeyheri	Alkaloids, flavonoids, glycosides, polyphenols, and steroids (Maroyi, 2019)	Anthelmintic, antibacterial, antioxidant, and toxicity activities (Lwalewa, 2007)
Boscia albitrunca	Tannin and phenol (Madibela et al., 2004)	Antifungal activities (Pendota et al., 2015)
Bridelia mollis	Lochidone (Ngueyem et al., 2009)	Antibacterial, antifungal antileishmanial, antioxidative, and antiplasmodial activities (Maroyi, 2019)
Ceropegia rendallii	Phenolics, proanthocyanidins, and flavonoids (Tselaesele et al., 2023)	Antioxidant activity (Tselaesele et al., 2023)

(*Continued*)

TABLE 15.4 (Continued)

Scientific Name	Phytochemicals	Biological Activities
Ceropegia spp.	Phenolics, flavonoids, alkaloids, triterpanes, ketos, tannins, saponins (Chavan et al., 2013)	Hepatoprotective, antipyretic, antiulcer, analgesic, mast cell stabilizing, tranquilizing, and hypotensive activities (Chavan et al., 2013)
Citrullus lanatus	Phenolics and flavonoids (Mogotlane, 2015)	Antimicrobial activity (Zamuz et al., 2021)
Cleome gynandra	Methyl-isothiocyanate, propyl-isothiocyanate, butyl-isothiocyanate, (E)-2-hexenal, hexanal, (Z)-3-hexenal, (E)-2-pentenal (Moyo & Aremu, 2022)	Anticancer and antioxidant (Bala et al., 2010)
Coccinia sessilifolia	Flavonoids, phenolic acids, terpenoids, alkaloids, sterols, and glycosides (Cai et al., 2023)	Antioxidant, antidiabetic, anti-inflammatory, and antidyslipidemic activities (Cai et al., 2023)
Corchorus olitorius	Phenols, flavonoids, sterols, lipids, and fatty acids (Ali et al., 2024)	Anticancer, antioxidant, analgesic, antiviral, and cardiovascular (Ali et al., 2024)
Cucurbita spp.	Carotenoids, tocopherols, phenols, terpenoids, saponins, sterols (Salehi et al., 2019)	Antidiabetic, antiulcer, analgesic, nephroprotective, and anticancer activities (Salehi et al., 2019)
Cyperus fulgens	Saponins, steroids, tannins, triterpenes (Urnukhsaikhan et al., 2021)	Antioxidant, anti-inflammatory, antimicrobial, and anticancer activities (Lipińska et al., 2023)
Cyperus papyrus	NF	Antifungal activity (Taha et al., 2019)
Diospyros lycioides	Alkaloids, anthraquinones, cardenolides, carotenoids, flavonoids, naphthoquinone, phenolics, polyphenols, saponins, steroids, tannins, and terpenoids (Maroyi, 2018a)	Antiadhesive, antibacterial, antifungal, and anti-inflammatory (Maroyi, 2018a)
Diospyros mespiliformis	Alkaloids, tannins, saponins, glycosides, and anthraquinones (Ebbo et al., 2020)	Antiparasitic and antiproliferative activities (Ramadwa & Meddows-Taylor, 2023b)
Echinocystis lobata	Phenolic and flavonoid (Vinogradova et al., 2021)	Antioxidant activity (Vinogradova et al., 2021)
Ehretia rigida	Flavonoids, phenolics, saponins, and tannins (Mnikathi, 2021)	Antimicrobial, antioxidant, anti-inflammatory, and antidiabetic activities (Mnikathi, 2021)
Englerophytum magalismontanum	Phenolics, flavonoids, triterpenes, alkaloids, tannins, glycosides, and coumarins (Papo, 2017)	Antidiabetic activity (Olaokun et al., 2022)
Ficus spp.	Terpenoids, polyphenols, alkaloids, organic acids, and saponins (Ashraf et al., 2021)	Antidiabetic activity (Ashraf et al., 2021)
Ficus sycomorus	Alkaloids, flavonoids, glycosides, tannins, and steroids (Ayuk et al., 2019)	Antifungal activity (Shinkafi & Abdullahi, 2018)
Fockea angustifolia	NF	NF
Garcinia livingstonei	Phenolics, steroids, alkaloids, terpenoids, saponins, flavonoids, anthraquinones, and tannins (Ramadwa & Meddows-Taylor, 2023a; Tabit et al., 2016)	Antifungal, cytotoxic, and antioxidant activities (Ramadwa & Meddows-Taylor, 2023a)
Grewia avellana	Triterpenoids, alkaloids, and sterols (Ullah et al., 2012)	Antimalarial, antimicrobial, and antiemetic (Ullah et al., 2012)
Grewia bicolor	Alkaloids, flavonoids, and phenols (Ullah et al., 2012)	Antioxidant, antibacterial, hepatoprotective, anti-inflammatory, antiemetic, antimalarial, analgesic, and antipyretic activities (Ullah et al., 2012)
Grewia flava	Glycoside, steroids, and flavonoids (Kumar et al., 2022)	Antimicrobial activity (Kumar et al., 2022)
Grewia flavescens	N-dimethylethanamine, acetic acid, 1,1-diethoxy ethane, pyridine, diethyl phthalate (Patel et al., 2023)	Antidiabetic, antiurolithiatic, anthelmintic, acute toxicity, anti-inflammatory, and antimicrobial activities (Patel et al., 2023)
Grewia retinervis	NF	NF

(Continued)

TABLE 15.4 (Continued)

Scientific Name	Phytochemicals	Biological Activities
Grewia schinzii	NF	NF
Hyphaene petersiana	NF	NF
Ipomea batatas	Phenolic, flavonoids, anthocyanins, carotenoids, and tannins (Amagloh et al., 2022)	Aphrodisiac and gonadoprotective activities (Suhendy et al., 2023)
Kalanchoe rotundifolia	NF	NF
Kigelia africana	Phenolic, coumarins, and sterols (Grace et al., 2002)	Antibacterial and antifungal activities (Hussain et al., 2016b)
Mentha aquatica	Flavonoid, terpenoids, and saponins (Thi et al., 2020)	Antioxidant activity (Thi et al., 2020)
Mimusops zeyheri	NF	NF
Momordica balsamina	Polyphenols, flavonoids, terpenes, and carotenoids (Thiaw et al., 2023)	Antioxidant, anti-inflammatory, antidiabetic, and anticarcinogenic (Thiaw et al., 2023)
Nymphaea nouchali	Phenols, flavonoids, tannins, glycosides, saponins, alkaloids, and steroids (Parimala & Shoba, 2013)	Antioxidant activity (Parimala & Shoba, 2013)
Nyphaeae Burm.f. var. *caerulea*	Alkaloids, monoterpenes, sesquiterpene lactones, diterpenes, triterpenes, and steroids (Selvakumari et al., 2016)	Antidiabetic and antimicrobial activities (Selvakumari et al., 2016)
Opuntia spp.	Phenolic, flavonoids, and betalains (Daniloski et al., 2022)	Antioxidant and anticancer activities (Daniloski et al., 2022)
Pappea capensis	Flavonoids (Tajuddeen et al., 2021)	Antiplasmodial activity (Tajuddeen et al., 2021)
Parinari curatellifolia	Anthraquinones, saponins, flavonoids, cardiac glycosides, and tannins (Halilu et al., 2008)	Antibacterial activities (Halilu et al., 2008)
Pentarrhinum insipidum	NF	NF
Phoenix reclinata	NF	NF
Rhus pyroides	Flavonoids, urushiols, and terpenoids (Opiyo et al., 2021)	Antimicrobial, cytotoxicity, anticancer, antioxidant, antiviral, anti-inflammatory, and antimalarial activities (Opiyo et al., 2021)
Rhus quartiniana	Flavonoids, urushiols, and terpenoids (Qiu et al., 2016)	Antimicrobial, cytotoxicity, anticancer, antioxidant, antiviral, anti-inflammatory, and antimalarial activities (Rayne & Mazza, 2007)
Rhus tenuinervis (synonym *Searsia tenuinervis*)	NF	NF
Sclerocarya birrea	Polyphenol, flavonoid, and tannin (Russo et al., 2018)	Anti-inflammatory, analgesic, and antimicrobial (Russo et al., 2018)
Stachys aethiopica	NF	NF
Stomatostemma monteiroae	Phenolic, tannin, and flavonoid (Masoko et al., 2019)	Antimycobacterial activity (Ramese, 2019)
Strychnos cocculoides Baker	NF	NF
Strychnos madagascariensis	Alkaloids, fatty acids, cardiac glycosides, flavonoids, saponins, steroids, tannins, terpenoids, triterpenoids (Maroyi, 2021)	Antibacterial, antifungal, antidiabetic, antioxidant, antiplasmodial, larvicidal, and toxicity activities (Maroyi, 2021)
Strychnos spinosa	Alkaloids, glycosides, tannins, saponins, anthraquinones, reducing sugar, steroids, and terpenoids (Ugoh & Bejide, 2013)	Antidiabetic, antifungal, anti-inflammatory, and antioxidant activities (Moyo & Aremu, 2022)
Strychnos spp.	Alkaloids, flavonoids, glycosides, phenols, saponins, and sterols (Thongphasuk et al., 2014)	Antidiabetic, anti-inflammatory, and antiulcerogenic activities (Yadav et al., 2014)
Trochomeria debilis	Flavonoids, saponins, tannins, and glycosides (Miro, 1995)	Antioxidant activity (Miro, 1995)

(Continued)

TABLE 15.4 (Continued)

Scientific Name	Phytochemicals	Biological Activities
Tylosema esculentum	Phenolics, flavonoids, saponins, and phytosterols (Omotayo & Aremu, 2021)	Antimicrobial (antibacterial) activities (Chingwaru et al., 2011)
Typha capensis	Alkaloids, sterols, and flavonoids (Musara & Aladejana, 2020)	Anti-inflammatory and antioxidant activities (Ondua et al., 2019)
Urochloa mosambicensis	NF	NF
Vangueria infausta	Alkaloids, anthraquinones, coumarins, glycosides, polyphenols, saponins, secoiridoids, steroids, tannins, and terpenoids (Nafuka, 2014)	Antibacterial and antifungal activities (de Boer et al., 2005)
Vangueriopsis lanciflora	NF	NF
Vigna unguiculata	Flavonoids, alkaloids, tannins, sterols, reducing sugar, terpenoids, and phenolic acids (Allah et al., 2017)	Antioxidant and antimicrobial activities (Hussain et al., 2016a)
Withania somnifera	Alkaloids, steroids, saponins, phenolics, flavonoids, phytophenols, and glycosides (Das & Afrin, 2019)	Antipyretic, analgesic, adaptogenic, and anti-inflammatory (Kalra & Kaushik, 2017)
Ximenia americana	Saponins, glycosides, flavonoids, tannins, phenolics, alkaloids, quinones, and terpenoids (Geyid et al., 2005; James et al., 2007)	Antimicrobial activities (Geyid et al., 2005; James et al., 2007)
Ximenia caffra	Flavonoids, phenolics, and tannins (Zhen et al., 2015)	Antimicrobial activity (Zhen et al., 2015)
Ziziphus mucronata	Phenolic, flavonoids, quercetin, terpenoids, quinones, and alkaloids (Nafuka & Mumbengegwi, 2013)	Antifungal, antibacterial, and antimycobacterial activities (Mongalo et al., 2020)

Note: *NF* = not found.

Additionally, these wild edible plants might possess other natural compounds that have not been studied yet, with properties such as fighting diverse microbes, lowering blood sugar, reducing inflammation, and promoting good health (Nishikito et al., 2023). More research is needed to fully understand how these compounds can benefit health. Generally, wild edible plants contain many natural compounds with various effects on the body. The inclusion of these plants in the diet can be a source of a wide range of nutrients that promote good health. Tannins, for example, possess properties that fight microbes by binding to and deactivating microbial enzymes or by disrupting their membranes. They can be harmful to certain fungi, yeast, and bacteria (Ejoh et al., 2007). Flavonoids can help protect against cancer, inflammation, and neurodegenerative conditions (Onyeka & Nwambekwe, 2007). Terpenoids have been proven to be effective in shielding the brain from health conditions, such as swelling in the nerves and damage caused by insufficient oxygen (Ruszkowski & Bobkiewicz-Kozlowska, 2014). The presence of saponins was indicated in some of the wild edible plants, such *as C. fulgens*, *B. discolor*, and *M. aquatica*. Saponins are known to help lower cholesterol levels by attaching to bile acids and cholesterol in the body, which stops them from being absorbed (James & Matemu, 2016).

15.3.6 Sustainable Harvesting Practices of Wild Edible Plants

The Batswana people have strategic ways for collecting and gathering parts from wild plants without harming them. For instance, when they collect leaves from plants such as *V. unguiculata*, *Z. mucronata*, and *Cucurbita* spp., they cut the leaves at the base with a special knife, leaving the roots unharmed. This helps the plants grow back, thereby ensuring their conservation and sustainability. Likewise, when they gather tubers from *P. reclinata*, they only take the older ones to ensure the survival of the mother plant (Tselaesele et al., 2023).

The traditional ways of gathering leaves are usually seen as sustainable. This method helps the leaves grow back after they are picked (Semenya & Mokgoebo, 2020). Similarly, when collecting twigs from *K. longiflora*, people cut them by hand in the wild. When it comes to getting tender leaves from plants such as *M. balsamina*, *M. boivinii*, and *V. unguiculata* to consume as vegetables, only a few leaves are taken so that the plants can maintain their vigor (Constant & Tshisikhawe, 2018). For fruits from *C. lanatus*, *C. zeyheri*, and *P. capensis*, people pick them by hand without breaking the branches. This method is also used by other native African groups (Legwaila et al., 2011). Additionally, *F. burkei* fruits are collected by throwing stones at the trees to knock the fruit down (Semenya & Maroyi, 2019). Fruits from *V. infausta* and *X. caffra* are only taken from the ground after they have naturally fallen off the tree (Rankoana, 2016).

15.3.7 Food and Nutrition Security and Economic Opportunities

Many people consume wild edible plants because most of them are readily available when there is no enough food or during extreme weather (Kanter et al., 2023). For example, during a big famine in Ethiopia, most people survived by eating wild plants, even though these plants are not always seen as very valuable (Guinand & Lemessa, 2000). Wild edible plants are important for ensuring people in rural areas have enough food and nutrients (Ju et al., 2013). The inclusion of these plants in diets can help achieve the goal of the United Nations to end hunger, UN SDG 2 (Singh et al., 2020). Some wild edible plants have higher nutritional value than the conventional foods, and more people are becoming more aware of the health benefits of these plants (Grace & Kayode, 2018; Mokganya & Tshisikhawe, 2019). Active utilization of wild plants ensures people in rural areas have a variety of foods to consume (Dejene et al., 2020; Mengistu & Hager, 2008), broadening the food base and serving as a strategy to enhance the nutritional value of diets (Ickowitz et al., 2016).

Additionally, these plants can generate income (Duguma, 2020). Fruits such as *G. bicolor*, *D. mespiliformis*, *G. flava*, and *B. discolor* are sold in Shorobe, Botswana. Similarly, households make money by selling *N. nouchali* (Neudeck et al., 2012). Evidence from Botswana suggests that wild edible plants can play a crucial role in reducing financial risk for farmers by providing alternative sources of income. Botswana faces recurrent droughts, which can devastate conventional crops. Wild edible plants which are adapted to local conditions can withstand drought better than many cultivated crops, providing a more reliable income source during periods of water scarcity and drought. By cultivating and selling wild edible plants, farmers can diversify their income sources. This reduces dependence on a single crop, mitigating risks associated with market fluctuations and price volatility. Wild edible plants often require minimal inputs, such as fertilizers and pesticides, making them more cost-effective to cultivate compared to conventional crops. This lowers production costs and improves profitability for farmers (Legwaila et al., 2011).

Farmers can add value to wild edible plants by processing them into products such as jams, teas, or herbal medicines. These value-added products attract higher prices in the market, increasing income and profitability for the farmers. Many wild edible plants have cultural significance in Botswana and are highly valued for their traditional uses. By cultivating and marketing these wild edible plants, farmers can tap into niche markets interested in preserving cultural heritage, thereby enhancing their income opportunities (Rashwan et al., 2023).

15.3.8 Challenges Associated with the Cultivation of Wild Edible Plants

A major problem associated with the cultivation of wild edible plants is inadequate knowledge on processing and value addition. Relative to conventional crops, there is limited knowledge on how to add value to wild edible plants. Given the limited valuable addition associated with these plants, the financial returns are low and attract less interest from a commercial perspective (Engels et al., 2002). Therefore, it is crucial to develop affordable crop processing methods tailored to the needs of rural farmers. This may entail providing training, information, and technology needed for food processing. Currently, many underutilized plants are processed by hand on farms, which can be time-consuming and can require intense efforts (Malkanthi et al., 2014). Developing quicker methods for processing these crops on a larger scale is important.

Also, limited access to affordable credit has been identified as a factor that is currently affecting the cultivation of wild edible plants. Even though there are loan programs for main crops such as rice and corn, it is difficult to get loans for wild edible plants (Williams & Haq, 2010). Many small-scale farmers and processors experience difficulties accessing loans, which inevitably affects the cultivation of wild edible plants.

15.4 CONCLUSION AND PROSPECTS

In conclusion, wild edible plants play a crucial role in ensuring food and nutrition security, as well as providing economic opportunities, particularly among the Batswana people in southern Africa. Despite being often overlooked, these plants are rich in nutrients, phytochemicals, and bioactive compounds, offering numerous health benefits. Moreover, they hold significant cultural and traditional values, serving various purposes beyond nutrition, including medicinal, ritual, and spiritual uses. Efforts to document and understand the diverse uses of wild edible plants among the Batswana have highlighted their importance in local diets and livelihoods. Sustainable harvesting practices, coupled with policies promoting crop diversification and the inclusion of underutilized plants, offer avenues for preserving traditional knowledge, enhancing food and nutrition security, and supporting rural economies. However, challenges such as limited knowledge on processing techniques and limited access to affordable credit hinder the full utilization of these valuable resources. Moving forward, concerted efforts are needed to address these challenges and promote the sustainable cultivation, utilization, and commercialization of wild edible plants. This requires collaboration among researchers, policymakers, and local communities to develop innovative strategies for value addition, market access, and capacity building. By harnessing the potential of wild edible plants, resilience, biodiversity conservation, and inclusive development in the face of environmental and socio-economic challenges can be fostered.

15.5 ACKNOWLEDGMENTS

This work is based on the research supported by the South Africa National Department of Agriculture, Land Reform, and Rural Development (DALRRD) and additional support from the National Research Foundation (UID: 150734), Pretoria, South Africa. We thank the North-West University, University of KwaZulu-Natal, and the Agricultural Research Council, South Africa, for institutional support. The opinions, findings, and conclusions or recommendations expressed are those of the authors alone, and the funders accept no liability whatsoever in this regard.

REFERENCES

Adenekan, O., Omoyajowo, K., Babalola, O., Amiolemen, J. O. S., Olaniyan, K., Akande, J., & Idowu, I. (2018). Comparative nutritional analysis of *Phoenix dactylifera* and *Phoenix reclinata* seeds. *Journal of Research and Review in Science*, *4*, 29–35.

Adesina, A., OlAleye, A., Popoola, O., Olatunya, A., Gbolagade, A., Idowu, K., & Ajakaye, A. (2022). Nutritional evaluation of leafy vegetables of *Corchorus olitorius* family from Ekiti State, Nigeria. *ChemSearch Journal*, *13*, 147–156.

Akubugwo, I., Obasi, N., Chinyere, G., & Ugbogu, A. (2007). Nutritional and chemical value of *Amaranthus hybridus* L. leaves from Afikpo, Nigeria. *African Journal of Biotechnology*, *6*, 2833–2839.

Alfa, A. A., Tijani, K. B., Omotoso, O., Junaidu, Y., & Sezor, A. A. (2020). Nutritional values and medicinal health aspects of brown, brown-black and white cowpea (*Vigna unguiculata* L. Walp.) grown in Okene, Kogi State, Nigeria. *Asian Journal of Advanced Research and Reports*, *14*, 114–124.

Ali, M. R., Ibrahim, H. H., & Salah-Eldin, A. A. (2024). Unveiling the chemical composition, bioactive profile and antioxidant capacity of dried Egyptian jew's mallow stems as a promising anticancer agent. *Molecules*, *29*, 1377.

Allah, N. S. K., Eltayeb, I. M., & Hamad, A. E. H. (2017). Phytochemical screening and hypolipidemic activity of extracts from seeds and leaves of *Vigna unguiculata* growing in Sudan. *Journal of Pharmacognosy and Phytochemistry*, *6*, 488–491.

Alshaikhi, A. I., Alzahrani, M. Y., Hazzazi, J. S., Kurdi, J. R., & Ramadan, M. F. (2024). Nutritional aspects, bioactive phytochemicals and biomedical traits of *Opuntia* spp.: Current trends and applications. *Journal of Umm Al-Qura University for Applied Sciences*, *10*, 367–378.

Amagloh, F. C., Kaaya, A. N., Yada, B., Chelangat, D. M., Katungisa, A., Amagloh, F. K., & Tumuhimbise, G. A. (2022). Bioactive compounds and antioxidant activities in peeled and unpeeled sweetpotato roots of different varieties and clones in Uganda. *Future Foods*, *6*, 100183.

Anand, A., Priyanka, U., Nayak, V. L., Zehra, A., Babu, K. S., & Tiwari, A. K. (2019). Nutritional composition and antioxidative stress properties in boiled tuberous rhizome of Neel Kamal (*Nymphaea nouchali* Burm. f.). *Indian Journal of Natural Products and Resources (IJNPR)*, [Formerly Natural Product Radiance (NPR)] *10*, 59–67.

Ashraf, K., Haque, M. R., Amir, M., Ahmad, N., Ahmad, W., Sultan, S., Shah, S. A. A., Alafeefy, A. M., Mujeeb, M., & Shafie, M. F. B. (2021). An overview of phytochemical and biological activities: *Ficus deltoidea*: Jack and other: *Ficus* spp. *Journal of Pharmacy and Bioallied Sciences*, *13*, 11–25.

Ayuk, E., Nwochi, S. P., & Ogugor, J. C. (2019). Investigation of the phytochemicals and some biological parameters of *Ficus sycomorus* leaf extracts. *IOSR Journal of Pharmacy and Biological Sciences*, *14*, 71–78.

Badimo, D., Lepetu, J., & Teketay, D. (2015). Utilization of edible wild plants and their contribution to household income in Gweta Village, central Botswana. *African Journal of Food Science and Technology*, *6*, 220–228.

Bala, A., Kar, B., Haldar, P. K., Mazumder, U. K., & Bera, S. (2010). Evaluation of anticancer activity of *Cleome gynandra* on Ehrlich's Ascites Carcinoma treated mice. *Journal of Ethnopharmacology*, *129*, 131–134.

Bhat, R., Rubuluza, T., & Jäger, A. (2002). The bio-diversity of traditional vegetables of the Transkei region in the Eastern Cape of South Africa. *South African Journal of Botany*, *68*, 94–99.

Bower, N., Hertel, K., Oh, J., & Storey, R. (1988). Nutritional evaluation of marama bean (*Tylosema esculentum*, Fabaceae): Analysis of the seed. *Economic Botany*, *42*, 533–540.

Braca, A., Sinisgalli, C., De Leo, M., Muscatello, B., Cioni, P. L., MiLella, L., OStuni, A., Giani, S., & Sanogo, R. (2018). Phytochemical profile, antioxidant and antidiabetic activities of *Adansonia digitata* L. (Baobab) from Mali, as a source of health-promoting compounds. *Molecules*, *23*, 3104.

Bultosa, G., Molapisi, M., Tselaesele, N., Kobue-Lekalake, R., Haki, G. D., Makhabu, S., Sekwati-Monang, B., Seifu, E., & Nthoiwa, G. P. (2020). Plant-based traditional foods and beverages of Ramotswa Village, Botswana. *Journal of Ethnic Foods*, *7*, 1.

Bvenura, C., & Afolayan, A. J. (2014). Ethnobotanical survey of wild vegetables in Mbashe and Nkonkobe municipalities, Eastern Cape Province, South Africa. *Acta Botanica Gallica*, *161*, 189–199.

Bvenura, C., & Afolayan, A. J. (2015). The role of wild vegetables in household food security in South Africa: A review. *Food Research International*, *76*, 1001–1011.

Cai, L., Pan, R., Zeng, Q., Zhang, X., Zeng, R., & Zhu, Q. (2023). The complete plastome sequence of *Momordica cochinchinensis* (Cucurbitaceae). *Mitochondrial DNA Part B*, *8*, 329–332.

Cámara-Leret, R., & Bascompte, J. (2021). Language extinction triggers the loss of unique medicinal knowledge. *Procedings of the National Academy of Sciences of the United States of America*, *118*, e2103683118.

Cartabiano-Leite, C. E., Porcu, O. M., & de Casas, A. F. (2020). Sweet potato (*Ipomoea batatas* L. Lam) nutritional potential and social relevance: A review. *International Journal of Engineering Research and Applications*, *11*, 23–40.

Chavan, J. J., Gaikwad, N. B., Kshirsagar, P. R., & Dixit, G. B. (2013). Total phenolics, flavonoids and antioxidant properties of three *Ceropegia* species from Western Ghats of India. *South African Journal of Botany*, *88*, 273–277.

Cheikhyoussef, A., & Maroyi, A. (2017). Traditional knowledge of wild edible fruits in Southern Africa: A comparative use patterns in Namibia and Zimbabwe. *Indian Journal of Traditional Knowledge*, *16*, 385–392.

Chen, C. C., Huang, M. Y., Lin, K. H., & Hsueh, M. T. (2022). The effects of nitrogen application on the growth, photosynthesis, and antioxidant activity of *Amaranthus viridis*. *Photosynthetica*, *60*, 420–429.

Chingwaru, W., Duodu, G., Van, Z., Yl, Y., Schoeman, C., Majinda, R. T., Yeboah, S. O., Jackson, J., Kapewangolo, P. T., Kandawa-Schulz, M., & Minnaar, A. (2011). Antibacterial and anticandidal activity of *Tylosema esculentum* (marama) extracts. *South African Journal of Science, 107*, Art. #366.

Chiteva, R., & Wairagu, N. (2013). Chemical and nutritional content of Opuntia ficus-indica (L.). *African Journal of Biotechnology, 12*, 3309–3312.

Constant, N. L., & Tshisikhawe, M. P. (2018). Hierarchies of knowledge: Ethnobotanical knowledge, practices and beliefs of the Vhavenda in South Africa for biodiversity conservation. *Journal of Ethnobiology and Ethnomedicine, 14*, 56.

Cornwell, R. (1988). African past:" origins" of the Sotho-Tswana peoples and the history of the Batswana. *Africa Insight, 18*, 96–103.

Cruz, M. P., Peroni, N., & Albuquerque, U. P. (2013). Knowledge, use and management of native wild edible plants from a seasonal dry forest (NE, Brazil). *Journal of Ethnobiology and Ethnomedicine, 9*, 79.

Daniloski, D., D'cunHa, N. M., Speer, H., McKune, A. J., Alexopoulos, N., Panagiotakos, D. B., Petkoska, A. T., & Naumovski, N. (2022). Recent developments on *Opuntia* spp., their bioactive composition, nutritional values, and health effects. *Food Bioscience, 47*, 101665.

Danley, K. (2006). Letters of the bush: A case study of traditional setswana herbal medicine. *Independent Study Project (ISP) Collection, 270.*

Das, R., & Afrin, N. (2019). *In vivo* study of anti-diabetic activity and safety profile analysis of ethanolic extract of root of *Withania somnifera* on alloxan-induced diabetic rats. *World Journal of Pharmaceutical Research, 8*, 318–328.

de Boer, H. J., Kool, A., Broberg, A., Mziray, W. R., Hedberg, I., & Levenfors, J. J. (2005). Anti-fungal and anti-bacterial activity of some herbal remedies from Tanzania. *Journal of Ethnopharmacology, 96*, 461–469.

Dejene, T., Agamy, M. S., AgúnDez, D., & Martin-Pinto, P. (2020). Ethnobotanical survey of wild edible fruit tree species in lowland areas of Ethiopia. *Forests, 11*, 177.

Duguma, H. T. (2020). Wild edible plant nutritional contribution and consumer perception in Ethiopia. *International Journal of Food Science, 20*, 16.

Earnest, E., Ilodigwe, E., & Ihekwereme, C. (2018). *Ficus Sycomorus* L (Moraceae): A review on its phytopharmacology and toxicity profile. *Discovery Phytomedicine, 5*, 64.

Ebbo, A. A., Sani, D., Suleiman, M. M., Ahmad, A., & Hassan, A. Z. (2020). Acute and sub-chronic toxicity evaluation of the crude methanolic extract of *Diospyros mespiliformis* hochst ex a. Dc (ebenaceae) and its fractions. *Toxicology Reports, 7*, 1138–1144.

Ejoh, R. A., Nkonga, D. V., Inocent, G. G., & Moses, M. C. (2007). Nutritional components of some non-conventional leafy vegetables consumed in Cameroon. *Pakistan Journal of Nutrition, 6*, 712–717.

Elhassan, G., & Yagi, S. (2010). Nutritional composition of *Grewia* species (*Grewia tenax* (Forsk.) Fiori, *G. flavescens* Juss and *G. villosa* Willd) fruits. *Advance Journal of Food Science and Technology, 2*, 159–162.

Engels, J., Rao, R., & Jackson, M. (2002). *Managing plant genetic diversity.* CABI Publishing.

Erarslan, Z., & Koçyiğit, M. (2019). The important taxonomic characteristics of the family Malvaceae and the herbarium specimens in ISTE. *Turkish Journal of Bioscience and Collections, 3*, 1–7.

Fentahun, M. T., & Hager, H. (2009). Exploiting locally available resources for food and nutritional security enhancement: Wild fruits diversity, potential and state of exploitation in the Amhara region of Ethiopia. *Food Security, 1*, 207–219.

Fiedor, J., & Burda, K. (2014). Potential role of carotenoids as antioxidants in human health and disease. *Nutrients, 6*, 466–488.

Geyid, A., Abebe, D., Debella, A., Makonnen, Z., Aberra, F., Teka, F., Kebede, T., Urga, K., Yersaw, K., Biza, T., Mariam, B. H., & Guta, M. (2005). Screening of some medicinal plants of Ethiopia for their anti-microbial properties and chemical profiles. *Journal of Ethnopharmacology, 97*, 421–427.

Grace, O., Light, M., Lindsey, K., Mulholland, D., van Staden, J., & Jager, A. (2002). Antibacterial activity and isolation of active compounds from fruit of the traditional African medicinal tree *Kigelia africana. South African Journal of Botany, 68*, 220–222.

Grace, O. O., & Kayode, J. (2018). Ethno-medicinal assessment of wild edible plants in Ijesa region, Osun State, Nigeria. *Bulletin of Pure & Applied Sciences-Botany, 37*, 89–107.

Gray, D. (2004). Herbal Remedies (CD-ROM) 5.0. In T. Brendler, J. Gruenwald, & C. Jaenicke (Eds.), *PhytoPharm consulting GmbH* (pp. 925–925). MedPharm GmbH Scientific Publishers.

Guimarães, A. C., Meireles, L. M., Lemos, M. F., Guimarães, M. C. C., Endringer, D. C., Fronza, M., & Scherer, R. (2019). Antibacterial activity of terpenes and terpenoids present in essential oils. *Molecules, 24*, 2471.

Guinand, Y., & Lemessa, D. (2000). *Wild food plants in Southern Ethiopia: Reflections on the role of 'famine foods' at the time of drought.* UNDP-EUE, UN-Emergencies Unit for Ethiopia.

Gundidza, M., & Sibanda, M. (1991). Antimicrobial activities of Ziziphus abyssinica and Berchemia discolor. *Central African Journal of Medicine*, *37*, 80–83.

Halilu, M., Akpulu, I., Agunu, A., Ahmed, A., & Abdurahman, E. (2008). Phytochemical and antibacterial evaluation of *Parinari curatetellifolia* Planch ex Benth (Chrysobalanaceae). *Nigerian Journal of Basic and Applied Sciences*, *16*, 281–285.

Hassan, L., & Umar, K. J. (2006). Nutritional value of balsam apple (*Momordica balsamina* L.) leaves. *Pakistan Journal of Nutrition*, *5*, 522–529.

Hettiarachchi, M., Liyanage, C., WickRemasinghe, R., Hilmers, D., & Abrams, S. (2006). Nutrient intake and growth of adolescents in Southern Sri Lanka. *The Ceylon Medical Journal*, *51*, 89–92.

Hiwilepo-van Hal, P. (2013). *Processing of marula (*Sclerocarya birrea *subsp.* caffra*) fruits: A case study on health-promoting compounds in marula pulp* [Doctoral thesis, Auspices of the Graduate School VLAG, Wageningen University, Wageningen, the Netherlands].

Hussain, M. S., Hossain, M. S., Amin, M. T., & Millat, M. S. (2016a). In vitro thrombolytic potentials of methanolic extract of *Vigna unguiculata* Linn (seed). *Journal of Pharmacognosy and Phytochemistry*, *5*, 129–131.

Hussain, T., FatiMa, I., Rafay, M., Shabir, S., AkraM, M., & Bano, S. (2016b). Evaluation of antibacterial and antioxidant activity of leaves, fruit and bark of *Kigelia africana*. *Pakistan Journal of Botany*, *48*, 277–283.

Ibrahim, K., Cox, N. J., Stevenson, J. M., Lim, S., Fraser, S. D. S., & Roberts, H. C. (2021). A systematic review of the evidence for deprescribing interventions among older people living with frailty. *BMC Geriatrics*, *21*, 258.

Ickowitz, A., Rowland, D., Powell, B., Salim, M. A., & Sunderland, T. (2016). Forests, trees, and micronutrient-rich food consumption in Indonesia. *PLoS One*, *11*, e0154139.

Ismail, B. B., Guo, M., Pu, Y., Çavuş, O., Ayub, K. A., Watharkar, R. B., Ding, T., Chen, J., & Liu, D. (2021). Investigating the effect of *in vitro* gastrointestinal digestion on the stability, bioaccessibility, and biological activities of baobab (*Adansonia digitata*) fruit polyphenolics. *LWT*, *145*, 111348.

James, A., & Matemu, A. (2016). Solar-drying of vegetables for micronutrients retention and product diversification. *American Journal of Research Communication*, *4*, 1–13.

James, D., Abu, E., Wurochekke, A., & Orji, G. (2007). Phytochemical and antimicrobial investigation of the aqueous and methanolic extracts of *Ximenia americana*. *Journal of Medical Sciences*, *7*, 284–288.

Ju, Y., Zhuo, J., Liu, B., & Long, C. (2013). Eating from the wild: Diversity of wild edible plants used by Tibetans in Shangri-la region, Yunnan, China. *Journal of Ethnobiology and Ethnomedicine*, *9*, 28.

Kalra, R., & Kaushik, N. (2017). *Withania somnifera* (Linn.) Dunal: A review of chemical and pharmacological diversity. *Phytochemistry Reviews*, *16*, 953–987.

Kanter, R., Kennedy, G., & Boza, S. (2023). Editorial: Local, traditional and indigenous food systems in the 21st century to combat obesity, undernutrition and climate change. *Frontiers in Sustainable Food Systems*, *7*, 1195741.

Karanja, J., Mugendi, B. J., Khamis, F., & Muchugi, A. (2013). Nutritional composition of the pumpkin (*Cucurbita* spp.) seed cultivated from selected regions in Kenya. *Journal of Horticulture Letters*, *3*, 17–22.

Karau, G. M., Njagi, E. N., Machocho, A. K., & Wangai, L. N. (2012). Phytonutrient, mineral composition and in vitro antioxidant activity of leaf and stem bark powders of *Pappea capensis* (L.). *Pakistan Journal of Nutrition*, *11*, 123.

Kaul, P. (2011). Nutritional potential, bioaccessibility of minerals and functionality of watermelon (*Citrullus vulgaris*) seeds. *LWT-Food Science and Technology*, *44*, 1821–1826.

Khan, H., Jan, S. A., Javed, M., Shaheen, R., Khan, Z., Ahmad, A., Safi, S. Z., & Imran, M. (2016). Nutritional composition, antioxidant and antimicrobial activities of selected wild edible plants. *Journal of Food Biochemistry*, *40*, 61–70.

Kobayashi, Y., Sato, H., Yorita, M., Nakayama, H., Miyazato, H., Sugimoto, K., & Jippo, T. (2016). Inhibitory effects of geranium essential oil and its major component, citronellol, on degranulation and cytokine production by mast cells. *Bioscience, Biotechnology, and Biochemistry*, *80*, 1172–1178.

Krishnamurthy, S., & Sarala, P. (2010). Proximate nutritive values and mineral components of *Withania somnifera* (Linn.) Dunal. *E-Journal of Chemistry*, *7*, 985–996.

Kumar, S., & Pandey, A. K. (2013). Chemistry and biological activities of flavonoids: An overview. *Scientific World Journal*, *2013*, 162750.

Kumar, S., Singh, B., & Bajpai, V. (2022). Traditional uses, phytochemistry, quality control and biological activities of genus *Grewia*. *Phytomedicine Plus*, *2*, 100290.

Legwaila, G., MojeRemane, W., Madisa, M., Mmolotsi, R., & Rampart, M. (2011). Potential of traditional food plants in rural household food security in Botswana. *Journal of Horticulture and Forestry, 3*, 171–177.

Lipińska, M. M., Haliński, Ł. P., Gołębiowski, M., & Kowalkowska, A. K. (2023). Active compounds with medicinal potential found in Maxillariinae Benth.(Orchidaceae Juss.) representatives – A review. *International Journal of Molecular Sciences, 24*, 739.

Liu, S., Huang, X., Bin, Z., Yu, B., Lu, Z., Hu, R., & Long, C. (2023). Wild edible plants and their cultural significance among the Zhuang ethnic group in Fangchenggang, Guangxi, China. *Journal of Ethnobiology and Ethnomedicine, 19*, 52.

Lwalewa. (2007). Inflammation: The foundation of diseases and disorders. A review of phytomedicines of South African origin used to treat pain and inflammatory conditions. *African Journal of Biotechnology, 6*, 2868–2885.

Madibela, O. R., Letso, M., MakoBa, B., & Seitshiro, O. (2004). Do indigenous browse trees influence chemical composition and in vitro dry matter digestibility of parasitic plants? *Animal Feed Science and Technology, 115*, 357–369.

Maiyo. (2009). Phytochemical constituents and antimicrobial activity of leaf extracts of three *Amaranthus* plant species. *African Journal of Biotechnology, 9*, 3178–3182.

Malkanthi, P., Karunaratne, A., Amuwala, S., & Silva, P. (2014). Opportunities and challenges in cultivating underutilized field crops in Moneragala district of Sri Lanka. *Asian Journal of Agriculture and Rural Development, 4*, 96–105.

Maluleke, M. K., Boitumelo, L., & Marokane-Radebe, C. K. (2024). Exploring the biochemical constituents, medicinal properties and potential commercialization of *Ximenia caffra* Sond. for enhancing human health: A comprehensive review. *Discover Applied Sciences, 6*, 21.

Mander, M., MaNder, J., Crouch, N., McKean, S., & Nichols, G. (1995). Catchment action: Growing and knowing muthi plants. *Share Net, 1998*.

Maroyi, A. (2018a). *Diospyros lycioides* Desf.: Review of its botany, medicinal uses, pharmacological activities and phytochemistry. *Asian Pacific Journal of Tropical Biomedicine, 8*, 130–136.

Maroyi, A. (2018b). Nutraceutical and ethnopharmacological properties of *Vangueria infausta* subsp. infausta. *Molecules, 23*, 1089.

Maroyi, A. (2019). Utilization of *Bridelia mollis* as herbal medicine, nutraceutical and functional food in Southern Africa: A review. *Tropical Journal of Pharmaceutical Research, 18*, 203.

Maroyi, A. (2021). Evaluation of medicinal uses, phytochemistry and pharmacological properties of Strychnos madagascariensis Poir. *Medicinal Plants-International Journal of Phytomedicines and Related Industries, 13*, 369–377.

Maroyi, A. (2023). A systematic review on biological and medicinal properties of *Ehretia rigida* (Thunb.) Druce (Ehretiaceae) in Southern Africa. *Plant Science Today, 10*, 74–82.

Mashamaite, C., Manyevere, A., & Chakauya, E. (2022). *Cleome gynandra*: A wonder climate-smart plant for nutritional security for millions in semi-arid areas. *Frontiers in Plant Science, 13*, 1003080.

Masoko, P., Ramese, N., & Adebayo, S. (2019). *Stomatostemma monteiroae*: A food plant with potential medicinal and therapeutic values. *South African Journal of Botany, 126*, 182–189.

Matenge, S. T. P. (2011). *Utilisation of traditional and indigenous foods in the North West Province of South Africa* [Doctoral dissertation, North-West University, South Africa].

Mengistu, F., & Hager, H. (2008). Wild edible fruit species cultural domain, informant species competence and preference in three districts of Amhara region, Ethiopia. *Ethnobotany Research and Applications, 6*, 487–502.

Miro, M. (1995). Cucurbitacins and their pharmacological effects. *Phytotherapy Research, 9*, 159–168.

Mngadi, S., Moodley, R., & Jonnalagadda, S. B. (2019). Elemental composition and nutritional value of the edible fruits of Transvaal red milkwood (*Mimusops zeyheri*) and impact of soil quality. *Environmental Monitoring and Assessment, 191*, 135.

Mnikathi, M. M. N. (2021). *A comparative evaluation of the biological activities and phytochemical properties in Ehretia obtusifolia and Ehretia rigida* [Doctoral dissertation, University of KwaZulu-Natal, South Africa].

Mogotlane, E. A. (2015). *Chemical composition of indiginous of watermelon (Citrullus lanatus Tthunb) Matsum. and Nakai)) landrace seeds from the Sekhukhune and Capricorn district in the Limpopo province* [Doctoral dissertation. University of Limpopo, South Africa].

Moher, D., Liberati, A., Tetzlaff, J., Altman, D. G., & Group, P. (2009). Preferred reporting items for systematic reviews and meta-analyses: The PRISMA statement. *PLoS Medicine, 151*, 264–269.

Mojeremane, W., & Tshwenyane, S. (2004). The resource role of morula (*Sclerocarya birrea*): A multipurpose indigenous fruit tree of Botswana. *Journal of Biological Sciences, 4*, 771–775.

Mokganya, M. G., & Tshisikhawe, M. P. (2019). Medicinal uses of selected wild edible vegetables consumed by Vhavenda of the Vhembe district municipality, South Africa. *South African Journal of Botany*, *122*, 184–188.

Mokganya, M. G., Tshisikhawe, M. P., Mushapi, L. F., Amoo, S., & Potgieter, M. (2021). Nutritional analysis of selected wild edible vegetables of the Vhembe region, South Africa. *Technium Social Sciences Journal*, *26*, 855.

Mongalo, N. I., Mashele, S. S., & Makhafola, T. J. (2020). *Ziziphus mucronata* Willd.(Rhamnaceae): It's botany, toxicity, phytochemistry and pharmacological activities. *Heliyon*, *6*, e03708.

Monteiro, S., Reboredo, F. H., Lageiro, M. M., Lourenço, V. M., Dias, J., Lidon, F., Abreu, M., Martins, A. P., & Alvarenga, N. (2022). Nutritional properties of baobab pulp from different Angolan origins. *Plants*, *11*, 2272.

Motlhanka, D. M. T., & Makhabu, S. W. (2011). Medicinal and edible wild fruit plants of Botswana as emerging new crop opportunities. *Journal of Medicinal Plants Research*, *5*, 1836–1842.

Motlhanka, D. M. T., Motlhanka, P., & Selebatso, T. (2008). Edible indigenous wild fruit plants of eastern Botswana. *International Journal of Poultry Science*, *7*, 457–460.

Moyo, M., & Aremu, A. O. (2022). Nutritional, phytochemical and diverse health-promoting qualities of *Cleome gynandra*. *Critical Reviews in Food Science and Nutrition*, *62*, 3535–3552.

Muhammad, A., Haruna, S. Y., Birnin-Yauri, A. U., Muhammad, A. H., & Elinge, C. M. (2019). Nutritional and anti-nutritional composition of *Ximenia americana* fruit. *American Journal Applied Chemistry*, *7*, 123–129.

Musara, C., & Aladejana, E. B. (2020). *Typha capensis* (Rohrb.) NE Br.(Typhaceae): Morphology, medicinal uses, biological and chemical properties. *Plant Science Today*, *7*, 578–583.

Nafuka, S. N. (2014). *In vitro antiplasmodial activity and phytochemicals screening of ethnomedicinal plants used to treat malaria associated symptoms* [Doctoral dissertation, Department of Biological Science, the University of Namibia, Namibia].

Nafuka, S. N., & Mumbengegwi, D. R. (2013). Phytochemical analysis and *in vitro* anti-plasmodial activity of selected ethnomedicinal plants used to treat malaria associated symptoms in Northern Namibia. *International Science and Technology Journal of Namibia*, *2*, 78–93.

Neudeck, L., Avelino, L., Bareetseng, P., Ngwenya, B. N., Teketay, D., & Motsholapheko, M. (2012). The contribution of edible wild plants to food security, dietary diversity and income of households in Shorobe village, Northern Botswana. *Ethnobotany Research and Applications*, *10*, 449–462.

Ngadze, R. T., Linnemann, A. R., Nyanga, L. K., Fogliano, V., & Verkerk, R. (2017). Local processing and nutritional composition of indigenous fruits: The case of monkey orange (*Strychnos* spp.) from Southern Africa. *Food Reviews International*, *33*, 123–142.

Ngueyem, T. A., Brusotti, G., Caccialanza, G., & Finzi, P. V. (2009). The genus *Bridelia*: A phytochemical and ethnopharmacological review. *Journal of Ethnopharmacology*, *12*(4), 339–349.

Nishikito, D. F., Borges, A. C. A., Laurindo, L. F., Otoboni, A., DiReito, R., Goulart, R. A., Nicolau, C. C. T., Fiorini, A. M. R., Sinatora, R. V., & Barbalho, S. M. (2023). Anti-inflammatory, antioxidant, and other health effects of dragon fruit and potential delivery systems for its bioactive compounds. *Pharmaceutics*, *15*, 159.

Nkafamiya, I. I., Osemeahon, S., Modibbo, U., & Aminu, A. (2010). Nutritional status of non-conventional leafy vegetables, *Ficus asperifolia* and *Ficus sycomorus*. *African Journal of Food Science*, *4*, 104–108.

Ntuli, N. R., Zobolo, A. M., Siebert, S. J., & Madakadze, R. M. (2012). Traditional vegetables of Northern KwaZulu-Natal, South Africa: Has indigenous knowledge expanded the menu. *African Journal of Agricultural Research*, *7*, 6027–6034.

Nyambe, M. M., Hakwenye, H., & Benyamen, M. (2019). Nutritional and anti-nutritional composition of Diospyros mespiliformis and Hyphaene petersiana fruits from Namibia. *International Science and Technology Journal of Namibia*, *13*, 2–11.

Oduor, F. O., Kaindi, D. W. M., Abong, G. O., Thuita, F., & Termote, C. (2023). Diversity and utilization of indigenous wild edible plants and their contribution to food security in Turkana County, Kenya. *Frontiers in Sustainable Food Systems*, *7*, 1113771.

Olaokun, O. O., Manonga, S. A., Zubair, M. S., Maulana, S., & Mkolo, N. M. (2022). Molecular docking and molecular dynamics studies of antidiabetic phenolic compound isolated from leaf extract of *Englerophytum magalismontanum* (Sond.) TD Penn. *Molecules*, *27*, 3175.

Olarewaju, O. O., Fajinmi, O. O., Arthur, G. D., Coopoosamy, R. M., & Naidoo, K. K. (2021). Food and medicinal relevance of Cucurbitaceae species in Eastern and Southern Africa. *Bulletin of the National Research Centre*, *45*, 208.

Omotayo, A. O., & Aremu, A. O. (2021). Marama bean [*Tylosema esculentum* (Burch.) A. Schreib.]: An indigenous plant with potential for food, nutrition, and economic sustainability. *Food and Function*, *12*, 2389–2403.

Omotayo, A. O., Ndhlovu, P. T., Tshwene, S. C., & Aremu, A. O. (2020). Utilization pattern of indigenous and naturalized plants among some selected rural households of North West Province, South Africa. *Plants*, *9*, 953.

Ondua, M., Njoya, E. M., Abdalla, M. A., & McGaw, L. J. (2019). Anti-inflammatory and antioxidant properties of leaf extracts of eleven South African medicinal plants used traditionally to treat inflammation. *Journal of Ethnopharmacology*, *234*, 27–35.

Onyeka, E. U., & Nwambekwe, I. O. (2007). Phytochemical profile of some green leafy vegetables in South East, Nigeria. *Nigerian Food Journal*, *25*, 67–76.

Opiyo, S. A., Njoroge, P. W., Ndirangu, E. G., & Kuria, K. M. (2021). A review of biological activities and phytochemistry of Rhus species. *American Journal of Chemistry*, *11*, 28–36.

Oseni, O. A., & Williams, O. D. (2018). In-vitro compositional investigations of antioxidants, phytochemicals, nutritional, and minerals in the fruit of *Kigelia africana* (Lam.) Benth. *International Journal of Contemporary Research and Review*, *9*, 20259–20268.

Papo, L. A. (2017). *The ethnobotanical, antimicrobial and phytochemical screening of selected medicinal plants from Ga-Mashashane, Limpopo Province, South Africa* [Master's thesis, Faculty of Science, University of Johannesburg, South Africa].

Parimala, M., & Shoba, F. G. (2013). Phytochemical analysis and In vitro antioxidant acitivity of hydroalcoholic seed extract of *Nymphaea nouchali* Burm. f. *Asian Pacific Journal of Tropical Biomedicine*, *3*, 887–895.

Patel, K., Kumar, V., Pant, V., & Ravikumar, R. (2023). Biological, nutritional, and therapeutic significance of Grewia flavescens Juss. *Medicinal Plants-International Journal of Phytomedicines and Related Industries*, *15*, 636–644.

Pavithra, M., Kandikere, R., Sridhara, A. G. A., & Tomita-Yokotani, K. (2016). Bioactive potential of the wild mushroom *Astraeus hygrometricus* in South-west India. *Mycologia*, *7*, 191–202.

Pendota, S. C., Aderogba, M. A., & Van Staden, J. (2015). In vitro antimicrobial activity of extracts and an isolated compound from *Boscia albitrunca* leaves. *South African Journal of Botany*, *96*, 91–93.

Pitso, F. S., & Lebese, M. R. (2014). Traditional uses of wild edible plants in arid areas of South Africa. *Journal of Human Ecology*, *48*, 23–31.

Potgieter, M. (2001). Nutritional value of wild, leafy plants consumed by the Vhavenda. *South African Journal of Science*, *97*, 51–54.

Potočnjak, I., Gobin, I., & Domitrović, R. (2018). Carvacrol induces cytotoxicity in human cervical cancer cells but causes cisplatin resistance: Involvement of MEK–ERK activation. *Phytotherapy Research*, *32*, 1090–1097.

Qiu, D. R., Wang, D. C., Yang, S. X., Zhang, Y. M., Wei, D. S., Zhang, M. Z., Sun, J. Z., Cong, J., Guo, J., He, S. L., & Qin, J. C. (2016). Chemical constituents from the fruits of *Rhus typhina* L. and their chemotaxonomic significance. *Biochemical Systematics and Ecology*, *69*, 261–265.

Rahul, J., Jain, M. K., Singh, S. P., Kamal, R. K., Anuradha, Naz, A., Gupta, A. K., & Mrityunjay, S. K. (2015). *Adansonia digitata* L. (baobab): A review of traditional information and taxonomic description. *Asian Pacific Journal of Tropical Biomedicine*, *5*, 79–84.

Ramadwa, T. E., & Meddows-Taylor, S. (2023a). *Garcinia livingstonei* T. Anderson (Clusiaceae): A comprehensive review of traditional uses, phytochemistry, pharmacology and toxicity. *South African Journal of Botany*, *162*, 767–782.

Ramadwa, T. E., & Meddows-Taylor, S. (2023b). Traditional uses, pharmacological activities, and phytochemical analysis of *Diospyros mespiliformis* Hochst. ex. A. DC (Ebenaceae): A review. *Molecules*, *28*, 7759.

Ramese, N. (2019). *Identification and characterisation of compounds with antimycobacterial activity from Stomatostemma monteiroae* [Doctoral dissertation, Faculty of Science and Agriculture, University of Limpopo, South Africa].

Rankoana, S. A. (2016). Sustainable use and management of indigenous plant resources: A case of mantheding community in Limpopo Province, South Africa. *Sustainability*, *8*, 221.

Rashwan, A., Bai, H., Osman, A., Eltohamy, K., Chen, Z., Younis, H. A., Alfatesh, A., Rooney, D., & Yap, P. S. (2023). Recycling food and agriculture by-products to mitigate climate change: A review. *Environmental Chemistry Letters*, *21*, 3351–3375.

Rayne, S., & Mazza, G. (2007). Biological activities of extracts from sumac (*Rhus* spp.): A review. *Plant Foods for Human Nutrition*, *62*, 165–175.

Russo, D., Miglionico, R., Carmosino, M., Bisaccia, F., Andrade, P. B., Valentão, P., MiLella, L., & Armentano, M. F. (2018). A comparative study on phytochemical profiles and biological activities of *Sclerocarya birrea* (A. Rich.) Hochst leaf and bark extracts. *International Journal of Molecular Sciences*, *19*, 186.

Ruszkowski, P., & Bobkiewicz-Kozlowska, T. (2014). Natural triterpenoids and their derivatives with pharmacological activity against neurodegenerative disorders. *Mini-Reviews in Organic Chemistry*, *11*, 307–315.

Salehi, B., CapaNoGlu, E., Adrar, N., Catalkaya, G., Shaheen, S., Jaffer, M., Giri, L., Suyal, R., Jugran, A. K., & Calina, D. (2019). Cucurbits plants: A key emphasis to its pharmacological potential. *Molecules*, *24*, 1854.

Selmi, M., Lassoued, L., Chbili, C., Fradj, M. B., & Charfeddin, R. (2022). Nutritional value and therapeutic effect of Mentha pulegium L. and Artemisia abrotanum L. *Journal of Medicinal Herbs and Ethnomedicine*, *8*, 16–20.

Selvakumari, E., Shantha, A., Kumar, C. S., & Prabhu, T. P. (2016). Phytochemistry and pharmacology of the genus *Nymphaea*. *Journal of Academia and Industrial Research*, *5*, 98–108.

Semenya, S. S., & Maroyi, A. (2019). Source, harvesting, conservation status, threats and management of indigenous plant used for respiratory infections and related symptoms in the Limpopo Province, South Africa. *Biodiversitas Journal of Biological Diversity*, *20*, 789–811.

Semenya, S. S., & Mokgoebo, M. J. (2020). The utilization and conservation of indigenous wild plant resources in the Limpopo Province, South Africa. *Natural Resources Management and Biological Sciences*, *317*. https://doi.org/10.5772/intechopen.89920

Shinkafi, S., & Abdullahi, H. (2018). Antifungal activity and phytochemical analysis of *Ficus sycomorus* leaf extract on Malassezia glubosa. *Advances in Plants & Agriculture Research*, *8*, 432–436.

Singh, A., Dubey, R. K., Bundela, A. K., & Abhilash, P. C. (2020). The trilogy of wild crops, traditional agronomic practices, and UN-sustainable development goals. *Agronomy*, *10*, 648.

Sinha, J., Purwar, S., Chuhan, S. K., & Rai, G. (2015). Nutritional and medicinal potential of *Grewia subinaequalis* DC.(syn. G. asiatica.)(Phalsa). *Journal of Medicinal Plants Research*, *9*, 594–612.

Steyn, C., Hawkins, P., Saito, T., Noakes, D. E., Kingdom, J. C. P., & Hanson, M. A. (2001). Undernutrition during the first half of gestation increases the predominance of fetal tissue in late-gestation ovine placentomes. *European Journal of Obstetrics & Gynecology and Reproductive Biology*, *98*, 165–170.

Suhendy, H., Fidrianny, I., & Insanu, M. (2023). Phytochemical compounds and pharmacological activities of *Ipomoea batatas* L.: An updated review. *Pharmacia*, *70*, 1283–1294.

Tabit, F. T., Komolafe, N. T., Tshikalange, T. E., & Nyila, M. A. (2016). Phytochemical constituents and antioxidant and antimicrobial activity of selected plants used traditionally as a source of food. *Journal of Medicinal Food*, *19*, 324–329.

Taha, A. S., Salem, M. Z., Elgat, W. A. A., Ali, H. M., Hatamleh, A. A., & Abdel-Salam, E. M. (2019). Assessment of the impact of different treatments on the technological and antifungal properties of papyrus (*Cyperus papyrus* L.) sheets. *Materials*, *12*, 620.

Tairo, V. E. (2021). Comparison of nutritional and anti-nutritional qualities of *Grewia forbesii* Hav. Ex Mast and *Grewia bicolor* Juss fruits from Kitapilimwa forest reserve in Iringa District. *Tanzania Journal of Science*, *47*, 1436–1441.

Tajuddeen, N., Swart, T., Hoppe, H. C., & van Heerden, F. R. (2021). Antiplasmodial and cytotoxic flavonoids from *Pappea capensis* (Eckl. & Zeyh.) leaves. *Molecules*, *26*, 3875.

Thi, N., Duc, L., Minh, L., & Tien, L. (2020). Phytochemicals and antioxidant activity of aqueous and ethanolic extracts of *Mentha aquatica* L. In *IOP conference series: Materials science and engineering* (p. 012027). IOP Publishing.

Thiaw, M., Samb, I., Genva, M., Gaye, M. L., & Fauconnier, M.-L. (2023). *Momordica balsamina* L.: A plant with multiple therapeutic and nutritional potential – A review. *Nutraceuticals*, *3*, 556–573.

Thongphasuk, P., Suttisri, R., Bavovada, R., & Verpoorte, R. (2014). Antioxidant lignan glucosides from *Strychnos* vanprukii. *Fitoterapia*, *75*, 623–628.

Tselaesele, N., Bultosa, G., Molapisi, M., Makhabu, S., Kobue-Lekalake, R., Haki, G. D., Sekwati-Monang, B., Seifu, E., Mokhawa, G., & Sonno, K. (2023). Plant-based traditional foods and beverages of Gumare Village, Botswana. *Food Production Processing and Nutrition*, *5*, 28.

Ugoh, S. C., & Bejide, O. S. (2013). Phytochemical screening and antimicrobial properties of the leaf and stem bark extracts of *Strychnos spinosa*. *Nature and Science*, *11*, 123–128.

Ullah, W., Uddin, G., & Siddiqui, B. S. (2012). Ethnic uses, pharmacological and phytochemical profile of genus *Grewia*. *Journal of Asian Natural Products Research*, *14*, 186–195.

Urnukhsaikhan, E., Bold, B.-E., Gunbileg, A., Sukhbaatar, N., & Mishig-Ochir, T. (2021). Antibacterial activity and characteristics of silver nanoparticles biosynthesized from *Carduus crispus*. *Scientific Reports*, *11*, 21047.

Van der Hoeven, M., Osei, J., Greeff, M., Kruger, A., Faber, M., & Smuts, C. M. (2013). Indigenous and traditional plants: South African parents' knowledge, perceptions and uses and their children's sensory acceptance. *Journal of Ethnobiology and Ethnomedicine*, *9*, 78.

Van Rayne, K. K., Adebo, O. A., Wokadala, O. C., & Ngobese, N. Z. (2023). The potential of *Strychnos* spp L. utilization in food insecurity alleviation: A review. *Food Reviews International*, *39*, 3400–3414.

van Rensburg, J. W., Van AveRbeke, W., Slabbert, R., Faber, M., Van Jaarsveld, P., Heerden, I., & Wenhold, F. (2009). African leafy vegetables in South Africa. *Water SA*, *33*, 317–326.

van Rensburg, J. W., Venter, S., Netshiluvhi, T., Van den, H. E., Vorster, H. J., Ronde, J. A., & Bornman, C. H. (2004). Role of indigenous leafy vegetables in combating hunger and malnutrition. *South African Journal of Botany*, *70*, 52–59.

Vinogradova, Y., ShelepOva, O., Vergun, O., Grygorieva, O., & Brindza, J. (2021). Phenlic content and antioxidant activity of *Echinocystis lobata* (Mich.) Torr.Et Gray (Cucurbitaceae). *Potravinarstvo Slovak Journal of Food Sciences*, *15*, 784–791.

Wang, C.-Y., Chen, Y.-W., & Hou, C.-Y. (2019). Antioxidant and antibacterial activity of seven predominant terpenoids. *International Journal of Food Properties*, *22*, 230–238.

Wehmeyer, A. S. (1986). *Edible wild plants of Southern Africa: Data on the nutrient contents of over 300 species*. The Council for Scientific and Industrial Research (CSIR) Report, Pretoria, South Africa.

Welcome, A. K., & Van Wyk, B. E. (2019). An inventory and analysis of the food plants of Southern Africa. *South African Journal of Botany*, *122*, 136–179.

Williams, J., & Haq, N. (2010). *Underutilized crops for food security and poverty alleviation*. University of Southampton.

WFO. (2024). *World flora online*. Retrieved January 18, 2024, from www.worldfloraonline.org

World Health Organization (WHO). (2004). *Vitamin and mineral requirements in human nutrition*. Report of a Joint FAO/WHO Expert Consultation, Bangkok, Thailand, September 21–30, 1998.

Yadav, K. N., Kadam, P. V., Patel, J. A., & Patil, M. J. (2014). *Strychnos potatorum*: Phytochemical and pharmacological review. *Pharmacognosy Reviews*, *8*, 61.

Yerima, B., & Markus, M. (2020). Chemical and nutritive composition of Buffalo Thorn (*Ziziphus mucronata*) seeds from Kwami LGA of Gombe State, Nigeria. *Chemistry Research Journal*, *5*, 256–265.

Yusuf, A. A., Lawal, B., Sani, S., Garba, R., Mohammed, B. A., Oshevire, D. B., & Adesina, D. A. (2020). Pharmacological activities of *Azanza garckeana* (Goron Tula) grown in Nigeria. *Clinical Phytoscience*, *6*, 27.

Zamuz, S., Munekata, P. E. S., Gullón, B., Rocchetti, G., Montesano, D., & Lorenzo, J. M. (2021). *Citrullus lanatus* as source of bioactive components: An up-to-date review. *Trends in Food Science & Technology*, *111*, 208–222.

Zhen, J., Guo, Y., Villani, T., Carr, S., Brendler, T., Mumbengegwi, D. R., Kong, A.-N. T., Simon, J. E., & Wu, Q. (2015). Phytochemical analysis and anti-inflammatory activity of the extracts of the African medicinal plant *Ximenia caffra*. *Journal of Analytical Methods in Chemistry*, *2015*, 948262.

16 Exploring Some Cape Wild Foods

Implications on Future Food and Nutrition Security

[1]Hildegard Witbooi, [2]Alfred Maroyi, and [1]Alen Manyevere*
[1]Department of Agronomy, University of Fort Hare, Alice, South Africa
[2]Department of Botany, University of Fort Hare, Alice, South Africa
*Corresponding author: hwitbooi@ufh.ac.za

16.1 INTRODUCTION

Southern Africa alone has more than 30,000 species of higher plants, with the Cape Floral Kingdom (CFK) possessing close to 9,000 of diverse temperate flora, of which 3,000 species are used as medicinal plants (UNESCO, 2024). One of the globe's most biodiverse nations, South Africa holds the third position in overall rankings. The clear evidence of its exceptional biodiversity lies in its rich CFK. Numerous beneficial plants remain undiscovered in this kingdom, with some unlikely to be explored in our lifetime. We can, however, take hints from our foreparents about the harvesting techniques and uses of specific terrestrial and marine plants and adapt it to our current knowledge for further exploration, particularly with some of the technologies now available to us. The use of wild native plants for nourishment, beverage, and pharmaceutical purposes dates back several decades. The use of indigenous wild food plants and their increasingly popular resurgence is gaining momentum. The Cape Coast of South Africa in particular provides an exciting and interesting range of wild or uncultivated indigenous food plant species. An added benefit of these selected plants is that they require little to no agronomical attention and could easily be adopted by marginalized farmers to generate income. Furthermore, they offer organic foods with minimal to no synthetic chemicals, yielding additional health benefits attributable to the accumulation of secondary metabolites. These compounds are synthesized by plants to enhance their competitiveness in challenging environments.

The sustainable production of food through the integration of indigenous knowledge systems by serving native communities carries significance for meeting the UN SDGs. An in-depth and multifaceted analysis of certain species, covering various aspects ranging from botanical characteristics and cultivation to diverse uses in ornamental, agricultural, culinary, and medicinal domains, needs to be explored. Based on the latest realities around climate warming, it is imperative to carry out a bibliometric analysis of wild foods of the Cape to harness the biological and economic potentials of the species documented in literature in the past ten years. The focus of this bibliometric study was to review the published outputs on the cultivation, phytochemical constituents, bioavailability, and health benefits of the two crops reported herein, *viz.*, *Citrullus lanatus* subsp. *lanatus* var. *citroides* (L. H. Bailey) Mansf. ex Greb. and *Harpephyllum caffrum* Bernh. This report will potentially assist in the exploration of key research outputs and highlight areas with knowledge gaps on cultivation, biological activities, and medicinal applications for commercial use.

Wild foods in the Cape of South Africa have been studied for many years and reported in the literature. Interestingly, while many species of edible wild plants have been extensively observed,

DOI: 10.1201/9781003469766-16

there is still a lack of basic knowledge regarding their cultivation, phytochemicals, and bioavailability. Challenges such as soil degradation, biodiversity loss, and the erosion of indigenous knowledge systems are significant in today's world. Major soil issues include poor fertility, shallow depth, and high degradability. The global demand for these edible plants for food security and agro-processing, as well as the support for indigenous knowledge systems, is considerable. Wild food species are featured in selected restaurants, where different plant parts are showcased in various dishes and often become the highlights of the meals. This study examines and discusses two species of wild foods in terms of their cultivation, phytochemical content, bioavailability, and some trade-offs associated with the consumption of the phytochemicals they produce in their current status.

16.2 *CITRULLUS LANATUS* SUBSP. *LANATUS* VAR. *CITROIDES*: BACKGROUND, CULTIVATION, AND USES

Citrullus lanatus subsp. *lanatus* var. *citroides* (L. H. Bailey) Mansf. ex Greb. is commonly known as *makataan*, *citron melon*, as well as *preserving melon* and belongs to the Cucurbitaceae family. This crop plant is underutilized and underinvestigated, with the potential to contribute to crop diversification in sub-Saharan Africa. This vine creeper is uncommonly planted due to its bitter fruits and unpalatability in its raw state. It occurs naturally near watercourses in drier bushlands and grasslands across Southern Africa (Vijayalakshmi, 2014).

This cucurbit crop is versatile, with multiple culinary uses (Table 16.1), and cultivated in many parts of Africa and serves people and animals particularly in times of drought. The crop's dietary value, coupled with its drought tolerance, makes it a crop of choice to achieve food and nutrition security. This crop is cultivated economically with an annual yield of 118 million tons on 3.5 million hectares globally (Yavuza, 2020).

Makataan can easily be grown from seed, and its seedlings thrive in fertile soils. Its roots grow deep, and the non-woody creeping stems spread 3 m wide, reaching a height of 0.2 m. The leaf stalks are hairy, and the leaves visibly three-lobed, with vigorous tendrils that split at the ends. Both female and male flowers occur on the same plant (monoecious) from January to May. These flowers are bright yellow and attract pollinators. The pollinated flowers mature to become fruits from February to May. The fruits can grow approximately 600 × 300 mm in size and can weigh up to 20 kg each. The rind is smooth, with pale green mottled and irregular bands down the length of the fruit. It is filled with maroon to light-brown seeds which turn dark brown with drying.

TABLE 16.1
Culinary Versatility of *C. lanatus* subsp. *Lanatus* var. *Citroides*

Plant Part	Preparation	Uses	References
Fresh and dried leaves	Cooked as a side dish	Food	Pasandide et al., 2018 Abiola et al., 2020
Fresh and dried vine material	Harvested	Fodder	Bultosa et al., 2020; Mashilo et al., 2021
Melon flesh below rind	Cooked with sugar and spice	"Makataankonfyt"; sweet preserve	N/A
Melon central, soft flesh	Cooked with sugar; traditionally cooked as a vegetable or porridge with maize meal	Jam or chutney; food	N/A Abiola et al., 2020
Rind	Extraction using hot water or solvents	Pectin	Pérez et al., 2022
Seeds	Harvest from fruit, sun dry, and roast; ground	Snack; flour	Mandizvo et al., 2022 Abiola et al., 2020

Source: N/A. Personal generational knowledge.

According to the Department of Agriculture, Forestry, and Fisheries (2013), the plant requires full sun and sandy soil conditions for growth and development. Soil preparation should be completed one month prior to planting for commercial or controlled cultivation. Furthermore, it is recommended that the field be deeply ploughed, and the ground levelled, using a harrow. Raised planting beds about 100 to 500 mm high and the removal of stones are recommended to assist in good drainage and rooting. All weeds, particularly perennial grasses, should be removed prior to planting. In addition, the soil may be fertilized with organic or inorganic material during soil preparation. It is planted in rows 2 to 2.5 m apart and spaced 0.5 to 2 m between the plants. Traditionally, this crop is often intercropped with maize and sorghum. Makataan prefers well-drained, loamy, and fertile soils. Also, the crop grows best on a rich, sandy loam medium, although it will grow in most soil types, provided the soil is well-drained. It also grows best at a soil pH of between 5.6 and 7.0 but will tolerate a pH as low as 5.0.

16.3 *C. LANATUS* PEST AND DISEASE MANAGEMENT

Makataan is susceptible to a wide range of diseases (Table 16.2), which do not reliably appear in the fields, but their occurrence should always be a cause for concern because of their ability to quickly spread and become uncontrollable, especially under suitable weather conditions. Viruses are transmitted by whiteflies, pimples, and ringspots (tobacco ringspot virus) (Sudha & Lakshmanan, 2007). Anthracnose, which is caused by the genus *Colletotrichum*, is seed-transmissible and must be controlled (Schippers, 2000). The melon fly has been reported to affect young fruit, while the major pests of stored seeds are *Tribolium castaneum* and *Lasioderma serricorne* (Schippers, 2000). Natural predators may be considered for successful control. These may include ladybirds and soldier beetles that can be used to feed on the eggs or adults of some pests. Further reports of natural repellents include plants such as broccoli and radish that can be intercropped with bitter watermelon to ward off some pests like cucumber beetles (Department of Agriculture, Forestry and Fisheries, 2013). Natural remedies are recommended – for example, mixing wood ash with hydrate lime or hot pepper and garlic in water to be sprayed on the leaves to control some of the pests (Department of Agriculture, Forestry and Fisheries, 2013). Natural cover crop repellents may be considered as a biofumigant, for example, mustards, broccoli, and cauliflower, which are used due to their high

TABLE 16.2
Pests and Diseases That Are Prone to Affect *C. lanatus*

Pests	Common Name	Fungal Diseases	Common Name
Meloidogyne incognita	Root knot nematodes	*Cercospora citrullina*	Cercospora leaf spot
Daucus sp.	Melon fly	*Pseudoperonospora cubensis*	Downy mildew
Aphidoidea	Aphids	*Alternaria cucumerina*	Alternaria leaf blight
Tetranychus urticae	Spider mites	*Didymella bryoniae*	Gummy stem blight
Epicauta fabricii	Blister beetles	*Acidovorax avenae* subsp. *citrulli*	Powdery mildew
Acalymma vittatum	Striped cucumber beetle	*Erwinia* sp.	Bacterial rind necrosis
Diabrotica undecimpunctata	Spotted cucumber beetle	*Rhizoctonia solani*	Belly rot
Thysanoptera	Thrips	*Macrophomina phaseolina*	Damping off
Diaphania nitidalis	Pickleworm		
Tribolium castaneum	Red flour beetle		
Lasioderma serricorne	Tobacco beetle		

Source: Department of Agriculture, Forestry and Fisheries (2013).

glucosinolate content. When these plants decompose, glucosinolates are broken down to volatile isothiocyanates, which become toxic to soil organisms, such as bacteria, fungi, and nematodes. Further control of diseases may also include the planting of disease-free seeds. Other conventional methods, such as good sanitation, crop rotation, soil fumigation, and mulching, are recommended to control these diseases.

16.4 *C. LANATUS* GENERAL AGRONOMIC PRACTICES

Makataan responds well to fertilizer. The required nutrition depends on the nutrient condition of the soil. Poultry or pig manure recommendation is 2 kg to cover an approximate area of 300 m^2. Similarly, kraal manure is recommended at 2 kg, covering 25 m^2. Chemical fertilizer recommendation ratio rate is 2:3:2 or 3:2:1 mixture of NPK at the general application of 40 g/m.

Weed infestation is not regarded as a management problem when the vine grows in its natural habitat; however, when needed, weed management can be done through cultivation by hand or hoeing. Weeding should occur twice or three times prior to the plants' natural growth pattern of a thick vegetative cover on the ground, six to eight weeks after sowing of seeds. Cultivation without herbicides is recommended; however, a pre- and post-emergence herbicide can be applied, and this should be done three days prior to sowing.

Makataan is adapted to survive in dry conditions with its deep roots but requires regular irrigation to avoid water stress. Due to the high rate of water loss through evaporation, supplementary irrigation is necessary, particularly when growing in light, sandy soils. Adequate water is required in the first four weeks, during which drip irrigation is recommended, which assists in water conservation and prevents the development of fungi-related diseases. Deep irrigation of approximately 5 L of water per day is required without leaching nutrients. Once the mature fruits are harvested, they are known to last for months thereafter, and this can be attributed to the layer of wax around the melon.

FIGURE 16.1 *Citrullus lanatus* subsp. *lanatus* var. *citroides*: (a) actively growing in the field; (b) harvested and matured; (c) matured melon ready for use; (d) seeds air-drying.

16.5 PHYTOCHEMICAL CONSTITUENTS OF MAKATAAN

Masoko et al. (2022) reported the presence of bioactive phytoconstituent groups responsible for the medical properties of makataan, which are primarily attributed to the basis for potential biological activity in plants. Total polyphenolic content was greatest in the seeds, followed by the flesh and rind, respectively. Potassium was the most abundant element present in the flesh, seeds, and untreated rind control. Masoko et al. (2022) further reported the flesh to have higher antioxidant activity than the rind and seeds, whilst Abiola (2020) reported the seeds to be extremely rich in antioxidant, vitamin C, minerals, fat, starch, and riboflavin.

16.6 HEALTH BENEFITS OF MAKATAAN

Although makataan is broadly used as a food source, some members of its family have been reported to possess medicinal properties, such as antioxidant, antidiabetic, antibacterial, anti-HIV, antipyretic, antidiarrhea, carminative, laxative, anthelmintic, antitubercular, purgative, and hepatoprotective (Rajasree et al., 2016). The matured yellow- and orange-fleshed fruits are a rich source of phytochemical compounds, including linoleic acid, palmitic acid, oleic acid, and stearic acid (Nkoana et al., 2021). Makataan has the potential to fulfill nutritional requirements by supplying biochemical compounds which carry health-promoting properties. Watermelon seeds which have the same derivative of makataan contain four identified proteinogenic amino acids (phenylalanine, threonine, tryptophan, and valine) and proteins which have higher accumulative quantities than those found in cereals (10–15%). Watermelon seeds are also reported to contain vitamins A, C, D, E, and K and several antioxidants, including flavonoids, among others (Guo et al., 2003). The seeds are harvested from the fruit, sun-dried, and roasted to enjoy as a snack, providing key nutritional macro- and micronutrients, such as calcium, magnesium, zinc, iron, and copper (Mandizvo et al., 2022). **Paliyath et al. (2011)** reported that populations consuming flavonoids-enriched foods have a low cancer frequency. The consumption of makataan seeds can produce higher-quality milk in lactating mothers, as seen in animal models fed with isoflavone-rich fodder (Zhengkang et al., 2006). Through their antioxidant activity and modulating multiple protein functions, flavonoids are known to inhibit degenerative diseases, like coronary heart disease, cancer, atherosclerosis, diabetes, and Alzheimer's disease (Pandey & Rizvi, 2009). An added advantage of makataan seeds is the absence of gluten, which offers alternative nourishment for the celiac population, and it has the potential to counteract the increasing problems of obesity in the developed world. In less-developed countries, makataan could significantly reduce famine and death by hunger.

16.7 *HARPEPHYLLUM CAFFRUM*: BACKGROUND, CULTIVATION, AND USES

Harpephyllum caffrum Bernh. is commonly known as wildepruim in Afrikaans and belongs to the Anarcadiaceae family. This evergreen medium to large tree stands out as a valuable fruit tree and source of traditional medicines. It occurs naturally in frost-free areas of the Eastern Cape, Kwazulu-Natal, and Limpopo provinces in South Africa. The fruits of *H. caffrum* are used as a snack, sweet preserve, and alcoholic and non-alcoholic beverage in Eswatini and South Africa (Maroyi, 2019).

16.8 PHYTOCHEMICAL CONSTITUENTS, BIOAVAILABILITY, AND HEALTH BENEFITS OF WILDEPRUIM

Analyses of the nutritional values of *H. caffrum* show that the fruits are a better source of energy; minerals, such as calcium, copper, iron, magnesium, manganese, phosphorus, potassium, sodium, and zinc; and classic nutrients, such as fiber, protein, thiamine, and vitamin C than a commercial fruit crop, mango, *Mangifera indica* L (Table 16.3). Furthermore, *H. caffrum* has been reported to contain abscisic acids and derivatives, alk(en)ylhydroxycyclohexanes, amino acids, cardanols, depsipeptides, fatty acyl glycosides, flavan-3-ol, flavone, flavonoids, gallate esters, hydrolyzable tannins, iridoid glycosides, lignan, lignan glycosides, methyl esters, pentagallic acid esters, phenolics, phenolic glycosides, proanthocyanidins, and triterpenoids (Table 16.4).

Some of these phytochemical compounds may be responsible for the biological activities associated with the species, which include acetylcholinesterase inhibitory, analgesic, antibacterial, antifungal, antimycobacterial, anti-HIV, anticonvulsant, antidiabetic, anti-inflammatory, antioxidant, antipyretic, antityrosinase, cytotoxicity, hepatoprotective, hypoglycemic, and hypotensive activities (Table 16.5). These results corroborate with the traditional uses of *H. caffrum* against gastrointestinal problems, such as diarrhea; respiratory infections, which include asthma, chest problems, and chronic cough; skin problems; and wounds (Maroyi, 2019).

TABLE 16.3
***Harpephyllum caffrum* and *Mangifera indica* Fruit Nutritional Constituents**

Nutritional Composition	*Harpephyllum caffrum*	*Mangifera indica* L.
Ash (g/100 g)	0.8	0.34–0.52
Calcium (mg/100 g)	47.0–115.8	7.0–21.0
Carbohydrates (g/100 g)	9.1	16.2–17.2
Chromium (mg/20 g dry mass)	0.1	–
Cobalt (mg/20 g dry mass)	0.003	–
Copper (mg/100 g)	0.1–0.4	0.04–0.32
Energy kJ/100 g	172.0	62.1–190.0
Fat (g/100 g)	0.2	0.3–0.53
Fiber (g/100 g)	1.7	0.9–1.1
Iron (mg/100 g)	0.6–2.9	0.09–0.9
Magnesium (mg/100 g)	23.7–26.4	8.0–38.0
Manganese (mg/100 g)	0.2	0.03–0.12
Moisture (g/100 g)	87.5	78.9–82.8
Nickel (mg/20 g dry mass)	0.08	–
Phosphorus (mg/100 g)	13.3	10.0–23.0
Potassium (mg/100 g)	254.0	120.0–617.0
Protein (g/100 g)	0.7	0.36–0.40
Sodium (mg/100 g)	5.7	0–4.0
Thiamine (mg/100 g)	0.1	0.01–0.04
Vitamin C (mg/100 g)	70.7	13.2–92.8
Zinc (mg/20 g dry mass)	0.1–0.3	0–0.15

Source: Wehmeyer (1986); Moodley et al. (2013); Maldonado-Celis et al. (2019).

TABLE 16.4
***Harpephyllum caffrum* Phytochemical Constituents**

Phytochemical Composition	Plant Part	Reference
Abscisic Acids and Derivatives		
(+)-Abscisyl beta-D-glucopyranoside	Fruit peel and pulp	Pfukwa et al., 2022a
Alk(en)ylhydroxycyclohexanes		
(1R,3R)-1,3-dihydroxy-3-[heptadec-12'(Z)-enyl]cyclohexane	Gum exudate	Bodede et al., 2022
(1R,2S,4R)-1,2,4-trihydroxy-4-[heptadec-12'(Z)-enyl]cyclohexane	Gum exudate	Bodede et al., 2022
(1S,2S,3S,4S,5R)-1,2,3,4,5-pentahydroxy-5-[octadec-13'(Z)-enyl] cyclohexane	Gum exudate	Bodede et al., 2022
Amino Acid		
S-methylcysteine sulfoxide	Fruits	Olofinsan et al., 2022b

(Continued)

TABLE 16.4 (Continued)

Phytochemical Composition	Plant Part	Reference
Cardanols		
1-hydroxy-3-[(Z)-12'-nonadecenyl] benzene	Stem bark	Moodley et al., 2014
1-hydroxy-3-[(Z)-12'-heptadecenyl] benzene	Stem bark	Moodley et al., 2014
1-hydroxy-3-heptadecanyl benzene	Stem bark	Moodley et al., 2014
1-hydroxy-3-pentadecanyl benzene	Stem bark	Moodley et al., 2014
Eicosanyl-trans-p-coumarate	Stem bark	Moodley et al., 2014
3-heptadec-12'-Z-enyl phenol	Gum exudate	Bodede et al., 2022
Depsipeptides		
Viridamide A	Fruit pulp	Pfukwa et al., 2022b
Fatty Acyl Glycosides		
7-Epi-12-hydroxyjasmonic acid glucoside	Fruit peel and pulp	Pfukwa et al., 2023
Flavan-3-ol		
(+)-catechin	Fruits and stem bark	Moodley et al., 2014
Flavone		
8-D-glucosyl-4′,5,7-trihydroxyflavone	Fruits	Olofinsan et al., 2022b
Apigenin-7-glucoside	Leaves	El-Sherbeiny & El-Ansari, 1976
Apigenin-7-O-α-glucoside	Leaves	Shabana et al., 2011
Nobiletin	Fruits	Olofinsan et al., 2022b
Puerarin	Fruits	Olofinsan et al., 2022b
Flavonoids		
4″-O-acetylmyricitrin	Fruit peel and pulp	Pfukwa et al., 2022a
Astragalin 7-rhamnoside	Fruit peel and pulp	Pfukwa et al., 2022b
Dihydroquercetin	Fruits	Olofinsan et al., 2022b
(-)-epigallocatechin	Fruit peel and pulp	Pfukwa et al., 2023
(-)-gallocatechin	Fruit peel and pulp	Pfukwa et al., 2022a
Mammeigin	Fruit peel and pulp	Pfukwa et al., 2022b
Myricetin 3-galactoside	Fruit peel and pulp	Pfukwa et al., 2023
Phellamurin	Fruits	Olofinsan et al., 2022b
Quercetin 3-galactoside	Fruit pulp	Pfukwa et al., 2022a
Quercetin 3-glucosyl-(1->2)-galactoside	Fruit peel and pulp	Pfukwa et al., 2022b
Rutin	Fruit peel and pulp	Pfukwa et al., 2023
Gallate Esters		
Ethyl gallate	Leaves	Shabana et al., 2011
Methyl gallate	Leaves	Shabana et al., 2011
Hydrolyzable Tannins		
Corilagin	Fruit peel and pulp	Pfukwa et al., 2022a
Beta-glucogallin	Fruit peel and pulp	Pfukwa et al., 2022b
Strictinin	Fruit peel and pulp	Pfukwa et al., 2023
Vescalagin	Fruit pulp	Pfukwa et al., 2022a
Hydroxycinnamic Acids		
3-O-caffeoyl-4-O-methylquinic acid	Fruit peel and pulp	Pfukwa et al., 2022b
3-O-p-coumaroylquinic acid	Fruit peel and pulp	Pfukwa et al., 2023
Chlorogenic acid	Fruit pulp	Pfukwa et al., 2022a
Cryptochlorogenic acid	Fruit peel and pulp	Pfukwa et al., 2022b
Neochlorogenic acid	Fruit peel and pulp	Pfukwa et al., 2023
Iridoid Glycosides		
Asperulosidic acid	Fruits	Olofinsan et al., 2022b
Lignan		
3,4-dimethyl-2,5-bis(3,4,5-trimethoxyphenyl) tetrahydrofuran	Fruits	Olofinsan et al., 2022b

(Continued)

TABLE 16.4 (Continued)

Phytochemical Composition	Plant Part	Reference
Lignan Glycosides		
(+)-7-epi-syringaresinol 4′-glucoside	Fruit peel and pulp	Pfukwa et al., 2022a
Methyl Esters		
Methyl linoleate	Leaves	Shabana et al., 2011
Pentagallic acid esters		
1,2,4,6-tetragalloylglucose	Fruits	Olofinsan et al., 2022b
Phenolics		
1,3-di-O-galloyl glucose	Leaves	Nawwar et al., 2011
2-methoxyhydroquinone	Bark	Olofinsan et al., 2022a
2,3-di-ogalloyl glucose	Leaves	Nawwar et al., 2011
3,3″-dimethoxy ellagic acid 4-O-glucosid	Leaves	Nawwar et al., 2011
3,3′,4-trimethoxyellagic acid	Leaves	Nawwar et al., 2011
3-methoxy ellagic acid 4-O-β-galactopyranoside	Leaves	Nawwar et al., 2011
3-methoxy gallic acid	Leaves	Nawwar et al., 2011
3-methoxy gallic acid 5-sodium sulfate	Leaves	Nawwar et al., 2011
10-undecenyl hexofuranoside	Bark	Olofinsan et al., 2022a
24,25-dihydroxyvitamin D	Bark	Olofinsan et al., 2022a
-(+)-ascorbic acid 2,6-dihexadecanoate	Bark	Olofinsan et al., 2022a
Catechol	Bark	Olofinsan et al., 2022a
Cyanidin 3,5-glucoside	Fruits	Pfukwa et al., 2022b
Cyanidin 3-galactoside	Fruits	Pfukwa et al., 2022b
Cyanidin 3-diglucoside 5-glucoside	Fruits	Pfukwa et al., 2022b
Cyanidin 3-O-galactoside	Fruits	Pfukwa et al., 2022b
Delphinidin 3-glucoside	Fruits	Pfukwa et al., 2022b
Gallic acid	Leaves	Nawwar et al., 2011
Gentisic acid 2-O-glucoside	Leaves	Nawwar et al., 2011
Gentisic acid 5-O-glucoside	Leaves	Nawwar et al., 2011
Hendecane	Leaves	Shabana et al., 2011
Hydrocortisone acetate	Bark	Olofinsan et al., 2022a
Hydroquinone	Bark	Olofinsan et al., 2022a
Hydroquinone acetate	Bark	Olofinsan et al., 2022a
p-hydroxybenzoic acid	Leaves	Nawwar et al., 2011
Kaempferol	Leaves	Shabana et al., 2011
Kaempferol 3-rhamnoside	Leaves	El-Sherbeiny & El-Ansari, 1976
Kaempferol 3-galactoside	Leaves	El-Sherbeiny & El-Ansari, 1976
Kaempferol 3-O-galactoside	Leaves	Nawwar et al., 2011
Kaempferol 3-O-rhamnoside	Leaves	Nawwar et al., 2011
Kaempferol-3-O-β-galactoside	Leaves	Shabana et al., 2011
Kaempferol-3-O-β-(2″-sulphatogalactopyranside)	Leaves	Nawwar et al., 2011
Kaempferol-3-O-α-rhamnoside	Leaves	Shabana et al., 2011
Maltol	Bark	Olofinsan et al., 2022a
Malvidin 3-glucoside	Fruits	Pfukwa et al., 2022b
Mequinol	Bark	Olofinsan et al., 2022a
Petunidin 3-glucoside	Fruits	Pfukwa et al., 2022b
Phytol acetate	Bark	Olofinsan et al., 2022a
Protocatechuic acid	Bark	Olofinsan et al., 2022a
Protocatechuic acid	Bark and leaves	Nawwar et al., 2011
Pyrogallol 1-methyl ether	Bark	Olofinsan et al., 2022a

(Continued)

TABLE 16.4 (Continued)

Phytochemical Composition	Plant Part	Reference
Quercetin	Leaves	Shabana et al., 2011
Quercetin 3-O-arabinopyranoside	Leaves	Nawwar et al., 2011
Quercetin 3-O-galactoside	Leaves	Nawwar et al., 2011
Quercetin 3-O-rhamnoside	Leaves	Shabana et al., 2011
Quercetin 3-O-β-(2″-sulphatogalactopyranoside)	Leaves	Nawwar et al., 2011
Quercetin-3-O-β-arabinoside	Leaves	Shabana et al., 2011
Quercetin 3-arahinoside	Leaves	El-Sherbeiny & El-Ansari, 1976
Quercetin 3-galactoside	Fruit peels	Pfukwa et al., 2023
Quercetin 3-glucoside	Leaves	El-Sherbeiny & El-Ansari, 1976
Quercetin 3-rhamnoside	Leaves	El-Sherbeiny & El-Ansari, 1976
Quercetin 3-rutinoside	Fruits	Olofinsan et al., 2022b
Retinal	Bark	Olofinsan et al., 2022a
Strophanthidol	Bark	Olofinsan et al., 2022a
Uncinoside A	Fruits	Pfukwa et al., 2022b
Phenolic Glycosides		
Methylsyringin	Fruit pulp	Pfukwa et al., 2022b
Picramnioside F	Fruit pulp	Pfukwa et al., 2022a
(-)-Picramnioside F	Fruits	Pfukwa et al., 2022b
Proanthocyanidins		
Procyanidin B1	Fruit peel and pulp	Pfukwa et al., 2023
Terpene Glycosides		
Cinnamoside	Fruit peel and pulp	Pfukwa et al., 2022a
Oleuropein	Fruit peel and pulp	Pfukwa et al., 2022b
Secologanin	Fruit peel and pulp	Pfukwa et al., 2023
Triterpenoids		
3-acetyl methyl betulinate	Leaves	Shabana et al., 2011
Betulonic acid	Leaves	Shabana et al., 2011
Lupeol	Fruits and leaves	Moodley et al., 2014
Lupenone	Leaves	Shabana et al., 2011
β-sitosterol	Fruits	Moodley et al., 2014

TABLE 16.5
***Harpephyllum caffrum* Different Extract Pharmacological Activities**

Activity Tested	Extract	Part/ Compound	Model	Effect
Acetylcholinesterase inhibitory	Dichloromethane (DCM) and methanol (MeOH)	Leaves	Microplate	Half-maximal inhibitory concentration (IC_{50}) = 0.12–0.17 mg/ml (Moyo et al., 2010)
	DCM	Stem bark	Microplate	IC_{50} = 0.02 mg/ml (Moyo et al., 2010)
Analgesic	Aqueous (Aq)	Stem bark	Acetic acid and hot plate	Dose-dependent activities (Ojewole & Amabeouku, 2007)
	Ethanol (EtOH)	Leaves	Electric current as noxious stimulus	Extract exhibited activities (Shabana et al., 2011)

(Continued)

TABLE 16.5 (Continued)

Activity Tested	Extract	Part/ Compound	Model	Effect
Antibacterial	Aq and hexane (He)	Bark	Disk diffusion	Minimum inhibitory concentration (MIC) = 0.8–1.6 mg/ml against *Bacillus subtilis* (*Bs*) (McGaw et al., 2000)
	EtOH	Bark	Disk diffusion	MIC = 0.1 mg/ml against *Bs*, *Escherichia coli* (*Ec*) (1.6 mg/ml), *Klebsiella pneumoniae* (*Kp*) (3.1 mg/ml), and *Staphylococcus aureus* (*Sa*) (3.1 mg/ml) (McGaw et al., 2000)
	Aq	Bark	Microdilution	MIC = 0.39 mg/ml against *Bs*, *Ec* (1.56 mg/ml), *Kp* (1.56 mg/ml), and *Sa* (0.39 mg/ml) (Buwa & Van Staden, 2006, 2007)
	EtOH	Bark	Microdilution	MIC = 0.1 mg/ml against *Bs*, *Ec* (0.78 mg/ml), *Kp* (1.56 mg/ml), and *Sa* (0.2 mg/ml) (Buwa & Van Staden, 2006, 2007)
	Ethyl acetate (Ea)	Bark	Microdilution	MIC = 3.13 mg/ml against *Bs*, *Ec* (3.13 mg/ml), *Kp* (3.13 mg/ml), and *Sa* (3.13 mg/ml) (Buwa & Van Staden, 2006, 2007)
	DCM	Leaves and twigs	Microdilution	MIC = 3.13 mg/ml against *Bs*, *Sa* (1.56 mg/ml), *Ec* (3.13 mg/ml), and *Kp* (3.13 mg/ml) (Moyo et al., 2011)
	EtOH	Leaves	Microdilution	MIC = 0.78 mg/ml against *Bs*, *Sa* (0.39 mg/ml), *Ec* (1.56 mg/ml), and *Kp* (1.56 mg/ml) (Moyo et al., 2011)
	EtOH	Twigs	Microdilution	MIC = 0.2 mg/ml against *Bs*, *Sa* (0.2 mg/ml), *Ec* (0.78 mg/ml), and *Kp* (0.78 mg/ml) (Moyo et al., 2011)
	Petroleum ether (PE)	Leaves and twigs	Microdilution assay	MIC = 3.13 mg/ml against *Bs*, *Sa* (3.13 mg/ml), *Ec* (6.25 mg/ml), and *Kp* (6.25 mg/ml) (Moyo et al., 2011)
	EtOH	Leaves	Disk agar diffusion	Zone of inhibition = 12.0–16.0 mm against *Bs*, *Enterococcus faecalis* (*Ef*), *Sa*, *Ec*, *Pseudomonas aeruginosa* (*Ps*), and *Neisseria gonorrhoeae* (*Ng*) (Shabana et al., 2011)
	Aq	Bark	Microdilution	MIC = 1.0 mg/ml against *Brevibacillus agri* (*Ba*), *Propionibacterium acnes* (*Pa*) (0.5 mg/ml), *Ps* (1.0 mg/ml), *Sa* (1.0 mg/ml), and *Staphylococcus epidermidis* (*Se*) (1.0 mg/ml) (Mabona et al., 2013)
	DCM–MeOH (1:1)	Bark	Microdilution	MIC = 1.0 mg/ml against *Ba*, *Pa* (1.0 mg/ml), *Ps* (0.25 mg/ml), *Sa* (0.4 mg/ml), and *Se* (0.5 mg/ml) (Mabona et al., 2013)
	EtOH	Leaves	Microdilution	MIC = 125.0 μg/ml against *Pa* (Sharma & Lall, 2014)
	Aq	Bark	Agar diffusion	MIC = 1.0 mg/ml against *Se*, *Bacillus cereus* (*Bc*) (10.0 mg/ml), *Ef* (1.0 mg/ml), *Ec* (1.0 mg/ml), *Ps* (1.0 mg/ml), *Shigella flexneri* (*Sf*) (10.0 mg/ml), *Kp* (1.0 mg/ml), and *Serratia marcescens* (*Sm*) (1.0 mg/ml) (Kambizi, 2016)

(Continued)

TABLE 16.5 (Continued)

Activity Tested	Extract	Part/ Compound	Model	Effect
Antimycobacterial	Acetone (Ac)	Leaves	Microdilution	MIC = 0.11–0.21 mg/ml against *Mycobacterium tuberculosis* (Kabongo-Kayoka et al., 2016)
Antifungal	Aq, Ea, and EtOH	Bark	Microdilution	MIC = 0.78 mg/ml to 3.13 mg/ml against *Candida albida* (*Ca*) (Buwa & Van Staden, 2006)
	Ac, DCM, He, and MeOH	Leaves	Microdilution	MIC = 0.02–2.5 mg/ml against *Aspergillus niger* (*An*), *Aspergillus parasiticus* (*Ap*), *Colletotricum gloeosporioides* (Cg), *Penicillium janthinellum* (*Pj*), *Penicillium expansum* (*Pe*), *Trichoderma harzianum* (Th), and *Fusarium oxysporum* (*Fo*) (Mahlo et al., 2010, 2016)
	DCM, EtOH, and PE	Leaves	Microdilution	MIC = 3.13 mg/ml against *Ca* (Moyo et al., 2011)
	DCM, EtOH, and PE	Twigs	Microdilution	MIC = 3.13–4.69 mg/ml against *Ca* (Moyo et al., 2011)
	Aq and DCM–MeOH (1:1)	Bark	Microdilution	MIC = 0.25–8.0 mg/ml against *Ca*, *Mc*, and *Tm* (Mabona et al., 2013)
	EtOH	Leaves	Disk agar diffusion	Zone of inhibition of 12 mm against *Ca* (Shabana et al., 2011)
	Aq and organic	Leaves	Microdilution	MIC = 0.02–4.7 mg/ml against *Aspergillus flavus* (Af), *Aspergillus ochraceous* (Ao), *Aspergillus parasiticus* (Ap), Fo, *Fusarium graminearum* (Fg), and *Furasium verticillioides* (*Fv*) (Mongalo et al., 2018)
Anti-HIV	DCM and MeOH	Leaves	Non-radioactive HIV-RT colorimetric ELISA	Inhibition of 81% at a concentration of 500.0 µg/ml (Mkhize, 2015)
Anticonvulsant	Aq	Stem bark	Pentylenetetrazole (PTZ)– and picrotoxin (PCT)–induced seizures	Dose-dependent activities delayed onset of seizures and antagonized PTZ- and PCT-induced seizures (Ojewole & Amabeouku, 2007)
Antidiabetic	EtOH	Stem bark	α-amylase and α-glucosidase inhibition	Inhibited α-amylase and α-glucosidase activities and increased glucose uptake in hepatic tissues (Olofinsan et al., 2022a)
Anti-inflammatory	EtOH	Leaves	Cyclooxygenase (COX) assay	Demonstrated inhibition of 93% (Jäger et al., 1996)
	DCM	Leaves and twigs	COX-1 and COX-2 assay	COX-1 inhibition of 90.7% to 99.8% and 69.0% to 92.6% against COX-2 (Moyo et al., 2011)
	EtOH	Leaves	Carrageenan-induced paw edema	Extract exhibited activities of 80% (Shabana et al., 2011)
	Ac	Leaves	Inhibition of 15-lipoxygenase (15-LOX) enzyme	IC_{50} = 40.0 µg/ml (Adebayo et al., 2015)
	EtOH	Leaves	COX-2 assay	COX-2 inhibition of 77.5% and IC_{50} = 6.4 µg/ml (Twilley et al., 2017)

(Continued)

TABLE 16.5 (Continued)

Activity Tested	Extract	Part/ Compound	Model	Effect
Antioxidant	MeOH	Floral parts, leaves, and stems	2,2-diphenyl-1-picryhydrazyl (DPPH) assay	Half-maximal effective concentration (EC_{50}) values of 5.0–6.9 μg/ml (Moyo et al., 2010)
	MeOH	Leaves and stem bark	Ferric-reducing power (FRAP) assay	Dose-dependent activities exhibited (Moyo et al., 2010)
	MeOH	Leaves and stem bark	β-carotene/linoleic acid model	Antioxidant activity of 84.1–93.7% (Moyo et al., 2010)
	MeOH	Leaves	DPPH assay	IC_{50} = 8.4 μg/ml (Nawwar et al., 2011)
	MeOH	Leaves	Oxygen radical absorbance capacity (ORAC) assay	Demonstrated activities (Nawwar et al., 2011)
	–	(+)-catechin	DPPH and FRAP assays	Exhibited activities (Moodley et al., 2014)
	EtOH	Leaves	DPPH assay	EC_{50} = 2.6 μg/ml (Sharma & Lall, 2014)
	MeOH	Leaves	DPPH assay	EC_{50} = 1.5 μg/ml (Makhafola et al., 2016)
	EtOH	Leaves	DPPH assay	IC_{50} = 2.4 μg/ml (Twilley et al., 2017)
	EtOH	Leaves	Nitric oxide (NO) radical assay	IC_{50} = 248.0 μg/ml (Twilley et al., 2017)
	EtOH	Fruits and stem bark	DPPH assay	IC_{50} = 40.0–45.2 μg/ml (Olofinsan et al., 2022a, 2022b)
	EtOH	Fruits and stem bark	FRAP assay	Reduced ferric iron in dose-dependent manner (Olofinsan et al., 2022a, 2022b)
	EtOH	Stem bark	Hydroxyl (OH°) assay	IC_{50} = 32.3 μg/ml (Olofinsan et al., 2022a, 2022b)
	EtOH	Fruits	FRAP assay	IC_{50} = 43.3 μg/ml (Olofinsan et al., 2022a, 2022b)
	EtOH	Fruits	DPPH assay	EC_{50} = 52.57–458.20 μg/ml (Pfukwa et al., 2022a)
Antipyretic	EtOH	Leaves	Yeast-induced hyperthermia model	Extract exhibited activities (Shabana et al., 2011)
Antityrosinase	EtOH	Bark and leaves	Tyrosinase enzyme assay	Inhibition of 90.0–92.0% at 0.25 mg/ml (Lall et al., 2007)
	EtOH	Bark and leaves	Tyrosinase enzyme assay	IC_{50} = 40.0–51.0 μg/ml (Mapunya et al., 2012)
	EtOH	Gum	Tyrosinase inhibitory assay	IC_{50} = 11.32 μg/ml (Bodede et al., 2022)
	EtOH	Phytochemical compounds isolated from the gum	Tyrosinase inhibitory assay	Compounds (1R,3R)-1,3-dihydroxy-3-[heptadec-12'(Z)-enyl]cyclohexane and (1S,2S,3S,4S,5R)-1,2,3,4,5-pentahydroxy-5-[octadec-13'(Z)-enyl]cyclohexane exhibited IC_{50} = 24.9 μg/ml and 26.99 μg/ml, respectively (Bodede et al., 2022)
Cytotoxicity	EtOH	Leaves	2,3-Bis-(2-methoxy-4-nitro-5-sulfophenyl]-2Htetrazolium-5-carboxyanilide salt (XTT) assay	IC_{50} = 62.5 μg/ml (cervical epithelial carcinoma, HeLa), epidermoid carcinoma (A431) (68.3 μg/ml), human melanoma (A375) (106.0 μg/ml), and human embryonic kidney cells (HEK-293) (135.0 μg/ml) (Twilley et al., 2017)

(Continued)

TABLE 16.5 (Continued)

Activity Tested	Extract	Part/ Compound	Model	Effect
	EtOH	Leaves	Sulforhodamine B stain (SRB) assay	IC_{50} = 1.2 µg/ml (liver carcinoma, HEPG2), larynx carcinoma (HEP2) (1.3 µg/ml), and colon carcinoma (HCT116) cell lines (1.3 µg/ml) (Shabana et al., 2011)
	EtOH	Bark	XTT assay	Cell viability >80% at a concentration of 100.0 µg/ml against melanocyte cells (Mapunya et al., 2012)
	DCM–MeOH (1:1)	Leaves	MTT assay	Median lethal dose (LD_{50}) = values of 0.2 mg/ml and 0.3 mg/ml against Bovine dermis and Vero cell lines, respectively (Mongalo et al., 2018)
Hepatoprotective	EtOH	Leaves	Assessing aspartate aminotransferase (AST), alanine aminotransferase (ALT), and alkaline phosphatase (ALP) enzymes serum levels	Extracts exhibited activities by normalizing the high enzyme parameters in a dose-dependent manner by 100% for AST, 64% for ALT, and 50% for ALP (Shabana et al., 2011)
Hypoglycemic	Aq	Stem bark	Streptozotocin (STZ)–induced diabetes mellitus model	Extract exhibited dose-dependent activities (Ojewole, 2006)
Hypotensive	Aq	Stem bark	Streptozotocin (STZ)–induced diabetes mellitus model	Extract exhibited dose-dependent activities (Ojewole, 2006)

16.9 CONCLUSION AND PROSPECTS

Globally, the predominant source of food supply is cereal crops, and these crops demonstrate a high level of sensitivity to fluctuations in weather and climate conditions.

Similarly, this holds true for other agricultural crops that have served as the foundation of food systems for over 10,000 years. This shift occurred as communities transitioned from gathering and hunting to the cultivation of food (Hueston & McLeod, 2012). Southern Africa has a rich and expansive plant diversity, and consequently, various cultures from different geographical locations have developed unique techniques toward the implementation and/or integration of plants for nutritional and medicinal purposes (Chinsamy and Koitsiwe, 2016). Therefore, climate-resilient underutilized indigenous crops may be a good option to be considered as potential crops for dietary diversification and human needs and for posterity. The Industrial Revolution, which commenced in Europe in the 1800s, played an important role in improving farming practices as well as the invention of new tools and technologies (Hueston & McLeod, 2012). Fast-forward to the mid-1900s, the world population was approximately 2.5 billion. Industrialized countries were largely food-secure, but Africa, Asia, and Latin America continued to suffer from continued hunger emanating from food and nutrition insecurity (Pingali, 2012). The green revolution led to high-yielding varieties of major cereal crops, such as wheat (*Triticum aestivum*), maize (*Zea mays*), and rice (*Oryza sativa*), all of which were responsive to additional agricultural inputs and water, resulting in the start of an agro-industrial food regime (Pingali, 2012; Evenson & Gollin, 2003). The rise of the agro-industrial food system subsequently gave rise to new challenges, including malnutrition (Gómez et al., 2013), loss of plant biodiversity (Kerr, 2012; Dudley & Alexander, 2017), and environmental pollution and degradation (Pingali, 2012; Kerr, 2012). It is now evident that sub-Saharan Africa was left behind during the first revolution, but there has been a concerted effort to enable a "new" green revolution (Dawson et al., 2016; Blaustein, 2008). Modern food choices are exceptionally vulnerable to economic and climate change pressures, and wild foods of the Cape are potential buffers against these risks. This is particularly applicable in developing economies that are experiencing climate change and variability (Gordon et al., 2017). Drimie and Pereira (2016) reported that South Africa is one of the few African countries that has been rooted in the global agro-industrial food system for decades. Despite the country's association in this regard, its food prices remain too high for most of its people. About 21.3% of the population does not have access to adequate food (Statistics South Africa, 2017). The current concerns regarding biodiversity loss, climate change vulnerability, and environmental degradation have stimulated the need to rethink South Africa's current standing with its food system (Drimie & Pereira, 2016). During an irrigation water deficit experiment in Turkey, makataan rootstock was tested on commercial watermelons, which demonstrated that the citron watermelons could be used commercially as watermelon rootstocks, and citron watermelon rootstocks could be used as watermelon rootstocks in breeding programs, to save water, especially in arid and semi-arid regions (Yavuz et al., 2020). Furthermore, makataan acts as a key crop for good fodder management, as each part of the plant is used. The seeds may be harvested and dried to either reseed for plantings or may be used as a snack. The preservatives made from melon serve as a perfect condiment on a cheese board or simply to enjoy the sweet chunks as is. The fresh and dried leaves are well used as a compliment side dish. Wildepruim is an evergreen tree, making for a perfect fast-growing shade tree in the garden, but it can also serve as an ideal alley tree, with several planted along a road. The fruit does attract birds, making it a messy affair when the fruits are ripe. The thick-skinned fruit contains a large seed and sweet and tangy mash. When a wildepruim tree thrives, it can produce hundreds of fruits. Harvesting and fermenting the entire fruit in a vessel are an excellent method to extract the juice or create an alcoholic beverage.

The utilization of plant biodiversity for ethnobotanical remedies often involves the wild harvesting of native species. However, for the extensive production of phytomedicines, cultivation presents numerous advantages as an ex situ conservation strategy, helping to mitigate biodiversity losses (Chen et al., 2016). For industrial production, environmental and biogeographical regional restrictions can be a limiting factor for the collection of plants from natural habitats. The domestication of wild food plants using conventional, in-field cultivation may be time-consuming, where agronomic requirements are unknown to generate high crop yields. Some species may be slow-growing, making it difficult and time-consuming to access crops of interest. To meet commercial demands for natural products derived from plants, *in vitro* and hydroponic cultivation systems for wild edible plants thus offer some advantages in producing yield with shorter plant life cycles. Unless agro-processing interventions are executed to increase the resilience of food systems, the impact will be more severe in the face of the next adverse events. Wild foods – in particular, the ones discussed in this chapter – are the ideal alternatives to narrow down the nutrition security gap and increase food system resilience. Commercial opportunities exist to adopt underutilized wild foods into the food systems in South Africa and Africa at large. Despite technological advances on the continent and beyond, the global food system is failing to meet the basic food requirements of the world's citizens equitably. The information presented here proposes and necessitates the need to identify the status of some Cape wild foods in the broader food system, an the limitations, and to identify opportunities that exist in already-functioning agro-systems to incorporate indigenous wild foods for environmental sustainability. Improved health outcomes, resilient agricultural systems, and agro-ecological biodiversity will follow the latter. A call for reinvigorating underutilized traditional and indigenous crops has been suggested by bringing them to the market and, in so doing, improving diets and making them more sustainable (Drimie & Pereira, 2016; Mbhenyane, 2017; Pereira, 2014). A better understanding of this process could allow researchers to develop cultivation strategies to optimize/maximize crop yield and meet the ever-rising global food demands.

16.10 ACKNOWLEDGMENTS

The authors are grateful to the Govan Mbeki Research and Development Centre (GMRDC) of the University of Fort Hare for funding this work. The authors further extend their gratitude to Mr. Andre Brand, a makataan farmer, for the photographs.

REFERENCES

Abiola, O. A. (2020). Vegetable production and nematodes infestation: Impacts on small-scale farming communities of South Africa. *World Journal of Advanced Research and Reviews*, *7*(2), 168–177.

Adebayo, S. A., Dzoyem, J. P., Shai, L. J., & Eloff, J. N. (2015). The anti-inflammatory and antioxidant activity of 25 plant species used traditionally to treat pain in Southern African. *BMC Complementary and Alternative Medicine*, *15*, 159.

Blaustein, R. J. (2008). The green revolution arrives in Africa. *AIBS Bulletin*, *58*, 8–14.

Bodede, O., More, G. K., Moodley, R., Steenkamp, P., Baijnath, H., Maharaj, V., & Prinsloo, G. (2022). New Alk(en)ylhydroxycyclohexanes with Tyrosinase inhibition potential from Harpephyllum caffrum Bernh. *Gum Exudate, Molecules*, *27*, 3839.

Bultosa, G., Molapisi, M., Tselaesele, N., Kobue-Lekalake, R., Desse Haki, G., Makhabu, S., SekwatiMonang, B., Seifu, E. P., & Nthoiwa, G. (2020). Plant-based traditional foods and beverages of Ramotswa Village, Botswana. *Journal of Ethnic Foods*, *7*. https://doi.org/10.1186/s42779-019-0041-3

Buwa, L. V., & Van Staden, J. (2006). Antibacterial and antifungal activity of traditional medicinal plants used against venereal diseases in South Africa. *Journal of Ethnopharmacology*, *103*, 139–142.

Buwa, L. V., & Van Staden, J. (2007). Effects of collection time on the antimicrobial activities of *Harpephyllum caffrum* bark. *South African Journal of Botany*, *73*, 242–247.

Chen, S. L., Yu, H., Luo, H. M., Wu, Q., Li, C. F., & Steinmetz, A. (2016). Conservation and sustainable use of medicinal plants: Problems, progress, and prospects. *Chinese Medicine*, *30*, 11–37.

Chinsamy, M., & Koitsiwe, M. (2016). Indigenous knowledge-based vegetable garden projects for poverty alleviation in South Africa. *Journal of Social Sciences*, *46*, 137–145.

Dawson, N., Martin, A., & Sikor, T. (2016). Green revolution in Sub-Saharan Africa: Implications of imposed innovation for the wellbeing of rural smallholders. *World Development*, *78*, 204–218.

Department of Agriculture, Forestry and Fisheries. (2013). *Citrullus lanatus (Thunberg) production guidelines Bitter watermelon*. Retrieved January 9, 2024. from https://old.dalrrd.gov.za/Portals/0/Brochures%20and%20Production%20guidelines/Production%20Guidelines%20Bitter%20Watermelon.pdf

Drimie, S., & Pereira, L. (2016). Advances in food security and sustainability in South Africa. In D. Barling (Ed.), *Advances in food security and sustainability* (pp. 1–31). Academic Press.

Dudley, N., & Alexander, S. (2017). Agriculture and biodiversity: A review. *Biodiversity*, *18*, 45–49.

El-Sherbeiny, A., & El-Ansari, M. (1976). The polyphenolics and flavonoids of *Harpephyllum caffrum*. *Planta Medica*, *29*, 129–132.

Evenson, R. E., & Gollin, D. (2003). Assessing the impact of the Green Revolution, 1960 to 2000. *Science*, *300*, 758–762.

Gómez, M. I., Barrett, C. B., Raney, T., Pinstrup-Andersen, P., Meerman, J., Croppenstedt, A., Carisma, B., & Thompson, B. (2013). Post-green revolution food systems and the triple burden of malnutrition. *Food Policy*, *42*, 129–138.

Gordon, L. J., Bignet, V., Crona, B., Henriksson, P. J. G., & Van Holt, T. (2017). Rewiring food systems to enhance human health and biosphere stewardship. *Environmental Research Letters*, *12*.

Guo, C. J., Yang, J. J., Wei, J. Y., Li, Y. F., Xu, J., & Jiang, Y. G. (2003). Antioxidant activities of peel, pulp and seed fractions of common fruits as determined by FRAP assay. *Nutrition Research*, *23*, 1719–1726.

Hueston, W., & McLeod, A. (2012). Overview of the global food system: Changes over time/space and lessons for future food safety. In *Improving food safety through a one health approach: Workshop summary* (p. 189). National Academies Press.

Jäger, A. K., Hutchings, A., & Van Staden, J. (1996). Screening of Zulu medicinal plants for prostaglandin-synthesis inhibitors. *Journal of Ethnopharmacology*, *52*, 95–100.

Kabongo-Kayoka, P. N., Eloff, J. N., Obi, C. L., & McGaw, L. J. (2016). Antimycobacterial activity and low cytotoxicity of leaf extracts of some African Anacardiaceae tree species. *Phytotherapy Research*, *30*, 2001–2011.

Kambizi, L. (2016). Indigenous plants for ethnoveterinary uses in the Pondoland, South Africa. *Acta Horticulturae*, *1125*, 309–314.

Kerr, R. B. (2012). Lessons from the old Green Revolution for the new: Social, environmental and nutritional issues for agricultural change in Africa. *Progress in Development Studies*, *12*, 213–229.

Lall, N., Mapunya, M. B., Nikolova, R. V., & Houghton, P. J. (2007). Anti-tyrosinase activity of South African plant extracts. *Planta Medica*, *73*, 464.

Mabona, U., Viljoen, A., Shikanga, E., Marston, A., & Van Vuuren, S. (2013). Antimicrobial activity of Southern African medicinal plants with dermatological relevance: From an ethnopharmacological screening approach to combination studies and the isolation of a bioactive compound. *Journal of Ethnopharmacology*, *148*, 45–55.

Mahlo, S. M., Chauke, H. R., McGaw, L., & Eloff, J. (2016). Antioxidant and antifungal activity of selected medicinal plant extracts against phytopathogenic fungi. *African Journal of Traditional, Complementary and Alternative Medicine*, *13*, 216–222.

Mahlo, S. M., McGaw, L. J., & Eloff, J. N. (2010). Antifungal activity of leaf extracts from South African trees against plant pathogens. *Crop Protection*, *29*, 1529–1533.

Makhafola, T. J., Elgorashi, E. E., McGaw, L. J., Verschaeve, L., & Elof, J. N. (2016). The correlation between antimutagenic activity and total phenolic content of extracts of 31 plant species with high antioxidant activity. *BMC Complementary Medicine and Therapies*, *16*, 490.

Maldonado-Celis, M. E., Yahia, E. M., Bedoya, R., Landázuri, P., Loango, N., Aguillón, J., Restrepo, B., Guerrero Ospina, J. C. (2019). Chemical composition of mango (*Mangifera indica* L.) fruit: Nutritional and phytochemical compounds. *Frontiers in Plant Science*, *10*, 1073.

Mandizvo, T., Odindo, A. O., & Mashilo, J. (2022). Nutrient composition and physical properties of citron watermelon (*Citrullus lanatus* var. *citroides* (L.H. Bailey) Mansf. ex Greb.) seeds are related to seed coat visual appearance. *South African Journal of Botany*, *145*, 405–419.

Mapunya, M. B., Nikolova, R. V., & Lall, N. (2012). Melanogenesis and antityrosinase activity of selected South African plants. *Evidence-Based Complementary and Alternative Medicine*, *2012*, Article 374017.

Maroyi, A. (2019). Medicinal uses, biological and chemical properties of wild plum (*Harpephyllum caffrum*): An indigenous fruit plant of Southern Africa. *Journal of Pharmacy and Nutrition Sciences*, *9*, 258–268.

Mashilo, J., Shimelis, H., & Ngwepe, R. M. (2021). Genetic resources of bottle gourd (*Lagenaria siceraria* (Molina) Standl.] and citron watermelon (*Citrullus lanatus* var. *citroides* (L.H. Bailey) Mansf. ex Greb.): Implications for genetic improvement, product development and commercialization: A review. *South African Journal of Botany*, *145*, 28–47.

Masoko, P., Matotoka, M. M., Maboko, S., & Mphosi, M. S. (2022). Phytochemical analysis and antibacterial activity of *Citrullus lanatus* var. *citroides* (Citron watermelon) fruit and the effect of temperature on the biological activity of the rind. *South African Journal of Botany*, *150*, 1111–1121.

Mbhenyane, X. G. (2017). Indigenous foods and their contribution to nutrient requirements. *South African Journal of Clinical Nutrition*, *30*, 5–7.

McGaw, L. J., Jäger, A. K., & Van Staden, J. (2000). Antibacterial, antihelmentic and anti-amoebic activity of South African medicinal plants. *Journal of Ethnopharmacology*, *72*, 247–263.

Mkhize, Z. (2015). *Structure and synthesis of bioactive natural products* [PhD thesis, University of KwaZulu-Natal].

Mongalo, N. I., Dikhoba, P. M., Soyingbe, S. O., & Makhafola, T. J. (2018). Antifungal, anti-oxidant activity and cytotoxicity of South African medicinal plants against mycotoxigenic fungi. *Heliyon*, *4*, e00973.

Moodley, R., Koorbanally, N. A., Islam, S., & Jonnalagadda, S. B. (2014). Structure and antioxidant activity of phenolic compounds isolated from the edible fruits and stem bark of *Harpephyllum caffrum*. *Journal of Environmental Science and Health*, *49*, 938–944.

Moodley, R., Koorbanally, N. A., & Jonnalagadda, S. B. (2013). Elemental composition and nutritional value of the edible fruits of *Harpephyllum caffrum* and impact of soil quality on their chemical characteristics. *Journal of Environmental Science and Health, 48*(7), 539–547.

Moyo, M., Finnie, J. F., & Van Staden, J. (2011). Antimicrobial and cyclooxygenase enzyme inhibitory activities of *Sclerocarya birrea* and *Harpephyllum caffrum* (Anacardiaceae) plant extracts. *South African Journal of Botany, 77*, 592–597.

Moyo, M., Ndhlala, A. R., Finnie, J. F., & Van Staden, J. (2010). Phenolic composition, antioxidant and acetylcholinesterase inhibitory activities of *Sclerocarya birrea* and *Harpephyllum caffrum* (Anacardiaceae) extracts. *Food Chemistry, 123*, 69–76.

Nawwar, M., Hussein, S., Ayoub, N., Hashim, A., El-Sharawy, R., Lindequist, U., Harms, M., & Wende, K. (2011). Constitutive phenolics of *Harpephyllum caffrum* (Anacardiaceae) and their biological effects on human keratinocytes. *Fitoterapia, 82*, 1265–1271.

Nkoana, D. K., Mashilo, J., Shimelis, H., & Ngwepe, R. M. (2021). Nutritional, phytochemical compositions and natural therapeutic values of citron watermelon (*Citrullus lanatus* var. *citroides*): A review. *South African Journal of Botany, 145*, 65–77.

Ojewole, J. A. O. (2006). Hypoglycaemic and hypotensive effects of *Harpephyllum caffrum* Bernh ex CF Krauss (Anacardiaceae) stem-bark aqueous extract in rats. *The Cardiovascular Journal of Africa, 17*, 67–72.

Ojewole, J. A. O., & Amabeouku, G. J. (2007). Anticonvulsant and analgesic effects of *Harpephyllum caffrum* Bernh ex CF Krauss (Anacardiaceae) stem-bark aqueous extract in mice. *International Journal of Pharmacology, 3*, 241–247.

Olofinsan, K. A., Erukainure, O. L., Brian, B. K., & Islam, M. S. (2022a). *Harpephyllum caffrum* stimulates glucose uptake, abates redox imbalance and modulates purinergic and glucogenic enzyme activities in oxidative hepatic injury. *Asian Pacific Journal of Tropical Biomedicine, 12*(1), 9–19.

Olofinsan, K. A., Salau, V. F., Erukainure, O. L., & Islam, M. S. (2022b). *Harpephyllum caffrum* fruit (wild plum) facilitates glucose uptake and modulates metabolic activities linked to neurodegeneration in isolated rat brain: An in vitro and in silico approach. *Journal of Food Biochemistry, 46*, e14177.

Paliyath, G., Bakovic, M., Shetty, K. (2011). *Functional foods, nutraceuticals, and degenerative disease prevention.* John Wiley & Sons.

Pandey, K. B., & Rizvi, S. I. (2009). Plant polyphenols as dietary antioxidants in human health and disease. *Oxidative Medicine and Cellular Longevity, 2*, 270–278.

Pasandide, B., Khodaiyan, F., Mousavi, Z., & Hosseini, S. S. (2018). Pectin extraction from citron peel: Optimization by Box-Behnken response surface design. *Food Science and Biotechnology, 27*, 997–1005.

Pereira, P. C. (2014). Milk nutritional composition and its role in human health, *Nutrition, 30*, 619–627.

Pérez, J., Karina Gómez, K., & Vega, L. (2022). Optimization and preliminary physicochemical characterization of pectin extraction from Watermelon Rind (*Citrullus lanatus*) with citric acid. *International Journal of Food Science, 2022*, 10.

Pfukwa, T. M., Chikwanha, O. C., Marais, J., Fawole, O. A., Manley, M., & Mapiye, C. (2023). Oxidative preservative and sensory effects of *Harpephyllum caffrum* (wild plum) peel extracts in fresh ground beef patties. *Food Bioscience, 53*, 102684.

Pfukwa, T. M., Fawole, O. A., Manley, M., & Mapiye, C. (2022b). Phenolic profiling and antioxidant evaluation of extracts from Southern African indigenous fruits byproducts. *Food Research International, 157*, 111388.

Pfukwa, T. M., Pheiffer, W., Fawole, O. A., Manley, M., & Mapiye, C. (2022a). Cellular antioxidant and viability efficacy of *Harpephyllum caffrum* peel and *Syzygium guineense* seed extracts. *Food Bioscience, 49*, 101934.

Pingali, P. L. (2012). Green revolution: Impacts, limits, and the path ahead. *Proceedings of the National Academy of Sciences of the United States of America, 109*, 12302–12308.

Rajasree, R. S., Sibi, P. I., Femi, F., & Helen, W. (2016). Phytochemicals of Cucurbitaceae family: A review. *International Journal of Pharmacognosy and Phytochemical Research, 8*(1), 113–123.

Schippers, R. R. (Ed.). (2000). *African indigenous vegetables. An overview of the cultivated species.* CAB International.

Shabana, M. M., El Sayed, A. M., Yousif, M. F., El Sayed, A. M., & Sleem, A. A. (2011). Bioactive constituents from *Harpephyllum caffrum* Bernh. and *Rhus coriaria* L. *Pharmacognosy Magazine, 7*(28), 298–306.

Sharma, R., & Lall, N. (2014). Antibacterial, antioxidant activities and cytotoxicity of plants against *Propionibacterium acnes. South African Journal of Science, 110*(11/12), Article #2013–0293.

Statistics South Africa. (2018). *General household survey 2017.* Author.

Sudha, A., & Lakshmanan, P. (2007). *Solanum nigrum*, a new host for powdery mildew disease of Capsicum annuum in the Madurai district of Tamil Nadu, India. *Australasian Plant Disease Notes*, *2*, 97–98.

Twilley, D., Langhansová, L., Palaniswamy, D., & Lall, N. (2017). Evaluation of traditionally used medicinal plants for anticancer, antioxidant, anti-inflammatory and anti-viral (HPV-1) activity. *South African Journal of Botany*, *112*, 494–500.

UNESCO. *World heritage convention*. Retrieved March 23, 2024, from https://whc.unesco.org/en/list/1007/

Vijayalakshmi, K. (2014). *Pharmacognostic, phytochemical and pharmacological evaluation of the leaves of Citrullus Lanatus (Thunb.) Matsum. & Nakai,(Cucurbitaceae)* [Doctoral dissertation, College of Pharmacy, Madurai Medical College].

Wehmeyer, A. S. (1986). *Edible wild plants of Southern Africa: Data on the nutrient contents of over 300 species*. CSIR.

Yavuza, D., Seymenb, M., Süheria, S., Yavuza, N., Türkmenb, Ö., & Kurtarb, E. S. (2020). How do rootstocks of citron watermelon (*Citrullus lanatus* var. *citroides*) affect the yield and quality of watermelon under deficit irrigation? *Agricultural Water Management*, *241*, 1063.

Zhengkang, H., Wang, G., Yao, W., & Zhu, W.-Y. (2006). Isoflavonic phytoestrogens-new prebiotics for farm animals: A review on research in China. *Current Issues in Intestinal Microbiology*, *7*, 53–60.

17 Potential for the Exploitation of Nutritional Traits of Sweet Sorghum (*Sorghum bicolor* (L.) Moench) in Food Systems

[1,6]Olivia Mukondwa, [2]Hapson Mushoriwa, [3]MacDonald Bright Jumbo, [4]Rogerio Marcos Chiulele, [5]Esther Kathini Muema, and [6]Eastonce Tendayi Gwata*

[1]Crop Breeding Institute, Sorghum and Millets Research, Matopos Research Station, P. Bag K5137, Bulawayo, Zimbabwe
[2]International Institute of Tropical Agriculture, PMB 5320, Oyo Road, Ibadan 200001, Oyo State, Nigeria
[3]International Crops Research Institute for Semi-Arid Tropics – Mali, BP 320 Bamako, Mali
[4]Centre of Excellence in Agri-Food Systems and Nutrition, 5th Floor, Rectory Building, 25th June Square, Maputo 1100, Mozambique
[5]Stellenbosch University, Faculty of Agri-Sciences, Department of Soil Science, Matieland, 7602, South Africa
[6]University of Venda, Faculty of Science, Engineering and Agriculture, Department of Plant and Soil Sciences, P. Bag X5050, Thohoyandou 0950, South Africa
*Corresponding author: ectgwata@gmail.com

17.1 INTRODUCTION

Sweet sorghum (*Sorghum bicolor* (L.) Moench) belongs to the grass family Poaceae, subfamily Panicoideae, tribe Andropogoneae, subtribe Sorghinae, subsp. *bicolor*, and genus *Sorghum* (Hariprasanna & Patil, 2015). Sweet sorghum is native to Africa, where several closely related wild species occur (Wu et al., 2022). There are four primary types of sorghum, namely, grain sorghum, sweet sorghum, forage sorghum, and biomass sorghum (Lekgari & Dweikat, 2014). Within each of the groups, genetic variants characterized by varying morphological traits and applications occur (Mullet et al., 2014). Sweet sorghum is a natural variant of the common grain sorghum with high content of sugars in the stem (Chen et al., 2019; Appiah-Nkansah et al., 2019). However, currently, all sweet sorghums are classified within *S. bicolor* ssp. *Bicolor*, and the term *S. saccharatum* is not recognized as an accepted synonym for sorghum. Studies involving molecular markers report that clusters developed by agronomic descriptors are unable to approximate the groupings produced by RAPD markers (Dahlberg et al., 2002). Using AFLP markers, the clusters obtained do not group clearly based on racial classification, suggesting that the sweet sorghums are interspersed amongst grain sorghum lines of different racial classifications, and they are of polyphyletic origin (Ritter et al., 2007). Moreover, the phytochemical compositions of the grain in both sorghums are similar.

 DOI: 10.1201/9781003469766-17

The term *sweet sorghum* refers to the high-content juicy stalks which accumulate 10–25% sugar near the time of grain maturity. Fructose and glucose are the predominant reducing sugars in the juice, and sucrose is the predominant disaccharide. The sweet sorghum varieties can be used to produce sorghum syrup or sugar (Hunter & Anderson, 1997), although many are used for forage. Sweet sorghums generally have lower grain yields and are often taller than modern grain sorghums. In many countries, grain sorghum is sold as F_1 hybrids as part of a commercial seed market. The grain color in sweet sorghum is based mainly on the pericarp and testa color (Choi et al., 2019). The color ranges include white to red types (Figure 17.1). In grain sorghums, the genotypes with a black pericarp show higher levels of flavan-4-ols and anthocyanins than the other varieties (Dykes et al., 2005). In a separate study, the antioxidant activity and the content of total phenolic compounds and anthocyanins were correlated with grain color in several grain sorghum types (Choi et al., 2019). This indicated that, likely, genes for plant color, pericarp thickness, and pigmented testa increase phenol and antioxidant activity.

Currently, there are no records of commercial production of sweet sorghum in most African countries, including Zimbabwe, where smallholder farmers occasionally cultivate the crop in fields alongside other varieties of sorghum or intercropped with maize (*Zea mays*), peanuts (*Arachis hypogaea*), and Bambara groundnut (*Vigna subterranea*) (Chiambiro et al., 2022). The crop can be cultivated in poor soils in tropical, subtropical, as well as temperate regions and requires low rainfall (300–380 mm) during the growing season (da Silva et al., 2020). Therefore, sweet sorghum fits well in smallholder dryland cropping systems in Southern Africa with limited access to crop production inputs, including chemical fertilizers and irrigation facilities (Nziguheba et al., 2016). The crop is highly tolerant to terminal drought stress partly due to its ability to maintain the stay-green trait and accumulation of high-hexose sugars at the expense of total soluble sugars and sucrose mainly at the median internode as an osmoregulation mechanism to cope with moisture stress (Tovignan et al., 2020). However, the crop responds favorably to increased soil moisture.

Currently, there are efforts by plant breeders to improve this small-grain cereal for food, feed, fuel, and fiber (Dube & Maphosa, 2022). In general, the target traits for genetic improvement include grain and stem quality, pest, and disease resistance, as well as grain yield performance and stability (Albahri et al., 2023). Therefore, this chapter explores the potential for exploiting the nutritional traits of sweet sorghum in food systems, particularly in semi-arid regions, such as Southern Africa. Inevitably, some of the examples in agronomic practices for the crop are borrowed from countries in the region, such as Zimbabwe, which is typically characterized by semi-arid conditions.

FIGURE 17.1 The creamy-white (left) and red (right) sweet sorghum grains.

17.2 IMPORTANT USES OF SWEET SORGHUM

Currently, sweet sorghum is primarily used for food and feed as well as organic fertilizer (Nasidi et al., 2019). In some countries, the stem juice is processed into syrup as well as ethanol. Sweet sorghum grain can be used also in the brewing industry (Mesta et al., 2018). Both the grain and the juice contain useful nutrients, including minerals and vitamins (Holou & Stevens, 2012; Mukabane et al., 2014).

17.2.1 Source of Food

Sweet sorghum grain can be utilized for human consumption and livestock feed. It can be used in the preparation of various food products, such as fermented or unfermented bread, snacks, cakes, and malted alcoholic beverages in many African and Asian countries (Adedugba et al., 2023). Sweet sorghum grain has been used successfully as cattle feed (Fernandes et al., 2014). The syrup from sweet sorghum, also known as sweet molasses, is widely used in the southern regions of America (Motsi et al., 2022a). Instead of using sugarcane, molasses is, in this case, manufactured by processing sweet sorghum cane. In the southern parts of the United States, the syrup is often consumed as a condiment alongside biscuits, hot cereals (grits), cornmeal mush, and pancakes (Dube & Maphosa, 2022; Naoura et al., 2020). In addition, the syrup can be used in cooking, and it has a similar sweetening effect to blackstrap molasses (Lopez et al., 2017). In some Asian countries, sweet sorghum syrup is advertised as a healthy snack (Andrzejewski et al., 2013). This is partly because most of the minerals and sugars in the syrup are retained when the juice is purified of contaminants before being concentrated by an open-pan evaporation process (Umakanth et al., 2018; Andrzejewski et al., 2013). The deep roots of sweet sorghum can access beneficial trace elements (Dalla Marta et al., 2014). Furthermore, the syrup is often used for sweetening gingerbread, popcorn balls, or sweet potatoes with apple slices (Dube & Maphosa, 2022; Andrzejewski et al., 2013).

In some sub-Saharan countries, sweet sorgum can also be chewed as a snack (Legwaila et al., 2004; Naoura et al., 2020). Frequently, markets for sweet sorghum chews are available, particularly in urban areas (Pfeiffer et al., 2010). In Zimbabwe, for instance, surplus fresh sticks of sweet sorghum are often sun-dried for consumption during the off-season period. On the other hand, sweet sorghum crop residue can be utilized for livestock feeds, particularly cattle (Fernandes et al., 2014). The bagasse from sweet sorghum is useful in ruminant nutrition partly due to the favorable fiber composition (Cifuentes et al., 2014). Sweet sorghum is also recognized as a source of renewable and sustainable biomass feedstock that can be annually produced.

17.2.2 Source of Biofuel

Fermented sweet sorghum sugars are used in the production of ethanol for utilization as automotive fuel in several countries, including India and the USA (Prasad et al., 2007; Ribeiro et al., 2020). The production process involves squeezing the juice from the stalks and fermenting into ethanol (Nazli, 2020; Chohnan et al., 2011). The biomass is converted into a variety of biofuels, such as bioethanol, biodiesel, biohydrogen, biogas, bio-oil, biochar, and syngas (Umakanth et al., 2018; Olukoya et al., 2015). In contrast, ethanol production from sweet sorghum is uncommon; however, its potential is significant. The crop offers many advantages over sugarcane because of its tolerance to drought, water logging, salinity, and alkalinity and grows rapidly (four-month cycle on average) (Table 17.1). The ethanol yield from sweet sorghum has been estimated at about 8,000 L/ha (Woods, 2001). In addition, it can be used in various processes for biofuel production, including lignocellulosic-to-biogas, starch-to-ethanol, as well as sugar-to-ethanol (Dalla Marta et al., 2014; Fernandes et al., 2014).

TABLE 17.1
Agronomic Advantages of Sweet Sorghum over Sugarcane

Agronomic Feature	Sugarcane	Sweet Sorghum
Crop duration	About 12–13 months	About 3 1/2 months
Growing season	Only one season	All season and summer
Soil requirements	Grows well in loamy soil	Most types of drained soil
Water management	Requires water throughout the year	Less water requirement; can be grown as rain-fed crop
Crop management	Requires good management	Little fertilizer required; less pests and disease complex; easy management
Yield per acre	25–30 t	20–25 t
Sugar content on weight basis	8–12%	8–10%
Sugar yield	2.5–4.8 t/ac	2–3 t/ac
Ethanol production directly from juice	1,700–2,700 L/ac	1,140–1,640 L/ac
Harvesting	Difficult and laborious	Easy, both manual and with the use of simple small mechanical machine

Source: Adapted from Prasad et al. (2007).

17.2.3 Drought Tolerance and Ratoon Ability

Sweet sorghum is highly tolerant to drought, thus making it a good candidate for securing household food and nutrition security, particularly in marginalized rural communities in sub-Saharan Africa (SSA) that are prone to frequent droughts (Anami et al., 2015b). This is partly because of its morphophysiological traits, which confer drought tolerance (Zegada-Lizarazu & Monti, 2013), and its C_4 photosynthetic system, which enables efficient CO_2 fixation and exceptional dry matter accumulation (Anami et al., 2015a). The C_4 photosynthetic pathway involves two carboxylation steps instead of only one, as in C_3 plants. Initially, CO_2 is fixed into C_4 acids by the phosphoenolpyruvate carboxylase in mesophyll cells, followed by their transportation into the bundle sheath cells, where they are decarboxylated and the CO_2 is refixed by RuBisCo (which suppresses photorespiration). This process is Kranz anatomy of C_4 leaves, which separates the two carboxylation steps (Leegood, 1997). The combination of these anatomical features and biochemistry produces a 3- to 20-fold increase in the CO_2 concentration in the bundle sheath cells, relative to that in the surrounding air, resulting in improved carboxylation efficiency (Jenkins, 1997; Laisk & Edwards, 1998).

Sweet sorghum is able to endure or tolerate extended dry spells or severe drought and resume growth when soil moisture becomes available (Kumar et al., 2011). Partly because of its ability to exhibit morphological changes, such as a deep and dense root system, reduced transpiration facilitated by leaf rolling and stomatal closure, as well as a reduction in metabolic activity to the point of near dormancy in response to terminal soil moisture stress (Devnarain et al., 2016). Both drought and heat stress adversely affect various physiological, genetic, and molecular changes that adversely affect sorghum growth, development, yield, and yield components (Chiambiro et al., 2022). Drought stress reduces tissue water content and the green leaf area as well as carbon assimilation, leading to reduced membrane stability, hence diminished growth, biomass, and yield (Hossain et al., 2022; Mathur et al., 2020). The key traits associated with drought tolerance include maintaining a green color, reduced canopy temperature, limited transpiration, higher reproductive success and root structure (Ngara et al., 2021).

Stress disrupts dry matter partitioning (Boyer & Westgate, 2004; Prasad et al., 2011), grain filling (Sehgal et al., 2017), and grain quality (Gooding et al., 2003; Britz et al., 2007). High-temperature stress also reduces membrane stability and carbon assimilation, floret fertility, seed number, size, and yield (Prasad et al., 2008; Shah & Paulsen, 2003; Sharkey, 2005). The main traits associated with heat stress include improved membrane stability, improved gamete viability, and reproductive success to produce the seed (Machado & Paulsen, 2001; Rajendra Prasad et al., 2021).

Sweet sorghum can be ratooned to enable multiple harvests per season (Lestari et al., 2022). *Ratooning* is a technique in which a single planting is harvested twice or more times during the growth season (Tsuchihashi & Goto, 2008). Genetic and environmental factors, both in the air and the soil, are critical since they determine cultivar productivity (Dahlberg, 2019). Different cultivars of sweet sorghum vary in their suitability for ratoon cropping (Paesal et al., 2021). In dry regions, ratoon cropping can be achieved by early planting on wetlands so that, after harvesting the first crop, the second crop (from the ratooned crop) can make use of residual moisture (Srinivasa Rao et al., 2013).

17.3 NUTRITIONAL VALUE AND BIOACTIVE COMPOUNDS

The nutritional value of sweet sorghum stems from both the vegetative and reproductive parts of the plant (Widowati & Luna, 2022). The nutritional composition compares well with that of other cereals (Table 17.2). The stalks accumulate a nutritious juice which is used for several products, including sugar for ethanol production (Dalla Marta et al., 2014; Fernandes et al., 2014) and syrup (Mukabane et al., 2014). The sweet sorghum biomass is suitable for livestock feed (Nasidi et al., 2019; Bonin et al., 2016).

The genotypic similarity between grain and sweet sorghum implies that some of their phenotypic and nutritional characteristics also exhibit considerable similarities. Both sorghums possess some common bioactive compounds, including phenolic acids, and flavonoids. The polyphenols act as metal chelators of transition metals, such as Fe^{2+}, thus preventing oxidation due to reactive hydroxyl radicals (Tsao, 2010). In general, the bioactive compounds vary in content, depending on the genotype and the plant part. For instance, the quantities of apigeninidin and luteolinidin were approximately 10- and 100-fold more in the seed than in the stalk dermal layer of sweet sorghum, respectively (Vanamala et al., 2018). Both apigeninidin and luteolinidin are not common in higher plants but unique to sorghum, where they occur in the glume, grain, leaf, and leaf sheath (Bröhan et al., 2011). The phenolic acids in sorghum include gentisic acid, cinnamic acid, salicylic acid, protocatechuic acid, caffeic acid, sinapic acid, ferulic acid, and p-coumaric acid, as well as the flavonoids eriodictyol and naringenin (Mawouma et al., 2022). This genotypic variation in the bioactive compounds presents opportunities for sorghum breeders to exploit during the genetic improvement of the crop for the aforementioned specific traits.

The small-molecule bioactive compounds which occur in the sorghums also have the potential to confer protective and preventive effects against chronic diseases, such as colon cancer, for instance (Amin et al., 2009). In addition, polyamines, which occur widely in sorghums, can play roles in the

TABLE 17.2
Nutritional Composition of Sweet Sorghum and Other Cereal Grain Crops

Crop	Moisture %	Protein %	Fat %	Carbohydrates	Ash
Milled rice	11.5	8.4	1.7	77.1	1.3
Yellow	11.4	9.4	7.3	69.1	2.4
Sorghum	11	11	3.3	73	1.7
Millet	11.9	9.7	3.5	73.4	1.5
Wheat	11.8	9	1	77.2	1
Sweet sorghum	56.23	8.36	3.21	27.13	1.6

Source: Adapted from Widowati and Luna (2022).

prevention of lipid peroxidation of membranes, regulation of inflammation, food allergy prevention, and antiglycation of flavonoids and phenolic acids (Rani et al., 2024). The predominant amines in sorghums include spermine, spermidine, putrescine, and cadaverine (Paiva et al., 2015).

The grain of sweet sorghum also has the potential to be used in food fortification efforts since it has a better nutritional and mineral profile and is also rich in dietary fiber (Adedugba et al., 2023). Sweet sorghum's superior nutritional value and better flavor make it appropriate for use in the creation of value-added food products (Appiah-Nkansah et al., 2019). In comparison to maize, sweet sorghum yields 23% more fermentable carbohydrates, requires 37% less nitrogen fertilizer, 17% less irrigation water, and may produce more ethanol in a dry year (Vlachos et al., 2020). Recent findings on sweet sorghum harvested in Asia and Africa showed a great deal of variation in grain production between sweet sorghum accessions, with some yielding up to 2.3 t ha^{-1} (Jebril et al., 2021).

The major antinutritional components present in sweet sorghum are tannins (polyphenolic compounds found in the grain) and dhurrin, a cyanogenic glucoside found primarily in the aerial shoot and sprouted seeds (Chamarthi et al., 2011). Beneath the pericarp layer of the sweet sorghum grain is the testa, followed by the endosperm. The testa contains tannins (Kumar et al., 2022; Redondo et al., 2014). Often, grains with pigmented testa or a reddish-colored pericarp contain condensed tannins which combine with proteins to form complex molecules and carbohydrate polymers that reduce digestibility (Baloch et al., 2023). On the other hand, tannins in sweet sorghum are found in small amounts in unpigmented grains (Etuk et al., 2012).

The starch, which is made up of amylose and amylopectin polysaccharides, is a major source of carbohydrates and has a higher amylopectin–amylose ratio than maize, resulting in slow digestibility of the starch compared to maize (Nasidi et al., 2019; Deenanath et al., 2012). The starch is also gluten-free and an inexpensive source of phenolics and dietary fiber (Nasidi et al., 2010). Similarly, phytic acid in the grain of sweet sorghum also reduces digestibility and forms insoluble compounds with minerals such as Ca, Mg, Fe, and Zn, among others, to decrease their bioavailability for metabolic activities (Nasidi et al., 2010). Moreover, polyphenolic compounds interact with the carbohydrate polysaccharides to form insoluble compounds that resist digestive enzyme activities, consequently leading to slower digestibility in comparison with maize or wheat (Nasidi et al., 2010). In general, slow digestibility reduces the buildup of calories and resultant energy supply over extended periods, thus contributing to the reduction of obesity. However, tannins deter birds and insect invasion due to the ability of the grains to impart astringent palatability qualities (Taylor, 2003; Hariprasanna et al., 2015). Tannins are also suitable for malting. In addition, the antinutritional effects of the tannins can be remedied by removing the testa through mechanical dehulling or chemical charcoal treatment (Johnston & Nghiem, 2018). The level of tannins can also be reduced by grinding and pelleting (Semere et al., 2023).

Tannins also decrease the digestibility of amino acids as well as palatability, which, in turn, affects feed intake and growth (Chamarthi et al., 2011). Therefore, the utilization of low-tannin cultivars (<0.5% tannin) is merited particularly in animal feeds (Semere et al., 2023; Ostmeyer et al., 2020; Barbanti et al., 2006). When using the grain in livestock feeds, chemical treatment is sometimes recommended so as to reduce the tannin concentration and increase digestibility (Etuk et al., 2012). In some cases, adding polyethylene glycol and micronizing high-tannin sorghum can increase its feeding value, but these methods tend to raise the total cost of feeding sorghum, making them unfeasible (Santos et al., 2008). Therefore, it is imperative to develop low-tannin, dual-purpose sweet sorghum varieties for use by farmers for food and feed (Mirfakhar et al., 2017). Tannin extracts from plants are also being used successfully as additives in poultry feeds to control diseases and to improve animal performance (Kumar et al., 2022; Redondo et al., 2014).

17.4 STALK JUICE

The sweet sorghum plant converts free atmospheric carbon dioxide to variable amounts of fermentable sugars, proteins, and starch which accumulate in the stem. It is estimated that <2.0% of these sugars accumulate in the leaves and panicles (Blanc, 2012). The fermentable sugars consist

TABLE 17.3
The Stalk Juice Contains Crude Protein and Sugars

Constituent	Concentration
Starch	0.3–1.2 g/L
Crude protein	1.1–1.9 g/L
Free amino acids	115.0–340.0 mg/L
Sucrose	36.0–117.0 g/L
Glucose	20.0–35.0 g/L
Fructose	10.0–15.0 g/L

Source: Adapted from Roland et al. (2012); Mukabane et al. (2014).

TABLE 17.4
Sweet Sorghum Juice Contains Higher Mineral and Vitamin Content Than Honey

Constituent	Sweet Sorghum	Sugarcane
***Mineral* (mg/100 g)**		
Calcium	441.12 ± 15.17	18.39 ± 0.26
Magnesium	88.28 ± 1.25	77.76 ± 1.01
Phosphorus	40.60 ± 0.51	61.02 ± 0.45
Iron	0.27 ± 0.04	0.16 ± 0.06
Manganese	0.54 ± 0.01	0.54 ± 0.00
Zinc	1.25 ± 0.08	0.08 ± 0.01
***Vitamin* (mg/100 g)**		
Riboflavin (vitamin B_2)	10.00	0.06
Vitamin C	11.50	5.00
Nicotinic acid	153.00	32.00

Source: Adapted from Asikin et al. (2018).

of mainly fructose, glucose, and sucrose, while other sugars, notably xylose, arabinose, galactose, and maltose, occur in low concentrations in the stalks (Holou & Stevens, 2012). The composition of the stalk juice (Table 17.3) varies with genotype, growth stage, and environmental factors (Blanc, 2012). During the juvenile vegetative growth stage, fructose is the most abundant sugar, but as the crop progresses to heading, sucrose becomes the predominant sugar. Sucrose continues to accumulate, reaching its maximum concentration when the grains reach the soft-dough stage (Blanc, 2012; Holou & Stevens, 2012). The syrup from sorghum tends to have more minerals as well as riboflavin than honey (Table 17.4) (Mukabane et al., 2014). The stalk juice also contains crude proteins, free amino acids, and minerals (Holou & Stevens, 2012). Therefore, sweet sorghum juice can positively contribute to human nutrition.

17.5 CROP RESIDUE, BIOMASS, AND BAGASSE

Sweet sorghum stover, which contains both lignin and hemicellulose, as well as cellulose, is a viable feedstock for ethanol production (Kanbar et al., 2021). Lignin does not contain any sugars, and it surrounds the hemicellulose and cellulose, making it difficult to convert them into ethanol and raising the amount of processing energy (Nazli, 2020). Significantly less lignin is present in the sweet sorghum stems and leaves of brown midrib types (Rivera-Burgos et al., 2019). In these types of sweet sorghum, enzymatic hydrolysis significantly increases (by about 50%) the production of fermentable sugars, thus lowering the cost of producing ethanol from biomass (Meena et al., 2022; Burks et al., 2015). The biomass yield can reach 60 t/ha, while the lignocellulosic parts of the bagasse typically contain 35–50% cellulose, 20–30% hemicellulose, and 15–25% lignin (Lolasi et al., 2018).

Bagasse is the fibrous material generated after stalk juice extraction, and its quantity as well as quality depend largely on the sweet sorghum variety and environmental factors. Essentially, bagasse is composed of lignin, cellulosic, and hemicellulosic biomass (Nasidi et al., 2010). The amount of lignin fraction contained in the lignocellulose biomass determines its digestibility efficiency; hence, effective utilization of this material as a feedstock source to produce bioethanol entails the reduction of the fibrous complex material into its separate basic components, namely, lignin, hemicellulose, and cellulose. This is achievable through a pre-treatment process, followed by hydrolysis of the polysaccharides (cellulose and hemicellulose), to break them down into their respective monomeric units, namely, glucose, pentose, and hexose sugars. The bagasse can also be potentially used as a feedstock for biochemical production through microbial conversion (Lo et al., 2020).

17.6 CONCLUSIONS AND PROSPECTS

The production of sweet sorghum for food and feed offers substantial opportunities, especially in dryland areas. Notable potential uses include stalk juices that can be processed industrially into several products. Sweet sorghum contains both desirable and negative nutritional components, particularly tannins (both the hydrolyzable types, for example, geraniin, which, when treated with acid, alkali, or enzymes, split into sugars and recognizable phenolic carboxylic acids and the condensed types that cannot be hydrolyzed into simpler components. Light-colored grains contain relatively low tannin content and, therefore, are associated with less bitterness and astringency.

In the future, there will be merit in enhancing sweet sorghum cultivars probably through the exploitation of the cytoplasmic male sterility system and mutation breeding. Moreover, there is already substantial diversity in the current sweet sorghum germplasm collections from various African countries. Mutation breeding, in conjunction with modern molecular tools, could be useful in future genetic enhancement efforts. The selection of high-grain and high-stalk-juice-yield genotypes could produce dual-purpose types for end users and other stakeholders in the sweet sorghum value chain. Perhaps an intensified utilization of sweet sorghum products, such as the syrup and feeds, could raise the demand for the crop, especially in Southern Africa. The new improved sweet sorghum varieties need to meet the technological demands of the sugar and beverage industries, as well as the needs of the farmers to facilitate adoption. The use of sweet sorghum as an ingredient in the food and pharmaceutical industries can also offer new opportunities in the value chain. Sweet sorghum varieties that possess unacceptable levels of condensed tannins in the grain may still be useful in the production of juice from their stalks. Moreover, such varieties can play a role in beer brewing, as well as other industrial applications. It is also important to note that from a carbon biogeochemical cycle perspective, sweet sorghum captures and converts atmospheric carbon dioxide to fermentable sugars, such as fructose, glucose, sucrose, xylose, arabinose, galactose, and maltose, as well as proteins and starch, which accumulate in the stem. These sugars can be exploited for economic benefits, provided deliberate future investments are made to support the industries that process these sugars. In addition, sweet sorghum breeders can continue to develop superior hybrid cultivars that accumulate these useful sugars. The crop is adapted to limited production inputs, including water and chemical fertilizers. Moreover, the production cycle is also markedly shorter than that for sugarcane, for instance. With the current efforts to increase sources of renewable energy, sweet sorghum, which accumulates sugar in the stalks while producing grain in the panicles, is ideal for bioethanol production without compromising food security. Biofuels are considered viable substitutes for fossil fuels. Moreover, both the biomass and grain of sweet sorghum can be used as animal feeds. Therefore, in our view, sweet sorghum has immense potential in future food systems. However, favorable policies that promote the sweet sorghum value chain are critical for deriving maximum economic benefits from the species.

17.7 ACKNOWLEDGMENTS

The authors wish to express their gratitude to all the colleagues who graciously reviewed and improved the original manuscript of this chapter.

REFERENCES

Adedugba, A. A., Adeyemo, O. A., Adetumbi, A. J., Amusa, O. D., & Ogunkanmi, L. A. (2023). Evaluation of genetic variability for major agro-morphological and stalk sugar traits in African sorghum genotypes. *Heliyon*, *9*(3).

Albahri, G., Alyamani, A. A., Badran, A., Hijazi, A., Nasser, M., Maresca, M., & Baydoun, E. (2023). Enhancing essential grains yield for sustainable food security and bio-safe agriculture through latest innovative approaches. *Agronomy*, *13*, 1709.

Amin, A. R., Kucuk, O., Khuri, F. R., & Shin, D. M. (2009). Perspectives for cancer prevention with natural compounds. *Journal of Clinical Oncology*, *27*, 2712–2725.

Anami, S. E., Zhang, L. M., Xia, Y., Zhang, Y. M., Liu, Z. Q., & Jing, H. C. (2015a). Sweet sorghum ideotypes: Genetic improvement of stress tolerance. *Food and Energy Security*, *4*, 3–24.

Anami, S. E., Zhang, L. M., Xia, Y., Zhang, Y. M., Liu, Z. Q., & Jing, H. C. (2015b). Sweet sorghum ideotypes: Genetic improvement of the biofuel syndrome. *Food and Energy Security*, *4*, 159–177.

Andrzejewski, B., Eggleston, G., & Powell, R. (2013). Pilot plant clarification of sweet sorghum juice and evaporation of raw and clarified juices. *Industrial Crops and Products*, *49*, 648–658.

Appiah-Nkansah, N. B., Li, J., Rooney, W., & Wang, D. (2019). A review of sweet sorghum as a viable renewable bioenergy crop and its techno-economic analysis. *Renewable Energy*, *143*, 121–1132.

Asikin, Y., Wada, K., Imai, Y., Kawamoto, Y., Mizu, M., Mutsuura, M., & Takahashi, M. (2018). Compositions, taste characteristics, volatile profiles, and antioxidant activities of sweet sorghum (*Sorghum bicolor* L.) and sugarcane (*Saccharum officinarum* L.) syrups. *Journal of Food Measurement and Characterization*, *12*, 884–891.

Baloch, F. S., Altaf, M. T., Liaqat, W., Bedir, M., Nadeem, M. A., Cömertpay, G., Çoban, N., Habyarimana, E., Barutçular, C., Cerit, I., Ludidi, N., Karaköy, T., Aasim, M., Chung, Y. S., Nawaz, M. A., Hatipoğlu, R., Kökten, K., & Sun, H. J. (2023). Recent advancements in the breeding of sorghum crop: Current status and future strategies for marker-assisted breeding. *Frontiers in Genetics*, *14*, 1–36.

Barbanti, L., Grandi, S., Vecchi, A., & Venturi, G. (2006). Sweet and fibre sorghum (*Sorghum bicolor* (L.) Moench), energy crops in the frame of environmental protection from excessive nitrogen loads. *European Journal of Agronomy*, *25*, 30–39.

Blanc, E. (2012). The impact of climate change on crop yields in Sub-Saharan Africa. *American Journal of Climate Change*, *1*, 1–13.

Bonin, C. L., Heaton, E. A., Cogdill, T. J., & Moore, K. J. (2016). Management of sweet sorghum for biomass production. *Sugar Technology*, *18*, 150–159.

Boyer, J. S., & Westgate, M. E. (2004). Grain yield with limited water. *Journal of Experimental Botany*, *55*, 2385–2394.

Britz, S. J., Prasad, P. V. V., Moreau, R. A., Allen, L. H., Jr., Kremer, D. F., & Boote, K. J. (2007). Influence of growth temperature on amounts of tocopherols, tocotrienols and γ-oryzanol in brown rice. *Journal of Agricultural Food Chemistry*, *55*, 7559–7565.

Bröhan, M., Jerkovic, V., & Collin, S. (2011). Potentiality of red sorghum for producing stilbenoid-enriched beers with high antioxidant activity. *Journal of Agricultural and Food Chemistry*, *59*, 4088–4094.

Burks, P. S., Kaiser, C. M., Hawkins, E. M., & Brown, P. J. (2015). Genome-wide association for sugar yield in sweet sorghum. *Crop Science*, *55*, 2138–2148.

Chamarthi, S. K., Sharma, H. C., Sahrawat, K. L., Narasu, L. M., & Dhillon, M. K. (2011). Physico-chemical mechanisms of resistance to shoot fly, Atherigona soccata in sorghum, Sorghum bicolor. *Journal of Applied Entomology*, *135*, 446–455.

Chen, B. R., Wand, C. Y., Ping, W. A. N. G., Zhu, Z. X., Ning, X. U., Shi, G. S., Miao, Y. U., Nai, W. A. N. G., Li, J. H., Hou, J. M., & Li, S. J. (2019). Genome-wide association study for starch content and constitution in sorghum (*Sorghum bicolor* (L.) Moench). *Journal of Integrative Agriculture*, *18*, 2446–2456.

Chiambiro, G., Madzimure, J., & Mpofu, I. D. T. (2022). Constraints and opportunities of sweet sorghum (*Sorghum bicolor* L.) silage production and utilisation by smallholder milk production sector in Zimbabwe. *Tropical Animal Health and Production*, *54*, 1–11.

Chohnan, S., Nakane, M., Rahman, M. H., Nitta, Y., Yoshiura, T., Ohta, H., & Kurusu, Y. (2011). Fuel ethanol production from sweet sorghum using repeated-batch fermentation. *Journal of Bioscience and Bioengineeringi*, *111*, 433–436.

Choi, S. C., Kim, J. M., Lee, Y. G., & Kim, C. (2019). Antioxidant activity and contents of total phenolic compounds and anthocyanins according to grain colour in several varieties of Sorghum bicolor (L.) Moench. *Cereal Research Communications*, *47*, 228–238.

Cifuentes, R., Bressani, R., & Rolz, C. (2014). The potential of sweet sorghum as a source of ethanol and protein. *Energy for Sustainable Development*, *21*, 13–19.

da Silva, M. J., de Olieveira Ribeiro, P. C., da Silva, R. A., da Silva, K. J., Schaffert, R. E., & da Costa Parrella, R. A. (2020). Multi-trait selection of sweet sorghum (*Sorghum bicolor* (L.) Moench) genotypes for bioenergy production. *Journal of Bioenergy and Food Science*, *7*, 1–13.

Dahlberg, J. (2019). The role of sorghum in renewables and biofuels. *Methods in Molecular Biology*, *1931*, 269–277.

Dahlberg, J. A., Zhang, X., Hart, G. E., & Mullet, J. E. (2002). Comparative assessment of variation among sorghum germplasm accessions using seed morphology and RAPD measurements. *Crop Science*, *42*, 291–296.

Dalla Marta, A., Mancini, M., Orlando, F., Natali, F., Capecchi, L., & Orlandini, S. (2014). Sweet sorghum for bioethanol production: Crop responses to different water stress levels. *Biomass and Bioenergy*, *64*, 211–219.

Datta Mazumdar, S., Poshadri, A., Srinivasa Rao, P., Ravinder Reddy, C. H., & Reddy, B. V. S. (2012). Innovative use of sweet sorghum juice in the beverage industry. *International Food Research Journal*, *19*, 1361–1366.

Deenanath, E. D., Iyuke, S., & Rumbold, K. (2012). The bioethanol industry in Sub-Saharan Africa: History, challenges, and prospects. *Journal of Biomedicine and Biotechnology*, *2012*, 1–11.

Devnarain, N., Crampton, B. G., Chikwamba, R., Becker, J. V. W., & O'Kennedy, M. M. (2016). Physiological responses of selected African sorghum landraces to progressive water stress and re-watering. *South African Journal of Botany*, *103*, 61–69.

Dube, F., & Maphosa, M. (2022). Significance of sweet sorghum as a multi-purpose crop for Sub-Saharan Africa. *African Crop Science Journal*, *30*, 235–243.

Dykes, L., Rooney, L. W., Waniska, R. D., & Rooney, W. L. (2005). Phenolic compounds and antioxidant activity of sorghum grains of varying genotypes. *Journal of Agricultural and Food Chemistry*, *53*, 6813–6818.

El-Razek, A. M. A., & Besheit, S. Y. (2009). Potential of some sweet sorghum (*Sorghum bicolor* L. Moench) varieties for syrup and ethanol production in Egypt. *Sugar Technology*, *11*, 239–245.

Etuk, E. B., Okeudo, N. J., Esonu, B. O., & Udedibie, A. B. I. (2012). Antinutritional factors in Sorghum: Chemistry, mode of action and effects on livestock and poultry. *Journal of Animal Science*, *2*, 113–119.

Fernandes, G., Braga, T. G., Fischer, J., Parrella, R. A. C., de Resende, M. M., & Cardoso, V. L. (2014). Evaluation of potential ethanol production and nutrients for four varieties of sweet sorghum during maturation. *Renewable Energy*, *71*, 518–524.

Gooding, M. J., Ellis, R. H., Shewry, P. R., & Schofield, J. D. (2003). Effects of restricted water availability and increased temperature on the grain filling, drying and quality of winter wheat. *Journal of Cereal Science*, *37*, 295–309.

Hariprasanna, K., Agte, V., Elangovan, M., Gite, S., & Kishore, A. (2015). Anti-nutritional factors and antioxidant capacity in selected genotypes of sorghum [*Sorghum bicolor* (L.) Moench]. *International Journal of Agriculture Sciences*, *7*, 620–625.

Hariprasanna, K., & Patil, J. V. (2015). Sorghum: Origin, classification, biology and improvement. In R. Madhusudhana, P. Rajendrakumar & J. Patil (Eds.), *Sorghum molecular breeding* (pp. 3–20). Springer.

Holou, R. A., & Stevens, G. (2012). Juice, sugar, and bagasse response of sweet sorghum (S orghum bicolor (L.) M oench cv. M81E) to N fertilization and soil type. *GCB Bioenergy*, *4*, 302–310.

Hossain, M. S., Islam, M. N., Rahman, M. M., Mostofa, M. G., & Khan, M. A. R. (2022). Sorghum: A prospective crop for climatic vulnerability, food and nutritional security. *Journal of Agriculture and Food Research*, *8*, 100300.

Hu, W., Zhou, L., & Chen, J. H. (2022). Conversion sweet sorghum biomass to produce value-added products. *Biotechnology for Biofuels and Bioproducts*, *15*, 1–11.

Hunter, E., & Anderson, I. (1997). Sweet sorghum. In J. Janick (Ed.), *Horticultural reviews* (pp. 73–104). John Wiley and Sons.

Jebril, J., Wang, D., Rozeboom, K., & Tesso, T. (2021). Grain sink removal increases stalk juice yield, sugar accumulation, and biomass in sweet sorghum [*Sorghum bicolor* (L.) Moench]. *Industrial Crops and Products, 173*, 114089.

Jenkins, C. L. (1997). The CO_2 concentrating mechanism of C_4 photosynthesis: Bundle sheath cell CO2 concentration and leakage. *Functional Plant Biology, 24*, 543–547.

Johnston, D. B., & Nghiem, N. (2018). Evaluation of sweet sorghum juice for the production of lysine using *Corynebacterium glutamicum*. *Fermentation, 4*, 29.

Kanbar, A., Shakeri, E., Alhajturki, D., Riemann, M., Bunzel, M., Morgano, M. T., Stapf, D., & Nick, P. (2021). Sweet versus grain sorghum: Differential sugar transport and accumulation are linked with vascular bundle architecture. *Industrial Crops and Products, 167*, 1–14.

Kumar, A. A., Reddy, B. V. S., Sharma, H. C., Hash, C. T., Rao, P. S., Ramaiah, B., & Reddy, P. S. (2011). Recent advances in sorghum genetic enhancement research at ICRISAT. *American Journal of Plant Sciences, 2*, 589–600.

Kumar, K., Sinha, R. K., Kumar, S., Nirala, R. K., Kumari, S., & Sahu, S. P. (2022). Significance of tannins as an alternative to antibiotic growth promoters in poultry production. *The Journal of Pharmaceutical Innovation, 11*, 1435–1440.

Laisk, A., & Edwards, G. E. (1998). Oxygen and electron flow in C 4 photosynthesis: Mehler reaction, photorespiration and CO_2 concentration in the bundle sheath. *Planta, 205*, 632–645.

Leegood, R. C. (1997). The regulation of C4 photosynthesis. In *Advances in botanical research* (pp. 251–316). Academic Press.

Legwaila, G., Balole, T., & Karikari, S. (2004). Review of sweet sorghum: A potential cash and forage crop in Botswana. *UNISWA Journal of Agriculture, 12*, 5–14.

Lekgari, A., & Dweikat, I. (2014). Assessment of genetic variability of 142 sweet Sorghum germplasm of diverse origin with molecular and morphological markers. *Open Journal of Ecology, 4*, 371–393.

Lestari, E. G., Dewi, I. S., Yunita, R., & Nur, A. (2022). Evaluation of ratooning ability in several sweet sorghum (Sorghum bicolor [L.] Moench) mutant lines. In *AIP conference proceedings* (p. 2462). AIP Publishing.

Lo, E., Brabo-Catala, L., Dogaris, I., Ammar, E. M., & Philippidis, G. P. (2020). Biochemical conversion of sweet sorghum bagasse to succinic acid. *Journal of Bioscience and Bioengineering, 129*, 104–109.

Lolasi, F., Amiri, H., Asadollahi, M. A., & Karimi, K. (2018). Using sweet sorghum bagasse for production of amylases required for its grain hydrolysis via a biorefinery platform. *Industrial Crops and Products, 125*, 473–481.

Lopez, J. R., Erickson, J. E., Asseng, S., & Bobeda, E. L. (2017). Modification of the CERES grain sorghum model to simulate optimum sweet sorghum rooting depth for rainfed production on coarse textured soils in a sub-tropical environment. *Agricultural Water Management, 181*, 47–55.

Machado, S., & Paulsen, G. M. (2001). Combined effects of drought and high temperature on water relations of wheat and sorghum. *Plant and Soil, 233*, 179–187.

Marta, A. D., Mancini, M., Orlando, F., Natali, F., Capecchi, L., & Orlandini, S. (2014). Sweet sorghum for bioethanol production: Crop responses to different water stress levels. *Biomass and Bioenergy, 64*, 211–219.

Mathur, S., Priyadarshini, S. S., Singh, V., Vashisht, I., Jung, K. H., Sharma, R., & Sharma, M. K. (2020). Comprehensive phylogenomic analysis of ERF genes in sorghum provides clues to the evolution of gene functions and redundancy among gene family members. *3 Biotech, 10*, 1–19.

Mawouma, S., Condurache, N. N., Turturică, M., Constantin, O. E., Croitoru, C., & Rapeanu, G. (2022). Chemical composition and antioxidant profile of sorghum (sorghum bicolor (L.) moench) and pearl millet (*Pennisetum glaucum* (L.) R. Br.) grains cultivated in the far-North region of Cameroon. *Foods, 11*, 2026.

Meena, K., Visarada, K. B. R. S., & Meena, D. K. (2022). Sorghum bicolor (L.) Moench a multifarious crop-fodder to therapeutic potential and biotechnological applications: A future food for the millennium. *Future Foods, 6*, 1–14.

Mesta, S., Geeta, G. S., & Ashwini, M. (2018). Standardization of fermentation parameters for beer production from sweet sorghum (*Sorghum bicolor* (L.) Moench) grains. *International Journal of Current Microbiology and Applied Sciences, 7*, 659–665.

Mirfakhar, M., Asadollahi, M. A., Amiri, H., & Karimi, K. (2017). Phenolic compounds removal from sweet sorghum grain for efficient biobutanol production without nutrient supplementation. *Industrial Crops and Products, 108*, 225–231.

Motsi, H., Molapo, M., & Phiri, E. E. (2022a). A review on sweet sorghum adaptive capacity on improving food security and poverty alleviation in Sub-Saharan Africa. *South African Journal of Botany, 150*, 323–329.

Mukabane, B. G., Thiongo, G., Gathitu, B., Murage, H., Ojijo, N. O., & Owino, W. O. (2014). Evaluating the potential of juice from some sweet sorghum varieties grown in Kenya to crystallize. *Food Science and Quality Management*, *30*, 31–39.

Mullet, J., Morishige, D., McCormick, R., Truong, S., Hilley, J., McKinley, B., Anderson, R., & Olson, S. N. R. W. (2014). Energy sorghum – a genetic model for the design of C4 grass bioenergy crops. *Journal of Experimental Botany*, *65*, 1–11.

Naoura, G., Emendack, Y., Baloua, N., Brocke, K., Hassan, M. A., Sawadogo, N., Nodjasse, A. D., Djinodji, R., Trouche, G., & Laza, H. E. (2020). Characterization of semi – arid Chadian sweet sorghum accessions as potential sources for sugar and ethanol production. *Scientific Reports*, *10*, 1–11.

Nasidi, M., Agu, R., Walker, G., & Deeni, Y. (2019). Sweet sorghum: Agronomic practice for food, animal feed and fuel production in Sub-Saharan Africa. Sweet sorghum: Characteristics, cultivation and uses. In L. Rogers & M. Wills (Eds.), *Sweet sorghum* (pp. 2–49). Nova Science Publishers Inc.

Nasidi, M., Akunna, J., Deeni, Y., Blackwood, D., & Walker, G. (2010). Bioethanol in Nigeria: Comparative analysis of sugarcane and sweet sorghum as feedstock sources. *Energy and Environmental Science*, *3*, 1447–1457.

Nazli, R. I. (2020). Evaluation of different sweet sorghum cultivars for bioethanol yield potential and bagasse combustion characteristics in a semiarid Mediterranean environment. *Biomass and Bioenergy*, *139*, 105624.

Ngara, R., Goche, T., Swanevelder, D. Z. H., & Chivasa, S. (2021). Sorghum's whole-plant transcriptome and proteome responses to drought stress: A review. *Life*, *11*, 704.

Nziguheba, G., Zingore, S., Kihara, J., Merckx, R., Njoroge, S., Otinga, A., Vandamme, E., & Vanlauwe, B. (2016). Phosphorus in smallholder farming systems of Sub-Saharan Africa: Implications for agricultural intensification. *Nutrient Cycling in Agroecosystems*, *104*, 321–340.

Olukoya, I. A., Bellmer, D., Whiteley, J. R., & Aichele, C. P. (2015). Evaluation of the environmental impacts of ethanol production from sweet sorghum. *Energy for Sustainable Development*, *24*, 1–8.

Ostmeyer, T., Bheemanahalli, R., Srikanthan, D., Bean, S., Peiris, K. H. S., Madasamy, P., Perumal, R., & Jagadish, S. V. K. (2020). Quantifying the agronomic performance of new grain sorghum hybrids for enhanced early-stage chilling tolerance. *Field Crops Research*, *258*, 1–11.

Paesal, S. S., & Aqil, M. (2021). Sorghum cultivation of the ratoon system for increased yields in dry land. *IOP Conference Series Earth and Environmental Science*, *911*, 012035.

Paiva, C. L., Evangelista, W. P., Queiroz, V. A. V., & Glória, M. B. A. (2015). Bioactive amines in sorghum: Method optimisation and influence of line, tannin and hydric stress. *Food Chemistry*, *173*, 224–230.

Pfeiffer, T. W., Bitzer, M. J., Toy, J. J., & Pedersen, J. F. (2010). Heterosis in sweet sorghum and selection of a new sweet sorghum hybrid for use in syrup production in appalachia. *Crop Science*, *50*, 1788–1794.

Prasad, P. V. V., Pisipati, S. R., Momcilovic, I., & Ristic, Z. (2011). Independent and combined effects of high temperature and drought stress during grain filling on plant yield and chloroplast protein synthesis elongation factor (EF-Tu) expression in spring wheat. *Journal of Agronomy and Crop Science*, *197*, 430–441.

Prasad, P. V. V., Pisipati, S. R., Mutava, R. N., & Tuinstra, M. R. (2008). Sensitivity of grain sorghum to high temperature stress during reproductive development. *Crop Science*, *48*, 1911–1917.

Prasad, S., Singh, A., & Joshi, H. C. (2007). Ethanol production from sweet sorghum syrup for utilization as automotive fuel in India. *Energy and Fuels*, *21*, 2415–2420.

Prasad, V. B. R., Govindaraj, M., Djanaguiraman, M., Djalovic, I., Shailani, A., Rawat, N., Lata Singla-Pareek, S., Pareek, A., & Vara Prasad, P. V. (2021). Molecular sciences drought and high temperature stress in sorghum: Physiological, genetic, and molecular insights and breeding approaches. *International Journal of Molecular Sciences*, *2021*, 9826.

Rani, R., Pandey, C., & Kapoor, N. (2024). Nurturing health through millet derived nutraceuticals. *Food and Humanity*, *100277*, 1–14.

Rao, P. S., Rathore, A., & Reddy, B. V. S. (2013). Interrelationship among biomass related traits and their role in sweet sorghum cultivar productivity in main and ratoon crops. *Sugar Technology*, *15*, 278–284.

Redondo, L. M., Chacana, P. A., Dominguez, J. E., & Miyakawa, M. E. F. (2014). Perspectives in the use of tannins as alternative to antimicrobial growth promoter factors in poultry. *Frontiers in Microbiology*, *5*, 118.

Ribeiro, P. M., Tiago, O., Isadora, R. R., Carneiro, P. O., & Aluízio, P. R. (2020). Insight into genetic potential of male sterile sweet sorghum a-lines for agro-industrial traits using tester R-lines. *Industrial Crops and Products*, *153*, 1–9.

Ritter, K. B., McIntyre, C. L., Godwin, I. D., Jordan, D. R., & Chapman, S. C. (2007). An assessment of the genetic relationship between sweet and grain sorghums, within *Sorghum bicolor* ssp. bicolor (L.) Moench, using AFLP markers. *Euphytica*, *157*, 161–176.

Rivera-Burgos, L. A., Volenec, J. J., & Ejeta, G. (2019). Biomass and bioenergy potential of brown midrib sweet sorghum germplasm. *Frontiers in Plant Science, 10*, 1–12.

Santos, J. E. P., Bilby, T. R., Thatcher, W. W., Staples, C. R., & Silvestre, F. T. (2008). Long chain fatty acids of diet as factors influencing reproduction in cattle. *Reproduction in Domestic Animals, 43*, 23–30.

Sehgal, A., Sita, K., Kumar, J., Kumar, S., Singh, S., Siddique, K. H., & Nayyar, H. (2017). Effects of drought, heat and their interaction on the growth, yield and photosynthetic function of lentil (Lens culinaris Medikus) genotypes varying in heat and drought sensitivity. *Frontiers in Plant Science, 8*, 1776.

Semere, T., Tsehaye, Y., Tareke, L., Westengen, O. T., & Fjellheim, S. (2023). Nutritional and antinutritional potentials of sorghum: A comparative study among different sorghum landraces of Tigray, Northern Ethiopia. *Agriculture, 13*, 1–22.

Shah, N., & Paulsen, G. M. (2003). Interaction of drought and high temperature on photosynthesis and grain-filling of wheat. *Plant Soil, 257*, 219–226.

Sharkey, T. D. (2005). Effect of moderate heat stress on photosynthesis: Importance of thylakoid reactions, rubisco deactivation, reactive oxygen species and thermotolerance provided by isoprene. *Plant Cell Environment, 28*, 269–277.

Taylor, J. R. N. (2003). Overview: Importance of sorghum in Africa. In: Afripro: Workshop on the proteins of sorghum and millets: Enhancing nutritional and functional properties for Africa. *Pretoria, 2*, 1–21.

Tovignan, T. K., Adoukonou-Sagbadja, H., Diatta, C., Clément-Vidal, A., Soutiras, A., Cisse, N., & Luquet, D. (2020). Terminal drought effect on sugar partitioning and metabolism is modulated by leaf stay-green and panicle size in the stem of sweet sorghum (Sorghum bicolor L. Moench). *CABI Agriculture and Bioscience, 1*, 1–11.

Tsao, R. (2010). Chemistry and biochemistry of dietary polyphenols. *Nutrients, 2*, 1231–1246.

Tsuchihashi, N., & Goto, Y. (2008). Year-round cultivation of sweet sorghum [*Sorghum bicolor* (L.) Moench] through a combination of seed and ratoon cropping in Indonesian Savanna. *Plant Production Science, 11*, 377–384.

Umakanth, A. V., Kumar, A. A., Vermerris, W., & Tonapi, V. A. (2018). Sweet sorghum for biofuel industry. *Breeding Sorghum for Diverse End Uses*, 255–270.

Vanamala, J. K. P., Massey, A. R., Pinnamaneni, S. R., Reddivari, L., & Reardon, K. F. (2018). Grain and sweet sorghum (*Sorghum bicolor* L. Moench) serves as a novel source of bioactive compounds for human health. *Critical Reviews in Food Science, 58*, 2867–2881.

Vlachos, C. E., Pavli, O. I., Flemetakis, E., & Skaracis, G. N. (2020). Exploiting pre-and post-harvest metabolism in sweet sorghum genotypes to promote sustainable bioenergy production. *Industrial Crops and Products, 155*, 112758.

Widowati, S., & Luna, P. (2022). Nutritional and functional properties of sorghum (Sorghum bicolor (L.) Moench)-based products and potential valorisation of sorghum bran. *IOP Conference Series: Earth and Environmental Science, 1024*, 012031.

Woods, J. (2001). The potential for energy production using sweet sorghum in Southern Africa. *Energy for Sustainable Development, 5*, 31–38.

Wu, X., Liu, Y., Luo, H., Shang, L., Leng, C., Liu, Z., Li, Z., Lu, X., Cai, H., Hao, H., & Jing, H. C. (2022). Genomic footprints of sorghum domestication and breeding selection for multiple end uses. *Molecular Plant, 15*, 537–551.

Zegada-Lizarazu, W., & Monti, A. (2013). Photosynthetic response of sweet sorghum to drought and re-watering at different growth stages. *Physiologia Plantarum, 149*, 56–66.

18 Nutraceutical, Agricultural, and Economic Potential of *Trachyandra ciliata* (Wild Cabbage), an Underutilized Halophyte from the Western Cape Province, South Africa

*Sihle Ngxabi,[1] Muhali Olaide Jimoh,[1, 2] Avela Sogoni,[1] Charles Petrus Laubscher,[1] and *Learnmore Kambizi[1]*

[1]Department of Horticultural Sciences, Faculty of Applied Sciences, Cape Peninsula University of Technology, PO Box 1905, Bellville, 7535 South Africa

[2]Department of Plant Science, Olabisi Onabanjo University, PMB 2002, Ago-Iwoye 120107, Nigeria

*Corresponding author: *KambiziL@cput.ac.za*

18.1 INTRODUCTION

The discussion over global water scarcity and food security has recently intensified, with uncertainties looming over the future demand for food and water (Hanjra & Qureshi, 2010). About 80% of the world's agricultural land – 1.1 billion ha out of 1.5 billion ha – is rain-fed, producing 60% of the world's staple foods without any irrigation systems (FAO, 2008; Schultz & De Wrachien, 2002). According to reports, 19% of all cultivated land is irrigated, which accounts for around 40% of all agricultural output worldwide. Surprisingly, although contributing relatively little to global food supply, irrigated agricultural land is accountable for 70% of water withdrawals from rivers (Molden et al., 2007).

By 2050, the Sustainable Development Goal 2 (SDG2) aims to eradicate hunger, boost food security and nutrition, and encourage the cultivation of crops sustainably. However, food production has been significantly impacted by climate change, increased soil salinization, and a lack of access to fresh water, notably in South Africa (Acquaah, 2009; Corwin, 2021). This makes it a challenge to reach the desired goal. Salinity restricts plant growth and causes a drop in overall yield, decreased leaf formation, a delay in flowering, and the abortion of flower buds. Therefore, ensuring sustainability in terms of the future supply of food for poor communities and the protection of water resources presents a significant problem for scholars. The use of wild salt-tolerant plants (halophytes) which possess considerable commercial value, such as green vegetables, feed crops, and medicinal properties, is one recently identified method to adapt to these ever-changing environmental conditions (Ventura & Sagi, 2013).

DOI: 10.1201/9781003469766-18

Water is crucial for future global food security because water shortage lowers agricultural output, which contributes to the rising trend of food security issues. Therefore, this means that there is a connection between water scarcity and food poverty. Researchers predict that the world's population will increase by nearly 2 billion people within the next two to three decades, necessitating an increased water supply for domestic, municipal, industrial, and environmental purposes (Colvin & Muruven, 2017; Condon et al., 2004). It is estimated that agricultural production must rise by 50 to 100% by 2050 to meet the demands of the world's constantly expanding population, but in reality, the present increase is just 1 to 5%, making it extremely difficult to ensure sustainable agriculture (Colvin & Muruven, 2017; Ngxabi et al., 2021b). As a result, government efforts to achieve high productivity are ongoing in an attempt to boost profitability and satisfy the nation's rising food demand in response to this issue.

The use of salt-tolerant crops can be a useful alternative to investing time and money in remediating the effect of water scarcity and rising soil salinity (Ventura et al., 2011). Halophytes are the ideal crop candidates for salty environments. Recent developments in the food science sector and consumer preferences for varied diets point to the consumption of wild edible plants as diet complements as well as nutritious and useful foods for specific conditions, making commercial cultivation of these plants very crucial (Nikalje et al., 2019; Petropoulos et al., 2018). However, in South Africa, most wild species are regarded as weeds for conventional crops, and farmers typically use chemical fertilizers to get rid of them (Tardío et al., 2006). Many of these wild edible species have been utilized by rural communities for many years as leafy greens and as the foundation of numerous traditional cuisines and regionally significant local recipes but neglected in the commercial vegetable production (Hamed & Custódio, 2019; Petropoulos et al., 2018). Therefore, the cultivation of wild halophytes presents a possible solution to support food and nutrition security while preserving water resources for domestic, municipal, and environmental uses. These developments need to be implemented at the national level to align countries to sustainable development goals (Ngxabi et al., 2021b).

Trachyandra ciliata is a wild underutilized halophyte that is endemic to the Western Cape coastal dunes in South Africa (Ngxabi et al., 2021a). The genus *Trachyandra* belongs to the Asphodelaceae (Aloe) family, which is widely known for its pharmaceutical importance. Studies on this plant are scarce, and there are currently no *Trachyandra* species in commercial cultivation (Manning, 2007). However, it is known that this plant is edible and that the Khoisan people who lived around the South African Cape Coast utilized it as food (De Vynck et al., 2016; Ngxabi et al., 2021a). We previously conducted the only study (Ngxabi, 2020) and focused on the vegetative growth of *T. ciliata* in response to different growth media and salinity to develop its cultivation protocol. Furthermore, we reported the presence of some antioxidants and phytochemicals in the plant, especially when subjected to salt stress. As such, the present study was conducted to explore the nutraceutical, agricultural, and economic potential of *Trachyandra ciliata* to support its cultivation and commercialization.

18.2 ACCUMULATION OF SALTS ON AGRICULTURAL LANDS

One of the main abiotic factors in agriculture that promotes deficiency, physiological abnormalities, and lower output of field crops around the world is salinity, along with drought (Ahanger & Agarwal, 2017; Ahmad et al., 2018). Salt stress has, over the years, become one of the most limiting factors in plant production, especially in arid or semi-arid zones (Silveira et al., 2009). The issue arises when soil ion concentrations (Na+, Cl, and other related salts) are higher than the normal levels. This disrupts osmotic processes first, which subsequently changes a number of metabolic processes necessary for healthy plant growth and development (Ahmad et al., 2018; Ho & Adams, 1995). Although the sensitivity of different plants to soil salinity varies, problems with respiration, photosynthesis, mineral uptake, and oxidative stress have a negative impact on most of their growth and production attributes (Ho & Adams, 1995; Petropoulos et al., 2017). Some plants, like halophytes, are not vulnerable to certain amounts of salts in their root zone; in fact, they prefer saline conditions for optimum growth (Ngxabi et al., 2021b). This therefore necessitates thorough research to determine the precise salt tolerance per species. If the problem is not properly addressed, decreased agricultural output due to salinity could pose serious concerns to global food security in the coming decades. This then calls for the cultivation of plants with high salt tolerance to mitigate the effect of salt toxicity in agricultural lands (Ngxabi et al., 2021a).

18.3 ADAPTATION OF HALOPHYTES TO SALINE CONDITIONS

The physiological processes within plants can often be substantially altered by unfavorable circumstances, like abiotic stressors (drought and salinity), which often reduce growth and production in plants (Ngxabi et al., 2021b). Plants known as *halophytes* are those that naturally thrive in salty areas and benefit from excess salt in their growing medium (Shabala, 2013). From coastal areas, salt marshes, and mudflats to inland deserts, salt flats, and steppes, halophytes can thrive in a range of saline habitats (Flowers & Colmer, 2008). Although wild cabbage is also classified as a halophyte, its adaptation mechanism to salinity is undocumented. Halophytes have developed a number of adaptations to withstand seawater and increased salinity concentrations. According to Shabala (2013), "[t]hese include adjustment of their internal water relations through ion compartmentation in cell vacuoles, the accumulation of compatible organic solutes, succulence, and salt secreting glands and bladders." Shabala (2013) argues that 50 mm of salt concentration is best to achieve optimum growth for monocotyledonous halophytes, and 100 mm for dicots. This author further reported that some halophytes do not show a significant yield decrease when irrigated with seawater. However, Ngxabi et al. (2021b) reported that 100 mm treatment recorded the highest values for vegetative growth in the cultivation of wild cabbage under hydroponics (Figure 18.1). This then makes halophytes the appropriate plants to be cultivated in land affected by salt toxicity.

Most halophytes have been discovered to use ion transport (Na+ and Cl–) as a salt-mitigation technique from roots to leaves, where they are contained in vacuoles (Munns & Tester, 2008). Studies

FIGURE 18.1 An experiment showing the response of *T. ciliata* to different salinity levels.

Source Ngxabi et al. (2021b).

by Souid et al. (2016) and Al Hassan et al. (2017) reported that ion exportation from roots to shoots keeps NaCl levels stable and minimizes the damaging effects of salt at the root level. González-Orenga et al. (2021) showed that excessive accumulation of inorganic ions in the vacuole is neutralized by the production of osmolytes or solutes in the cytoplasm. They further argued that even at high intracellular levels, these organic non-toxic osmolytes do not affect cellular metabolism, but they serve as "reactive oxygen species" (ROS) scavengers and play critical roles in the osmotic adjustment and protein stabilization processes that occur in plants in response to abiotic stresses.

Plants benefit a lot from reactive oxygen species (ROS) because they promote cellular growth, biological processes, and stability. Even though ROS are necessary for life, an excessive amount in a cell can harm lipids, proteins, DNA, and nucleic acids (Mittler, 2017; Ngxabi et al., 2021b). In a case of excessive concentration of ROS, plants address this issue by the production of antioxidant enzymes. These antioxidant enzymes aid in the removal of excess ROS and contribute to the preservation of healthy cellular pH (Souid et al., 2016). Furthermore, halophytes' adaptive defense mechanism against oxidative stress is said to be highly dependent on antioxidant enzymes (Mittler, 2017). This is supported by the study of Jeeva (2020), who reported that *Tetragonia tetragonioides* (a well-recognized halophyte) stimulated superoxide dismutase (SOD) enzyme's activity to reduce cell membrane damage in saline conditions.

It is also reported that succulents and other salt-tolerant species have developed a system for excreting salt through specialized organs, such as salt glands, salt hairs, or salt bladders, which are primarily located on the leaf and epidermis. These specific structures move ions from the leaf tissues to the surface of the leaf, where the saline solution crystallizes and is quickly blown or washed away by wind or rain (Caperta et al., 2020). The wild cabbage's tolerance mechanisms to salinity remain largely unknown, unlike the wealth of information available for other halophytes. This underscores the necessity to unravel the precise mechanisms enabling these plants to thrive in saline environments.

18.4 THE SIGNIFICANCE OF HALOPHYTES IN FOOD SECURITY

Water scarcity and rising soil salinity are significant global issues, particularly in agricultural regions (Hanjra & Qureshi, 2010; Tug & Yaprak, 2019). Therefore, using salt-tolerant crops can be a useful alternative to investing time and money in remediating such areas (Ventura et al., 2011). Halophytes are the ideal crop candidates for salty environments. Recent developments in the food science sector and consumer preferences for varied diets point to the consumption of wild edible plants as diet complements as well as nutritious and useful foods for specific conditions, making commercial cultivation of these plants crucial to preventing threats from irrational gathering and genetic erosion (Nikalje et al., 2019; Petropoulos et al., 2018). Many of these wild edible species have been utilized by rural communities for generations as leafy greens and as the foundation of numerous traditional cuisines and regionally significant local recipes, and they still exist throughout Southern Africa and, indeed, the world at large (Petropoulos et al., 2018).

Wild halophytes typically possess higher nutritional and bioactive contents and flavors, like those of common salad crops (Petropoulos et al., 2016). In addition to hand-harvested wild greens, modern human lifestyles and these species' seasonality have created a niche in the market for commercialized production of several species (Ruiz-Rodríguez et al., 2011). With wild plants becoming an essential part of the food supply on a worldwide scale, in both developed and developing countries, recent trends in food science and technology mandate the manufacture of unique and nutritious food products and the integration of creative processing techniques (Petropoulos et al., 2018). In addition, many halophytes have been confirmed to possess properties that treat many chronic diseases (Ksouri et al., 2011). Therefore, consumers' newfound interests compel a return to previous eating patterns, with a focus on balanced nutrition and self-medication with healthy foods, nutritional supplements, herbal medicines, and functional foods (Łuczaj et al., 2012).

Most wild species are regarded as weeds for conventional crops, and farmers typically use chemical fertilizers to get rid of them. However, this practice has the potential for long-term negative effects, including the extinction of threatened species, the resulting genetic erosion, as well as the drastic changes in the biodiversity of ecosystems (Tardío et al., 2006). Hence, it is crucial to suggest substitute crop species within the context of saline agriculture that can adapt under difficult conditions and are suitable candidates as prospective cash food and therapeutic crops (Panta et al., 2014). This therefore challenges researchers to conduct extensive studies on the edibility of halophytes and present them for commercial vegetable production, especially in areas that are struggling with salinization of agricultural lands.

18.5 HEALTH BENEFITS OF PLANT-BASED FOODS

The food industry has encountered new obstacles in the last few decades, forcing it to create new food products and diets that can halt the progression of lifestyle and/or chronic illnesses (Fehér et al., 2020). Additionally, the health of people has deteriorated in recent decades, a trend that can be linked to bad lifestyle choices, an imbalanced diet, and excessive intake of indulgent foods and beverages (Mariotti, 2023). Consequently, each year, 63% of all deaths worldwide are caused by malignant tumors, diabetes, cardiovascular illnesses, respiratory conditions, and obesity, all of which are lifestyle-linked (Fehér et al., 2020). In conjunction with the swift spread of chronic illnesses, the population in developed and semi-developed nations is aging at a rapid pace; as a result, the number of individuals who are inactive and in need of medical attention is rising (Mullins & Arjmandi, 2021). On the other hand, some regions like Asia still boast aging populations, and that is partly related to the lifestyle and dietary choices of consuming plant-based foods (Tokudome et al., 2016). Studies on selected Asian countries revealed that people who followed a prudent diet, which is characterized by consuming a lot of whole grains, legumes, vegetables, seafood, and seaweed, had longer life expectancy, while no significant correlation was found between Western diet (which is characterized by consuming a lot of refined grains, red or processed meat, and sweetened carbonated drinks) and prolonged life expectancy (Crous-Bou et al., 2019; Gong et al., 2018).

A *plant-based diet* is a dietary trend in which animal products, including eggs, meat, dairy products, and eggs, are consumed in moderation, while highly processed foods like sugar, wheat flour, and oil are kept to a minimum. Most of these diets consist mostly of plant-based foods, such cereals, nuts, seeds, tubers, legumes, vegetables grains, and fruit that are raw, unprocessed, or just lightly processed foods (Bvenura & Afolayan, 2015; Leitzmann, 2014). Lichtenstein et al. (2021) linked predominantly plant-based diets to improved cardiovascular risk factors, decreased incidence, and slower progression of coronary heart disease (CHD). Randomized controlled trials and epidemiological research have demonstrated that plant-based diets lead to a considerable reduction in CHD occurrences, lowering risk factors like diabetes and hypertension, as well as symptomatic and scintigraphic myocardial ischemia and coronary artery disease (Brennan et al., 2012; Fraser, 2009; Israeli, 2013). Research shows that a plant-based diet can potentially reduce blood pressure, cholesterol, HbA1C, body mass index, and consequently, costs associated with health care (Fehér et al., 2020; Fraser, 2009). Additionally, such a diet may lessen mortality rates from ischemic heart disease and decrease the amount of medication required to manage chronic illnesses (Israeli, 2013). Furthermore, plants are the only source of phytochemicals, like carotenoids, glucosinolates, and flavonoids, which perform multiple health functions in the human body, including neutralizing free radicals, anti-inflammation, immunity enhancement, optimization of serum cholesterol, and protection against cancer (Hever, 2016). Although studies on plant-based diets have been an area of interest globally, South African wild edible plants continue to be foraged without sufficient knowledge on their nutritional and medicinal benefits. This then necessitates extensive studies on the nutritional and medicinal properties of these plants to support their consumption and commercialization.

Table 18.1 shows the ethnomedicinal uses of some wild edible species that thrive in similar conditions as *T. ciliata* and some species in the Asphodelaceae family, which are mostly indigenous to Southern Africa, with numerous species traditionally used as laxatives or analgesics for pain relief. Many of these species are widely used by humans for the treatment of skin, cancer, kidney,

TABLE 18.1
Some Popular Wild Edible Plants Used in Ethnomedicine That Thrive in Similar Conditions as *T. ciliata*

Species	Part Used	Mode of Application	References
Tulbaghia violacea Harv.	Leaves and bulb	Decoctions of leaves are used to treat cancer of the esophagus. Bulb is used as a remedy for pulmonary tuberculosis and to destroy intestinal worms.	(Amoo et al., 2014)
Pelargonium culallatum	Leaves	For stomach disorders, sores, and wounds.	(Amoo et al., 2014)
Aloe cooperi Baker subsp. *cooperi* Bak.	Roots, leaves, flowers	Roots infusion for easy delivery; flowers cooked and eaten as vegetable.	(Malmir et al., 2018)
Aloe ecklonis Salm-Dyck	Flowers	Flowers cooked and eaten as vegetable.	(Amoo et al., 2014)
Agathosma apiculata	Leaves	Source of essential oils and flavor. Leaf decoctions are used for colds and flu.	(Amoo et al., 2014)
Coleonema pulchellum	Leaves	Rich source of essential oils and flavor. Can treat colds and flu.	(Bodede & Prinsloo, 2020)
Artemisia afra	Leaves	An addition to herbal drinks and iced tea. Used to treat colds, sore throat, coughs, asthma, and headaches.	(Bodede & Prinsloo, 2020)
Carissa macrocarpa	Fruits	Rich source of antioxidants, vitamin C, and pectin.	(Asong et al., 2019)
Carpobrotus edulis	Leaves and fruits	Leaves are used to treat cold sores, bruises, bee stings, rashes, and sunburns.	(Asong et al., 2019)
Mentha longifolia	Leaves	Rich source of essential oils. Tea made from leaves used for stomach disorders.	(Rankoana, 2021)
Salvia africana-lutea	Leaves	Tea made from the leaves good for colds, coughs, and stomach ailments.	
Portulacaria afra		Leaves chewed by lactating mothers to stimulate production of milk. Also, it is high in phytochemicals.	(Dlamini et al., 2010)
Trachyandra falcata Kunth.	Flowers	Cooked as vegetable.	(Tshayingwe et al., 2023)
Trachyandra divaricata Kunth.	Flowers	Cooked as vegetable.	(Bulawa et al., 2022)

bladder, and stomach problems (Malmir et al., 2018). Furthermore, some well-known and unspecified species have been used by women as oral contraceptives and during birth, while some edible *Portulacaria* species have been traditionally used by lactating mothers to stimulate milk production. Some of the species are consumed as vegetables and fruits, added in salads, and some are used to make tea, although their edibility and nutrition are yet to be recorded (Amoo et al., 2014). Some species are widely used to treat skin-related problems, while they are also edible (Asong et al., 2019). Few *Trachyandra* species have been reported only as edible vegetable plants, although thorough research is still needed on this genus (Bulawa et al., 2022; Nortje & van Wyk, 2019).

18.6 *ASPHODELACEAE* FAMILY

Asphodelaceae is a family of flowering plants which includes 12 genera and over 1060 species and is indigenous to Africa, the Mediterranean basin, the Arabian Peninsula, West and Central Europe, Madagascar, Central Asia, Australia, and New Zealand (Chase et al., 2016; S. Ngxabi et al., 2021a). The nomenclature of this family dates back to 1789, when it was first scientifically published in Jussieu's *Genera Plantarum* (Ngxabi et al., 2021a). Asphodeloideae and Alooideae are the two subfamilies found under this family (Chase et al., 2000). The Asphodelaceae family has a highly diverse range of members; however, there are a few traits they all share. They can

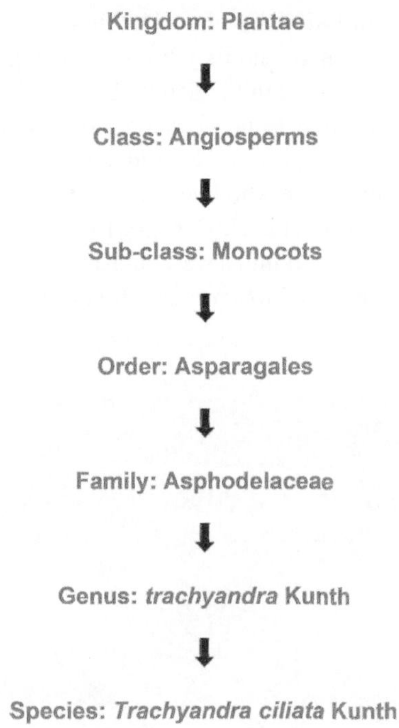

FIGURE 18.2 Classification of *Trachyandra ciliata.*

be small- to medium-sized plants, which are frequently succulent, herbs, or enormous trees with terminal rosettes of leaves instead of succulent stems (Ngxabi, 2020). According to Smith and Van Wyk (1998): "Leaves dorsiventral, lanceolate-acuminate, linear or subulate, terete, often succulent and thickly conical, spirally arranged or distichous as in some species of Alooideae, amplexicaul, margins toothed, serrate or entire, sharply pointed, parallel veins often obscure."

Trachyandra is a genus under Asphodelaceae that is made up of 50 more species distributed throughout the sub-Saharan Africa. *Trachyandra* species are distinguished from the species within the family by their soft-textured leaves and white to pink flowers that are short-lived, with a deciduous perianth and rough filaments (Ngxabi et al., 2021a; Smith & Van Wyk, 1998). *Trachyandra* species can either be woody or herbaceous, and some species can have granular pubescence. They are deciduous plants with fibrous or spindle-shaped root systems that occasionally swell toward the terminals to produce thin tubers.

The upright, woody, or herbaceous stems are typically naked but occasionally coated with old leaf bases. Rhizomes are most often established as vertical but can also be horizontal. The bottoms of the leaves, which are tube-like and frequently resemble a sheath of fibers, are either homogeneous or have two unique morphologies. On aerial stems, the leaves are grouped in a rosette formation, infrequently in alternating form, or in dense spirals. The lamina is flat, cylindrical, sharply three-angled, hairy, or smooth, or glandular pubescent, but occasionally spirally twisted and folded (Obermeyer, 1962). According to our earlier work (Ngxabi et al., 2021a):

> The inflorescence is axillary with single or branched raceme unusually sub-umbellate, naked or with sterile bracts. The inflorescence has single pedicels that change their position during the flowering stage and when the capsule ripens. These species have flowers that are either scentless or characterised by a strong scent that usually open during the day and at night. Plants in this genus have six stamens that are attached to the base of the perianth.

The term *Trachyandra*, which refers to the hairy filaments, is derived from the Greek words *trachy*, which means "rough," and *andro*, which indicates "male." *Trachyandra ciliata* is a spreading perennial herbaceous plant that can reach a height of 0.5 m. The mature leaves are linear, 100

× 4 cm in size, and succulent, with hairy margins. The flowers are grouped in crowded racemes and have white tepals with a pink midrib, and they last only a single day. These snake-like flowers distinguish this species from other species in the genus (De Vynck et al., 2016).

The literature on this plant is limited to non-existent, and there are currently no *Trachyandra* species under commercial cultivation (Manning & Goldblatt, 2007). However, it is known that this plant is edible and that the Khoisan people who lived around the South African Cape Coast utilized it as food (De Vynck et al., 2016; Ngxabi et al., 2021a, 2021b). The inflorescence, just like asparagus, is steamed and eaten as a vegetable, added into a stew, and added into salads. Only two other species in the *Trachyandra* genus are edible (*T. divaricata* and *T. falcata*) and are also eaten as a vegetable or added into a stew.

FIGURE 18.3 Edible inflorescence of *T. ciliata*.

Source: Ngxabi et al. (2021a).

FIGURE 18.4 Veldkool cooked with lamb stew.

Source: Ngxabi et al. (2021a).

18.7 DISTRIBUTION OF *TRACHYANDRA* SPECIES IN SOUTHERN AFRICA

Trachyandra is primarily a South African genus, with majority of species occurring in the Southwestern Cape. *Trachyandra* species can be found all over Southern Africa, with most of them native to the Southwestern Cape's winter rainfall region. Others can be found in southern Zimbabwe, Angola, Kenya, with one occurring in Ethiopia (Smith & Van Wyk, 1998). Given that this plant is found naturally on the coastal sand dunes of the Western Cape, it suggests a strong tolerance to salinity, making it a potentially beneficial addition to coastal gardens (Ngxabi et al., 2021a).

18.8 SIGNIFICANCE AND POTENTIAL OF ASPHODELACEAE

Recent research examining the potential of indigenous South African species as food crops has revealed that several species could be profitable products for the food and beverage industries, although Asphodelaceae plants have always been known for their effective medicinal capacity (Van Wyk, 2011). The use of the inner, non-bitter part of some *Aloe* species as food supplements is a potentially new frontier in research. The presence of flavonoids in some *Aloe* species has been reported, while some species in this family are used as food preservatives by local Cape farmers (Cock, 2015). Ngxabi (2020) and Ngxabi et al. (2021b) also reported the presence of antioxidants and phytochemicals in *T. ciliata*, especially when subjected to salt stress.

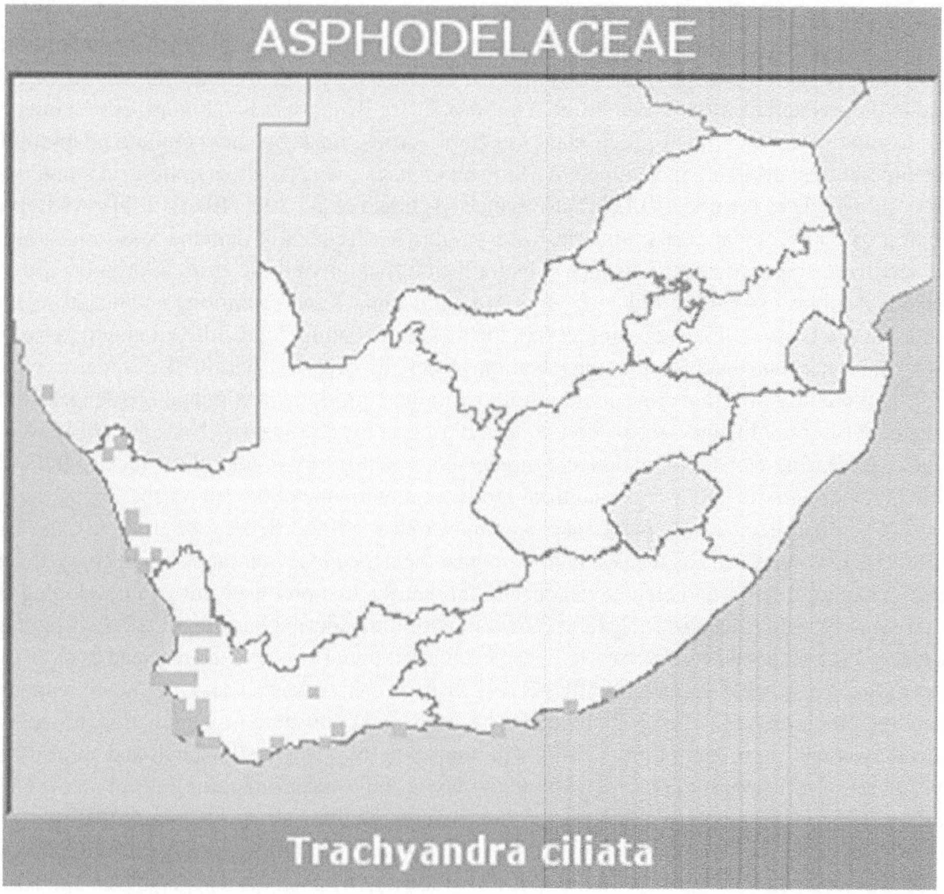

FIGURE 18.5 Distribution of *T. ciliata* in South Africa.

Source: http://redlist.sanbi.org/species.php?species=2203-18.

Researchers have identified the health advantages of food and drink items made from the leaf parenchyma of *Aloe* species, and the use of *Aloe* extracts as direct food additives for human consumption has been approved by the Food and Drug Administration (Salehi et al., 2018). Some *Aloe* leaf gels have been found to include flavonols as well as some vital vitamins (vitamins C, B_1, B_2, B_6, B_{12}, and E) that are essential for the effectiveness of the human body (Cock, 2015). These authors also reported iron, calcium, manganese, potassium, phosphorous, copper, magnesium, zinc, molybdenum, and sodium as essential minerals present on the gel of *Aloe* species. Ngxabi et al. (2021a) reported that *Trachyandra ciliata*, together with two other species in the genus *Trachyandra* (*T. Divaricata* and *T. falcata*), is edible and was used by the original Khoisan people who used to live in the Western Cape coastal areas. With little to no literature available about this plant, more research should be conducted to evaluate its nutrition.

18.9 POTENTIAL VALUE OF *TRACHYANDRA CILIATA*

18.9.1 Economic Potential as a Commercial Edible Vegetable

Due to their saltiness and crispiness, edible halophytes are used in a variety of green foods served in restaurants and have become popular in several cultures worldwide, contributing to the preservation of traditional food systems and culinary heritage (Bvenura & Afolayan, 2015; S. Ngxabi et al., 2021a). Many underutilized vegetables have been reported to be rich in phytochemicals and essential nutrients, such as vitamins, dietary fiber, proteins, calcium, iron, and magnesium, all of which are essential for human and animal health and nutrition. Therefore, promoting their consumption can contribute to improved nutrition and public health outcomes, potentially reducing health-care costs related to diet-related illnesses (Bvenura & Afolayan, 2015; Bvenura & Sivakumar, 2017; Gong et al., 2018). Compared to traditional green leafy vegetable crops, these species' climate adaptability to high temperatures, drought, and saline environments makes them excellent options in some regions that are vulnerable to the effects of climate change (Agudelo et al., 2021; Ebert, 2014). Additionally, compared to conventional crops, underutilized halophytes frequently demand less inputs, such as water, fertilizer, and pesticides. This then calls for the cultivation of these crops to promote more sustainable agricultural practices, reducing environmental impact and promoting ecological resilience (Bvenura & Afolayan, 2015; Zuccarini, 2008). Furthermore, commercial cultivation of underutilized crops by farmers contributes to the conservation of genetic diversity within crop species, which is crucial for breeding programs and maintaining resilience amid environmental challenges, such as climate change, drought, and salinization of agricultural lands (Galluzzi & Noriega, 2014).

Due to increasing global population and threatened food security in the future, recent studies have reported the necessity to enhance agricultural production, with one of the approaches being the introduction of underutilized edible plant species (Bulawa et al., 2022; S. Ngxabi et al., 2021b; Sogoni et al., 2021). This approach has the potential to reduce the dependency on mainstream crops, diversify farmers' income streams, and create resilience to market fluctuations, especially in developing countries (Bvenura & Afolayan, 2015; Ngxabi, 2020). In addition, Aderibigbe et al. (2022) highlighted the potential of exploring wild edible crops to open new market opportunities that may lead to more export opportunities, since these species usually possess distinct characteristics and are mostly endemic to specific regions, which may be profitable to the farmers and contribute to job creation, especially in disadvantaged and rural communities. This will contribute to poverty alleviation and rural development. Other market opportunities for wild edible crops may include tapping into processed value-added products, such as jams, pickles, sauces, and beverages (Bvenura & Sivakumar, 2017).

Trachyandra ciliata is less studied, with the only existing literature mainly based on its taxonomy. However, it has been documented that it was traditionally eaten as a vegetable by the Khoisan people in the past, but that information was eradicated due to colonization and migration. Our previous study (Ngxabi et al., 2021b) reported that maximum yield was achieved when irrigated with saline water. We further reported increased phytochemical contents under saline conditions.

Further studies on closely related species (*Trachyandra divaricata* Kunth) reported the presence of some significant nutrients, such as Ca, Mg, P, N, K, Na, Cu, Zn, and Fe, which support the need for more studies on this genus (Bulawa et al., 2022; Tshayingwe et al., 2023).

Trachyandra remains an underexploited genus, with none of its species under known cultivation. People using this plant as food access it through foraging, which is not a recommended practice because it tempers with the diversity and sustainability of wild ecosystems (Ngxabi et al., 2021a). This therefore makes *T. ciliata* an appropriate candidate as a commercial vegetable crop. This means there is a need to further study its nutritional contents and attempt to introduce it in the commercial vegetable market.

18.9.2 Potential Medicinal Value of Halophytes

Halophytes have been used medicinally in various cultures around the world and have been shown to be beneficial in treating and managing chronic diseases that trouble modern worlds (Ksouri et al., 2011). According to Ksouri et al. (2011), "[c]urrently, an increasing interest is granted to these species because of their high content in bioactive compounds (primary and secondary metabolites) such as polyunsaturated fatty acids, carotenoids, vitamins, sterols, essential oils (terpenes), polysaccharides, glycosides, and phenolic compounds." These authors further alluded that these bioactive chemicals exhibit strong antioxidant, antimicrobial, anti-inflammatory, and antitumor properties, and as a result, they serve as important antiaging and disease-prevention agents. This phenomenon in South Africa is, however, underexplored, with very few halophytes having been studied for medicinal purposes (Ngxabi et al., 2021b). *T. ciliata* is therefore no exception, and its medicinal potential can be inferred from its relatives in the Asphodelaceae family.

The demand for new drugs with robust activity against these eukaryotic microorganisms has significantly increased because of a rise in the number of immune-deficient and immunocompromised individuals suffering from parasitic illnesses and fungi (Era et al., 2014). In Southern Africa, the vast majority (85%) of usage records pertaining to *Aloe* species highlight their pharmaceutical benefits (Grace et al., 2008). The use of the species of *Bulbine* (Asphodelaceae) dates to the eighteenth century, when British and Dutch settlers in South Africa used the species to treat a variety of illnesses, including parasitic infections, diarrhea, as well as stomachaches, urinary tract infections, viral infections like HIV, chicken pox, shingles, and sexually transmitted diseases (Bodede & Prinsloo, 2020; van Wyk, 2008). Halophytes are useful for boosting the immune system, lowering inflammation, and guarding against cancers, hypertension, and atherosclerosis (Im et al., 2003). In recent years, there has been a rising research interest worldwide in investigating the medicinal properties of halophytes, as they demonstrate significant potential as a source of secondary metabolites. However, South African halophytes, including *T. ciliata*, remain undervalued as potential sources of useful bioactive compounds.

18.9.3 Desalinization

Around 80 million hectares of agricultural lands are affected by soil salinity worldwide, and this will potentially increase (Flowers et al., 2010; Rabhi et al., 2009). Such soils might be recovered using a variety of techniques, including hydraulic, physical, chemical, and biological approaches (Rabhi et al., 2009; Shiyab et al., 2003). However, in developing and semi-developed countries, hydraulic, physical, and chemical approaches make it difficult to explore these measures due to high costs involved (Flowers et al., 2010; Pinto & Marques, 2017). This then calls for more biological approaches to be adopted as sustainable and environmentally friendly measures for desalinization.

Crop rotation and the use of salt-tolerant crops to absorb excess salts in agricultural lands are gaining momentum worldwide. Particularly for some poor nations, where chemical additives are becoming increasingly expensive, this plant-based approach is of paramount importance. Shabala (2013) stated that for revegetation and rehabilitation of salt-affected land, halophytes are the best plant species. The study of Watson and O'leary (1993) showed that *Atriplex nummularia* (old man saltbush) produced 20 to 30 t/ha of biomass per year and accumulated between 20 and 40% NaCl in its dry matter when irrigated with saline water. *Suaeda fruticosa* (sea blite), on the other hand, may remove more than 2.5 t of salt per hectare in a single harvest of the plant's aerial portions each year (Shabala, 2013). In the context of African halophytes, this phenomenon is underexplored and more studies should be conducted to explore such avenues.

18.9.4 Phytoremediation Using Halophytes

Global industrialization over the past 200 years has significantly increased the production, usage, and environmental release of heavy metals (Flowers & Colmer, 2015). Due to the buildup of harmful ions brought by using chemical fertilizers in agricultural fields, as well as climate change, soil fertility has been lost, and it has also caused salinization and desertification, making it difficult to cultivate crops (Yasin et al., 2019). Farmers have mainly been adopting chemical amendments for the improvement of salty and sodic soils to combat this devastating situation. Nonetheless, this use of chemicals has become expensive, especially for developing countries, and it can also exert adverse effects on the environment (Hasanuzzaman et al., 2014). This therefore challenges researchers to find cost-effective and environmentally friendly methods to reduce the catastrophic effects of salinity in agricultural lands. Halophytes are frequently recommended for phytoremediation (phytostabilization and phytoextraction) because tolerance to salt and heavy metals depends, at least in part, on similar physiological mechanisms (Manousaki & Kalogerakis, 2011; Shabala, 2013; Thomas et al., 1998). *Phytostabilization* occurs when a metal-tolerant non-accumulator plant can withstand metals in its aboveground parts without transferring or accumulating them. These plants offer soil stabilization and prevent metals from being mobilized or leached into the groundwater, and they also provide vegetation cover (Manousaki & Kalogerakis, 2011). *Phytoextraction* is the process of removing harmful metals from the soil by building up in the plant's top layers so they can be collected and removed. *Trachyandra ciliata* is one of the unexplored halophytes from South Africa, which makes it one of the options to address the issue of contaminated soils.

18.9.5 Companion Planting/Intercropping

Utilizing halophytic plants in crop rotations or mixed cropping systems may be a promising management strategy to reduce salt stress–related yield losses because halophytes are able to accumulate high quantities of NaCl in their tissues (Karakas et al., 2016). Intercropping is mostly used in developing nations to increase crop productivity (De La Fuente et al., 2014; Karakas et al., 2016). In South Africa, crop rotation stands out as the most commonly adopted practice, particularly among small-scale farmers, aimed at averting the risk of crop failure by alternating different plant species. Intercropping using halophytic plants in South Africa remains unexplored, although the country is one of the countries threatened by soil salinization and drought (Bantie et al., 2014). In other countries, there are methods that have been adopted to reduce salinity in the soil, such as physical reclamation (deep ploughing), chemical reclamation, plant control, and water-based approaches. However, these methods are cost- and labor-intensive and therefore not sustainable or feasible for many developing countries (Karakas et al., 2016; Qadir et al., 2007; Singh et al., 2012).

In other countries, there are some studies that have been conducted to test the performance of normal cash crops when cultivated with halophytes under saline conditions. Halophytes take up salt from the soil and store it in their tissues, preventing crop plants from getting to harmful ions (Zuccarini, 2008). The study of Colla et al. (2006) investigated the effectiveness of *Salsola soda* (halophyte) as a companion plant for peppers. According to these authors, the presence of *S. soda* decreased the EC value by 43% and increased total crop, marketable crop, and total biomass values by 26, 32, and 22%, respectively. In addition, a study conducted by Zuccarini (2008) found that tomato plants intercropped with halophytes (purslane and garden orache) had lower NaCl concentration and a yield increase of 44% compared to those that were not intercropped. These studies reveal that there is a potential for cultivating traditional food crops in saline soils by intercropping with halophytes. In the case of *T. ciliata*, there is a need to examine its potential as a companion plant because of its edibility. Studies to examine the potential of South African halophytes as companion plants to traditional food crops are desperately needed.

18.10 CONCLUSION AND PROSPECTS

Water scarcity and high salinity in agricultural lands cause restrictions in plants, which decrease overall output and leaf production, delay flowering, and cause flower bud abortion. To cater for the constantly increasing population, researchers predict that overall global food production needs to increase by 50 to 100% by 2050, although the current increase is around 1 to 1.5% annually. This then necessitates the cultivation of more salt- and drought-tolerant plant species (halophytes) for food and medicine to substitute the traditional vegetable crops that are currently under cultivation. In South Africa, native wild halophytes have been neglected and underresearched, although they have been gaining increasing attention in the past decade. Previous studies have identified *T. ciliata* as an edible wild halophyte that was used for food by the Khoisan people that used to live in the Western Cape coastal areas before colonization and massive industrialization of the province. Previous research on this plant that focused on determining its cultivation protocol also showed that, indeed, the plant is more productive under saline conditions, although its anatomic salt tolerance mechanisms, nutritional profile, seed germination protocol, medicinal properties, and biochemical and physiological properties remain unknown. Further research is recommended to study its leaf micromorphological properties and chemical composition to understand its precise salt tolerance and changes in its anatomical structures that are vital for biochemical processes. Further studies are recommended to determine the plant's nutritional profile in the form of macronutrients, micronutrients, and antinutrients to promote its consumption and commercialization. Recent studies including this research recommend the commercial production and promotion of wild edible crops to keep up with the increasing demand for food due to population growth. Other production techniques, like hydroponics and vertical cultivation, are also recommended to grow these plants, especially in urban areas and private homes, to promote healthy dietary lifestyles, more especially that these techniques use less water than field cultivation, contributing to sustainable water usage. To promote the commercial cultivation and popularity of these crops, more resources need to be made available for marketing in the form of education and awareness campaigns to capacitate the farmers and communities about the economic potential of these crops. Capacity building seminars and market days in local and rural communities should also include the benefits of cultivating wild edible crops, which may lead to the emergence of more culinary innovations that can be sold to prominent food markets and wholesalers. Findings from this study serve as points of reference to commercial farmers, communities, scholars, policymakers, and aspiring researchers whose interests are on the potential of easily accessible underutilized wild edible crops to develop new strategies to address global issues.

18.11 ACKNOWLEDGMENTS

This research was supported by the National Research Foundation (grant number MND210624615237).

REFERENCES

Acquaah, G. (2009). *Principles of plant genetics and breeding*. Blackwell Publishing.

Aderibigbe, O. R., Ezekiel, O. O., Owolade, S. O., Korese, J. K., Sturm, B., & Hensel, O. (2022). Exploring the potentials of underutilized grain amaranth (Amaranthus spp.) along the value chain for food and nutrition security: A review. *Critical Reviews in Food Science and Nutrition*, *62*, 656–669. https://doi.org/10.1080/10408398.2020.1825323

Agudelo, A., Carvajal, M., & Martinez-Ballesta, M. D. C. (2021). Halophytes of the mediterranean basin – underutilized species with the potential to be nutritious crops in the scenario of the climate change. *Foods*, *10*. https://doi.org/10.3390/foods10010119

Ahanger, M. A., & Agarwal, R. M. (2017). Salinity stress induced alterations in antioxidant metabolism and nitrogen assimilation in wheat (Triticum aestivum L) as influenced by potassium supplementation. *Plant Physiology and Biochemistry*, *115*, 449–460. https://doi.org/10.1016/j.plaphy.2017.04.017

Ahmad, P., Alyemeni, M. N., Ahanger, M. A., Egamberdieva, D., Wijaya, L., & Alam, P. (2018). Salicylic Acid (SA) induced alterations in growth, biochemical attributes and antioxidant enzyme activity in Faba Bean (Vicia faba L.) seedlings under NaCl toxicity. *Russian Journal of Plant Physiology*, *65*, 104–114. https://doi.org/10.1134/S1021443718010132

Ahmad, Z., Waraich, E. A., Akhtar, S., Anjum, S., Ahmad, T., Mahboob, W., Hafeez, O. B. A., Tapera, T., Labuschagne, M., & Rizwan, M. (2018). Physiological responses of wheat to drought stress and its mitigation approaches. *Acta Physiologiae Plantarum*, *40*. https://doi.org/10.1007/s11738-018-2651-6

Al Hassan, M., Estrelles, E., Soriano, P., López-Gresa, M. P., Bellés, J. M., Boscaiu, M., & Vicente, O. (2017). Unraveling salt tolerance mechanisms in halophytes: A comparative study on four mediterranean Limonium species with different geographic distribution patterns. *Frontiers in Plant Science*, *8*. https://doi.org/10.3389/fpls.2017.01438

Amoo, S. O., Aremu, A. O., & Van Staden, J. (2014). Unraveling the medicinal potential of South African aloe species. *Journal of Ethnopharmacology*, *153*, 19–41. https://doi.org/10.1016/j.jep.2014.01.036

Asong, J. A., Ndhlovu, P. T., Khosana, N. S., Aremu, A. O., & Otang-Mbeng, W. (2019). Medicinal plants used for skin-related diseases among the Batswanas in Ngaka Modiri Molema district municipality, South Africa. *South African Journal of Botany*, *126*, 11–20. https://doi.org/10.1016/j.sajb.2019.05.002

Bantie, Y. B., Abera, F. A., & Woldegiorgis, T. D. (2014). Competition indices of Intercropped Lupine (Local) and small cereals in additive series in West Gojam, North Western Ethiopia. *American Journal of Plant Sciences*, *5*, 1296–1305. https://doi.org/10.4236/ajps.2014.59143

Bodede, O., & Prinsloo, G. (2020). Ethnobotany, phytochemistry and pharmacological significance of the genus Bulbine (Asphodelaceae). *Journal of Ethnopharmacology*, *260*. https://doi.org/10.1016/j.jep.2020.112986

Brennan, N., Bartee, S., Wheatcroft, G., & Krometis, J. (2012). *Chronic conditions among medicare beneficiaries*. CMS, Office of Information Products and Data Analysis.

Bulawa, B., Sogoni, A., Jimoh, M. O., & Laubscher, C. P. (2022). Potassium application enhanced plant growth, mineral composition, proximate and phytochemical content in Trachyandra divaricata Kunth (Sandkool). *Plants*, *11*. https://doi.org/10.3390/plants11223183

Bvenura, C., & Afolayan, A. J. (2015). The role of wild vegetables in household food security in South Africa: A review. *Food Research International*, *76*, 1001–1011. https://doi.org/10.1016/j.foodres.2015.06.013

Bvenura, C., & Sivakumar, D. (2017). The role of wild fruits and vegetables in delivering a balanced and healthy diet. *Food Research International*, *99*, 15–30. https://doi.org/10.1016/j.foodres.2017.06.046

Caperta, A. D., Róis, A. S., Teixeira, G., García-Caparros, P., & Flowers, T. J. (2020). Secretory structures in plants: Lessons from the Plumbaginaceae on their origin, evolution and roles in stress tolerance. *Plant Cell Environment*, *43*, 2912–2931. https://doi.org/10.1111/pce.13825

Chase, M. W., Christenhusz, M. J. M., Fay, M. F., Byng, J. W., Judd, W. S., Soltis, D. E., Mabberley, D. J., Sennikov, A. N., Soltis, P. S., Stevens, P. F., Briggs, B., Brockington, S., Chautems, A., Clark, J. C., Conran, J., Haston, E., Möller, M., Moore, M., Olmstead, R., & Perret, M., Skog, L., Smith, J., Tank, D., Vorontsova, M., & Weber, A. (2016). An update of the Angiosperm Phylogeny Group classification for the orders and families of flowering plants: APG IV. *Botanical Journal of the Linnean Society*, *181*, 1–20. https://doi.org/10.1111/BOJ.12385

Chase, M. W., De Bruijn, A. Y., Cox, A. V., Reeves, G., Rudall, P. J., Johnson, M. A. T., & Eguiarte, L. E. (2000). Phylogenetics of asphodelaceae (Asparagales): An analysis of plastid rbcL and trnL-F DNA sequences. *Annals of Botany*, *86*, 935–951. https://doi.org/10.1006/anbo.2000.1262

Cock, I. E. (2015). The genus aloe: Phytochemistry and therapeutic uses including treatments for gastrointestinal conditions and chronic inflammation. In K. D. Rainsford, M. C. Powanda, & M. W. Whitehouse (Eds.), *Progress in drug research* (pp. 179–235). Springer.

Colla, G., Rouphael, Y., Fallovo, C., Cardarelli, M., & Graifenberg, A. (2006). Use of salsola soda as a companion plant to improve greenhouse pepper (Capsicum annuum) performance under saline conditions. *New Zealand Journal of Crop and Horticultural Science*, *34*, 283–290. https://doi.org/10.1080/01140671.2006.9514418

Colvin, C., & Muruven, D. (2017). *Scenarios for the future of water in South Africa*. The Boston Consulting Group. www.wwf.org.za

Condon, A. G., Richards, R. A., Rebetzke, G. J., & Farquhar, G. D. (2004). Breeding for high water-use efficiency. *Journal of Experimental Botany*, 2447–2460. https://doi.org/10.1093/jxb/erh277

Corwin, D. L. (2021). Climate change impacts on soil salinity in agricultural areas. *European Journal of Soil Science*. https://doi.org/10.1111/ejss.13010

Crous-Bou, M., Molinuevo, J. L., & Sala-Vila, A. (2019). Plant-rich dietary patterns, plant foods and nutrients, and telomere length. *Advances in Nutrition*. https://doi.org/10.1093/advances/nmz026

De La Fuente, E. B., Suárez, S. A., Lenardis, A. E., & Poggio, S. L. (2014). Intercropping sunflower and soybean in intensive farming systems: Evaluating yield advantage and effect on weed and insect assemblages. *NJAS – Wageningen Journal of Life Sciences*, *70*, 47–52. https://doi.org/10.1016/j.njas.2014.05.002

De Vynck, J. C., Van Wyk, B. E., & Cowling, R. M. (2016). Indigenous edible plant use by contemporary Khoe-San descendants of South Africa's Cape South Coast. *South African Journal of Botany*, *102*, 60–69. https://doi.org/10.1016/j.sajb.2015.09.002

Dlamini, N. R., Moroka, T., Mlotshwa, L., Reddy, J., & Botha, G. (2010). Indigenous edible plants as sources of nutrients and health benefitting components (nutraceuticals). In *Science real and relevant conference*. CSIR.

Ebert, A. W. (2014). Potential of underutilized traditional vegetables and legume crops to contribute to food and nutritional security, income and more sustainable production systems. *Sustainability (Switzerland)*, *6*, 319–335. https://doi.org/10.3390/su6010319

Era, V., Isca, M. S., Seca, A. M. L., Pinto, D. C. G. A., & Silva, A. M. S. (2014). An overview of Salícornia Genus: The phytochemical and pharmacological profile. In V. K. Gupta (Ed.), *Natural products: Research reviews* (pp. 145–164). Daya Publishing House.

FAO. (2008). *The state of food and agriculture*. Food and Agriculture Organization of the United Nations.

Fehér, A., Gazdecki, M., Véha, M., Szakály, M., & Szakály, Z. (2020). A comprehensive review of the benefits of and the barriers to the switch to a plant-based diet. *Sustainability (Switzerland)*. https://doi.org/10.3390/su12104136

Flowers, T. J., & Colmer, T. D. (2008). Salinity tolerance in halophytes. *New Phytologist*, *179*, 945–963. https://doi.org/10.1111/j.1469-8137.2008.02531.x

Flowers, T. J., & Colmer, T. D. (2015). Plant salt tolerance: Adaptations in halophytes. *Annals of Botany*, *115*, 327–331. https://doi.org/10.1093/aob/mcu267

Flowers, T. J., Gaur, P. M., Gowda, C. L. L., Krishnamurthy, L., Samineni, S., Siddique, K. H. M., Turner, N. C., Vadez, V., Varshney, R. K., & Colmer, T. D. (2010). Salt sensitivity in chickpea. *Plant Cell Environment*, *33*, 490–509. https://doi.org/10.1111/j.1365-3040.2009.02051.x

Fraser, G. E. (2009). Vegetarian diets: What do we know of their effects on common chronic diseases? *The American Journal of Clinical Nutrition*, *89*, 1607–1612.

Galluzzi, G., & Noriega, I. L. (2014). Conservation and use of genetic resources of underutilized crops in the Americas-A continental analysis. *Sustainability (Switzerland)*, *6*, 980–1017. https://doi.org/10.3390/su6020980

Gong, Y., Tian, G., Xue, H., Zhang, X., Zhao, Y., & Cheng, G. (2018). Higher adherence to the 'vegetable-rich' dietary pattern is related to longer telomere length in women. *Clinical Nutrition*, *37*, 1232–1237. https://doi.org/10.1016/j.clnu.2017.05.005

González-Orenga, S., Grigore, M. N., Boscaiu, M., & Vicente, O. (2021). Constitutive and induced salt tolerance mechanisms and potential uses of limonium mill. *Species Agronomy*. https://doi.org/10.3390/agronomy11030413

Grace, O. M., Simmonds, M. S. J., Smith, G. F., & van Wyk, A. E. (2008). Therapeutic uses of Aloe L. (Asphodelaceae) in Southern Africa. *Journal of Ethnopharmacology*, *119*, 604–614. https://doi.org/10.1016/j.jep.2008.07.002

Hamed, K., & Custódio, L. (2019). How could halophytes provide a sustainable alternative to achieve food security in marginal lands? In M. Hasanuzzaman, K. Nahar, & M. Öztürk (Eds.), *Ecophysiology, abiotic stress responses and utilization of halophytes* (pp. 259–270). Springer.

Hanjra, M. A., & Qureshi, M. E. (2010). Global water crisis and future food security in an era of climate change. *Food Policy*, *35*, 365–377. https://doi.org/10.1016/j.foodpol.2010.05.006

Hasanuzzaman, M., Nahar, K., Alam, M. M., Bhowmik, P. C., Hossain, M. A., Rahman, M. M., Prasad, M. N. V., Ozturk, M., & Fujita, M. (2014). Potential use of halophytes to remediate saline soils. *BioMed Research International*, 1–12. https://doi.org/10.1155/2014/589341

Hever, J. (2016). Plant-based diets: A physician's guide. *The Permanente Journal*, *20*, 15–82.

Ho, L. C., & Adams, P. (1995). Nutrient uptake and distribution in relation to crop quality. *Acta Horticulturae*, 33–44.

Im, S. A., Kim, G. W., & Lee, C. K. (2003). Immunomodulatory activity of Salicornia herbacea L. Components. *Natural Product Sciences*, *9*, 273–277.

Israeli, E. (2013). Nocturnal glucose control with an artificial pancreas at a diabetes camp. *Israel Medical Association Journal, 15*, 255. https://doi.org/10.1056/NEJMOA1206881/SUPPL_FILE/NEJMOA1206881_DISCLOSURES.PDF

Jeeva, S. (2020). Studies on the performance and phytoremediation effect of underutilized leafy vegetables in salt affected soils. *International Journal of Chemical Studies, 8*, 1762–1764. https://doi.org/10.22271/chemi.2020.v8.i2aa.9015

Karakas, S., Cullu, M. A., Kaya, C., & Dikilitas, M. (2016). Halophytic companion plants improve growth and physiological parameters of tomato plants grown under salinity. *Pakistan Journal of Botany, 48*, 21–28.

Ksouri, R., Ksouri, W. M., Jallali, I., Debez, A., Magné, C., Hiroko, I., & Abdelly, C. (2011). Medicinal halophytes: Potent source of health promoting biomolecules with medical, nutraceutical and food applications. *Critical Reviews in Biotechnology, 32*, 289–326.

Leitzmann, C. (2014). Vegetarian nutrition: Past, present, future, in: American Journal of Clinical Nutrition. *American Society for Nutrition, 100*, 496–502. https://doi.org/10.3945/ajcn.113.071365

Lichtenstein, A. H., Appel, L. J., Vadiveloo, M., Hu, F. B., Kris-Etherton, P. M., Rebholz, C. M., Sacks, F. M., Thorndike, A. N., Van Horn, L., & Wylie-Rosett, J. (2021). Dietary guidance to improve cardiovascular health: A scientific statement from the american heart association. *Circulation, 144*, E472–E487. https://doi.org/10.1161/CIR.0000000000001031

Łuczaj, Ł., Pieroni, A., Tardío, J., Pardo-De-Santayana, M., Sõukand, R., Svanberg, I., & Kalle, R. (2012). Wild food plant use in 21st century Europe: The disappearance of old traditions and the search for new cuisines involving wild edibles. *Acta Societatis Botanicorum Poloniae, 81*, 359–370. https://doi.org/10.5586/asbp.2012.031

Malmir, M., Serrano, R., Caniça, M., Silva-Lima, B., & Silva, O. (2018). A comprehensive review on the medicinal plants from the genus Asphodelus. *Plants, 7*, 20. https://doi.org/10.3390/plants7010020

Manning, J. (2007). *Field guide to fynbos* (1st ed.). Struik Nature.

Manning, J., & Goldblatt, P. (2007). Asphodelaceae. *Bothalia*, 27–31.

Manousaki, E., & Kalogerakis, N. (2011). Halophytes – an emerging trend in phytoremediation. *International Journal of Phytoremediation, 13*, 959–969.

Mariotti, F. (2023). Nutritional and health benefits and risks of plant-based substitute foods. *Proceedings of the Nutrition Society*. https://doi.org/10.1017/S0029665123004767

Mittler, R. (2017). ROS are good. *Trends in Plant Science*. https://doi.org/10.1016/j.tplants.2016.08.002

Molden, D., Frenken, K., Barker, R., De Fraiture, C., Mati, B., Svendsen, M., Sadoff, C., & Finlayson, C. M. (2007). *Changing diets lead to changing water use in agriculture*. Trends in Water and Agricultural Development. https://horizon.documentation.ird.fr/exl-doc/pleins_textes/divers15-08/010050288.pdf

Mullins, A. P., & Arjmandi, B. H. (2021). Health benefits of plant-based nutrition: Focus on beans in cardiometabolic diseases. *Nutrients, 13*, 519. https://doi.org/10.3390/nu13020519

Munns, R., & Tester, M. (2008). Mechanisms of salinity tolerance. *Annual Review of Plant Biology, 59*, 651–681. https://doi.org/10.1146/annurev.arplant.59.032607.092911

Ngxabi, S. (2020). *The effect of different growth media and salinity on the vegetative growth of Trachyandra ciliata (Wild cabbage) in hydroponics* [MSc, Cape Peninsula University of Technology].

Ngxabi, S., Jimoh, M. O., Kambizi, L., & Laubscher, C. P. (2021a). Growth characteristics, phytochemical contents, and antioxidant capacity of trachyandra ciliata (L.f) kunth grown in hydroponics under varying degrees of salinity. *Horticulturae, 7*. https://doi.org/10.3390/horticulturae7080244

Ngxabi, S., Jimoh, M. O., Kambizi, L., & Laubscher, C. P. (2021b). Edibility of Trachyandra ciliata (L.f.) kunth-an underutilized vegetable from South African Western Cape Coast: A review. *African Journal of Food, Agriculture, Nutrition and Development, 21*, 18644–18661. https://doi.org/10.18697/AJFAND.104.20810

Nikalje, G. C., Bhaskar, S. D., Yadav, K., & Penna, S. (2019). Halophytes: Prospective plants for future. In M. Hasanuzzaman, K. Nahar, & M. Öztürk (Eds.), *Ecophysiology, abiotic stress responses and utilization of halophytes* (pp. 221–234). Springer.

Nortje, J. M., & van Wyk, B. E. (2019). Useful plants of Namaqualand, South Africa: A checklist and analysis. *South African Journal of Botany, 122*, 120–135. https://doi.org/10.1016/J.SAJB.2019.03.039

Obermeyer, A. A. (1962). A revision of the south african species of anthericum, chlorophytum and trachyandra. *Bothalia, 7*, 669–767. https://doi.org/10.4102/abc.v7i4.1679

Panta, S., Flowers, T., Lane, P., Doyle, R., Haros, G., & Shabala, S. (2014). Halophyte agriculture: Success stories. *Environmental and Experimental Botany, 107*, 71–83. https://doi.org/10.1016/j.envexpbot.2014.05.006

Petropoulos, S. A., Karkanis, A., Martins, N., & Ferreira, I. C. F. R. (2018). Edible halophytes of the Mediterranean basin: Potential candidates for novel food products. *Trends in Food Science and Technology, 74*, 69–84. https://doi.org/10.1016/j.tifs.2018.02.006

Petropoulos, S. A., Levizou, E., Ntatsi, G., Fernandes, Â., Petrotos, K., Akoumianakis, K., Barros, L., & Ferreira, I. C. F. R. (2017). Salinity effect on nutritional value, chemical composition and bioactive compounds content of Cichorium spinosum L. *Food Chemistry*, *214*, 129–136. https://doi.org/10.1016/j.foodchem.2016.07.080

Petropoulos, S., Karkanis, A., Martins, N., & Ferreira, I. C. F. R. (2016). Phytochemical composition and bioactive compounds of common purslane (Portulaca oleracea L.) as affected by crop management practices. *Trends in Food Science and Technology*, *55*, 1–10. https://doi.org/10.1016/j.tifs.2016.06.010

Pinto, F. S., & Marques, R. C. (2017). Desalination projects economic feasibility: A standardization of cost determinants. *Renewable and Sustainable Energy Reviews*, *78*, 904–915. https://doi.org/10.1016/j.rser.2017.05.024

Qadir, M., Oster, J. D., Schubert, S., Noble, A. D., & Sahrawat, K. L. (2007). Phytoremediation of sodic and saline-sodic soils. *Advances in Agronomy*, *96*, 197–247. https://doi.org/10.1016/S0065-2113(07)96006-X

Rabhi, M., Hafsi, C., Lakhdar, A., Hajji, S., Barhoumi, Z., Hamrouni, M. H., Abdelly, C., & Smaoui, A. (2009). Evaluation of the capacity of three halophytes to desalinize their rhizosphere as grown on saline soils under nonleaching conditions. *African Journal of Ecology*, *47*, 463–468. https://doi.org/10.1111/j.1365-2028.2008.00989.x

Rankoana, S. A. (2021). Indigenous plant foods of Dikgale community in South Africa. *Journal of Ethnic Foods*, *8*. https://doi.org/10.1186/s42779-021-00080-w

Ruiz-Rodríguez, B. M., Morales, P., Fernández-Ruiz, V., Sánchez-Mata, M. C., Cámara, M., Díez-Marqués, C., Pardo-de-Santayana, M., Molina, M., & Tardío, J. (2011). Valorization of wild strawberry-tree fruits (Arbutus unedo L.) through nutritional assessment and natural production data. *Food Research International*, *44*, 1244–1253. https://doi.org/10.1016/j.foodres.2010.11.015

Salehi, B., Albayrak, S., Antolak, H., Kręgiel, D., Pawlikowska, E., Sharifi-Rad, M., Uprety, Y., Fokou, P. V. T., Yousef, Z., Zakaria, Z. A., Varoni, E. M., Sharopov, F., Martins, N., Iriti, M., & Sharifi-Rad, J. (2018). Aloe genus plants: From farm to food applications and phytopharmacotherapy. *International Journal of Molecular Sciences*, *19*, 2843. https://doi.org/10.3390/ijms19092843

Schultz, B., & De Wrachien, D. (2002). Irrigation and drainage systems research and development in the 21st century. *Irrigation and Drainage*. https://doi.org/10.1002/ird.67

Shabala, S. (2013). Learning from halophytes: Physiological basis and strategies to improve abiotic stress tolerance in crops. *Annals of Botany*, *112*, 1209–1221. https://doi.org/10.1093/aob/mct205

Shiyab, S. M., Shibli, R. A., & Mohammad, M. M. (2003). Influence of sodium chloride salt stress on growth and nutrient acquisition of sour orange in vitro. *Journal of Plant Nutrition*, *26*, 985–996. https://doi.org/10.1081/PLN-120020070

Silveira, J. A. G., Araújo, S. A. M., Lima, J. P. M. S., & Viégas, R. A. (2009). Roots and leaves display contrasting osmotic adjustment mechanisms in response to NaCl-salinity in Atriplex nummularia. *Environmental and Experimental Botany*, *66*, 1–8. https://doi.org/10.1016/j.envexpbot.2008.12.015

Singh, K., Pandey, V. C., Singh, B., & Singh, R. R. (2012). Ecological restoration of degraded sodic lands through afforestation and cropping. *Ecological Engineering*, *43*, 70–80. https://doi.org/10.1016/J.ECOLENG.2012.02.029

Smith, G. F., & Van Wyk, B.-E. (1998). Asphodelaceae. In K. Kubitzki (Ed.), *Flowering plants Monocotyledons* (pp. 130–140). Springer. https://doi.org/10.1007/978-3-662-03533-7_16/COVER

Sogoni, A., Jimoh, M. O., Kambizi, L., & Laubscher, C. P. (2021). The impact of salt stress on plant growth, mineral composition, and antioxidant activity in tetragonia decumbens mill.: An underutilized edible halophyte in South Africa. *Horticulturae*, *7*. https://doi.org/10.3390/horticulturae7060140

Souid, A., Gabriele, M., Longo, V., Pucci, L., Bellani, L., Smaoui, A., Abdelly, C., & Ben Hamed, K. (2016). Salt tolerance of the halophyte Limonium delicatulum is more associated with antioxidant enzyme activities than phenolic compounds. *Functional Plant Biology*, *43*, 607–619. https://doi.org/10.1071/FP15284

Tardío, J., Pardo-De-Santayana, M., & Morales, R. (2006). Ethnobotanical review of wild edible plants in Spain. *Botanical Journal of the Linnean Society*, *152*, 27–71. https://doi.org/10.1111/j.1095-8339.2006.00549.x

Thomas, J. C., Malick, F. K., Endreszl, C., Davies, E. C., & Murray, K. S. (1998). Distinct responses to copper stress in the halophyte Mesembryanthemum crystallinum. *Physiologia Plantarum*, *102*, 360–368. https://doi.org/10.1034/j.1399-3054.1998.1020304.x

Tokudome, S., Hashimoto, S., & Igata, A. (2016). Life expectancy and healthy life expectancy of Japan: The fastest graying society in the world. *BMC Research Notes*, *9*, 1–6. https://doi.org/10.1186/s13104-016-2281-2

Tshayingwe, A., Jimoh, M. O., Sogoni, A., Wilmot, C. M., & Laubscher, C. P. (2023). Light intensity and growth media influence growth, nutrition, and phytochemical content in Trachyandra divaricata Kunth. *Agronomy*, *13*. https://doi.org/10.3390/agronomy13010247

Tug, G. N., & Yaprak, A. E. (2019). Halophytes as a potential food source Gül Nilhan TUG potansiyel gida kaynaği olarak halofitler. *ANADOLU Journal of Aegean Agricultural Research Institute*, *27*, 78–81.

Van Wyk, B. E. (2008). A review of Khoi-San and Cape Dutch medical ethnobotany. *Journal of Ethnopharmacol*, *119*, 331–341. https://doi.org/10.1016/j.jep.2008.07.021

Van Wyk, B. E. (2011). The potential of South African plants in the development of new medicinal products. *South African Journal of Botany*, *77*, 812–829. https://doi.org/10.1016/J.SAJB.2011.08.011

Ventura, Y., & Sagi, M. (2013). Halophyte crop cultivation: The case for salicornia and sarcocornia. *Environmental and Experimental Botany*, *92*, 144–153. https://doi.org/10.1016/j.envexpbot.2012.07.010

Ventura, Y., Wuddineh, W. A., Myrzabayeva, M., Alikulov, Z., Khozin-Goldberg, I., Shpigel, M., Samocha, T. M., & Sagi, M. (2011). Effect of seawater concentration on the productivity and nutritional value of annual Salicornia and perennial Sarcocornia halophytes as leafy vegetable crops. *Scientia Horticulturae*, *128*, 189–196. https://doi.org/10.1016/j.scienta.2011.02.001

Watson, M. C., & O'leary, J. W. (1993). Performance of Atriplex species in the San Joaquin Valley, California, under irrigation and with mechanical harvests. *Ecosystems and Environment*, *43*, 255–266.

Yasin, N. A., Khan, W. U., Ahmad, S. R., Ahmad, A., Akram, W., & Ijaz, M. (2019). Role of acinetobacter sp. CS9 in improving growth and phytoremediation potential of Catharanthus longifolius under cadmium stress. *Polish Journal of Environmental Studies*, *28*, 435–443. https://doi.org/10.15244/pjoes/80806

Zuccarini, P. (2008). Ion uptake by halophytic plants to mitigate saline stress in Solanum lycopersicon L., and different effect of soil and water salinity. *Soil & Water Research*, *3*, 62–73.

19 Exploring the Nutritional Wealth of Edible Members of the Genus *Vitex*

Insights into Health Benefits and Nutrient Bioavailability

**Paul Oluwatimilehin Ogunbamowo, Linda Ifeanyichukwu Sowunmi, and Yetunde Oyinkansade Babalola*
Biomedicinal Research Centre, Forestry Research Institute of Nigeria, Oyo State, Nigeria
*Corresponding author: olutimmy7@gmail.com

19.1 INTRODUCTION

In recent times, plants have increasingly been utilized for their nutraceutical properties, contributing significantly to the creation of food- and plant-derived phytocomplexes with medicinal attributes. These natural resources are harnessed for their potential in addressing health conditions, serving as both preventive and curative remedies (Daliu et al., 2019; Souto et al., 2020).

Underutilized plants, including those in the *Vitex* genus, hold immense potential in diversifying agricultural landscapes and fortifying cropping systems against diverse stresses. Despite their abundance of vitamins, minerals, and essential nutrients, these plants have often been overlooked or underutilized within modern agricultural practices (Okigbo & Anyaegbu, 2021). Furthermore, even though they possess the potential to provide valuable sources of diverse phytochemicals and secondary metabolites, underutilized wild herbs among the *Vitex* sp. have received minimal attention from both the commercial and scientific communities. Expanding the utilization of underutilized plants offers a wider array of options for establishing resilient cropping systems, thereby enhancing agricultural resilience.

The genus *Vitex*, also known as the chaste tree, consists of approximately 270 species, primarily trees and shrubs with woody stems that can grow up to around 3 m tall. These species are widely distributed in tropical and subtropical regions globally, including Brazil, Nigeria, Turkey, Thailand, Vietnam, Algeria, and South Africa (Mohd Zaki & Wan Salleh, 2020). Among the most renowned and extensively studied species within the *Vitex* genus are *Vitex agnus-castus*, *Vitex negundo*, *Vitex trifolia*, *Vitex rotundifolia*, and *Vitex doniana*. Although they are potentially rich in nutritional content, many members of these species are underutilized. However, in recent years, there has been a growing recognition of the diverse range of health-promoting compounds and the potential therapeutic properties that they possess. Following the growing recognition of the diverse range of health-promoting compounds and potential therapeutic properties within the *Vitex* genus, scientific evidence has highlighted the efficacy of these plants in traditional and modern medicine. To mention a few, studies have shown that various species within the *Vitex* genus possess bioactive compounds known for their anti-inflammatory, antioxidant, antimicrobial, and hormone-regulating properties (Rani & Sharma, 2013). For instance, *V. agnus-castus* has been extensively researched

DOI: 10.1201/9781003469766-19

for its effectiveness in addressing conditions such as PMS (premenstrual syndrome) and menopausal symptoms (Rani & Sharma, 2013; Niroumand et al., 2018; Naseri et al., 2019). Similarly, *V. negundo* has demonstrated promising results in the treatment of ailments like arthritis and skin conditions due to its anti-inflammatory and analgesic properties (Dharmasiri et al., 2003; Kulkarni et al., 2008). Furthermore, *V. grandifolia* presents a rich ethnobotanical history. The plant's leaves are commonly acknowledged as a wild vegetable, yet they hold significance in various ethnomedicinal practices. Traditional remedies utilize the bark of this plant to address stomachaches, diarrhea, bronchial complaints, rickets, soreness, and fever. Moreover, the fruits of *V. grandifolia* are not only edible but also incorporated into the production of alcoholic beverages, adding to the plant's diverse utility (Bello et al., 2018b). Thus, the leaves, flowers, and fruits of *Vitex* plants stand out due to their potential nutritional and therapeutic value, underscoring the importance of exploring the ethnopharmacological applications of these species within the *Vitex* genus. The scientific community's curiosity in the therapeutic capabilities of these *Vitex* species is ever-growing, highlighting their significant role in promoting overall health and well-being.

Exploring the nutritional value of the *Vitex* genus is of paramount importance as it presents a valuable source of essential nutrients, bioactive compounds, and health-promoting phytochemicals (Pío-León et al., 2014). Identifying and harnessing the potential of underutilized plants within the *Vitex* genus can play a significant role in diversifying food sources and fostering innovative nutritional interventions.

A comprehensive understanding of the phytochemical compositions of any potential nutraceuticals is an essential initial step in unlocking their beneficial health properties. Therefore, the primary objective of this chapter is to comprehensively explore the nutritional value of some members of the *Vitex* genus. The chapter aims to showcase the macronutrient and micronutrient profiles, phytochemical compositions, and potential health-promoting properties found within various species of the *Vitex* genus. Furthermore, the chapter seeks to explore the bioavailability of nutrients present in the *Vitex* genus, shedding light on their absorption and utilization in the human body.

Although many studies and reviews have highlighted in detail the pharmacological activities and ethnobotany of many *Vitex* species, the significance of investigating the nutritional potential of plants in the *Vitex* genus lies in the capacity to enrich and diversify our food systems while simultaneously addressing prevalent nutritional deficiencies. By tapping into the nutritional reservoirs of underutilized plants in the genus *Vitex* which are readily available in many parts of the world, we not only expand the range of available food resources but also bolster dietary diversity and resilience against food insecurity.

This chapter aims to bridge the gap between traditional knowledge and modern science, unlocking the nutritional potential of the *Vitex* genus and facilitating its integration into mainstream dietary practices. Through a comprehensive exploration encompassing botanical, nutritional, and medicinal aspects, we endeavor to provide insights that stimulate further research, shape policy decisions, and foster innovative approaches to addressing food security and promoting nutrition.

19.2 LITERATURE SEARCH AND DATA COMPILATION

The information used in this review was obtained by searching databases like Google Scholar, PMC, and other databases using the keywords "genus vitex." Research and review articles on the subject matter published between 2010 and 2023 were considered for the review.

19.3 BOTANICAL DESCRIPTION OF THE *VITEX* GENUS

Plants within the *Vitex* genus exhibit a diverse array of characteristics and features that contribute to their botanical significance. While specific details may vary across different species, a general overview of *Vitex* plants can be gleaned from the collective attributes observed within the genus. The genus *Vitex*, once classified under the Verbenaceae family, is now acknowledged as a prominent group within the Lamiaceae family, boasting a diverse range of over 200 species (Govaerts, 2017). *Vitex* species are shrubs and trees ranging from 1 to 35 m in height (Bello et al., 2018b); members of the genus are often

FIGURE 19.1 Different shades of *Vitex* species: (a) *V. agnus-castus* with fruits; (b) *V. trifolia* with fruits; (c) *V. negundo*; (d) *V. doniana*; (e) *V. ferruginea* with fruits; (f) *V. payos* with fruits; (g) *V. doniana* fruits.

Source: (a) www.alamy.com/vitex-agnus-castus-branch-with-fruits-image; (b) cabicompendium.56555, CABI International (2022); (c) Singh et al. (2020); (d) Dadjo (2014); (e) Bingham et al. (2024); (f) Günter Baumann (www.eastafricanplants.org).

recognized by their opposite, and usually compound, leaves with five to seven leaflets, typically arranged palmately, which may vary in size and shape across different species (Figure 19.1). These leaves typically exhibit serrated edges, contributing to the distinctive appearance of the plant; the flowers of *Vitex* plants are known for their vibrant colors, with clusters of violet or blue flowers being a common sight among various species within the genus, also, the fruits of various *Vitex* species exhibit comparable morphology, often presenting challenges in differentiation (Pío-León et al., 2014), and many *Vitex* species bear edible fruits which are black or reddish, adding to their ecological and potentially nutritional value within their ecosystems (Rani & Sharma, 2013). For example, *V. grandifolia*, also known as the large-leaved chaste tree, is an annual evergreen shrub. It is characterized by its distinctive large, compound leaves with serrated edges and clusters of violet or blue flowers (Ali et al., 2020). It has edible fruits, and the plant typically reaches heights of up to 20 m and is known for its ornamental and medicinal properties. The trunk of the tree typically exhibits sinuous patterns and a slight fluting at its base. It generally reaches a diameter of up to 60 cm, with certain instances recorded at 120 cm. Moreover, it can remain without branches for heights of up to 15 m, although this is typically shorter (Tropical Plants Database, 2024).

19.4 HABITAT AND GEOGRAPHICAL DISTRIBUTION OF SELECTED SPECIES

The habitat and geographic distribution of various *Vitex* species revealed a diverse array of ecological landscapes worldwide. From the lush tropical forests of South Asia to the arid plains of Africa, *Vitex* species flourish in a wide range of habitats, each contributing uniquely to their growth and adaptation. The distribution of *V. agnus-castus* spans across Mediterranean Europe, Southern Europe, Central Asia, and North Africa, making it the most extensively studied and recognized plant within the *Vitex* genus (Kamal et al., 2022; Islam et al., 2024). *V. negundo* is predominantly found in India, Bangladesh,

Afghanistan, Pakistan, Thailand, Sri Lanka, Nepal, Eastern Africa, China, and Malaysia (Singh et al., 2020; Islam et al., 2024). *V. rotundifolia* is distributed across the Mediterranean region, East Asia, Central Asia, and along the seacoast from South to North China. *V. trifolia* is present in various Asian countries, notably in Vietnam, southwest of China, and the Taiwan province (Sun et al., 2019). Among the predominant *Vitex* species across the African continent are *V. doniana*, *V. payos*, *V. fischeri*, *V. grandifolia*, *V. simplicifolia*, *V. madiensis*, *V. mombassae*, *V. ferruginea*, and *V. pooara* (Pío-León et al., 2014). *V. doniana*, the most widespread species in Africa, spans Senegal to Somalia, to South Africa, and is widely distributed in Nigeria (Adoukonou-Sagbadja et al., 2023), while *V. grandifolia* is native to various regions of Africa, where it thrives in diverse ecological habitats. It is commonly found in tropical and subtropical areas, including forest margins, savannas, and riverbanks. The plant demonstrates a degree of adaptability to different soil types and climatic conditions, contributing to its wide distribution across countries such as Nigeria, Cameroon, Ethiopia, and India (Bello et al., 2019).

19.5 CULTIVATION PRACTICES

The cultivation of *Vitex* typically involves propagation through seeds or stem cuttings, with careful attention to soil preparation and planting techniques. The plant thrives in well-drained, fertile soils with appropriate level of moisture and sunlight. Agronomic considerations for successful cultivation include appropriate spacing between plants, irrigation management, and potential pest and disease control measures. Studies on the propagation methods of *Vitex* species vary with the technique employed. *Vitex* species exhibit diverse propagation methods tailored to their unique characteristics. For instance, *V. negundo* thrives in various soil types with moderate water needs and can be efficiently propagated vegetatively through stem cuttings, facilitating rapid multiplication (Patel, 2016). Similarly, *V. doniana* has shown successful vegetative propagation outcomes using root segment cuttings (Mapongmetsem et al., 2016). Biotechnological approaches such as micropropagation are recommended for the mass production of different *Vitex* plants to support their conservation and sustainable utilization (Ahmed & Anis, 2014; Ahmad et al., 2015) because the conventional method of propagation, which is through seeds or root suckers, has poor seed germination potential (Groach et al., 2014). Moreover, *V. doniana* and *V. agnus-castus* have demonstrated promising results with seed propagation, provided proper treatments are employed to overcome dormancy (N'Danikou et al., 2014: Nor Hisham Shah & Bridgen, 2016; Haruna et al., 2024). A recent study conducted in the Democratic Republic of Congo focused on four versatile *Vitex* species (*V. doniana*, *V. madiensis* subsp. *milanjiensis*, *V. fischeri*, and *V. mombassae*), highlighting their varied responses to different propagation techniques for nursery production. This research sheds light on the importance of understanding the cultivation requirements of different *Vitex* species. The study identified four main propagation methods: sowing, root cuttings, stem cuttings, and aerial layering. Each species displayed distinct responses to these techniques, underscoring the necessity for species-specific protocols. For example, *V. doniana* demonstrated a high success rate with scarified seeds and root cuttings, while *V. fischeri* exhibited favorable results with aerial layering (Numbi Mujike et al., 2023).

19.6 NUTRITIONAL COMPOSITIONS

Within the *Vitex* genus, while specific literature evaluating the nutritional components of individual species remains limited, the collective body of research on various Vitex species offers valuable insights into their nutritional richness. This section aims to provide insights into the nutritional composition of *Vitex* plants by drawing upon studies conducted on species within the *Vitex* genus. The nutritional composition of certain plants within the *Vitex* genus has been well-documented, showcasing their rich content of essential nutrients, vitamins, and bioactive compounds. For example, *V. negundo* has been reported to contain essential nutrients, such as carbohydrates, proteins, calcium, and iron (Rana et al., 2019; Koirala et al., 2020). Similarly, *V. agnus-castus* has been recognized for its nutritional value, containing minerals and bioactive compounds with potential health benefits

(Boujbiha et al., 2023). Among the diverse species encompassed by the *Vitex* genus, certain members, particularly those indigenous to West Africa, have attracted attention for their nutritional significance. Investigations into the nutritional profiles of these species have unveiled a diverse array of essential nutrients, including vitamins, minerals, and phytochemicals. Particularly, *V. doniana* stands out as one of the most studied species in this context, reflecting its prevalence in West Africa. Studies by Bello et al. (2014) have delved into the nutritional content of *V. doniana* leaves, revealing varying percentages consistent with those reported in similar species and other *Vitex* species. For instance, the ascorbic acid content was found to be 24.6%, contributing to a nutritive value of 361.7 kcal/g. Results from the proximate composition of *V. doniana* fruits from two locations in Tanzania suggest that the fruits contain carbohydrates in the range of 8.72–23.98%, crude protein 7.13–21.73%, crude fat 1.90–2.40%, moisture content 64.67–65.79%, and ash content 4.20–5.35% (Charles & Mgina, 2023). Similarly, the fruits of *V. doniana* from Nigeria were reported to contain carbohydrates (28.40%), crude protein (8.24%), crude fat (34.62), moisture (16.66%), and ash (11.5%) (Vunchi et al., 2011). The mineral content shows K (15.70 mg/100 g), Na (10.40 mg/100 g), Ca (30.27 mg/100 g), P (16.5 mg/100 g), Mg (20.10 mg/100 g), Fe (5.20 mg/100 g), and Cu (2.70 mg/100 g), while the vitamin contents were reported as vitamin A (0.27 mg/100 g), vitamin B_1 (18.33 mg/100 g), vitamin B_2 (4.80 mg/100 g), vitamin B_6 (20.45 mg/100 g), and vitamin C (35.58 mg/100 g) (Vunchi et al., 2011). Additionally, research on the effect of processing methods on the proximate composition of *V. doniana* leaves demonstrated that oven-dried leaves exhibited higher levels of carbohydrate, crude protein, fats, fiber, and ash content compared to raw leaves, except for moisture content (Osum et al., 2013). Adejumo et al. (2013) reported similar findings on the proximate composition of raw *V. doniana* leaves, along with significant concentrations of vitamin C and various macro- and micronutrients, such as calcium, potassium, magnesium, sodium, iron, manganese, zinc, and copper. The proximate composition of different documented species within the *Vitex* genus is summarized in Table 19.1.

TABLE 19.1
Proximate Compositions (%) of *Vitex* Genus

Species	Location	Plant Part	CHO	Protein	Ash	Fat	Fiber	Moisture	Reference
V. doniana	Nigeria	Leaves	41.31	12.83	21.16		15.73	11.83	Abdullahi et al. (2022)
	Nigeria	Fruit	67.84	0.006	21.50	0.75	Nd	9.90	Imoisi et al. (2021)
	Kenya	Fruit	29.57	0.85	3.41	2.44	11.48	39.42	Ochieng and Nandwa (2010)
	Nigeria	Fruit	43.20	0.60	18.00	23.7	4.5	10.00	Aiwonegbe et al. (2018)
	Nigeria	Fruit	28.40	8.24	11.50	34.62	0.58	16.66	Vunchi et al. (2011)
	Nigeria	Seed	17.64	27.57	5.18	36.52	4.84	8.25	Amah and Okogeri (2019)
	Nigeria	Leaves	4.02	15.46	6.50	0.80	14.67	58.55	Ifeanacho and Ogunwa (2021)
	Nigeria	Leaves	57.50	17.29	6.42	1.25	6.29	11.25	Osum et al. (2013)
	Nigeria	Leaves	70.20	8.75	7.92	5.10	15.58	8.04	Bello et al. (2014)
	Nigeria	Raw leaves	7.57	8.10	1.63	2.92	2.75	77.03	Adejumo et al. (2013)
	Nigeria	Fruit pulp	73.10	10.80	2.00	3.40	0.30	10.40	Adeleke et al. (2023)
	Benin	Leaves	7.74[a]	6.12[a]	1.68[a]	2.39[a]	1.96[a]	80.56[a]	Dadjo (2014)
V. keniensis	Kenya	Fruit	28.04	0.87	3.40	2.35	10.42	40.56	Ochieng and Nandwa (2010)
V. fischeri	Kenya	Fruits	27.28	0.98	3.66	2.66	12.38	37.74	Ochieng and Nandwa (2010)
V. negundo	Nepal	Bark	51.39[+]	9.69	7.86	6.19	6.52	18.35	Koirala et al. (2020)
	India	Leaves	8.50	13.70	5.4	7.00	28.02	16.50	Kumar et al. (2013)
	Bangladesh	Leaves	37.75	19.27	20.15	3.13	19.70	82.78	Rana et al. (2019)

Note:

[+]Obtained by difference.

[a]Average value from three regions.

19.6.1 Mineral and Vitamin Composition

The mineral composition of the *Vitex* genus exhibits a diverse array of micronutrients essential for human health and well-being. While specific data on mineral composition may vary among different species, certain members of the genus are recognized for their significant mineral content. Analysis of their macronutrient profiles reveals certain levels of essential minerals such as calcium, iron, and potassium, which play crucial roles in various physiological functions. Furthermore, the presence of vitamins, including vitamin C and vitamin A, further enhances the nutritional value of these *Vitex* species, contributing to their potential as valuable food and medicinal resources. *Vitex* species, including *V. doniana*, exhibit significant mineral content, contributing to their nutritional value and potential health benefits. *V. doniana*, for instance, was observed to contain essential minerals, such as calcium (3.36%), potassium (1.13%), and iron (0.12 mg/kg). Importantly, the levels of antinutrients found in *V. doniana* leaves, including tannins, phytate, saponin, alkaloid, and oxalate, were determined to be below permissible limits, suggesting no adverse nutritional effects on human health. This underscores the potential of these vegetables to complement conventional ones, contributing to the enhancement of food security and the establishment of sustainable livelihoods (Bello et al., 2014). The mineral and vitamin composition of *Vitex* species, as summarized in Tables 19.2 and 19.3, respectively, highlights their nutritional richness and diversity across different locations and plant parts.

This summary provides insights into the mineral content of *Vitex* species across different regions and plant parts, emphasizing their potential contribution to human nutrition and health. *V. doniana*, a species native to Nigeria, demonstrates variability in mineral composition across its various plant parts. While its fruits and fruit pulp exhibit significant levels of potassium and calcium, its leaves contain significant concentrations of calcium, iron, and phosphorus. Similarly, *V. agnus-castus*, found in Tunisia, showcases substantial levels of potassium, calcium, and magnesium in its fruits (Boujbiha et al., 2023). In contrast, *V. negundo*, prevalent in India and Bangladesh, presents high

TABLE 19.2
Mineral Contents (mg/100 g) of Some Members of the *Vitex* Genus

Species	Location	Plant Part	K	Na	Ca	Mg	Fe	Cu	P	Reference
	Nigeria	Leaves	13.89	–	23.53	12.71	18.87	0.92	–	
V. doniana	Nigeria	Fruits	15.7	10.4	30.27	20.10	5.20	2.70	16.50	Vunchi et al. (2011)
	Nigeria	Fruits	166.72	30.48	96.19	58.35	–	–	–	Aiwonegbe et al. (2018)
	Nigeria	Leaves	3.80	0.18	7.3	4.1	2.4	0.19	5.1	Ifeanacho and Ogunwa (2021)
	Benin	Leaves		–	190.52[a]	61.32[a]	2.48[a]	–	–	Dadjo (2014)
	Nigeria	Fruit pulp	26.26	2.26	0.18	26.66	2.22	–	24.83	Adeleke et al. (2023)
	Nigeria	Leaves	–	1.25	51.70	–	17.29	–	–	Osum et al. (2013)
	Nigeria	Leaves	1,130	–	3,360	–	0.012	ND	–	Bello et al. (2014)
	Nigeria	Raw leaves	260	0.19	730	410	2.42	0.62	–	Adejumo et al. (2013)
V. agnus-castus	Tunisia	Fruits	1,428.84	23.28	781.29	190.34	5.89	0.44	–	Boujbiha et al. (2023)
	Turkey	Seeds	–	–	–	–	0.94	0.3	–	Ozkaya et al. (2013)
V. negundo	India	Leaves	–	–	5,267.13	–	9.4		680.30	Devi et al. (2016)
	India	Leaves	16,180	–	14,340	–	533	20	–	Kumar et al. (2013)
	Bangladesh	Leaves	3,345.20	577.82	1,786.24	315.15	62.05	1.08	–	Rana et al. (2019)

Note:
[a]Average value from 3 regions.
–not determined.

TABLE 19.3
Vitamins Compositions (mg/100 g) of Some *Vitex* Genus

Species	Location	Part	A	B_1	B_2	B_6	C	D	K	E	Reference
V. doniana	Nigeria	Fruits	0.27	18.33	4.80	20.45	35.58	–	–	–	Vunchi et al. (2011)
	Benin	Leaves	3.49[a]	–	–	–	16.09[a]	–	–	–	Dadjo (2014)
	Nigeria	Leaves	0.09	0.0001	0.012	0.0005	2.76	0.002	0.0001	0.41	Ifeanacho and Ogunwa (2021)
	Nigeria	Fruit pulp	–	14.22	2.11	16.22	31.23	0.00003	–	0.71	Adeleke et al. (2023)
	Nigeria	Leaves	1.07[b]	–	9.13	–	20.83	–	–	53.20	Osum et al. (2013)
V. agnus-castus	Turkey	Seeds	–	–	–	–	–	–	2.5	1.82	Ozkaya et al. (2013)

Note:
[a]Average value from three regions.
[b] – converted from IU.

concentrations of calcium, iron, and phosphorus in its leaves (Kumar et al., 2013; Devi et al., 2016; Rana et al., 2019). These findings highlight the nutritional richness of *Vitex* species and their potential to contribute to human health and well-being.

19.7 PHYTOCONSTITUENTS

Numerous phytochemicals have been identified from various parts of the *Vitex* genus. The phytochemical composition of *Vitex* species encompasses a diverse range of bioactive compounds, including flavonoids, tannins, saponins, alkaloids, cardiac glycosides, terpenoids, and phenolic acids, which contribute to their potential therapeutic and health-promoting properties. Some of these fundamental phytochemicals have been reported and quantified in certain species, like *V. doniana* (Dauda et al., 2011; Ifeanacho et al., 2019) and *V. grandifolia* (Epidi et al., 2016). While comprehensive phytochemical analyses of these plants remain limited, existing studies on species within the same genus indicate a substantial presence of phytoconstituents. These findings underscore the genus's capacity to serve as a repository of natural bioactive compounds, showcasing a wide array of biological activities.

19.7.1 Identification and Characterization of Key Phytoconstituents

The *Vitex* genus is known for its rich phytochemical composition, with various compounds contributing to its medicinal properties. One of the common classes of compounds found in *Vitex* is iridoid. In *V. grandifolia*, key phytoconstituents have been identified, including two iridoid glucosides, agnuside and bartsioside, isolated from the methanol leaves extract (Bello et al., 2018a). These iridoids, characterized as acetal derivatives of iridodial, are a type of monoterpenoids (Dong & Zhang, 2004). Furthermore, the inherent instability of their C1-OH group often leads to reactions with sugars, resulting in the formation of glycosides (Wang et al., 2020). Additionally, *V. grandifolia* contains three known flavonoid compounds – isoorientin, orientin, and isovitexin – isolated and structurally elucidated from the methanol fraction of the leaves (Bello et al., 2019). Alongside these compounds, other terpenoids, such as diterpenes and triterpenoids, have also been identified in the plant. Table 19.4 highlights some of the phytoconstituents isolated and characterized from the edible parts of various *Vitex* species. Flavonoids constitute a class of polyphenolic hydroxylated compounds characterized by the benzo-π-pyrone backbone. They are widely distributed in plants and synthesized through the phenylpropanoid pathway in plant organisms (Achika et al., 2023).

TABLE 19.4
A Summary of the Compounds and Their Biological Activities Found in *Vitex* Species

Compound	Structural Class	Vitex Species (Part)	Biological Activity	Reference
Agnuside and bartsioside	Iridoid glycoside	*V. grandifolia* (leaves)	Agnuside showed significant activity against the NF-kB assay, with an IC_{50} of 8.9 µg/ml, while bartsioside exhibited an IC_{50} of 12 µg/ml. Both compounds also displayed moderate activity against the iNOS assay, with agnuside having an IC_{50} of 18 µg/ml and bartsioside with 24 µg/ml.	Bello et al. (2018a)
Isoorientin, orientin, and isovitexin	Flavonoids	*V. grandifolia* (leaves)	The compounds demonstrate in vitro inhibition of monoamine oxidase and anti-inflammatory activity.	Bello et al. (2019)
2-(4-hydroxyphenyl)-3,5,7-trihydroxy chromen-4-one; 2-(5-methoxyphenyl)-3,40,5,7,8-trihydroxychroman-4-one; 2-(50-methoxyphenyl) 40,5,7-trihydroxy-3-methoxychromen-4-one; 2-(30,40-dimethoxyphenyl)-7-hydroxychromen-4-one; 2-(40-hydroxyphenyl)-5-hydroxy 3,7-dimethoxy chromen-4-one	Flavonoids	*V. simplicifolia* (leaves)	All compounds displayed moderate trypanocidal activity against *T. brhodesiense*.	Nwodo et al. (2015)
4,5-diethyl-30-ethoxy-pyro flavone	Flavonoids	*V. negundo* L. (leaves)	Demonstrated a dose-dependent antifilarial activity.	Rana (2018)
Vitex negheteroins K and L	Iridoid glycoside	*V. negundo* var. *heterophylla* (leaves)	Iridoid glycosides displayed relatively weaker antioxidant properties, as evidenced by IC_{50} values exceeding 20 µM.	Hu et al. (2017)
10-p-hydroxybenzoyl-6*b*-hydroxyiridoid 1-O-*b*-D-(60-O-p-hydroxybenzoyl) glucopyranoside	Iridoid glycoside	*V. negundo* var. *heterophylla* (leaves)	No observed inhibition of nitric oxide production.	Qiu et al. (2017)
Nishindacin A and isonishindacin A	Iridoids	*V. negundo* var. *heterophylla* (leaves)	Both compounds exhibited weak radical-scavenging effects on stable free radicals.	Niu et al. (2020)
(1S, 5S,6R,9R)-10-O-p-hydroxybenzoyl-5,6*b*-dihydroxy iridoid 1-O-*b*-D-glucopyranosidem	Iridoid glycoside	*V. trifolia* L. (leaves)	The compound demonstrated moderate inhibitory effects on nitric oxide production induced by LPS-stimulated RAW 264.7 macrophages.	Bao et al. (2018)
Viteagnuside A	Diterpenoids	*V. agnus-castus* L. (fruits)	Not tested.	Ono et al. (2011)
Vitexlactam B and C	Diterpenoids	*V. agnus-castus* L. (fruits)	Vitexlactam C exhibited a moderate effect on the induction activity of NADP(H)-quinone oxidoreductase type 1 (QR1).	Li et al. (2013)
Vitextrifolin A-G	Diterpenoids	*V. trifolia* L. (fruits)	The compounds showed no activity against A549, HCT116, HL60, and ZR-75–30 cell lines.	Zheng et al. (2013)

(Continued)

TABLE 19.4 (Continued)

Compound	Structural Class	Vitex Species (Part)	Biological Activity	Reference
6*a*, 7a-diacetoxy-13-hydroxy-8(9),14-labdadien; and 9-hydroxy-13(14)-labden-15, 16-olide	Diterpenoids	*V. trifolia* L. (leaves)	Exhibits activity against *M. tuberculosis*, with a minimum inhibitory concentration (MIC) value of 100 μg/ml.	Tiwari et al. (2013)
Viteagnusin I	Diterpenoids	*V. agnus-castus* L. (fruits)	No activity was observed in assays targeting opioid receptors DOR and MOR.	Chen et al. (2011)
1α, β-dihydroxybauer-7-en-28-oic acid; 2β,3β,19α, 24-tetrahydroxy-23-norus-12-en-28-oic acid	Triterpenoid	*V. doniana* (fruits)	Not determined.	Mohammed et al. (2016)

Source: Adapted from Kamal et al. (2022).

Studies on the essential oils of some *Vitex* species have revealed some key compounds that have been demonstrated to exhibit biological activity. In the only study documenting the chemical composition of the essential oils obtained from *V. grandifolia*, the leaves and stem of the plant exhibited distinct major constituents. The stem oil extract showed the presence of 9-octadecanoic acid ethyl ester (32.46%) as the major constituent, accompanied by hexadecanoic acid (1.68%) and nerolidol (0.90%). In contrast, the major constituents in the leaf oil extract included E-9-octadecenoic acid ethyl ester (35.39%) and hexadecanoic acid (20.78%), with limonene, eucalyptol, and caryophyllene present as minor constituents (Odusina & Yusuf-Mikail, 2022). The major compound in the leaf essential oil, E-9-octadecenoic acid ethyl ester, derived from another plant, has been reported to attenuate inflammatory responses by regulating MAPKs and NF-κB signaling pathways in LPS-induced RAW 264.7 macrophages (Xie et al., 2022). Furthermore, a report on the leaf essential oil constituents in another member of the *Vitex* species, *V. doniana* from Nigeria, revealed that β-phellandrene (31.13%), phytol (28.3%), and β-caryophyllene (12.6%) are the major constituents (Sonibare et al., 2009). Additionally, a study on the essential oil from *V. doniana* in Ghana disclosed that the fruit essential oil possessed 3-phenylpropanol (11.58%), hexyl acetate (10.46%), 3-decanoic acid, (E) (9.70%), octanoic acid (5.16%), ethyl cinnamate (4.96%), and methyl cinnamate (4.91%) as the major constituents, while n-hexadecanoic acid (15.90%), octacosane (8.45%), tetradecane (5.87%), hexadecane (5.56%), and oleic acid (4.74%) were the major compounds in the leaf essential oil (Odoom et al., 2023). Another study on the fruit essential oil of *V. doniana* identified eucalyptol (34.86%), 3-decanoic acid (16.91%), n-octanoic acid (14.86%), terpinene-4-ol (8.01%), and alpha-terpineol acetate (4.29%) as the major constituents (Chinyere et al., 2021). Remarkably, hexadecenoic acid is identified as a major constituent of the essential oils of the leaves of *V. grandifolia* and *V. doniana*. This compound has been reported to have anti-inflammatory, antibacterial, antioxidant, and hypocholesterolemic properties (Aparna et al., 2012).

Given the close phylogenetic relationship among various species within the *Vitex* genus, it is reasonable to speculate that these species may share similar nutritional characteristics. The documented nutritional wealth of other members of the genus suggests the potential for *Vitex* species to be valuable sources of vitamins, minerals, and phytochemicals. However, further research is essential to elucidate the specific nutritional composition of different *Vitex* species. Comprehensive studies involving proximate analysis, mineral profiling, and phytochemical screening are crucial to uncovering their nutritional potential. Moreover, investigations into the bioavailability and functional properties of their nutrients can provide valuable insights into their dietary significance and health-promoting effects.

19.8 NUTRIENT BIOAVAILABILITY

The concept of *bioavailability* pertains to the quantity of a nutrient that is absorbed from the gastrointestinal tract and subsequently utilized for regular physiological processes or stored within the body (Schonfeldt et al., 2016). The significance of nutrient bioavailability cannot be overstated, as the extent to which consumed nutrients can be absorbed directly influences their beneficial impact on overall health. In cases where a nutrient lacks bioavailability, its absorption is compromised, potentially leading to insufficiency or even deficiency over time (Kiani et al., 2022). The assessment of nutrient bioavailability in *V. grandifolia* remains an area requiring further investigation, as specific studies on this aspect of the plant are limited. However, insights from related *Vitex* species provide a foundation for understanding potential nutrient bioavailability in *V. grandifolia* while considering various factors that influence nutrient absorption and utilization. Some antinutritional factors, such as phytate, tannins, saponins, etc., are known to inhibit nutrient bioavailability and absorption (Samtiya et al., 2020). Adejumo et al. (2013) revealed that *V. doniana* contains low levels of antinutrients, which can be further minimized through different processing techniques to enhance the availability of nutrients. Furthermore, the antinutrient analysis conducted by Bello et al. (2014) indicated that the levels of tannin, phytate, saponin, alkaloid, and oxalate in *V. doniana* were below the permissible limit, thus not posing any concerns regarding nutrient bioavailability. Furthermore, the bioavailability of phytoconstituents in *Vitex* species is a crucial aspect to consider in understanding their potential health benefits and therapeutic effects. Various phytoconstituents have been identified, including flavonoids, terpenoids, phenolic acids, and iridoid glycosides (Table 19.4), all of which have shown variable bioavailability based on the different outcomes in the bioactivities assessed. Iridoids are the most widely detected and reported in many studies isolating and characterizing bioactive compounds in *Vitex* species. These bioactive compounds contribute to the pharmacological properties of *Vitex* species and may play a role in their bioavailability. However, the specific bioavailability of these metabolites in *Vitex* species remains to be fully elucidated. Factors such as plant genetics, phytochemical composition, environmental conditions, processing methods, and interactions within the gastrointestinal tract play pivotal roles in determining nutrient bioavailability (Rein et al., 2013).

19.9 POTENTIAL HEALTH IMPLICATIONS AND BIOLOGICAL ACTIVITIES

The health benefits of *Vitex* species have been substantiated through a growing body of scientific evidence, providing valuable insights into its potential therapeutic applications and traditional medicinal uses. Through systematic reviews, in vitro and in vivo studies, and clinical trials, the scientific evidence has underscored the significant health-promoting potential of *Vitex*, positioning it as a valuable natural resource for addressing various health conditions and promoting overall well-being. Phytoconstituents within *Vitex* species offer a diverse range of biological activities that may hold promise for human health. The presence of phytoconstituents in various *Vitex* species holds significant health implications, with documented biological activities that encompass a wide range of potential therapeutic benefits. Many studies have elucidated the plant's diverse pharmacological properties, including antioxidant, anti-inflammatory, antimicrobial, anticancer, and neuroprotective activities, which collectively contribute to its potential health benefits. Some of the compounds identified in various *Vitex* plants, along with their tested biological activities, are summarized in Table 19.4. To recap, some of these activities include antioxidant properties that protect against oxidative stress–related diseases, anti-inflammatory effects that may alleviate inflammatory conditions, antimicrobial actions that combat pathogens, and potential anticancer properties. Many of the compounds reported in various *Vitex* species have been characterized for their pharmacological activities, such as anti-inflammatory, antimicrobial, and anticancer properties, contributing to the plant's potential therapeutic relevance. A comprehensive review by Wang et al. (2020) and Kamal et al. (2022) has detailed the various biological activities attributed to iridoid compounds and other classes of compounds found in various *Vitex* species. Results from the study of Bello et al. (2018a) showed the effectiveness of the iridoid glycosides isolated from the leaves of *V. grandifolia* in acting as an anti-inflammatory agent in the various anti-inflammatory assays. Specifically, agnuside was found to exhibit significant activity against the nuclear factor kappa-light-chain-enhancer of activated B cells (NF-kB) assay with IC_{50} of 8.9 μg/ml, while bartsioside showed an IC_{50} of 12 μg/ml in the inhibition of NF-kB, an in vitro anti-inflammatory assay that measured the ability of a substance or compound to inhibit the activity of NF-κB, which plays a crucial role in regulating the immune response, inflammation, cell proliferation, and survival. Therefore, the assay assesses the impact of the isolated iridoid glycosides on the activity of NF-κB, which is essential for understanding its potential role in modulating various cellular processes and its potential therapeutic implications (Olivera et al., 2012). Furthermore, agnuside and bartsioside were found to display moderate activity against the inhibition of inducible nitric oxide (iNOS) assay, with an IC_{50} of 18 μg/ml and 24 μg/ml, respectively. In the synthase iNOS assay, this assay measured the ability of the iridoid glycosides to inhibit the activity of iNOS. iNOS is an enzyme that plays a key role in the production of nitric oxide (NO), which is involved in various physiological and pathological processes, including immune response, inflammation, and cardiovascular regulation (Balaji & Kripa, 2021). The iridoid glucosides extracted from the ethyl acetate extracts of *V. altissima* leaves exhibited remarkable antioxidant activity, as evidenced by their effectiveness in both the superoxide nitro blue tetrazolium (NBT) riboflavin photoreduction, free radical scavenging, and 2,2-diphenyl-1-picrylhydrazyl (DPPH) radical scavenging methods (Sridhar et al., 2004). Moreover, the anti-inflammatory and antioxidant activities of various leaves and bark extract fractions of *V. grandifolia* were reported in a study from Ghana (Amo-Mensah et al., 2020). The flavonoids isolated from the leaves of the plant have been shown to elicit *in vitro* monoamine oxidase inhibition and anti-inflammatory activity (Bello et al., 2019). Only one *Vitex* species has passed through clinical trials to assess its health-promoting properties in clinical subjects. Evidence from the health-promoting effect of *V. agnus-castus* L has been established through several clinical trials. In one study, the effect of the plant in alleviating menopausal symptoms was evaluated, and it was reported that after the intervention, the *Vitex* group showed a reduction in vasomotor dysfunction, anxiety, and total menopausal disorder (Naseri et al., 2019).

The exploration of the bioactivities of compounds found in the essential oils obtained from the leaves and stems of *V. grandifolia* in the study by Odusina and Yusuf-Mikail (2022) suggests that they exhibit no significant antimicrobial activities; however, some of these compounds have been reported in other *Vitex* species, and the compounds have been shown to exhibit antimicrobial activities (Dahham et al., 2015). Similarly, the biological activities of the essential oil from the *V. doniana* showed significant antioxidant activity as tested using the phosphomolybdenum, hydrogen peroxide scavenging, inhibition of lipid peroxidation, and DPPH assays, while the essential oils from the fruits and leaves showed better activity against gram-positive bacteria when compared with the gram-negative bacteria used in this study (Odoom et al., 2023).

In addition to scientific evidence, insights into the traditional medicinal uses of *Vitex* species provide valuable cultural and historical perspectives on its health benefits. Traditional medicinal practices, especially in regions where the plants are indigenous, have long recognized its therapeutic properties and incorporated it into remedies for various ailments. Traditional uses include the treatment of fevers, gastrointestinal disorders, respiratory conditions, and wound healing, reflecting the plant's historical significance in indigenous healing systems. Understanding the traditional medicinal uses of *Vitex* provides valuable insights into its historical context, cultural relevance, and traditional knowledge systems, enriching the understanding of its health benefits and medicinal potential (Das et al., 2022; Kamal et al., 2022).

19.10 TOXICITY ASSESSMENTS

To date, there have been no reports of toxicity associated with the utilization of the plant in humans, as well as with other *Vitex* species. A study presented findings on the toxicity potential of *V.*

grandifolia, while other *Vitex* species have also undergone toxicity evaluations in experimental animals. In a specific investigation focusing on the potential toxicity of the aqueous extract of *V. grandifolia* on liver and kidney function in rats (Owolabi et al., 2010), the administration of aqueous extracts at doses ranging from 0.5 to 2 g/kg body weight (b. wt.) resulted in a significant, dose-dependent increase in serum electrolytes, creatinine, and liver function enzymes compared to the control group. Furthermore, structural alterations were observed in the liver, kidney, and lungs of the treated groups in comparison to the control. Observed clinical signs, including polydipsia, polyuria, hair puffiness, and sedation, were evident in the treated groups, with the severity of these manifestations correlating with the dosage of the extract. These findings underscore the potential toxicity associated with prolonged administration of the aqueous leaf extracts of *V. grandifolia* at the specified doses in this investigation. Therefore, cautious consideration is recommended regarding the utilization of this plant in traditional medicinal practices. In a separate investigation focusing on the toxicity of *V. simplicifolia* ethyl acetate leaf extract in Wistar rats (Salim & Imam, 2016), it was observed that the extract was well tolerated at the high dose of 5,000 mg/kg body weight. Furthermore, the sub-chronic toxicity study revealed no significant increase in the liver function indices and some kidney function indices. Additionally, the results from the study of Nguyen et al. (2024) examining the toxicity of *V. negundo* indicate a favorable safety profile for oral administration. In the acute toxicity study conducted in mice, the highest tolerable dose of *V. negundo* leaf extract, at 620.25 g/kg/day, did not induce any signs of oral toxicity. Similarly, in the sub-chronic toxicity study involving rabbits, two groups were orally administered *V. negundo* at doses of 3.2 g/kg/day and 9.6 g/kg/day for eight weeks. Throughout the study period, no evidence of general toxicity was observed. There were no notable changes in growth parameters, hematological parameters, liver and kidney functions, and no discernible effects on the macroscopic and microscopic morphology of the liver and kidneys in the treated animals. These findings suggest that *V. negundo* exhibits a low risk of toxicity when administered orally at the specified doses for both acute and sub-chronic duration. The toxicity profile of *V. doniana* fruits as evaluated in the study by Adjei et al. (2021) indicates that the fruit of this plant is safe for consumption.

19.11 CONCLUSION AND PROSPECTS

In this chapter, we have delved into the nutritional composition of edible members of the genus *Vitex*, encompassing proximate analysis, vitamins, minerals, and bioactive phytoconstituents. Our exploration has unveiled the rich nutritional wealth present in various *Vitex* species, shedding light on their potential health benefits and nutrient bioavailability. Moving forward, there are promising avenues for future research and application. Firstly, further investigations are warranted to comprehensively elucidate the nutritional profiles of additional *Vitex* species, thereby expanding our understanding of their dietary significance and therapeutic potential. Additionally, exploring the bioavailability and functional properties of nutrients within *Vitex* species will enhance their utilization as dietary supplements, functional food ingredients, or medicinal resources. Furthermore, leveraging this knowledge on the nutritional wealth of *Vitex* species can play a crucial role in addressing food security challenges. By harnessing the nutritional potential of these underutilized plant species, we can diversify food sources, enhance dietary diversity, and mitigate malnutrition. Moreover, integrating *Vitex* species into agricultural systems and promoting their cultivation can contribute to sustainable livelihoods and support economic development in regions where they are indigenous.

19.12 ACKNOWLEDGMENTS

The authors appreciate the leadership of Biomedicinal Research Centre, Forestry Research Institute of Nigeria, for supporting this work.

REFERENCES

Abdullahi, S., Dankat, C. B., Ayuba, A., Auta, C., & Danjuma, I. (2022). Assessment of proximate, minerals and amino acids content of black plum (*Vitex doniana*) young leaves as dietary vegetable substitute. *International Journal of Home Economics, Hospitality and Allied Research*, *1*(2), 170–177. https://doi.org/10.57012/ijhhr.v1n2.001

Achika, J. I., Yusuf, A. J., Ayo, R. G., & Liman, D. U. (2023). Flavonoids from Nigerian indigenous medicinal plants and their bioactivities: A review. *Phytomedicine Plus*, *3*, 100405. https://doi.org/10.1016/j.phyplu.2023.100405

Adejumo, A. A., Alaye, S. A., Ajagbe, R. O., Abi, E. A., & Adedokun, F. T. (2013). Nutritional and anti-nutritional composition of black plum (Vitex doniana). *Journal of Natural Sciences Research*, *3*(12), 144–148.

Adeleke, A. E., Isola, O. E., Onifade, A. P., Oke, A. M., Olanipekun, I. M., & Aghanti, S. S. (2023). Proximate, minerals and nutritive composition of *Vitex doniana* flour and oil for industrial applications. *Adeleke University Journal of Engineering and Technology*, *6*(1), 222–230.

Adjei, S., Amponsah, I. K., Bekoe, S. O., Harley, B. K., Mensah, K. B., Mensah, A. Y., Baah, M. K., & Fosu-Mensah, G. (2021). Fruits of *Vitex doniana* sweet: Toxicity profile, anti-inflammatory and antioxidant activities, and quantification of one of its bioactive constituents oleanolic acid. *Heliyon*, *7*(9), e07910. https://doi.org/10.1016/j.heliyon.2021.e07910

Adoukonou-Sagbadja, H., Ahoyo, D., Tovignan, T. K., Zavinon, F., & Ahoton, L. (2023). Conservation status and phenotypic diversity of natural populations of *Vitex doniana* sweet in Southern Benin assessed using quantitative morphometric traits. *South African Journal of Botany*, *156*, 286–297. https://doi.org/10.1016/j.sajb.2023.03.023

Ahmad, N., Shahid, A., Javed, S. B., Khan, M. I., & Anis, M. (2015). Micropropagation of *Vitex* spp. through *in vitro* manipulation: Current status and future prospectives. *Journal of Applied Research on Medicinal and Aromatic Plants*, *2*(4), 114–123. https://doi.org/10.1016/j.jarmap.2015.09.001

Ahmed, R., & Anis, M. (2014). Rapid in vitro propagation system through shoot tip cultures of *Vitex trifolia* L.-an important multipurpose plant of the Pacific traditional Medicine. *Physiology and Molecular Biology of Plants: An International Journal of Functional Plant Biology*, *20*(3), 385–392. https://doi.org/10.1007/s12298-014-0233-7

Aiwonegbe, A. E., Iyasele, J. U., & Izevbuwa, N. O. (2018). Proximate composition, phytochemical and antimicrobial screening of the methanol and acetone extracts of *Vitex doniana* fruit pulp. *Ife Journal of Science*, *20*(2), 317–323.

Ali, T., Bello, O. M., & Ogbesejana, A. B. (2020). Flavonoids with monoamine oxidase a and b inhibitory and anti-inflammatory effects from *Vitex grandif.olia*. *Malasian Journal of Analytical Sciences*, *24*(6), 1035–1044.

Amah, U. J., & Okogeri, O. (2019). Nutritional and phytochemical properties of wild black plum (Vitex doniana) seed from Ebonyi state. *International Journal of Horticulture, Agriculture and Food Science*, *3*(1), 32–36.

Amo-Mensah, J., Darko, G., & Borquaye, L. S. (2020). Anti-inflammatory and antioxidant activities of the root and bark extracts of *V. grandifolia* (Verbanaceae). *Scientific African*, *10*, e00586. https://doi.org/10.1016/j.sciaf.2020.e00586

Aparna, V., Dileep, K. V., Mandal, P. K., Karthe, P., Sadasivan, C., & Haridas, M. (2012). Anti-inflammatory property of n-hexadecanoic acid: Structural evidence and kinetic assessment. *Chemical Biology and Drug Design*, *80*(3), 434–439.

Balaji, S., & Kripa, K. G. (2021). Inhibition of inducible nitric oxide production by *Caryota urens* and its active constituents umbelliferone and rutin. *Journal of Ayurveda and Integrative Medicine*, *12*(2), 369–372. https://doi.org/10.1016/j.jaim.2020.09.002

Bao, F., Tang, R., Cheng, L., Zhang, C., Qiu, C., Yuan, T., Zhu, L., Li, H., & Chen, L. (2018). Terpenoids from *Vitex trifolia* and their anti-inflammatory activities. *Journal of Natural Medicine*, *72*, 570–575. https://doi.org/10.1007/s11418-018-1178-x

Bello, O. M., Asubonteng, K., Sodamade, A., & Adeniyi, S. (2014). Nutrient potentials of two lesser-known leafy vegetables (*Vitex doniana* L. and *Sesamum indicum* L.). *International Food Research Journal*, *21*(5), 1993–1999.

Bello, O. M., Ogbesejana, A. B., Adetunji, C. O., & Oguntoye, S. O. (2019). Flavonoids isolated from *V. grandifolia*, an underutilized vegetable, exert monoamine a & b inhibitory and anti-inflammatory effects and their structure-activity relationship. *Turkish Journal of Pharmaceutical Science*, *16*(4), 437–443.

Bello, O. M., Ogbesejana, A. B., & Osibheme, M. (2018a). Iridoid glycosides from *V. grandifolia* displayed anti-inflammatory and antileishmanial effects and structure-activity relationship. *Journal of Applied Science and Environmental Management*, *22*(3), 373–378.

Bello, O. M., Zaki, A. A., Aloko, S., Fasinu, P. S., Bello, E. O., Ajao, U. L., & Oguntoye, O. S. (2018b). The genus Vitex. An overview of iridoids as chemotaxonomic markers. *Beni-Suef University Journal of Basic and Applied Science*, *7*, 414–419.

Bingham, M. G., Willemen, A., Wursten, B. T., Ballings, P., & Hyde, M. A. (2024). *Flora of Zambia: Species information: Individual images*. Vitex Ferruginea. Retrieved April 14, 2024, from www.zambiaflora.com/speciesdata/image-display.php?species_id=148890&image_id=7

Boujbiha, M. A., Chahdoura, H., Adouni, K., Ziani, B. E. C., Snoussi, M., Chakroun, Y., Ciudad-Mulero, M., Fernández-Ruiz, V., Achour, L., Selmi, B., Morales, P., Flamini, G., & Mosbah, H. (2023). Wild *Vitex agnus-castus* L.: Phytochemical characterization, acute toxicity, and bioactive properties. *Molecules (Basel, Switzerland)*, *28*(13), 5096. https://doi.org/10.3390/molecules28135096

Charles, D., & Mgina, C. (2023). Proximate composition of Vitex doniana and Saba comorensis fruits. *Scientific Report*, *13*, 20553. https://doi.org/10.1038/s41598-023-46874-7

Chen, S. N., Friesen, J. B., Webster, D., Nikolic, D., van Breemen, R. B., Wang, Z. J., Fong, H. H. S., Farnsworth, N. R., & Pauli, G. F. (2011). Phytoconstituents from *Vitex agnus-castus* Fruits. *Fitoterapia*, *82*, 528–533.

Chinyere, I., Julius, I. U., & Samuel, O. E. (2021). Determination of the flavoring components in Vitex doniana fruit following hydrodistillation extraction. *American Journal of Food Science and Nutrition*, *9*(2), 69–75.

Dadjo, C. (2014). *Assessment of nutritional variation and establishment of an efficient tissue culture propagation protocol for Vitex doniana sweet (Lamiaceae) from Benin (Western Africa)* [MSc thesis, Submitted to Pan African University Institute for Basic Sciences Technology and Innovation].

Dahham, S. S., Tabana, Y. M., Iqbal, M. A., Ahamed, M. B. K., Ezzat, M. O., Majid, A. S. A., & Majid, A. M. S. A. (2015). The anticancer, antioxidant and antimicrobial properties of the sesquiterpene β-caryophyllene from the essential oil of *Aquilaria crassna*. *Molecules*, *20*, 11808–11829.

Daliu, P., Santini, A., & Novellino, E. (2019). From pharmaceuticals to nutraceuticals: Bridging disease prevention and management. *Expert Review in Clinical Pharmacology*, *12*, 1–7.

Das, N., Salgueiro, A. C. F., Choudhury, D. R., Mandal, S. K., Logesh, R., Hassan, M. M., & Devkota, H. P. (2022). Traditional uses, phytochemistry, and pharmacology of genus Vitex (Lamiaceae). *Phytotherapy Research: PTR*, *36*(2), 571–671. https://doi.org/10.1002/ptr.7330

Dauda, B. E. N., Oyeleke, S. B., Jigam, A. A., Salihu, S. O., & Balogun, M. M. (2011). Phytochemical and in-vitro antibacterial investigation of *Vitex doniana* leaves, stem bark and root bark extracts. *Australian Journal of Basic and Applied Sciences*, *5*(7), 523–528.

Devi, N. T., Rani, P. R. J., Theivaprakasham, H., Arumugam, S., & Nachiappan, R. M. V. (2016). Estimation of micronutrients in *Vitex negundo* L. (Karunochi) Leaves. *Food Science: Indian Journal of Research in Food Science and Nutrition*, *3*(1), 13–17.

Dharmasiri, M. G., Jayakody, J. R., Galhena, G., Liyanage, S. S., & Ratnasooriya, W. D. (2003). Anti-inflammatory and analgesic activities of mature fresh leaves of *Vitex negundo*. *Journal of Ethnopharmacology*, *87*(2–3), 199–206. https://doi.org/10.1016/s0378-8741(03)00159-4

Dong, J. E., & Zhang, J. (2004). Advances in the research of iridoids occurring in plants. *Journal of Northwest Forestry College*, *19*, 131–135.

Epidi, J. O., Izah, S. C., Ohimain, E. I., & Epidi, T. T. (2016). Antibacterial and synergistic potency of tissues of *V. grandifolia*. *Biotechnological Research*, *2*(2), 69–76.

Govaerts, R. (Ed.). (2017). WCSP: World Checklist of Selected Plant families (version Aug 2017). In Y. Roskov, G. Ower, T. Orrell, D. Nicolson, N. Bailly, P. M. Kirk, T. Bourgoin, R. E. DeWalt, W. Decock, E. vanNieukerken, J. Zarucchi, & L. Penev (Eds.), *Species 2000 & ITIS catalogue of life, 2019 annual checklist* (Digital resource at www.catalogueoflife.org/annual-checklist/2019. Species 2000). Naturalis.

Groach, R., Yadev, K., & Singh, N. (2014). In vitro studies on Vitex negundo, a potent medicinal plant. *Environmental and Experimental Biology*, *12*, 149–153.

Haruna, M., Nakhooda, M., & Shaik, S. (2024). Development of seed germination and in vitro propagation protocols for *Vitex doniana*. *South African Journal of Botany*, *167*, 399–409. https://doi.org/10.1016/j.sajb.2024.02.025

Hu, P., Li, D. H., Jia, C. C., Liu, Q., Wang, X. F., Li, Z. L., & Hua, H. M. (2017). Bioactive constituents from *Vitex Negundo Var. Heterophylla* and their antioxidant and α-Glucosidase inhibitory activities. *Journal of Functional Foods*, *35*, 236–244.

Ifeanacho, M., & Ogunwa, S. (2021). nutritional and bioactive potentials of an underutilized vegetable – Vitex doniana. *Food and Nutrition Sciences*, *12*, 978–995. https://doi.org/10.4236/fns.2021.1210072

Ifeanacho, M., Ogunwa, S. C., & Amadi, P. U. (2019). Phytochemical composition of *Vitex doniana*. *Analytical Chemistry Letters*, *9*(6), 863–875.

Imoisi, C., Iyasele, J., & Okhale, S. (2021). Proximate and acute toxicity profile of Vitex doniana (Black Plum) fruit. *Journal of Chemical Society of Nigeria*, *46*(2). https://doi.org/10.46602/jcsn.v46i2.597

Islam, Z., Caldeira, G. I., Canica, M., Islam, N., & Silva, O. (2024). Vitex genus as a source of antimicrobial agents. *Plants*, *13*, 401. https://doi.org/10.3390/plants13030401

Kamal, N., Mio Asni, N. S., Rozlan, I. N. A., Mohd Azmi, M. A. H., Mazlan, N. W., Mediani, A., Baharum, S. N., Latip, J., Assaw, S., & Edrada-Ebel, R. A. (2022). Traditional medicinal uses, phytochemistry, biological properties, and health applications of *Vitex* sp. *Plants (Basel, Switzerland)*, *11*(15), 1944. https://doi.org/10.3390/plants11151944

Kiani, A. K., Dhuli, K., Donato, K., Aquilanti, B., Velluti, V., Matera, G., Iaconelli, A., Connelly, S. T., Bellinato, F., Gisondi, P., & Bertelli, M. (2022). Main nutritional deficiencies. *Journal of Preventive Medicine and Hygiene*, *63*(2, Suppl. 3), E93–E101. https://doi.org/10.15167/2421-4248/jpmh2022.63.23.2752

Koirala, N., Dhakal, C., Munankarmi, N. N., Ali, S. W., Hameed, A., Martins, N., Sharifi-Rad, J., Imran, M., Arif, A. M., Hanif, M. S., Basnyat, R. C., & Salehi, B. (2020). *Vitex negundo* Linn.: Phytochemical composition, nutritional analysis, and antioxidant and antimicrobial activity. *Cellular and Molecular Biology*, *66*(4), 1–7. https://doi.org/10.14715/cmb/2020.66.4.1

Kulkarni, R. R., Virkar, A. D., & D'mello, P. (2008). Antioxidant and antiinflammatory activity of Vitex negundo. *Indian Journal of Pharmaceutical Sciences*, *70*(6), 838–840. https://doi.org/10.4103/0250-474X.49140

Kumar, M., Dandapat, S., Kumar, A., & Sinha, M. P. (2013). Determination of nutritive value and mineral elements of five-leaf chaste tree (*Vitex negundo* L.) and malabar nut (*Adhatoda vasica* Nees). *Academic Journal of Plant Sciences*, *6*(3), 103–108.

Li, S., Qiu, S., Yao, P., Sun, H., Fong, H. H. S., & Zhang, H. (2013). Compounds from the fruits of the popular European medicinal plant *Vitex agnus-castus* in chemoprevention via NADP(H): quinone oxidoreductase type 1 induction. *Evidence Based Complementary and Alternative Medicine*, *2013*, Article 432829. https://doi.org/10.1155/2013/432829

Mapongmetsem, P. M., Fawa, G., Noubissie-Tchiagam, J. B., Nkongmeneck, B. A., Biaou, S. S. H., & Bellefontaine, R. (2016). Vegetative propagation of *Vitex doniana* sweet from root segments cuttings. *Bois et forêts des tropiques*, *327*(1), 29–37.

Mohammed, M., Danmallam, A., Jajere, U. M., Kolo, M. T., Abubakar, A., & Babakano, J. M. (2016). Three triterpenoids from the leaf extract of Vitex doniana (Verbenaceae). *Journal of Pharmaceutical Research International*, *12*(5), 1–8. https://doi.org/10.9734/BJPR/2016/24335

Mujike, D. N., Meerts, P., Lokoto, B. M., & Shutcha, M. N. (2023). Four multipurpose species of the genus *Vitex* (Lamiaceae) in the democratic republic of the Congo show different responses to propagation techniques for nursery production. *Bois et Forets des Tropiques*, *357*, 43–56. https://doi.org/10.19182/bft2023.357.a36943

N'Danikou, S., Achigan-Dako, E. G., Tchokponhoué, D. A., Assogba Komlan, F., Gebauer, J., Vodouhè, R. S., & Ahanchédé, A. (2014). Enhancing germination and seedling growth in *Vitex doniana* sweet for horticultural prospects and conservation of genetic resources. *Fruits*, *69*(4), 279–291.

Naseri, R., Farnia, V., Yazdchi, K., Alikhani, M., Basanj, B., & Salemi, S. (2019). Comparison of *Vitex agnus-castus* extracts with placebo in reducing menopausal symptoms: A randomized double-blind study. *Korean Journal of Family Medicine*, *40*(6), 362–367.

Nguyen, T. T., Do, P. T., Nguyen, L. T. N., & Pham, A. T. (2024). Acute and sub-chronic toxicity studies of aqueous ethanol leaf extract of Vitex negundo L in experimental animals. *Tropical Journal of Natural Product Research (TJNPR)*, *8*(1), 5806–5810. www.doi.org/10.26538/tjnpr/v8i1.14

Niroumand, M. C., Heydarpour, F., & Hosein, F. M. (2018). Pharmacological and therapeutic effects of *Vitex agnus-castus* L.: A review. *Pharmacognosy Reviews*, *12*(23), 103–114.

Niu, Y. X., Wang, D., Chu, X. Y., Gao, S. Y., Yang, D. X., Chen, L. X., & Li, H. (2020). Iridoids from *Vitex negundo Var. Heterophylla* and their antioxidant activities. *Phytochemistry Letters*, *35*, 186–190. https://doi.org/10.1016/j.phytol.2019.11.018

Nor Hisham Shah, N. K. A., & Bridgen, M. (2016). Seed germination studies of *Vitex agnus-castus*©. *Acta Horticulture*, *1140*, 299–302. https://doi.org/10.17660/ActaHortic.2016.1140.67

Nwodo, N., Okoye, F., Lai, D., Debbab, A., Kaiser, M., Brun, R., & Proksch, P. (2015). Evaluation of the in vitro trypanocidal activity of methylated flavonoid constituents of *Vitex Simplicifolia* leaves. *BMC Complementary and Alternative Medicine*, *15*, 82. https://doi.org/10.1186/s12906-015-0562-2

Ochieng, C. O., & Nandwa, B. O. (2010). Proximate composition, phenolic content and antioxidant activities of three black plum (Vitex sp.) fruits: Preliminary results. *Journal of Food Technology*, *8*(3), 118–125.

Odoom, J. F., Aboagye, C. I., Acheampong, P., Asiamah, I., Darko, G., & Borquaye, L. S. (2023). Chemical composition, antioxidant, and antimicrobial activities of the leaf and fruit essential oils of the West African Plum, *Vitex doniana*. *Journal of Chemistry*, *2023*, Article 9959296. https://doi.org/10.1155/2023/9959296

Odusina, B. O., & Yusuf-Mikail, R. A. (2022). Chemical constituents and antibacterial activities of essential oils from V. grandifolia Gurke. *Bells University Journal of Applied Sciences and Environment*, *3*(2), 1–5.

Okigbo, R. N., & Anyaegbu, C. F. (2021). Underutilized plants of Africa. *Journal of Biology and Nature*, *13*(2), 34–49.

Olivera, A., Moore, T. W., Hu, F., Brown, A. P., Sun, A., Liotta, D. C., Snyder, J. P., Yoon, Y., Shim, H., Marcus, A. I., Miller, A. H., & Pace, T. W. (2012). Inhibition of the NF-κB signalling pathway by the curcumin analogue, 3,5-Bis(2-pyridinyl methylidene)-4-piperidone (EF31): Anti-inflammatory and anti-cancer properties. *International Immunopharmacology*, *12*(2), 368–377. https://doi.org/10.1016/j.intimp.2011.12.009

Ono, M., Eguchi, K., Konoshita, M., Furusawa, C., Sakamoto, J., Yasuda, S., Ikeda, T., Okawa, M., Kinjo, J., Yoshimitsu, H., & Nohara, T. (2011). A new diterpenoid glucoside and two new diterpenoids from the fruit of Vitex agnus-castus. *Chemical & Pharmaceutical Bulletin*, *59*(3), 392–396. https://doi.org/10.1248/cpb.59.392

Osum, F. I., Okonkwo, T. M., & Okafor, G. I. (2013). Effect of processing methods on the chemical composition of *Vitex doniana* leaf and leaf products. *Food Science & Nutrition*, *1*(3), 241–245. https://doi.org/10.1002/fsn3.31

Owolabi, M. A., Abass, M. M., Emeka, P. M., Jaja, S. I., Nnoli, M., & Dosa, B. O. (2010). Biochemical and histologic changes in rats after prolonged administration of the crude aqueous extract of the leaves of *V. grandifolia*. *Pharmacognosy Research*, *2*(5), 273–278. https://doi.org/10.4103/0974-8490.72322

Ozkaya, A., Ciftci, H., Yilmaz, O., Tel, A. Z., Cil, E., & Cevrimli, B. S. (2013). *Itamin, trace element, and fatty acid levels of Vitex agnus-castus L., Juniperus oxycedrus L., and Papaver somniferum L.* Plant Seeds.

Patel, D. K. (2016). Vitex negundo Linn: Stem cutting propagation in Herbal Garden. *International Journal of Applied Research*, *2*(1), 1–5.

Pío-León, J. F., Montes-Avila, J., Díaz-Camacho, S. P., & Delgado-Vargas, F. (2014). Chapter 4 – Biological activities and phytochemicals of the fruits of Vitex Plants. In V. K. Gupta (Ed.), *Bioactive phytochemicals: Perspectives for modern medicine* (Vol. 2). Daya Publishing House.

Qiu, C., Tong, L., Yuan, T., Wang, F., Zhao, F., & Chen, L. (2017). Constituents from Vitex negundo var. heterophylla and their inhibition of nitric oxide production. *Journal of Natural Medicines*, *71*(1), 292–298. https://doi.org/10.1007/s11418-016-1032-y

Rana, G. (2018). Inhibition efficiency of a newly isolated flavonoid compound from Vitex Negundo L. leaves against cattle-endosymbiont Setaria Cervi: Phytomedicine for Lymphatic Filariasis. *Parasite Epidemiology and Control*, *3*(2), 88–95. https://doi.org/10.1016/j.parepi.2018.03.002

Rana, Z. H., Alam, M. K., & Akhtaruzzaman, M. (2019). Nutritional composition, total phenolic content, antioxidant and α-amylase inhibitory activities of different fractions of selected wild edible plants. *Antioxidants*, *8*, 203. https://doi.org/10.3390/antiox8070203

Rani, A., & Sharma, A. (2013). The genus Vitex: A review. *Pharmacognosy Review*, *7*(14), 188–198.

Rein, M. J., Renouf, M., Cruz-Hernandez, C., Actis-Goretta, L., Thakkar, S. K., & da Silva Pinto, M. (2013) Bioavailability of bioactive food compounds. A challenging journey to bioefficacy. *British Journal of Clinical Pharmacology*, *75*(3), 588–602. https://doi.org/10.1111/j.1365-2125.2012.04425.x

Salim, M. A., & Imam, A. A. (2016). Hepato-renal toxicological studies of ethyl acetate leaf extract of *Vitex simplicifolia* in Wistar rats. *Bayero Journal of Pure and Applied Sciences*, *9*(2), 76–81.

Samtiya, M., Aluko, R. E., & Dhewa, T. (2020). Plant food anti-nutritional factors and their reduction strategies: An overview. *Food Production Processing and Nutrition*, *2*, 6. https://doi.org/10.1186/s43014-020-0020-5

Schonfeldt, H., Pretorius, B., & Hall, N. (2016). Bioavailability of nutrients. In *Encyclopedia of food and health*. https://doi.org/10.1016/B978-0-12-384947-2.00068-4

Singh, Y., Mishra, P., & Kannojia, P. (2020). Morphology, phytochemistry and pharmacological activity of vitex negundo: An overview. *Journal of Drug Delivery and Therapeutics*, *10*(3-s), 280–285. https://doi.org/10.22270/jddt.v10i3-s.4173

Sonibare, O. O., Effiong, I., Oladosu, I. A., & Ekundayo, O. (2009). Chemical constituents and antimicrobial activity of the essential oil of Vitex doniana sweet (Verbernaceae). *Journal of Essential Oil-Bearing Plants*, *12*(2), 185–188.

Souto, E. B., Durazzo, A., Nazhand, A., Lucarini, M., Zaccardelli, M., Souto, S. B., Silva, A. M., Severino, P., Novellino, E., & Santini, A. (2020). *Vitex agnus-castus* L.: Main features and nutraceutical perspectives. *Forests*, *11*(7), 761. https://doi.org/10.3390/f11070761

Sridhar, C., Subbaraju, G. V., Venkateswarlu, Y., & Venugopal, R. T. (2004). New acylated iridoid glucosides from *Vitex altissima*. *Journal of Natural Product*, *67*, 2012–2016.

Sun, Y., Yang, H., Zhang, Q., Qin, L., Li, P., Lee, J., Chen, S., Rahman, K., Kang, T., & Jia, M. (2019). Genetic diversity and its conservation implications of *Vitex rotundifolia* (Lamiaceae) populations in East Asia. *Peer J*, *7*, e6194. https://doi.org/10.7717/peerj.6194

Tiwari, N., Yadav, A. K., & Gupta, M. M. (2013). Validated HPTLC method for the simultaneous quantification of diterpenoids in *Vitex trifolia L. Journal of Separation Science*, *36*, 2373–2378.

Tropical Plants Database. (2024). *Ken fern* (tropical.theferns.info. 2024-03-07). tropical.theferns.info/view-tropical.php?id=Vitex+grandifolia

Vunchi, M. A., Umar, A. N., King, M. A., Liman, A. A., Jeremiah, G., & Aigbe, C. O. (2011). Proximate, vitamins and mineral composition of Vitex doniana (black plum) fruit pulp. *Nigerian Journal of Basic and Applied Science*, *19*(1), 97–101.

Wang, C., Gong, X., Bo, A., Zhang, L., Zhang, M., Zang, E., Zhang, C., & Li, M. (2020). Iridoids: Research advances in their phytochemistry, biological activities, and pharmacokinetics. *Molecules (Basel, Switzerland)*, *25*(2), 287. https://doi.org/10.3390/molecules25020287

Xie, C., Wang, S., Cao, M., Xiong, W., & Wu, L. (2022). (E)-9-Octadecenoic acid ethyl ester derived from lotus seedpod ameliorates inflammatory responses by regulating MAPKs and NF-κB signalling pathways in LPS-induced RAW264.7 macrophages. *Evidence-Based Complementary and Alternative Medicine: eCAM*, *2022*, 6731360. https://doi.org/10.1155/2022/6731360

Zaki, F. L. M., & Wan Salleh, W. M. N. H. (2020). Essential oils and biological activities of the genus Vitex (Lamiaceae) – a review. *Natural Volatiles and Essential Oils*, *7*(2), 1–9. https://doi.org/10.37929/nveo.743784

Zheng, C. J., Zhu, J. Y., Yu, W., Ma, X. Q., Rahman, K., & Qin, L. P. (2013). Labdane-type diterpenoids from the fruits of *Vitex trifolia*. *Journal of Natural Products*, *76*, 287–291.

20 Exploring the Nutritional Profile of Tepary Bean (*Phaseolus acutifolius*) and Evaluating Seed Protein Alterations Caused by Chemical Mutagenesis

*[1]Dipoo Charity Mashifane, [2]Dorcus Maja, [3]Esther Kathini Muema, [4]Addmore Shonhai, and *[1]Eastonce Tendayi Gwata*

[1]University of Venda, Faculty of Science, Engineering, and Agriculture, Department of Plant and Soil Sciences, Private Bag X5050, Thohoyandou, 0950, South Africa

[2]North-West University, Faculty of Natural and Agricultural Sciences, Food Security and Safety Focus Area, Private Bag X2046, Mmabatho 2735, South Africa

[3]Stellenbosch University, Faculty of Agri-Sciences, Department of Soil Science, Matieland, 7602, South Africa

[4]University of Venda, Faculty of Science, Engineering, and Agriculture, Department of Biochemistry and Microbiology, Private Bag X5050, Thohoyandou, 0950, South Africa

*Corresponding author: ectgwata@gmail.com

20.1 INTRODUCTION

Tepary bean (*Phaseolus acutifolius*) is a self-pollinating diploid (2n = 2x = 22) leguminous species originating from the semi-arid region of northwestern Mexico and southwestern United States (Nabhan & Felger, 1978). Tepary bean was introduced in West Africa in the 1930s and into the Congo about the same time. The production area of the legume was expanded to Southern and Eastern Africa as well as Australia and the Caribbean region (Nong et al., 2023; Chivenge et al., 2015; Crews et al., 2004; Miklas & Santiago, 1996). Currently, traditional varieties of the crop are cultivated by smallholder growers in several African countries, including Botswana, Kenya, Malawi, South Africa, Uganda, and Zimbabwe, as well as Benin in West Africa (Thangwana et al., 2021; Dinghani et al., 2018; Molosiwa et al., 2014; Mogotsi, 2006; Shisanya, 2002). It is an annual summer crop which possesses unique genetic attributes, such as tolerance to drought and heat, making it suitable for cultivation in harsh agro-ecological conditions where other grain legumes often

 DOI: 10.1201/9781003469766-20

fail. It is often classified as a climate-smart crop partly because of its ability to withstand extremely harsh environmental conditions and can complete its life cycle rapidly within about 90 days.

Tepary bean is cultivated primarily for human consumption and is often consumed as sprouts or green beans. The leaves are also consumed while haulms are used as animal feed (Porch et al., 2017). The grain of tepary bean is a rich source of affordable protein particularly for rural communities in developing countries. The grain contains high protein levels (25.0%); essential mineral elements, such as calcium (Ca), iron (Fe), copper (Cu), and zinc (Zn); as well as considerable amounts of fatty acids and antioxidants (Heredia-Rodríguez et al., 2019; Salas-Lopez et al., 2018; Bhardwaj & Hamama, 2005). Tepary bean also contributes to soil fertility improvement by fixing atmospheric nitrogen (Shisanya, 2004). However, despite its valuable potential, tepary bean remains underutilized particularly in Africa and Asia. Consequently, in these regions, there is a limited number or no improved varieties for commercial production. However, the current extreme weather conditions, characterized by increased frequency of drought, heat, and floods, pose a serious threat to crop production as well as food and nutrition security (Lesk et al., 2016). Leguminous plants such as tepary bean that are nutrient-dense and tolerant to drought present opportunities for diversified, secure, nutritious, and climate-resilient food systems for sustainable crop production in many parts of the world (Dwivedi et al., 2023; Abberton et al., 2022; Heredia-Rodríguez et al., 2019).

20.2 MAJOR USES OF TEPARY BEAN

The grain is used mainly for human consumption. In some countries, the grain is commonly used as supplementary feed for poultry and other livestock (Bhardwaj & Hamama, 2005). The grain is also potentially useful in chemotherapy due to the presence of lectins and other compounds (De Mejía & Prisecaru, 2005). The lectins exhibit antitumor properties by inducing programmed cell death (apoptosis) and autophagy (Konozy & Osman, 2022; Yau et al., 2015). Traditionally, the grain of tepary bean is ideal for people who are diabetic because of its low glycemic index (GI) resulting from its high fiber levels, which consequently release sugar slowly, thus enabling diabetic patients to retain protein in the body without risking an increase in body sugar levels (Jenkins et al., 2012; No?thlings et al., 2008).

Tepary bean is used also in agroforestry systems as a cover crop and in intercropping due to its ability to improve soil fertility through biological nitrogen fixation, up to 260 kg N/ha (Shisanya, 2003). In the process, tepary bean translocates ureides in the xylem sap as the dominant products to various parts of the plant, where they are subsequently catabolized and synthesized into biological molecules, such as amino acids (Pratelli & Pilot, 2014). Since amino acids are the protein building blocks, there is a direct relationship between root nodulation and protein production in legumes.

Tepary bean is highly tolerant to drought and high temperatures (Nong et al., 2023). It can grow in areas where annual rainfall is less than 450 mm. The drought tolerance mechanism in the species has been attributed to sensitive stomata which close at high-water potentials (Medina et al., 2017). Tepary bean is a source of useful genes for disease resistance, insect pests, and drought tolerance. For instance, it harbors genes for resistance to some diseases, such as bacterial blight (*Xanthomonas campestris* pv. Phaseoli), fusarium wilt (*Fusarium oxysporum*), bean golden mosaic virus (BGMV – family Geminiviridae, genus *Begomovirus*), and insect pests, such as bruchids (or weevils) (*Acanthoscelides obtectus*) (Jiménez et al., 2017; Miklas et al., 2006; Salgado et al., 1994). The resistance to insect pests is partly due to the presence of lectin-like seed storage proteins and α-amylase inhibitor (Velten et al., 2007; Sales et al., 2000). Because of these desirable traits, tepary bean was used in the introgression of useful genes into common bean (*Phaseolus vulgaris*) (Pratt et al., 1990).

The seed also contains high levels of antioxidant, mineral, and anticancerous properties, which potentially minimize the risks of various ailments, such as colon cancer, coronary heart diseases, and diabetes (Valadez-Vega et al., 2011). Furthermore, its "slow sugar release" quality enables diabetic patients to retain protein in the body without risking a surge in body sugar levels.

20.3 NUTRITIVE VALUE OF TEPARY BEAN

Tepary bean consists of a range of nutrients, with protein serving as the primary component. The other nutrients in tepary bean include minerals, crude fat (vegetable oil), and carbohydrates (Table 20.1). The nutrients occur in variable quantities largely due to genetic and environmental factors (González de Mejía et al., 2003; Bhardwaj, 2013). The health benefits associated with the consumption of plant-based proteins have been recently elucidated (López-Ibarra et al., 2021; Moreno-Valdespino et al., 2020; Tran et al., 2020). For instance, the functional effects of the lectin protein fraction of tepary bean on cancer cells were demonstrated both *in vitro* and *in vivo* (García-Gasca et al., 2012; Moreno-Celis et al., 2017, 2020). The seed storage proteins in tepary bean include albumins (83.2%), globulins (13.7%), glutelins (2.6%), and prolamins (0.5%) (Idouraine et al., 1994). The seed storage proteins are important in determining the total protein content in the seed, as well as functional qualities, particularly in cellular metabolic activities.

20.3.1 Mineral and Vitamin Content

The range of minerals in tepary bean grain compares well with those reported for most common tropical legumes (Table 20.2). However, the concentrations are generally lower, particularly Fe (Celmeli et al., 2018; Gouveia et al., 2014). Moreover, a significant negative relationship between Fe and P was observed in forage tepary bean (Bhardwaj, 2013). In soybean, the Fe content is approximately double that in tepary bean (Etiosa et al., 2018; Sharma et al., 2014). Iron is critical in hemoglobin synthesis and oxygen transport (Aisen et al., 2021; Lieu et al., 2021). Similarly, Ca content has been found relatively inferior in comparison with other common legumes, averaging about 28.0 mg/100 g, which is tenfold less than in soybean. In a study involving a tepary bean landrace, Fe and Mn were low, showing 0.6 mg/100 g and 2.3 mg/100 g, respectively (Dinghani et al., 2018). However, in the same study, Ca and Mg contents were 48 mg/100 g and 51.8 mg/100 g, respectively. Ca is important for skeletal tissue development and metabolic regulation of biomolecules, such as coenzymes. Similarly, Mg is involved in regulating blood sugar levels, blood pressure, and cardiovascular disease (Chiuve et al., 2013; Vormann, 2003; Rude, 1998). The intake of 625 mg daily in the diet has been shown to reduce hypertension in high blood pressure individuals (Montoyama, 1989), while about 500 mg of Mg daily was shown to improve systolic and diastolic blood pressure significantly. Magnesium deficiency can lead to back, neck, and joint pain, as well as muscle tension and spasms (Faryadi, 2012). This macronutrient is also important for the maintenance of electrical balance in the human body and protects the nerve cells in the brain, enhancing the transmission

TABLE 20.1
Tepary Bean Proximate Constituents

Name of Legume	Carbohydrate	Fat	Protein	Calcium	Phosphorus	Iron	Reference
		(%)			(mg/100 g)		
Tepary bean (*Phaseolus acutifolius*)	65.0–69.0	0.9–1.2	21.0–25.0	28.0	450.0	7.1–8.3	Porch et al. (2017)
Common bean (*Phaseolus vulgaris*)	55.0–65.0	0.3–1.3	18.5–29.6	75.0–83.0	300.0–750.0	4.1–10.0	Celmeli et al. (2018); Gouveia et al. (2014)
Cowpea (*Vigna unguiculata*)	50.0–60.0	5.4–11.2	23.0–32.0	85.0–93.0	438.0–498.0	10.0–11.0	Enyiukwu et al. (2020); Jayathilake et al. (2018)
Soybean (*Glycine max*)	30.2	31.0–71.0	36.5–44.3	277.0–300.0	695.0–704.0	15.7–16.4	Etiosa et al. (2018); Sharma et al. (2014)

TABLE 20.2
Tepary Bean Mineral Constituents

Mineral	Range (mg/100 g of Dry Grain)
Calcium	126.3–445.8
Magnesium	94.8–191.9
Phosphorus	60–451.0
Manganese	0.9–3.6
Zinc	1.7–6.1
Potassium	1501.0–1,607.8
Iron	0.2–12.6

Source: Adapted from Dinghani et al. (2018); Bhardwaj and Hamama (2004).

of electrical impulses. In another study, in which Mg intake was maintained, diabetes was reduced considerably (Rosique-Esteban et al., 2018). It is also important to point out that serum Mg represents only a minimal proportion of the total Mg present in the entire human body; hence, intracellular [Mg^{2+}] is regarded as a more accurate method reflecting the Mg status, albeit with additional difficulties to be measured (Jahnen-Dechent & Ketteler, 2012).

However, tepary bean's and other leguminous crops' mineral availability is potentially limited by the occurrence of phytic acid (phytin), which contains a P molecule and an affinity to bind with either protein or certain minerals. Domesticated teparies generally exhibit less phytin content than wild types, with brown and white teparies averaging 0.61 and 0.73% of this compound, respectively (Nabhan & Felger, 1978). In other legumes, such as common bean, phytin content has been recorded to range between 0.54 and 1.58% and correlates positively with total P content (Lolas & Markakis, 1975). Nonetheless, the phytic acid in most legume seeds is generally reduced by soaking and cooking processes (Reddy, 2001). The phytic acid contents of both wild and domesticated teparies are 30.0–50.0% less than those attributed to soybean and resemble values reported for common beans. Domesticates exhibit less phytin content than wild beans, with brown and white teparies averaging 0.61 and 0.73% of this compound. In legumes, both starch and protein digestibility can also be limited by the presence of tannins and α-amylase (Alonso et al., 2000). However, digestibility can be improved by various domestic processes, such as cooking, germination, microwave, and gamma irradiation (Tresina et al., 2017; Bhagyawant et al., 2015).

20.3.2 Protein Content and Quality

Protein content in tepary bean can be found ranging between 17.19 and 26.26% (Narina et al., 2014; Bhardwaj & Hamama, 2004). In addition, the beans contain carbohydrates, dietary fiber, vitamins, and about 15 essential minerals. Higher protein levels (21.0–31.9%) have also been reported (Porch et al., 2017). However, the protein content varies widely with domesticated cultivars, averaging about 25% (Narina et al., 2014; González de Mejía et al., 2003). In the 1980s, the protein content in white-seeded and brown-seeded teparies varied from 18.9 to 27.3% and from 18.1 to 27.1%, respectively (Nabhan et al., 1980). In another study, the mean protein content in wild teparies was significantly greater than that of the improved cultivars (Waines, 1978). The diversity in protein content is attributed largely to genetic and environmental factors (Nabhan et al., 1980).

Protein quality can also be determined chemically by analyzing the amino acid constituents as well as the digestibility of the proteins *in vivo* using rats in the laboratory or other animals (Queiroz Mendes et al., 2016; Deglaire & Moughan, 2012). In a study that determined both mineral and amino acid content, sulfur correlated with proline, glycine, alanine, cysteine, valine, and lysine, as well as with Mg (Porch et al., 2017). However, phenylalanine was significantly higher in wild teparies in comparison with the domesticated species (Table 20.3) (Nabhan & Felger, 1978).

TABLE 20.3
Wild and Domesticated Tepary Bean Essential Amino Acids

Essential Amino Acid	Tepary Bean	
	Wild (mg/g)	Improved (mg/g)
Isoleucine	4.9–5.3	4.1–4.7
Leucine	9.0–10.1	7.2–12.4
Lysine	7.1–8.3	6.1–7.6
Methionine	0.8	0.9–1.2
Phenylalanine	6.1–6.8	3.0–3.6
Threonine	4.3–4.5	3.9–5.0
Tryptophan	–	0.4–0.7
Valine	5.9–6.7	5.2–5.6
Tyrosine	9.7–10.7	8.0–9.7

Source: Adapted from Nabhan and Felger (1978).

20.3.3 Lectins in Tepary Bean Seed

Lectins are glycoproteins capable of agglutinating red blood cells *in vitro* as well as *in vivo* by interacting with specific carbohydrate residues on cell membrane structures (Santos et al., 2014; Sharon & Lis, 1995). Wild teparies possess a higher content of lectins than the domesticated types (Sotelo et al., 1995). Lectins are also thermolabile and can distinguish between different cell types, including malignant and normal cells, such that they have numerous applications in experimental medical science (Schwartz et al., 1999).

The family of legume lectins, present in tepary bean, is one of the seven families of proteins that are related structurally and evolutionarily (Pino et al., 2006). They are generally classified according to their binding specificity to arabinose (Wang & Ng, 2005), lactose (Liu et al., 2008), mannose (Francis et al., 2011), N-acetyl-Dgalactosamine (Chumkhunthod et al., 2006), or xylose (Zheng et al., 2007). Four main categories of plant lectins, namely, the merolectins, hololectins, chimerolectins, and superlectins, are recognized. The *merolectins* have one carbohydrate-binding domain and are small monovalent proteins with a single polypeptide. This group is unable to precipitate glycoconjugates or agglutinating cells. The *hololectins* have carbohydrate-binding domains, in which two or more of them are similar or identical, and they exhibit multiple binding sites. They can agglutinate cells or precipitate glycoconjugates. The third group belongs to the *chimerolectins*, which arise from protein fusion and possess one carbohydrate-binding domain in addition to an unrelated domain acting independently. The fourth group, namely, *superlectins*, is made up of molecules consisting of two or more distinct carbohydrate-binding domains. Plant lectins can also be classified according to the specificity of their interactions with mannose/maltose, mannose/glucose, galactose/N-acetylgalactosamine, N-acetylglucosamine/N-acetylglucosamine, galactose, fucose, mannose, and sialic acid, among others (Bari et al., 2013; Wong et al., 2006).

The major functions of plant lectins include acting as a nitrogen reserve; protein-specific recognition factors; and defense mechanisms against pathogenic microorganisms, insects, nematodes, as well as viruses, largely by interfering with glycoconjugates in the growth, development, and physiology of these organisms (Ripoll et al., 2003). Among their several other functions, they actively engage in the transport of carbohydrates and their establishment in plant tissues. Lectin molecules have a high affinity to oligosaccharides, which are either unusual or absent in plants. For example, chitin-binding lectins recognize the typical fungal cell wall as well as the exoskeleton of invertebrates, while sialic acid ligand lectins recognize and bind to a carbohydrate that is absent in plants but is the major component of animal glycoproteins (Fonseca et al., 2022). Moreover, plant lectins

TABLE 20.4
Biological Activities of Lectins in Various Plant Tissues of Tepary Bean

Plant Tissue	Biological Activity	Reference
Seed	Anticoagulant and antiplatelet aggregating properties; coagulant, mitogenic, antibacterial, antifungal, and antitumor activities	Silva et al.,2012, 2014 Sitohy et al., 2007
Stem	Antiviral and apoptosis-inducing activities	Peng et al., 2009
Leaves	Antiviral, antibacterial, and antifungal activities	Ooi et al., 2004
Roots	Antifungal and termiticidal activities	Souza et al., 2011

Source: Adapted from Dinghani et al. (2018); Bhardwaj and Hamama (2004).

are highly stable under unfavorable conditions, such as changes in temperature or pH or exposure to proteases of animals (or insects). Consequently, they are found preferentially in certain plant tissues, such as in storage organs and seeds, which are susceptible to attack by other organisms (Peumans, 1995) (Table 20.4).

In a study involving the toxic effects of lectins and lectin-related glycopolypeptides in experimental animals, the most affected target organs were the small intestines, spleen, and thymus (Reynoso-Camacho et al., 2003). In a related study involving experimental pigs, the detrimental effects of bean lectins on the small intestines affected other digestive organs, such as the pancreas, in which exocrine pancreatic acini hypertrophy was observed (Alatorre-Cruz et al., 2018). The small intestine closely communicates with the pancreas through intestinal hormones, such as cholecystokinin and secretin, such that the down-regulation feedback of cholecystokinin secretion leads to an increased pancreatic secretion of trypsin, chymotrypsin, and amylase in a compensatory manner (Grant et al., 2000). Moreover, other studies reported pancreas weight changes in rats caused by some lectins and attributed the alteration to the low digestibility of dietary lipids and protein in the intestine, due to disrupted permeability and villous atrophy (Ferriz-Martínez et al., 2015; Zhao et al., 2011; Kelsall et al., 2002; Radberg et al., 2001; Pusztai et al., 1995). Nonetheless, lectins from tepary bean exhibit differential cytotoxicity on cancer cell lines by apoptosis or autophagy (Yau et al., 2015; Wu et al., 2014; García-Gasca et al., 2012; Petrossian et al., 2007; Ferríz-Martínez et al., 2010). These findings demonstrate that tepary bean grain is potentially useful in the treatment of cancer, due to the presence of lectin proteins (González de Mejía & Prisecaru, 2005).

20.4 POTENTIAL FOR WIDENING THE GENETIC BASE FOR SEED PROTEIN THROUGH INDUCED CHEMICAL MUTATION

Breeding efforts for tepary bean improvement are hampered by a narrow genetic pool of the species. As an alternative approach to widen the genetic base for improvement, chemical mutagenesis with varying doses of ethyl methanesulfonate (EMS) has previously been applied on tepary bean genotypes (Thangwana et al., 2021). The protein content of the seed storage proteins was determined, and the electrophoretic profiles were separated by sodium dodecyl sulphate–polyacrylamide gel electrophoresis (SDS-PAGE), among early mutant generations (M_2 to M_4) (Mashifane, 2018). The results showed variability in the content of seed storage protein among the mutant genotypes that were used in the study (Table 20.5). Similarly, genotype–dose interaction effects were significant ($p < 0.05$) for crude protein accumulation. The mean crude protein content among the genotypes was 16.93% (Table 20.5). "Gen-4" showed consistently lower protein content compared to the other genotypes, but "Gen-6" revealed about 20.0% higher crude protein accumulation when treated with 0.5% EMS compared to the control (0%). Generally, tepary bean mutants showed a significant improvement in protein content accumulation in comparison with the control (untreated) genotypes (Mashifane, 2018).

TABLE 20.5
EMS Dose Effects on Crude Seed Protein Content of Tepary Bean Genotypes over Three Successive Mutant Generations

Mutant Generation	EMS Dose (%)	% Crude Protein		
		Gen-3	Gen-4	Gen-6
2	0.0	$18.04^{bc} \pm 0.37$	$17.32^{a} \pm 0.27$	$14.56^{b} \pm 0.47$
	0.5	$20.88^{a} \pm 0.90$	$14.98^{a} \pm 1.61$	$16.39^{a} \pm 0.17$
	1.0	$18.05^{bc} \pm 0.51$	$17.69^{a} \pm 0.76$	$12.20^{c} \pm 0.33$
	1.5	$20.18^{ab} \pm 0.73$	$16.08^{a} \pm 0.62$	$16.83^{a} \pm 1.16$
	2.0	$17.71^{c} \pm 0.52$	$18.59^{a} \pm 1.06$	$12.37^{c} \pm 0.67$
Mean		18.97	16.93	14.47
3	0.0	$16.39^{b} \pm 1.02$	$18.68^{a} \pm 0.36$	$12.91^{b} \pm 1.38$
	0.5	$18.73^{ab} \pm 0.77$	$14.66^{d} \pm 0.37$	$17.38^{a} \pm 0.73$
	1.0	$21.08^{a} \pm 0.90$	$17.19^{ab} \pm 0.26$	$16.45^{ab} \pm 0.88$
	1.5	$18.79^{ab} \pm 0.28$	$16.53^{bc} \pm 0.24$	$18.73^{a} \pm 0.82$
	2.0	$18.25^{ab} \pm 0.13$	$15.01^{cd} \pm 0.87$	$15.18^{ab} \pm 1.15$
Mean		18.65	16.41	16.13
4	0.0	$18.61^{c} \pm 0.65$	$16.42^{ab} \pm 0.07$	$13.51^{a} \pm 0.79$
	0.5	$21.96^{ab} \pm 0.59$	$15.92^{ab} \pm 0.05$	$16.61^{a} \pm 0.99$
	1.0	$19.16^{c} \pm 1.08$	$17.05^{a} \pm 0.49$	$15.11^{a} \pm 1.62$
	1.5	$24.23^{a} \pm 0.76$	$14.19^{c} \pm 0.11$	$13.72^{a} \pm 0.63$
	2.0	$20.10^{bc} \pm 0.23$	$15.06^{bc} \pm 0.45$	$12.44^{a} \pm 0.59$
Mean		20.81	15.73	14.28
Grand Mean		**19.48**	**16.36**	**14.96**

Source: Mashifane (2018).

Note: Means followed by the same letter in a column are not significantly different ($p \leq 0.05$). The mutant generations (M_2–M_4) were derived from each of three improved tepary bean genotypes, namely, "Gen-3," "Gen-4," and "Gen-6," using varying doses of EMS.

The electrophoretic patterns of the seed storage proteins showed highly significant ($p \leq 0.01$) differences in crude protein accumulation among the mutant tepary bean genotypes (Figures 20.1 and 20.2) (Mashifane, 2018). The findings demonstrated that EMS mutagenesis could generate potential useful mutant variants for improving the nutritional quality of tepary bean, while the presence of distinct protein subunits could be used in varietal identification.

Seed storage proteins were resolved into 17 protein subunits with molecular weights ranging from 10 kDa to 170 kDa. The differences in seed crude protein profiles were predominantly observed in the small protein subunits (<35 kDa). For instance, the 0.5% EMS dose showed the presence of a unique protein band corresponding to a unique 20 kDa subunit which was absent in the rest of the EMS doses observed in "Gen-3." In the M_2 generation, the protein-banding pattern in "Gen-4" showed two extra protein subunits (at 40 kDa and 100 kDa) that were absent in the control (Figure 20.3). The main genotypic differences in seed storage protein profiles were observed in the small and relatively large protein subunits, with the 1.5% EMS dose in "Gen-6" showing a unique and distinct 15 kDa protein subunit. In similar studies, electrophoretic analysis of tepary bean lectins revealed the presence of a 33 kDa protein subunit in tepary bean which was associated with resistance to bruchids (Li et al., 2017; Kusolwa & Myers, 2012). Several similar studies determined that the protein bands corresponding to 25–40 kDa and 50–55 kDa regions were mainly tepary bean lectins (Idouraine et al., 1994).

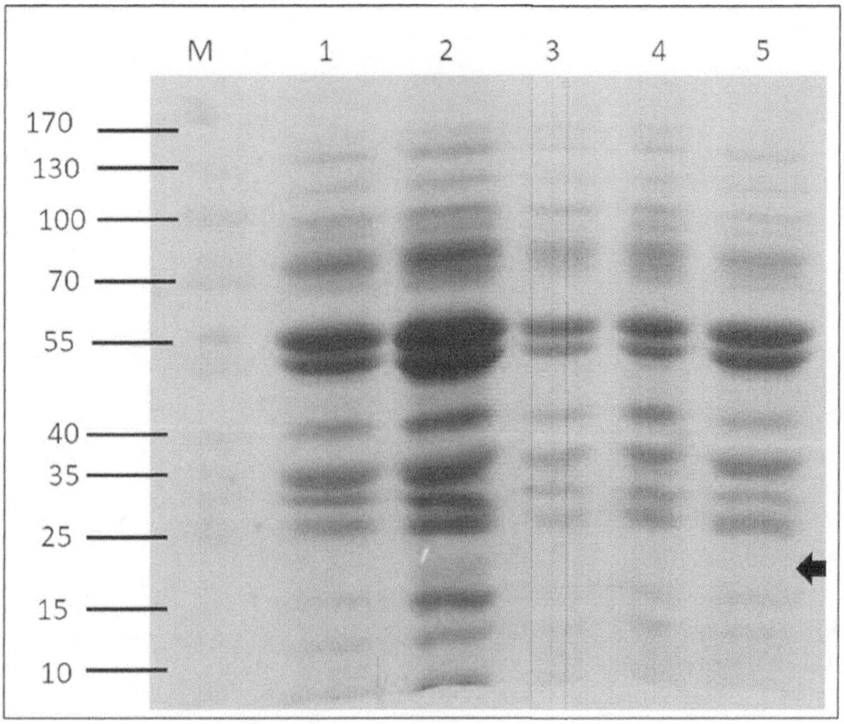

FIGURE 20.1 Electrophoretic banding pattern of EMS-derived seed storage proteins in M_2 tepary bean "GEN-3." Lanes: *M* = protein marker; *1* = control (0.0%); *2* = 0.5%; *3* = 1.0%; *4*= 1.5%; and *5* = 2.0% EMS dose. Arrow represents the 20 kDa subunit that was present only in lane 2.

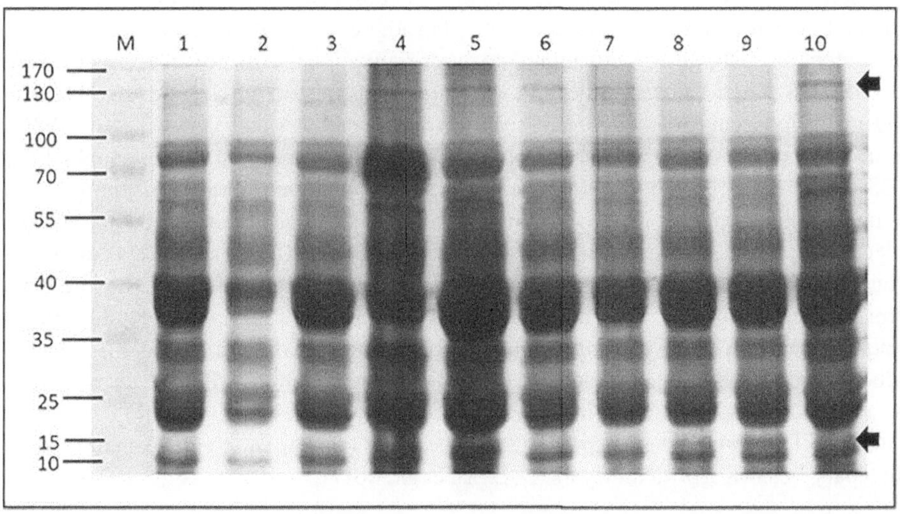

FIGURE 20.2 Electrophoretic banding pattern of seed storage proteins induced by EMS mutagenesis between M_4 tepary bean "Gen-4" and "Gen-6." Lanes: M = protein marker. "Gen-4": *1* = control (0.0%); *2* = 0.5%; *3* = 1.0%; *4* = 1.5%; and *5* = 2.0% EMS dose. "Gen-6" = control (0.0%); 7 = 0.5%; 8 = 1.0%; 9 = 1.5%; and 10 = 2.0% EMS dose. Top and bottom arrows represent the absence/presence of a 130 kDa and 15 kDa protein subunits, respectively.

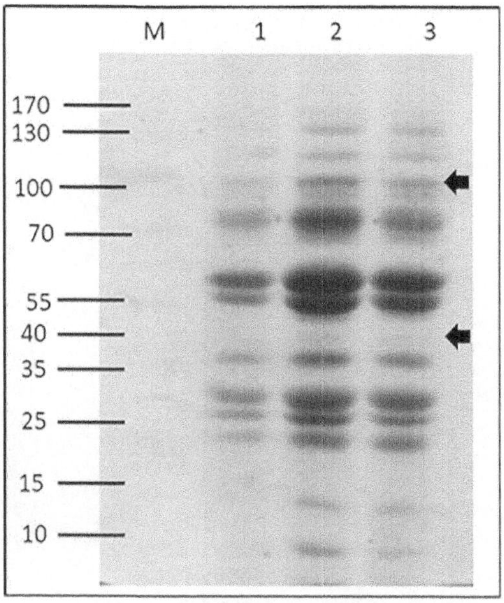

FIGURE 20.3 Electrophoretic banding pattern of EMS-derived seed storage proteins in M_2 tepary bean "GEN-4." Lanes: *M* = protein marker; *1* = control (0.0%); *2* = 0.5%; and *3* = 2.0% EMS dose. Top and bottom arrows highlight protein species of approximately 40 kDa and 100 kDa, respectively.

In other leguminous species, high-crude-seed-protein genotypes of lentil (Shahwar et al., 2022), soybean (Imsande, 2001), and green gram (Samiullah & Wani, 2006) derived through mutation breeding were reported (Shahwar et al., 2022; Imsande, 2001; Samiullah & Wani, 2006). The increase in seed crude protein was attributed to an increase in auxin levels and modification of biochemical pathways for protein synthesis (Jain & Suprasanna, 2011). Enhanced protein content accumulation over successive mutant generations was observed also in studies involving cowpea, soybean, and peanut (Adekola & Oluleye, 2007; Pavadai et al., 2010; Chen et al., 2020). The high protein content in mutants was attributed to micro-mutations with positive effects (Kamble & Petkar, 2015). In general, chemically induced mutagenesis results in chromosome breakages and rearrangements, consequently accounting for the unique gene products (Prasad et al., 1986).

The unique protein profiles observed on account of the varied EMS doses applied suggested that SDS-PAGE could be relied on to resolve the phenotypic variations, in support of previous studies investigating similar phenomenon in different bean species, such as cowpea (Gaafar et al., 2016), lentil (Laskar et al., 2018), groundnut (Muniappan et al., 2022), and black gram (Goyal et al., 2021) mutants. However, there is a need to identify the various protein species using appropriate proteomics-based techniques, such as LC-MS (Alves et al., 2019; Schmelter et al., 2018). Ideally, the identities of the informative subunits that could be utilized to determine protein functionality or as genetic markers in tepary bean selection programs could be merited. For instance, lectins, which possess anticancer properties (García-Gasca et al., 2012; Wu et al., 2014), could be closely associated with a 33 kDa subunit which is produced by the genes that confer tolerance against seed storage pests in tepary bean (Kusolwa & Myers, 2012; Li et al., 2017).

20.5 CONCLUSION AND PROSPECTS

Tepary bean is a highly nutritious grain legume containing a wide range of minerals, vitamins, and proteins. It also contains antinutritional factors, notably various classes of lectins that hinder bioavailability of the minerals. However, the lectins act as a reserve of N and offer defense mechanisms against pathogenic microorganisms, viruses, and insects, largely by interfering with glycoconjugates in the growth, development, and physiology of these organisms The lectins are also involved in the transport of carbohydrates and their establishment in plant tissues and have a high affinity to oligosaccharides that are either unusual or absent in plants. Tepary bean mutants demonstrated the variability in seed protein content induced by EMS. A significant improvement in protein content accumulation of mutants in comparison with the wild teparies suggested that there is potential to increase the seed protein to supplement the dietary needs of communities that face low-protein diets, for instance, in low-income rural communities which rely largely on starch-based diets with low protein intake. Protein-rich legume-based diets could be a viable, sustainable, and healthy option in preventing malnutrition from developing. The presence of distinct protein subunits in the germplasm could be used in varietal identification of this crop. The findings demonstrated that chemically induced mutagenesis could be exploited in the genetic improvement of tepary bean in the future. Tepary bean also has many advantages over similar species, such as the common bean (*Phaseolus vulgaris*), for instance, which augurs well for the future. Tepary bean can mature within 60 days versus 90 days for the common bean. In hot environments, it outperforms common bean in grain yield. Therefore, it has the potential to benefit growers as a complementary source of affordable protein and contribute significantly to household food security in drought-prone regions. However, bean breeders may consider the development of cultivars that require a short cooking time to encourage a wider adoption of the crop by growers. In addition, the species needs to be promoted as a complementary, thus diversifying the farm portfolio.

20.6 ACKNOWLEDGMENTS

We wish to acknowledge the smallholder growers from Sekhukhune District (Limpopo, South Africa) and Prof. I. K. Mariga for the original tepary bean germplasm that was subsequently used in several scientific investigations, particularly by the principal author of this chapter.

REFERENCES

Abberton, M., Paliwal, R., Faloye, B., Marimagne, T., Moriam, A., & Oyatomi, O. (2022). Indigenous African orphan legumes: Potential for food and nutrition security in SSA. *Frontiers in Sustainable Food Systems*, *6*, 83–102.

Adekola, O. F., & Oluleye, F. (2007). Induction of genetic variation in cowpea (*Vigna unguiculata*) by gamma irradiation. *Asian Journal of Plant Sciences*, *6*, 869–873.

Aisen, P., Enns, C., & Wessling-Resnick, M. (2021). Chemistry and biology of eukaryotic iron metabolism. *International Journal of Biochemistry and Cell Biology*, *33*, 940–959.

Alatorre-Cruz, J. M., Pita-López, W., López-Reyes, R. G., Ferriz-Martínez, R. A., Cervantes-Jiménez, R., Carrillo, M. D. J. G., Vargas, P. J. A., López-Herrera, G., Rodríguez-Méndez, A. J., Zamora-Arroyo, A., & Gutiérrez-Sánchez, H. (2018). Effects of intragastrically-administered tepary bean lectins on digestive and immune organs: Preclinical evaluation. *Toxicology Reports*, *5*, 56–64.

Alonso, R., Aguirre, A., & Marzo, F. (2000). Effects of extrusion and traditional processing methods on antinutrients and in vitro digestibility of protein and starch in faba and kidney beans. *Food Chemistry*, *68*, 159–165.

Alves, T. O., D'Almeida, C. T. S., Scherf, K. A., & Ferreira, M. S. L. (2019). Modern approaches in the identification and quantification of immunogenic peptides in cereals by LC-MS/MS. *Frontiers in Plant Sciences*, *10*, 1470.

Bari, A. U., Silva, H. C., Silva, M. T., Pereira Junior, F. N., Cajazeiras, J. B., Sampaio, A. H., Leal, R. B., Teixeira, E. H., Rocha, B. A., Nascimento, K. S., & Nagano, C. S. (2013). Purification and partial characterization of a new mannose/glucose-specific lectin from *Dialium guineense* wild seeds that exhibits toxic effect. *Journal of Molecular Recognition*, *26*, 351–356.

Bhagyawant, S. S., Gupta, N., & Shrivastava, N. (2015). Biochemical analysis of chickpea accessions vis-a-vis; zinc, iron, total protein, proline and antioxidant activity. *American Journal of Food Technology*, 158–162.

Bhardwaj, H. L. (2013). Preliminary evaluation of tepary bean (*Phaseolus acutifoilus*) as a forage crop. *Journal of Agricultural Sciences*, *5*, 160–166.

Bhardwaj, H. L., & Hamama, A. A. (2004). Protein and mineral composition of tepary bean seed. *HortScience, 39*, 1363–1365.

Bhardwaj, H. L., & Hamama, A. A. (2005). Oil and fatty acid composition of tepary bean seed. *HortScience, 40*, 1436–1438.

Celmeli, T., Sari, H., Canci, H., Sari, D., Adak, A., Eker, T., & Toker, C. (2018). The nutritional content of common bean (*Phaseolus vulgaris* L.) landraces in comparison to modern varieties. *Agronomy, 8*, 166. https://doi.org/10.3390/agronomy8090166

Chen, T., Huang, L., Wang, M., Huang, Y., Zeng, R., Wang, X., Wang, L., Wan, S., & Zhang, L. (2020). Ethyl methyl sulfonate-induced mutagenesis and its effects on peanut agronomic, yield and quality traits. *Agronomy, 10*, 655.

Chiuve, S. E., Sun, Q., Curhan, G. C., Taylor, E. N., Spiegelman, D., Willett, W. C., Manson, J. E., Rexrode, K. M., & Albert, C. M. (2013). Dietary and plasma magnesium and risk of coronary heart disease among women. *Journal of the American Heart Association, 2*, 1–10.

Chivenge, P., Mabhaudhi, T., Modi, A., & Mafongoya, P. (2015). The potential role of neglected and underutilised crop species as future crops under water scarce conditions in Sub-Saharan Africa. *International Journal of Environmental Research and Public Health, 12*, 5685–5711.

Chumkhunthod, P., Rodtong, S., Lambert, S. J., Fordham-Skelton, A. P., Rizkallah, P. J., Wilkinson, M. C., & Reynolds, C. D. (2006). Purification and characterization of an N-acetyl-D-galactosamine-specific lectin from the edible mushroom *Schizophyllum commune. Biochimica et Biophysica Acta (BBA)-General Subjects, 1760*, 326–332.

Crews, T. E., Brockwell, J., & Peoples, M. B. (2004). Host–rhizobia interaction for effective inoculation: Evaluation of the potential use of the ureide assay to monitor the symbiotic performance of tepary bean (*Phaseolus acutifolius* A. Gray). *Soil Biology and Biochemistry, 36*, 1223–1228.

Deglaire, A., & Moughan, P. J. (2012). Animal models for determining amino acid digestibility in humans–A review. *British Journal of Nutrition, 108*, S273–S281.

De Mejía, E. G., & Prisecaru, V. I. (2005). Lectins as bioactive plant proteins: A potential in cancer treatment. *Critical Reviews in Food Science and Nutrition, 45*(6), 425–445.

de Mejía, E. G., Martínez-Resendiz, V., Castaño-Tostado, E., & Loarca-Piña, G. (2003). Effect of drought on polyamine metabolism, yield, protein content and in vitro protein digestibility in tepary (*Phaseolus acutifolius*) and common (*Phaseolus vulgaris*) bean seeds. *Journal of the Science of Food and Agriculture, 83*, 1022–1030.

Dinghani, D. Y., Yu, H., Zeng, X., Bo, L., Leo, S. F., & Okorley, B. A. (2018). Some nutritional characteristics of a local landrace of tepary bean seed from the republic of Benin. *Food Science and Quality Management, 73*, 4–46.

Dwivedi, S. L., Chapman, M. A., Abberton, M. T., Akpojotor, U. L., & Ortiz, R. (2023). Exploiting genetic and genomic resources to enhance productivity and abiotic stress adaptation of underutilized pulses. *Frontiers in Genetics, 14*, 1193780.

Enyiukwu, D. N., Chukwu, L. A., & Bassey, I. N. (2020). Nutrient and anti-nutrient compositions of cowpea (Vigna unguiculata) and mung bean (Vigna radiata) seeds grown in humid Southeast Nigeria: A comparison. *International Journal of Tropical Drylands, 4*, 42–45.

Etiosa, O., Chika, N., & Benedicta, A. (2018). Mineral and proximate composition of soyabean. *Asian Journal of Physical and Chemical Sciences, 4*, 1–6.

Faryadi, Q. (2012). The magnificent effect of magnesium to human health: A critical review. *International Journal of Applied Science and Technology, 2*, 118–126.

Ferríz-Martínez, R. A., Torres-Arteaga, I. C., Blanco-Labra, A., & García-Gasca, T. (2010). The role of plant lectins in cancer treatment. In *The new approaches in the treatment of cancer* (pp. 71–90). Nova Science Publishers.

Ferriz-Martínez, R., García-García, K., Torres-Arteaga, I., Rodriguez-Mendez, A. J., Guerrero-Carrillo, M. J., Moreno-Celis, U., Ángeles-Zaragoza, M. V., Blanco-Labra, A., Gallegos-Corona, M. A., Robles-Álvarez, J. P., Mendiola-Olaya, E., Andrade-Montemayor, H. M., Garcia, O. P., & García-Gasca, T. (2015). Tolerability assessment of a lectin fraction from Tepary bean seeds (*Phaseolus acutifolius*) orally administered to rats. *Toxicology Reports, 2*, 63–69.

Fonseca, V. J. A., Braga, A. L., Filho, R. J., Teixeira, C. S., da Hora, G. C., & Morais-Braga, M. F. B. (2022). A review on the antimicrobial properties of lectins. *International Journal of Biological Macromolecules, 195*, 163–178.

Francis, F., Jaber, K., Colinet, F., Portetelle, D., & Haubruge, E. (2011). Purification of a new fungal mannose-specific lectin from *Penicillium chrysogenum* and its aphicidal properties. *Fungal Biology, 115*, 1093–1099.

Gaafar, R. M., Hamouda, M., & Badr, A. (2016). Seed coat color, weight and eye pattern inheritance in gamma rays induced cowpea M_2 mutant line. *Journal of Genetic Engineering and Biotechnology, 14*, 61–68.

García-Gasca, T., García-Cruz, M., Hernandez-Rivera, E., López-Matínez, F. J., Castañeda-Cuevas, A. L., & Yllescas-Gasca, L. (2012). Effects of tepary bean protease inhibitor and semi-pure lectin fractions on cancer cells. *Nutrition and Cancer, 64*, 1269–1278.

Gouveia, C. S., Freitas, G., Brito, J. H. D., Slaski, J. J., & Carvalho, M. Â. (2014). Nutritional and mineral variability in 52 accessions of common bean varieties (Phaseolus vulgaris L.) from Madeira Island. *Agricultural Sciences, 5*, 317–329.

Goyal, S., Wani, M. R., Raina, A., Laskar, R. A., & Khan, S. (2021). Quantitative assessments on induced high yielding mutant lines in urdbean (*Vigna mungo*). *Legume Science, 4*, e125.

Grant, G., Alonso, R., Edwards, J. E., & Murray, S. (2000). Dietary soya beans and kidney beans stimulate secretion of cholecystokinin and pancreatic digestive enzymes in 400-day-old Hooded-Lister rats but only soya beans induce growth of the pancreas. *Pancreas, 20*, 305–312.

Heredia-Rodríguez, L., Gaytán-Martínez, M., Morales-Sánchez, E., de Jesús Garza-Juárez, A., Urias-Orona, V., González-Martínez, B. E., Lomelí, M. L. C., & Vázquez-Rodríguez, J. A. (2019). Nutritional and technological properties of Tepary bean (Phaseolus acutifolius) cultivated in Mexican Northeast. *Czech Journal of Food Sciences, 37*, 62–68.

Idouraine, A., Yensen, S. B., & Weber, C. W. (1994). Fractionation and partial characterization of tepary bean (*Phaseolus acutifolius*) proteins. *Food Chemistry, 50*, 13–18.

Imsande, J. (2001). Selection of soybean mutants with increased concentrations of seed methionine and cysteine. *Crop Science, 41*, 510–515.

Jahnen-Dechent, W., & Ketteler, M. (2012). Magnesium basics. *Clinical Kidney Journal, 5*, i3–i14.

Jain, S. M., & Suprasanna, P. (2011). Induced mutations for enhancing nutrition and food production. *Geneconserve, 40*, 201–215.

Jayathilake, C., Visvanathan, R., Deen, A., Bangamuwage, R., Jayawardana, B. C., Nammi, S., & Liyanage, R. (2018). Cowpea: An overview on its nutritional facts and health benefits. *Journal of the Science of Food and Agriculture, 98*, 4793–4806.

Jenkins, D. J., Kendall, C. W., Augustin, L. S., Mitchell, S., Sahye-Pudaruth, S., Mejia, S. B., Chiavaroli, L., Mirrahimi, A., Ireland, C., Bashyam, B., Vidgen, E., de Souza, R. J., Sievenpiper, J. L., Coveney, J., Leiter, L. A., & Josse, R. G. (2012). Effect of legumes as part of a low glycemic index diet on glycemic control and cardiovascular risk factors in type 2 diabetes mellitus: A randomized controlled trial. *Archives of Internal Medicine, 26*, 1653–1660.

Jiménez, J. C., de la Fuente, M., Ordás, B., Domínguez, L. E. G., & Malvar, R. A. (2017). Resistance categories to *Acanthoscelides obtectus* (Coleoptera: Bruchidae) in tepary bean (*Phaseolus acutifolius*), new sources of resistance for dry bean (*Phaseolus vulgaris*) breeding. *Crop Protection, 98*, 255–266.

Kamble, G. C., & Petkar, H. J. (2015). Some biochemical and biotechnological perspectives in wild chickpea and its induced mutants. *American Journal of Pharmacology and Pharmacotherapeutic, 2*, 41–49.

Kelsall, A., Fitzgerald, A. J., Howard, C. V., Evans, R. C., Singh, R., Rhodes, J. M., & Goodlad, R. A. (2002). Dietary lectins can stimulate pancreatic growth in the rat. *International Journal of Experimental Pathology, 83*, 203–208.

Konozy, H. E., & Osman, M. E. M. (2022). Plant lectin: A promising future anti-tumor drug. *Biochimie, 202*, 136–145.

Kusolwa, P. M., & Myers, J. R. (2012). Peptide sequences from seed storage proteins of tepary bean (*Phaseolus acutifolius*) accession G40199 demonstrate the presence of multiple variants of APA proteins. *International Journal of Biochemistry and Biotechnology, 1*, 12–18.

Laskar, R. A., Laskar, A. A., Raina, A., Khan, S., & Younus, H. (2018). Induced mutation analysis with biochemical and molecular characterization of high yielding lentil mutant lines. *International Journal of Biological Macromolecules*, 109, 16–179.

Lesk, C., Rowhani, P., & Ramankutty, N. (2016). Influence of extreme weather disasters on global crop production. *Nature, 529*, 84–87.

Li, M., Liu, Q., Cui, Y., Li, D., Wang, H., & Tzi, B. N. (2017). Isolation and characterization of a *Phaseolus vulgaris* trypsin inhibitor with antiproliferative activity on leukemia and lymphoma cells. *Molecules, 22*, 187–193.

Lieu, P. T., Heiskala, M., Peterson, P. A., & Yang, Y. (2021). The roles of iron in health and disease. *Molecular Aspects of Medicine, 2*, 1–87.

Liu, C., Zhao, X., Xu, X. C., Li, L. R., Liu, Y. H., Zhong, S. D., & Bao, J. K. (2008). Hemagglutinating activity and conformation of a lactose-binding lectin from mushroom *Agrocybe cylindracea*. *International Journal of Biological Macromolecules, 42*, 138–144.

Lolas, G. M., & Markakis, P. (1975). Phytic acid and other phosphorus compounds of beans (*Phaseolus vulgaris* L.). *Journal of Agricultural and Food Chemistry, 23*, 13–15.

López-Ibarra, C., Ruiz-López, F. J., Bautista-Villarreal, M., Báez-González, J. G., Romero, B. A. R., González-Martínez, B. E., López-Cabanillas, L. M., & Vázquez-Rodríguez, J. A. (2021). Protein concentrates on Tepary Bean (*Phaseolus acutifolius* Gray) as a functional ingredient: In silico docking of tepary bean lectin to peroxisome proliferator-activated receptor gamma. *Frontiers in Nutrition, 8*, 661463.

Mashifane, D. C. (2018). *The effect of chemomutagenesis on root nodulation and seed protein in tepary bean (Phaseolus acutifolius)* [MSc thesis, Department of Crop Science, University of Venda, 1–69].

Medina, V., y Teran, J.C.B.M., Gepts, P. & Gilbert, M.E. (2017). Low stomatal sensitivity to vapor pressure deficit in irrigated common, lima and tepary beans. *Field Crops Research, 206*, 128–137.

Miklas, P. N., Kelly, J. D., Beebe, S. E., & Blair, M. W. (2006). Common bean breeding for resistance against biotic and abiotic stresses: From classical to MAS breeding. *Euphytica, 147*, 105–131.

Miklas, P. N., & Santiago, J. (1996). Reaction of select tepary bean to bean golden mosaic virus. *HortScience, 31*, 430–432.

Mogotsi, K. K. (2006). *Phaseolus acutifolius* A. Gray. In M. Brink & G. Belay (Eds.), *Plant resources of tropical Africa 1. cereals and pulses* (pp. 133–137). Prota Foundation.

Molosiwa, O. O., Kgokong, S. B., Makwala, B., Gwafila, C., & Ramokapane, M. G. (2014). Genetic diversity in tepary bean landraces grown in Botswana. *Journal of Plant Breeding and Crop Science, 6*, 194–199.

Montoyama, T., Sano, H., & Fukuzaki, H. (1989). Oral magnesium supplements in patients with essential hypertension. *Hypertension, 13*, 227–232.

Moreno-Celis, U., López-Martínez, F. J., Cervantes-Jiménez, R., Ferríz-Martínez, R. A., Blanco-Labra, A., & García-Gasca, T. (2020). Tepary bean (*Phaseolus acutifolius*) lectins induce apoptosis and cell arrest in G0/G1 by P53(Ser46) phosphorylation in colon cancer cells. *Molecules, 25*, 1021.

Moreno-Celis, U., López-Martínez, J., Blanco-Labra, A., Cervantes-Jiménez, R., Estrada-Martínez, L. E., García-Pascalin, A. E., Guerrero-Carrillo, M. J., Rodríguez-Méndez, A. J., Mejía, C., Ferríz-Martínez, R. A., & García-Gasca, T. (2017). *Phaseolus acutifolius* Lectin fractions exhibit apoptotic effects on colon cancer: Preclinical studies using dimethilhydrazine or azoxi-methane as cancer induction agents. *Molecules, 22*, 1670.

Moreno-Valdespino, C. A., Luna-Vital, D., Camacho-Ruiz, R. M., & Mojica, L. (2020). Bioactive proteins and phytochemicals from legumes: Mechanisms of action preventing obesity and type-2 diabetes. *Food Research International, 130*, 108905.

Muniappan, V., Manivel, P., Prabakaran, V., & Palanivel, S. (2022). Effect of diethyl sulphate on M_1 attributes and sds-page analysis in groundnut. *Plant Archives, 22*, 15–18.

Nabhan, G. P., & Felger, R. S. (1978). Teparies in Southwestern North America. A biogeographical and ethnohistorical study of *Phaseolus acutifolius. Economic Botany, 32*, 3–19.

Nabhan, G., Berry, J., Anson, C., & Weber, C. (1980). Papago Indian floodwater fields and tepary bean protein yields. *Ecology of Food and Nutrition, 10*(2), 71–78.

Narina, S. S., Pathak, S. C., & Bhardwaj, H. L. (2014). Chlorophyll fluorescence to evaluate pigeon-pea breeding lines and mung bean for drought tolerance. *Journal of Agricultural Science, 6*, 11–16.

Nong, R. A., Gerrano, A. S., & Gwata, E. T. (2023). A preliminary evaluation of phenotypic traits of tepary bean (*Phaseolus acutifolius* A. Gray). *Journal of Underutilised Crops Research*, 2(1), a5. https://doi.org/10.4102/jucr.v2i1.5

No?thlings, U., Schulze, M. B., Weikert, C., Boeing, H., Van Der Schouw, Y. T., Bamia, C., Benetou, V., Lagiou, P., Krogh, V., Beulens, J. W., & Peeters, P. H. (2008). Intake of vegetables, legumes, and fruit, and risk for all-cause, cardiovascular, and cancer mortality in a European diabetic population. *Journal of Nutrition, 138*, 775–781.

Ooi, L. S., Sun, S. S., & Ooi, V. E. (2004). Purification and characterization of a new antiviral protein from the leaves of *Pandanus amaryllifolius* (Pandanaceae). *International Journal of Biochemistry and Cell Biology, 36*, 1440–1446.

Pavadai, P., Girija, M., & Dhanavel, D. (2010). Effect of gamma rays, EMS, DES and COH on protein and oil content in soybean. *Journal of Ecobiotechnology, 2*, 47–50.

Peng, H., Lv, H., Wang, Y., Liu, Y. H., Li, C. Y., Meng, L., Chen, F., & Bao, J. K. (2009). Clematis montana lectin, a novel mannose-binding lectin from traditional Chinese medicine with antiviral and apoptosis-inducing activities. *Peptides, 30*, 1805–1815.

Petrossian, K., Banner, L., & Oppenheimer, S. (2007). Lectin binding and effects in culture on human cancer and non-cancer cell lines: Examination of issues of interest in drug design strategies. *Acta Histochemistry, 109*, 491–500.

Peumans, W. P. (1995). Lectin as plant defense proteins. *Plant Physiology, 109*, 347–352.

Pino, A. G., Buts, L., Wyns, L., & Loris, R. (2006). Interplay between metal binding and cis/trans isomerization in legume lectins: Structural and thermodynamic study of *P. angolensis* lectin. *Journal of Molecular Biology*, *361*, 153–671.

Porch, T. G., Cichy, K., Wang, W., Brick, M., Beaver, J. S., Santana-Morant, D., & Grusak, M. A. (2017). Nutritional composition and cooking characteristics of tepary bean (*Phaseolus acutifolius*) in comparison with common bean (*Phaseolus vulgaris*). *Genetic Resources and Crop Evolution*, *64*, 935–953.

Prasad, A. B., Verma, N. P., & Jha, A. M. (1986). Seed protein content and protein pattern in gamma-ray induced *Phaseolus* mutants. In A. B. Prassad (Ed.), *Mutagenesis basics and applied* (pp. 158–184). Print House.

Pratelli, R., & Pilot, G. (2014). Regulation of amino acid metabolic enzymes and transporters in plants. *Journal of Experimental Botany*, *65*, 5535–5556.

Pratt, R. C., Singh, N. K., Shade, R. E., Murdock, L. L., & Bressan, R. A. (1990). Isolation and partial characterization of a seed lectin from tepary bean that delays bruchid beetle development. *Plant Physiology*, *93*, 1453–1459.

Pusztai, A., Grant, G., Duguid, T., Brown, D. S., Peumans, W. J., Van Damme, E. J., & Bardocz, S. (1995). Inhibition of starch digestion by a-amylase inhibitor reduces the efficiency of utilization of dietary proteins and lipids and retards the growth of rats. *Journal of Nutrition*, *125*, 1554–1562.

Queiroz Mendes, F., De Almeida Oliveira, M. G., Brunoro Costa, N. M., Vieira Pires, C., & Passos, F. R. (2016). Capability of in vitro digestibility methods to predict in vivo digestibility of vegetal and animal proteins. *Archivos Latinoamericanos de Nutrición*, *66*, 005–016.

Radberg, K., Biernat, M., Linderoth, A., Zabielski, R., Pierzynowski, S. G., & Westrom, B. R. (2001). Enteral exposure to crude red kidney bean lectin induces maturation of the gut in suckling pigs. *Journal of Animal Science*, *79*, 2669–2678.

Reddy, N. R. (2001). Occurrence, distribution, content, and dietary intake of phytate. In N.R. Reddy & S.K. Sathe (Eds.), *Food phytates* (pp. 41–68). CRC Press.

Reynoso-Camacho, R., & Mejia, E. (2003). Purification and acute toxicity of a lectin extracted from Tepary bean (*Phaseolus acutifolius*). *Food and Chemical Toxicology*, *41*, 21–27.

Ripoll, C., Favery, B., Lecomte, P., Van Damme, E., Peumans, W., Abad, P., & Jouanin, L. (2003). Evaluation of the ability of lectin from snowdrop (*Galanthus nivalis*) to protect plants against root-knot nematodes. *Plant Science*, *164*(4), 517–523.

Rosique-Esteban, N., Guasch-Ferré, M., Hernández-Alonso, P., & Salas-Salvadó, J. (2018). Dietary magnesium and cardiovascular disease: A review with emphasis in epidemiological studies. *Nutrients*, *10*, 168–189.

Rude, R. K. (1998). Magnesium deficiency: A cause of heterogeneous disease in Humans. *Journal of Bone Mineral Research*, *13*, 749.

Salas-Lopez, F., Gutierrez-Dorado, R., Milan-Carrillo, J., Cuevas-Rodriguez, E. O., Canizalez-Roman, V. A., Leon-Sicairos, C. D. R., & Reyes-Moreno, C. (2018). Nutritional and antioxidant potential of a desert underutilized legume–tepary bean (*Phaseolus acutifolius*). Optimization of germination bioprocess. *Food Science and Technology*, *38*, 254–262.

Sales, M. P., Gerhardt, I. R., Grossi-de-Sá, M. F., & Xavier-Filho, J. (2000). Do legume storage proteins play a role in defending seeds against bruchids? *Plant Physiology*, *124*, 515–522.

Salgado, M. O., Schwartz, H. F., Brick, M. A., & Pastor-Corrales, M. A. (1994). Resistance to *Fusarium oxysporum* f.sp. phaseoli in tepary bean (*Phaseolus acutifolius*). *Plant Disease*, *78*, 357–360.

Samiullah, K., & Wani, M. R. (2006). Induced mutations for yield contributing traits in greengram. *International Journal of Agriculture and Biology*, *8*, 528–530.

Santos, A. F., Da Silva, M. D. C., Napoleão, T. H., Paiva, P. M. G., Correia, M. D. S., & Coelho, L. C. B. B. (2014). Lectins: Function, structure, biological properties and potential applications. *Current Topics in Peptide and Protein Research*, *15*, 41–62.

Schmelter, C., Funke, S., Treml, J., Beschnitt, A., Perumal, N., & Manicam, C. (2018). Comparison of two Solid-Phase Extraction (SPE) methods for the identification and quantification of porcine retinal protein markers by LC-MS/MS. *International Journal of Molecular Science*, *19*, 3847.

Schwartz, R. E., Wojciechowicz, D. C., Picon, A. I., Schwarz, M. A., & Paty, P. B. (1999). Wheatgerm agglutinin-mediated toxicity in pancreatic cancer cells. *British Journal of Cancer*, *80*, 1754–1762.

Shahwar, D., Ansari, M. Y., & Park, Y. (2022). Physio-biochemical analysis and molecular characterization of induced lentil mutant lines. *PLoS One*, *17*, e0274937.

Sharma, S., Kaur, M., Goyal, R., & Gill, B. S. (2014). Physical characteristics and nutritional composition of some new soybean (*Glycine max* (L.) Merrill) genotypes. *Journal of Food Science and Technology*, *51*, 551–557.

Sharon, N., & Lis, H. (1995). Lectins–proteins with a sweet tooth: Functions in cell recognition. *Essays in Biochemistry*, *30*, 59–75.

Shisanya, C. A. (2002). Improvement of drought adapted tepary bean (*Phaseolus acutifolius* A. Gray var. latifolius) yield through biological nitrogen fixation in semi-arid SE-Kenya, *European Journal of Agronomy*, *16*, 13–24.

Shisanya, C. A. (2003). Yield and nitrogen fixation response by drought tolerant tepary bean (*Phaseolus acutifolius*) in sole and maize intercrop systems in semi-arid Kenya. *Journal of Agronomy*, *2*, 126–137.

Shisanya, C. A. (2004). Improvement of grain legume production in semi-arid Kenya through biological nitrogen fixation: The experience with tepary bean (*Phaseolus acutifolius* A. Gray var. Latifolius). In R. Dris & S. M. Jain (Eds.), *Production Practices and Quality Assessment of Food Crops* (pp. 163–188), Kluwer Academic Publisher.

Silva, M. C., de Paula, C. A., Ferreira, J. G., Paredes-Gamero, E. J., Vaz, A. M., Sampaio, M. U., Correia, M. T. S., & Oliva, M. L. V. (2014). *Bauhinia forficata* lectin (BfL) induces cell death and inhibits integrin-mediated adhesion on MCF7 human breast cancer cells. *Biochimica et Biophysica Acta (BBA)-General Subjects*, *1840*, 2262–2271.

Silva, M. C., Santana, L. A., Mentele, R., Ferreira, R. S., de Miranda, A., Silva-Lucca, R. A., Sampaio, M. U., Correia, M. T., & Oliva, M. L. (2012). Purification, primary structure and potential functions of a novel lectin from *Bauhinia forficata* seeds. *Process Biochemistry*, *47*, 1049–1059.

Sitohy, M., Doheim, M., & Badr, H. (2007). Isolation and characterization of a lectin with antifungal activity from Egyptian *Pisum sativum* seeds. *Food Chemistry*, *104*, 971–979.

Sotelo, A., Sousa, H., & Sanchez, M. (1995). Comparative study of the chemical composition of wild and cultivated beans (*Phaseolus vulgaris*). *Plant Foods for Human Nutrition*, *47*, 93–100.

Souza, J. D., Silva, M. B., Argolo, A. C., Napoleão, T. H., Sá, R. A., Correia, M. T., Paiva, P. M., Silva, M. D., & Coelho, L. C. (2011). A new *Bauhinia monandra* galactose-specific lectin purified in milligram quantities from secondary roots with antifungal and termiticidal activities. *International Biodeterioration and Biodegradation*, *65*, 696–702.

Thangwana, A., Gwata, E. T., & Zhou, M. M. (2021). Impact of chemical mutagenesis using ethyl methane sulphonate on tepary bean seedling vigour and adult plant performance, *Heliyon*, *7*(1), e06103.

Tran, E., Dale, H. F., Jensen, C., & Lied, G. A. (2020). Effects of plant-based diets on weight status: A systematic review. *Diabetes, Metabolic Syndrome and Obesity: Targets and Therapy*, *13*, 3433–3348.

Tresina, P. S., Paulpriya, K., Mohan, V. R., & Jeeva, S. (2017). Effect of gamma irradiation on the nutritional and antinutritional qualities of *Vigna aconitifolia* (Jacq.) Marechal: An underutilized food legume. *Biocatalysis and Agricultural Biotechnology*, *10*, 30–37.

Valadez-Vega, C., Alvarez-Manilla, G., Riverón-Negrete, L., García-Carrancá, A., Morales-González, J. A., Zuñiga-Pérez, C., Madrigal-Santillán, E., Esquivel-Soto, J., Esquivel-Chirino, C., Villagómez-Ibarra, R., & Bautista, M. (2011). Detection of cytotoxic activity of lectin on human colon adenocarcinoma (Sw480) and epithelial cervical carcinoma (C33-A). *Molecules*, *16*(3), 2107–2118.

Velten, G., Rott, A. S., Cardona, C., & Dorn, S. (2007). The inhibitory effect of the natural seed storage protein arcelin on the development of *Acanthoscelides obtectus*. *Journal of Stored Products Research*, *43*, 550–557.

Vormann, J. (2003). Magnesium: Nutrition and metabolism. *Molecular aspect of Medicine*, *23*, 27–37.

Waines, J. G. (1978). Protein contents, grain weights, and breeding potential of wild and domesticated tepary beans. *Crop Science*, *18*, 587–589.

Wang, H., & Ng, T. B. (2005). First report of an arabinose-specific fungal lectin. *Biochemical and Biophysical Research Communications*, *337*, 621–625.

Wong, J. H., Wong, C. C., & Ng, T. B. (2006). Purification and characterization of a galactose-specific lectin with mitogenic activity from pinto beans. *Biochimica et Biophysica Acta (BBA)-General Subjects*, *1760*, 808–813.

Wu, J., Zhu, J., Yin, H., Buckanovich, R., & Lubman, D. M. (2014). Analysis of glycan variation on glycoproteins from serum by the reverse lectin-based ELISA assay. *Journal of Proteome Research*, *13*, 2197–2204.

Yau, T., Dan, X., Ng, C. C., & Ng, T. B. (2015). Lectins with potential for anti-cancer therapy. *Molecules*, *20*, 3791–3810.

Zhao, Y., Qin, G., Sun, Z., Che, D., Bao, N., & Zhang, X. (2011). Effects of soybean agglutinin on intestinal barrier permeability and tight junction protein expression in weaned piglets. *International Journal of Molecular Sciences*, *12*, 8502–8512.

Zheng, S., Li, C., Ng, T. B., & Wang, H. X. (2007). A lectin with mitogenic activity from the edible wild mushroom Boletus edulis. *Process Biochemistry*, *42*, 1620–1624.

21 Exploring the Food Security and Nutrition Potential of *Digitaria exilis*

*Yetunde Oyinkansade Babalola, Linda Ifeanyichukwu Sowunmi, and Paul Oluwatimilehin Ogunbamowo
Biomedicinal Research Centre, Forestry Research Institute of Nigeria, Oyo State, Nigeria
*Corresponding author: babalolayetundeo209@gmail.com

21.1 INTRODUCTION

In recent times, global challenges in ensuring food security have intensified, primarily attributed to the expanding population and diminishing crop yields, posing increased challenges in adequately nourishing the growing. Food security is achieved when there is universal physical, financial, and social access to an ample supply of nutritious food for everyone (Liese, 2018; Wani & Sawargaonkar, 2018). Approximately 850 million people worldwide suffer from extreme hunger, which creates obstacles for the economy and makes it difficult to meet the Sustainable Development Goals (SDG) (Kousar et al., 2021).

Climate change, the economy, the environment, underutilization of some crops with essential values, seed quality, and soil nutrient status, among others, are factors that impact food security (Wang et al., 2015; Szabo, 2016; Weih et al., 2017; Abiad & Meho, 2018; Chauhan et al., 2018; Richardson et al., 2018; Begna, 2021; Rahman et al., 2021; Setsoafia et al., 2022). Food insecurity can be significantly reduced by using underutilized or orphan crops, particularly cereals like fonio. This cereal, possessing a high nutrient content, exhibits resilience against droughts and, crucially, demonstrates a significant potential for nitrogen fixation, thereby enhancing plant growth, crop yield, and overall food production (Bano & Iqbal, 2016).

According to some authors, *underutilized food plants* are crops that are well-known in their home countries due to their economic, cultural, or agronomic values but were previously disregarded by plant breeders, policymakers, consumers, agricultural researchers, donors, extension services, and technology providers (Mabhaudhi et al., 2017a, 2017b; Popoola et al., 2019). This lessens the amount of information that is available about them and their overall acceptability, making it difficult to fully realize their potential.

Researchers all over the world are beginning to understand the potential contribution of biodiversity to increased dietary diversity and quality, which in turn can lead to food and nutrition security (Johns & Eyzaguirre, 2006; Lutaladio et al., 2010). In order to provide and increase sustainable production levels as well as nutritional diversity, plant genetic diversity is essential (Lutaladio et al., 2010). In many parts of the world, traditional or indigenous foods are important sources of nutrition during times of food scarcity (de Merode et al., 2003; Shackleton & Shackleton, 2004), particularly for populations with limited resources. These species can also be used as a major supplement to cereal-based diets (Grivetti & Ogle, 2000; Bharucha & Pretty, 2010).

DOI: 10.1201/9781003469766-21

Compared to popular and often-used foods, several of these lesser-known foods and variants offer superior micronutrients (Lutaladio et al., 2010). They also provide valuable genetic traits for developing new crops through breeding and selection (Ford-Lloyd et al., 2011).

Cereal grains make up more than 60% of the world's food production and, when combined with oilseeds and pulses, supply most of the world's population – both in the poor world and globally – with dietary proteins, calories, vitamins, and minerals, including iron and zinc (Betschart, 1982; Salunkhe & Desai, 1986). Across sub-Saharan Africa, locally grown traditional cereal grains, such as sorghum (*Sorghum bicolor*), amaranth seeds (*Amaranthus* sp.), fonio (*Digitaria* sp.), teff (*Eragrostis tef*), and millets (*Pennisetum* sp.) (CCP/GR, 1995), may prove to be a potent tool in the fight against hunger on the continent (Kuta et al., 2004).

The potential nutritional values of underutilized species, as well as how to include those biological food resources into initiatives aimed at reducing malnutrition, have been widely reported (Moron, 2006; Toledo & Burlingame, 2006). Many previously disregarded species and varieties have since been given new consideration as viable means of promoting better nutrition and serving as practical means of boosting food supply and income for the population groups most affected by food insecurity (CCP/GR, 1995; Moron, 2006; Toledo & Burlingame, 2006). Their contribution to food requirements can be substantial in many developing nations because, unlike cash crops, they are subsistence-oriented food staples that are accessible, available in rural areas, and suitable for family gardening to combat nutrition insecurity (CCP/GR, 1995). Literature search and information obtained in this work were generated from searches on the subject matter from current literature conducted in the databases of ResearchGate, Google, and Google Scholar. The search terms, amongst others, included *description and distribution of fonios, nutritional information on fonio, African diets and health, constraints and prospects of growing fonios*, etc. Published papers and other useful materials, presented on some URLs, were also considered for inclusion.

21.1.1 Literature Search and Data Compilation

Literature searches were conducted on online academic databases, including Google Scholar, Science Direct, and other relevant sources. Keywords, such as *Digitaria exilis, fonio, nutritional value, cultivation*, and additional pertinent terms, were used to gather information for this review.

21.2 BOTANICAL DESCRIPTION

Due to their long history of cultivation, indigenous crops are highly suited to the climate and soil types of the region. In contrast to introduced cultivars, these crops frequently demand less inputs, like water, fertilizer, and pesticides. They therefore have the potential to lessen resource scarcity and lessen the negative effects of agriculture on the environment. A vital resource for food security in areas vulnerable to water scarcity are millets, a family of drought-resistant grains that thrive in desert climates.

There are about 300 species belonging to the genus *Digitaria*; however, not all of them are fonio. According to reports, *Digitaria* spp., also known as fonio, acha, or hungry rice in Nigeria, is the oldest domesticated cereal crop in West Africa. It is grown in several countries in the region from Cape Verde, Lake Chad, in the Sudan-Sahel, and to the edges of tropical rainforest zones (Vodouhè & Achigan-Dako, 2006; Bosch & Borus, 2007). Fonio is a member of the genus *Digitaria*, tribe Paniceae, subfamily Paniceae, and family Poaceae. The sole crop in the genus with any real economic importance is fonio; however, there are two recognized cultivated species in the genus *Digitaria*, viz., *D. exilis* Stapf and *D. iburua* Stapf (Taylor, 2019). West Africans cultivate *D. exilis* extensively, although *D. iburua* is reportedly limited to Nigeria and the northern parts of Benin and Togo. *D. exilis* is also known as white fonio, while *D. iburua* is also known as black fonio (Taylor, 2019).

FIGURE 21.1 The fonio plant and grain.

Source: Courtright (2017); Sylvia (2015).

Fonios belong to the grass family and are herbaceous, free-tillering, carbon-4 (C4) plants with upright, thin, and glabrous culms (Figure 21.1). Black fonio only reaches a maximum height of 1.0 m, whereas white fonio reaches a height of 1.5 m. The crop has simple, alternating leaves with upward-tapering linear blades and finger-shaped terminal panicles with two to five digits that are 5 to 10 cm long. The crop's ellipsoidal grains, which have dimensions of 0.5–1.0 mm in diameter and 0.75–2.0 mm in length, are incredibly colored and encased in two brown husks which, in the case of the black fonio kinds, may be darker. Approximately 0.5–0.6 g of these tiny grains makes up to a thousand pieces. The crop typically matures in the field in 60–150 days, with the gestation period being longer in humid rainforests than in tropical arid zones (Gomez & Gupta, 2003; Vodouhè & Achigan-Dako, 2006; Bosch & Borus, 2007; Adoukonou-Sagbadja, 2010) and shorter (60–70 days) for certain early-maturing varieties in the arid zones (Del Coro, 2019). Warm-season cereals, such as fonios, thrive in tropical climates with 900–1,000 mm of annual rainfall. Given the difficulties posed by climate extremes, its fast growth and drought tolerance may make it more significant as a possible component of tropical farming systems for food and nutrition security (Animasaun et al., 2014).

21.3 CULTIVATION REQUIREMENTS

Fonio is cultivated in tropical climates with a marked dry season, average temperatures of 25–30°C, and between 900 and 1,000 mm rainfall. It is a fast-growing, drought-tolerant cereal, a factor which could make it assume greater importance as a potential component of tropical farming systems for food and nutrition security, given the challenges of extremes of environmental conditions due to climate change (Animasaun et al., 2014). This crop is also tolerant to marginal or acidic soils occasioned by aluminum ion toxicity, as in the Fouta Djallon area of Senegal (Del Coro, 2019). Generally, it takes between 60 and 150 days for the crop to attain full maturity in the field, with the gestation period being longer in humid rainforests than in the arid zones of the tropics (Gomez & Gupta, 2003; Vodouhè & Achigan-Dako, 2006; Bosch & Borus, 2007; Adoukonou-Sagbadja, 2010), and a shorter period (60–70 days) for some early-maturing varieties in the arid zones (Del Coro, 2019). The crop produces small-sized, good-tasting, low–glycemic index, sulfur- and amino acid–rich grains (Del Coro, 2019).

21.4 POTENTIAL CONTRIBUTION OF FONIO TO NUTRIENT INTAKE

The remarkable nutritional value of native and/or underutilized crops is one of their main advantages. Staples such as rice, wheat, and maize are important sources of calories, but they could be deficient in certain vital elements. These voids can potentially be filled by native crops. Quinoa, a staple grain in the Andes, is full of vital amino acids, vitamins, and minerals, making it a complete protein source. Including these crops in the diet can potentially help fight malnutrition and offer a wider range of nutrients that are vital to human health.

Due to their subsistence-oriented nature, accessibility in remote areas, and suitability for family gardening as a means of addressing nutrition insecurity – a feature that sets them apart from income crops – they can play a key role in meeting food requirements in many developing nations. By addressing some of these characteristics in underutilized crops, constraints to their wider adoption are eliminated, thus making a useful contribution to food and nutrition security. At the same time, optimization of agronomy and management techniques tailored to the crop and location can make a real contribution to yields and quality of the product.

The information on dietary compositions that is now available, including expanding knowledge about the nutritional varieties found within these species, makes a compelling argument for the contributions of biodiversity to nutrition and health (Burlingame et al., 2009). Large gaps exist in our understanding of the composition and nutritional contribution of traditional or underutilized foods to human nutrition, however, because these foods have received little professional attention; food composition database compilers and researchers that create and carry out food consumption surveys have either ignored or discounted the potential contribution of underutilized crops (Burlingame, 2004). Therefore, it is challenging to evaluate these foods' actual and potential contributions to improved nutrition and health. Furthermore, given the prominence of these staple foods in the diets of resource-poor populations, it is quite likely that enhancing the nutritional value of these traditional cereals could aid in improving nutrition (Bouis, 2003). Fonio is one of these ancient and underappreciated cereals that hold great potential in sub-Saharan Africa (Vall et al., 2011). It is currently being "rediscovered" and improved as a cultivated species (Ibrahim, 2001; Morales-Payán et al., 2002). Like many other cereals, fonio is high in carbohydrate energy content and nutritious, making it a useful component of dietary and nutritional balance in foods. It can be taken as a whole dish in the form of porridge, non-fermented steamed and granulated dumplings, sourdough fonio jollof, and others, or as a side dish in the form of bread, biscuits, and cookies, among others.

21.5 INDIGENOUS PATTERN OF FONIO CONSUMPTION

Previous research on the grain's involvement in West African diets revealed that fonio is mostly cultivated in rural regions for domestic use (Vall et al., 2011; Morales-Payán et al., 2002; Vietmeyer et al., 1996a; Cruz et al., 2011). All age groups from these regions consume fonio about three times a day in a variety of recipes (Konkobo-Yameogo et al., 2004). About 54% of fonio consumers in urban West Africa reportedly consume this crop less than once a month on average (Konkobo-Yameogo et al., 2004). About 4% of the population in Burkina Faso and Guinea consumes fonio on a weekly basis (Konkobo-Yameogo, 2004). Fonio makes up less than 1% of the cereals consumed in Mali's cities, with an average consumption of 0.5–1.0 kg per capita annually. It is mainly consumed as a porridge, couscous, traditional/alcoholic beverage, and fatty fonio, or in salads, or as cakes, doughnuts, cookies, and bread. Furthermore, it is used to make pudding, kunu acha, and tuwo in a variety of African contexts (Jideani, 2012).

The consumption of fonio in West African diets is frequently associated with sociocultural and religious activities and beliefs (Vietmeyer et al., 1996b; Adoukonou-Sagbadja et al., 2004; Cruz, 2004; Konkobo-Yameogo et al., 2004). Fonio is specifically saved for chiefs, nobility, and important occasions, like Ramadan in Senegal, Mali, Guinea, Burkina Faso, and Togo (Adoukonou-Sagbadja et al., 2006). Fonio is the most significant dish served at women's initiation rituals, at weddings, and in customary baptism of newborns in various Togolese traditional communities. Fonio is traditionally recommended for nursing mothers to increase milk supply (Vodouhe et al., 2003), and it is thought to help reduce blood clotting in women postpartum (Adoukonou-Sagbadja et al., 2006). However, the majority of data on fonio usage comes from anecdotal reports found in descriptive socio-economic research (Konkobo-Yameogo et al., 2004). Furthermore, no thorough research on the intake of fonio has been conducted thus far in the field of food consumption. An important first step in determining whether fonio and its derivatives can enhance nutritional intake is to assess the amount of fonio consumed in West African cities.

21.6 NUTRITIONAL VALUES AND POTENTIAL HEALTH BENEFITS

Fonio is a staple food of about 4 million people in some West African communities; it is therefore not surprising that members of this family are ranked sixth among the world's most significant grains in the millets family (Taylor, 2019). They are prized for their flavor and reported to taste better than any cereal in the world (NAP, 2016). They can be made into porridge, combined with other cereals, cooked, and consumed like rice with stews, used to make couscous, and occasionally brewed into local beers and non-alcoholic drinks (Brink, 2006; Adoukonou-Sagbadja, 2010).

Fonio is abundant in a variety of vitamins, minerals, and nutrients. The nutritional values of 100 g include the following nutrients: protein (8.9 g), fat (3.0 g), carbohydrate (75.6 g), fiber (6.2 g), P (234 mg), and Fe (10.0 mg) (Irving & Jideani, 1997; Brink, 2006). According to some studies, fonio has a lower glycemic index than sorghum.

Foods with a low glycemic index (GI) release glucose into the bloodstream more slowly, resulting in a gradual and steady increase in blood sugar levels. This helps prevent rapid spikes and crashes in blood sugar, which can contribute to insulin resistance over time. Thus, to make bread and pastries, confectioners employ fonios (Bashiri et al., 2010). Diabetics can consume fonio starch since it has a lower glycemic index than other cereals because of its branched chain carbohydrates, which release glucose units gradually into the bloodstream to maintain a stable blood sugar level (Ballougou et al., 2013; Koreissi, 2015). Because of its high fiber content, low fat content, and ease of digestion, fonio is thought to offer a healthy nutrition profile. Because the sugars are assimilated more slowly than in other grains, this fonio has a lower GI (Fliedel et al., 2003; Besançon, 2000). It may be a good food for diabetics since it includes branched chain carbs, which release gradually to maintain a stable blood sugar level in the body (Besançon, 2000). This, along with its high mineral and amino acid content, has led to health professionals in some countries, such as Nigeria, recommending fonio for diabetics and nursing mothers (Bailey, 2011; FFBPR, 2013). According to Ballougou et al. (2013), grains are high in fiber and possess very little (approximately 1%) soluble sugars, such as fructose, saccharose, stachyose, and raffinose. Non-starchy, non-digestible polysaccharides make up fiber. Foods high in fiber may help lower the risk of diabetes, colon cancer, and cardiovascular disorders. A systematic review and meta-analysis by Reynolds et al. (2020) examined the role of high-fiber diets on mortality and the impact of increasing fiber intake on glycemic control and other cardiometabolic risk factors in adults with prediabetes or diabetes. The study found that increased fiber intake reduced glycated hemoglobin (HbA1c), fasting plasma glucose, and insulin resistance, among other benefits. Similarly, fiber helps promote regular bowel movements, and it may also have protective effects on the colon. In another systematic review and dose-response meta-analysis of prospective studies, Aune et al. (2011) investigated the association between dietary fiber, whole grains, and the risk of colorectal cancer. The analysis concluded that a high intake of dietary fiber, particularly cereal fiber, and whole grains was associated with a reduced risk of colorectal cancer. Also, maintaining a healthy weight is crucial for managing blood pressure. A low-GI diet, by promoting weight loss or preventing weight gain, can contribute to overall weight management. Excess body weight, especially around the waist, is a risk factor for hypertension. Furthermore, high-fiber foods help lower cholesterol levels by binding to cholesterol molecules and promoting their excretion from the body. A systematic review and meta-analysis conducted by Threapleton et al. (2013)

explored the association between dietary fiber intake and the risk of cardiovascular disease. The study found that total dietary fiber intake was inversely associated with the risk of both cardiovascular disease and coronary heart disease, with a risk ratio of 0.91 per 7 g/day increase in fiber.

Significant levels of vitamins and minerals can be found in fonios. The mean mineral value of 1.4 g found in fonio is less than the mean mineral contents of 4.4 g and 3.3 g found in barnyard and foxtail millets, respectively (Jideani et al., 2014). However, it compares favorably with values of 2.7 g, 1.9 g, and 1.5 g per 100 g of finger, prosso, and little millet, respectively. When compared to other West African grains, fonios have higher levels of magnesium, zinc, iron, and phosphorus. Despite having greater calcium and potassium values than white rice (Ballougou et al., 2013), they are thought to have lower Sn, Mn, and Ca contents than other varieties of millet. According to TACAM, fonio ranks second in terms of iron (8.5 mg/100 g dry matter) and zinc (1.5 mg/100 g dry matter) concentrations among cereals, with sorghum having the highest concentrations (2.1 mg/100 g dry matter for zinc and 11 mg/100 g dry matter for iron) (Barikmo et al., 2004). According to some studies based on composite samples obtained from retailers in various markets (Barikmo et al., 2004), most of the iron and zinc contents come from decorticated grain of the Fingoloni variety (Fliedel et al., 2003). However, soil iron may also have contaminated these samples. Minerals such as magnesium, zinc, iron, phosphorus, calcium, and potassium are essential for a wide range of physiological functions in the human body. Magnesium is crucial for over 300 enzymatic reactions, including muscle contraction and blood pressure regulation (Al Alawi et al., 2018). Zinc is vital for immune function, DNA synthesis, and cell division, affecting growth and development (Chasapis et al., 2020). Iron is necessary for oxygen transport in the blood and is a key component of hemoglobin, with deficiencies leading to anemia (Charlebois & Pantopoulos, 2023). Phosphorus plays a role in energy production, bone health, and cellular function, with imbalances affecting cardiovascular health (Takeda et al., 2012). Calcium is important for bone health, muscle function, and nerve signaling and may protect against hypertension and certain cancers (Tai et al., 2015). Potassium is key for cardiovascular health, renal function, and bone health, with high dietary intake offering protection against various conditions (McCormick et al., 2023). These minerals are integral to maintaining health and preventing disease.

Fonio is a significant source of glucidic energy and starch, but in comparison to other common cereals, like rice and maize, it has a relatively low protein and fat content; however, among cereals, fonio is the cereal with the second-highest protein concentration (7.1 mg/100 g dry matter) according to the TACAM, after sorghum (11 mg/100 g dry matter) (Barikmo et al., 2004). Compared to sorghum and millet, the proteins in fonio grains are more easily digestible and difficult to extract. According to Ayo and Nkama (2004), fonio's significant residue protein levels could possess vital functional implications. Essential amino acids such as methionine (4.5 mg/100 h-g), cystine (2.5 mg/100 g), and leucine (10.5 mg/100 g) are abundant in fonio (Fliedel et al., 2003). Given that fonio has twice as much methionine as the protein in eggs, it may be valuable as a supplement to regular diets in addition to serving as a means of survival. Compared to sorghum, millet, and rice, milled fonio has less polyunsaturated fatty and saturated fatty acids (Fliedel et al., 2003). Like most cereals, fonio is low in linolenic acid, and it has a very small amount of total pentosanase content when compared to sorghum or pearl millet.

21.6.1 Some Nutritional Challenges of Fonio

On the other hand, fonios have very small amounts of some non-nutritional compounds (NNCs), like tannins, phytate, oxalate, and cyanide. It has been observed that processing fonios, for example, into akamu, fufu, tuwo, or kunu, significantly reduces these NNCs, increasing the grains' mineral and proximate contents (Echendu et al., 2009; Ukim et al., 2013; Olu-Owolabi et al., 2014). Due to their lack of gluten (Taylor & Emmabux, 2008), fonio grains can be used as a complement to or replacement for wheat in baking sourdough and bread for the gluten-intolerant community, as this improves the fermentation process and bread's sensory qualities (Bailey, 2011).

21.7 ECONOMIC PROSPECTS OF FONIOS

Indigenous crops are closely linked to the customs and cultural identity of numerous people worldwide. By retaining their traditional knowledge and practices, their cultivation and preservation can potentially benefit local communities and cultural heritage. The biodiversity of traditional food systems and their sustainability, by encouraging the use of native crops, can be preserved. Their health-promoting qualities, particularly concerning their antidiabetic activities, warrant a strong research focus on the crop. They are resilient, tolerant to drought and flooding, marginal and acidic soils, insects, diseases, and mycotoxin contamination (Enyiukwu & Bassey, 2020; Tabassum et al., 2024). The commercialization and cultivation of this indigenous and underutilized crop can generate income for small-scale farmers because these crops often have niche markets and can command higher prices due to their unique qualities. Encouraging their cultivation can help improve rural community economies.

The increasing market demand for fonio is driven by its nutritional value, its adaptability to diverse growing conditions, and its status as a culturally important grain. According to Taylor and Emmabux (2008), fonios are a versatile and easy-to-cultivate crop. They may thrive in acidic and marginal soils, which are too poor for sorghum and pearl millet production. They are highly digestible cereals and grains packed with nutrients. Although it is generally acknowledged that fonios are nutritious, they have for long been disregarded in many agricultural programs. As gluten-free cereal rich in amino acids, fiber, and micronutrients, fonio appeals to health-conscious consumers and those with specific dietary needs, including the growing global market for gluten-free products (Xhakollari et al., 2018). This niche market provides an opportunity for farmers to tap into higher-value markets, enhancing their income streams.

Furthermore, cultivating fonio can serve as a risk mitigation strategy for farmers, offering an alternative source of income amid the uncertainties of climate change and market fluctuations (Barnaud et al., 2013). This is because, unlike many other cereal crops, fonio is highly drought-resistant and can thrive in poor soils with minimal inputs (Bano & Iqbal, 2016). This resilience makes it an attractive option for farmers in semi-arid regions who face the threat of crop failure with more traditional cereals. Diversifying crops to include fonio can therefore reduce dependency on a single crop, spreading risk and increasing the stability of farmers' incomes.

In the same vein, the value addition through the processing of fonio into various products represents another promising economic prospect. Fonio can be transformed into a range of value-added products, such as biscuits, cakes, bread, beverages, and specialty foods (Jideani & Jideani, 2011); many other value-added products derived from fonio are well documented in the review by Zhu (2020). This not only enhances its market appeal but also provides opportunities for small- and medium-sized enterprises to thrive. The development of these value-added products can create jobs and stimulate local economies.

Also, policy frameworks play a crucial role in the promotion and support of underutilized crops like fonio. Governments in West Africa have increasingly recognized the importance of crop diversification as a strategy for food security and economic resilience. Policies aimed at promoting the cultivation of underutilized crops often include research and development funding, extension services to educate farmers, and infrastructure development to improve market access (Aboagye et al., 2007: Mabhaudhi et al., 2017a, 2017b). However, the level of government support can vary significantly across countries, and the effectiveness of these policies depends on their implementation and the availability of resources.

21.8 CONCLUSION AND PROSPECTS

Indigenous and underutilized crops like fonio are an invaluable asset in the quest for increased food and nutrition security and better human health. These crops provide diverse dietary profiles, climatic resilience, and economic prospects while maintaining cultural legacies. Incorporating them into conventional agriculture and diets honors the resourcefulness of the communities that have cultivated them for generations and promotes a more resilient and sustainable food system. Investment in the preservation and promotion of these food resources is crucial to ensuring a healthy, food-secure future. Our exploration into fonio's cultivation, nutritional and health benefits, and economic potential reveals a promising future for this resilient cereal. Fonio's adaptability to various climatic conditions, particularly its ability to thrive in poor soils and withstand drought, makes it an invaluable crop amid the current climate crisis. As climate change continues to threaten global food security, fonio offers a sustainable solution for regions prone to erratic weather patterns, providing a reliable food source and reducing the risk of crop failure for smallholder farmers. Nutritionally, fonio is a powerhouse. Rich in essential amino acids, fiber, vitamins, and minerals, it addresses several dietary needs and health issues, including gluten intolerance and malnutrition. Integrating fonio into diets can combat food insecurity and promote better health outcomes, particularly in West Africa, where malnutrition rates are high. Its health benefits are equally compelling for global markets increasingly oriented toward health-conscious consumers. Economically, fonio presents numerous opportunities. Its growing market demand, driven by its nutritional benefits and cultural significance, can enhance farmers' incomes and stimulate rural economies. Looking ahead to the next 10 to 15 years, fonio's prospects are promising but contingent on several factors. Increased investment in agricultural research and development is essential to improve fonio yields, develop better processing technologies, and create efficient value chains. Raising awareness about fonio's benefits among consumers and farmers through targeted education and marketing campaigns is crucial for expanding its cultivation and consumption. Policy support will play a pivotal role in fonio's mainstreaming. Governments need to implement and strengthen policies that promote crop diversification, support smallholder farmers, and invest in agricultural infrastructure. Such policies should include subsidies for fonio production, the development of processing facilities, and the improvement of transportation networks to facilitate market access. International collaboration and support from organizations focused on food security and climate resilience can further bolster these efforts. To ensure fonio becomes a staple not only in West Africa but also in other regions facing climate challenges, concerted efforts are required. Integrating fonio into global agricultural policies, promoting it through international development programs, and encouraging private sector investment can drive its adoption. Research into adapting fonio to different agro-ecological zones beyond West Africa will be key to its global dissemination.

21.9 ACKNOWLEDGMENTS

The authors appreciate the leadership of the Biomedicinal Research Centre, Forestry Research Institute of Nigeria, for supporting this work.

REFERENCES

Abiad, M. G., & Meho, L. I. (2018). Food loss and food waste research in the Arab world: A systematic review. *Food Security, 10*(2), 311–322.

Aboagye, L. M., Obirih-Opareh, N., Amissah, L., & Adu-Dapaah, H. (2007). *Analysis of existing policies that enable or inhibit the wider use of underutilized plant species in Ghana.* Underutilized species and strategies, Global facilitation unit for underutilized species. (Via dei Tre Denari, 472/a, Maccarese (Fiumicino), Italy).

Adoukonou-Sagbadja, H. (2010). *Genetic characteristics of traditional fonio millets (Digitaria exilis, D. iburua Stapf) landraces from West Africa: Implications for conservation and breeding* [PhD dissertation, Justus-Liebig-University, p. 199]. https://jlupub.ub.uni-giessen.de/items/a6052c6c-1576-4f69-aa79-e98de2d68760

Adoukonou-Sagbadja, H., Dansi, A., Vodouhè, R., & Akpagana, K. (2004). Collecting fonio (Digitaria exilis Kipp. Stapf, D. Iburua Stapf) landraces in Togo. *Plant Genetic Resources Newsletter, 139*, 63–67.

Adoukonou-Sagbadja, H., Dansi, A., Vodouhè, R., & Akpagana, K. (2006). Indigenous knowledge and traditional conservation of fonio millet (Digitaria exilis Stapf, Digitaria iburua Stapf) in Togo. *Biodiversity & Conservation, 15*(8), 2379–2395.

Al Alawi, A. M. Majoni, S. W., & Falhammar, H. (2018). Magnesium and human health: Perspectives and research directions. *International Journal of Endocrinology, 2018*, Article 9041694. https://doi.org/10.1155/2018/9041694

Animasaun, D. A., Morakinyo, J. A., & Mustapha, O. T. (2014). Assessment of the effect of gamma radiation on the growth and yield of Digitaria exilis (Hailer). *Journal Applied Biosciences, 75*, 6164–6172.

Aune, D., Chan, D. S. M., Lau, R., Vieira, R., Greenwood, D. C., Kampman, E., & Norat, T. (2011). Dietary fibre, whole grains, and risk of colorectal cancer: Systematic review and dose-response meta-analysis of prospective studies. *BMJ, 343*. https://doi.org/10.1136/bmj.d6617

Ayo, J., & Nkama, I. (2004). Effect of acha (Digitaria exilis) grain flour on the physico-chemical and sensory properties of bread. *International Journal of Food Properties, 7*(3), 561–569.

Bailey, E. M. (2011). Fonio (acha and iburu) an anciet African grain and next quinoa? *Delicious Living News*. Retrieved September 16, 2024, from www.deliciousliving.com/health/fonio-acha-and-iburu-ancient-african-grain-and-next-quinoa/

Ballougou, V. Y., Soumanou, M. M., Toukourou, F., & Hounhouigan, J. D. (2013). Structure and nutritional composition of fonio (Digitaria exilis) grains: A review. *International Research Journal of Biological Sciences, 2*(1), 73–79.

Bano, S. A., & Iqbal, S. M. (2016). Biological nitrogen fixation to improve plant growth and productivity. *International Journal of Agriculture Innovations and Research, 4*, 596–569. https://ijair.org/administrator/components/com_jresearch/files/publications/1_IJAIR_1732_Final.pdf

Barikmo, I., Ouattara, F., & Oshaug, A. (2004). *Food composition table for Mali*. Research Series No. 9. Arkhesus University College, Norway Publications Series.

Barnaud, A., Vigouroux, Y., Barry, B., Beavogui, F., Camara, M., Billot, C., Noyer, J. L., Pham, J. L., & Bakasso, Y. (2013). From advanced to underutilized crops: Making fonio benefit from research on major cereals in Africa. *Acta Horticulturae, 979*, 421–430. https://doi.org/10.17660/ActaHortic.2013.979.45

Bashiri, M., Mustafa, Z., & Anthony, G. (2010). Development of an acha hulling machine. *Journal Research Natural Development, 8*(1), 1–7.

Begna, T. (2021). Role and economic importance of crop genetic diversity in food security. *International Journal of Agriculture and Food Science, 7*(1), 164–169. https://doi.org/10.17352/2455-815X.000104

Besançon, S. (2000). *Etude de l'influence de la consommation de fonio dans le traitement du diabète sucré au Mali*. Mémoire DESS Université de Montpellier II.

Betschart, A. A. (1982). World food and nutrition problems: The problem. *Cereal Foods World (USA), 27*, 562–1982. (21).

Bharucha, Z., & Pretty, J. (2010). The roles and values of wild foods in agricultural systems. *Philosophical Transactions of the Royal Society B: Biological Sciences, 365*(1554), 2913–2926.

Bosch, C. H., & Borus, D. J. (Eds.). (2007). *Cereals and pulses of Tropical Africa. Conclusions and recommendations based on PROTA 1: 'Cereals and pulses'*. PROTA Foundation.

Bouis, H. E. (2003). Micronutrient fortification of plants through plant breeding: Can it improve nutrition in man at low cost? *Proceedings of the Nutrition Society, 62*(2), 403–411.

Brink, M. (2006). Digitaria iburua Stapf. In M. Brink & Bailey (Eds.), *PROTA 1: Cereals, Pulses/Cereals et legumes Sec*. (CD-ROM) PROTA.

Burlingame, B. (2004). Fostering quality data in food composition databases: Visions for the future. *Journal of Food Composition and Analysis, 17*(3), 251–258.

Burlingame, B., Charrondiere, R., & Mouille, B. (2009). Food composition is fundamental to the cross-cutting initiative on biodiversity for food and nutrition. *Journal of Food Composition and Analysis, 22*(5), 361–365.

CCP/GR. (1995). *Development strategy for traditional grains*. Committee on Commodity Problems Intergovernmental Group on Grains.

Charlebois, E., & Pantopoulos, K. (2023). Nutritional aspects of iron in health and disease. *Nutrients, 15*, 2441. https://doi.org/10.3390/nu15112441

Chasapis, C. T., Ntoupa, P. S. A., Spiliopoulou, C. A., & Stefanidou, M. E. (2020). Recent aspects of the effects of zinc on human health. *Archives of Toxicology, 94*, 1443–1460. https://doi.org/10.1007/s00204-020-02702-9

Chauhan, N., Vaidya, D., & Pandit, A. (2018). Underutilized grains of Himalayan region: A mini review. *Journal of Pharmacognosy and Phytochemistry, 7*(1), 1044–1047. https://www.phytojournal.com/archives/2018/vol7issue1/PartN/6-4-484-543.pdf

Courtright, J. (2017, November 27). The Fonio Revival: West Africa's indigenous superfood is coming back from obscurity and making an appearance on the world stage. *Slate News*. https://roadsandkingdoms.com/2017/the-fonio-revival/

Cruz, J. F. (2004). Fonio: A small grain with potential. *LEISA*, *20*, 16–17.

Cruz, J. F., Béavogui, F., & Dramé, D. (2011). *Fonio, an African cereal (Le fonio, une céréale africaine). Collection agricultures tropicales en poche* (p.175). Editions Quae Versailles Cedex, Presses Agronomiques de Gembloux.

Del Coro. (2019). *Fonio: Nutrition facts* (calories, carbohydrates and health benefits of fonio). Retrieved February 22, 2020, from www.verywellfit.com/fonio-nutrition-facts-4693487

de Merode, M., Homewood, K., & Cowlishaw, G. (2003). The value of bushmeat and other wild foods to rural households living in extreme poverty in Democratic Republic of Congo. *Biological Conservation*, *118*(5), 573–581.

Echendu, C. A., Obiora, I. C., & Anyika, J. U. (2009). Changes in chemical composition of treated and untreated hungry rice 'acha' (Digitaria exilis). *Pakistan Journal Nutrition*, *8*(1). https://doi.org/10.39923/pjn.2009.1770-1785

Enyiukwu, D. N., & Bassey, I. N. (2020). Fonio (Digitaria spp.): The good tasting, potential food security cereal for Africa begs for research attention. *Direct Research Journal of Agriculture and Food Science*, *8*(7), 231–238. https://doi.org/10.26765/DRJAFS86631905

FFBPR (Fibre, Food and Beauty for Poverty Reduction). (2013). *Nutritional poperties of fonio: Health claims and evidences.* Retrieved June 6, 2024, from www.pan.uk.orrg/fonio

Fliedel, G., Ouattara, M., Grabulos, J., Dramé, D., & Cruz, J. (2003). Effet du blanchiment mécanique sur la qualité technologique, culinaire et nutritionnelle du fonio, céréale d'Afrique de l'Ouest. In *Voies alimentaires d'amélioration des situations nutritionnelles en Afrique de l'Ouest: Le rôle des technologues alimentaires et des nutritionnistes* (pp. 23–28). Actes du 2e Atelier international.

Ford-Lloyd, B. V., Schmidt, M., Armstrong, S. J., Barazani, O., Engels, J., Hadas, R., Hammer, K., Kell, S. P., Kang, D., Khoshbakht, K., Li, Y., Long, C., Lu, B-R., Ma, K., Nguyen, V. T., Qiu, L., Ge, S., Wei, W., Zhang, Z., & Maxted, N. (2011). Crop wild relatives undervalued, underutilized and under threat? *BioScience*, *61*(7), 559–565.

Gomez, M. I., & Gupta, S. C. (2003). Millets. In L. Trugo & P. M. Finglas (Eds.), *Encyclopeadia of life science and nutrition* (2nd ed., pp. 3974–3979). Elsevier.

Grivetti, L. E., & Ogle, B. M. (2000). Value of traditional foods in meeting macro- and micronutrient needs: The wild plant connection. *Nutrition Research Reviews*, *13*(1), 31–46.

Ibrahim, A. (2001). Hungry rice (Fonio): A neglected cereal crop. *NAQAS Newsletter*, *1*(4), 4–5.

Irving, D. W., & Jideani, I. A. (1997). Main structure and composition of Digitaria exilis Stapf (acha): A potential crop. *Cereal Chemistry*, *74*, 224–228.

Jideani, I. A. (2012). Digitaria exilis (acha, fonio), Digitaria iburrua (iburu, fonio) Eleusine corocana (finger millet, tanba) – non-conventional grains with potentials. *Scientific Research Essays*, *7*(45), 3834–3843.

Jideani, I. A., & Jideani, V. A. (2011). Developments on the cereal grains Digitaria exilis (acha) and Digitaria iburua (iburu). *Journal of Food Science and Technology*, *48*(3), 251–259. https://doi.org/10.1007/s13197-010-0208-9

Jideani, I. A., Salami, R. A., & Salami, I. A. (2014). Effects of Irish potato starch, yeast and sprouted soybean flour on quality of acha bread. *British Food Journal*, *110*(3), 271–282.

Johns, T., & Eyzaguirre, P. B. (2006). Linking biodiversity, diet and health in policy and practice. *Proceedings of the Nutrition Society*, *65*(2), 182–189.

Konkobo-Yameogo, C., Chaloub, Y., Kergna, A., Bricas, N., Karimou, R., & Ndiaye, J. L. (2004). Urban consumption of fonio, a traditional cereal in West Africa. (La consommation urbaine d'une céréale traditionnelle en Afrique de l'Ouest: le fonio). *Cahiers Agricultures*, *13*, 125–128.

Koreissi, Y. (2015). *Fonio (Digitaria exilis) in West Africa: Towards improving nutrient quality* [PhD dissertation, Wageningen University Netherlands, p. 180].

Kousar, S., Ahmed, F., PervAiz, A., & Bojnec, S. (2021). Food insecurity, population growth, urbanization and water availability: The role of government stability. *Sustainability*, *13*(22), 12336. https://doi.org/10.3390/su132212336

Kuta, D. D., Kwon-Ndung, E., Dachi, S., Ukwungwu, M., & Imolehin, E. D. (2004). Potential role of biotechnology tools for genetic improvement of "lost crops of Africa": The case of fonio (Digitaria exilis and Digitaria iburua). *African Journal of Biotechnology*, *2*(12), 580–585.

Liese, A. (2018). Food security. In *The Oxford handbook of governance limited statehood* (pp. 459–478). Oxford University Press.

Lutaladio, N., Burlinghame, B., & Crews, J. (2010). Horticulture, biodiversity and nutrition. *Journal of Food Composition and Analysis*, *23*(6), 481–485.

Mabhaudhi, T., Chimonyo, V. G., Chibarabada, T. P., & Modi, A. T. (2017a). Developing a roadmap for improving neglected and underutilized crops: A case study of South Africa. *Frontiers in Plant Science*, *8*, 2143. https://doi.org/10.3389/fpls.2017.02143

Mabhaudhi, T., Chimonyo, V. G. P., & Modi, A. T. (2017b). Status of underutilised crops in South Africa: Opportunities for developing research capacity. *Sustainability*, *9*(9), 1569. https://doi.org/10.3390/su9091569

McCormick, J. A., Topf, J., Tomacruz, I. D., & Grimm, P. R. (2023). A new understanding of potassium's influence upon human health and renal physiology. *Advances in Kidney Disease and Health*, *30*(2), 137–147. https://doi.org/10.1053/j.akdh.2023.01.001

Morales-Payán, J. P., Ortiz, J. R., Cicero, J., & Taveras, F. (2002). Digitaria exilis as a crop in the Dominican Republic. In *Supplement to: Trends in new crops and new uses*. ASHS Press

Moron, C. (2006). Food-based nutrition interventions at community level. *British Journal of Nutrition*, *96*(S1), S20–S22.

NAP (The National Academic Press). (2016). *Fonio (Acha) lost crops of Africa* (Vol. 1). Grains. (Online). Retrieved September 14, 2024, from https://nap.nationalacademies.org/read/2305/chapter/6

Olu-Owolabi, B. I., Olayinka, O. O., Adegbemile, A. A., & Adebowale, K. O. (2014). Comparison of the functional properties between native and chemically modified starches from acha (Digitaria spp. Stapf) grains. *Food and Nutritional Science*, *5*(2), 222–230.

Pandey, A., Tomer, A. K., Bhandari, D. C., & Pareek, S. K. (2008). Towards collection of wild relatives of crop plants in India. *Genetic Resources and Crop Evolution*, *55*(2), 187–202.

Popoola, J., Ojuederie, O., Omonhinmin, C., & Adegbite, A. (2019). Neglected and underutilized legume crops: Improvement and future prospects. In *Recent advances in grain crops research*. IntechOpen.

Rahman, M. S., Toiba, H., & Huang, W.-C. (2021). The impact of climate change adaptation strategies on income and food security: Empirical evidence from small-scale fishers in Indonesia. *Sustainability*, *13*(14), 7905. https://doi.org/10.3390/su13147905

Reynolds, A. N., Akerman, A. P., & Mann, J. (2020). Dietary fibre and whole grains in diabetes management: Systematic review and meta-analyses. *PLoS Medicine*, *17*(3), e1003053. https://doi.org/10.1371/journal.pmed.1003053

Richardson, K. J., Lewis, K. H., Krishnamurthy, P. K., Kent, C., Wiltshire, A. J., & Hanlon, H. M. (2018). Food security outcomes under a changing climate: Impacts of mitigation and adaptation on vulnerability to food insecurity. *Climatic Change*, *147*(1), 327–341. https://doi.org/10.1007/s10584-018-2137-y

Salunkhe, D. K., & Desai, B. B. (1986). *Postharvest biotechnology of oilseeds* (p. 264). CRC Press.

Setsoafia, E. D., Ma, W., & Renwick, A. (2022). Effects of sustainable agricultural practices on farm income and food security in Northern Ghana. *Agricultual and Food Economics*, *10*(1), 1–15. https://doi.org/10.1186/s40100-022-00216-9

Shackleton, C., & Shackleton, S. (2004). The importance of non-timber forest products in rural livehood security and as safety nets: A review of evidence from South-Africa. *South African Journal of Science*, *100*(11 & 12), 658–664.

Sylvia. (2015, July 13). Health benefits of Fonio. *Health Benefits Times*. www.healthbenefitstimes.com/health-benefits-of-fonio/

Szabo, S. (2016). Urbanisation and food insecurity risks: Assessing the role of human development. *Oxford Development Studies*, *44*(1), 28–48. https://doi.org/10.1080/13600818.2015.1067292

Tabassum, N., Ahmed, H. I., Parween, S., Sheikh, A. H., Saad, M. M., Krattinger, S. G., & Hirt, H. (2024). Host genotype, soil composition, and geo-climatic factors shape the fonio seed microbiome. *Microbiome*, *12*(11). https://doi.org/10.1186/s40168-023-01725-5

Tai, V., Leung, W., Grey, A., Reid, I. R., & Bolland, M. J. (2015). Calcium intake and bone mineral density: Systematic review and meta-analysis. *BMJ*, *351*, h4183. https://doi.org/10.1136/bmj.h4183

Takeda, E., Yamamoto, H., Yamanaka-Okumura, H., & Taketani, Y. (2012). Dietary phosphorus in bone health and quality of life. *Nutrition Reviews*, *70*(6), 311–321. https://doi.org/10.1111/j.1753-4887.2012.00473.x

Taylor, J. R. N. (2019). Sorghum and millets. In *Chemistry, taxonomy and nutritional attributes* (2nd ed., pp. 1–21). ScienceDirect.

Taylor, J. R. N., & Emmabux, M. N. (2008). Gluten-free foods and beverages from millets. In *Food science and technology* (pp. 119–148). ScienceDirect.

Threapleton, D. E., Greenwood, D. C., Evans, C. E. L., Cleghorn, C. L., Nykjaer, C., Woodhead, C., Cade, J. E., Gale, C. P., & Burley, V. J. (2013). Dietary fibre intake and risk of cardiovascular disease: Systematic review and meta-analysis. *BMJ*, *347*, f6879. https://doi.org/10.1136/bmj.f6879

Toledo, A., & Burlingame, B. (2006). Biodiversity and nutrition: A common path toward global food security and sustainable development. *Journal of Food Composition and Analysis, 19*(6), 477–483.

Ukim, C. I., Ojewola, G. S., & Obun, C. O. (2013). Nutrition and replacement value of hungry rice acha (Digitaria exilis) grain for maize grain in broiler starter chicks. *Global Journal of Agricultural* Sciences, *12*(1), 55–61. https://doi.org/10.4814/gjass.v1211.8

Vall, E., Andrieu, N., Beavogui, F., & Sogodogo, D. (2011). Les cultures de soudure comme stratégie de lutte contre l'insécurité alimentaire saisonnière en Afrique de l'Ouest: le cas du fonio (Digitaria exilis Stapf). *Cahiers Agricultures, 20*(4), 294–300.

Vietmeyer, N., Borlaugh, N., Axtell, J., Burton, G., Harlan, J., & Rachie, K. (1996a). Chapter 3 Fonio (Acha), in NRC/BSTID ed lost crops of Africa. In Board on Science and Technology for International Development, National Research Council (Eds.), *Chapter 3 Grains* (Vol. I, pp. 59–76). National Academy Press. (33).

Vietmeyer, N., Borlaugh, N., Axtell, J., Burton, G., Harlan, J., & Rachie, K. (1996b). Fonio (Acha), in lost crops of Africa. In Board on Science and Technology for International Development, National Research Council (Eds.), *Chapter 3 Grains* (Vol. I, pp. 59–76), National Academy Press.

Vodouhè, S. R., & Achigan-Dako, E. G. (2006). *Digitaria exilis (Kippist) Stapf, in MBaG Belay ed plant resources of tropical Africa* (Vol. 1, pp. 59–63). PROTA Foundation; Backhuys Publishers.

Vodouhe, S. R., Zannou, A., & Achigan Dako, E. G. (2003). *Proceedings of the first workshop on genetic diversity of fonio (Digitaria exilis) in West Africa* (Premier Atelier sur la Diversité Génétique du Fonio en Afrique de l'Ouest, 4–6 August 1998, pp. 36–41). IPGRI.

Wang, Z.-G, Bao, X.-G., Li, X.-F, Jin, X., Zhao, J.-H, Sun, J.-H., et al. (2015). Intercropping maintains soil fertility in terms of chemical properties and enzyme activities on a timescale of one decade. *Plant Soil, 391*(1), 265–282. https://doi.org/10.1007/s11104-015-2428-2

Wani, S. P., & Sawargaonkar, G. L. (2018). Future smart crops for paddy fallow agri-food systems in southeast Asia. In *Future smart food-rediscovering hidden treasures of neglected underutilized species for zero hunger in Asia* (pp. 61–78). FAO.

Weih, M., Westerbergh, A., & Lundquist, P.-O. (2017). Role of nutrient-efficient plants for improving crop yields: Bridging plant ecology, physiology, and molecular biology. In *Plant macronutrient use efficiency* (pp. 31–44). Academic Press.

Xhakollari, V., Canavari, M., & Osman, M. (2018). Factors affecting consumers adherence to gluten-free diet, A systematic review. *Trends in Food Science & Technology, 85*. https://doi.org/10.1016/j.tifs.2018.12.005

Zhu, F. (2020). Fonio grains: Physicochemical properties, nutritional potential, and food applications. *Comprehensive Reviews in Food Science and Food Safety, 2020*, 1–25. https://doi.org/10.1111/1541-4337.12608

22 Leveraging Underutilized Food Plants for Food and Nutrition Security

Selected Nutritional Attributes in Pigeonpea (Cajanus cajan) Seed

[1]Eastonce Tendayi Gwata, [1]Dipoo Charity Mashifane, [1]Tsakani Maria Maphosa, [2]Liana Jank, and [3]Hapson Mushoriwa*

[1]University of Venda, Faculty of Science, Engineering, and Agriculture, Department of Plant and Soil Sciences, Private Bag X5050, Thohoyandou, 0950, South Africa
[2]Embrapa Beef Cattle, Embrapa, Rua Radio Maia, 830, 79106–550, Campo Grande, MS, Brazil
[3]International Institute of Tropical Agriculture, PMB 5320, Oyo Road, Ibadan 200001, Oyo State, Nigeria
*Correspondence: ectgwata@gmail.com

22.1 INTRODUCTION

Pigeonpea (*Cajanus cajan*) is an important grain legume for food and feed value chains, particularly in parts of Africa and Asia. Based mainly on morpho-biochemical and cytogenetical analyses, pigeonpea evolved through genetic mutations in the species *Atylosia cajanifolia* (Debata et al., 2012; van der Maesen, 1980). However, pigeonpea–type segregants were produced from a cross (*A. lineata* × *A. scarabaeoides*), which suggested a polyphyletic origin of the species from several *Atylosia* species. Other similar studies revealed a close affinity of pigeonpea with both *A. lineata* and *A. scarabaeoides* (Debata et al., 2012; Pundir & Singh, 1985). Currently, pigeonpea is cultivated by farmers in many parts of the world, including sub-Saharan Africa, Asia, the Americas, and the Caribbean islands (Gomes et al., 2021; Beekham et al., 2010; Gwata et al., 2006; Sugui et al., 2000).

The wide ecological plasticity of pigeonpea is largely enhanced by three main factors. Firstly, the pigeonpea plant can nodulate promiscuously (that is, form effective nodules with naturally occurring soil rhizobia) (Bopape et al., 2022, 2023). Secondly, there are several maturity groups that are adapted to low- and high-latitude areas (Gwata & Siambi, 2009; Gwata & Silim, 2009; Gwata et al., 2007). For instance, the short-duration types mature within 90 days or less, while the long-duration types require up to 180 days or more (Silim et al., 2006). Thirdly, pigeonpea is also highly tolerant to drought stress, thus enabling it to grow in harsh, semi-arid areas (Gwata & Siambi, 2009). In general, it is a low-input requirement crop, thus making it suitable for cultivation by resource-poor smallholder farmers.

DOI: 10.1201/9781003469766-22

22.2 USES OF PIGEONPEA

Pigeonpea is a multipurpose crop which is cultivated primarily for human food. The seeds are highly nutritious, and the foliage, especially the green leaves and tops, can serve dual purposes as green manure as well as for livestock feeds, including rearing silkworms (Gowda et al., 2016). The tender leaves can be used as forage for supplementing protein in areas that have poor-quality pastures (Berardo et al., 1997). In Brazil, some of the crop–livestock systems utilize forage and pigeonpea in the production of corn silage (Gomes et al., 2021; Jank et al., 2014). The pigeonpea plant is used also in rotations with cereals and other crops because of its ability to improve soil fertility through biological nitrogen (N_2) fixation (Bopape et al., 2020, 2021; Mapfumo et al., 1999). Many smallholder farmers also derive household income from pigeonpea, especially in Africa. Pigeonpea stalks and leaves can be used in the cultivation of mushrooms (*Pleurotus* spp.) (Mane et al., 2007). Pigeonpea is also used as fuelwood, as a soil erosion management tool, and in traditional medicine (Robertson et al., 2005). Pigeonpea plays a significant ethnopharmacological role in many parts of China, Bangladesh, India, and the Caribbean for the treatment of several ailments (Pal et al., 2011; Lans, 2007). For example, young stems are used for curing gingivitis and stomatitis, while the leaf juice is used as a laxative in some parts of India (Ganesan, 2008). Pigeonpea is also used for the treatment of diabetes, dysentery, hepatitis, and jaundice (Zu et al., 2010).

These medicinal properties are partly due to the presence of numerous bioactive compounds in various parts of the plant. The leaves are rich in flavonoids, hexadecanoic acid, α-amyrin, β-sitosterol, and pinostrobin, as well as longistylin A ($C_{20}H_{22}O_2$) (Figure 22.1), which is associated with antibacterial activity against methicillin-resistant *Staphylococcus aureus* (Wu et al., 2020). Longistylin is also associated with anticancer activity (Pal et al., 2011). In addition, antiplasmodic activity is attributed to the presence of cajaninstilbene acid, pinostrobin, vitexin, and orientin in pigeonpea leaves (Pal et al., 2011). The average contents of pinostrobin and cajaninstilbene acid in pigeonpea are 3.35 mg/g and 2.49 mg/g (Gai et al., 2021). The plant also contains two globulins, cajanin (Figure 22.2) and concajanin (Pal et al., 2011).

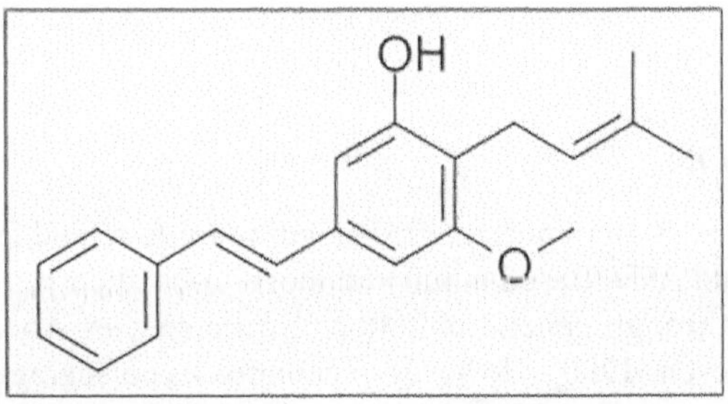

FIGURE 22.1 Chemical structure of longistylin A.

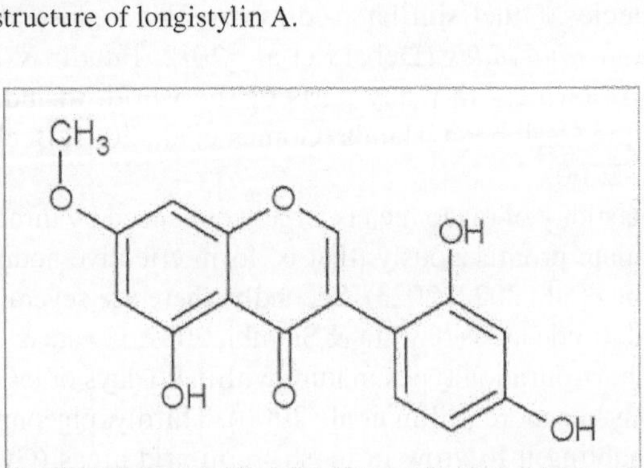

FIGURE 22.2 Chemical structure of cajanin.

22.3 MINERAL AND PROTEIN COMPOSITION

Several studies have reported the chemical composition of pigeonpea seed (Eltayeb et al., 2010; Oloyo, 2002). In a recent study involving 14 pigeonpea landraces, grain mineral and protein revealed highly significant ($p \leq 0.001$) differences for most of the mineral elements, including Ca, Cu, Mg, P, and Zn (Mashifane et al., 2024). In the study, the mean P content (4,031.94 mg/kg) was more than double that of both Ca (1,257.46 mg/kg) and Mg (1,409.94 mg/kg). Among the micronutrients, the Fe and Zn content ranged between 32.99 and 66.05 mg/kg and 20.48 and 38.56 mg/kg, respectively. However, the mean Fe content (45.24 mg/kg) was threefold more than that of Cu, but the mean crude protein content was only 21.50%. The study also reported a highly significant ($p \leq 0.01$) negative association between Cu and Mn (Table 22.1). In contrast, a highly significant ($p \leq 0.01$) strong positive linear relationship was observed between K and Cu, suggesting that breeding for an elevated content of either mineral could result in a concomitant improvement of the other mineral.

Further investigation using a biplot analysis technique revealed that protein as well as K and Cu were closely associated with five landraces ("G-01," "G-02," "G-06," "G-08," and "G-10") that were in the top left quadrant, but <30.0% of the landraces were closely associated with both Fe and Zn (Figure 22.3). Moreover, a negative relationship was detected between protein and each of these two microelements, as indicated by the obtuse angle between the vector for protein and that of Fe and Zn. Furthermore, two pigeonpea genotypes ("G-09" and "G-14") that were used in the study were positioned distantly from the origin, suggesting that they possessed unique alleles for the grain mineral elements and protein that were evaluated (Mashifane et al., 2024). Similarly, a study which examined the relationships among trace elements in pigeonpea seed indicated a highly significant ($p < 0.01$) positive linear relationship between Pb and each of Se and V (Table 22.2). However, Ba was negatively ($p < 0.00$) correlated with both Se and Ni but strongly positively correlated with Co. These studies demonstrate significant genotypic variability for mineral elements and protein, suggesting that there is potential for genetic improvements and selections for these traits.

TABLE 22.1
Pearson's correlation coefficients (*r*) indicating the association between selected grain mineral and protein content in pigeonpea landraces

Variable	Ca	Cu	Fe	K	Mg	Mn	P	Zn	Protein
Ca	1.000								
Cu	−0.541*	1.000							
Fe	−0.148	0.157	1.000						
K	−0.139	0.704**	0.074	1.000					
Mg	0.557*	−0.573*	0.021	−0.523	1.000				
Mn	0.424	−0.646**	0.011	−0.592*	0.507	1.000			
P	0.348	−0.514	−0.275	−0.579*	0.647*	0.377	1.000		
Zn	−0.105	0.246	0.502	−0.171	0.223	0.287	0.043	1.000	
Protein	−0.086	0.347	−0.575*	0.332	−0.04	−0.532*	0.104	−0.226	1.000

Source: Adapted from Mashifane et al. (2024).

**,*Highly significant and significant at the 1.0% and 5.0% probability levels, respectively.

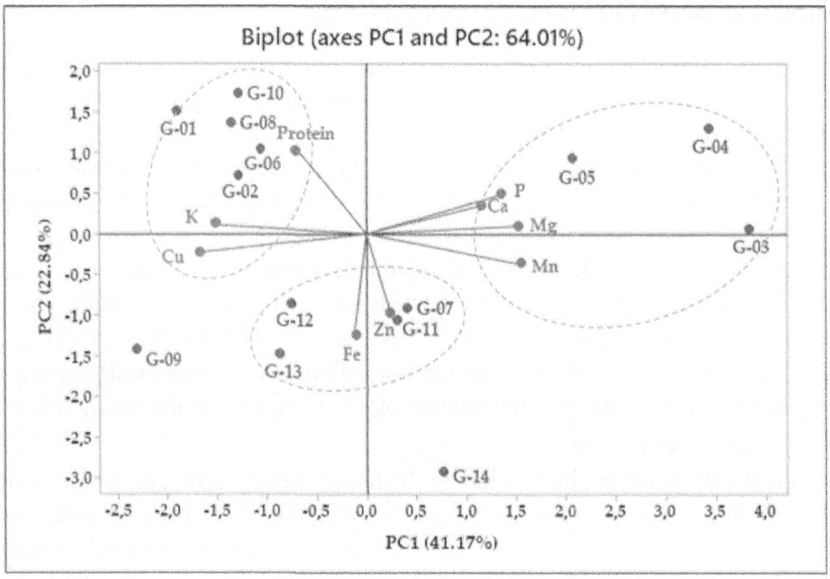

FIGURE 22.3 Biplot analysis for seed mineral and protein content among pigeonpea landraces.
Source: Adapted from Mashifane et al. (2024).

TABLE 22.2
Relationships among Selected Trace Elements in Pigeonpea Landraces

Variable	Ba	Co	Ni	Pb	Sb	Se	V
Ba	1.000						
Co	0.656***	1.000					
Ni	−0.478***	−0.279	1.000				
Pb	−0.318	−0.303	0.289	1.000			
Sb	0.064	−0.085	−0.237	0.184	1.000		
Se	−0.452**	−0.331	0.307	0.541***	0.297	1.000	
V	0.041	0.050	0.062	0.684***	0.080	0.223	1.000

Source: Adapted from Mashifane et al. unpublished.
***, **, *Very highly significant, highly significant, and significant at the 0.1%, 1.0%, and 5.0% probability levels, respectively.

22.4 SEED STORAGE PROTEINS

In general, the seed storage proteins from legumes contribute a significant proportion of human protein consumption. Their classification is based on their solubility and extraction in solvents such as water (albumins), dilute saline (globulins), alcohol/water mixtures (prolamins), and dilute acid or alkaline (glutelins) (Shewry et al., 1995). In legumes such as pigeonpea, the salt-soluble globulins are the predominant group, and they are further classified into two categories based on their sedimentation coefficients. The 7S globulins consist of vicilin and convicilin, while the 11S globulins consist of legumin, glycinin, and legumin-like proteins (Müntz, 1998).

On average, the seed protein content in legumes is twofold higher than in cereal seed. In pigeonpea, studies on protein fractionation showed differential accumulation in the seed coat, embryo, and cotyledons (Figure 22.4) (Singh & Jambunathan, 1982). Overall, the embryo contains the highest

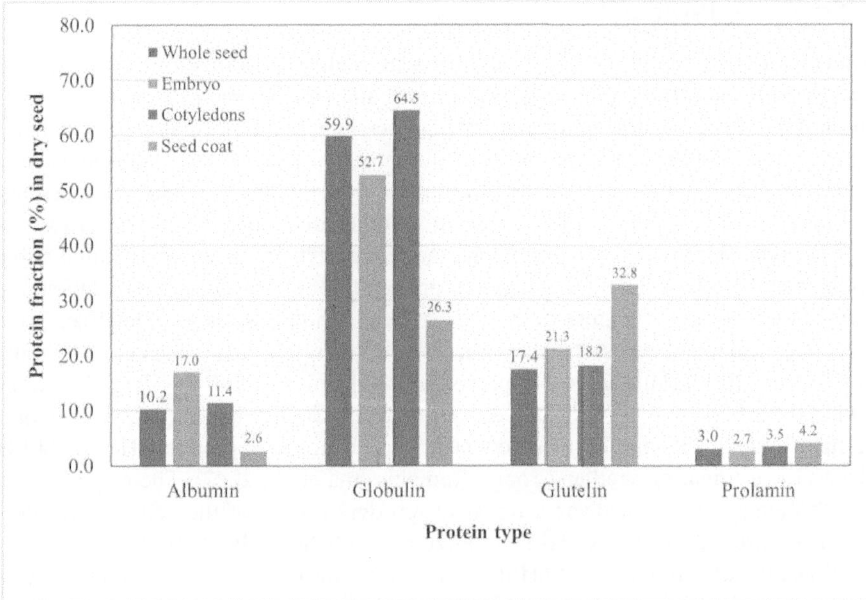

FIGURE 22.4 Distribution of protein fraction in pigeonpea dry seed.

Source: Adapted from Singh and Jambunathan (1982).

TABLE 22.3
Distribution of Selected Nutrients in Pigeonpea Seed

Constituent	Whole Seed	Cotyledons	Embryo	Seed Coat
Carbohydrates (%)	64.2	66.7	31.0	58.7
Protein (%)	20.5	22.2	49.6	4.9
Fiber (%)	5.0	0.4	1.4	31.9
Amino acid (g/100 g protein)				
Lysine	6.8	7.1	7.0	3.9
Threonine	3.8	4.3	4.7	2.5
Methionine	1.0	1.2	1.4	0.7
Cystine	1.2	1.3	1.7	–

Source: Adapted from Faris and Singh (1990).

amount of protein. Globulin was the major fraction of the embryos and cotyledons of these legumes, while the albumin fraction of cotyledons contained the highest concentration of two sulfur amino acids, namely, cystine and methionine (Table 22.3). Similarly, glutelin contained a relatively higher concentration of methionine and cystine than globulin.

The protein content in pigeonpea seed is rich in lysine (an essential amino acid), which is relatively deficient in cereals. However, the protein quality is limited by the deficiency of sulfur-containing amino acids, such as methionine and tryptophan. Moreover, pigeonpea is associated with low digestibility and the presence of antinutritional factors. However, a combined diet of cereal and pigeonpea can provide a good balance of amino acids, since cereals contain adequate methionine and tryptophan while lysine is supplied by the legume (Mandal & Mandal, 2000).

22.4.1 Specialized Proteins

Most biochemical reactions in a cell are regulated by highly specialized proteins, which are the prime mediators of the cellular phenotype. Therefore, the identification, quantitation, and characterization of all proteins in a cell are necessary to understand the molecular processes that mediate cellular metabolism and physiology, as well as the integration of internal molecular and external signals. Each protein is synthesized by translating the information encoded in a messenger RNA molecule to a polypeptide chain, which adopts a specific functional structure. For example, both heat shock proteins (HSPs) and proteases have been reported in pigeonpea (Ramakrishna et al., 2022; Godbole et al., 1994). The HSPs regulate protein folding, assembly, degradation, and translocation in cellular processes (Figure 22.5). Under stress conditions, these specialized proteins can stabilize proteins and membranes as well as assist in protein refolding (Wang et al., 2004; Frydman, 2001; Easton et al., 2000; Hartl, 1996). In pigeonpea seed, various HSP families, including HSP70, HSP90, and HSP100, were reported (Danekar et al., 2014; Sri Devi et al., 1999). In another study involving the root and leaf samples of pigeonpea, small heat shock protein (sHSPs) (14 kDa) sizes were induced in germinating seed and roots (Ramakrishna et al., 2022). The response of the HSP under elevated temperatures involves a transient reprogramming of the cellular activities to discontinue normal protein synthesis and synthesize a set of the HSPs (Schoffl et al., 1998). Under such conditions, it was estimated that sHSPs can account for up to 1% of the total leaf or root cell proteins (Hsieh et al., 1992; DeRocher et al., 1991). Nonetheless, these sHSPs, which consist of conserved N-terminal domains, can form high-molecular-weight (200–400 kDa) oligomers that are often referred to as heat shock granules through the C-terminal interactions (Jaya et al., 2009; Basha et al., 2006). The N-terminal domain is responsible for substrate binding (Figure 22.6). In contrast, these sHSPs are produced strictly at specific developmental stages of the plant when there is no stress.

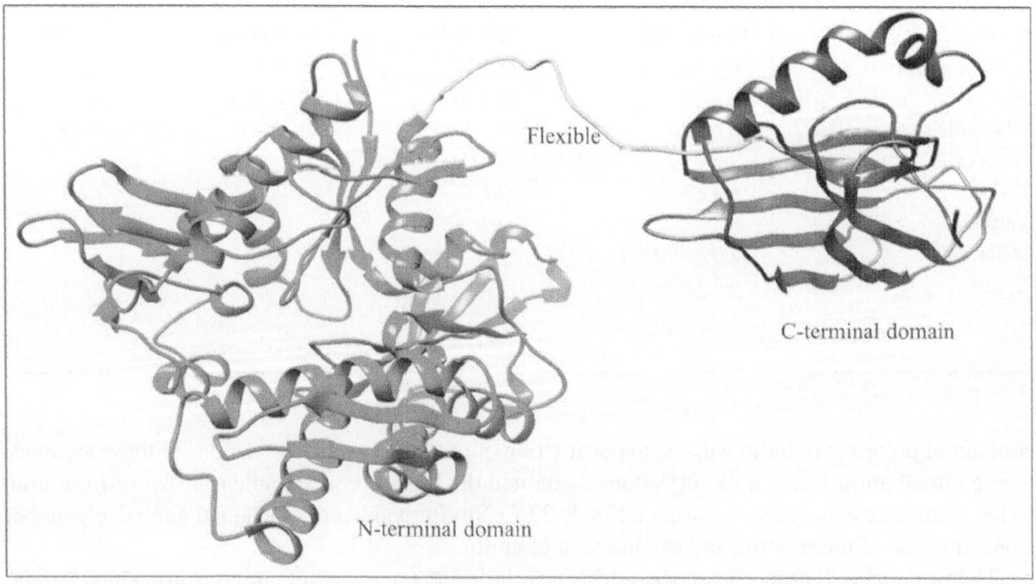

FIGURE 22.5 Structural features of a model heat shock protein (HSP70) indicating the C-terminal substrate-binding domain (blue) and the N-terminal binding domain (gold) and the bound peptide (red). The C-terminal domain is involved in the activation of ATPase (adenosine 5'-triphosphatase, EC 3.6.1.3)

22.5 PHENOLIC COMPOUNDS

Phenolic compounds possess various health benefits, such as antiproliferative, antimicrobial, anti-inflammatory, antioxidant, antiviral, anticancer, insecticidal, and antitumor properties (Al-Saeedi & Hossain, 2015; Maneechai & Rinthong, 2018). The phenolic compounds in pigeonpea are also associated with its medicinal value, as indicated earlier (Talari & Shakappa, 2018). In the plant, the phenolics have diverse biological functions. They also neutralize reactive oxygen species (ROS) and reactive nitrogen species (RNS) through a process known as radical scavenging (Nimse & Pal, 2015). The main phenolic compounds found in pigeonpea are flavonoids, isoflavonoids, and stilbenes (Nix et al., 2015; Wei et al., 2015). Vitexin, isovitexin, genistin, pinostrobin, and cajaninstilbene acid are frequently used as target bioactive compounds for extraction and separation in pigeonpea (Kong et al., 2011). The flavonoids play an important role in maintaining plant health, development, and growth. For instance, they function as phytoalexins, photoprotectors (against ultraviolet radiation), and nod inducers for N_2-fixing bacteria (Julkunen-Tiitto et al., 2014). The isoflavonol cajanol is associated with antifungal, antioxidant, and anticancer properties (Luo et al., 2010; Zhao et al., 2013). Similarly, genistein and genistin (Figure 22.6) possess activity against cancer (Cui et al., 2015). In addition, cajaninstilbene acid has hypoglycemic and hypocholesterolemic activities and the potential for the treatment of postmenopausal osteoporosis (Luo et al., 2008; Kadioglu et al., 2016; Wu et al., 2011). Isovitexin and vitexin possess hypotensive, antispasmodic, and radioprotective effects (Fu et al., 2007). The variations in total phenolic content are influenced by storage conditions, maturity period, genetic, and environmental factors (Panche et al., 2016).

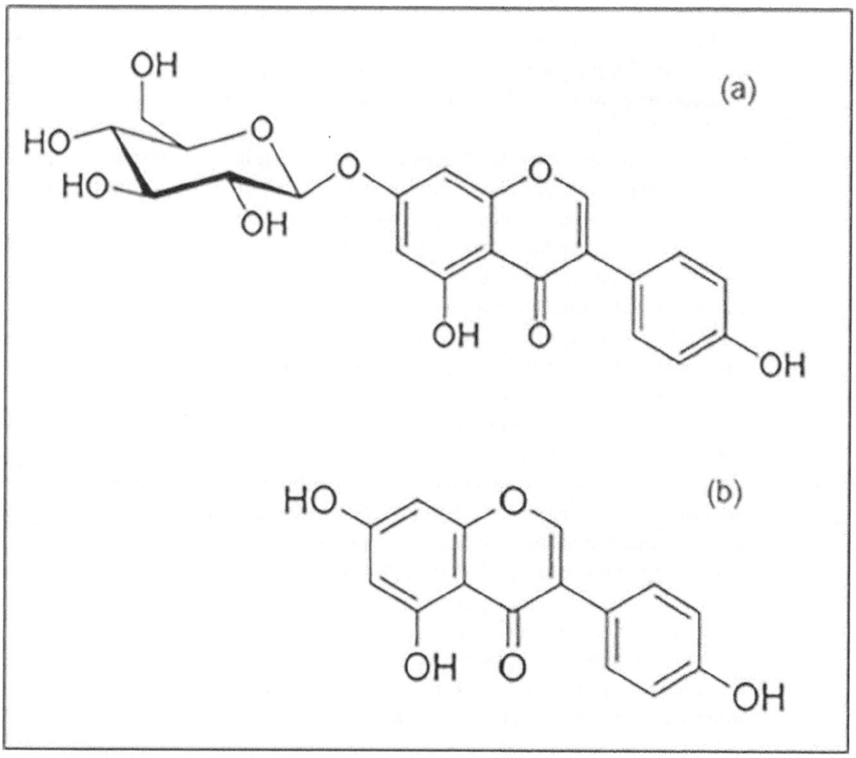

FIGURE 22.6 The structure of (a) genistin and (b) genistein. Upon ingestion, the isoflavone genistin is converted to the aglycone form (genistein) by enzymes in the small intestine and liver.

22.6 ANTINUTRITIONAL FACTORS

The main antinutritional factors in pigeonpea include total phenols, tannins, trypsin inhibitor, chymotrypsin inhibitor, amylase inhibitor, raffinose, stachyose, verbascose, cyanogenic glycosides, hemagglutinin, phytates, and saponins (Table 22.4 and 22.5) (Devindra & Aruna, 2017; Nwosu et al., 2013). They occur in variable content levels depending on several factors, such as the genotype, growth stage, environmental stresses, and plant part (Faris & Singh, 1990). The seed coat is rich in antinutritional factors; hence, whole-seed consumers tend to prefer white-seeded types more than brown-seeded pigeonpea cultivars that contain about three times higher levels of polyphenols as well as an elevated enzyme inhibition activity. Anaemene (2020) found that antinutritional factors in pigeonpea varied from 61.50 to 303.50 mg/100 g for tannins, from 802.50 to 2,149.65 mg/100 g for phytates, and from 659.00 to 1,191.50 mg/100 g for oxalates. (However, boiling or removal of the seed coat to produce *dhal* – or removing seed coat and splitting the cotyledons or decorticating – depletes the tannins from the colored pigeonpea seed. The antinutritional factors can also be reduced through chemical soaking, fermentation, germination, and parching, or a combination of these methods (Duhan et al., 2004; Onwuka, 2006; Ukom et al., 2023).

TABLE 22.4
Mode of Action of Antinutritional Phenolic Compounds That Occur in Pigeonpea Grain

	Compound	Mode of Action	Reference
1	Total phenols	Neutralize reactive oxygen species and reactive nitrogen species through radical scavenging; inhibit the production of pro-inflammatory cytokines and enzymes such as cyclooxygenase-2 (COX-2) and nitric oxide synthase (iNOS); induce apoptosis in cancer cells, inhibiting tumor growth, metastasis, and suppressing angiogenesis; prevents the oxidation of low-density lipoprotein (LDL) cholesterol.	Nimse & Pal, 2015; Zeb, 2020
2	Tannins	Inhibit lipid peroxidation and lipoxygenases; scavenge radicals (hydroxyl, superoxide, and peroxyl) which are important in cellular pro-oxidant state; inhibit inflammatory pathways and reduce production of inflammatory cytokines; inhibit growth of bacteria, fungi, and viruses by disrupting their cell membrane activities, growth, and replication; inhibit the growth and proliferation of cancer cells, inducing apoptosis, and prevent formation of blood vessels that supply tumors with nutrients by modulating the signaling pathways involved in cell proliferation and cancer development; inhibit the activity of pro-inflammatory cyclooxygenase (COX) and lipoxygenase (LOX).	Scalbert, 1991; Sekhon et al., 2017; Gbenga-Fabusiwa et al., 2021
3	Phytic acid	Forms complexes with minerals such as iron, calcium, magnesium, and zinc, and binds minerals in the digestive tract, reducing their absorption; scavenges free radicals and reduces oxidative stress; promotes apoptosis and inhibits tumor angiogenesis and metastasis; modulates the production of pro-inflammatory mediators, such as cytokines and chemokines; modulates gut microbiota by serving as a substrate for beneficial bacteria in the gut, leading to the production of short-chain fatty acids and other metabolites that support gut health and immune function.	Kumar et al., 2010; Igbedioh et al., 1994; Balogun et al., 2021
4	Oxalate	Chelates minerals such as iron, magnesium, and calcium to form oxalate-insoluble complexes, thus reducing their availability for absorption, and prevents absorption of excess iron in individuals with hemochromatosis; neutralizes free radicals and reduces oxidative stress.	Gautam et al., 2020; Petroski & Minich, 2020; Shi et al., 2018

Antinutritional factors, through their metabolic products, inhibit bioavailability and nutrient utilization, causing serious health issues, such as reduction in nutrient intake, digestion, and absorption and potential carcinogenic effects (Lawrence et al., 2023). For instance, antinutritional factors negatively affect the bioavailability of protein, Ca, Fe, Zn, and carbohydrate digestibility by forming complexes with protein and minerals (Kaushik et al., 2018). Pigeonpea contains about 0.23 g/100 g tannin, 6.35 g/100 g saponin, 3.82 g/100 g phenolics, 2.11 g/100 g flavonoids, and 2.65 g/100 g alkaloids (Harris et al., 2014). In another study, pigeonpea contained 7.18 g/100 g hydrogen cyanide, 1.02 g/100 g phytate, 0.14 g/100 g oxalate, 12.42 mg/100 g cyanogenic glycoside, and 4.75 mg/100 g anthocyanin (Nwosu

TABLE 22.5
Mode of Action of Three Antinutritional Enzyme Inhibitors That Occur in Pigeonpea Grain

	Compound	Mode of Action	Reference
1	Amylase inhibitor	Binds to active site of amylase, inhibiting its activity, particularly the breaking down of starch molecules into sugars, leading to delayed digestion and absorption of carbohydrates in the body; regulates blood sugar levels by slowing down the rate at which carbohydrates are converted into glucose.	Giri & Kachole, 1998; Li et al., 2021; Ambekar et al., 1996
2	Trypsin inhibitor	Inhibits the activity of the trypsin enzyme, regulating protein absorption and digestion; inhibits the growth and proliferation of cancer cells (breast, lung, and prostate) by inducing apoptosis and inhibition of signaling pathways involved in cell growth and survival; reduces the levels of pro-inflammatory cytokines and mediators; binds to the active site of trypsin, preventing the enzyme from binding to its substrate and carrying out its proteolytic function.	Ambekar et al., 1996; Duhan et al., 2001; Chinma et al., 2022
3	Chymotrypsin inhibitor	Forms a reversible, non-covalent complex with chymotrypsin and binds to the active site of the enzyme, preventing it from cleaving its substrate proteins, thereby regulating protein digestion in the digestive system; binds to serine, histidine, and aspartate residues via hydrogen bonds and hydrophobic interactions, effectively blocking the access of substrates to the enzyme's active site.	Obizoba, 1991; Kollipara et al., 1994; Kumar et al., 2021
4	Raffinose	Scavenges free radicals and protects cells from oxidative damage; α-galactosidase breaks down raffinose into galactose, glucose, and fructose, which can be used as an energy source for cell metabolism; maintains turgor pressure in seedlings, helping to regulate water uptake, retention, and cell integrity; acts as a prebiotic, promoting growth of beneficial bacteria in the gut, such as *Bifidobacteria* and *Lactobacilli*	Cirunay et al., 2020; Gopal et al., 2001
5	Stachyose	Serves as an energy source for cells lining the colon with short-chain fatty acids (SCFAs) produced from fermentation of stachyose, enhancing the integrity of the intestinal barriers, thereby reducing the risk of colon cancer and modulating immune responses; acts as a prebiotic, promoting the growth of beneficial bacteria in the gut, such as *Bifidobacteria* and *Lactobacilli*.	Aruna & Devindra, 2016
6	Verbascose	On fermentation by gut bacteria, produces short-chain fatty acids (SCFAs), such as acetate, butyrate, and propionate, acting as sources of energy for the cells lining the colon, reducing inflammation in the gut; the SCFAs enhance glucose tolerance and insulin sensitivity and reduce inflammation in adipose tissues as well regulation of the immune system.	Aruna & Devindra, 2016; Raja et al., 2016; Duszkiewicz-Reinhard et al., 1994

et al., 2013). The phytates chelate cations such as Ca, Mg, Zn, and Fe (Sandberg, 2002). Similarly, the presence of tannins in food lowers feed efficiency, depresses growth, decreases Fe absorption, and increases the excretion of proteins and essential amino acids (Gan et al., 2016; Ferreira et al., 2004). The phytolectins interact with glycoproteins on the surface of red blood cells, causing them to agglutinate. Therefore, it is essential to remove antinutritional factors to utilize pigeonpea effectively.

22.7 DIGESTION OF DIETARY CARBOHYDRATES

The dietary carbohydrate component goes through colonic bacterial fermentation, leading to the production of beneficial short-chain fatty acids, such as acetic, propionic, and n-butyric acid, in the colon (Hosseini et al., 2011). Propionate is an important microbial fermentation metabolite in the human gut, with putative health benefits that extend beyond the gut epithelium (Connolly et al., 2010; Blaut, 2002). It lowers lipogenesis and serum cholesterol levels as well as carcinogenesis in other tissues. Therefore, stimulating microbial propionate production through diet can increase health benefits from microbial carbohydrate fermentation (van Hylckama et al., 2011; Macfarlane & Macfarlane, 2011). Some of the potential substrates of this metabolite include arabinoxylan, polydextrose, and L-rhamnose (Hosseini et al., 2011). In legumes such as pigeonpea, both stachyose and verbascose occur among the non-digestible polysaccharides (Aruna & Devindra, 2016). On the other hand, phytic acid, which is a 6-phosphate ester of inositol (or hexahydroxycylohexane), is not digested in the upper intestinal region because phytase is absent there. However, it manifests no colloidal physical properties of non-digestible carbohydrates, hence is not classified as a fiber. Nonetheless, it is found in close association with fiber in legumes, and it exhibits chelating and antioxidant properties. In general, the dietary fibers are intrinsically part of the cell wall material or the inner layer of the legume seed (Augustin et al., 2020).

The formation of propionate involves the microbial fermentation of indigestible carbohydrates in the gut rather than from dietary intake (Vogt et al., 2004; Cheng & Lai, 2000). Two major biochemical pathways, each involving multiple enzymes, lead to propionate production in the intestine (Table 22.6) (Hosseini et al., 2011). In the succinate pathway, oxaloacetate resulting from the action of pyruvate carboxylase on pyruvate, is reduced to malate and dehydrated to fumarate by malate

TABLE 22.6
The Multiple Enzymes That Are Involved in Propionate Production

Propionate Pathway		Enzyme Involved
1. Succinate	(i)	Pyruvate carboxylase
	(ii)	Malate dehydrogenase
	(iii)	Fumarate hydratase
	(iv)	Succinate dehydrogenase
	(v)	Succinyl-CoA synthetase
	(vi)	Methylmalonyl-CoA mutase
	(vii)	Methylmalonyl-CoA epimerase
	(viii)	Methyl malonyl-CoA decarboxylase
	(ix)	Propionate CoA-transferase
2. Acrylate	(i)	L-lactate dehydrogenase
	(ii)	Propionate CoA-transferase
	(iii)	Lactoyl-CoA dehydratase
	(iv)	Acyl-CoA dehydrogenase
	(v)	Phosphate acetyltransferase
	(vi)	Propionate kinase

Source: Adapted from Hosseini et al. (2011).

dehydrogenase and fumarate hydratase, respectively (Hosseini et al., 2011). The succinate produced from this phase is further metabolized to propionate through the decarboxylation of methylmalonyl-CoA (Macy et al., 1978). The second distinct route for propionate production is defined as the acrylate pathway (Hosseini et al., 2011). Unlike the first pathway, in which fumarate and succinate are symmetrical intermediates, it contains no symmetrical intermediates (Léger et al., 2001; Tholozan et al., 1992). Pyruvate is reduced to lactate (by lactate dehydrogenase), which is subsequently converted to lactyl-CoA, then dehydrated to acrylyl-CoA, followed by reduction to propionyl-CoA and phosphorylation, resulting in propionyl phosphate, which is finally converted to propionate by propionate kinase (Macy et al., 1978). While the propionate pathway involves gut microbes such as *Bacteroides fragilis* (an important anaerobe in the human intestine) and *Propionibacterium shermanii* and *Veillonella* spp., the acrylate pathway typically utilizes the *Megasphaera elsdenii* (Prabhu et al., 2012; Hino et al., 1994). Other notable bacterial genera that are common in the human intestine include *Lactobacillus* and *Bacteroides* (Wang et al., 2021; Varyukhina et al., 2012). In addition, *Escherichia coli*, *Prevotella ruminicola*, and *Selenomonas ruminantium* are also involved in the production of propionate in the gut (De Filippis et al., 2019; Wang et al., 1996). All these findings demonstrate the importance and implications of pigeonpea in the human diet.

22.8 CONCLUSIONS AND PROSPECTS

The consumption of pigeonpea in the human diet has tangible health benefits, which include significant ethnopharmacological effects. The seed contains a wide spectrum of beneficial mineral elements. The seed storage proteins occur mainly in four distinct classes, in which the globulin types are predominant. Protein stability is facilitated by the presence of specialized proteins, such as HSPs. Modern proteomic analytical approaches present realistic opportunities for determining protein functionality in the seed of pigeonpea and other underutilized legume crops. The antinutritional factors that are associated with pigeonpea can be reduced by using a variety of methods which range from boiling to parching. In addition, carbohydrate fermentation results in beneficial short-chain fatty acids, such as propionate, in the gut.

In our opinion, pigeonpea remains highly underutilized despite its potential to ameliorate food and nutrition insecurity and enhance health. Moreover, it is a low-input, climate-smart crop. One of the major challenges centers on dietary preferences that are often influenced by sociocultural dimensions, among other factors. For instance, pigeonpea recipes are preferred largely by consumers of Asian ancestry. Other communities probably lack sufficient knowledge about the immense nutritional benefits that are associated with pigeonpea. Perhaps promoting diversified culinary uses and preparation techniques of pigeonpea in other communities outside Asia may also increase consumption of this pulse. In addition, farmers, particularly in many parts of sub-Saharan Africa, require exposure to the superlative ameliorative capabilities of pigeonpea in the cropping systems to broaden their choices of the legumes to adopt. Potentially, adoption of pigeonpea will increase the production of the crop as well as improve soil fertility, thus contributing to environmental and agricultural sustainability. Over and above these efforts, value addition of pigeonpea will be critical for improving its utilization. Some of the value-added products include canned pigeonpea, flour for biscuits and allied confectionaries, plywood bonding glue, pulp and paper from the stalks. We believe that if effort is exerted across these multipronged solutions, pigeonpea utilization can improve significantly in many parts of the world and possibly mainstream the crop. Apart from future genetic improvement of pigeonpea for nutritional attributes for human consumption, the crop has considerable potential in crop–livestock systems. This is partly because of the perennial nature of the crop, implying that it can provide feeds during the off-season. Studies involving beef cattle, for instance, observed that there is potential to produce more meat and milk without increasing the area through intensifying the utilization of better forages and better management of integrated crop–livestock–forage systems. Both the pigeonpea foliage and grain can be used in formulating supplementary livestock feeds. The genotypic variation in fodder production and the mineral content in pigeonpea auger well for the genetic improvement of fodder varieties, particularly in dryland farming systems, where crude protein for ruminants is low. Perennial pigeonpea types can be useful also in agroforestry systems, which utilize forest tree species as forage for various types of livestock to optimize the exploitation of agricultural land. The integration of pigeonpea in such systems provides high-protein fodder as well as organic nitrogenous fertilizers. Nonetheless, the relatively high lignin content associated with pigeonpea foliage tends to depress its digestibility in ruminants. Overall, pigeonpea utilization is likely to increase worldwide because of multiple factors, as highlighted in this chapter.

22.9 ACKNOWLEDGMENTS

The authors wish to thank the National Research Foundation (South Africa) for funding part of the work which is reported in this chapter.

REFERENCES

Al-Saeedi, A. H., & Hossain, M. A. (2015). Total phenols, total flavonoids contents and free radical scavenging activity of seeds crude extracts of pigeonpea traditionally used in Oman for the treatment of several chronic diseases. *Asian Pacific Journal of Tropical Disease*, *5*(4), 316–321.

Ambekar, S. S., Patil, S. C., Giri, A. P., & Kachole, M. S. (1996). Proteinaceous inhibitors of trypsin and of amylases in developing and germinating seeds of pigeonpea (*Cajanus cajan*). *Journal of the Science of Food and Agriculture*, *72*(1), 57–62.

Anaemene, D. I. (2020). A comparative evaluation of the nutrient and antinutrient compositions of four pigeonpea (*Cajanus cajan*) varieties. *Anchor University Journal of Science and Technology*, *1*(1), 102–109.

Aruna, T., & Devindra, S. (2016). Nutritional and anti-nutritional characteristics of two varieties of red gram (Cajanus cajan, L) Seeds. *International Journal of Scientific and Research Publications*, *6*(9), 89–95.

Augustin, L. S., Aas, A. M., Astrup, A., Atkinson, F. S., Baer-Sinnott, S., Barclay, A. W., Brand-Miller, J. C., Brighenti, F., Bullo, M., Buyken, A. E., & Ceriello, A. (2020). Dietary fibre consensus from the international carbohydrate quality consortium (ICQC). *Nutrients*, *12*(9), 2553.

Balogun, M. A., Ahmed, R. N., Akintayo, O. A., Aruna, T. E., Omovbude, M. O., & Shittu, T. (2021). Microbial, chemical and sensory evaluation of pigeonpea condiment from wild and controlled fermentation. *Ceylon Journal of Science*, *50*(3), 269–277.

Basha, E., Friedrich, K. L., & Vierling, E. (2006). The N-terminal arm of small heat shock proteins is important for both chaperone activity and substrate specificity. *Journal of Biological Chemistry*, *281*(52), 39943–39952.

Beekham, A., Umaharan, P., & Wilson, L. (2010). Consumer acceptance of vegetable pigeonpea (*Cajanus cajan*) in Trinidad and Tobago with observations for farmers and processors. *Tropical Agriculture*, *87*(1), 21–29.

Berardo, N., Dzowela, B. H., Hove, L., & Odoardi, M. (1997). Near infrared calibration of chemical constituents of *Cajanus cajan* (pigeonpea) used as forage. *Animal Feed Science and Technology*, *69*, 201–206.

Blaut, M. (2002). Relationship of prebiotics and food to intestinal microflora. *European Journal of Nutrition*, *41*, 111–116.

Bopape, F. L., Beukes, C. W., Katlego, K., Hassen, A. I., Steenkamp, E. T., & Gwata, E. T. (2022). Symbiotic performance and characterization of pigeonpea (*Cajanus cajan*) rhizobia occurring in South African soils. *Agriculture*, *13*(1), 30.

Bopape, F. L., Gwata, E. T., Hassen, A. I., & Zhou, M. M. (2021). Symbiotic efficiency of pigeonpea (*Cajanus cajan* (L.) Huth) with different sources of nitrogen. *Plant Genetic Resources: Characterization and Utilization*, *19*(4), 312–315.

Bopape, F. L., Hassen, A. I., Shonhai, A., Chiulele, R. M., & Gwata, E. T. (2023). The genome of a pigeonpea compatible rhizobial strain '10ap3' lacks common nodulation genes. *Genes*, *14*(5), 1084.

Bopape, F. L., Hassen, A. I., Swanevelder, Z. H., & Gwata, E. T. (2020). Draft genome sequence of *Rhizobium tropici* SARCC-755, a free-living Rhizobium that nodulated and promoted growth in pigeonpea [*Cajanus cajan* (L.) Millsp]. *Microbiology Resource Announcements*, *9*(2), e01122–19.

Cheng, H. H., & Lai, M. H. (2000). Fermentation of resistant rice starch produces propionate reducing serum and hepatic cholesterol in rats. *The Journal of Nutrition*, *130*(8), 1991–1995.

Chinma, C. E., Abu, J. O., Adedeji, O. E., Aburime, L. C., Joseph, D. G., Agunloye, G. F., Adebo, J. A., Oyeyinka, S. A., Njobeh, P. B., & Adebo, O. A. (2022). Nutritional composition, bioactivity, starch characteristics, thermal and microstructural properties of germinated pigeonpea flour. *Food Bioscience*, *49*, 101900.

Cirunay, A. R. T., Mopera, L. E., Sumague, M. J. V., & Bautista, J. A. N. (2020). In vitro fermentation and prebiotic potential of pigeonpea (*Cajanus cajan*) flour. *Food Research*, *5*(1), 174–184.

Connolly, M. L., Lovegrove, J. A., & Tuohy, K. M. (2010). In vitro evaluation of the microbiota modulation abilities of different sized whole oat grain flakes. *Anaerobe*, *16*(5), 483–488.

Cui, Q., Peng, X., Yao, X. H., Wei, Z. F., Luo, M., Wang, W., Zhao, C. J., Fu, Y. J., & Zu, Y. G. (2015). Deep eutectic solvent-based microwave-assisted extraction of genistin, genistein and apigenin from pigeonpea roots. *Separation and Purification Technology*, *150*, 63–72.

Danekar, P., Tyagi, A., Mahto, A., Krishna, K. G., Singh, A., Raje, R. S., Gaikwad, K., & Singh, N. K. (2014). Genome wide characterization of Hsp 100 family genes from pigeonpea. *Indian Journal of Genetics and Plant Breeding*, *74*(03), 325–334.

Debata, A. K., Mukherjee, A. K., & Panda, P. C. (2012). Phylogenetic relationships among pigeonpea (*Cajanus cajan*) and its wild relatives as revealed by RAPD and ISSR markers. *Plant Science Research*, *34*, 72–92.

De Filippis, F., Pasolli, E., Tett, A., Tarallo, S., Naccarati, A., De Angelis, M., Neviani, E., Cocolin, L., Gobbetti, M., Segata, N., & Ercolini, D. (2019). Distinct genetic and functional traits of human intestinal *Prevotella copri* strains are associated with different habitual diets. *Cell Host and Microbe*, *25*(3), 444–453.

DeRocher, A. E., Helm, K. W., Lauzon, L. M., & Vierling, E. (1991). Expression of a conserved family of cytoplasmic low molecular weight heat shock proteins during heat stress and recovery. *Plant Physiology*, *96*(4), 1038–1047.

Devindra, S., & Aruna, T. (2017). Effect of chemical soaking, toasting and crude α-galactosidase enzyme treatment on the oligosaccharide content of red gram flour. *Journal of Food Processing and Preservation*, *41*(3), e12922.

Duhan, A., Khetarpaul, N., & Bishnoi, S. (2001). Saponin content and trypsin inhibitor activity in processed and cooked pigeonpea cultivars. *International Journal of Food Sciences and Nutrition*, *52*(1), 53–59.

Duhan, A., Khetarpaul, N., & Bishnoi, S. (2004). HCl-extractability of zinc and copper as affected by soaking, dehulling, cooking and germination of high yielding pigeonpea cultivars. *Journal of Food Composition and Analysis*, *17*(5), 597–604.

Duszkiewicz-Reinhard, W., Gujska, E., & Khan, K. (1994). Reduction of stachyose in legume flours by lactic acid bacteria. *Journal of Food Science*, *59*(1), 115–117.

Easton, D. P., Kaneko, Y., & Subjeck, J. R. (2000). The Hsp110 and Grp170 stress proteins: Newly recognized relatives of the Hsp70s. *Cell Stress and Chaperones*, *5*(4), 276.

Eltayeb, A. R. S. M., Ali, A. O., & Haron, R. (2010). The chemical composition of pigeonpea (*Cajanus cajan*) seed and functional properties of protein isolate. *Pakistan Journal of Nutrition*, *9*(11), 1069–1073.

Faris, D. G., & Singh, U. (1990). Pigeonpea: Nutrition and products. In Y. L. Nene, S. D. Hall, & V. K. Sheila (Eds.), *The Pigeonpea* (pp. 401–434). CAB International.

Ferreira, E. C., Nogueira, A. R. A., Souza, G. B., & Batista, L. A. (2004). Effect of drying method and length of storage on tannin and total phenol concentrations in pigeonpea seeds. *Food Chemistry*, *86*(1), 17–23.

Frydman, J. (2001). Folding of newly translated proteins in vivo: The role of molecular chaperones. *Annual Review of Biochemistry*, *70*, 603–647.

Fu, Y., Zu, Y., Liu, W., Hou, C., Chen, L., Li, S., Shi, X., & Tong, M. (2007). Preparative separation of vitexin and isovitexin from pigeonpea extracts with macroporous resins. *Journal of Chromatography A*, *1139*(2), 206–213.

Gai, Q. Y., Jiao, J., Wang, X., Fu, Y. J., Lu, Y., Liu, J., Wang, Z. Y., & Xu, X. J. (2021). Simultaneous quantification of eleven bioactive phenolic compounds in pigeonpea natural resources and in vitro cultures by ultra-high performance liquid chromatography coupled with triple quadrupole mass spectrometry (UPLC-QqQ-MS/MS). *Food Chemistry*, *335*, 127602.

Gan, R. Y., Deng, Z. Q., Yan, A. X., Shah, N. P., Lui, W. Y., Chan, C. L., & Corke, H. (2016). Pigmented edible bean coats as natural sources of polyphenols with antioxidant and antibacterial effects. *LWT-Food Science and Technology*, *73*, 168–177.

Ganesan, S. (2008). Traditional oral care medicinal plants survey of Tamil Nadu. *Natural Product Radiance*, *7*(2), 166–172.

Gautam, A. K., Sharma, D., Sharma, J., & Saini, K. C. (2020). Legume lectins: Potential use as diagnostics and therapeutics against the cancer. *International Journal of Biological Macromolecules*, *142*, 474–483.

Gbenga-Fabusiwa, F. J., Oladele, E. P., Oboh, G., Adefegha, S. A., Fabusiwa, O. F., & Osho, P. O. (2021). Modulatory effect of pigeonpea wheat biscuits on lipid profile, lipid peroxidation level, α-glucosidase, and butyrylcholinesterase activities in type-2 diabetic patients. *Journal of Food Biochemistry*, *45*(4), e13658.

Giri, A. P., & Kachole, M. S. (1998). Amylase inhibitors of pigeonpea (*Cajanus cajan*) seeds. *Phytochemistry*, *47*(2), 197–202.

Godbole, S. A., Krishna, T. G., & Bhatia, C. R. (1994). Purification and characterisation of protease inhibitors from pigeonpea (*Cajanus cajan*) seeds. *Journal of the Science of Food and Agriculture*, *64*(1), 87–93.

Gomes, V. C., de Lima Meirelles, P. R., Costa, C., da Silva Barros, J., de Castilhos, A. M., de Souza, D. M., Tardivo, R., & Pariz, C. M. (2021). Production and quality of corn silage with forage and pigeonpeas in a crop-livestock system. *Semina: Ciências Agrárias Londrina*, *42*, 861–876.

Gopal, P. K., Sullivan, P. A., & Smart, J. B. (2001). Utilisation of galacto-oligosaccharides as selective substrates for growth by lactic acid bacteria including *Bifidobacterium lactis* DR10 and *Lactobacillus rhamnosus* DR20. *International Dairy Journal*, *11*, 19–25.

Gowda, C. L. L., Saxena, K. B., Srivastava, R. K., Upadhyaya, H. D., & Silim, S. N. (2016). Pigeonpea: From an orphan to a leader in food legumes. In P. Gepts, T. R. Famula, & R. L. Bettinger (Eds.), *Biodiversity in agriculture: Domestication, evolution, and sustainability* (pp. 361–373). Cambridge University Press.

Gwata, E. T., Mligo, J. K., & Silim, S. N. (2007). Registration of pigeonpea cultivar Tumia. *Crop Science*, *47*(1), 436–436.

Gwata, E. T., & Siambi, M. (2009a). Genetic enhancement of pigeonpea for high latitude areas in Southern Africa. *African Journal of Biotechnology*, *8*(18), 4413–4417.

Gwata, E. T., & Silim, S. N. (2009b). Utilization of landraces for the genetic enhancement of pigeonpea in Eastern and Southern Africa. *Journal of Food, Agriculture and Environment*, *7*(2), 803–806.

Gwata, E. T., Silim, S. N., & Mgonja, M. (2006). Impact of a new source of resistance to Fusarium wilt in pigeonpea. *Journal of Phytopathology*, *154*(1), 62–64.

Harris, K. K., Sahu, M., & Verma, D. (2014). Phytochemical analysis of the leaf, stem and seed extracts of *Cajanus cajan*. *World Journal of Pharmaceutical Sciences*, *3*, 694–733.

Hartl, F. U. (1996). Molecular chaperones in cellular protein folding. *Nature*, *381*, 571–580.

Hino, T., Shimada, K., & Maruyama, T. (1994). Substrate preference in a strain of *Megasphaera elsdenii*, a ruminal bacterium, and its implications in propionate production and growth competition. *Applied and Environmental Microbiology*, *60*(6), 1827–1831.

Hosseini, E., Grootaert, C., Verstraete, W., & Van de Wiele, T. (2011). Propionate as a health-promoting microbial metabolite in the human gut. *Nutrition Reviews*, *69*(5), 245–258.

Hsieh, M. H., Chen, J. T., Jinn, T. L., Chen, Y. M., & Lin, C. Y. (1992). A class of soybean low molecular weight heat shock proteins: Immunological study and quantitation. *Plant Physiology*, *99*(4), 1279–1284.

Igbedioh, S. O., Olugbemi, K. T., & Akpapunam, M. A. (1994). Effects of processing methods on phytic acid level and some constituents in bambara groundnut (*Vigna subterranea*) and pigeonpea (*Cajanus cajan*). *Food Chemistry*, *50*(2), 147–151.

Jank, L., Barrios, S. C., do Valle, C. B., Simeão, R. M., & Alves, G. F. (2014). The value of improved pastures to Brazilian beef production. *Crop and Pasture Science*, *65*(11), 1132–1137.

Jaya, N., Garcia, V., & Vierling, E. (2009). Substrate binding site flexibility of the small heat shock protein molecular chaperones. *Proceedings of the National Academy of Sciences*, *106*(37), 15604–15609.

Julkunen-Tiitto, R., Nenadis, N., Neugart, S., Robson, M., Agati, G., Vepsäläinen, J., Zipoli, G., Nybakken, L., Winkler, B., & Jansen, M. A. K. (2014). Assessing the response of plant flavonoids to UV radiation: An overview of appropriate techniques. *Phytochemistry Reviews*, *14*(2), 273–297.

Kadioglu, O., Fu, Y., Wiench, B., Zu, Y., & Efferth, T. (2016). Synthetic cajaninstilbene acid derivatives inhibit c-MYC in breast cancer cells. *Archives of Toxicology*, *90*(3), 575–588.

Kaushik, G., Singhal, P., & Chaturvedi, S. (2018). Food processing for increasing consumption: The case of legumes. In *Food processing for increased quality and consumption* (pp. 1–28). Academic Press.

Kollipara, K. P., Singh, L., & Hymowitz, T. (1994). Genetic variation of trypsin and chymotrypsin inhibitors in pigeonpea (*Cajanus cajan*) and its wild relatives. *Theoretical and Applied Genetics*, *88*(8), 986–993.

Kong, Y., Wei, Z. F., Fu, Y. J., Gu, C. B., Zhao, C. J., Yao, X. H., & Efferth, T. (2011). Negative-pressure cavitation extraction of cajaninstilbene acid and pinostrobin from pigeonpea (*Cajanus cajan*) leaves and evaluation of antioxidant activity. *Food Chemistry*, *128*(3), 596–605.

Kumar, V., Sinha, A. K., Makkar, H. P. S., & Becker, K. (2010). Dietary roles of phytate and phytase in human nutrition: A review. *Food Chemistry*, *120*(4), 945–959.

Kumar, Y., Basu, S., Goswami, D., Devi, M., Shivhare, U. S., & Vishwakarma, R. K. (2021). Anti-nutritional compounds in pulses: Implications and alleviation methods. *Legume Science*, *4*(2), e111.

Lans, C. (2007). Comparison of plants used for skin and stomach problems in Trinidad and Tobago with Asian ethnomedicine. *Journal of Ethnobiology and Ethnomedicine*, *3*, 22–28.

Lawrence, I. G., Akande, E. A., & Oke, O. M. (2023). Effects of fermentation and steaming on the proximate and antinutritional properties of pigeonpea seed flour. *Asian Journal of Food Research and Nutrition*, *2*(4), 298–312.

Léger, C., Heffron, K., Pershad, H. R., Maklashina, E., Luna-Chavez, C., Cecchini, G., Ackrell, B. A., & Armstrong, F. A. (2001). Enzyme electrokinetics: Energetics of succinate oxidation by fumarate reductase and succinate dehydrogenase. *Biochemistry*, *40*(37), 11234–11245.

Li, H., Zhou, H., Zhang, J., Fu, X., Ying, Z., & Liu, X. (2021). Proteinaceous α-amylase inhibitors: Purification, detection methods, types and mechanisms. *International Journal of Food Properties*, *24*(1), 277–290.

Luo, M., Liu, X., Zu, Y., Fu, Y., Zhang, S., Yao, L., & Efferth, T. (2010). *Cajanol*, a novel anticancer agent from pigeonpea (*Cajanus cajan*) roots, induces apoptosis in human breast cancer cells through a ROS-mediated mitochondrial pathway. *Chemico-Biological Interactions*, *188*(1), 151–160.

Luo, Q. F., Sun, L., Si, J. Y., & Chen, D. H. (2008). Hypocholesterolemic effect of stilbenes containing extract fraction from *Cajanus cajan* on diet-induced hypercholesterolemia in mice. *Phytomedicine*, *15*(11), 932–939.

Macfarlane, G. T., & Macfarlane, S. (2011). Fermentation in the human large intestine: Its physiologic consequences and the potential contribution of prebiotics. *Journal of Clinical Gastroenterology*, *45*, S120–S127.

Macy, J. M., Ljungdahl, L. G., & Gottschalk, G. (1978). Pathway of succinate and propionate formation in *Bacteroides fragilis*. *Journal of Bacteriology*, *134*(1), 84–91.

Mandal, S., & Mandal, R. K. (2000). Seed storage proteins and approaches for improvement of their nutritional quality by genetic engineering. *Current Science*, *79*(5), 576–589.

Mane, V. P., Patil, S. S., Syed, A. A., & Baig, M. (2007). Bioconversion of low quality lignocellulosic agricultural waste into edible protein by *Pleurotus sajor-caju* (Fr.) singer. *Journal of Zhejiang University Science*, *B8*(10), 745–751.

Maneechai, S., & Rinthong, P. (2018). Total phenolic content and tyrosinase inhibitory potential of extracts from *Cajanus cajan*. *Pharmacognosy Journal*, *10*(6), 109–112.

Mapfumo, P., Giller, K. E., Mpepereki, S., & Mafongoya, P. L. (1999). Dinitrogen fixation by pigeonpea of different maturity types on granitic sandy soils in Zimbabwe. *Symbiosis*, *27*, 305–318.

Mashifane, D. C., Chiulele, R. M. & Gwata, E. T. (2024). Diversity in selected grain mineral and protein among pigeonpea landraces. *Applied Sciences*, *14*, 573.

Müntz, K. (1998). Globulins from legume seeds: Structure and function during storage and reactivation. In *Plant proteins from european crops: Food and non-food applications* (pp. 3–12). Springer.

Nimse, S. B., & Pal, D. (2015). Free radicals, natural antioxidants, and their reaction mechanisms. *Royal Society of Chemistry Advances*, *5*(35), 27986–28006.

Nix, A., Paull, C. A., & Colgrave, M. (2015). The flavonoid profile of pigeonpea, *Cajanus cajan*: A review. *SpringerPlus*, *4*(1), 125.

Nwosu, J. N., Ojukwu, M., Ogueke, C. C., Ahaotu, I., & Owuamanam, C. I. (2013). The antinutritional properties and ease of dehulling on the proximate composition of pigeonpea (*Cajanus cajan*) as affected by malting. *International Journal of Life Sciences*, *2*(2), 60–67.

Obizoba, I. C. (1991). Effect of sprouting on the nitrogenous constituents and mineral composition of pigeonpea (*Cajanus cajan*) seeds. *Plant Foods for Human Nutrition*, *41*(1), 21–26.

Oloyo, R. A. (2002). Processing effects on the chemical composition and nutritional potential of the pigeonpea (*Cajanus cajan* L.). *La Rivista Italiana delle Sostanze Grasse*, *79*, 273–276.

Onwuka, G. I. (2006). Soaking, boiling and antinutritional factors in pigeonpeas (*Cajanus cajan*) and cowpeas (*Vigna unguiculata*). *Journal of Food Processing and Preservation*, *30*(5), 616–630.

Pal, D. K., Mishra, P., Sachan, N., & Ghosh, A. K. (2011). Biological activities and medicinal properties of Cajanus cajan (L) Millsp. *Journal of Advanced Pharmaceutical Technology and Research*, *2*(4), 207–214.

Panche, A. N., Diwan, A. D., & Chandra, S. R. (2016). Flavonoids: An overview. *Journal of Nutritional Science*, *5*, 1–15.

Petroski, W., & Minich, D. M. (2020). Is there such a thing as "anti-nutrients"? a narrative review of perceived problematic plant compounds. *Nutrients*, *12*(10), 2929.

Prabhu, R., Altman, E., & Eiteman, M. A. (2012). Lactate and acrylate metabolism by *Megasphaera elsdenii* under batch and steady-state conditions. *Applied and Environmental Microbiology*, *78*(24), 8564–8570.

Pundir, R. P. S., & Singh, R. B. (1985). Biosystematic relationships among Cajanus, Atylosia, and Rhynchosia species and evolution of pigeonpea (Cajanus cajan). *Theoretical and Applied Genetics*, *69*, 531–534.

Raja, R. B., Agasimani, S., Varadharajan, A., & Ram, S. G. (2016). Natural variability and effect of processing techniques on raffinose family oligosaccharides in pigeonpea cultivars. *Legume Research*, *39*(4), 528–532.

Ramakrishna, G., Singh, A., Kaur, P., Yadav, S. S., Sharma, S., & Gaikwad, K. (2022). Genome wide identification and characterization of small heat shock protein gene family in pigeonpea and their expression profiling during abiotic stress conditions. *International Journal of Biological Macromolecules*, *197*, 88–102.

Sandberg, A. S. (2002). In vitro and in vivo degradation of phytate. In N. R. Reddy & S. K. Sathe (Eds.), *Food phytates* (pp. 139–155). CRC Press.

Scalbert, A. (1991). Antimicrobial properties of tannins. *Phytochemistry*, *30*(12), 3875–3883.

Schoffl, F., Prandl, R., & Reindl, A. (1998). Regulation of the heat-shock response. *Plant Physiology*, *117*(4), 1135–1141.

Sekhon, J., Grewal, S. K., Singh, I., & Kaur, J. (2017). Evaluation of nutritional quality and antioxidant potential of pigeonpea genotypes. *Journal of Food Science and Technology*, *54*(11), 3598–3611.

Shewry, P. R., Napier, J. A., & Tatham, A. S. (1995). Seed storage proteins: Structures and biosynthesis. *The Plant Cell*, *7*(7), 945–956.

Shi, L., Arntfield, S. D., & Nickerson, M. (2018). Changes in levels of phytic acid, lectins and oxalates during soaking and cooking of Canadian pulses. *Food Research International*, *107*, 660–668.

Silim, S. N., Coe, R., Omanga, P. A., & Gwata, E. T. (2006). The response of pigeonpea genotypes of different duration types to variation in temperature and photoperiod under field conditions in Kenya. *Journal of Food, Agriculture and Environment*, *4*(1), 209–214.

Singh, U., & Jambunathan, R. (1982). Distribution of seed protein fractions and amino acids in different anatomical parts of chickpea (*Cicer arietinum* L.) and pigeonpea (*Cajanus cajan* L.). *Qualitas Plantarum Plant Foods for Human Nutrition*, *31*(4), 347–354.

Sri Devi, V., Satyanarayana, N., & Rao, K. M. (1999). Induction of heat shock proteins and acquisition of thermotolerance in germinating pigeonpea seeds. *Biologia Plantarum*, *42*, 589–597.

Sugui, F. P., Rasalan, R. E., Tadena, D. A., & Sugui, C. C. (2000). Evaluation of vegetable pigeonpea lines in the Philippines. *International Chickpea and Pigeonpea Newsletter*, *7*, 45–46.

Talari, A., & Shakappa, D. (2018). Role of pigeonpea (*Cajanus cajan*) in human nutrition and health: A review. *Asian Journal of Dairy and Food Research*, *37*(3), 212–220.

Tholozan, J. L., Touzel, J. P., Samain, E., Grivet, J. P., Prensier, G., & Albagnac, G. (1992). *Clostridium neopropionicum* sp. nov., a strict anaerobic bacterium fermenting ethanol to propionate through acrylate pathway. *Archives of Microbiology*, *157*, 249–257.

Ukom, A., Odoabuchi, E., Udofia, P., & Ejinkeonye, U. (2023). Variations in antioxidant activity, physical, chemical and selected antinutritional factors of two pigeonpea (*Cajanus cajan*) varieties as affected by processing methods. *Food Science and Engineering*, *4*(1), 62–74.

van der Maesen, L. J. G. (1980). *India is the native home of pigeonpea* (H. C. D. de Wit, J. C. Arends, G. Boelema, C. T. de Groot, & A. J. M. Leeuwenberg, Eds., pp. 257–262). Agricultural University Wageningen.

van Hylckama, V. J. E., Veiga, P., Zhang, C., Derrien, M., & Zhao, L. (2011). Impact of microbial transformation of food on health – from fermented foods to fermentation in the gastro-intestinal tract. *Current Opinion in Biotechnology*, *22*(2), 211–219.

Varyukhina, S., Freitas, M., Bardin, S., Robillard, E., Tavan, E., Sapin, C., Grill, J. P., & Trugnan, G. (2012). Glycan-modifying bacteria-derived soluble factors from *Bacteroides thetaiotaomicron* and *Lactobacillus casei* inhibit rotavirus infection in human intestinal cells. *Microbes and Infection*, *14*(3), 273–278.

Vogt, J. A., Ishii-Schrade, K. B., Pencharz, P. B., & Wolever, T. M. (2004). L-rhamnose increases serum propionate after long-term supplementation, but lactulose does not raise serum acetate. *The American Journal of Clinical Nutrition*, *80*(5), 1254–1261.

Wang, C., Zhao, J., Zhang, H., Lee, Y. K., Zhai, Q., & Chen, W. (2021). Roles of intestinal bacteroides in human health and diseases. *Critical Reviews in Food Science and Nutrition*, *61*(21), 3518–3536.

Wang, R. F., Cao, W. W., & Cerniglia, C. E. (1996). PCR detection and quantitation of predominant anaerobic bacteria in human and animal fecal samples. *Applied and Environmental Microbiology*, *62*(4), 1242–1247.

Wang, W., Vinocur, B., Shoseyov, O., & Altman, A. (2004). Role of plant heat shock proteins and molecular chaperones in the abiotic stress response. *Trends in Plant Science*, *9*(5), 244–252.

Wei, Z., Qi, X., Li, T., Luo, M., Wang, W., Zu, Y., & Fu, Y. (2015). Application of natural deep eutectic solvents for extraction and determination of phenolics in *Cajanus cajan* leaves by ultra performance liquid chromatography. *Separation and Purification Technology*, *149*, 237–244.

Wu, J., Li, B., Xiao, W., Hu, J., Xie, J., Yuan, J., & Wang, L. (2020). Longistylin A, a natural stilbene isolated from the leaves of *Cajanus cajan*, exhibits significant anti-MRSA activity. *International Journal of Antimicrobial Agents*, *55*(1), 105821.

Wu, N., Kong, Y., Fu, Y., Zu, Y., Yang, Z., Yang, M., Peng, X., & Efferth, T. (2011). In vitro antioxidant properties, DNA damage protective activity, and xanthine oxidase inhibitory effect of cajaninstilbene acid, a stilbene compound derived from pigeonpea (*Cajanus cajan*) leaves. *Journal of Agricultural and Food Chemistry*, *59*(1), 437–443.

Zeb, A. (2020). Concept, mechanism, and applications of phenolic antioxidants in foods. *Journal of Food Biochemistry*, *44*(9), 1–22.

Zhao, J., Li, C., Wang, W., Zhao, C., Luo, M., Mu, F., Fu, Y., Zu, Y., & Yao, M. (2013). *Hypocrea lixii*, novel endophytic fungi producing anticancer agent *cajanol*, isolated from pigeonpea (*Cajanus cajan*). *Journal of Applied Microbiology*, *115*(1), 102–113.

Zu, Y., Liu, X., Fu, Y., Wu, N., Kong, Y., & Wink, M. (2010). Chemical composition of the SFE-CO_2 extracts from *Cajanus cajan* (L.) Huth and their antimicrobial activity in vitro and in vivo. *Phytomedicine*, *17*(14), 1095–1101.

23 Organic Medium–Enclosed Trough System

A Climate-Smart Technology for Enhancing Indigenous Vegetable Productivity and Nutritional Constituents

**Semakaleng Mpai and Ashwell Rungano Ndhlala*
University of Limpopo, Department of Plant Production, Soil Science, and Agricultural Engineering, Green Biotechnologies Research Centre of Excellence, Private Bag X1106, Sovenga, 0727, South Africa
*Corresponding author: semakaleng.mpai@ul.ac.za

23.1 INTRODUCTION

Global warming and drought are negatively affecting crop production and productivity due to their detrimental effects, such as excessive heat stress, water scarcity, greenhouse gas, and carbon dioxide buildup. These effects contribute to climate change, which has recently emerged as a direct or indirect threat for poverty and hunger reduction in sub-Saharan Africa (SSA). Therefore, the effects of climate change caused by global warming contribute to environmental, social, and economic livelihoods in SSA. This is because crop production is the cornerstone for ensuring accessibility of adequate, safe, and nutritious food, consequently contributing to food security through subsistence farming and income generation. Crop production for common grains, such as maize, beans, sorghum, and other African staple crops, used to depend on rain-fed agriculture and predictable warm climate in the tropical areas. Yet today's climate is characterized by increased frequency, duration, and intensity of heat waves and increased daytime temperatures, reduced nighttime cooling, and higher air pollution levels associated with heat events. These changes come at a time when the global human population is estimated to reach 9.8 billion in 2050. Thus, increased pressure on food production lies within the crop production sector. In response to the aforementioned factors, different innovations on cropping systems are being developed to improve crop yields.

Therefore, the present chapter introduces the organic medium–enclosed trough (OMET) growing technique as a tool to mitigate climate change on crop production of some indigenous food crops. OMET is a climate-smart and sustainable crop production method aimed at reducing production inputs, such as labor and irrigation water, to increase yields. In addition, this method has the potential of growing more crops on a relatively small area in comparison to conventional seedbeds. It is designed to allow plant waste to be incorporated in the growth medium to feed the growing seedling, thus reducing fertilizer requirements. Furthermore, the plastic sheet mulch used in the OMET system virtually reduces evaporation and drainage to allow for a lengthy utility of the available water and nutrients by the plant. In cases of heavy rainfall, OMET reduces detrimental effects

DOI: 10.1201/9781003469766-23

by keeping the growth medium's top layer intact (Ferreira, 2013). Its smooth surface deters crawling insects and prevents weed growth, minimizing the need for weeding or herbicides application.

Okra (*Abelmoschus esculentus* (L.) Moench) has gained interest in developing cropping systems that improve their yield and nutrients. Amaranth belongs to the Amaranthaceae family, and it has over 60 different species which differ based on many attributes, including morphological and nutritional aspects. In South Africa, tender and young leaves are cooked as a component of daily meals. Leaves of different *A. cruentus* accessions are rich in dietary fiber, protein, fat, carbohydrates, energy, and ash (Manyelo et al., 2020). In fact, amaranth has sulfur-containing amino acids, which are normally limited in other protein-rich crops, such as grain legumes, thus making amaranth ideal for supporting human dietary needs. Moreover, Mateos-Maces et al. (2020) reported that phytochemical compositions in *A. cruentus* were predominated by total phenolic compounds equipped with phenolic acids (hydroxycinnamic and hydroxybenzoic acid) and flavonoids.

Abelmoschus esculentus (okra), belonging to the Malvaceae family, has been indigenized in many African countries, where it is consumed as a relish in daily meals. The consumption of one cup of okra pods contains dietary fiber (3.2 g), carbohydrate (91.2 g), protein (2 g), and vitamin C (21.1 mg) (Sanchez et al., 2022). These contents contribute 10, 28, 200, and 23% of the daily recommend intake. In addition, okra contains noticeable amounts of amino acids and mineral elements, such as iron, potassium, and calcium (Salameh, 2014). The aim of this chapter is to contribute to a consensus of information on the effect of OMET on some growth, yield, and nutritional components in three amaranth species and okra.

23.2 MATERIALS AND METHODS

23.2.1 OMET Growth Medium Establishment

The OMET system was prepared by demarcating an area of, for example, 200 × 200 m space. An underlying plastic was set underneath at approximately 40 cm height down the surface in such a way that it can cover a 30 cm growing plastic pot filled with the growth medium, as shown in Figure 23.1. Before seedling transplanting, the plastic was carefully closed, and holes were created on top of the plastic with a diameter of approximately 90 mm to allow a seedling intra-spacing of 40 × 40 cm. Growth medium consisting of Hygromix® growing medium, pasteurized (300°C for 45 min) loam soil, and fine sand at a 2:1:1 ratio was used. The non-OMET system was prepared by planting the plants in 30 cm pots with the same spacing but without the use of any plastic (Mopai et al., 2023; Mokgalabone et al., 2023).

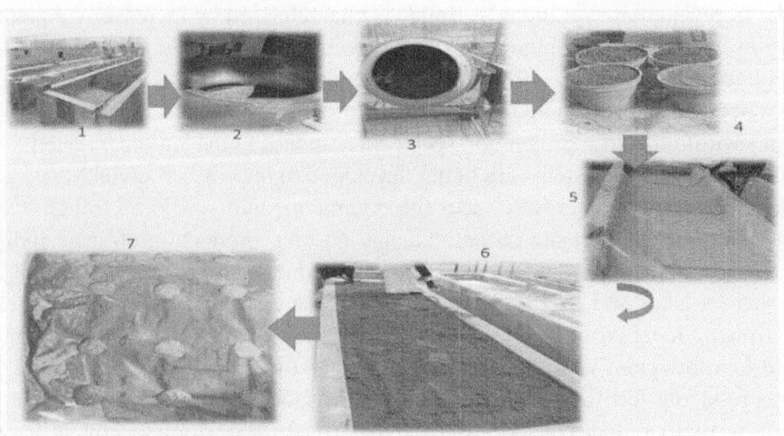

FIGURE 23.1 Step-by-step OMET establishment.

23.3 DATA COLLECTION

23.3.1 Growth and Yield Parameters

Stem diameter and plant height were collected once per week during the growth period. The stem diameter was measured 5 cm above the soil surface using a digital Vernier caliper, while plant height was measured using a tape measurer and/or meter ruler.

23.3.2 Minerals

About 10 g of dried leaf and pod materials was digested in 40 ml of 4% nitric acid (HNO_3), followed by placing the container on a vortex to allow for the complete wetting of the mixture. The materials were magnetically stirred and, thereafter, incubated in a 95°C water bath for 90 min before being allowed to cool down at room temperature, filtered, and decanted into 50 ml tubes which were covered with foil, and then selected nutrient elements were analyzed using the Inductively Coupled Plasma Optical Emission Spectrometry (ICPE-9000). The conditions of the analysis and the development of the mineral standard curve were similar to those reported by Mafokoane et al. (2019).

23.3.3 Amino Acids

The dried leaf and pod samples were prepared for amino acid determination using an Amino Acid Analysis Application Solution (AccQTag) derivatization kit following the methods described in Mpai et al. (2018). The analysis of individual amino acids, including arginine (Arg), serine (Ser), glycine (Gly), aspartate (Asp), glutamate (Glu), alanine (Ala), proline (Pro), histidine (His), threonine (Thr), methionine (Met), lysine (Lys), tyrosine (Tyr), leucine (Leu), phenylalanine (Phe), asparagine (Asn), and glutamine (Gln), was performed using the ultra-performance liquid chromatography analysis following the methods described by Mpai et al. (2018)

23.3.4 Irrigation Water

Irrigation water was calculated based on the sum of the total volume per treatment. Three marked pots were evaluated for moisture status using a moisture tester (Xlus, model: 10.43 × 1.18 × 2.17 in, South Africa), and 250 ml of water was supplied when the moisture status was "dry." Therefore, the frequency of irrigation needed and the irrigation volume constituted the overall irrigation of water over time.

23.4 RESULTS AND DISCUSSION

23.4.1 Effect of OMET on Amaranth and Okra Growth and Yield

From a crop production perspective, growth is measured based on primary objective indices, including leaf length, root length, plant height, stem diameter, and number of leaves per plant (Taia et al., 2021). These growth parameters are often associated with the yield components measured as dry or wet biomass. In fact, these growth attributes play a key role in the absorption and transportation of water and nutrients from the ground to various parts of the plant. Improved leaf length or surfaces are generally associated with improved gaseous exchange and photosynthesis rate, which ultimately improves the plant biomass and yield (Chen et al., 2021). Similarly, thick long stem diameter and roots are associated with effective transportation of water and nutrients from the roots to the leaves, leading to well-nourished plants (Peng et al., 2019).

23.4.1.1 Amaranth

The OMET system improved the growth and yield of three green-pigmented amaranth species, including *Amaranthus tricolor*, *Amaranthus caudatus*, and *Amaranthus cruentus*. The results, illustrated in Figure 23.2a–c, demonstrate that there were significant ($p \leq 0.05$) differences in stem diameter and plant height growth among the three amaranth species cultivated under OMET and non-OMET system. *A. cruentus* stem diameter and plant height grown under OMET and non-OMET system outperformed all other treatments over the growth period (Figure 23.2a).

However, stem diameter and plant height in *A. caudatus* grown under the OMET system were higher than from the non-OMET and *A. tricolor* species grown in both OMET and non-OMET (Figure 23.2b and c).

FIGURE 23.2A Effect of OMET on plant height and stem diameter in *A. cruentus*.

FIGURE 23.2B Effect of OMET on plant height and stem diameter in *A. caudatus*.

FIGURE 23.2C Effect of OMET on plant height and stem diameter in *A. tricolor.*

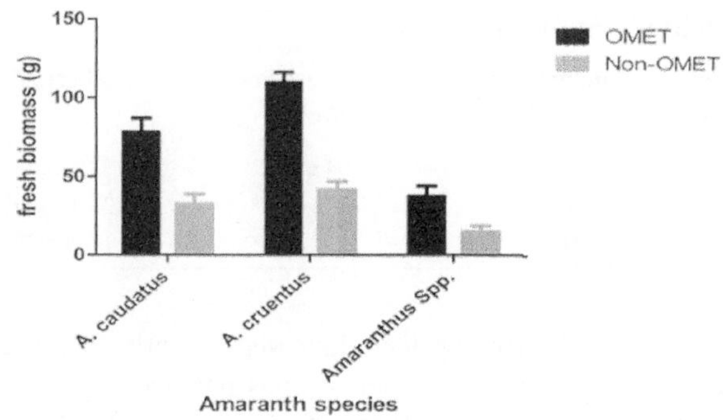

FIGURE 23.3 Effect of OMET on plant biomass of three amaranth species.

When it comes to yield attributes, OMET enhanced fresh biomass in amaranth species compared to those grown under the non-OMET system. The fresh biomass of each amaranth species (*A. caudatus*, *A. cruentus*, and *A. tricolor.*) grown under OMET was twofold higher than the one grown under the non-OMET (Figure 23.3).

The amaranth species growth and yield results may be attributed to species variation and OMET system influence. These results are comparable to those of a previous mulching study. Saeed and Ahmad (2009) reported that mulch is effective for vegetable growth and yield by improving moisture content of soil and heat energy. Furthermore, reasons related to the potential of OMET to conserve water and nutrients, which causes moderate availability of water, resulting in good nutrient dissolution and absorption by the plants, which could also be attributed to the results obtained (Ferrini et al., 2008; Ngala et al., 2019).

23.4.1.2 Okra

On the other hand, OMET-grown okra resulted in a gradual increase in stem diameter from week 3 up to week 8, thus suggesting a twofold thicker stem diameter in OMET than in the control plants. Furthermore, there was a gradual increase in plant height which was twofold taller than the control (Figure 23.4).

FIGURE 23.4 Effect of OMET on plant height and stem diameter in okra.

TABLE 23.1
Effect of OMET on Yield Components of Okra

	Number of Branches	Number of Pods per Plant	Fresh Pod Weight (g)	Pod Length (mm)	Pod Diameter (mm)
OMET	8.00±0.24[a]	16.00±0.40[a]	43.64±2.02[a]	172.43±6.2[a]	27.26±0.8[a]
Non-OMET	4.00±0.18[b]	8.00±0.29[b]	20.30±0.87[b]	130.18±2.8[b]	28.65±0.8[a]
F-statistics	1.231**	0.193**	2.078*	4.569**	1.886 ns

On the other hand, the OMET growing technique significantly increased the yield attributes as follows: number of branches by 50%, number of pods per plant by 49%, fresh pod weight by 53%, and pod length by 68% (Table 23.1). Pod diameter was unaffected by the OMET system. The increase in okra pods grown under OMET may be attributed to the increased number of branches, stem diameter, and plant height.

23.4.2 Effect of OMET on Amaranth and Okra Protein Content

Food insecurity is associated with the consumption of imbalanced diets lacking primary macro- and micronutrients (Wekeza et al., 2022). Macronutrients contain nutrients required in large quantities (greater than 100 mg/dl) for daily body functioning in comparison to their counterparts, that is, micronutrients (Eruvbetine, 2003; Heber et al., 2016; Rosborg & Kozisek, 2016; Drago, 2022). The consumption of any immoderate contents of these nutrients may lead to health undernourishment (Wekeza et al., 2022). Therefore, food security can be improved by the consumption of diets rich in fruits and vegetables, some rich sources of nutrients. Furthermore, macronutrients, such as carbohydrates, amino acids, fatty acids, and organic acids, are involved in growth and development, respiration and photosynthesis, and hormone and protein synthesis in plants (Khan et al., 2020). In this study, OMET has been used to improve some of these nutritional components.

23.4.2.1 Amaranth

The protein constituents of the three amaranth species and okra leaves and pods are shown in Table 23.3. The highest protein content was recorded in OMET-grown *A. cruentus* (28.6%) and *A. caudatus* (21.4%) and non-OMET-grown *A. cruentus* (24.1%). The lowest protein content was

TABLE 23.2
Effect of OMET on Protein Content in Three *Amaranthus* Species and Okra

Species and Treatment		Protein (%)
A. caudatus	OMET	24.1±0.039b
	Non-OMET	21.4±0.044d
A. cruentus	OMET	28.6±0.46a
	Non-OMET	24.1±0.040b
A. tricolor	OMET	22.7±0.036c
	Non-OMET	20.3±0.042d
Okra leaf	OMET	3.1±0.22a
	Non-OMET	2.70±0.72b
Okra pod	OMET	2.4±0.40a
	Non-OMET	2.1±0.01a

TABLE 23.3
Effect of OMET on Macroelemental Nutrients in Three *Amaranthus* Species and Okra Leaf and Pod

Amaranth Species and Treatment		Macroelements (mg/kg DW)			
		Ca	Mg	K	P
A. caudatus	OMET	104±0.043b	77.3±0.032b	251±0.041c	29.7±0.045b
	Non-OMET	104±0.043b	62.6±0.030d	244±0.038d	21.8±0.037c
A. cruentus	OMET	130±0.038a	63.3±0.029d	254±0.046b	34.7±0.043a
	Non-OMET	88.3±0.033d	56.6±0.025e	217±0.033f	22.2±0.039c
A. tricolor	OMET	104±0.039b	82.5±0.038a	276±0.05a	30.3±0.041b
	Non-OMET	90.5±0.041c	66.2±0.032c	226±0.035e	21.5±0.036c
Okra leaf	OMET	237.0±9.91[a]	163.0±7.66[a]	110.0±6.23[a]	6.6±1.21[a]
	Non-OMET	129.0±5.63[b]	110.0±4.92[b]	53.1±4.47[b]	5.9±0.92[b]
Okra pod	OMET	55.6±3.44[a]	40.3±2.87[a]	170.0±8.11[a]	4.8±1.63
	Non-OMET	49.0 ± 3.11[a]	36.6±1.63[b]	156.0±6.80[b]	4.8±1.63

recorded in non-OMET-grown *A. tricolor* (20.3%), although this was not significantly different from that recorded in non-OMET-grown *A. caudatus* (21.4%) (Table 23.2).

23.4.2.2 Okra

Okra grown under OMET contained higher protein levels in comparison to that grown on the non-OMET. Furthermore, the levels of protein in the leaves were more than those detected in the pods (Table 23.2). Thus, the OMET potentially enhances okra dietary value.

23.4.3 Effect of OMET on Three Amaranth Species and Okra Minerals

23.4.3.1 Amaranth

Table 23.3 shows the macroelements (Ca, Mg, K, and P) and microelements (Cu, Mn, Fe, Se, and Zn) in three amaranth species and okra leaves and pods grown under the OMET and non-OMET growing systems. The OMET-grown *A. cruentus* had the highest Ca (130 mg/kg) and P (34.7 mg/kg) contents, whilst the highest Mg (82.5 mg/kg) and K (276 mg/kg) contents were found in the OMET-grown *A. tricolor*. Non-OMET-grown *A. cruentus* recorded the lowest concentrations of Ca

(88.3 mg/kg), Mg (56.6 mg/kg), and K (217 mg/kg) in comparison to all other studied amaranth species grown under both OMET and non-OMET. The same trend was observed in OMET-grown *A. tricolor*, which had the highest concentration of P (30.3 mg/kg). In contrast, the non-OMET-grown *A. tricolor* had the lowest concentration of P (21.4 mg/kg) among the other species under both treatments. There was no significant difference in Ca (104 mg/kg DW) reported in *A. caudatus* in both treatments and OMET-grown *A. tricolor* (104 mg/kg).

The following microelements – Cu, Mn, Fe, Se, and Zn – were quantified (Table 23.3). It was revealed that the OMET-grown *A. cruentus* had the highest concentration of Cu (1.04 mg/kg), Se (8.13 mg/kg), and Zn (1.66 mg/kg), whereas the *A. tricolor* grown under the OMET system was found to contain the highest concentration of Mn (2.16 mg/kg) and Fe (3.41 mg/kg) among the rest of the species grown under the same conditions. The OMET-grown *A. caudatus* contained the second highest concentration of Fe (2.62 mg/kg) and Se (8.03 mg/kg). The non-OMET-grown *A. tricolor* measured the lowest Cu (0.78 mg/kg DW), Se (6.36 mg/kg DW), and Zn (0.95 mg/kg DW) among all the other amaranth species grown under both treatments. There was no significant difference in Fe measured in all amaranth species grown under both treatments, except for OMET-grown *A. tricolor*, which had an outstanding Fe concentration (3.41 mg/kg DW). The highest Fe concentration of *A. tricolor* (3.41 mg/kg DW) was higher than the Fe concentration reported by Yahaya et al. (2012) in *A. caudatus* (1.06 mg/kg DW), Roselle (1.2 mg/kg DW), and Kenaf (1.17 mg/kg DW). The present results are similar to those reported by Maseko et al. (2019), who stated that severe drought conditions, characterized by a 30% reduction in water application increased Ca and Mg in *Amaranthus cruentus* species. The same study reported an increase in the contents of Zn, Na, P, and K when water availability was reduced to 60%. According to Kanda et al. (2020), moderate deficit irrigation on cowpea increased the carbohydrate and fiber contents due to water availability around the root surface. Furthermore, minerals are essential for plant growth, development, reproduction, and seed quality. Mineral deficiency and poor mineral uptake and transport due to abiotic stress, such as drought, especially at the reproductive stage, result in yield loss and poor seed quality (Bellaloui et al., 2012). Lowered absorption of the minerals could be due to reduced transpiration flow, limited availability of energy for assimilation, and interference in the unloading mechanism (Farooq et al., 2009; Rouphael et al., 2012). The OMET system has successfully increased the concentration of the mineral elements for amaranth species in comparison to the non-OMET system. This could probably be due to its potential to eliminate leaching of nutrients, conserving and reserving them for uptake by the plant roots. The less to non-fluctuating temperatures of the OMET growing condition was conducive for efficient nutrient absorption. The root exudates become useful for the plant since they are not leached. The non-OMET system, on the other hand, had little influence on the accumulation of mineral elements, probably due to its inability to conserve nutrients. The nutrients are leached during irrigation through a deep drainage or seepage, and since there is no underlying plastic to inhibit drainage, the nutrients become out of reach for plant root absorption. The mineral accumulation is also influenced by the degree of soil moisture stress. Both macro- and microelements were found to be higher in amaranth species grown under organic mulch with enhanced technology (OMET) and subjected to the highest soil moisture levels compared to amaranth species not grown under. It is therefore plausible that optimum moisture content positively enhances elemental composition in OMET-grown *Amaranthus* species.

23.4.3.2 Okra

The results of the present study showed that elemental nutrients, including calcium (Ca), potassium (K), magnesium (Mg), and phosphorus (P), were predominant attributes for the nutritional status of okra leaves and pods. The highest contents of Ca (237 mg/kg), Mg (163 mg/kg), K (110 mg/kg), and P (6.6 mg/kg) were recorded in OMET-grown leaves, while the lowest contents for Ca (49 mg/kg), Mg (36.6 mg/kg), and P (4.8 mg/kg) were recorded in okra pods from the non-OMET system, except for K (110 mg/kg), which recorded the lowest content in the non-OMET leaves (Table 23.4) – thus suggesting the positive impact of OMET in enhancing macronutrient contents. These macronutrients

TABLE 23.4
Effect of OMET on Microelemental Nutrients in Three *Amaranthus* Species and Okra Leaf and Pod

		Microelements (mg/kg DW)				
		Cu	Mn	Fe	Se	Zn
A. caudatus	OMET	0.89±0.023c	1.73±0.046b	2.62±0.05b	8.03±0.049a	1.11±0.024c
	Non-OMET	0.84±0.021d	1.46±0.043d	2.5±0.050c	7.01±0.048d	1.09±0.036c
A. cruentus	OMET	1.04±0.034a	1.78±0.048b	2.52±0.04c	8.13±0.050a	1.66±0.048a
	Non-OMET	0.89±0.023c	1.17±0.041e	2.51±0.04c	7.69±0.05b	1.28±0.045b
A. tricolor	OMET	0.93±0.025b	2.16±0.05a	3.41±0.39a	7.28±0.05c	1.27±0.047b
	Non-OMET	0.78±0.022e	1.64±0.04c	2.47±0.05c	6.36±0.048e	0.95±0.039d
Okra leaf	OMET	–	2.4±0.41[a]	3.8±0.80[a]	12.8±2.73[a]	6.9±1.32[a]
	Non-OMET	–	1.9±0.11[a]	6.1±1.31[b]	11.8±2.11[a]	6.7±1.17[a]
Okra pod	OMET	–	1.0±0.22[a]	1.5±0.34[a]	12.8±2.73[a]	2.9±0.70[a]
	Non-OMET	–	1.0±0.22[a]	1.6±0.33[a]	11.8±2.11[a]	2.6±0.52[a]

(calcium, potassium, and magnesium) were found in okra, and they potentially contributed 8, 9, and 14%, respectively, of the daily required nutritional value (Maseko et al., 2019).

The results indicated that OMET-grown pods and leaves exhibited comparable and high contents of Se (12.2 and 12.8 mg/kg) and Na (6.2 and 6.8 mg/kg). These levels were slightly higher than the selenium found in OMET leaves (11.8 mg/kg) and approximately twofold higher than the Na content (3.7 mg/kg) found in non-OMET leaves. OMET-grown leaves maintained higher contents of Zn, Fe, and Mn and were up to threefold, sixfold, and twofold, respectively, higher than the mean values reported in non-OMET pods, as shown in Table 23.4. These reports document the impact of water availability on the availability of okra nutrients. On the other hand, mulching using plant materials from *Pueraria phaseoloides*, *Mucuna pruriens*, *Pennisetum purpureum*, and *Panicum maximum* increased N, P, K, Ca, and Mg nutrient contents in okra leaves (Adekiya et al., 2017). Therefore, different cropping systems affected the nutritional compositions. According to Fawibe et al. (2022), plastic mulch under the *Elaeis guineensis* canopy reduced the ash, crude protein, and carbohydrates as compared to the canopy without mulching. In addition, different mulching materials, including *Pueraria phaseoloides*, *Mucuna pruriens*, *Pennisetum purpureum*, and *Panicum maximum*, impacted the mineral element compositions such that nitrogen, phosphorus, potassium, calcium, and magnesium were increased in comparison to the no-mulch treatment (Fawibe et al., 2022). In the current study, the OMET growing technique maintained a higher contents of minerals than the non-OMET system. Under non-OMET conditions, samples were often under dry conditions caused by aeration, water seepage, and trans-evaporation, which often lead to moderate drought conditions (Sehgal et al., 2018). These drought conditions were reported to reduce elemental nutrients, such as protein, Zn, Fe, P, and N (Sehgal et al., 2018). Furthermore, it is plausible that consumption of approximately 50 g of okra pods and 25 g of leaves grown under OMET and cultivated in micro-plot conditions can ensure 1,000 mg for adults' adequate intake of Ca according to the National Academy of Medicine (2016).

23.4.4 Effect of OMET on Amino Acids in Three Amaranth Species and Okra

23.4.4.1 Amaranth

A total of 16 free amino acids was quantified in the three amaranth species grown under the OMET and non-OMET system (Table 23.5). The results revealed that amaranth species contained both essential and non-essential amino acids. There was a significant difference ($p \leq 0.05$) between amino acid concentration among the same Amaranth species grown under the OMET and non-OMET system. Essential amino acids threonine, lysine, valine, isoleucine, leucine, and phenylalanine were detected in all amaranth species grown under both the OMET and non-OMET system. The levels of phenylalanine in OMET-grown *A. cruentus* (2.54 mg/kg) were higher than the amount reported in the leaves of *A. cruentus* (0.66 mg/kg) (Manyelo et al., 2020). The OMET-grown *A. cruentus* also showed a slightly higher amount of leucine (2.21 mg/kg) than the amount reported in *A. cruentus* (1.55 mg/kg) (Manyelo et al., 2020). The highest concentration of valine (1.45 mg/kg) detected in OMET-grown *A. cruentus* was shown to be lower than the amount reported in *A. cruentus* (1.51 mg/kg) (Manyelo et al., 2020). Moreover, the highest quantity of isoleucine (1.29 mg/kg) and threonine (1.15 mg/kg) detected in OMET-grown *A. cruentus* showed to be both higher than the amount reported in *A. cruentus* leaves isoleucine (0.83 mg/kg) and threonine (0.85 mg/kg) (Manyelo et al., 2020). The lysine concentration was 2.37 and 2.25 mg/kg in OMET-grown *A. cruentus* and *A. caudatus*, respectively. This was higher than the concentration reported in the leaves of *A. cruentus* (1.73 mg/kg) (Manyelo et al., 2020).

Non-essential amino acids serine, arginine, glycine, aspartate, and glutamate were detected in three amaranth species. Glycine, aspartate, and glutamate were identified as the predominant non-essential amino acids. The highest concentration of glycine in non-OMET-grown *A. cruentus* (1.46 mg/kg) was much higher than the amount reported by Manyelo et al. (2020) in *A. cruentus* (0.94 mg/kg). Aspartate content was highest in non-OMET-grown *A. cruentus* (2.02 mg/kg). Non-OMET-grown *A. cruentus* was predominated by glutamate (2.55 mg/kg). Serine was highest in non-OMET-grown *A. cruentus* (1.16 mg/kg) but higher than the amount reported in *A. cruentus* (0.90 mg/kg) by Manyelo et al. (2020). Moreover, the lowest concentration of arginine (0.53 mg/kg) was detected in non-OMET-grown *A. cruentus*, but lower than the concentration reported in early harvested *A. cruentus* (0.90 mg/kg) grown in a greenhouse condition (Manyelo et al., 2020).

According to the carbon–nitrogen theory (Kazimierczak et al., 2015), nitrogen fertilizer enhances the accumulation of nitrogen containing metabolites, such as amino acids and protein, while reducing the production of compounds such as polyphenols (phenols and flavonoids) (Coley

TABLE 23.5
Essential and Non-Essential Amino Acids (mg/kg) Found in Three Amaranth Species Grown Both under OMET and Non-OMET Systems

		Essential Amino Acids					
		Threonine	Valine	Isoleucine	Lysine	Leucine	Phenylalanine
A. caudatus	OMET	0.92±0.024b	1.09±0.043c	0.97±0.031b	2.25±0.05a	1.66±0.044b	2.36±0.05a
	non-OMET	0.66±0.025d	0.84±0.027d	0.76±0.026c	1.3±0.034b	1.25±0.038c	1.14±0.035c
A. cruentus	OMET	1.15±0.034a	1.45±0.041a	1.29±0.027a	2.37±0.05a	2.21±0.042a	2.54±0.051a
	non-OMET	0.93±0.036b	1.22±0.042b	1.12±0.039a	2.19±0.050a	1.92±0.051b	1.95±0.043b
A. tricolor	OMET	0.88±0.021c	1.08±0.034c	0.98±0.028b	1.62±0.036b	1.66±0.041b	1.47±0.038c
	non-OMET	0.43±0.021e	0.52±0.026d	0.46±0.022d	0.8±0.012c	0.74±0.024d	0.71±0.023d

		Non-Essential Amino Acids				
		Arginine	Serine	Glycine	Aspartate	Glutamate
A. caudatus	OMET	1.12±0.047b	0.97±0.034b	1.37±0.05b	1.2±0.041c	1.52±0.05c
	non-OMET	0.86±0.031c	0.7±0.030c	0.95±0.041c	0.96±0.042c	1.29±0.049c
A. cruentus	OMET	1.37±0.048a	0.98±0.036b	1.33±0.044b	1.33±0.044b	1.75±0.05b
	non-OMET	1.54±0.05a	1.16±0.045a	1.46±0.048a	2.02±0.05a	2.55±0.05a
A. tricolor	OMET	1.12±0.049b	0.91±0.042b	1.21±0.047b	1.53±0.049b	1.95±0.05b
	non-OMET	0.53±0.028d	0.46±0.022c	0.59±0.029d	0.6±0.030d	0.73±0.032d

et al., 1985). On this basis, the OMET-grown amaranth had fewer bioactive compounds and higher contents of amino acids and proteins. Minor drought stress is associated with major causes for crop loss and poor performance. In this case, amaranth species grown under the non-OMET system created a localized minor drought, leading to a significant reduction in photosynthesis (da Silva Ferreira et al., 2017).

23.4.4.2 Okra

The effect of the OMET growth technique significantly ($p \leq 0.05$) affected okra nutritional levels. A total of 16 amino acid compounds were quantified in okra pods and leaves grown under the OMET and non-OMET growing techniques. The essential amino acid compounds included: Arg, Ser, Gly, Asp, Glu, Ala, and Pro and are illustrated in Figure 23.5. On the other hand, non-essential amino acids, such as His, Thr, Met, Lys, Tyr, Leu, Phe, Asn, and Gln, were quantified, as shown in Figure 22.6. Quantitatively, all essential amino acid compounds were 79 (pods) and 74% (leaf) higher than non-essential amino acids in samples grown under OMET and non-OMET conditions, respectively.

Glutamine (Glu) and aspartate (Asp) amino acids were the most predominant compounds in both pods and leaves when grown under OMET and non-OMET conditions. The Glu contents ranged between 1.00 and 1.11 mg/kg in the pods and between 2.06 and 2.59 mg/kg in the leaves, respectively, in non-OMET and OMET systems, while Asp content ranged between 3.64 and 4.04 mg/kg and 2.38

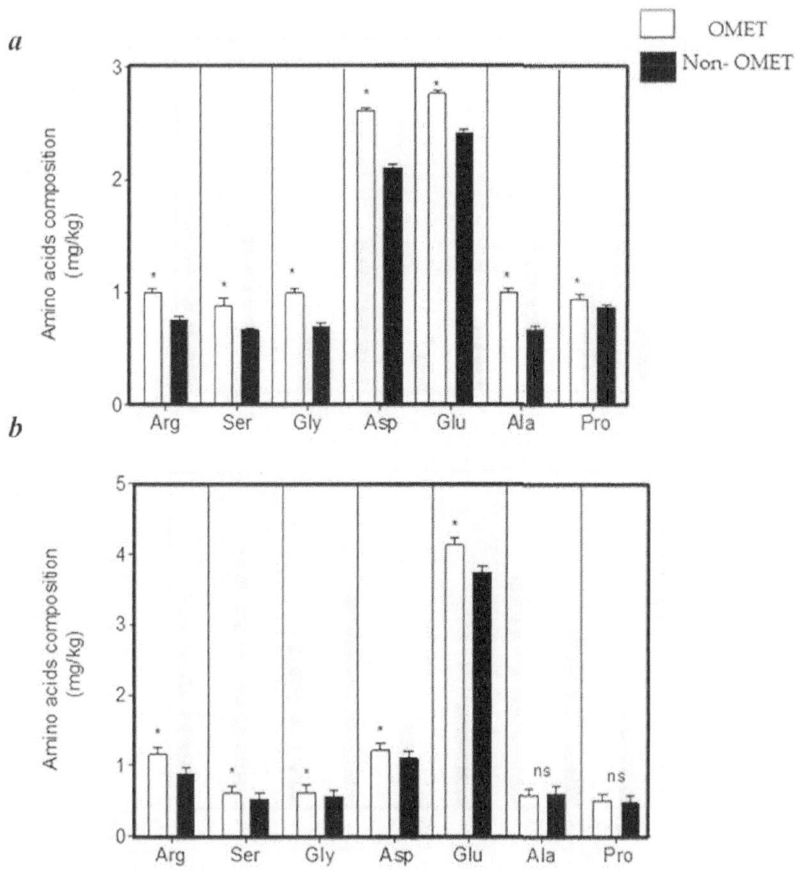

FIGURE 23.5 Total contents of essential amino acids in (a) okra leaves and (b) pods, respectively. Significantly different at * $p \leq 0.05$, while ns = not significant. Mean separation was conducted using a t-test. *Arg* = arginine, *Ser* = serine, *Gly* = glycine, *Asp* = aspartate, Glu-glutamate, Ala = Alanine and Pro = Proline

and 2.73 mg/kg, respectively. Overall, the trend of the essential amino acids in okra pods grown under the OMET growing technique exhibited higher contents of Arg (1.06 mg/kg), Ser (0.51 mg/kg), Gly (0.52 mg/kg), Asp (1.11 mg/kg), and Glu (4.04 mg/kg) than those grown under the non-OMET growing technique in both okra pods and leaves (Figure 23.3). However, a non-significant effect between OMET and the non-OMET growing technique was observed for Ala and Pro amino acids in okra pods.

All the quantified non-essential amino acids, which included Arg, Ser, Gly, Glu, Ala, Pro, and Asp, were higher in OMET-grown leaf samples, as shown in Figure 23.3a and b. On the other hand, essential amino acids, including His, Lys, Met, Leu, Tyr, and Phe, were higher in okra pods grown under OMET. Yet Thr (0.38 mg/kg), Val (0.49 mg/kg), and ILe (0.36 mg/kg) were unaffected by the OMET growing technique. Okra leaves grown under the OMET growing technique exhibited higher contents of His, Tyr, Lys, Thr, Met, Val, ILe, Leu, and Phe in comparison to non-OMET. In this study, the growth of okra under the OMET growth technique affected the nutritional compositions based on protein basic units: amino acids. It was interesting to observe that Glu and Asp were predominant amino acids in okra, which were confirmed by the study of Sami et al. (2013). Glu and Asp were found to predominate soybean genotypes, which had the highest preferred sensory score concerning good taste (Kumar et al., 2011; Carrera et al., 2011).

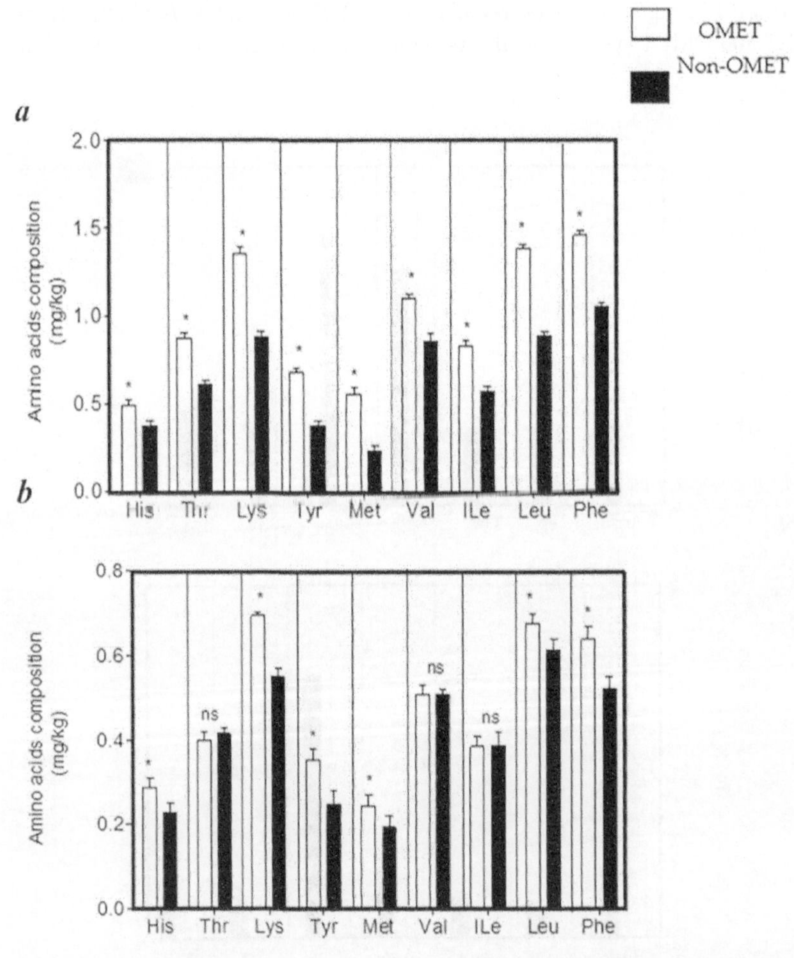

FIGURE 23.6 Non-essential amino acids in okra (a) leaves and (b) pods, respectively. Significantly different at * $p \leq 0.05$, while ns = not significant. Mean separation was conducted using a t-test. *His* = histidine, *Thr* = threonine, *Lys* = lysine, *Tyr* = tryptophan, *Met* = methionine, *Val* = valine, *Leu* = leucine, *Phe* = phenylalanine.

In addition, the OMET growth technique directly brought changes to primary factors, such as soil temperature, light, water availability, and evapotranspiration, which affected amino acid compositions. Amino acid accumulation was directly proportional to increasing environmental temperatures (El-Kereamy et al., 2012; Viana et al., 2022). The study of Viana et al. (2022) revealed that Asp increased linearly with temperature, and this agrees with our results. Amino acids, including Ile, Met, Phe, Ala, Glu, and Pro, increased with the temperature up to a certain threshold, which, when surpassed, led to a drop in their contents (El-Kereamy et al., 2012). For example, Pro showed a decrease at 22°C, followed by Phe at 21°C, Glu at 20°C, Met at 19°C, and Ile at 16°C (El-Kereamy et al., 2012). This evidence encompasses findings from this study, showing that the OMET growing technique consistently maintains higher amino acid levels compared to other methods. The OMET system could have adjusted the root surface temperature to be more favorable compared to the non-OMET conditions. Consequently, under the OMET setup, there could have been reductions in trans-evaporation, irrigation intervals, and water seepage while also providing a shade for the rooting zone.

On the other hand, amino acids are defense mechanisms against abiotic conditions (Khan et al., 2020). Pro is well-known for its potential to regulate osmotic adjustment substances in response to cold resistance (Khan et al., 2020). However, the content of Pro was moderately upregulated in the OMET growth technique and thus suggested the optimum growth temperature for okra production. This was further authenticated by the higher contents of Glu, a precursor for Pro, which becomes depleted in cold environments. Under limited water availability conditions, the report of Khan et al. (2020) demonstrated the translocation and exudation of amino acids such as Pro, Ser, and Ala from the shoot down to the roots, which likely affected the soil's water presence.

23.4.5 Effect of OMET on Irrigation Water

The OMET growing technique significantly and positively affected the amount of irrigation water, as shown by an increase in water used to irrigate the non-OMET-grown okra ($p \leq 0.05$). The cumulative irrigation water increased throughout the weeks, as shown in Figure 23.6, demonstrating that the OMET system improved water retention. Hence, it enhanced growth and yield in the disposal of lower irrigation water. The results may be attributed to the fact that the OMET growing technique was practiced under an enclosed trough which eliminated water seepage, thus advocating for climate-smart agriculture. The present results are similar to those of Ferreira (2013), who reported improved Swiss chard growth and yield under OMET and decreased irrigation. The top plastic of the OMET system reduced the evaporation rate.

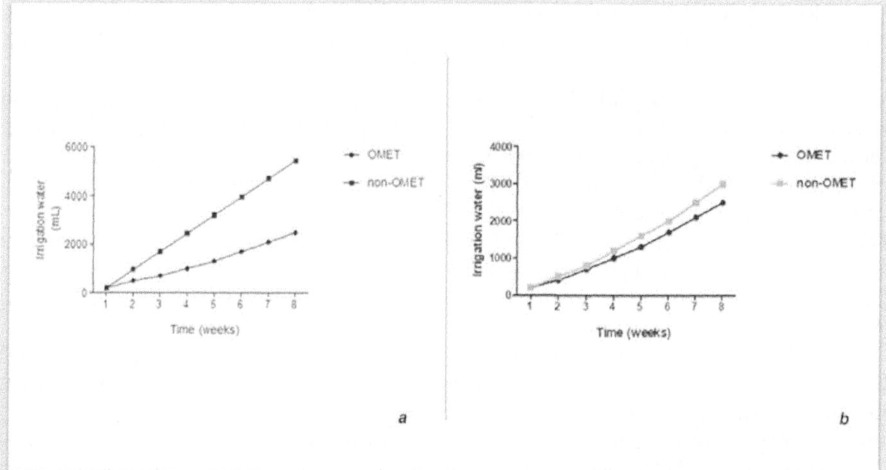

FIGURE 23.7 Effect of OMET on irrigation water on (a) *Amaranthus* species and (b) okra.

23.5 CONCLUSION AND PROSPECTS

Climate change lowers crop yields and nutritional quality due to drought, heat waves, and flooding, as well as increased pest and plant disease incidences, leading to high rates of malnutrition and food insecurity. The good news from this chapter is that OMET has shown the potential to mitigate some of these challenges. In this study, the use of OMET improved the growth and yield attribute of both amaranth species and okra. These results were in concomitant to increased protein, some minerals, and amino acids content in OMET as compared to the non-OMET. This was achieved by using the lowest level of water than in non-OMET conditions. Since OMET techniques advocate for organic farming, and given the rise in trend of preference for organically produced food, OMET has the potential to be recognized by policymakers around the production of organic foods. Furthermore, these results could lead to improved cultivation of amaranth species at small scale and, at a later stage, the commercial level. Although there are some commercializations of fresh okra products from local markets, the outcome of this study will close a gap of nationalizing its availability throughout the year. With the results, an OMET growth condition is recommended for small-scale farming and commercial level to produce amaranth and okra crops with enhanced phytonutrients. Furthermore, investigations in phytochemical compositions for the development of nutraceutical products are essential. The inclusion of the amaranth species and okra in the food database could be a valuable source of nutrition as well supplementary income for the South African population, more so the marginalized and resource-poor.

23.6 ACKNOWLEDGMENTS

Authors would like to thank the National Research Foundation (NRF, Pretoria) (grant number 12612) and the Department of Science and Innovations (DSI) of the South African Government (grant number DSI/CON C2235/2021) for funding the projects.

REFERENCES

Adekiya, A. O., Agbede, T. M., Aboyeji, C. M., & Dunsin, O. (2017). Response of okra (*Abelmoschus esculentus* (L.) Moench) and soil properties to different mulch materials in different cropping seasons. *Scientia Horticulturae*, *217*, 209–216.

Bellaloui, N., Mengistu, A., Fisher, D. K., & Abel, C. A. (2012). Soybean seed composition constituents as affected by drought and phomopsis in phomopsis susceptible and resistant genotypes. *Journal of Crop Improvement*, *26*(3), 428–453.

Carrera, C. S., Reynoso, C. M., Funes, G. J., Martínez, M. J., Dardanelli, J., & Resnik, S. L. (2011). Amino acid composition of soybean seeds as affected by climatic variables. *Pesquisa Agropecuária Brasileira*, *46*, 1579–1587.

Chen, G., Yang, K., Zhang, J., Wang, M., Wang, L., & Xian, J. (2021). Effects of succession stages and altitudinal gradient on leaf surface area and biomass allocation of typical plants in the subalpine of Eastern Tibetan Plateau. *Global Ecology and Conservation*, *27*, e01590. https://doi.org/10.1016/j.gecco.2021.e01590

Coley, P. D., Bryant, J. P., & Chapin, F. S., III. (1985). Resource availability and plant antiherbivore defense. *Science*, *230*(4728), 895–899.

da Silva Ferreira, V., & Sant'Anna, C. (2017). Impact of culture conditions on the chlorophyll content of microalgae for biotechnological applications. *World Journal of Microbiology and Biotechnology*, *33*(1), 20.

Drago, S. R. (2022). Minerals. In *Nutraceutical and functional food components* (pp. 153–181). Academic Press.

El-Kereamy, A., Bi, Y. M., Ranathunge, K., Beatty, P. H., Good, A. G., & Rothstein, S. J. (2012). The rice R2R3-MYB transcription factor OsMYB55 is involved in the tolerance to high temperature and modulates amino acid metabolism. *PLoS One*, *7*(12), e52030.

Eruvbetine, D. (2003). *Canine nutrition and health: A paper presented at the seminar organized by Kensington Pharmaceuticals Nig. Ltd.* Lagos.

Farooq, M., Wahid, A., Kobayashi, N. S. M. A., Fujita, D. B. S. M. A., & Basra, S. M. (2009). Plant drought stress: Effects, mechanisms and management. *Sustainable Agriculture*, 153–188.

Fawibe, O. O., Bankole, A. O., Mustafa, A. A., & Ewajane, O. N. (2022). Agronomic performance and nutritional content of Okra [Abelmoschus esculentus (L.) Moench] as influenced by plastic-film mulch under Elaeis guineensis (Jacq.) canopy. *Nigeria Agricultural Journal*, *53*(1), 212–222.

Ferreira, J. (2013, September 3). OMET farming better than hydroponics? *Farmers Weekly Publication*. Retrieved from www.farmersweekly.co.za/crops/field-crops/omet-farming-better-than-hydroponics/

Ferrini, F., Fini, A., Frangi, P., & Amoroso, G. (2008). Mulching of ornamental trees: Effects on growth and physiology. *Arboric Urban Forum*, *34*, 157.

Heber, D., Henning, S. M., & Li, Z. (2016). Energy balance, epigenetics, and prostate cancer. In *Epigenetics, Energy Balance, and Cancer* (pp. 235–250). Springer.

Kanda, E. K., Senzanje, A., Mabhaudhi, T., & Mubanga, S. C. (2020). Nutritional yield and nutritional water productivity of cowpea (Vigna unguiculata L. Walp) under varying irrigation water regimes. *Water SA*, *46*(3), 410–418.

Kazimierczak, R., Hallmann, E., & Rembiałkowska, E. (2015). Effects of organic and conventional production systems on the content of bioactive substances in four species of medicinal plants. *Biological Agriculture & Horticulture*, *31*(2), 118–127.

Khan, N., Ali, S., Zandi, P., Mehmood, A., Ullah, S., Ikram, M., & Babar, M. A. (2020). Role of sugars, amino acids and organic acids in improving plant abiotic stress tolerance. *Pakistan Journal of Botany*, *52*(2), 355–363.

Kumar, V., Rani, A., Goyal, L., Pratap, D., Billore, S. D., & Chauhan, G. S. (2011). Evaluation of vegetable-type soybean for sucrose, taste-related amino acids, and isoflavones contents. *International Journal of Food Properties*, *14*(5), 1142–1151.

Mafokoane, A. M., Mphosi, M. S., & Shadung, K. G. (2019). Effect of time-based oven-drying on the essential mineral elements of cowpea (*Vigna unguiculata*) leaves. *Research on Crops*, *20*(1), 50–54.

Manyelo, T. G., Sebola, N. A., & Mabelebele, M. (2020). Nutritional and phenolic profile of early and late harvested amaranth leaves grown under cultivated conditions. *Agriculture*, *10*(10), 432.

Maseko, I., Ncube, B., Mabhaudhi, T., Tesfay, S., Chimonyo, V. G. P., Araya, H. T., & Du Plooy, C. P. (2019). Nutritional quality of selected African leafy vegetables cultivated under varying water regimes and different harvests. *South African Journal of Botany*, *126*, 78–84.

Mateos-Maces, L., Chávez-Servia, J. L., Vera-Guzmán, A. M., Aquino-Bolaños, E. N., Alba-Jiménez, J. E., & Villagómez-González, B. B. (2020). Edible leafy plants from Mexico as sources of antioxidant compounds, and their nutritional, nutraceutical and antimicrobial potential: A review. *MDPI*, *9*(6), 41.

Mokgalabone, T. T., Mpai, S., & Ndhlala, A. R. (2023). Organic medium enclosed trough growing technique improves *Abelmoschus esculentus* (Okra) growth, yield and some nutritional components. *Applied Sciences*, *13*(9), 5645.

Mopai, M. G., Mpai, S., & Ndhlala, A. R. (2023). Practices of organic medium enclosed trough and amaranth species variation enhanced growth, nutritional composition and bioactive compounds. *Applied Sciences*, *13*(23), 12574.

Mpai, S., Du Preez, R., Sultanbawa, Y., & Sivakumar, D. (2018). Phytochemicals and nutritional composition in accessions of Kei-apple (*Dovyalis caffra*): Southern African indigenous fruit. *Food Chemistry*, *253*, 37–45.

National Academy of Medicine. (2016). Retrieved January 31, 2023, from https://nam.edu/event/nam-annual-meeting-2016-scientific-program/

Ngala, J. M., Ndiso, J. B., & Mundi, E. M. (2019). Effects of selected organic mulches on growth and yield of Amaranth in Kilifi country. *International Journal of Agriculture, Environment and Bioresearch*, *4*(6).

Peng, H., Yan, Z., Chen, Y., Zhao, X., & Han, W. (2019). Effects of body size and root to shoot ratio on foliar nutrient resorption efficiency in *Amaranthus* mangostanus. *American Journal of Botany*, *106*(3), 363–370. https://doi.org/10.1002/ajb2.1246

Rosborg, I., & Kozisek, F. (2016). *Drinking water minerals and mineral balance*. Springer International Publication.

Rouphael, Y., Cardarelli, M., Schwarz, D., Franken, P., & Colla, G. (2012). Effects of drought on nutrient uptake and assimilation in vegetable crops. In *Plant responses to drought stress: From morphological to molecular features* (pp. 171–195). Springer.

Saeed, R., & Ahmad, R. (2009). Vegetative growth and yield of tomato as affected by the application of organic mulch and gypsum under saline rhizosphere. *Pakistan Journal of Botany*, *41*(6), 3093–3105.

Salameh, N. M. (2014). Genetic diversity of okra (*Abelmoschus esculentus* L.) landraces from different agro-ecological regions revealed by AFLP analysis. *American-Eurasian Journal of Agricultural & Environmental Sciences*, *14*(2), 155–160.

Sami, R., Lianzhou, J., Yang, L., Ma, Y., & Jing, J. (2013). Evaluation of fatty acid and amino acid compositions in okra (Abelmoschus esculentus) grown in different geographical locations. *BioMed Research International*, *2013*.

Sanchez, D. T., Peconcillo, L. B., Jr., & De Vera, J. V. (2022). Effects of natural and home-made bio-inoculants on containerized okra plants (Abelmoschus Esculentus). *Current Agriculture Research Journal*, *10*(1). (Review. *Antioxidants*, *9*(6), 541).

Sehgal, A., Sita, K., Siddique, K. H., Kumar, R., Bhogireddy, S., Varshney, R. K., & Nayyar, H. (2018). Drought or/and heat-stress effects on seed filling in food crops: Impacts on functional biochemistry, seed yields, and nutritional quality. *Frontiers in Plant Science*, *9*, 388913.

Taia, W. K., Shehata, A. A., Ibrahim, M. M., & El-shamy, I. M. (2021). Vegetative morphological variations within some Egyptian. *Jordan Journal of Biological Sciences*, *14*(1), 137–146.

Viana, V. E., Aranha, B. C., Busanello, C., Maltzahn, L. E., Panozzo, L. E., de Oliveira, A. C., & Pegoraro, C. (2022). Metabolic profile of canola (Brassica napus L.) seedlings under hydric, osmotic and temperature stresses. *Plant Stress*, *3*, 100059.

Wekeza, S. V., Sibanda, M., & Nhundu, K. (2022). Prospects for organic farming in coping with climate change and enhancing food security in Southern Africa: A systematic literature review. *Sustainability*, *14*(20), 13489.

Yahaya, Y., Birnin-Yauri, U. A., Bagudo, B. U., & Noma, S. S. (2012). Quantification of macro and micro elements in selected green vegetables and their soils from Aliero agricultural fields in Aliero, Kebbi State, Nigeria. *Journal of Soil Science and Environmental Management*, *3*(8), 207–215.

24 A Survey of Goitrogenic Compounds in Selected Millets and Cruciferous Vegetables

Nutritional Downside

[1]Eastonce Tendayi Gwata, [2]Oluwatoyin Oladayo Onipe, [1]Odeshnee Naicker, [1]Tsakani Maria Maphosa, and [1]Dipoo Charity Mashifane*

[1]University of Venda, Faculty of Science, Engineering, and Agriculture, Department of Plant and Soil Sciences, Private Bag X5050, Thohoyandou, 0950, South Africa

[2]University of Johannesburg, Faculty of Science, Department of Biotechnology and Food Technology, Doornforntein 2028, South Africa

*Corresponding author: ectgwata@gmail.com

24.1 INTRODUCTION

Human nutrition can derive immense health benefits from cereals in their natural form. Small-grain cereals such as millets are generally rich in calcium, polyphenols, proteins, and dietary fiber (Devi et al., 2011). The gluten-free grain of millets also contains vitamin B, antioxidants, and minerals (manganese, magnesium, phosphorus, and iron), essential amino acids (methionine and cysteine), as well as essential fatty acids (linoleic, oleic, palmitic, monogalactosul, phosphatidyl choline, diacylglycerols, phosphatidylserine, digalactosyl diacylglycerols, and phosphatidylethanolamine) (Bagdi et al., 2011). In addition, fermented millet products provide natural probiotics that improve digestion (Amadou et al., 2013; Lei et al., 2006). The grains also exhibit prebiotic activity, which helps increase the population of probiotics in the human gut (Palaniappan et al., 2017). The high fiber which is present in cereals is important for managing weight gain and heart disease (Smith & Tucker, 2011). Moreover, human diets that contain both cereal grains and vegetables provide additional nutrition and fiber.

The *Brassicaceae* (cruciferous) species also possess beneficial effects in human health, such as antioxidant, anti-inflammatory, and cardioprotective activities, partly due to the synergistic action of glucosinolates, polyphenols, and triterpenes that occur widely in the vegetables (Šamec & Salopek-Sondi, 2019). Among other roles in the human body, glucosinolates are associated with reducing various types of cancer. Numerous studies have reported that the consumption of cruciferous vegetables can reduce the risks of breast cancer (Zhang et al., 2018) and prostate cancer (Steinbrecher et al., 2010), as well as lung and ovarian cancer (McManus et al., 2018; Chuang et al., 2007). Furthermore, tocopherols, carotenoids, folic acid, and vitamins C and E are common vitamins found in high quantities in cruciferous vegetables (Singh et al., 2007).

DOI: 10.1201/9781003469766-24

However, both the cereal grains and vegetables also contain biological compounds that can impact negatively on human health (Skibola & Smith, 2000). Among such biological compounds, goitrogens occur in some cereal grains and vegetables. Goitrogens are compounds that interfere with or impede the thyroid gland from producing the hormones that are needed by the body for normal metabolic functions. The association of goitrogens with the thyroid function was first observed when enlargement of the thyroid gland (or goiter) occurred in rabbits that were on a diet consisting of fresh cabbage (McCarrison, 1933).

In cruciferous vegetables such as cabbage (*Brassica oleracea*) and turnip greens (*Brassica rapa*), three main types of goitrogens are recognized, namely, goitrins, thiocyanates, and flavonoids (Petroski & Minich, 2020). Glucosinolates (Figure 24.1), which consist of thioglucosides, in which the glucose molecule is linked to an O-sulfated (Z)-thiohydroximate, occur widely in cruciferous vegetables (Di Dalmazi & Giuliani, 2021; Hertog et al., 1992; Carlson et al., 1987). The indole glucosinolates in these vegetables degrade into the goitrin metabolite, thiocyanate (Figure 24.2), which inhibits iodine uptake by the thyroid gland, hence inhibiting the synthesis of thyroid hormones (Felker et al., 2016). Among several other roles, thyroid hormones are critical for fetal brain development, skeletal maturation, increasing serum glucose, decreasing serum cholesterol, as well as the conversion of carotene to vitamin A (Babiker et al., 2020).

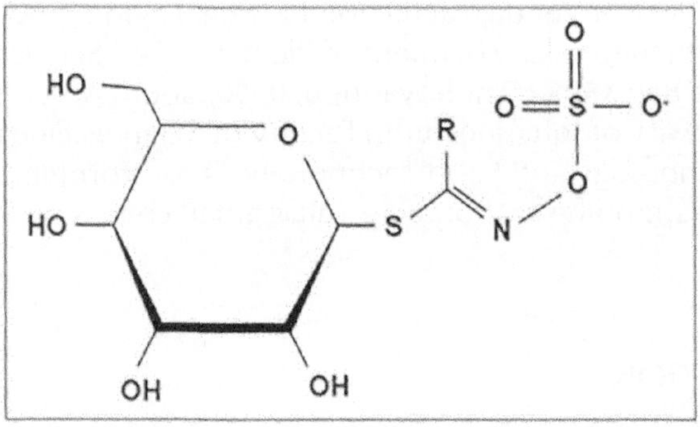

FIGURE 24.1 The general structure of a glucosinolate. (*R* = aglycone side chain.)

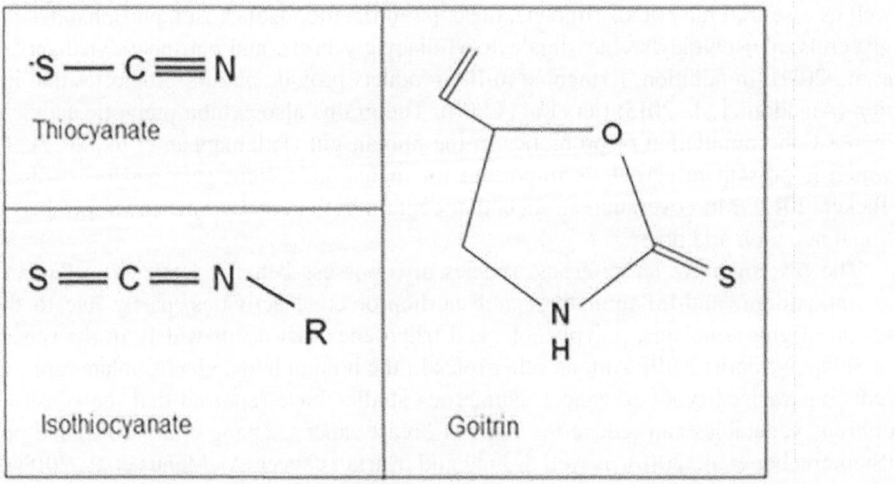

FIGURE 24.2 The chemical structure of thiocyanate, isothiocyanate, and goitrin. (*R* = an alkyl or aryl group.)

24.2 TYPES AND MAJOR SOURCES OF GOITROGENIC COMPOUNDS

Goitrogenic compounds belong to a broad range of chemical groups, such as the sulfonated organics (thiocyanate, isothiocyanate, goitrin, and disulfides), flavonoids (polyphenols), polyhydroxyphenols, and phenol derivatives, pyridines, phthalate esters and metabolites, polychlorinated or polybrominated biphenyls, organochlorines, polycyclic aromatic hydrocarbons, and inorganic iodine (Muzzaffar et al., 2022; Gaitan, 1988). They are divided into two major groups, namely, those that act directly and those that act indirectly on the thyroid gland (Gaitan, 1988). The direct group consists of the following:

1. Inhibiting transport of iodide into the thyroid (for instance, thiocyanate and isothiocyanate)
2. Acting on the intrathyroidal oxidation and organic binding process of iodide (for example, disulfides and goitrin)
3. Interfering with proteolysis, dehalogenation, and hormone release

In contrast, the indirect goitrogens essentially increase the rate of thyroid hormone metabolism (for example, 2,4-dinitrophenol). The goitrogenic compounds that are found in millets and cruciferous vegetables are broadly categorized into two groups, namely, flavonoid glycosides and cyanogenic glycosides. Thiocyanate and rhamnosyl vitexin occur in finger millet and pearl millet, respectively (Table 24.1). In white cabbage, goitrogenic compounds, including sinigrin, gluconapin, progoitrin, and glucoiberin, were detected in varying amounts ranging between 1.8 and 17.28 mg/kg wet weight (Clark, 2010; Šamec et al., 2017).

Major glucosinolates also include thiocyanate precursors glucobrassicin, glucoraphanin, and neoglucobrassicin (Table 24.2). Several findings showed >20.00 μmol/g dw content for both glucobrassicanapin and gluconapin, but the indolyl glucosinolates were generally low (<2.00–20.00 μmol/g dw) in turnip greens (Bonnema et al., 2019; Klopsch et al., 2018; Lee et al., 2013; Francisco et al., 2010; Padilla et al., 2007). They are mainly metabolized to reactive isothiocyanates (or mustard oils) that impart garlic-like odors which are associated with mustard and horseradish (Burrows & Tyrl, 2013). Additional metabolites include goitrins, nitriles, sulfates, and thiocyanates (Burrows & Tyrl, 2013). Isothiocyanates are depicted within a wider framework of flavor chemicals, which also have a role in the pungent taste and odor of cruciferous plants (Wieczorek et al., 2018). The combined effect of goitrogens, linoleic acid, polyphenols, and enzymatic activities renders pearl millet unsuitable for consumption after ten days of storage due to rancid odor (Ali et al., 2023). Glucobrassicin (which is a 3-indolylmethyl glucosinolate found in almost all cruciferous vegetables) is hydrolyzed to an unstable isothiocyanate, while glucoraphanin (4-methylsulfinylbutyl glucosinolate) is hydrolyzed to the isothiocyanate sulforaphane, which is associated with tumor prevention properties by blocking the cell cycle and promoting apoptosis (Wagner et al., 2013). Progoitrin is hydrolyzed to goitrin, which is named for its antithyroid effects (Burrows & Tyrl, 2013). Similarly, sinigrin (which is a prop-3-enyl glucosinolate) is hydrolyzed to allyl isothiocyanate, which can inhibit microRNA expression as well as activate the transcription factor Nrf2, a key modulator of the cellular antioxidant response (Gerstgrasser et al., 2017; Wagner et al., 2013).

TABLE 24.1
Goitrogenic Compound Containing Millets

Name of Millet	Goitrogenic Compound	Reference
Pearl millet	C-glucosyl flavones (vitexin, glucosyl vitexin, and glucosyl orientin)	Gaitan et al., 1989
	Vitexin, orientin, rhamnosyl vitexin, glucosyl orentin, glucosyl vitexin, rhamnosyl orientin	Boncompagni et al., 2018
Finger millet	Thiocyanate, C-glycosyl flavones	Chebet et al., 2018

TABLE 24.2
Glucosinolates and Their Derivatives in Turnip Greens

Class of Glucosinolates	Name of Compound	Concentration (μmol/g dw)	Reference
Aliphatic			
1	2-hydroxy-4-pentenyl/gluconapoleiferin	0.15–0.3	Bonnema et al., 2019; Klopsch et al., 2018
2	2-(R)-2-hydroxy-3-butenyl (progoitrin)	0.32–0.8	Bonnema et al., 2019; Klopsch et al., 2018; Francisco et al., 2011
3	4-pentenyl/glucobrassicanapin	0.69–22.16	Klopsch et al., 2018; Lee et al., 2013; Padilla et al., 2007
4	3-butenyl (gluconapin)	2.04–26.93	Bonnema et al., 2019; Francisco et al., 2010
Indolyl			
1	1-methoxy-3-indolylmethyl/neoglucobrassicin	0.04–1.06	Klopsch et al., 2018; Francisco et al., 2011
2	3-indolylmethyl (glucobrassicin)	0.03–1.71	Klopsch et al., 2018; Lee et al., 2013; Padilla et al., 2007
3	4-hydroxy-3-indolylmethyl/4-hydroxyglucobrassicin	0.04–0.98	Klopsch et al., 2018; Padilla et al., 2007
4	4-methoxy-3-indolylmethyl/4-methoxyglucobrassicin	0.01	Klopsch et al., 2018; Padilla et al., 2007
Aromatic			
1	2-phenylethyl (gluconasturtiin)	0.09–2.68	Klopsch et al., 2018; Lee et al., 2013
Isothiocyanate			
1	Goitrin	1.32–2.55	Vieites-Outes et al., 2016
2	Napin	0.84–1.06	Vieites-Outes et al., 2016

24.2.1 Goitrogenic Compounds in Millets

The existence of goitrogenic compounds in pearl millet has been well documented, specifically derivatives of phenolic flavonoids referred to as C-glycosyl flavones (C-GFs), which, together with their by-products, cause the development of unpleasant odors in flours during storage (Taylor & Duodu, 2015; Priya et al., 2023). C-glycosyl flavones in pearl millet are predominant in the bran fraction (Hassan et al., 2021) and cause inflammation of the thyroid glands when the bran is fed to rats (Gaitan et al., 1989). Glucosyl vitexin, the primary C-glycosylflavone found in millet, exhibits similar effects to 1 μmol/L MMI when present at a dose of 60 μmol/L (Gaitan et al., 1989). Moreover, the goitrogens inhibit thyroid peroxidase (the enzyme that induces thyroid hormone production) by up to 85%. A recent study which quantified the distribution of six C-GFs in 96 pearl millet varieties reported their presence in varying proportions ranging between 7 and 29% (Figure 24.3) (Boncompagni et al., 2018).

Several studies were conducted in Sudan since the 1980s, where pearl millet is a staple for some of the communities. When thiocyanate concentration exceeded 10 μmol/100 ml, it inhibited the incorporation of iodide into thyroglobulin by competing with iodide at the thyroid peroxidase level, thus exacerbating iodine deficiency (Fukayama et al., 1992). Other inconclusive studies involving human subjects who frequently consumed pearl millet in Western Sudan and experimental rats have also been reported (Elnour et al., 2000; Gaitan et al., 1989; Osman et al., 1983). The researchers postulated that, probably, a goitrogenic substance associated with pearl millet was synergistically contributing to iodine deficiency and, hence, goiter developments in the population. In another study involving feeding 1 g/kg/day pearl millet to goat kids in Egypt, the animals developed goiters (Gadir & Adam, 2007). A subsequent study in which potassium iodate was incorporated at 50 parts per million (ppm) failed to protect animals against the goitrogens in pearl millet (Elnour et al., 1997).

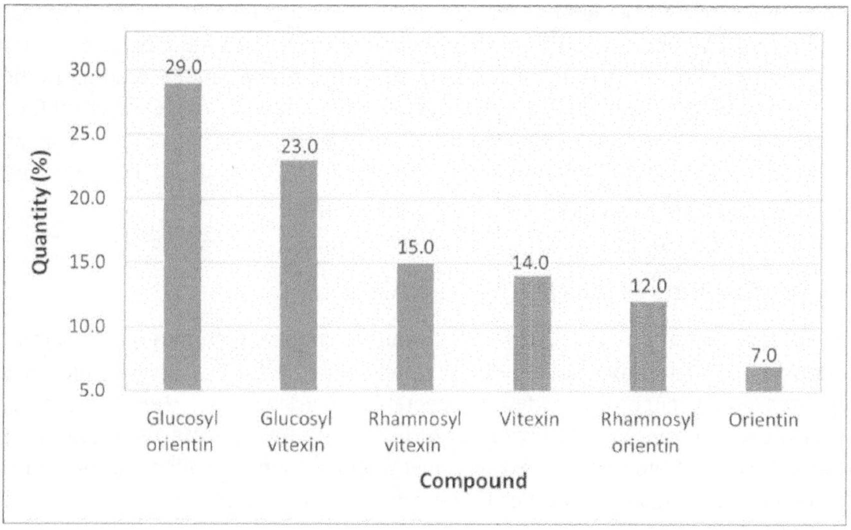

FIGURE 24.3 The distribution of six C-glycosyl flavones in pearl millet.

Source: Adapted from Boncompagni et al. (2018).

Apparently, the goitrogenic substance in pearl millet was suspected to be thiocyanate, which has a molecular size equal to that of iodide (Osman et al., 1983; Chandra, 2015). Further studies identified another compound, thionamide, which was attributed to the inhibition of iodine organification as well as the coupling of iodotyrosines, thus impeding the synthesis of hormones (Osman et al., 1983). Perhaps one of the interesting findings of related studies is that the goitrogenic compound found in fonio millet is distinct from the one found in pearl millet (Sartelet et al., 1996; Gaitan et al., 1989), while traditional fermentation increases the goitrogenic effect of pearl millet (Elnour et al., 1997).

In finger millet, the goitrogenic compounds are mostly phenolic flavonoid chemicals, particularly C-glycosyl flavones, which are linked to antithyroid and goitrogenic effects. They can hinder the function of the enzyme thyroid peroxidase (TPO) and disrupt the absorption of iodine in the body. Moreover, the level of these compounds in the grain fractions is directly linked to their goitrogenic and antithyroid activities. Hence, it is important to consider the goitrogenic qualities of finger millet, particularly for those with thyroid issues or those susceptible to goiter (Gaitan et al., 1989).

Flavonoids, which also occur widely in plant material, are generally considered as healthy antioxidants, but some of them can be converted into goitrogenic compounds by the gut bacteria (Dos Santos et al., 2011; Vanderpas, 2006). In millets, such as finger millet and pearl millet (*Pennisetum glaucum*), goitrogenic compounds have been detected (Chebet et al., 2018). Finger millet contains cyanogenic glycosides that can be converted to thiocyanate, which reduces the absorption of iodine and eventually causes goiter. In a study conducted in Kenya, thiocyanate content in the red finger millet ranged between 43.48 and 4.28 mg/kg (Chebet et al., 2018). The researchers observed that while the occurrence of these compounds was within the recommended rates (100 mg/kg), the frequency of intake could still elevate the risk to health and recommended processing the finger millet (for instance, through cooking) to reduce the thiocyanate content. The researchers also observed that the presence of goitrogenic compounds *per se* in finger millet was not a cause for worry, unless ingested in excessive amounts. Moreover, processing reduces thiocyanate in finger millet. Similarly, the goitrogenic compounds C-glycosyl flavones present in pearl millet, such as glucosyl vitexin, glucosyl orientin, and vitexin, can potentially contribute to some health problems since they are biologically active in dietary components (Gaitan et al., 1989, 1995). Orientin and vitexin are produced from aglycone forms (either apigenin or luteolin) and further glycosylated to produce the corresponding glucosylated or rhamnosylated forms (Boncompagni et al., 2018). Essentially, the C-glycosyl flavones can interfere with many enzymatic activities in the human body, particularly those that are involved in regulating the synthesis of the thyroid hormone, leading to the enlargement of the thyroid gland (or goiter) (Gonçalves et al., 2017). Nonetheless, C-glycosyl flavones have been implicated in some beneficial activities, which include antiproliferative effects on thyroid cancer cells (Benvenga et al., 2020; Montané et al., 2020; Sharifi-Rad et al., 2020) and thyroid autoimmunity (Khan et al., 2020; Schmeltz et al., 2014; Gaitan, 1990).

24.2.2 Glucosinolates in Cruciferous Vegetables

Depending on their amino acid precursor, glucosinolates are classified into three chemical classes, namely, aliphatic, aromatic, and indole glucosinolates (Figure 24.4) (Vieites-Outes et al., 2016). They occur in a broad range of cruciferous vegetables (Carlson et al., 1987). If the vegetable is consumed raw, the glucosinolates are hydrolyzed in the proximal part of the gastrointestinal tract by myrosinase to several bioactive breakdown metabolites, such as isothiocyanates, nitriles, thiocyanates, epithionitriles, and oxazolidines (Vieites-Outes et al., 2016).

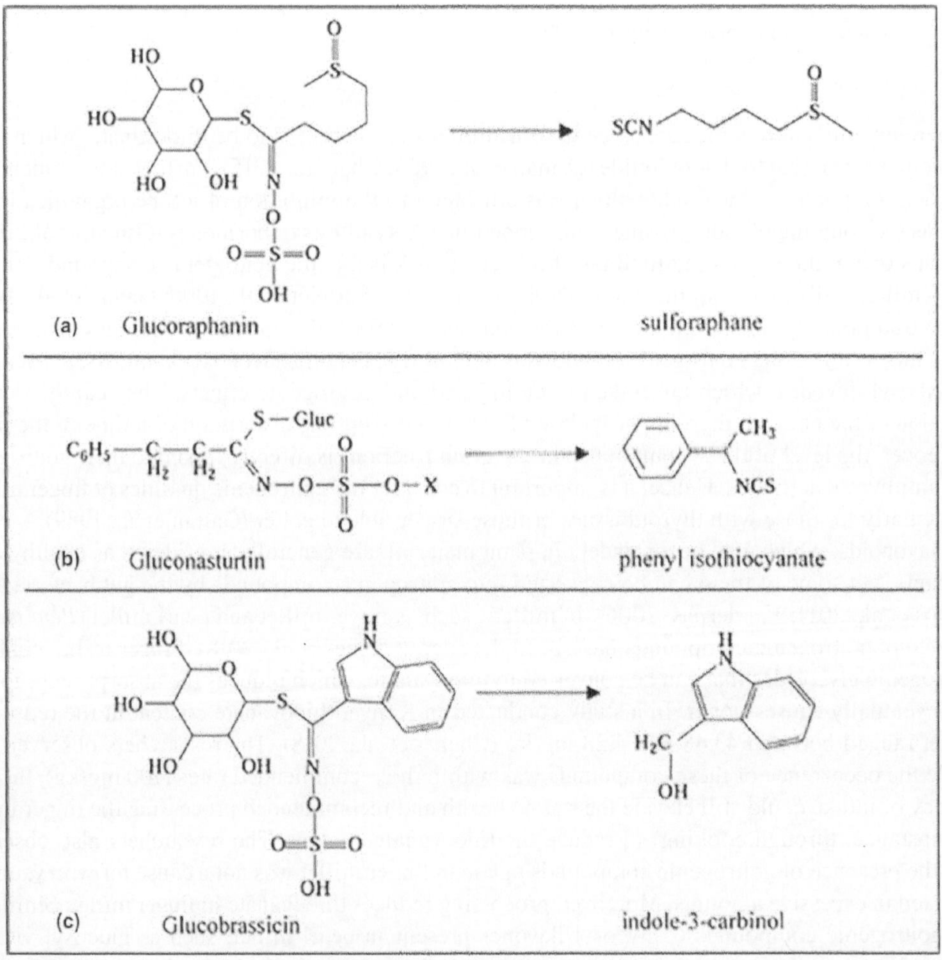

FIGURE 24.4 Examples of glucosinolates and their hydrolysis products implicated in human nutrition in each chemical class: (a) aliphatic, (b) aromatic, (c) indole.

In turnip greens, for instance, a broad range of glucosinolates, including neoglucobrassicin and 4-methoxyglucobrassicin, was detected in numerous studies (Table 24.2). Turnip is an important vegetable which is cultivated partly for its tops, fructiferous stems with the flower buds, and leaves that are consumed while still green before opening (Paul et al., 2019; Francisco et al., 2011). The leaves possess a bitter taste, which differentiates them from other *Brassica* vegetables, such as cabbage (Francisco et al., 2009). In Chinese cabbage (*Brassica rapa* L. ssp. *pekinensis*), five aliphatic glucosinolates (progoitrin, sinigrin, glucoalyssin, gluconapin, and glucobrassicanapin) and four indolyl glucosinolates (4-hydroxyglucobrassicin, glucobrassicin, 4-methoxyglucobrassicin, and neoglucobrassicin) were reported with glucobrassicanapin and gluconapin averaging 4.52 and 3.72 μmol/g DW, respectively (Lee et al., 2014). In raw cabbage (cv. Marathon), sinigrin accounted for 41.3% of the total glucosinolates, with a mean concentration of 6.98 μmol/g DW (Rungapamestry et al., 2006).

24.3 METABOLIC EFFECTS OF GLUCOSINOLATES ON THE THYROID GLAND FUNCTION

Upon ingestion, glucosinolates undergo enzymatic degradation by myrosinase (thioglucohydrolase; E.C.3.2.1.147), and some of them generate sulforaphane, phenethyl, and indolylic isothiocyanates, which exhibit anticarcinogenic properties. On the other hand, progoitrin and indolylic glucosinolates break down into goitrin and thiocyanate, respectively, and can potentially reduce the production of thyroid hormones.

After consumption, flavonoid glycosides are broken down into flavonoid aglycones by glycosidases produced by intestinal microbes. The bacteria with β-glucosidase activity dominate the human intestinal microbiota, including *Bacteroides ovatus*, *Bacteroides uniformis*, *Bifidobacterium adolescentis*, *Bifidobacterium longum*, *Enterococcus faecalis*, and *Parabacteroides distasonis* (Xiong et al., 2023). Also, the deglycosylation of flavonoids involves α-L-rhamnosidases and β-glucosidases (Hsiao & Hsieh, 2018; Braune & Blaut, 2016; Bokkenheuser et al., 1987). Larger molecules that reach the colon are degraded further by bacteria. The sugar moiety of the flavonoid glycosides is critical in determining their bioavailability, and dimerization tends to reduce bioavailability (Walle, 2004; Hollman et al., 1999).

Generally, glucosinolates are thermally and chemically stable. They are only broken down primarily through hydrolysis by myrosinase (β-thioglucosidase) found in plants alongside glucosinolates (Figure 24.5). The reaction products, such as thiocyanates, nitriles, and isothiocyanates, exhibit bioactive functions that protect plants against pathogens and have cancer-preventative potential in humans (Shakour et al., 2022). Disruption of the plant tissue leads to the hydrolytic reaction between myrosinase and glucosinolate, resulting in the breakdown of the glycosidic bond to glucose and aglycone, which is unstable.

The aglycans can either be absorbed and metabolized by mammalian tissues or undergo additional modifications (by gut microbiota), including hydroxylation of the B ring and fusion of the middle ring, producing different compounds, such as gallic acid, phenolic acid, and phloroglucinol (Chandra, 2010). These compounds have antithyroid activity. A study conducted on fonio millet (*Digitaria exilis* Stapf), a staple grain in West Africa, revealed that luteolin and apigenin were the primary flavones (Sartelet et al., 1996). Furthermore, when present at a concentration of 10 μM, these flavones effectively impeded the release of thyroid hormones in a culture of pig thyroid cells (Di Dalmazi & Giuliani, 2021).

From a dietary perspective, the metabolic effects of glucosinolates may also vary depending on the quantity of these vegetables consumed, potentially resulting in either beneficial or harmful outcomes (Polat, 2010; Boncompagni et al., 2018). The biosynthesis of glucosinolates occurs from amino acids. Each of the three subtypes of glucosinolates is associated with a specific precursor. When the vegetables are consumed raw, myrosinase (which is triggered in injured plant tissue and created by microbes in the human gut) hydrolyzes the glucosinolates in the upper part of the

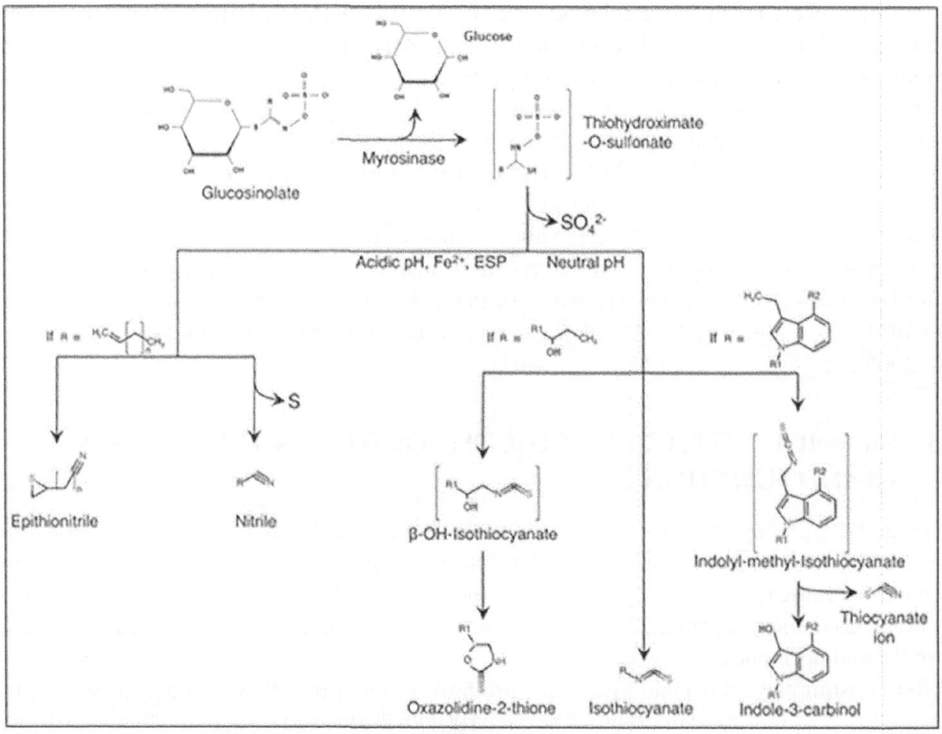

FIGURE 24.5 Breakdown products of glucosinolates.

Source: Adapted from Barba et al. (2016); Carlson et al. (1987).

digestive system into different by-products, such as sulforaphane, oxazolidine-2-thiones, nitriles, thiocyanates, indole-3-carbinols, and isothiocyanates. However, myrosinase is inactivated through cooking or thermal treatment. Because of their hydrophilic characteristics, glucosinolates (thioglucose and sulphate groups) pass to the colon, where they are broken down by bacteria in the intestines after myrosinase is no longer active (Barba et al., 2016). In general, the glucosinolates undergo many phases and states within the gastrointestinal tract, including absorption in the intact form in the stomach, followed by breakdown by myrosinase in the small intestine to form isothiocyanates, which are subsequently absorbed (Michaelsen, 1994).

24.4 METABOLISM AND ABSORPTION OF DIETARY FLAVONOIDS

The metabolism of flavonoids is catabolized by microbes in the lower intestinal tract, but it is recognized that while the flavones themselves are not goitrogens, some of the phenolic compounds, such as p-hydroxybenzoic acid, resorcinol, or phloroglucinol, that are produced during their degradation may actually be goitrogenic. The absorption of dietary flavonoids from the small intestine or colon also depends on their physicochemical properties, such as configuration, lipophilicity, molecular size, and solubility, as well as the structure of flavonoid, that is, whether it is a glycoside or an aglycone (Zhang et al., 2005). With the exception of the subclass of catechins, most flavonoids are bound to sugars as glycosides that need to be converted into aglycones (Figure 24.6) that can be absorbed easily by the small intestines (Hollman et al., 1999). Therefore, the absorption of flavonoid glycosides requires hydrolyzing their sugar group into aglycones, which are more hydrophobic and can be easily absorbed through passive diffusion across the epithelial cells (Chen et al., 2011). The microflora from the colon are key sources of hydrolases, such as α-arabinofuranosidase,

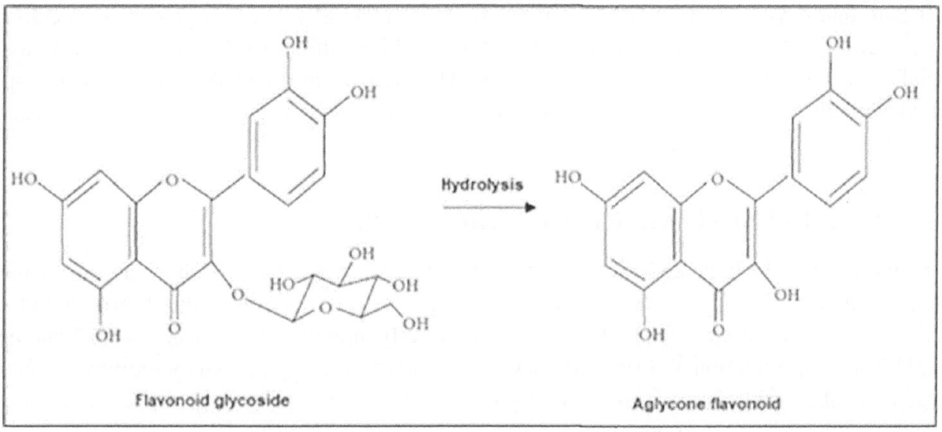

FIGURE 24.6 Hydrolytic conversion of flavonoid glycoside to an aglycone flavonoid, which is more hydrophobic and can be easily absorbed by the epithelial cells through passive diffusion. The conversion can also be achieved through the action of gut microbes.

Source: Chen et al. (2011); Heinonen et al. (2004); Crespy et al. (2001); Sesink et al. (2001).

α-fucosidase, β-fucosidase, β-glucosidase, β-glucuronidase, and α-rhamnosidase, for converting the flavonoid glycosides (Hur et al., 2000). It has been speculated that in an alternative mechanism, flavonoid glucosides are hydrolyzed by lactase-phlorizin hydrolase, which is a ??-glucosidase, on the outside of the brush border membrane of the small intestine, such that the liberated aglycone is subsequently absorbed across the small intestine (Day et al., 2000).

24.5 IMPLICATIONS OF PROCESSING ON GOITROGENIC COMPOUNDS

Roasting is one of the traditional operations used in rural households to process finger millet. It is achieved by exposing the dry grain to heat (70–80°C) treatment without applying water or oil on the grain. The process decreases the content of antinutrients, such as goitrogenic chemicals, cyanides, alkaloids, and saponins (Abioye et al., 2022). Roasting can also enhance the longevity of the grain during storage. However, roasting at 70°C reduced the antioxidant activity of finger millet, which was attributed partly to the capacity of phenols to form complexes with proteins, thus diminishing the quantity of phenols that could be extracted from the grains during thermal processing (Singh et al., 2018).

When finger millet flour was cooked at 100°C for 5, 10, and 30 min, about 95% thiocyanate content was eliminated in comparison with fresh (uncooked) flour (Chebet et al., 2018). Similarly, the products of aliphatic glucosinolate hydrolysis showed 5–12% reduction in steamed, 18–23% reduction in pressure-cooked, and 37–45% reduction in boiled vegetable extracts (Vieites-Outes et al., 2016). In a study involving turnip leaves, steaming resulted in an increase (+17%) of the amount of total glucosinolates, but boiling decreased (–50%) the total glucosinolates (Vieites-Outes et al., 2016).

Fermentation, which is a food preservation technique that utilizes endogenous microorganisms in food material and the environment to hydrolyze the starch into simple sugars and alcohol, lowers the pH of the fermented substrate. With regards to goitrogens, an increase in thyroid activity in rats fed with fermented millet was reported and attributed to the leaching out of dietary magnesium (Elnour et al., 1998). Fermentation at room temperature for 72 h had no significant effect on the thiocyanate content of finger millet flour (Chebet et al., 2018).

Mechanical processing of finger miller also affects the polyphenol content since about 90% of these compounds are concentrated in the seed coat. Therefore, removal of the seed coat would impact the polyphenols in the resulting flour (Chethan & Malleshi, 2007). For instance, clean extraction of the seed coat using hydrothermal treatment combining hydration at 33% moisture, steaming for 20 min, and drying at 40°C to 12% moisture consequentially lowered goitrogens (Shobana & Malleshi, 2007). Also, sprouting the seed at 30°C caused 27% thiocyanate reduction, while soaking reduced thiocyanate by 75% (Chebet et al., 2018). The increase in thiocyanate levels during sprouting could be attributed to the increased enzymatic activity in the shoots during the early seedling growth stages (Chebet et al., 2018).

24.6 RECOMMENDED MITIGATION METHODS

The maintenance of a balanced diet is necessary for attaining a healthy thyroid gland function. Although iodine deficiency (or excess) is a major factor in the function of the thyroid gland, there are other goitrogenic compounds that directly or indirectly influence the thyroid gland. Nonetheless, thyroid hormone production and function depend on micronutrients, primarily iodine and selenium (Triggiani et al., 2009). Apart from reducing goitrogens during processing, supplementation is an alternative mitigation method opted for by many nations.

Perhaps it is also important to point out that apart from goitrogenic compounds in millets and cruciferous vegetables, there are other sources or reasons, including genetic factors (Radkay et al., 2014; Hegedus et al., 2009), exposure to environmental pollutants (such as benzene, formaldehyde, pesticides, bisphenol A, polychlorinated biphenyl, polyhalogenated aromatic hydrocarbons, and polybrominated diphenyl ether) (Pellegriti et al., 2013), and smoking (Ericsson & Lindgarde, 1991), that can contribute to goiter development. In addition, other foods, such as cassava (*Manihot esculenta* Crantz.), may contain cyanogenic glucosides (linamarin, linustatin, and lotaustralin) that can be detoxified to thiocyanate, which can exacerbate iodine deficiency (Cressey & Reeve, 2019; Haque & Bradbury, 2002).

To mitigate the effect of goitrogens from plant foods, it is necessary to understand that these compounds, including thiocyanates, disrupt iodine metabolism in the thyroid gland by reducing iodide absorption, the stimulation of iodide efflux, or the substitution of iodide with thiocyanate. Moreover, these goitrogens primarily impede the iodide-concentrating process of the thyroid gland by hindering the one-way removal of iodide from the gland (Di Dalmazi & Giuliani, 2021).

24.7 CONCLUSIONS AND PROSPECTS

Goitrogenic compounds occur in both millets and cruciferous vegetables. However, for finger millet, there is no unequivocal evidence based on, for instance, metabolomic profiling and genetic mapping of goitrogenic substances and a detailed examination of their effects on thyroid activity. Similarly, the association of pearl millet with goiter prevalence is not compelling but somewhat fragmented, inconclusive, and not strong enough to assert that when consumed as part of a balanced diet, the millet has goitrogenic effects in a population. Therefore, more scientific research is needed. Moreover, the goitrogenic compounds that are present in both millets and cruciferous vegetables can be handled by the processing industry to produce safe products for human consumption. At the household level, there are several methods for reducing the goitrogenic content in the vegetables and millet flour (Kaushik et al., 2021).

Moving forward, it will be necessary to increase awareness in communities about the risks of environmental pollutants that lead to interference with the functioning of the thyroid gland and the development of goiters. Equally important will be the need to access a balanced diet to maintain good health. From a crop improvement perspective, millet breeders will need to consider selecting for low content of compounds that are associated with goitrogenic effects. There are excellent prospects of combining conventional approaches with speed breeding (Ghosh et al., 2018) and sophisticated breeding tools that utilize precise knowledge of plant genomes to develop high-quality end use products in the form of crop varieties (Ahmar et al., 2020). These technologies include the exploitation of zinc-finger nucleases (Wright et al., 2005), transcription activator–like effector nucleases (Gaj et al., 2013), and the clustered regularly interspaced short palindromic repeats (Butt et al., 2020; Puchta, 2017).

24.8 ACKNOWLEDGMENTS

We would like to sincerely acknowledge all the colleagues who reviewed and improved the original draft manuscript of this chapter.

REFERENCES

Abioye, V. F., Babarinde, G. O., Ogunlakin, G. O., Adejuyitan, J. A., Olatunde, S. J., & Abioye, A. O. (2022). Varietal and processing influence on nutritional and phytochemical properties of finger millet: A review. *Heliyon*, *8*(12), e12310.

Ahmar, S., Gill, R. A., Jung, K. H., Faheem, A., Qasim, M. U., Mubeen, M., & Zhou, W. (2020). Conventional and molecular techniques from simple breeding to speed breeding in crop plants: Recent advances and future outlook. *International Journal of Molecular Science*, *21*, 2590.

Ali, A., Kumar, R. R., Bansal, N., Bollinedi, H., Singh, S. P., Satyavathi, C. T., Praveen, S., & Goswami, S. (2023). Characterization of biochemical indicators and metabolites linked with rancidity and browning of pearl millet flour during storage. *Journal of Plant Biochemistry and Biotechnology*, *32*(1), 121–131.

Amadou, I., Gounga, M. E., & Le, G. W. (2013). Millets: Nutritional composition, some health benefits and processing-A review. *Emirates Journal of Food and Agriculture*, *25*(7), 501–508.

Babiker, A., Alawi, A., Al Atawi, M., & Al Alwan, I. (2020). The role of micronutrients in thyroid dysfunction. *Sudanese Journal of Paediatrics*, *20*(1), 13–19.

Bagdi, A., Balázs, G., Schmidt, J., Szatmári, M., Schoenlechner, R., Berghofer, E., & Tömösközia, S. J. A. A. (2011). Protein characterization and nutrient composition of Hungarian proso millet varieties and the effect of decortication. *Acta Alimentaria*, *40*, 128–141.

Barba, F. J., Nikmaram, N., Roohinejad, S., Khelfa, A., Zhu, Z., & Koubaa, M. (2016). Bioavailability of glucosinolates and their breakdown products: Impact of processing. *Frontiers in Nutrition*, *3*, 24.

Benvenga, S., Ferrari, S. M., Elia, G., Ragusa, F., Patrizio, A., Paparo, S. R., Camastra, S., Bonofiglio, D., Antonelli, A., & Fallahi, P. (2020). Nutraceuticals in thyroidology: A review of in vitro, and in vivo animal studies. *Nutrients*, *12*(5), 1337.

Bokkenheuser, V. D., Shackleton, C. H., & Winter, J. (1987). Hydrolysis of dietary flavonoid glycosides by strains of intestinal bacteroides from humans. *Biochemistry Journal*, *248*, 953–956.

Boncompagni, E., Orozco-Arroyo, G., Cominelli, E., Gangashetty, P. I., Grando, S., Kwaku Zu, T. T., Daminati, M. G., Nielsen, E., & Sparvoli, F. (2018). Antinutritional factors in pearl millet grains: Phytate and goitrogens content variability and molecular characterization of genes involved in their pathways. *PLoS One*, *13*(6), e0198394.

Bonnema, G., Lee, J. G., Shuhang, W., Lagarrigue, D., Bucher, J., Wehrens, R., De Vos, R., & Beekwilder, J. (2019). Glucosinolate variability between turnip organs during development. *PLoS One*, *14*(6), e0217862.

Braune, A., & Blaut, M. (2016). Bacterial species involved in the conversion of dietary flavonoids in the human gut. *Gut Microbes*, *7*, 216–234.

Burrows, G. E., & Tyrl, R. J. (2013). *Toxic plants of North America* (p. 1383). Wiley-Blackwell.

Butt, H., Zaidi, S. S. A., Hassan, N., & Mahfouz, M. (2020). (CRISPR-based directed evolution for Cartea, M. E., de Haro, A., Obregón, S., Soengas, P. and Velasco, P. (2012)). Glucosinolate variation in leaves of *Brassica rapa* crops. *Plant Foods for Human Nutrition*, *67*, 283–288.

Carlson, D. G., Daxenbichler, M. E., van Etten, C. H., Kwolek, W. F., & Williams, P. H. (1987). Glucosinolates in crucifer vegetables: Broccoli, brussels sprouts, cauliflower, collards, kale, mustard greens, and kohlrabi. *Journal of the American Society for Horticultural Science*, *112*(1), 173–178.

Chandra, A. K. (2010). Goitrogen in food: Cyanogenic and flavonoids containing plant foods in the development of goiter. In *Bioactive foods in promoting health* (pp. 691–716). Academic Press.

Chandra, A. K. (2015). Iodine, thiocyanate and the thyroid. *Biochemistry and Pharmacology: Open Access*, *4*(3), 171.

Chebet, S. J., Nawiri, M., & Murungi, J. (2018). Assessment of the rates of thiocyanate in treated and untreated red and brown finger millet (*Eleusine coracana*) cultivated in Mogotio Area, Baringo County, Kenya. *International Journal of Tropical Drylands*, *2*(1), 28–34.

Chen, Y., Wang, J., Jia, X., Tan, X., & Hu, M. (2011). Role of intestinal hydrolase in the absorption of prenylated flavonoids present in Yinyanghuo. *Molecules*, *16*, 1336–1348.

Chethan, S., & Malleshi, N. G. (2007). Finger millet polyphenols: Optimization of extraction and the effect of pH on their stability. *Food Chemistry*, *105*(2), 862–870.

Chuang, L. T., Moqattash, S. T., Gretz, H. F., Nezhat, F., Rahaman, J., & Chiao, J. W. (2007). Sulforaphane induces growth arrest and apoptosis in human ovarian cancer cells. *Acta Obstetricia et Gynecologica Scandinavica*, *86*(10), 1263–1268.

Clarke, D. B. (2010). Glucosinolates, structures and analysis in food. *Analytical Methods*, *2*(4), 310–325.

Crespy, V., Morand, C., Besson, C., Manach, C., Demigne, C., & Remesy, C. (2001). Comparison of the intestinal absorption of quercetin, phloretin and their glucosides in rats. *Journal of Nutrition*, *131*, 2109–2114.

Cressey, P., & Reeve, J. (2019). Metabolism of cyanogenic glycosides: A review. *Food and Chemical Toxicology*, *125*, 225–232.

Day, A. J., Cañada, F. J., D??az, J. C., Kroon, P. A., Mclauchlan, R., Faulds, C. B., Plumb, G. W., Morgan, M. R., & Williamson, G. (2000). Dietary flavonoid and isoflavone glycosides are hydrolysed by the lactase site of lactase phlorizin hydrolase. *FEBS Letters*, *468*, 166–170.

Devi, P. B., Vijayabharathi, R., Sathyabama, S., Malleshi, N. G., & Priyadarisini, V. B. (2011). Health benefits of finger millet (*Eleusine coracana*) polyphenols and dietary fiber: A review. *Journal of Food Science and Technology*, *51*(6), 1021–1040.

Di Dalmazi, G., & Giuliani, C. (2021). Plant constituents and thyroid: A revision of the main phytochemicals that interfere with thyroid function. *Food and Chemical Toxicology*, *152*, 112158.

Dos Santos, M. C. D. S., Gonçalves, C. F. L., Vaisman, M., Ferreira, A. C. F., & de Carvalho, D. P. (2011). Impact of flavonoids on thyroid function. *Food and Chemical Toxicology*, *49*(10), 2495–2502.

Elnour, A., Hambraeus, L., Eltom, M., Dramaix, M., & Bourdoux, P. (2000). Endemic goiter with iodine sufficiency: A possible role for the consumption of pearl millet in the etiology of endemic goiter. *American Journal of Clinical Nutrition*, *71*, 59–66.

Elnour, A., Liedén, S. A., Bourdoux, P., Eltom, M., Khalid, S. A., & Hambraeus, L. (1997). The goitrogenic effect of two Sudanese pearl millet cultivars in rats. *Nutrition Research*, *17*, 533–546.

Elnour, A., Liedén, S. A., Bourdoux, P., Eltom, M., Khalid, S. A., & Hambraeus, L. (1998). Traditional fermentation increases goitrogenic activity in pearl millet. *Annals of Nutrition and Metabolism*, *42*(6), 341–349.

Ericsson, U. B., & Lindgarde, F. (1991). Effects of cigarette smoking on thyroid function and the prevalence of goiter, thyrotoxicosis and autoimmune thyroiditis. *Journal of Internal Medicine*, *229*, 67–71.

Felker, P., Bunch, R., & Leung, A. M. (2016). Concentrations of thiocyanate and goitrin in human plasma, their precursor concentrations in brassica vegetables, and associated potential risk for hypothyroidism. *Nutrition Reviews*, *74*(4), 248–258.

Francisco, M., Cartea, M. E., Soengas, P., & Velasco, P. (2011). Effect of genotype and environmental conditions on health-promoting compounds in *Brassica rapa*. *Journal of Agricultural and food Chemistry*, *59*(6), 2421–2431.

Francisco, M., Moreno, D. A., Cartea, M. E., Ferreres, F., Garcia-Viguera, C., & Velasco, P. (2009). Simultaneous identification of glucosinolates and phenolic compounds in a representative collection of vegetable *Brassica rapa*. *Journal of Chromatography*, *1216*(38), 6611–9.

Francisco, M., Velasco, P., Moreno, D. A., García-Viguera, C., & Cartea, M. E. (2010). Cooking methods of *Brassica rapa* affect the preservation of glucosinolates, phenolics and vitamin C. *Food Research International*, *43*(5), 1455–1463.

Fukayama, H., Nasu, M., Murakami, S., & Sugawara, M. (1992). Examination of antithyroid effects of smoking products in cultured thyroid follicles: Only thiocyanate is a potent antithyroid agent. *Acta Endocrinology*, *127*, 520–525.

Gadir, A. W. S., & Adam, S. E. (2007). Effect of pearl millet (*Pennisetum typhoides*), potassium iodate or their mixture on Nubian goats. *Journal of Pharmacology and Toxicology*, *2*, 183–189.

Gaitan, E. (1988). Goitrogens. *Baillière's Clinical Endocrinology and Metabolism*, *2*(3), 683–702.

Gaitan, E. (1990). Goitrogens in food and water. *Annual Review of Nutrition*, *10*(1), 21–37.

Gaitan, E., Cooksey, R. C., Legan, J., & Lindsay, R. H. (1995). Antithyroid effects in vivo and in vitro of vitexin: A C-glucosylflavone in millet. *Journal of Clinical Endocrinology and Metabolism*, *80*(4), 1144–1147.

Gaitan, E., Lindsay, R. H., Reichert, R. D., Ingbar, S. H., Cooksey, R. C., Legan, J., Meydrech, E. F., Hill, J., & Kubota, K. (1989). Antithyroid and goitrogenic effects of millet: Role of C-glycosylflavones. *Journal of Clinical Endocrinology and Metabolism*, *68*(4), 707–714.

Gaj, T., Gersbach, C. A., & Barbas, C. F. (2013). ZFN, TALEN, and CRISPR/Cas-based methods for genome engineering. *Trends in Biotechnology*, *31*, 397–405.

Gerstgrasser, A., Melhem, H., Leonardi, I., Atrott, K., Schäfer, M., Werner, S., Rogler, G., & Frey-Wagner, I. (2017). Cell-specific activation of the Nrf2 antioxidant pathway increases mucosal inflammation in acute but not in chronic colitis. *Journal of Crohn's and Colitis*, *11*(4), 485–499.

Ghosh, S., Watson, A., Gonzalez-Navarro, O. E., Ramirez-Gonzalez, R. H., Yanes, L., Mendoza-Suárez, M., Simmonds, J., Wells, R., Rayner, T., Green, P., & Hafeez, A. (2018). Speed breeding in growth chambers and glasshouses for crop breeding and model plant research. *Nature Protocols, 13*(12), 2944–2963.

Gonçalves, C. F. L., de Freitas, M. L., & Ferreira, A. C. F. (2017). Flavonoids, thyroid iodide uptake and thyroid cancer – A review. *International Journal of Molecular Sciences, 18*(6), 1247.

Haque, M. R., & Bradbury, J. H. (2002). Total cyanide determination of plants and foods using the picrate and acid hydrolysis methods. *Food Chemistry, 77*(1), 107–114.

Hassan, Z. M., Sebola, N. A., & Mabelebele, M. (2021). The nutritional use of millet grain for food and feed: A review. *Agriculture and Food Security, 10*, 1–14.

Hegedus, L., Brix, T. H., & Paschke, R. (2009). Etiology of simple goiter. *Thyroid, 19*, 209–211.

Heinonen, S. M., Wahala, K., Liukkonen, K. H., Aura, A. M., Poutanen, K., & Adlercreutz, H. (2004). Studies of the *in vitro* intestinal metabolism of isoflavones aid in the identification of their urinary metabolites. *Journal of Agriculture and Food Chemistry, 52*, 2640–2246.

Hertog, M. G., Hollman, P. C., & Katan, M. B. (1992). Content of potentially anticarcinogenic flavonoids of 28 vegetables and 9 fruits commonly consumed in the Netherlands. *Journal of Agricultural and Food Chemistry, 40*(12), 2379–2383.

Hollman, P. C. H., Buijsman, M. N. C. P., van Gameren, Y., Cnossen, P. J., de Vries, J. H. M., & Katan, M. B. (1999). The sugar moiety is a major determinant of the absorption of dietary flavonoid glycosides in man. *Free Radical Research, 31*(6), 569–573.

Hsiao, Y. H., & Hsieh, J. F. (2018). The conversion and deglycosylation of isoflavones and anthocyanins in black soymilk process. *Food Chemistry, 261*, 8–14.

Hur, H. G., Lay, J. O., Jr., Beger, R. D., Freeman, J. P., & Rafii, F. (2000). Isolation of human intestinal bacteria metabolizing the natural isoflavone glycosides daidzin and genistin. *Archives of Microbiology, 174*, 422–428.

Kaushik, N., Yadav, P., Khandal, R. K., & Aggarwal, M. K. (2021). Review of ways to enhance the nutritional properies of millets for their value addition. *Journal of Food Processing and Preservation, 45*(4), 213–222.

Khan, M. S., Qadri, Q., Makhdoomi, M. J., Wani, M. A., Malik, A. A., Niyaz, M., Masoodi, S. R., Andrabi, K. I., Ahmad, R., & Mudassar, S. (2020). RET/PTC gene rearrangements in thyroid carcinogenesis: Assessment and clinico-pathological correlations. *Pathology and Oncology Research, 26*, 507–513.

Klopsch, R., Witzel, K., Artemyeva, A., Ruppel, S., & Hanschen, F. S. (2018). Genotypic variation of glucosinolates and their breakdown products in leaves of Brassica rapa. *Journal of Agricultural and Food Chemistry, 66*(22), 5481–5490.

Lee, J. G., Bonnema, G., Zhang, N., Kwak, J. H., de Vos, R. C., & Beekwilder, J. (2013). Evaluation of glucosinolate variation in a collection of turnip (*Brassica rapa*) germplasm by the analysis of intact and desulfo-glucosinolates. *Agricultural and Food Chemistry, 61*(16), 3984–3993.

Lee, M. K., Chun, J. H., Byeon, D. H., Chung, S. O., Park, S. U., Park, S., Arasu, M. V., Al-Dhabi, N. A., Lim, Y. P., & Kim, S. J. (2014). Variation of glucosinolates in 62 varieties of Chinese cabbage (*Brassica rapa* L. ssp. *pekinensis*) and their antioxidant activity. *Food Science and Technology, 58*(1), 93–101.

Lei, V., Friis, H., & Michaelsen, K. F. (2006). Spontaneously fermented millet product as a natural probiotic treatment for diarrhea in young children: An intervention study in Northern Ghana. *International Journal of Food Microbiology, 110*, 246–253.

McCarrison, R. (1933). A paper on food and goitre. *British Medical Journal, 2*(3797), 671–675.

McManus, H., Moysich, K. B., Tang, L., Joseph, J., & McCann, S. E. (2018). Usual cruciferous vegetable consumption and ovarian cancer: A case-control study. *Nutrition and Cancer, 70*(4), 678–683.

Michaelsen, S., Otte, J., Simonsen, L. O., & Sørensen, H. (1994). Absorption and degradation of individual intact glucosinolates in the digestive tract of rodents. *Acta Agriculturae Scandinavica, Section A-Animal Science, 44*(1), 25–37.

Middleton, E., Kandaswami, C., & Theoharides, T. C. (2000). The effects of plant flavonoids on mammalian cells: Implications for inflammation, heart disease, and cancer. *Pharmacological Reviews, 52*(4), 673–751.

Montané, X., Kowalczyk, O., Reig-Vano, B., Bajek, A., Roszkowski, K., Tomczyk, R., Pawliszak, W., Giamberini, M., Mocek-Płóciniak, A., & Tylkowski, B. (2020). Current perspectives of the applications of polyphenols and flavonoids in cancer therapy. *Molecules, 25*(15), 3342.

Muzzaffar, S., Nazir, T., Bhat, M. M., Wani, I. A., & Masoodi, F. A. (2022). Goitrogens. In *Handbook of plant and animal toxins in food* (pp. 125–154). CRC Press.

Osman, A. K., Basu, T. K., & Dickerson, J. W. (1983). A goitrogenic agent from millet (*Pennisetum typhoides*) in Darfur Province, Western Sudan. *Annals of Nutrition and Metabolism, 27*(1), 14–18.

Padilla, G., Cartea, M. E., Velasco, P., de Haro, A., & Ordás, A. (2007). Variation of glucosinolates in vegetable crops of *Brassica rapa*. *Phytochemistry*, *68*(4), 536–545.

Palaniappan, A., Yuvaraj, S. S., Sonaimuthu, S., & Antony, U. (2017). Characterisation of xylan from rice bran and finger millet seed coat for functional food applications. *Journal of Cereal Science*, *75*, 296–305.

Paul, S., Geng, C. A., Yang, T. H., Yang, Y. P., & Chen, J. J. (2019). Phytochemical and health-beneficial progress of turnip (*Brassica rapa*). *Journal of Food Science*, *84*(1), 19–30.

Pellegriti, G., Frasca, F., Regalbuto, C., Squatrito, S., & Vigneri, R. (2013). Worldwide increasing incidence of thyroid cancer: Update on epidemiology and risk factors. *Journal of Cancer Epidemiology*, *2013*, 965212.

Petroski, W., & Minich, D. M. (2020). Is there such a thing as "anti-nutrients"? A narrative review of perceived problematic plant compounds. *Nutrients*, *12*(10), 2929.

Polat, U. (2010). The effects on metabolism of glucosinolates and theirs hydrolysis products. *Journal of Biological and Environmental Sciences*, *4*(10), 39–42.

Priya Verma, R. K., Lakhawat, S., Yadav, V. K., Gacem, A., Abbas, M., Yadav, K. K., Park, H. K., Jeon, B. H., & Mishra, S. (2023). Millets: A sustainable treasure house of bioactive components. *International Journal of Food Properties*, *26*(1), 1822–1840.

Puchta, H. (2017). Applying CRISPR/Cas for genome engineering in plants: The best is yet to come. *Current Opinion in Plant Biology*, *36*, 1–8.

Radkay, L. A., Chiosea, S. I., Seethala, R. R., Hodak, S. P., LeBeau, S. O., Yip, L., McCoy, K. L., Carty, S. E., Schoedel, K. E., Nikiforova, M. N., & Nikiforov, Y. E. (2014). Thyroid nodules with KssssRAS mutations are different from nodules with NRAS and HRAS mutations with regard to cytopathologic and histopathologic outcome characteristics. *Cancer Cytopathology*, *122*(12), 873–882.

Rungapamestry, V., Duncan, A. J., Fuller, Z., & Ratcliffe, B. (2006). Changes in glucosinolate concentrations, myrosinase activity, and production of metabolites of glucosinolates in cabbage (*Brassica oleracea* var. *capitata*) cooked for different durations. *Journal of Agricultural and Food Chemistry*, *54*(20), 7628–7634.

Šamec, D., Pavlović, I., & Salopek-Sondi, B. (2017). White cabbage (Brassica oleracea): Botanical, phytochemical and pharmacological overview. *Phytochemistry Reviews*, *16*, 117–135.

Šamec, D., & Salopek-Sondi, B. (2019). Cruciferous (Brassicaceae) vegetables. In S.M. Nabavi & T. Sanches Silva. (Eds.), *Nonvitamin and nonmineral nutritional supplements* (pp. 195–202), Academic Press.

Sartelet, H., Serghat, S., Lobstein, A., Ingenbleek, Y., Anton, R., Petitfrere, E., Aguie-Aguie, G., Martiny, L., & Haye, B. (1996). Flavonoids extracted from fonio millet (*Digitaria exilis*) reveal potent antithyroid properties. *Nutrition*, *12*(2), 100–106.

Schmeltz, L. R., Blevins, T. C., Aronoff, S. L., Ozer, K., Leffert, J. D., Goldberg, M. A., Horowitz, B. S., Bertenshaw, R. H., Troya, P., Cohen, A. E., & Lanier, R. K. (2014). Anatabine supplementation decreases thyroglobulin antibodies in patients with chronic lymphocytic autoimmune (Hashimoto's) thyroiditis: A randomized controlled clinical trial. *Journal of Clinical Endocrinology and Metabolism*, *99*(1), E137–E142.

Sesink, A. L. A., O'Leary, K. A., & Hollman, P. C. H. (2001). Quercetin glucuronides but not glucosides are present in human plasma after consumption of quercetin-3-glucoside or quercetin-4'-glucoside. *Journal of Nutrition*, *131*, 1938–1941.

Shakour, Z. T., Shehab, N. G., Gomaa, A. S., Wessjohann, L. A., & Farag, M. A. (2022). Metabolic and biotransformation effects on dietary glucosinolates, their bioavailability, catabolism and biological effects in different organisms. *Biotechnology Advances*, *54*, 107784.

Sharifi-Rad, J., Rajabi, S., Martorell, M., López, M. D., Toro, M. T., Barollo, S., Armanini, D., Fokou, P. V. T., Zagotto, G., Ribaudo, G., & Pezzani, R. (2020). Plant natural products with anti-thyroid cancer activity. *Fitoterapia*, *146*, 104640.

Shobana, S., & Malleshi, N. G. (2007). Preparation and functional properties of decorticated finger millet (*Eleusine coracana*). *Journal of Food Engineering*, *79*(2), 529–538.

Singh, J., Upadhyay, A. K., Prasad, K., Bahadur, A., & Rai, M. (2007). Variability of carotenes, vitamin C, E and phenolics in *Brassica* vegetables. *Journal of Food Composition and Analysis*, *20*, 106–112.

Singh, N., David, J., Thompkinson, D. K., Seelam, B. S., Rajput, H., & Morya, S. (2018). Effect of roasting on functional and phytochemical constituents of finger millet (*Eleusine coracana* L.). *Pharma Innovation Journal*, *7*(4), 414–418.

Skibola, C. F., & Smith, M. T. (2000). Potential health impacts of excessive flavonoid intake. *Free Radical Biology and Medicine*, *29*, 375–383.

Smith, C. E., & Tucker, K. L. (2011). Health benefits of cereal fibre: A review of clinical trials. *Nutrition Research Reviews*, *24*(1), 118–131.

Steinbrecher, A., Rohrmann, S., Timofeeva, M., Risch, A., Jansen, E., & Linseisen, J. (2010). Dietary glucosinolate intake, polymorphisms in selected biotransformation enzymes, and risk of prostate cancer. *Cancer epidemiology, Biomarkers and Prevention*, *19*(1), 135–143.

Taylor, J. R., & Duodu, K. G. (2015). Effects of processing sorghum and millets on their phenolic phytochemicals and the implications of this to the health-enhancing properties of sorghum and millet food and beverage products. *Journal of the Science of Food and Agriculture*, *95*(2), 225–237.

Triggiani, V., Tafaro, E., Giagulli, V. A., Sabbà, C., Resta, F., Licchelli, B., & Guastamacchia, E. (2009). Role of iodine, selenium and other micronutrients in thyroid function and disorders. *Endocrine, Metabolic and Immune Disorders-Drug Targets*, *9*(3), 277–294.

Vanderpas, J. (2006). Nutritional epidemiology and thyroid hormone metabolism. *Annual Review of Nutrition*, *26*(1), 293–322.

Vieites-Outes, C., López-Hernández, J., & Lage-Yusty, M. A. (2016). Modification of glucosinolates in turnip greens (*Brassica rapa*) subjected to culinary heat processes. *CyTA-Journal of Food*, *14*(4), 536–540.

Wagner, A. E., Terschluesen, A. M., & Rimbach, G. (2013). Health promoting effects of brassica-derived phytochemicals: From chemopreventive and anti-inflammatory activities to epigenetic regulation. *Oxidative Medicine and Cellular Longevity*, *2013*, 1–12.

Walle, T. (2004). Serial review: Flavonoids and isoflavones (phytoestrogens): Absorption, metabolism, and bioactivity. *Free Radical Biology and Medicine*, *36*, 829–837.

Wieczorek, M. N., Walczak, M., Skrzypczak-Zielińska, M., & Jeleń, H. H. (2018). Bitter taste of brassica vegetables: The role of genetic factors, receptors, isothiocyanates, glucosinolates, and flavour context. *Critical Reviews in Food Science and Nutrition*, *58*(18), 3130–3140.

Wright, D. A., Townsend, J. A., Winfrey, R. J., Irwin, P. A., Rajagopal, J., Lonosky, P. M., Hall, B. D., Jondle, M. D., & Voytas, D. F. (2005). High-frequency homologous recombination in plants mediated by zinc-finger nucleases. *Plant Journal*, *44*, 693–705.

Xiong, H. H., Lin, S. Y., Chen, L. L., Ouyang, K. H., & Wang, W. J. (2023). The interaction between flavonoids and intestinal microbes: A review. *Foods*, *12*(2), 320.

Zhang, L., Lin, G., Chang, Q., & Zuo, Z. (2005). Role of intestinal firstpass metabolism of baicalein in its absorption process. *Pharmaceutical Research*, *22*, 1050–1058.

Zhang, Z., Bergan, R., Shannon, J., Slatore, C. G., Bobe, G., & Takata, Y. (2018). The role of cruciferous vegetables and isothiocyanates for lung cancer prevention: Current status, challenges, and future research directions. *Molecular Nutrition and Food Research*, *62*(18), 1700936.

25 The Untapped Potential of Wild and Underutilized Fruits in the Development of Novel Commercial Products in Africa

*Makomborero Nyoni, Cyprian Mahuni, Charity Ngirande, Melisa Nomalanga Sithole, Faith Tariro Mukoko, Elizabeth Vimbai Tavengana, Janet Chiedza Zambuko, Shantel Moyo, Rumbidzai Lesley Chitsiku, Lissette Mupanda, Blessing Mwanalota, Munyaradzi Mativavarira, Prisca Nonceba Ncube, Rorisang Michelle Moyo, Tariro Sarah Zimuto, Tatenda Clive Murashiki, Rose Tafadzwa Masekesa, and Reward Muzerengwa

National Biotechnology Authority, 21 Princess Drive, Newlands, Harare, Zimbabwe

*Corresponding author: makomboreronyoni@gmail.com

25.1 INTRODUCTION

Underutilized wild fruits refer to a variety of fruit plant species that grow naturally in the wild and have not been extensively cultivated or commercialized (Bhat & Karim, 2009; Hunter et al., 2019). These fruit plants often thrive in diverse ecosystems and exhibit unique resistance to pests, diseases, and adverse climatic conditions (Peduruhewa et al., 2021). They are usually indigenous to specific regions and have been common in these local communities for their nutritional and medicinal benefits (Ford-Lloyd et al., 2011). Despite their potential, these fruits have not received much attention from the mainstream market due to limited knowledge and awareness (Peduruhewa et al., 2021). However, they are promising for their richness in phytochemicals, antioxidants, vitamins, and minerals, as well as their unique flavors and textures (Dandin et al., 2019). Moreover, underutilized wild fruits contribute to sustainable agriculture and biodiversity conservation and provide economic opportunities for rural communities (Li et al., 2016; Gyanaranjan et al., 2021).

They offer a diverse range of flavors, textures, and colors, providing a valuable dietary source (Bvenura & Sivakumar, 2017). Wild fruits have evolved over time in response to various environmental challenges, including extreme weather conditions and pests. By integrating wild fruit species into agricultural practices, farmers can enhance the resilience of their food systems in the face of climate change (Ford-Lloyd et al., 2011). Promoting the cultivation, processing, and marketing of wild fruits results in the improvement of livelihoods and rural development (Gyanaranjan et al., 2021). Wild and underutilized fruits often hold cultural and traditional importance for Indigenous communities as they are deeply rooted in local knowledge systems, folklore, and culinary traditions (Bvenura & Sivakumar, 2017). Diversifying food sources through the inclusion of wild and

 DOI: 10.1201/9781003469766-25

underutilized fruits can enhance food security. In times of crop failures or disruptions, these fruits can serve as alternative food sources, reducing dependence on staple crops (Sahoo et al., 2021).

To fully harness the potential of wild and underutilized fruits, efforts are needed to raise awareness, conduct research on their nutritional composition and agronomic practices (Singh et al., 2014), establish sustainable value chains (Aschemann-Witzel et al., 2023), and develop policies that support their conservation and utilization (Tata Ngome et al., 2017). Incorporating wild and underutilized fruits in the food industry expands the range of available products, leading to greater diversification in flavors, textures, and nutritional profiles (Gyanaranjan et al., 2021). Wild and underutilized fruits often possess high nutritional value, including vitamins, minerals, fiber, and antioxidants (Li et al., 2016).

Many wild fruits contain pigments that can be used as natural colorants in food products. Mazhanje (*Uapaca kirkiana*) is a popular wild fruit in Zimbabwe. Its bright orange flesh contains carotenoid pigments, such as beta-carotene, which can be used as natural yellow and orange colorants in food and beverage applications. These natural colorants offer an alternative to artificial food dyes, which are often associated with health concerns (Singh & Sudha, 2024).

With respect to the food industry, some wild fruits have thickening or gelling properties, making them suitable to produce jams, jellies, and sauces (Imeson, 2012). Examples of these underutilized fruits with thickening or gelling properties include kei apple, cape gooseberry, marula, and monkey orange, which are widely found in the Southern African region (Ngemakwe et al., 2017). Others may possess antimicrobial or antioxidant properties, which can extend the shelf life of food products and enhance their safety (Li et al., 2016). Instead of relying solely on traditional commercial crops, the food industry can diversify its supply chains and reduce pressure on land, water, and other resources (FAO, 2015). Consumers are increasingly interested in novel and sustainable food options, and products featuring wild fruits can capture their attention and loyalty (Gyanaranjan et al., 2021). The involvement of small-scale farmers, foragers, and Indigenous communities in the supply chain can make the food industry contribute to poverty alleviation and inclusive development (Chamberlain et al., 2020).

To fully realize the potential benefits, collaboration between the food industry, farmers, researchers, and policymakers is crucial. Investing in research and development, improving post-harvest technologies, and establishing sustainable value chains are key steps in harnessing the potential of wild and underutilized fruits in the food industry (Mabhaudhi et al., 2017).

25.2 OBJECTIVES OF THE CHAPTER

The chapter identifies wild and underutilized fruits in Africa that have not received significant attention in mainstream food systems but possess unique flavors, nutritional profiles, and potential health benefits. The chapter aims to explore the development of innovative food products from underutilized fruits. The chapter also emphasizes conducting sensory assessments of the developed food products. The chapter further addresses the importance of assessing the shelf life of the novel food products. The chapter discusses marketing strategies and potential collaborations with local farmers, processors, and food industry stakeholders to promote these products. The inclusion criteria for the selected underutilized fruits were based on commercialization potential and abundance. Comprehensive literature search was conducted on Scopus, Google Scholar, PubMed, PlantZAfrica, and Useful Tropical Plants. There was no consistency regarding the existing information on the selected underutilized fruits.

25.3 EXPLORATION AND IDENTIFICATION OF UNDERUTILIZED FRUITS

25.3.1 Definition and Characteristics of Wild and Underutilized Fruits

Despite many fruits that have been identified, there remains quite a number of wild fruits whose purpose has not yet been identified and, hence, are underutilized. The enhanced use of wild underutilized fruits can bring about better nutrition and fight hidden hunger and other dietary deficiencies, particularly among the rural poor and the more vulnerable social groups in developing countries (Kour et al., 2018).

Underutilized fruits or fruit plants are generally characterized as those that meet a certain criterion. The fruit must have a scientific or ethnobotanical proof of food value. It should also have been cultivated, either in the past or only being cultivated in a specific geographical area, and must be cultivated less than other conventional fruit plants. The plant or fruit must have weak or no formal seed supply system and should have indigenous uses in localized areas. It should also have received little attention from research, extension services, farmers, policy, and decision-makers, as well as technology providers. Additionally, the fruit should be highly nutritious and possess therapeutic medicinal as well as multiple uses (Kour et al., 2018). The criteria of Kour et al. (2018) have been adopted to identify these species.

25.3.2 Methods for Identifying and Selecting Potential Fruits

Different methods can be utilized in identifying and selecting potential wild underutilized fruits for various purposes, as stated earlier. These methods can be summarized as highlighted in Table 25.1. Combining these methods can help in identifying and selecting fruits with untapped potential. It is therefore imperative to consider a holistic approach that considers not only the botanical and nutritional aspects but also the cultural, environmental, and market factors to ensure successful utilization and cultivation of these fruits.

TABLE 25.1
Methods Used for Identifying and Selecting Potential Fruits

Method	Description
Botanical research	Studying plant taxonomy, distribution, and characteristics through literature review, online databases, and consultation with experts.
Ecological surveys	Botanical surveys carried out to identify and record all plant species in a certain area.
Indigenous knowledge/expert knowledge	To acquire indigenous knowledge, that is, understandings and philosophies developed by local communities with long histories and experiences of interaction with the underutilized plant species.
Ethnobotanical studies	Studying the interrelations between humans and plants, including plants used as food, as medicines, and for other economic applications, with the aim of exploring the importance of plants that are used as food, clothing, shelter, fodder, fuel, furniture, and medicine. These studies are useful in identifying, disseminating, and documenting indigenous knowledge and the application of plant diversity for human and livestock ailments (Mekonnen et al., 2022).
Nutritional analysis	Detection of the exact nutritional value of the identified underutilized fruits by determining the percentage of macro- and micronutrients present in the fruit, as well as the presence of inhibitors, toxic chemicals, or any other new component.
Flavor and taste evaluation	Determining the acceptability of the fruits before commercialization.
Market and demand trends	Identifying patterns in the sales of the identified fruits before commercialization. Market trends can be used to predict future demand for the fruits or products from these fruits.
Agronomic feasibility	Determining the practicality of farming the identified underutilized fruits in a different area. It provides a thorough and comprehensive study covering all potential development aspects for the given land.
Field observation and focus group discussion	Repeated field observations conducted in study sites by walking transect where most of the underutilized fruits are grown/cultivated (Abdullah et al., 2020).

25.3.3 Challenges and Considerations in Fruit Exploration

Several challenges have been associated with the full utilization and adaptation of wild underutilized fruits. There is still scarcity of research investment and development for the improvement of underutilized fruit trees. This is because most of these fruits are still sourced from the wild and natural environments, which have generally limited their potential for higher yield and growth (Omotayo & Aremu, 2020). Some of the challenges associated with fruit exploration and utilization include inadequate baseline data on the nutritional properties of individual indigenous fruit trees, indiscriminate and illegal logging, low level of acceptability and accessibility of indigenous fruits, insidious domination, and preference for the exotic fruit species.

25.4 NUTRITIONAL AND FUNCTIONAL PROPERTIES OF WILD AND UNDERUTILIZED FRUITS

25.4.1 Overview of Nutritional Compositions

Malnutrition and undernutrition continue to be a threat to public health in sub-Saharan Africa, with children and the poor being the most affected (Bvenura & Sivakumar, 2017). Wild fruits, often found growing in natural environments, are known for their diverse flavors, vibrant colors, and unique nutrient profiles (Table 25.2). Most of the health benefits and nutrition claims made for wild fruits occurring in sub-Saharan Africa have, however, not been validated by scientific research, with only a few, such as *Solanum nigrum* and *Trilepisium madagascariense*, having undergone detailed studies. More research is needed to fully understand the nutritional profiles and health impacts of these and other wild fruits in the region (Yimer et al., 2023). Some of these fruits also contain antinutritional factors. Examples of these fruits include kei apple, baobab fruit, monkey orange, marula, and cape gooseberry (Ngemakwe et al., 2017)

TABLE 25.2
Nutritional Composition of Selected Southern African Indigenous Fruits

Scientific Name	Energy	Water	Protein (g/100 g)	Fat (g/100 g)	Carbohydrate (g/100 g)	Fiber (g/100 g)	Ash (g/100 g	References
Carissa macrocarpa	100.3 [kcal/100 g fruit weight (FW)]	78.83	0.74	3.53	16.40		0.50	Souilem et al. (2019)
Carpobrotus edulis	1,370 kJ/100 g dry weight (DW)	77.6 wet mass (WM)	23.5 (DW)	2 (DW)	64.4 (DW)	–	10.1 (DW)	Broomhead et al. (2020)
	8.53 (kJ/fruit)	80.56	12	–	68	–	–	Vilà and D'Antonio (1998)
Dovyalis caffra	–	15.5 DW	4 DW	3 DW	54.05 DW	16.3 DW	7.45 DW	Taher et al. (2018)
Harpephyllum caffrum	–	87.5	0.7	0.2	9.1	1.7	0.8	Maroyi (2019)
Sclerocarya birrea	–	86.3	0.7	0.5	–	–	–	Stadlmayr et al. (2013)
	–	–	1.4	0.9	–	–	3.0	Magaia & Skog (2017)
	–	85–87	–	13.5	0.7–1.2	2.9	10–20	Hiwilepo-Van Hal et al. (2013)
Syzygium guineense	295 kJ/100 g	81.5	1.6	0.7	13.5	1.8	1.0	Stadlmayr et al. (2013)
	1,096 kJ/100 g	–	10.1	4	48.5	30.3	7.1	

25.4.2 Health Benefits and Bioactive Compounds

Bioactive molecules play a crucial role in our overall well-being by exerting positive influences on physiological processes within the body. This chapter explores the medical and nutritional value of bioactive molecules, highlighting their importance in promoting health and preventing diseases, as shown in Table 25.3. One notable group of bioactive molecules are antioxidants, such

TABLE 25.3
Bioactive Ingredients and Medicinal Properties of Selected Indigenous Tropical Fruits

Fruit	Bioactives	Medical Properties	Experimental Models
Averrhoa bilimbi	Flavonoids, saponins, and triterpenoids	Antihypercholesterolemic	Triton-induced hypercholesterolemic rats
		Antibacterial	Disk diffusion method: gram-positive and gram-negative bacteria
		Antidiabetes	Streptozotocin-induced diabetic rats
Baccaurea motleyana	Phenolic compounds	Antimicrobial (peel)	Disk diffusion method: gram-positive and gram-negative bacteria, fungus, and yeast
Canarium odontophyllum	Flavonoids and anthocyanins	Antiatherosclerosis	Cholesterol-induced hypercholesterolemic rabbits
Cynometra cauliflora	Phenolic compounds	Antileukemic	Human promyelocytic leukemia HL-60 and normal mouse fibroblast NIH/3T3 cell cultures
Durio kutejensis	Not reported	Antimelanogenesis effect	Tyrosinase assay and melanin inhibition in B16 melanoma cell cultures
Garcinia hombroniana	Phenolic compounds	Antimicrobial	Well diffusion method: pathogenic and nonpathogenic bacteria
Mangifera pajang	Phenolic compounds and carotenoids	Antihypercholesterolemic and antiatherosclerotic	Cholesterol-induced hypercholesterolemic rabbit model
		Anticancer (kernel)	MTT assay: HepG2, HT-29, and Caov3 cultures
		Hepatoprotective effect	HepG2 cell culture and western blot method
Phyllanthus emblica	Phenolic compounds	Gastric ulcer healing effect	Indomethacin-induced ulceration of rats
		Anticlastogenicity	Cochran–Armitage trend test: bone marrow cells of Swiss albino mice treated with lead and aluminum
		Antiproliferative	MTT assay: MCF-7 tumor cell culture
		Antimicrobial	TLC-bioautographic method: drug-resistant bacteria and yeast
		Anticancer	In vitro cytotoxicity assays: human lung carcinoma (A549) and HepG2 cell lines
		Antiaging effect	In vitro MMP-1, MMP-2, and elastase inhibition assays: inhibitions of collagenase and elastase
		Chondroprotection	In vitro enzymatic assays: explant cultures of cartilage from osteoarthritis patients
Syzygium jambos	Phenols, tannins, alkaloids, and flavonoids	Antifungal (seed)	Disk diffusion method: microbroth dilution technique (*Microsporum gypseum*, *Microsporum canis*, and *Candida albicans*)
Syzygium malaccense	Phenolic compounds and terpenes	Antimicrobial	Disk diffusion method: test bacteria on Mueller–Hinton agar, and yeast on potato dextrose agar
Ziziphus mauritiana	Phenolic compounds and saponin	Antihyperglycemic, antidiarrheal, and hepatoprotective	Glucose-overloaded hyperglycemic rats, castor oil–induced diarrhea in mice, and tetrachloromethane-induced liver damage in rats, respectively
		Anticancer	Neutral red assay: cytotoxicity of various cultivars of jujube against different cancer cell lines
			Apoptosis detection by flow cytometry

Note: The medicinal properties are reported based on in vitro and in vivo animal studies, as well as human intervention trials.

as vitamins C and E, carotenoids, and polyphenols. Antioxidants neutralize harmful free radicals, reducing oxidative stress and lowering the risk of chronic diseases, including cardiovascular diseases, cancer, and neurodegenerative disorders. Another group of bioactive molecules with significant medical value are phytochemicals. These compounds possess anti-inflammatory, anticancer, and immune-boosting properties. For example, curcumin, a bioactive molecule present in turmeric, has shown potential in preventing/treating various cancers, including breast, colon, and prostate cancer.

25.5 PROCESSING TECHNIQUES FOR WILD AND UNDERUTILIZED FRUITS

25.5.1 Post-Harvest Treatment Techniques

Once harvested, fruits remain alive, and their metabolic processes and functions continue. Some years back, it was not possible to improve the post-harvest quality of the fruits through any post-harvest technology, except for fruit color. Currently, depending on fruit crops, different post-harvest strategies are commercially practiced to adapt ripening to industry needs, delay senescence, maintain fruit quality attributes, and thus, prolong shelf life (Pott et al., 2020). In the last few decades, besides the traditional approaches, advanced post-harvest physical and chemical treatments (active packaging, dipping, vacuum impregnation, conventional heating, pulsed electric field, high hydrostatic pressure, and cold plasma) and biocontrol techniques have been implemented to preserve the nutritional value and safety of fresh produce (Palumbo et al., 2022).

25.5.1.1 Temperature Control

Refrigeration is one of the most common effective methods of preserving fruit quality, including attributes like texture, nutrition, aroma, and flavor. However, some fruits are sensitive to chilling injury when they are stored below their critical temperature (10°C) for even short durations and may result in pitting, uneven ripening, and fungal infestation of stored fruits. Examples of these fruits include *topi tambo* or *leren* (*Calathea allouia*), which is a very popular tuber in Caribbean Latin American countries, and palm fruit peewah (*Bactris gasipaes)* (Bridgemohan & Isaac, 2017). The required optimum temperatures (10–15°C at 85–95% relative humidity) can be achieved by using less-expensive methods of cooling, such as an evaporative cooling system.

The marula fruit can be refrigerated, but there is no longevity, as it is a seasonal fruit which comes once in a season; hence, there might be a need to also use drying as a complementary method to enhance its shelf life, where it can later be processed into other products, such as marula jam. Baobab fruit is normally sun-dried as it is first harvested while it is in dry form, and to enhance its longevity, a combination of methods, such as refrigeration and chemical preservatives, can be applied (Nemapare et al., 2023).

25.5.1.2 Heat Treatment

Post-harvest heat treatments using hot air and heated water have been reported to reduce chilling injuries in fruits like mangoes and oranges. Heat treating (37–42°C) prior to cold storage can slow down ripening while increasing pathogenic resistance and enhances or causes no change in some quality traits, particularly in modified atmosphere storage systems. Heat treatment before cold storage at 14°C can also increase total soluble solids (TSS) and titratable acids (TA) when fruits ripen as compared to the untreated fruits (Fawole et al., 2020). For monkey oranges, thermal processing is of utmost importance as all its known products undergo some form of thermal treatment (Ngadze et al., 2017a)

Borochov-Neori et al. (2008) found that marula juice had 56 mg/100 ml of pyrogallol equivalence of phenols and an antioxidant capacity of 382 mg/100 ml of vitamin C equivalence. Antioxidant stability toward heat treatment is important and is yet to be fully studied with regards to the marula fruit, even though thermal treatment is always applied in the food industry. Sun-dried monkey orange pulp can be stored from two months to five years (Shackleton et al., 2000), which makes thermal drying an ideal preservation method due to its affordability for rural communities and to secure continuous fruit availability into the next season. However, studies on the species' suitability for dried products, the drying kinetics, and the characteristics of the commonly consumed monkey oranges need to be conducted to avail suitable drying techniques that fit local conditions and offer the possibility of commercialization.

25.5.1.3 Storage

Packaging materials with a predetermined composition of gases (O_2 and CO_2) are used in modified atmosphere packaging (MAP). Examples of wild fruits that have used the MAP method include the baobab fruit, marula fruit, mongongo fruits, tamarind fruit, and wild berries. On the traditional storage methods, unripe wild fruits are usually stored and ripened by incubating them in a shallow pit in the ground and covering with soil, storing in tightly closed plastic containers (Chadzimura, 2016). Burying of unripe fruits is also a common practice in Malawi, Zambia, and Namibia and is used both as a preservation and fruit ripening method (Kadzere et al., 2007). Another storage approach employed by fruit collectors at the homestead or market is to keep ripe fruits in a cool dry area, such as a shade, at ambient temperatures (Kadzere et al., 2004). The marula fruit can be stored using the previously stated traditional method. Desert date (bushman's date), which is native to the Sahara Desert, can be stored in oil or honey (Kadzere et al., 2007). When picked, ripe monkey orange can be stored for up to two weeks under ambient conditions, mainly in shade, before they spoil (Ruffo et al., 2002)

25.5.1.4 Chemical Treatment

Administering 1-methylcyclopropene (1-MCP) can suppress and reduce the effects of ethylene production in harvested climacteric fruit. This can prolong storage life and prevent abscission in fruits (Passam, 2007). The marula fruit, baobab fruit, and tamarind have benefitted from this method. Use of nitric oxide, salicylic acid, and oxalic acid can also be employed on underutilized fruits to extend their shelf through slowing down their metabolism, as mentioned earlier (Passam, 2007)

The use of calcium chloride ($CaCl_2$) before and after harvesting has been proven to enhance the longevity and quality of various fruits. This is accomplished by slowing the ripening and ageing processes, decreasing respiration, preserving firmness, and preventing physiological disorders. $CaCl_2$ has been used as a preservative in monkey orange (Ngadze et al., 2017a). This method is yet to be fully applied in fruits like marula and baobab.

25.5.1.5 Edible Coatings

Edible coatings are semipermeable membranes on the fruit skin, creating a modified internal atmosphere, decreasing moisture loss and respiration rate (Bourtoom, 2008). Increased total sugars, extended shelf life, and retained firmness have been found using *aloe vera* as an edible coating material (Abd El-Gawad et al., 2019). Chitosan edible coating has been associated with inhibited microbial growth and decreased weight loss (Zhu et al., 2008). The gum arabic (GA) coating improves the antioxidant pool (Khaliq et al., 2016a), which scavenges excess reactive oxygen species (ROS), thus decreasing oxidative damage to the fruit. Carboxymethyl cellulose (CMC) coating can form a semipermeable barrier, thereby modifying the internal fruit atmosphere, limiting the exchange of gases into and out of the fruit, and hence reducing respiration rate and delaying senescence (Gava et al., 2018). There is potential to use edible coatings like chitosan in the marula fruit, where the chitosan can be sprayed on the fruit or the fruits can be dipped in chitosan solutions (Xing et al., 2016)

TABLE 25.4
Harvesting and Post-Harvesting Techniques of Underutilized Fruits

Harvesting	Careful handling during harvest to reduce bruising, scratching, and punctures; harvesting during the cooler hours of the day (e.g., the early morning); shading crops once harvested.
Handling	Protecting the crops from injury can minimize pest attacks and physiological and dehydration damage.
Sorting and cleaning	Sorting and cleaning can increase shelf life considerably. By separating higher- and lower-quality crops, the risk that fungi or bacteria spread from damaged crops to others is reduced. Quality parameters like size and color can be determined through the use of visual charts and allows the crops to be targeted to appropriate markets to maximize revenue.
Packaging	Proper packaging to maintain freshness prevents quality deterioration, as well as protecting against physical damage during transportation. Clean, smooth, and ventilated containers are key, but the specific type depends on the crop.
Transportation	Use of clean, cool, ventilated, and covered vehicles for the transport of perishable crops, with transport during the colder hours of the day, is advised. The smoothness of the road is also important, as excessive vibrations and movement can degrade crop quality. Avoid watering the produce before transport, as this increases decay. Care during loading and unloading is a simple yet effective way to reduce loss.
Storage	Only crops that meet specific quality standards should be stored (correct level of maturity, undamaged). Optimal temperatures for each commodity should be known and used, as shelf life is longer when stored in optimal temperature conditions.
Processing	Processing allows producers to stabilize the produce and diversify the food supply for enhanced nutrition throughout the year, which generates employment. Drying, salting, fermenting, and pickling are among the simpler processing technologies.

Source: Adapted from Kitinoja & Tokala (2018).

25.6 PRODUCT DEVELOPMENT FROM WILD AND UNDERUTILIZED FRUITS – VALUE ADDITION

Integrating the use of wild and underutilized fruits into the food industry is an innovative way to promote biodiversity, sustainable agriculture, and nutrition. These fruits, commonly found in natural ecosystems, have long been neglected due to their limited commercial value and lack of cultivation. However, their unique flavor, nutritional profile, and elastic properties make them an excellent choice for inclusion in various culinary applications. There are several ways to incorporate wild and less commonly used fruits into food and beverage products. Selected African underutilized edible fruits in Africa, parts of fruit used as ingredients, indigenous products made, and their preservation methods are summarized in Table 25.5. One popular method is to use them as an ingredient in juices, smoothies, and fruit drinks. With respect to beverages, wild and underutilized fruits can be incorporated into a wide range of food products (Jimenez, 2024). By exploring the culinary potential of these fruits, we can tap into nature's rich diversity and create a more sustainable and delicious food system in Africa.

25.7 SENSORY EVALUATION AND CONSUMER ACCEPTANCE

25.7.1 Importance of Sensory Evaluation in Product Development

Sensory evaluation involves the systematic analysis and interpretation of human senses, such as taste, smell, texture, appearance, and sound, to assess the qualities and attributes of a product (Mihafu et al., 2020). By conducting sensory tests and gathering feedback from target consumers, companies can identify the sensory attributes that drive consumer preference and incorporate them into their product formulations (Viejo et al., 2019).

TABLE 25.5
Methods of Sensory Evaluation in Product Development

Method	Description
Descriptive analysis	It involves a trained panel of sensory experts who systematically evaluate and describe the sensory attributes of a product. In the case of wild and underutilized fruits, a panel would assess attributes such as flavor, aroma, texture, appearance, and overall acceptability. The panelists use a standardized vocabulary and scoring system to objectively describe and rate each attribute. This method provides detailed information about the sensory profile of the fruits, enabling product developers to understand their unique characteristics and potential applications.
Hedonic testing	It focuses on assessing consumer preferences and acceptance of products. It involves a larger group of consumers who evaluate the sensory attributes of different products made from wild and underutilized fruits. Participants rate their liking or preference for each product using a scale, providing insights into consumer preferences. This method helps determine which sensory attributes are most appealing to consumers and can guide the development of products that align with their tastes.
Triangle test	The triangle test is a discrimination test used to determine whether there is a perceptible difference between two samples. In the context of wild and underutilized fruits, this test can be used to evaluate if there are sensory differences between different varieties or preparations of the fruits. Panelists are presented with three samples, two of which are identical, while the third is different. Panelists must identify the odd sample. The test helps in evaluating subtle sensory differences and identifying preferred varieties or processing methods.
Ranking test	It involves presenting panelists with multiple samples and asking them to rank them based on specific sensory attributes or overall preference. This method can be used to compare different varieties or preparations of wild and underutilized fruits based on their sensory qualities. By analyzing the rankings, product developers can gain insights into the relative sensory attributes of the samples and make informed decisions about which varieties or preparations are more desirable.
Time–intensity analysis	The method measures the intensity of specific sensory attributes over time. Panelists evaluate the samples and continuously rate the intensity of a particular attribute, such as sweetness or tartness, at regular intervals. This analysis allows for the understanding of the temporal perception of sensory attributes and their duration. It is particularly useful when evaluating the flavor profiles of wild and underutilized fruits, as it provides insights into the dynamic sensory experiences they offer.
Acceptance testing	It involves presenting consumers with samples of products made from wild and underutilized fruits and gathering their overall acceptance/liking scores. This method helps assess the market potential and consumer acceptance of the products. It can be conducted using various scales or questionnaires to capture consumers' sensory preferences, purchase intent, and willingness to pay.

The sensory methods have been used in Africa, and narrowing it down to Zimbabwe, these methods have been used through unrecorded surveys and unpublished data. In Zimbabwe, these surveys were used to develop marula fruit products mainly through the use of indigenous knowledge systems (IKS) in Mwenezi District, Zimbabwe. Masawu fruit was also used to develop food products through sensory methods that were not published. Research is still underway regarding the product development of these underutilized crops as there are plans of commercializing the products.

25.7.2 Consumer Preference

Conducting sensory tests with target consumer groups can make developers understand which sensory attributes are most appealing and use this knowledge to optimize product formulations. Acceptance and preference of the sensory properties of foods have been and are still among the most important criteria determining food choice (Lawless et al., 2010). There is much empirical research showcasing the effect that our senses have on our perception, affective response to food products, and our food choices (Köster, 2009). This effect of the senses is of course also affected by both the aforementioned intrinsic food product factors as well as extrinsic factors in a multitude of manners, both independently and in synergy (Wang et al., 2019).

25.7.3 Methods of Sensory Analysis

In product development involving wild and underutilized fruits, several methods of sensory analysis can be used to evaluate and understand the sensory attributes of these fruits. Here are some commonly employed methods, as shown in Table 25.5.

The mentioned sensory analysis methods provide valuable information about the sensory attributes, consumer preferences, and market potential of products developed from wild and underutilized fruits. By employing a combination of these methods, product developers can gain comprehensive insights into the sensory properties of these fruits and use that knowledge to create appealing and successful products.

25.8 CASE STUDIES AND SUCCESS STORIES

African indigenous fruits have been successfully utilized in product development and subsequent commercialization. African underutilized edible fruits with potential for development of commercial products are listed in Table 25.6. Most underutilized indigenous fruits that have gained recognition include baobab and marula. Baobab and marula are native to sub-Saharan Africa. Across the region, baobab products are being exported to the EU, UK, and South African markets. Products from baobab include the baobab powder, paste, tea, and coffee. The baobab product market is estimated to reach 10.1 billion by 2027 (Chebet, 2022).

Zimbabwe is the leading producer of baobab with most exports to European markets. One of the largest companies that supplies baobab products is B'Ayoba, a natural ingredients company in Binga under the PhytoTrade Africa Southern Africa's natural products trade association. B'Ayoba's baobab products include baobab powder, baobab paste, and red baobab tea. These products have been certified to EU and NOP organic standards. The FairWild certification scheme which was initiated by the FairWild Foundation in partnership with TRAFFIC helps ensure sustainability to collectors of wild plants ingredients as well as the implementation of the resource management program, as well as ensuring fair trade. Through this initiative, B'Ayoba has partnered with the Organic Herb Trading company, which is a major EU distributor (Traffic, 2017).

Essential Delights, a baobab fruit juice company in Beitbridge, Zimbabwe, produces a wide range of products and supplies them to hotels and retail supermarkets. These include juice, mahewu, opaque beer, yoghurt, energy drink, and instant porridge (Chapanduka, 2023). Other Zimbabwean companies that have produced baobab beverages include Dairibord Ltd., Top-One (Pvt.) Ltd., and Ekhaya (Pvt.) Ltd. The Southern Liquor Company of South Africa was the first to produce marula fruit spirit in 1983 as Amarula and commercialized in 1989. In 2009, Amarula gained international recognition after winning a gold medal at the San Francisco World Spirits Competition (Japhe, 2023).

In Zimbabwe, a marula processing and value addition plant to produce marula wine was launched in 2021. The wine has been commercialized and is successfully penetrating the market. Through the National Biotechnology Authority of Zimbabwe (NBA), more products are being developed using the marula fruit.

A case study of baobab fruit in Malawi focused on the promotion of baobab fruit as a sustainable source of nutrition and income for rural communities. Efforts were made to raise awareness about the nutritional benefits of baobab fruit while developing value-added products and creating market linkages for smallholder farmers (Darr et al., 2022). In Nigeria, a community-led initiative was launched that aimed to unlock the economic potential of the tamarind fruit. By establishing tamarind processing facilities and promoting tamarind-based products, the initiative has created employment opportunities and improved the livelihoods of local communities (Toungos, 2019).

TABLE 25.6
African Underutilized Indigenous Edible Fruits, Products, and Preservation Methods

Underutilized Indigenous Fruits	Common Name	Parts Used as Ingredients	Products Made from Fruit	Modes of Preparation and Preservation	References
Uapaca kirkiana	Wild loquat	Pulp	Porridge, cakes, sweet wine	To make porridge, pulp is mixed with water, maize, millet, or sorghum meal. Fruit is pounded during the peeling of skins. To make an opaque or gray sweet wine, pulp is mixed with cold water and fermented.	Chawafambira et al. (2020); Ngadze (2018)
Berchemia discolor	Brown ivory	Pulp, seeds, peels	Juice and beer	Fresh fruit soaked in water and pounded during skin peeling or to expose the pulp. The pulp is fermented into beer overnight. Pulp is filtered to make juice. Seeds, peel, and pulp are ground and dried under the sun to make a powder for consumption.	Du Preez et al. (2013)
Ximenia caffra	Blue sour plum	Pulp and seeds	Juice, sweets, oil	To make juice, the pulp is typically mixed with water and then filtered. Seeds are pressed to make oil.	Du Preez et al. (2013)
Parinari curatellifolia	Mobola plum	Pulp and seeds	Porridge, beer, juice, dried food, oil, jam	To brew beer, fruits are pounded, and juice is extracted by squeezing. Juice is boiled, left to stand overnight, boiled again, cooled, and consumed. To make porridge, pulp juice is mixed with millet or sorghum and roasted to make a dry powder. Powder is dried under the sun for 3 to 4 days, which helps enhance its sweetness.	Du Preez et al. (2013); Gadaga et al. (1999)
Strychnos spinosa	Monkey apple	Pulp	Jam, jellies, fruit wines, dry fruit rolls	To make jam, pulp is mixed with sugar. Pulp can be sun-dried and fried to make dry fruit rolls.	Van Wyk (2011)
Strychnos cocculoides	Monkey orange		Juice, porridge, cake, bread, beer, jam, fermented liquid mealie-meal porridge	The fruit is cracked open and the pulp is soaked in water for 1–2 days and the pulp is separated from the seeds by filtration. To make porridge, the pulp is mixed with maize meal and boiled. To prepare fermented liquid mealie-meal porridge, the pulp is mixed with stiff maize meal and malted flour. The mixture is left to ferment until a slurry is formed. When preparing jam, the pulp is mixed with sugar and lemon and cooked until thick. When making juice, mix the pulp with an equal amount of water. Add lemon for a refreshing drink. To make cakes, the pulp is sieved and mixed with flour and baked. The pulp may be sun- or fire-dried by first treating the pulp with salt or sugar.	Van Wyk (2011); Bille et al. (2013); Ngadze et al. (2017b)
Opuntia ficus-indica	Sweet prickly pear	Pulp	Juice, beer	To make juice, the fruit is mixed with water, beaten to a pulp, and juice is extracted through filtration using a cloth. To brew beer, the juice is left to ferment for hours or overnight.	Rampedi and Olivier (2013)

(Continued)

TABLE 25.6 (Continued)

Underutilized Indigenous Fruits	Common Name	Parts Used as Ingredients	Products Made from Fruit	Modes of Preparation and Preservation	References
Sclerocarya birrea	Marula	Pulp and seeds	Juice, wine liquor, flavorant, sweets	Spontaneous fermentation of a mash obtained from the fruits. Ripe fruits washed, skins removed. Fleshy stones soaked in cold water for 3 days, removed, and juice allowed to ferment for 2 days. Juice is filtered to give wine, or ripe fruits are left to soak in cold water or ground to remove skins and stones. The pulp is mixed with an equal volume of water in a pot left to ferment overnight and is ready for drinking the next day.	Van Wyk (2011)
Azanza garckeana	Snot apple	Pulp	Jam	To make jam, fruit can be dried and preserved by boiling in salty water; the fruit pulp can be mixed with sugar and other ingredients.	Shava (2005)
Adansonia digitata	Baobab	Pulp/seeds/leaves	Dried powder, porridge, health drinks, leaves used as relish, seeds as coffee	Pulp can be dried, pounded, and added to water or milk to serve as a refreshing drink. The pulp can be sieved and added to a warm millet/sorghum gruel with water or milk. Fruit powder can be mixed with water, sweetened, and frozen to be consumed as an ice lollipop. Fruit powder can be added to sauces and soups as a flavorant. Seeds can be roasted and kernel eaten as a snack. Seeds can be pounded into flour which is mixed with milk and sweetened with honey to make a refreshing drink. Ground and roasted seeds can be used as coffee substitutes.	Kabore (2011); Shava (2005)
Ficus capensis	Cape fig	Figs	Dessert	Turned frequently and flattened by hand before sun-drying. To make jam, clean figs are boiled until soft, and sieved, then pulp is re-boiled with equal quantities of sugar until the jam sets. To make fig pudding, chopped figs and breadcrumbs are mixed and served with honey.	Tredgold (1986)
Piliostigma thonningii	Camel's foot/ monkey bread	Pods/seeds	Porridge	To make porridge, dried fruit is peeled and pounded into mealie-meal which is mixed with water and then boiled.	Tredgold (1986)
Ziziphus mauritiana	Buffalo thorn	Pulp	Juice, alcoholic beverages or spirits, sweets	To make alcoholic beverages or spirits, fruits are crushed, soaked in water for hours, and allowed to ferment. Liquid is distilled into a potent spirit. To make sweets, fruits are pounded, dried, mixed with water, stuck together with sugar, and dried.	Van Wyk (2011); Nyanga et al. (2008)

In Botswana, there has been successful commercialization of monkey orange fruit. Through value chain development, product diversification, and branding strategies, the project has transformed monkey orange fruit into a popular and profitable commodity, both domestically and internationally (Leakey et al., 2022).

The Swaziland Indigenous Fruit Value Chain Development Project was launched in Swaziland, which sought to promote the sustainable use of indigenous fruits, including marula, sour plum, and wild medlar. The project focused on training local communities in sustainable harvesting practices, processing techniques, and value addition. It also raised awareness about the nutritional value and economic potential of underutilized wild fruits in Swaziland (Maziya-Dixon & Dlamini, 2012).

In Namibia, the Namibian Indigenous Fruit Processing Initiative sought to promote the utilization of indigenous fruits, such as wild melon and nara melon. The project focused on establishing community-based processing facilities, training local women in fruit processing techniques, and creating market linkages for fruit products. The initiative led to the establishment of successful community-owned fruit processing enterprises, providing employment and income opportunities for women in rural areas. It also contributed to the conservation of indigenous fruit species and the promotion of sustainable harvesting practices (Shackleton & Shackleton, 2004).

25.9 IMPORTANT CONTRIBUTIONS OF WILD AND UNDERUTILIZED FRUIT TO THE FOOD INDUSTRY

The African continent is rich in biodiversity, which includes a wide variety of wild and underutilized fruit species, and Zimbabwe is no exception. The important contributions of these fruits to the food industry in Africa stem from their nutritional value, the economic prospects, and their potential for sustainable development. For example, the baobab fruit is known for its high vitamin C content and is a potent source of calcium, magnesium, and potassium. The marula fruit contains high levels of vitamins C and E and essential fatty acids. The African star apple (*Chrysophyllum albidum*) has been found to have anti-inflammatory, antidiabetic, and antimicrobial properties. The tamarind fruit is known for its antioxidant and anti-inflammatory effects. By harnessing their benefits, the food industry can foster inclusive growth, improve food security, and promote sustainable practices, ultimately contributing to the well-being of African populations and the preservation of its rich biodiversity (Ngemakwe et al., 2017)

25.10 CONCLUSIONS AND PROSPECTS

Research and development focused on value addition, breeding, and utilization of underutilized wild fruits in Africa hold immense potential for improving food security, promoting sustainable agriculture, and boosting economic growth. It can provide an enabling route for Africa to harness the full potential of its rich biodiversity, enhance the nutritional value of its food systems, and create economic opportunities for local communities. It is imperative that further research be conducted on the identification and characterization of underutilized wild fruits, genetic improvement and breeding programs, value addition and processing techniques, market research and consumer preferences, capacity building and knowledge transfer, policy support and institutional frameworks, and promotion of sustainable agriculture practices.

25.11 ACKNOWLEDGMENTS

We appreciate the participation of African rural communities in providing us with information around wild and underutilized fruits through IKS, with specific mention to the Mwenezi Rural Community, Zimbabwe, which was engaged in the launch of the Mapfura Project as part of giving value to the marula fruit in Zimbabwe. Special acknowledgments are extended to the volunteers who participated in sensory evaluations of some of the food products mentioned as part of product development. We also acknowledge the governments of various African countries for facilitating the recognition and utilization of the wild and underutilized fruits discussed in this chapter.

REFERENCES

Abd El-Gawad, M. G., Zaki, Z. A., & Ekbal, Z. A. (2019). Effect of some postharvest treatments on quality of "Alphonse" mango fruits during cold storage. *Middle East Journal*, *8*(4), 1067–1079.

Abdullah, M. R., Haque, M. E., Sarwar, A. K. M. G., Ashrafuzzaman, M., & Rahman, M. M. (2020). Diversity of underutilized fruits and their uses in Karnaphuli range, Rangamati, Bangladesh. *International Journal for Ecology & Environment*, *1*, 10–20.

Aschemann-Witzel, J., Bizzo, H. R., Doria Chaves, A. C. S., Faria-Machado, A. F., Gomes Soares, A., de Oliveira Fonseca, M. J., Kidmose, U., & Rosenthal, A. (2023). Sustainable use of tropical fruits? Challenges and opportunities of applying the waste-to-value concept to international value chains. *Critical Reviews in Food Science and Nutrition*, *63*(10), 1339–1351.

Bhat, R., & Karim, A. A. (2009). Exploring the nutritional potential of wild and underutilized legumes. *Comprehensive Reviews in Food Science and Food Safety*, *8*(4), 305–331.

Bille, P. G., Shikongo-Nambabi, M., & Cheikhyoussef, A. (2013). Value addition and processed products of three indigenous fruits in Namibia. *African Journal of Food Agriculture Nutrition & Development*, *13*(1), 7192–7212.

Borochov-Neori, H., Judeinstein, S., Greenberg, A., Fuhrman, B., Attias, J., Volkova, N., Hayek, T., & Aviram, M. (2008). Phenolic antioxidants and antiatherogenic effects of marula (*Sclerocarrya birrea* subsp. *caffra*) fruit juice in healthy humans. *Journal of Agricultural and Food Chemistry*, *56*, 9884–9891.

Bourtoom, T. (2008). Edible films and coatings: Characteristics and properties. *International Food Research Journal*, *15*(3), 237–248.

Bridgemohan, P., & Isaac, W. A. P. (2017). Postharvest handling of indigenous and underutilized fruits in Trinidad and Tobago. *Postharvest Handling*, *165*.

Broomhead, N. K., Moodley, R., & Jonnalagadda, S. B. (2020). Chemical and elemental analysis of the edible fruit of five Carpobrotus species from South Africa: Assessment of nutritional value and potential metal toxicity. *International Journal of Environmental Health Research*, *30*(4), 357–371.

Bvenura, C., & Sivakumar D. (2017). The role of wild fruits and vegetables in delivering a balanced and healthy diet. *Food Research International*, *99*(Pt. 1), 15–30.

Chadzimura, J. (2016). *Analysis of uses, marketing and contribution of Uapaca Kirkiana to household income in Goromonzi district, Zimbabwe* [MSc dissertation, Bindura University of Science and Education].

Chamberlain, J. L., Darr, D., & Meinhold, K. (2020). Rediscovering the contributions of forests and trees 'to transition global food systems. *Forests*, *11*(10), 1098.

Chapanduka, C. (2023, March 15). Baobab fruit juice firm sweetens lives. *The Herald*. www.herald.co.zw/baobab-fruit-juice-firm-sweetens-lives/

Chawafambira, A., Sedibe, M. M., Mpofu, A., Achilonu, M., & Yildiz, F. (2020). Uapaca kirkiana, an indigenous fruit tree in Sub-Saharan Africa: A comprehensive review. *Cogent Food and Agriculture Journal*, *6*(1), 1766735.

Chebet, C. (2022, November 28), Global baobab powder market will reach Sh1.2tr in five years, *The Standard*. www.standardmedia.co.ke/health/health-science/article/2001461836/global-baobab-powder-market-will-reach-sh12tr-in-five-years

Dandin, S. B., Kencharaddi, H. G., Kumar, N. K., & Chikkanna, G. S. (2019). Genetic resources, diversity, conservation and utilization of underutilized fruit crops. *Exploration of Processing and Value Addition Prospects of Underutilized Fruits*, *1*, 1–10.

Darr, D., Dumenu, W. K., Gebauer, J., Kasulo, V., Kleinke, M., Meinhold, K., Munthali, C., & Wichern, F. (2022). The potential of non-timber forest products to contribute to the bioeconomy transition: the example of baobab (Adansonia digitata L.) in Malawi. In Carsten Smith-Hall & James Chamberlain (Eds.), *The bioeconomy and non-timber forest products* (pp. 148–163). Routledge.

Du Preez, R. J., De Jager, K., Welgemoed, C. P., Maphanga, O., Mhlophe, S. D., & Hansmann, C. (2013). Preliminary investigation into the potential for commercialization of indigenous fruits in rural communities: Survey, fruit composition, propagation and product development. *Acta Horticulturae*, *1007*, 613–626.

FAO. (2015). The state of food insecurity in the world 2015. In *Meeting the 2015 international hunger targets: Taking stock of uneven progress* (pp. 1–54). Food and Agriculture Organization of the United Nation.

Fawole, O. A., Julian, A., Ebrahiema, A., & Umezuruike, O. O. (2020). Postharvest physiological responses of pomegranate fruit (cv. Wonderful) to exogenous putrescine treatment and effects on physico-chemical and phytochemical properties. *Food Science and Human Wellness*, *9*(2), 146–161.

Ford-Lloyd, B. V., Schmidt, M., Armstrong, S. J., Barazani, O. Z., Engels, J., Hadas, R., Hammer, K., Kell, S. P., Kang, D., Khoshbakht, K., & Li, Y. (2011). Crop wild relatives – undervalued, underutilized and under threat? *BioScience*, *61*(7), 559–565.

Gadaga, T. H., Mutukumira, A. N., Narvhus, J. A., & Feresu, S. B. (1999). A review of traditional fermented foods and beverages of Zimbabwe. *International Journal of Food Microbiology*, *53*(1), 1–11.

Gava, C. A. T., De Castro, A. P. C., Pereira, C. A., & Fernandes-Júnior, P. I. (2018). Isolation of fruit colonizer yeasts and screening against mango decay caused by multiple pathogens. *Biological Control*, *117*, 137–146.

Gyanaranjan, S., Swamy, S. L. R., Wani, S., & Mishra, A. A. (2021). Exploitation of wild leafy vegetables and underutilised fruits: Consequences for food and nutritional security. *Annals of the Romanian Society for Cell Biology*, *25*, 5656–5668.

Hiwilepo-Van Hal, P., Robben, J., Verkerk, R., & van Boekel, M. A. J. (2013). Optimising the juice yield and quality of marula fruit (Sclerocarya birrea subsp. Caffra) with pectolytic enzymes by a response surface method. *Processing of Marula (Sclerocarya Birrea Subsp Caffra) Fruits: A Case Study on Health-Promoting Compounds in Marula Pulp*, *51*, 49–51.

Hunter, D., Borelli, T., Beltrame, D. M., Oliveira, C. N., Coradin, L., Wasike, V. W., Wasilwa, L., Mwai, J., Manjella, A., Samarasinghe, G. W., & Madhujith, T. (2019). The potential of neglected and underutilized species for improving diets and nutrition. *Planta*, *250*, 709–729.

Imeson, A. P. (2012). *Thickening and gelling agents for food*. Springer; Science & Business Media.

Japhe, B. (2023, March 26). How Amarula became one of the world's most popular cream liqueurs. *Forbes*. www.forbes.com/sites/bradjaphe/2023/03/26/how-did-amarula-become-the-worlds-most-popular-cream-liqueur/?sh=6b09be1310a9

Jimenez, M. J. A., Ferreira, V. S., Ampese, L. C., Marinho, L., Rostagno, M. A., & Carneiro, T. F. (2024). An overview of the ellagic acid and proanthocyanidins' polyphenols from cambuci (Campomanesia Phaea Berg): Myrtaceae's family. *European Food Research and Technology*, *250*(3), 859–876.

Kabore, D., Sawatogo-Lingan, H., Diawara, B., Campaworee, C. S., Dicko, M. H., & Jakobsen, M. (2011). A review of baoba (Adansonia digitata) products: Effects of processing techniques, medicinal properties and uses. *African Journal of Food Science*, *5*(16), 833–844.

Kadzere, I., Hove, L., Gatsi, T., Masarirambi, M. T., Tapfumaneyi, L., Maforimbo, E., & Magumise, I. (2004). Current status of post-harvest handling and traditional processing of indigenous fruits in Zimbabwe. In M. R. Rao & K. Wesiga (Eds.), *Proceeding of regional agroforestry conference on agroforestry impacts on livelihoods in Southern Africa: Putting research into practice* (pp. 353–363). World Agroforestry Centre (ICRAF).

Kadzere, I., Watkins, C. B., Merwin, I. A., Akinnifesi, F. K., & Saka, J. D. K. (2007). Harvest date affects colour and Soluble Solids Concentrations (SSC) of Uapaca Kirkiana (Muell. Arg.) fruits from natural woodlands. *Journal of Agroforestry Systems*, *69*(2), 167–173.

Khaliq, G., Mohamed, M. T. M., Ghazali, H. M., Ding, P., & Ali. (2016a). Influence of gum arabic coating enriched with calcium chloride on physiological, biochemical and quality responses of mango (Mangifera indica L.) fruit stored under low temperature stress. *Postharvest Biology and Technology*, *111*, 362–369.

Kitinoja, L., & Tokala, V. Y. (2018). Postharvest education, training and capacity building for reducing losses in plant-based food crops: A critical review (2010–2017). *Postharvest Extension and Capacity Building for the Developing World*, *2018*, 29–40.

Köster, E. P. (2009). Diversity in the determinants of food choice: A psychological perspective. *Food Quality and Preference*, *20*, 70–82.

Kour, S., Bakshi, P., Sharma, A., Wali, V. K., Jasrotia, A., & Kumari, S. (2018). Strategies on conservation, improvement and utilization of underutilized fruit crops. *International Journal of Current Microbiology and Applied Sciences*, *7*(3), 638–650.

Lawless, H. T., & Heymann, H. (2010). Introduction. In H. T. Lawless & H. Heymann (Eds.), *Sensory evaluation of food: Principles and practices* (2nd ed., pp. 1–2). Springer Science and Business Media.

Leakey, R. R., Tientcheu Avana, M. L., Awazi, N. P., Assogbadjo, A. E., Mabhaudhi, T., Hendre, P. S., Degrande, D., Hlahla, S., & Manda, L. (2022). The future of food: Domestication and commercialization of indigenous food crops in Africa over the third decade (2012–2021). *Sustainability*, *14*(4), 2355.

Li, Y., Zhang, J. J., Xu, D. P., Zhou, T., Zhou, Y., Li, S., & Li, H. B. (2016). Bioactivities and health benefits of wild fruits. *International Journal of Molecular Sciences*, *17*, 1258.

Mabhaudhi, T., Chimonyo, V. G., & Modi, A. T. (2017). Status of underutilised crops in South Africa: Opportunities for developing research capacity. *Sustainability*, *9*(9), 1569.

Magaia, T. L. J., & Skog, K. (2017). Composition of amino acids, fatty acids and dietary fibre monomers in kernels of Adansonia digitate and Sclerocarya birrea. *African Journal of Food, Agriculture, Nutrition and Development*, *17*(3), 12441–12454.

Maroyi, A. (2019). Medicinal uses, biological and chemical properties of Wild Plum (Harpephyllum caffrum): An indigenous fruit plant of Southern Africa. *Journal of Pharmacy and Nutrition Sciences*, *9*(5), 258–268.

Maziya-Dixon, B., & Dlamini, N. R. (2012). Indigenous fruit trees in Swaziland: Contribution to improved household food security and nutrition. In F. K. Akinnifesi, R. R. B. Leakey, O. C. Ajayi, G. Sileshi, Z. Tchoundjeu, P. Matakala, & F. Kwesiga (Eds.), *Indigenous fruit trees in the tropics: Domestication, utilization and commercialization*. CABI.

Mekonnen, A. B., Mohammed, A. S., & Tefera, A. K. (2022). Ethnobotanical study of traditional medicinal plants used to treat human and animal diseases in sedie Muja District, South Gondar, Ethiopia. *Evidence-based complementary and alternative medicine: eCAM*, *2022*, 7328613. https://doi.org/10.1155/2022/7328613

Mihafu, F. D., Issa, J. Y., & Kamiyango, M. W. (2020). Implication of sensory evaluation and quality assessment in food product development: A review. *Current Research in Nutrition and Food Science Journal*, *8*(3), 690–702.

Nemapare, P., Gadaga, T. H., & Mugadza, D. T. (2023). Edible indigenous fruits in Zimbabwe: A review on the post-harvest handling, processing, and commercial value. *Cogent Social Sciences*, *9*(1).

Ngadze, R. T. (2018). *Value addition of Southern African monkey Orange (Strychnos Spp.): Composition, utilization and quality* [PhD thesis, Wageningen University]. https://doi.org/10.18174/459128

Ngadze, R. T., Linnemann, A. R., Nyanga, L. K., Fogliano, V., & Verkerk, R. (2017b). Local processing and nutritional composition of indigenous fruits: The case of monkey orange (*Strychnos* spp.) from Southern Africa. *Food Reviews International*, *33*(2), 123–142.

Ngadze, R. T., Verkerk, R., Nyanga, L. K., Fogliano, V., & Linnemann, A. R. (2017a). Improvement of traditional processing of local monkey orange (Strychnos Spp.) fruits to enhance nutrition security in Zimbabwe. *Food Security Journal*, *9*(3), 621–633.

Ngemakwe, P. N., Remize, F., Thaoge, M. L., & Sivakumar, D. (2017). Phytochemical and nutritional properties of underutilised fruits in the southern African region. *South African Journal of Botany*, *113*, 137–149.

Nyanga, L., Nout, M. J. R., Gadaga, T. H., Boekhout, T., & Zwietering, M. H. (2008). Traditional processing of masau fruits (Ziziphus Mauritiana) in Zimbabwe. *Ecology of Food and Nutrition Journal*, *47*(1), 95–107.

Omotayo, A. O., & Aremu, A. O. (2020). Underutilized African indigenous fruit trees and food–nutrition security: Opportunities, challenges, and prospects. *Food and Energy Security*. https://doi.org/10.1002/fes3.220

Palumbo, M., Attolico, G., Capozzi, V., Cozzolino, R., Corvino, A., de Chiara, M. L. V., Pace, B., Pelosi, S., Ricci, I., Romaniello, R., & Cefola, M. (2022, December). Emerging postharvest technologies to enhance the shelf-life of fruit and vegetables: An overview. *Foods*, *11*(23), 3925.

Passam, H. C., Karapanos, I. C., Bebeli, P. J., & Savvas, D. (2007). A review of recent research on tomato nutrition, breeding and post-harvest technology with reference to fruit quality. *The European Journal of Plant Science and Biotechnology*, *1*, 1–21.

Peduruhewa, P. S., Jayathunge, K. G. L. R., & Liyanage, R. (2021). Potential of underutilised wild edible plants as the food for the future–A review. *Journal of Food Security*, *9*(4), 136–147.

Pott, D. M., Vallarino, J. G., & Osorio, S. (2020, May 8). Metabolite changes during postharvest storage: Effects on fruit quality traits. *Metabolites*, *10*(5), 187.

Rampedi, I. T., & Olivier, J. (2013). Traditional beverages derived from wild food plant species in the Vhembe District, Limpopo Province in South Africa. *Ecology of Food and Nutrition*, *52*(3), 203–222.

Ruffo, C. K., Birnie, A., & Tengnäs, B. (2002). *Edible wild plants of Tanzania* (Vol. 27). Regional Land Management Unit/Sida.

Sahoo, G., Swamy, S., Rout, S., Wani, A., & Mishra, A. (2021). Exploitation of wild leafy vegetables and underutilised fruits: Consequences for food and nutritional security. *Annals of the Romanian Society for Cell Biology*, *25*(6), 5656–5668.

Shackleton, C., & Shackleton, S. (2004). The importance of non-timber forest products in rural livelihood security and as safety nets: A review of evidence from South Africa. *South African Journal of Science*, *100*(11–12), 658–664.

Shackleton, C. M., Dzerefos, C. M., Shackleton, S. E., & Mathabela, F. R. (2000). The use of and trade in indigenous edible fruits in the Bushbuckridge savanna region, South Africa. *Ecology of Food and Nutrition*, *39*(3), 225–245.

Shava, S. (2005). Research on indigenous knowledge and its application: A case study of food plants of Zimbabwe. *Southern African Journal of Environmental Education*, *22*, 73–86.

Singh, K. K., Singh, M., & Joshi, S. C. (2014). Phenolic content and antioxidant activity of some underutilised wild edible fruits of the Sikkim Himalaya. *SMU Medical Journal*, 283–293.

Singh, N., & Sudha, M. L. (2024). Natural food flavours: A healthier alternative for bakery industry – A review. *Journal of Food Science and Technology*, *61*(4), 642–650.

Souilem, F., Dias, M. I., Barros, L., Calhelha, R. C., Alves, M. J., Harzallah-Skhiri, F., & Ferreira, I. C. (2019). Phenolic profile and bioactive properties of Carissa macrocarpa (Eckl.) A. DC.: An in vitro comparative study between leaves, stems, and flowers. *Molecules*, *24*(9), 1696.

Stadlmayr, B., Charrondière, U. R., Eisenwagen, S., Jamnadass, R., & Kehlenbeck, K. (2013). Nutrient composition of selected indigenous fruits from sub-Saharan Africa. *Journal of the Science of Food and Agriculture*, *93*(11), 2627–2636.

Taher, M. A., Tadros, L. K., & Dawood, D. H. (2018). Phytochemical constituents, antioxidant activity and safety evaluation of Kei-apple fruit (Dovyalis caffra). *Food Chemistry*, *265*, 144–151.

Tata Ngome, P. I., Shackleton, C., Degrande, A., & Tieguhong, J. C. (2017). Addressing constraints in promoting wild edible plants' utilization in household nutrition: Case of the Congo Basin forest area. *Agriculture & Food Security*, *6*, 1–10.

Traffic. (2017, August 9). *New fairwild certification project brings sustainable and ethical baobab to the market.* www.traffic.org/news/new-study-finds-chinas-ivory-market-shrinking-ahead-of-incoming-domestic-ivory-ban/

Toungos, M. D. (2019). Tamarind (Tamarindus indicus L) fruit of potential value but underutilized in Nigeria. *International Kournal of Innovative Food, Nutrition and Sustainable Agriculture*, *7*(1), 1–10.

Tredgold, M. H. (1986). *Food plants of Zimbabwe: With old and new ways of preparation.* Mambo Press.

Van Wyk, B.-E. (2011). The potential of South African plants in the development of new food and beverage products. *South African Journal of Botany*, *77*(4), 857–868.

Viejo, C. G., Fuentes, S., Howell, K., Torrico, D. D., & Dunshea, F. R. (2019). Integration of non-invasive biometrics with sensory analysis techniques to assess acceptability of beer by consumers. *Physiology & Behaviour*, *200*, 139–147.

Vilà, M., & D'Antonio, C. M. (1998). Hybrid vigor for clonal growth in Carpobrotus (Aizoaceae) in coastal California. *Ecological Applications*, *8*(4), 1196–1205.

Wang, Q. J., Mielby, L. A., Junge, J. Y., Bertelsen, A. S., Kidmose, U., Spence, C., & Byrne, D. V. (2019). The role of intrinsic and extrinsic sensory factors in sweetness perception of food and beverages: A review. *Foods*, *8*, 211.

Xing, Y., Xu, Q., Li, X., Chen, C., Ma, L., Li, S., Che, Z., & Lin, H. (2016). Chitosan-based coating with antimicrobial agents: Preparation, property, mechanism, and application effectiveness on fruits and vegetables. *International Journal of Polymer Science*, Article 4851730.

Yimer, A., Forsido, S. F., Addis, G., & Ayelign, A. (2023). Nutritional composition of some wild edible plants consumed in Southwest Ethiopia. *Heliyon*, *9*(6), e16541.

Zhu, X., Wang, Q., Cao, J., & Jiang, W. (2008). Effects of chitosan coating on postharvest quality of mango (Mangifera Indica L. cv. Tainong) fruits. *Journal of Food Processing and Preservation*, *32*, 770.

26 Sustainable Food Production and Prospects

A Fundamental Guide to Hydroponic Agriculture

[1]Muhali Olaide Jimoh, Jesulewami Precious Adeoye,[1] Comfort Titilayomi Senjobi,[1] Karimat Adeyinka Ahmed,[1] Serifat Olatundun Salami,[3] and Charles Petrus Laubscher[2]*

[1]Department of Plant Science, Olabisi Onabanjo University, PMB 2002, Ago-Iwoye 120107, Nigeria
[2]Department of Horticultural Sciences, Cape Peninsula University of Technology, Symphony Way, Bellville 7535, South Africa
[3]Department of Agricultural Education, Federal College of Education (Special), Oyo, Oyo State, Nigeria
*Corresponding author: moajay006@gmail.com/muhali.jimoh@oouagoiwoye.edu.ng

26.1 INTRODUCTION

Food security presents a global challenge exacerbated by factors such as population growth, climate change, and biodiversity resource depletion. Moreover, overreliance of traditional agriculture on fertile and ample water availability, which is decreasing, suggests the need for innovative strategies to meet the increasing global food demand. Hydroponics, a soilless plant cultivation method, emerges as a potential solution which provides plants with essential nutrients through water-based solutions (Savvas et al., 2017). In this system, plants are anchored to the nutrient solution using inert media like rockwool, coco coir, or perlite, which provide physical support. These substrates are chosen due to their lack of nutrient-holding properties, ensuring a complete reliance on the provided solution, thus allowing precise control over nutrient levels needed for optimized plant growth and yield (Awad et al., 2017; Inden & Torres, 2001).

Hydroponics provides a viable answer to agricultural difficulties, such as water scarcity, limited land availability, and harsh temperatures. Hydroponics saves more water than traditional soil-based agriculture because it recirculates water in a closed system, reducing waste. This is especially helpful in dry areas, where water is scarce (Rodríguez-Delfín et al., 2017). Furthermore, hydroponic systems can be developed in several situations, including urban areas and non-arable land, overcoming the limits of restricted land availability. Moreover, hydroponics provides for fine control over the growth environment, allowing crops to be grown in harsh conditions that would otherwise be unsuitable for traditional agricultural methods. Unlike traditional agriculture, which is limited by seasons, hydroponic systems can operate year-round in controlled environments, ensuring a consistent source of fresh produce, even in harsh climates (Rufí-Salís, 2020). This system of cultivation significantly reduces water usage by recirculating the nutrient solution in most setups. This approach is particularly beneficial in drought-prone or arid regions with limited water resources (Jones, 2020). By amalgamating these advantages,

DOI: 10.1201/9781003469766-26

hydroponics offers several benefits for food and nutrition security, including increased production in resource-limited areas, reduced reliance on unpredictable weather patterns, enhanced access to fresh produce in urban areas or challenging climates, and more sustainable production practices with diminished environmental impact (Naresh et al., 2024). Thus, this chapter aims at assessing the suitability of food production utilizing various hydroponic systems for food security in changing global climates.

26.1.1 Sustainability of Food Production in Hydroponics

Hydroponics presents a promising prospect for sustainable agriculture, offering benefits such as decreased water usage, enhanced yields, and reduced dependence on pesticides. Its sustainability is evident in its capacity to utilize up to 90% less water compared to traditional soil-based farming (Specht et al., 2014), achieved through closed-loop systems that recycle water, mitigating runoff and evaporation. Furthermore, hydroponics allows for denser planting and precise management of nutrients and environmental factors, potentially resulting in increased crop yields (Alnaqbi & Alami, 2023). The controlled environment and the absence of soil in hydroponic systems minimize issues related to pests and weeds, potentially diminishing the necessity for pesticides. Particularly noteworthy is the flexibility of hydroponic farms, as they can be established anywhere and are not subject to weather conditions, seasonality, hence facilitating year-round food production (Viljoen et al., 2021).

Compared to conventional soil-based techniques, hydroponics, a land-efficient agricultural method, can produce more food per unit of land (Rahman et al., 2023). By vertically stacking plants, hydroponic systems maximize efficiency and enable more growth in limited spaces. Also, hydroponic systems can be installed in metropolitan areas, lowering transportation costs and carbon emissions from food production and distribution (Araya et al., 2024). It assists in maximizing output in constrained areas, like indoors or on balconies or rooftops, and gives room for regulation of light, temperature, and humidity, giving the plant conditions for optimal growth. Fertilizers and pesticides are frequently used by soil-based farmers to increase crop yield, leading to inorganic, contaminated, and inferior produce. In hydroponics, this issue does not arise, as hydroponic farming limits fertilizer consumption and increases fertilizer use efficiency as crops obtain minimum minerals required from the nutrient-rich water for optimum growth, hence making hydroponically cultivated vegetables taste better. Therefore, hydroponics provides a sustainable and effective solution to food production in the face of global environmental challenges, and it assures cleaner environments (Das et al., 2024). Through hydroponics, food yields can be augmented, as hydroponic systems deliver 20–50% more yield compared to the traditional cultivating methods (Specht et al., 2014).

26.1.2 Development of Cultivation Protocols for Underutilized Wild Vegetables in Hydroponics

Underutilized vegetables are neglected edible crops having the potential of cushioning hidden hunger and malnutrition (Bvenura & Sivakumar, 2017; Salami et al., 2022). Several reasons were adduced to their less patronage, among which are overdependence on exotic plants, dearth of information on their nutritional potentials, and a stark regard of wild edible plants as food for the poor. However, these crops have diverse nutrients and novel metabolites that can be used to elevate the nutritional values of food, formulate drugs and drug precursors, and may be consumed directly as herbal remedies (Jimoh et al., 2023).

Several studies have been conducted to develop suitable protocols for cultivating underutilized wild plants in hydroponics. Findings from such studies supported the suitability of hydroponics for the cultivation of underutilized vegetables. For instance, hydroponic studies of Ngxabi et al. (2021), Bulawa et al. (2022), and Tshayingwe et al. (2023) have reignited interest in the cultivation, consumption, pharmacological benefits, and reintroduction of edible *Trachyandra* species (*T. ciliata* and *T. divaricata*) as edible vegetables. These plants were reportedly consumed by the Indigenous Khoi-San people of the Western Cape province of South Africa (Van Wyk & Gorelik, 2017; Ngxabi et al., 2021). Also, species of *Mesembryanthemum* and *Tetragonia* were cultivated successfully in hydroponics under different fertigation regimes (Sogoni et al., 2021; Mndi et al., 2023; Cebani et al., 2024), suggesting the suitability of hydroponic systems for cultivating, domesticating, and reintroducing wild underutilized edibles.

26.1.3 Water Conservation in Hydroponics

Hydroponics provides a significant edge in water conservation compared to traditional soil-based agriculture, attributable to several pivotal factors (Kannan et al., 2022). In contrast to conventional methods, where water infiltration into the soil and evaporation are common, hydroponics employs closed-loop systems. These systems capture and recycle the nutritional solution delivering nutrients to plant roots, thereby curbing water waste. In other hydroponic setups, plants are either suspended or placed in a medium devoid of soil, substantially reducing water loss via soil surface evaporation, a prominent issue in traditional farming. Conversely, traditional methods may result in overwatering or underwatering, leading to water wastage in both scenarios. Hydroponics ensures precise water delivery directly to the roots, optimizing efficiency and minimizing waste (Savvas et al., 2017). Research indicates that hydroponics can achieve up to a 90% reduction in water usage compared to soil-based farming. This remarkable decrease is attributed to the combined effects of closed-loop systems, minimized evaporation, and precise water delivery (Pomoni et al., 2023).

26.1.4 Integrated Pest and Disease Management in Hydroponics

Integrated pest management (IPM) is a critical strategy for hydroponic growers aiming to uphold plant health while minimizing dependency on chemical control methods. The core principle of IPM revolves around thwarting pest and disease outbreaks. Hence, sustaining a hygienic growing environment and routinely sterilizing tools, equipment, and the hydroponic system aid in reducing potential harborage points for pathogens (Gruda & Bragg, 2021). A diversified strategy is needed to overcome obstacles in hydroponics, especially to manage pests and diseases without the use of soil. Conventional pest management techniques are frequently inappropriate in soilless systems, requiring the use of new tactics. Integrated pest management is a sustainable approach that integrates environmental management, close observation, and biological control agents like beneficial insects. As hydroponics eliminates soil, it minimizes the risk of soilborne diseases (Barzman et al., 2015).

Prior to introducing new plants into the hydroponic system, thorough inspection for signs of pests or diseases is crucial to prevent inadvertent pathogen introduction. Consistently monitoring plants for early indications of pest or disease presence is an effective management practice. Observable signs such as chewed leaves, unusual markings, or wilting and the presence of pests on plants may serve as an indicator of potential disease infestation. Integrating beneficial insects like ladybugs or lacewings into the hydroponic system aids in natural pest control (Sigognault Flochlay et al., 2017). Furthermore, organic pest control methods like insecticidal soap or neem oil sprays offer viable alternatives to synthetic pesticides, particularly for managing smaller outbreaks (Weinzierl & Henn, 2020).

The implementation of integrated pest management in hydroponics results in decreased reliance on chemicals, thereby reducing harm to beneficial insects and contamination of produce (Van Lenteren & Nicot, 2020). By fostering a healthy growing environment supportive of beneficial organisms, IPM promotes stronger, more resilient plants that are less susceptible to pest and disease attacks. This approach aligns with sustainable farming practices, minimizing environmental impacts from chemical usage and fostering a more balanced ecosystem within the hydroponic system (Folorunso et al., 2021).

26.1.5 Improved Crop Returns in Hydroponics

Hydroponics presents a distinctive approach to food cultivation, fostering enhancements in both quality and safety. Hydroponically grown food is generally regarded as nutritionally comparable to traditionally grown options, and some studies suggest potential advantages in nutrient content

(Croft et al., 2017). Operating as enclosed environments, hydroponic systems restrict pest intrusion and propagation (Jones, 2020), enabling growers to significantly reduce or eliminate pesticide usage, yielding cleaner and potentially healthier produce for consumers (Sharma et al., 2022). In contrast to traditional soil-based techniques, hydroponics offers direct oversight of the nutrient solution provided to plants. Growers can customize or regulate the nutrients to optimize plant growth, potentially augmenting the essential nutrients, like vitamins, minerals, and antioxidant content of the food (Croft et al., 2017).

Hydroponic cultivation often accelerates plant growth due to optimal growing conditions and the availability of nutrients in the solution (Asaduzzaman et al., 2022), resulting in expedited harvests and fresher produce reaching consumers. Unlike conventional agriculture, which carries the risk of soilborne diseases and pathogens, hydroponics mitigates this concern as controlled environment and absence of soil diminish the likelihood of contamination, potentially yielding safer food (Treftz & Omaye., 2016).

26.1.6 Efficient Nutrient System Delivery and Management in Hydroponics

Hydroponics presents a distinct advantage in plant nutrition by offering precise control over nutrient delivery (Jones, 2020). Various hydroponic systems employ different methods, such as ebb and flow, drip systems, aeroponics, and fertilizer injectors, to deliver the nutrient solution to plant roots. By integrating a balanced nutrient solution with an efficient delivery system, hydroponics fosters optimal nutrient uptake by plants. This direct access to essential nutrients leads to accelerated growth compared to soil-based cultivation (Asaduzzaman et al., 2022), translating to healthier plants with enhanced fruit or vegetable production. Additionally, precise control over nutrients enables the optimization of factors like flavor, size, and nutritional value of the produce (Jones, 2020). However, efficient nutrient delivery in hydroponics necessitates ongoing monitoring and management practices, as follows:

1. The pH and electrical conductivity (EC) of the nutrient solution must be tested regularly to maintain optimal nutrient availability for the specific crop being grown (Nkcukankcuka et al., 2021).
2. The nutrient solution needs to be adjusted based on monitoring results by adding concentrated nutrients or adjusting the water-to-nutrient ratio (Jones, 2020).
3. The reservoir needs to be routinely cleaned and disinfected to prevent the buildup of pathogens and organic matter that could obstruct drippers or affect nutrient stability (Kannan et al., 2022).
4. Clean water needs to be used for the nutrient solution to prevent the introduction of impurities that may harm plants or clog systems.
5. All system components, including pumps, filters, and drippers, need to be regularly inspected and maintained to ensure proper operation and prevent malfunctions that could disrupt nutrient delivery.

26.2 DIFFERENT HYDROPONIC CULTIVATION TECHNIQUES

26.2.1 Deep Water Culture (DWC)

In deep water culture (Figure 26.1), roots of plants are suspended in nutrient-rich water, and air is provided directly to the roots by an air stone (Nursyahid et al., 2021). The hydroponics bucket system is a classic example. Plants are placed in net pots, and roots are suspended in the nutrient solution, where they grow quickly in a large mass. It is important to monitor the oxygen and nutrient concentrations, salinity, and pH to ensure optimal conditions for plant growth (Domingues et al., 2012), as algae and molds can grow rapidly in the reservoir. Maintenance involves replenishing the nutrient solution, cleaning the reservoir, and checking for any root or plant diseases.

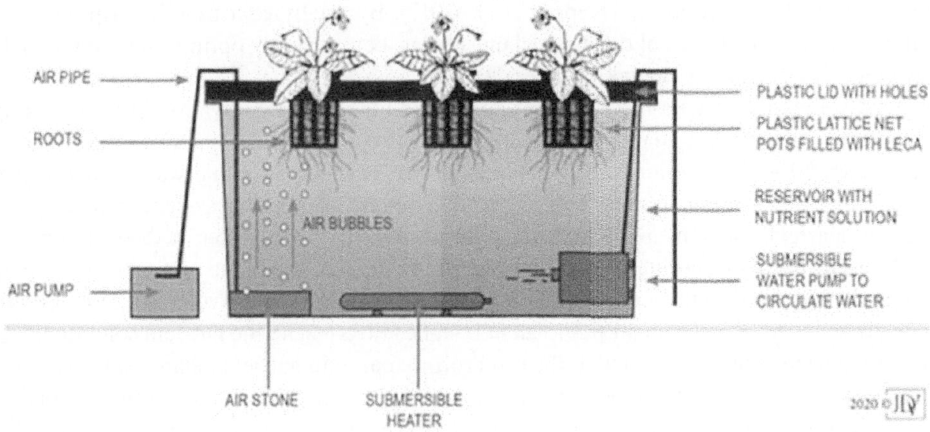

FIGURE 26.1 Typical deep water culture (DWC) setup.

Source: Viljoen et al. (2021).

This system works well for larger plants that produce fruits, especially cucumber and tomato. The main components of a DWC system include:

1. *Reservoir.* This container holds the nutrient solution and is usually made from non-toxic, light-resistant materials to prevent algae growth.
2. *Air pump and air stone.* These devices are essential for oxygenating the nutrient solution. The air pump pushes air through the air stone, which diffuses it into small bubbles, increasing oxygen absorption by the roots (Nursyahid et al., 2021).
3. *Net pots.* These pots allow the roots to grow through them and into the nutrient solution while supporting the plant above.
4. *Growth medium.* Materials like clay pellets or rockwool support the plant and roots, providing stability and aiding in moisture retention.
5. *Nutrient solution.* A balanced mix of water and essential nutrients tailored to the specific plants being grown (Savvas et al., 2017).

Principal benefits of DWC include accelerated plant growth, increased yield, space and water use efficiency, ease of maintenance, and reduced pests and diseases. The DWC also allows plants to focus energy on growth rather than searching for nutrients, leading to healthier and more robust plants (Guruchandran et al., 2024). Despite the afore-stated benefits, the initial cost of setting up a DWC, the routine maintenance to avoid the risk of waterborne diseases, the reliance on electrically powered air pumps, and the regular monitoring of nutrient level, pH, and electrical conductivity of the nutrient solution constitute some constraints (Gillani et al., 2023).

26.2.2 Drip Hydroponics

In drip hydroponic systems, the nutrient solution is set apart in a reservoir, and the plants are grown separately in a soilless medium (Figure 26.2). Water or nutrient solution from the reservoir is provided to individual plant roots in appropriate proportion with the help of a pump (Abedin et al., 2021). Drip systems dispense nutrients through nozzles, at a very slow rate, and the extra solutions can be collected and recirculated, or even allowed to drain out. With this system, it is possible to simultaneously grow several kinds of plants. This method conserves water and nutrients by minimizing evaporation and runoff, making it an efficient choice for both small-scale gardeners and large-scale agricultural producers (Kapoor et al., 2017). By briefly adjusting the drip rate, growers can cater to the specific needs of different plant species, consequently optimizing growth and yield (Yadav et al., 2020; Ani & Gopalakirishnan, 2020).

Originally, the drip irrigation system was developed for outdoor cultivation in Israel to improve water efficiency; it was later adapted successfully to hydroponics (Fishelson & Rymon, 1989; Garb & Friedlander, 2014). The emitters release liquid slowly, ensuring minimal water usage while providing precise control over nutrients. The essential components of such a system include a reservoir to hold the nutrient solution, a submersible pump to circulate the water, a distribution network consisting of drip lines or emitters which deliver the solution to the plants, a timer to automate the watering schedule, growth trays or pots to hold the plants above the reservoir, a support structure to hold the growth trays or pots, an air pump and air stone to oxygenate the nutrient solution, PVC tubing which connects return lines back to the reservoir to maintain nutrient balance and prevent water loss, and various monitoring equipment, such as pH and EC meters, to ensure the optimal conditions for plant growth (Camp & Lamm, 2003; Li et al., 2015). Crops suitable for drip hydroponics include tomatoes, lettuce, peppers, basil, mint, and parsley, among other herbs. Drip irrigation systems are suitable for various hydroponic setups, such as small-scale home gardens, large-scale commercial farms, greenhouse systems, vertical farming, and balcony gardens. Their flexibility and adaptability make them a versatile choice for different spaces (Ben-Gal et al., 2004).

The benefits of drip hydroponics include water usage efficiency, optimal nutrient delivery, reduced disease risk, higher yields, space saving, flexibility, reduced labor, and environmental friendliness (Li et al., 2015). However, shortcomings include system complexity, electricity dependency, high cost of setting up and maintenance, root zone issues due to poor drainage, uneven nutrient distribution, and requirements of a significant amount of water to maintain the system effectively (Sharma et al., 2018; Kumar et al., 2021).

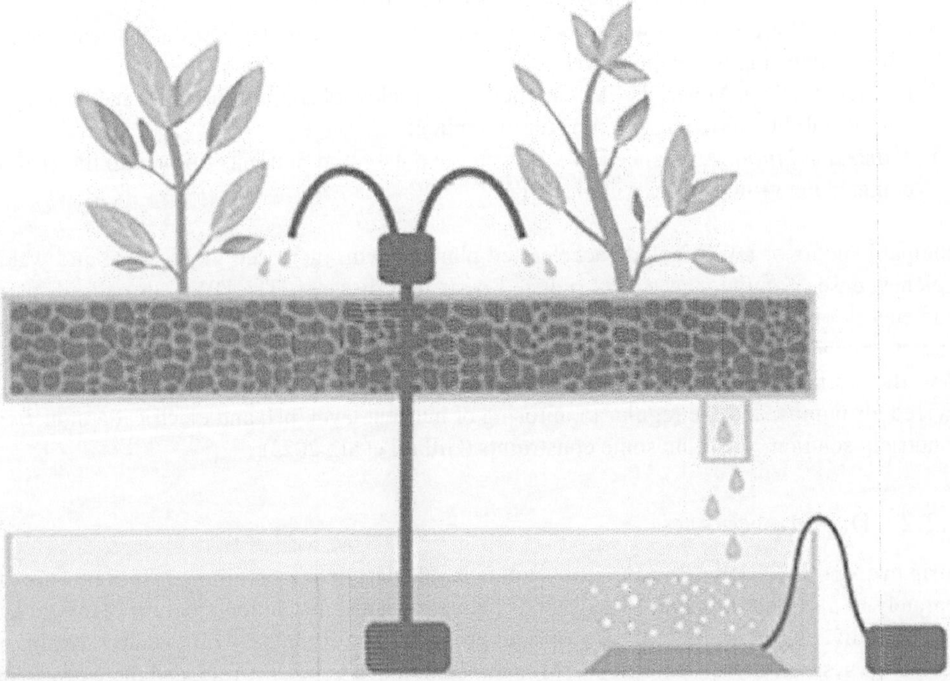

FIGURE 26.2 Typical drip hydroponic system setup.

Source: https://growwithoutsoil.com.

26.2.3 Aeroponics

Aeroponics is an advanced form of hydroponics which involves the growing of plants in an air or mist environment without the use of soil or an aggregate medium (Figure 26.3). This innovative cultivation method allows plant roots to absorb nutrients directly from a fine mist sprayed onto them, maximizing oxygen exposure and nutrient uptake (Lakhiar et al., 2018). The increased air exposure promotes rapid root development and offers greater control over the plant's environment. Aeroponics systems can lead to faster growth rates and higher yields and can be more water-efficient than traditional farming methods. Additionally, because aeroponics does not rely on soil, it eliminates the risk of soilborne diseases and pests, making it an ideal choice for space-limited urban farming and potentially for extraterrestrial agriculture (Eldridge et al., 2020). Components of aeroponics include aeroponic chamber, misting nozzles, nutrient reservoir, submersible pump, cycle timer, root supports, pressure gauge, drainage system, air filtration system, lighting system, and environmental controls.

26.2.4 Nutrient Film Techniques (NFT)

The nutrient film technique (Figure 26.4) was developed in the mid-1960s in England by Dr. Alen Cooper to overcome the shortcomings of the ebb-and-flow system. Similar to aeroponics, the NFT is the most popular hydroponic system. In this method, a nutrient solution is pumped constantly through channels in which plants are placed (Domingues et al., 2012). When the nutrient solutions reach the end of the channel, they are sent back to the beginning of the system. This makes it a recirculating system, but unlike DWC, the plant roots are not completely submerged, which is the main reason for naming this method NFT. This system is highly efficient, allowing for greater oxygenation of roots and optimal nutrient uptake (Ngxabi et al., 2021).

The NFT systems are scalable and can be used for a variety of crops, making them popular among commercial and hobbyist growers alike. The key to success with NFT is maintaining the proper flow rate of the nutrient solution to prevent drying out or oversaturation of plant roots (Graves, 1983; Wang et al., 2022). Additionally, regular monitoring of pH and nutrient levels is

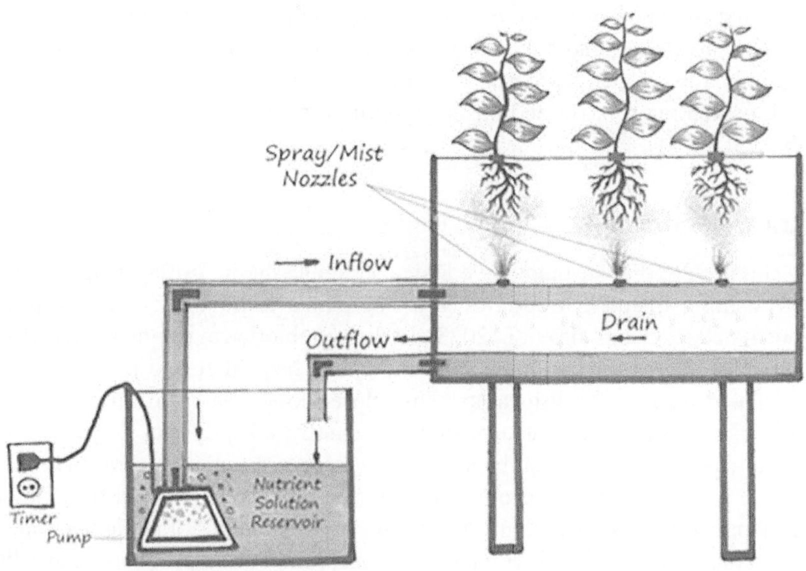

FIGURE 26.3 Typical aeroponics system setup.

Source: https://medium.com/krishi-wise.

FIGURE 26.4 Nutrient film technique hydroponic system setup.

Source: Ngxabi et al. (2021).

crucial to ensure plant health and maximize yield. The key components of an NFT system include a reservoir to hold the nutrient solution, a submersible pump to circulate the solution, channels or tubes for plant placement, and a recovery system to recirculate unused nutrients back into the reservoir. Additionally, proper slope and flow rate are crucial to ensure a thin film of solution is maintained, preventing root rot and ensuring optimal plant health (Castillo-Castellanos et al., 2016; Ngxabi et al., 2021).

26.2.5 Aquaponics

Aquaponics (Figure 26.5) is an innovative and sustainable method of agriculture that combines aquaculture (raising fish) and hydroponics (growing plants without soil) into one integrated system (Zhang et al., 2021; Okomoda et al., 2023). In this symbiotic environment, fish waste provides an organic nutrient source for the plants, and the plants help filter and purify the water, which is then recirculated back to the fish tanks. This closed-loop system mimics natural ecological cycles, resulting in a water-efficient and environmentally friendly way to grow food (Baganz et al., 2022). Aquaponics systems can be set up anywhere, are scalable from small to large sizes, and can produce a diverse range of crops, including vegetables, herbs, and fruits, as well as fish species like tilapia or trout (Zhang et al., 2022). The key advantages of aquaponics are water conservation, the elimination of chemical fertilizers, sustainable food production, and the potential for higher yield in a smaller footprint compared to traditional farming methods (Sonneveld et al., 2017; Maucieri et al., 2018).

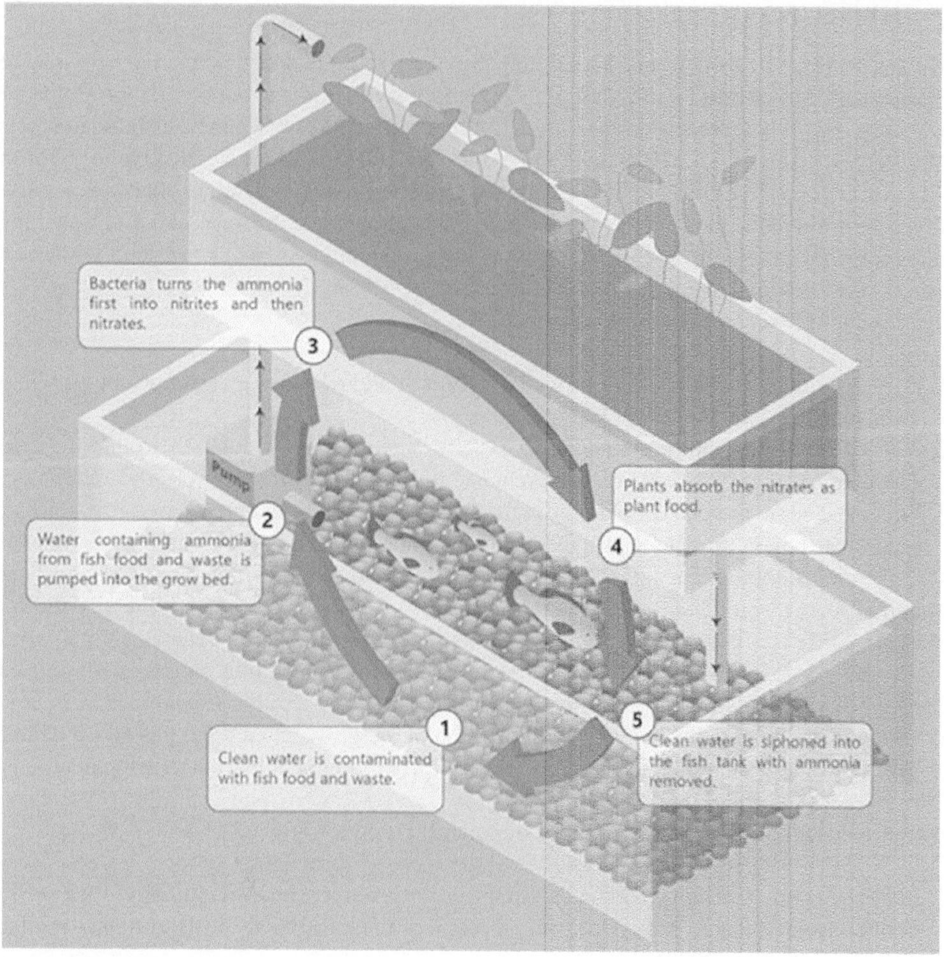

FIGURE 26.5 Typical aquaponics setup.

Source: www.btlliners.com.

26.2.6 Cosmosponics

One of the strategic plans of the United States National Aeronautics and Space Administration (NASA) is hydroponics in space (*cosmosponics*). This was made a matter of priority to complement long-term plans to occupy planetary surfaces and taking advantage of the complementary association between plants and animals (Ferl et al., 2002). The role of plants in bioconversion of animal wastes to essential nutrients for humans and animals and phytoremediation is very critical to life support systems in space. The biological effects of ionizing radiation and microgravity on chromosomal DNA damage have been well reported in literature. This may limit human exploration of space due to aging facilitated by DNA damage (Yatagai et al., 2019). This makes imperative the creation of life support ecosystems in extraterrestrial environments. Therefore, NASA has developed a capillary fluidics hydroponic system suitable for crop production in low-gravity environments (Mungin et al., 2019). In the model, a conduit irrigation and passive surface tension were employed to perform the role of gravity, with peculiar considerations for irrigation, priming, nutrient supply, aeration, germination, and root development. Hence, with hydroponics, food production is possible for weeks and months in extraterrestrial environments (Wright et al., 1988; Mungin et al., 2019).

26.3 HYDROPONIC MEDIA

Unlike soil, which fulfills both nutritional and structural roles, hydroponic media primarily serves as a framework for the plant's root system. Contemporary media has the potential to impact variables such as root zone temperature, aeration, and nutrient uptake efficiency, as indicated by studies of Ortega Martínez et al. (2016) and Vinci and Rapa (2019). There is a diverse array of media choices available, each possessing distinct characteristics tailored to different plant species and cultivation setups (Sánchez-Gutiérrez et al., 2015; Mavrogianopoulos, 2016). While traditional materials like rockwool and plastic are prevalent, ongoing research delves into eco-friendly alternatives, such as coconut coir and biochars (Sepúlveda-Vera et al., 2019).

Hydroponic media serves several key functions:

1. *Physical support.* The medium provides stability and facilitates proper upright growth by anchoring the plant (Gruda & Bragg, 2021).
2. *Airflow.* Healthy plant development necessitates adequate air circulation around the roots. The medium provides sufficient air exchange to prevent root rot (Savvas et al., 2017).
3. Water retention. Crucial for plant growth, the growing medium must possess the ability to retain water and nutrients within the root zone, thus creating a readily accessible moisture reservoir for uptake. The physical attributes of the medium, including pore size and distribution, play a significant role in determining its water-holding capacity (Sepúlveda-Vera et al., 2019). Additionally, combining different media with diverse water retention properties can prove to be an effective approach. For instance, blending biochar with coir can enhance water-holding capacity while ensuring adequate aeration (Sepúlveda-Vera et al., 2019). It is crucial to optimize the composition of the growing medium to strike a balance between water retention and air circulation for promoting healthy root development.

Hydroponic media can be either inert or organic. Inert options include:

1. *Rockwool.* Recognized for its exceptional water retention and air porosity, rockwool is derived from volcanic rock. However, it requires preadjustment of pH and may produce dust (Nerlich et al., 2022).
2. *Perlite.* Lightweight and possessing good drainage properties, perlite is a sterile volcanic glass material. Nevertheless, it retains less water compared to some alternatives (Gruda & Bragg, 2021).
3. *Expanded clay pebbles (LECA clay).* These lightweight fired clay pebbles offer effective drainage and aeration. They are reusable, pH-neutral, and widely favored for hydroponic systems (Savvas et al., 2017).

Organic media include the following:

1. *Coco coir.* An eco-friendly and biodegradable alternative crafted from coconut husks. Coco coir exhibits commendable water retention capabilities and natural buffering properties. However, prior rinsing may be advisable to eliminate excess salts (Vinci & Rapa, 2019).
2. *Biochar.* This charcoal-like substance, derived from pyrolysis (the combustion of organic material in a low-oxygen environment), contributes to soil health and fertility improvement. In hydroponic applications, biochar can be blended with other media to augment water retention and aeration and to potentially offer slow-release nutrients (Awad et al., 2017).
3. *Vermicompost.* Vermicompost, the nutrient-rich castings generated by worms digesting organic matter, represents a valuable reservoir of beneficial microbes and nutrients. Similar to composted manure, attention to hygiene and potential salt content is necessary. When integrated with other media, vermicompost can enhance microbial activity within the root zone and potentially boost nutrient absorption by plants (Weber, 2016).

26.3.1 Optimizing Quality of Hydroponically Cultivated Food

Hydroponics creates a controlled environment conducive to cultivating high-quality food. Urban farmers can enhance the taste, nutritional value, and overall quality of their hydroponically grown crops by focusing on several critical aspects. These include conducting research on optimal levels of macro- and micronutrients for each plant variety (Rouphael et al., 2018), tailoring the nutrient solution to maximize nutritional content for specific crops and regularly monitoring and adjusting pH, electrical conductivity (EC), and nutrient levels to maintain the right balance throughout the growth cycle (Savvas et al., 2017). Exploring the integration of organic or slow-release fertilizers into the hydroponic system can potentially enhance flavor and provide a more natural nutrient source for plants (Roy et al., 2023). Enhancing the growing environment involves providing the correct light spectrum and intensity tailored to the chosen crop, as inadequate light can stunt growth and affect taste and nutrient production (Rouphael et al., 2018). Additionally, maintaining consistent temperature and humidity levels within the ideal range for specific plants is essential, as fluctuations can stress plants and impact flavor and quality. Proper airflow is also critical, as adequate ventilation within the hydroponic system prevents stagnant air and moisture buildup, which can encourage disease (Peiro et al., 2020).

26.3.2 Climate-Resilient Hydroponics Farming

One of the primary factors contributing to the resilience of hydroponics in the face of environmental challenges is its reduced vulnerability to water scarcity (Verner et al., 2022). A fundamental advantage of hydroponics lies in its water efficiency, with hydroponically grown plants utilizing up to 90% less water compared to traditional soil-based agriculture. This is particularly beneficial in regions prone to drought, where water availability is of significant concern. Hydroponic systems are designed as circulating systems that capture, filter, and reuse nutrient solutions, thus minimizing water usage and maximizing resource utilization (Son et al., 2020).

Unlike traditional agriculture, where water use is dictated by weather conditions, hydroponic growers can closely monitor and adjust water usage based on the specific needs of crops and real-time environmental conditions. This allows for optimized water utilization and minimizes wastage. Furthermore, hydroponic systems provide a controlled environment unlike open fields, enabling growers to manipulate temperature using cooling systems, heating, and ventilation (Dev, 2013). This flexibility allows them to tailor the environment to suit the requirements of individual crops, thereby mitigating the impact of extreme heat or cold. Hydroponic systems are often housed within greenhouses, which offer additional temperature control and protection from adverse weather conditions, safeguarding plants from unseasonal frosts or extreme heat (Verner et al., 2022; Goddek et al., 2023).

26.3.3 Automation, Monitoring, and Climate Control System in Hydroponics

The integration of technology in hydroponics plays a pivotal role in enhancing the efficiency and productivity of these soilless farming systems. Automation stands at the forefront, allowing for precise control over environmental factors, such as lighting, temperature, and nutrient delivery, which are critical for plant growth (Shetty et al., 2021). Monitoring technologies provide real-time data on these parameters, enabling quick adjustments and ensuring optimal growing conditions. Control systems bring together automation and monitoring, facilitating seamless operation and management of the hydroponic setup, leading to higher yields and better resource utilization (Palande et al., 2018). By integrating advanced tools and techniques, hydroponics can achieve optimal water use, nutrient management, and crop yield. Sensors monitor pH levels, electrical conductivity, and nutrient availability to ensure that automated dosing systems carefully control nutrient concentrations and plant growth circumstances are ideal (Jung et al., 2019; Chowdhury et al., 2020).

26.4 CASE STUDY OF SUCCESSFUL HYDROPONIC PROJECTS IN DIFFICULT ENVIRONMENTS AROUND THE WORLD

Hydroponics has grown in favor as a sustainable farming alternative, particularly in areas where traditional agriculture is challenged. Notable among these is the Eden Project in Cornwall, UK, which has successfully grown tropical plants using hydroponic systems within enormous biomes (Prance, 2002; Imhof et al., 2018). Hydroponics has helped desert locations, such as the Middle East, to reach agricultural output; for example, Bustanica in Dubai, the world's largest vertical farm, produces leafy greens with 95% less water than conventional farming. Similarly, in Antarctica, researchers have used hydroponic technology to grow fresh veggies despite the hard cold, providing a healthy food source for scientists at research outposts (Brown-Paul, 2012). These case studies demonstrate how hydroponic technology can overcome environmental limits and contribute to global food security efforts.

26.4.1 Personalized Nutrition Approach

Poor nutrition has been identified as a major cause of chronic diseases globally due to choices of diets and lifestyles. With personalized nutrition, individuals can determine specific nutrient needs and deficiencies through consultations with nutritionists to create suitable personalized diet plans with measurable health benefits (Adams et al., 2020). Based on specific needs, hydroponic growers can formulate and adjust the nutrient solution in their systems to cultivate microgreens and vegetables enhanced with the identified vitamins and minerals. By incorporating these microgreens into their diets, individuals can access a concentrated source of the nutrients their bodies require (Bhaswant et al., 2023). This personalized nutrition approach offers several advantages which target individual nutrient deficiencies; provides a fresh, locally sourced, concentrated nutrient supply; and may be integrated seamlessly with urban living due to the space-efficient nature of hydroponics.

26.4.2 Hydroponics and Personalized Nutrition with Microgreens

Microgreens, the youthful seedlings of vegetables, herbs, and other edible plants, are emerging as a potent tool for tailored nutrition, particularly with the aid of hydroponics. Research indicates that microgreens boast higher concentrations of vitamins, minerals, and antioxidants compared to their fully grown counterparts (Bhaswant et al., 2023). The precise nutrient control afforded by hydroponics enables growers to produce microgreens which may be tailored toward specific age group or populations or customized to meet distinct individual nutritional needs based on dietary recommendations (D'Imperio et al., 2024).

26.4.3 Social and Economic Impact of Hydroponics

The societal impacts of hydroponics encompass enhanced food security, promotion of urban agriculture, and a year-round harvest. Hydroponics enables cultivation in areas with limited arable land or challenging climates, potentially bolstering food security, particularly in developing regions (Jones, 2020). Hydroponic systems can be established in urban settings, such as rooftops or warehouses, fostering local food production, reducing transportation expenses, and bringing fresh produce closer to consumers (Magwaza et al., 2020). Since hydroponic farming is not affected by seasonal variations, it ensures a consistent supply of fresh food throughout the year (Buwalda et al., 2013; Hasan et al., 2018).

The economic advantages of hydroponic farming entail increased crop yields, water conservation, and enhanced product quality. Compared to traditional farming methods, hydroponics often yields higher crop outputs, attributed to controlled environments and efficient nutrient delivery (Garcia & Hernandez, 2019; Hikmah & Susanti, 2023). It also utilizes significantly less water than conventional soil-based agriculture, making them well-suited for regions facing water scarcity (Jones, 2020). Also, hydroponically grown crops are typically less vulnerable to pests and diseases, resulting in higher-quality produce that commands premium prices (Garcia & Hernandez, 2019).

26.4.4 Market Opportunities in Hydroponics

The increasing trend of urbanization, combined with limited land availability, is fueling the demand for space-efficient growing solutions. Hydroponics thrives in vertical farms and urban environments, offering a means to grow fresh produce closer to consumers (Hikmah & Susanti, 2023). There is a growing consumer interest in organic and locally sourced food. Hydroponics enables year-round cultivation regardless of the season, ensuring a steady supply of local produce. The market for high-value crops, such as herbs, medicinal plants, vegetables, and microgreens, is experiencing significant growth. Hydroponics provides optimal conditions for cultivating these delicate or high-demand crops (Suryaningprang et al., 2021). Given the increasing concerns over water scarcity, the water efficiency of hydroponics has become a key selling point. The development of closed-loop systems for water recycling can further enhance the environmental sustainability of hydroponics (Sharma et al., 2023).

26.4.5 Empowerment Potential of Hydroponics Farming

Establishing hydroponic farms in urban areas, particularly in food deserts, offers residents reliable access to fresh, nutritious produce, enhancing dietary quality and overall health (Havens et al., 2020). In contrast to traditional farming methods, hydroponics ensures consistent yields regardless of the season, guaranteeing a continuous supply of nutritious food and reducing reliance on external sources, thereby promoting dietary diversity (Warren et al., 2015). Hydroponic farms can create new employment avenues within communities, especially benefiting youth and individuals exploring alternative paths in agriculture (Goodman & Minner, 2019). Training programs can equip individuals with the requisite skills for success. Hydroponics also lends itself well to small-scale operations, fostering local entrepreneurship and instilling a sense of community ownership in the food system (Bob et al., 2021).

26.4.6 Empowerment through Education and Skill Development

The integration of hydroponic projects into educational curricula can potentially provide students with valuable hands-on experience in science, technology, and sustainable agriculture (Gareau & Moscovitz, 2021). Hydroponic farms can serve as focal points for community engagement, offering knowledge-sharing platforms, workshops, and training programs. This empowers individuals to become growers of underutilized vegetables, self-sufficient food producers, and they can share their expertise with others. This could potentially foster personal growth and community development, ease the tension of food insecurity, and create additional streams of income (Currey et al., 2018).

26.4.7 Empowerment through Sustainability

Notably, hydroponic systems employ less water than conventional farming, rendering them suitable for regions grappling with water scarcity. This fosters conscientious water stewardship and alleviates strain on essential resources (Jones, 2020). Hydroponics has the potential to decrease pesticide application and mitigate nutrient deprivation from degraded soils in contrast to conventional practices. This fosters the development of a more sustainable food system characterized by reduced environmental impact (Sousa et al., 2024).

26.4.8 Economic and Environmental Advantages of Hydroponics for Sustainable Agriculture

Hydroponics offers several economic and environmental benefits consistent with the concepts of sustainable agriculture (Das et al., 2024). Hydroponic systems save up to 90% more water than standard agricultural methods since water can be recycled inside the system. This not only conserves a valuable resource but also considerably reduces farmers' water costs. Furthermore, hydroponics provides for larger yields in a smaller area because plants may be grown closer together and vertically, increasing space efficiency and lowering land expenses (Jat et al., 2015). Environmentally, hydroponics eliminates the need for herbicides and pesticides that are commonly used in soil-based agriculture, resulting in cleaner produce and less chemical runoff into neighboring water bodies. Phytoremediation techniques that are coupled to hydroponic systems are novel means of achieving effluent and sewage degradation and utilizing agricultural runoff water for crop production (Kim et al., 2021; Kumar et al., 2024). Hydroponic systems' regulated atmosphere also allows for year-round plant growth, consequently minimizing the carbon footprint involved with transporting out-of-season products over large distances. By embracing hydroponics, an agricultural system that is both economically feasible and environmentally responsible can be created (Kumar & Cho, 2014; Kim et al., 2021).

26.5 CONCLUSION AND PROSPECTS

Hydroponics offers a promising future for sustainable food production due to its flexibility in allowing plants to grow more quickly and healthily than in conventional farming. In addition to saving water and land, the soilless farming technique offers a workable way to produce food in places with bad soil or limited land availability and extreme climatic conditions. Hydroponics stands out as a crucial component in advancing food security and environmental preservation as we aspire to a sustainable future. However, addressing challenges such as cost, energy efficiency, and the development of user-friendly systems suitable for small-scale and developing world applications requires further research and development. To address this concern, governments may set up special intervention funds to subsidize the initial cost of setting up hydroponic systems for households and commercial growers, to ensure food security while minimizing environmental impact.

26.6 ACKNOWLEDGMENTS

Authors wish to acknowledge the Olabisi Onabanjo University, Ago-Iwoye, Nigeria, and Cape Peninsula University of Technology for supporting this study with library resources.

REFERENCES

Abedin, T., Yamamoto, A., Hayashi, T., & Hosokawa, M. (2021). Drip fertigation enhances the growth of hydroponic lettuce (Lactuca sativa) using polyester fiber substrate. *Scientia Horticulturae*, *276*, 09604.

Adams, S. H., Anthony, J. C., Carvajal, R., Chae, L., Khoo, C. S. H., Latulippe, M. E., Matusheski, N. V., McClung, H. L., Rozga, M., Schmid, C. H., & Wopereis, S. (2020). Perspective: Guiding principles for the implementation of personalized nutrition approaches that benefit health and function. *Advances in Nutrition*, *11*(1), 25–34.

Alnaqbi, S. A., & Alami, A. H. (2023). Sustainability and renewable energy in the UAE: A case study of Sharjah. *Energies*, *16*(20), 7034.

Ani, A., & Gopalakirishnan, P. (2020). Automated hydroponic drip irrigation using big data. In *2020 Second international conference on inventive research in computing applications (ICIRCA)* (pp. 370–375). IEEE.

Araya, N. A., Moremi, M. G., Mokgehle, S., Mofokeng, M. M., Malaka, M. J., Makgato, M., Araya, H. T., & Mampholo, B. (2024). Sustainable soilless recirculating hydroponics for productive use of marginal lands: A South African context. In *The marginal soils of Africa: Rethinking uses, management and reclamation* (pp. 279–306). Springer Nature.

Asaduzzaman, M., Niu, G., & Asao, T. (2022). Nutrients recycling in hydroponics: Opportunities and challenges toward sustainable crop production under controlled environment agriculture. *Frontiers in Plant Science*, *13*, 845472.

Awad, Y. M., Lee, S. E., Ahmed, M. B. M., Vu, N. T., Farooq, M., Kim, I. S., Kim, H. S., Vithanage, M., Usman, A. R. A., Al-Wabel, M., & Meers, E. (2017). Biochar, a potential hydroponic growth substrate, enhances the nutritional status and growth of leafy vegetables. *Journal of Cleaner Production*, *156*, 581–588.

Baganz, G. F., Junge, R., Portella, M. C., Goddek, S., Keesman, K. J., Baganz, D., Staaks, G., Shaw, C., Lohrberg, F., & Kloas, W. (2022). The aquaponic principle—It is all about coupling. *Reviews in Aquaculture*, *14*(1), 252–264.

Barzman, M., Bàrberi, P., Birch, A. N. E., Boonekamp, P., Dachbrodt-Saaydeh, S., Graf, B., Hommel, B., Jensen, J. E., Kiss, J., Kudsk, P., & Lamichhane, J. R. (2015). Eight principles of integrated pest management. *Agronomy for Sustainable Development*, *35*, 1199–1215.

Ben-Gal, A., Lazorovitch, N., & Shani, U. (2004). Subsurface drip irrigation in gravel-filled cavities. *Vadose Zone Journal*, *3*(4), 1407–1413.

Bhaswant, M., Shanmugam, D. K., Miyazawa, T., Abe, C., & Miyazawa, T. (2023). Microgreens—A comprehensive review of bioactive molecules and health benefits. *Molecules*, *28*(2), 867.

Bob, F., Fitriani, R., & Muhamad, D. J. (2021). Hydroponic Training to improve community living standards in densely populated and weak economies. *Budapest International Research and Critics Institute (BIRCI-Journal)*, *4*(1), 1569–1577.

Brown-Paul, C. (2012). Aquaponics in Antarctica. *Practical Hydroponics and Greenhouses*, *124*, 30–34.

Bulawa, B., Sogoni, A., Jimoh, M. O., & Laubscher, C. P. (2022). Potassium application enhanced plant growth, mineral composition, proximate and phytochemical content in Trachyandra divaricata Kunth (Sandkool). *Plants*, *11*(22), 3183.

Buwalda, F., van Os, E. A., Giacomelli, G., Samperio Ruiz, G., Vermeulen, T., van Weel, P. A., & Ruijs, M. N. A. (2013, June). Hydroponic systems: Hype or new perspective. *International Symposium on Growing Media and Soilless Cultivation*, *1034*, 201–207.

Bvenura, C., & Sivakumar, D. (2017). The role of wild fruits and vegetables in delivering a balanced and healthy diet. *Food Research International*, *99*, 15–30.

Camp, C. R., & Lamm, F. R. (2003). Irrigation systems, subsurface drip. *Encyclopedia Water Science*, 560–564.

Castillo-Castellanos, D., Zavala-Leal, I., Ruiz-Velazco, J. M. J., Radilla-García, A., Nieto-Navarro, J. T., Romero-Bañuelos, C. A., & González-Hernández, J. (2016). Implementation of an experimental nutrient film technique-type aquaponic system. *Aquaculture International*, *24*, 637–646.

Cebani, S., Jimoh, M. O., Sogoni, A., Wilmot, C. M., & Laubscher, C. P. (2024). Nutrients and phytochemical density in Mesembryanthemum crystallinum L. cultivated in growing media supplemented with dosages of nitrogen fertilizer. *Saudi Journal of Biological Sciences*, *31*(1), 103876.

Chowdhury, M., Ali, M., Rasool, K., Jeong, J. H., Choi, C. H., Han, M. W., Ko, H. J., & Chung, S. O. (2020, December). Identification of PID parameters for system-specific nutrient mixing control for ISE-based hydroponic nutrient management. In *III Asian Horticultural Congress-AHC2020 1312* (pp. 567–574).

Croft, M. M., Hallett, S. G., & Marshall, M. I. (2017). Hydroponic production of vegetable Amaranth (Amaranthus cruentus) for improving nutritional security and economic viability in Kenya. *Renewable Agriculture and Food Systems*, *32*(6), 552–561.

Currey, C. J., VanDerZanden, A. M., & Mitchell, J. J. (2018). Increasing student awareness and understanding of food security by integrating a service-learning project into an undergraduate hydroponic food production course. *HortTechnology*, *28*(4), 543–547.

Das, P. K., Sahu, K., Das, B. K., Das, B. P., & Dash, P. (2024). Hydroponics: A significant method for phytoremediation. In *Hydroponics and environmental bioremediation: Wastewater treatment* (pp. 1–25). Springer Nature.

Dev, P. K. (2013). Waterlogging through soil-less agriculture as a climate resilient adaptation option. *Climate Change and Disaster Risk Management*, 681–692.

D'Imperio, M., Bonelli, L., Mininni, C., Renna, M., Montesano, F. F., Parente, A., & Serio, F. (2024). Soilless cultivation systems to produce tailored microgreens for specific nutritional needs. *Journal of the Science of Food and Agriculture*, *104*(6), 3371–3380.

Domingues, D. S., Takahashi, H. W., Camara, C. A., & Nixdorf, S. L. (2012). Automated system developed to control pH and concentration of nutrient solution evaluated in hydroponic lettuce production. *Computers and Electronics in Agriculture*, *84*, 53–61.

Eldridge, B. M., Manzoni, L. R., Graham, C. A., Rodgers, B., Farmer, J. R., & Dodd, A. N. (2020). Getting to the roots of aeroponic indoor farming. *New Phytologist*, *228*(4), 1183–1192.

Ferl, R., Wheeler, R., Levine, H. G., & Paul, A. L. (2002). Plants in space. *Current Opinion in Plant Biology*, *5*(3), 258–263.

Fishelson, G., & Rymon, D. (1989). Adoption of agricultural innovations: The case of drip irrigation of cotton in Israel. *Technological Forecasting and Social Change*, *35*(4), 375–382.

Folorunso, E. A., Roy, K., Gebauer, R., Bohatá, A., & Mraz, J. (2021). Integrated pest and disease management in aquaponics: A metadata-based review. *Reviews in Aquaculture*, *13*(2), 971–995.

Garb, Y., & Friedlander, L. (2014). From transfer to translation: Using systemic understandings of technology to understand drip irrigation uptake. *Agricultural Systems*, *128*, 13–24.

Garcia, L., & Hernandez, J. (2019). Hydroponics for higher yields and improved quality. *Agriculture*, *9*(7), 147.

Gareau, T. P., & Moscovitz, A. (2021). An overview of urban agriculture youth programs in major cities of the US and the integration of STEM curriculum and activities. In L. T. Esters, A. Patchen, I. DeCoito, & N. Knobloch (Eds.), *Research approaches in urban agriculture and community contexts* (pp. 165–183). Urban Agriculture, Springer.

Gillani, S. A., Abbasi, R., Martinez, P., & Ahmad, R. (2023). Comparison of energy-use efficiency for lettuce plantation under nutrient film technique and deep-water culture hydroponic systems. *Procedia Computer Science*, *217*, 11–19.

Goddek, S., Körner, O., Keesman, K. J., Tester, M. A., Lefers, R., Fleskens, L., Joyce, A., van Os, E., Gross, A., & Leemans, R. (2023). How greenhouse horticulture in arid regions can contribute to climate-resilient and sustainable food security. *Global Food Security*, *38*, 100701.

Goodman, W., & Minner, J. (2019). Will the urban agricultural revolution be vertical and soilless? A case study of controlled environment agriculture in New York City. *Land Use Policy*, *83*, 160–173.

Graves, C. J. (1983). The nutrient film technique. *Horticultural Reviews*, *5*, 1–44.

Gruda, N. S., & Bragg, N. (2021). Developments in alternative organic materials for growing media in soilless culture systems. In *Advances in horticultural soilless culture* (pp. 73–106). Burleigh Dodds Science Publishing.

Guruchandran, S., Muninathan, C., & Ganesan, N. D. (2024). Novel strategy for effective utilization of anaerobic digestate as a nutrient medium for crop production in a recirculating deep water culture hydroponics system. *Biomass Conversion and Biorefinery*, *14*(8), 9491–9503.

Hasan, M., Sabir, N., Singh, A. K., Singh, M. C., Patel, N., Khanna, M., & Pragnya, P. (2018). Hydroponics technology for horticultural crops. *TechBull TB-ICN*, *188*(2018), 30.

Havens, S., Marks, D., Sandusky, M., Hedrick, A., Johnson, M., Robertson, M., & Trujillo, E. (2020). Automated Water Supply Model (AWSM): Streamlining and standardizing application of a physically based snow model for water resources and reproducible science. *Computers & Geosciences*, *144*, 104571.

Hikmah, N., & Susanti, A. (2023). Analysis of economic opportunities for farming with hydroponic systems in Makassar city. *Jurnal Manajemen Bisnis*, *10*(2), 392–405.

Imhof, B., Schlacht, I. L., Waclavicek, R., Schubert, D., Zeidler, C., Vrakking, V., Hoheneder, W., & Hogle, M. (2018, October 1–18). Eden Iss: A simulation testbed to an advanced xploration design concept for a greenhouse for moon and mars. *69th International Astronautical Congress (IAC)*.

Inden, H., & Torres, A. (2001, September). Comparison of four substrates on the growth and quality of tomatoes. *International Symposium on Growing Media and Hydroponics*, *644*, 205–210.

Jat, R. K., Singh, M., Kumar, A., & Singh, S. K. (2015). Hydroponics: A modern technique for efficient food production. *Journal of Experimental Biology and Agriculture Sciences*, *4*(1), 142–146.

Jimoh, M. A., Jimoh, M. O., Saheed, S. A., Bamigboye, S. O., Laubscher, C. P., & Kambizi, L. (2023). Commercialization of medicinal plants: Opportunities for trade and concerns for biodiversity conservation. In *Sustainable Uses and Prospects of Medicinal Plants* (pp. 309–332). CRC, Taylor and Francis.

Jones, M. (2020). The role of hydroponics in improving food security. *Journal of Sustainable Agriculture*, *14*(2), 123–132.

Jung, D. H., Kim, H. J., Cho, W. J., Park, S. H., & Yang, S. H. (2019). Validation testing of an ion-specific sensing and control system for precision hydroponic macronutrient management. *Computers and Electronics in Agriculture*, *156*, 660–668.

Kannan, M., Elavarasan, G., Balamurugan, A., Dhanusiya, B., & Freedon, D. (2022). Hydroponic farming–A state of art for the future agriculture. *Materials Today: Proceedings*, *68*, 2163–2166.

Kapoor, R., Pandey, S., Singh, A., & Singh, S. K. (2017). Soilless culture techniques for vegetable production under protected environment: A review. *Agricultural Reviews*, *36*(4), 311–322.

Kim, T. H., An, B. R., & Clementi, M. (2021). Phytoremediation as adaptive design strategy to improve indoor air quality. Experimental results relating to the application of a vertical hydroponic biofilter. In *Sustainability in energy and buildings 2020* (pp. 479–489). Springer.

Kumar, A., Mukherjee, G., & Gupta, S. (2024). Environmental remediation using hydroponics. In *Hydroponics and environmental bioremediation: Wastewater treatment* (pp. 115–141). Springer Nature.

Kumar, R. R., & Cho, J. Y. (2014). Reuse of hydroponic waste solution. *Environmental Science and Pollution Research, 21*, 9569–9577.

Kumar, S., Singh, M., Yadav, K. K., & Singh, P. K. (2021). Opportunities and constraints in hydroponic crop production systems: A review. *Environment Conservation Journal, 22*(3), 401–408.

Lakhiar, I. A., Gao, J., Syed, T. N., Chandio, F. A., & Buttar, N. A. (2018). Modern plant cultivation technologies in agriculture under controlled environment: A review on aeroponics. *Journal of Plant Interactions, 13*(1), 338–352.

Li, Y., Song, P., Pei, Y., & Feng, J. (2015). Effects of lateral flushing on emitter clogging and biofilm components in drip irrigation systems with reclaimed water. *Irrigation Science, 33*, 235–245.

Magwaza, S. T., Magwaza, L. S., Odindo, A. O., & Mditshwa, A. (2020). Hydroponic technology as decentralised system for domestic wastewater treatment and vegetable production in urban agriculture: A review. *Science of the Total Environment, 698*, 134154.

Maucieri, C., Nicoletto, C., Junge, R., Schmautz, Z., Sambo, P., & Borin, M. (2018). Hydroponic systems and water management in aquaponics: A review. *Italian Journal of Agronomy, 13*(1), 1–11.

Mavrogianopoulos, G. N. (2016). Irrigation dose according to substrate characteristics, in hydroponic systems. *Open Agriculture, 1*(1), 1–6.

Mndi, O., Sogoni, A., Jimoh, M. O., Wilmot, C. M., Rautenbach, F., & Laubscher, C. P. (2023). Interactive effects of salinity stress and irrigation intervals on plant growth, nutritional value, and phytochemical content in mesembryanthemum crystallinum L. *Agriculture, 13*(5), 1026.

Mungin, R., Weislogel, M. M., Hatch, T. R., & McQuillen, J. B. (2019, July). Omni-gravity hydroponics for space exploration. *International Conference on Environmental Systems (ICES)* (No. GRC-E-DAA-TN66314).

Naresh, R., Jadav, S. K., Singh, M., Patel, A., Singh, B., Beese, S., & Pandey, S. K. (2024). Role of hydroponics in improving water-use efficiency and food security. *International Journal of Environment and Climate Change, 14*(2), 608–633.

Nerlich, A., Karlowsky, S., Schwarz, D., Förster, N., & Dannehl, D. (2022). Soilless tomato production: Effects of hemp fiber and rock wool growing media on yield, secondary metabolites, substrate characteristics and greenhouse gas emissions. *Horticulturae, 8*(3), 272.

Ngxabi, S., Jimoh, M. O., Kambizi, L., & Laubscher, C. P. (2021). Growth characteristics, phytochemical contents, and antioxidant capacity of Trachyandra ciliata (Lf) Kunth grown in hydroponics under varying degrees of salinity. *Horticulturae, 7*(8), 244.

Nkcukankcuka, M., Jimoh, M. O., Griesel, G., & Laubscher, C. P. (2021). Growth characteristics, chlorophyll content and nutrients uptake in Tetragonia decumbens Mill. cultivated under different fertigation regimes in hydroponics. *Crop and Pasture Science, 73*(2), 67–76.

Nursyahid, A., Setyawan, T. A., Sa'diyah, K., Wardihani, E. D., Helmy, H., & Hasan, A. (2021, March). Analysis of Deep Water Culture (DWC) hydroponic nutrient solution level control systems. In *IOP Conference Series: Materials Science and Engineering* (Vol. 1108, No. 1, p. 012032). IOP Publishing.

Okomoda, V. T., Oladimeji, S. A., Solomon, S. G., Olufeagba, S. O., Ogah, S. I., & Ikhwanuddin, M. (2023). Aquaponics production system: A review of historical perspective, opportunities, and challenges of its adoption. *Food Science & Nutrition, 11*(3), 1157–1165.

Ortega Martínez, L. D., Martínez Valenzuela, C., Ocampo Mendoza, J., Sandoval Castro, E., & Pérez Armendáriz, B. (2016). Efficiency of substrates in soil and hydroponic system for greenhouse tomato production. *Revista Mexicana de Ciencias Agrícolas, 7*(3), 643–653.

Palande, V., Zaheer, A., & George, K. (2018). Fully automated hydroponic system for indoor plant growth. *Procedia Computer Science, 129*, 482–488.

Peiro, E., Pannico, A., Colleoni, S. G., Bucchieri, L., Rouphael, Y., De Pascale, S., Paradiso, R., & Gòdia, F. (2020). Air distribution in a fully-closed higher plant growth chamber impacts crop performance of hydroponically-grown lettuce. *Frontiers in Plant Science, 11*, 537.

Pomoni, D. I., Koukou, M. K., Vrachopoulos, M. G., & Vasiliadis, L. (2023). A review of hydroponics and conventional agriculture based on energy and water consumption, environmental impact, and land use. *Energies, 16*(4), 1690.

Prance, G. T. (2002). A paradise for economic botanists: The Eden Project. *Economic Botany, 56*(3), 226–230.

Rahman, S. I., Tahalil Azim, M., Fardin Hossain, M., Mahmud, S., Sajid, S., & Motaharul Islam, M. (2023, June). Green IoT-based automated door hydroponics farming system. In *International conference on iot based control networks and intelligent systems* (pp. 507–521). Springer Nature.

Rodríguez-Delfín, A., Gruda, N., Eigenbrod, C., Orsini, F., & Gianquinto, G. (2017). Soil based and simplified hydroponics rooftop gardens. In C. Aubry, É. Duchemin & J. Nasr (Eds.), *Rooftop urban agriculture* (pp. 61–81). Springer Nature.

Rouphael, Y., Eid, M., Flores-Serrano, F. Z., Hussain, A. R., Diaz-Vivancos, P., Rico, M. J., Xing, Y., Liu, J., Cruz-Rus, E., & Bautista, M. J. (2018). Salinity and water stress in vegetable production using hydroponics: A review. *Agronomy*, *8*(3), 39.

Roy, A. S., de Beer, M., Pillai, S. K., & Ray, S. S. (2023). Application of layered double hydroxides as a slow-release phosphate source: A comparison of hydroponic and soil systems. *ACS Omega*, *8*(17), 15017–15030.

Rufí-Salís, M., Petit-Boix, A., Villalba, G., Ercilla-Montserrat, M., Sanjuan-Delmás, D., Parada, F., Arcas, V., Muñoz-Liesa, J., & Gabarrell, X. (2020). Identifying eco-efficient year-round crop combinations for rooftop greenhouse agriculture. *The International Journal of Life Cycle Assessment*, *25*, 564–576.

Salami, S. O., Adegbaju, O. D., Idris, O. A., Jimoh, M. O., Olatunji, T. L., Omonona, S., Orimoloye, I. R., Adetunji, A. E., Olusola, A., Maboeta, M. S., & Laubscher, C. P. (2022). South African wild fruits and vegetables under a changing climate: The implications on health and economy. *South African Journal of Botany*, *145*, 13–27.

Sánchez-Gutiérrez, M., Romero-Troxler, C., Ruiz-Bejarano, B., Camacho-Ferre, F., & Martínez-Gaitán, Y. (2015). Comparison of deep flow technique (DFT) and Ebb and flow technique (EFT) for basil (Ocimum basilicum L.production. In *Advances in controlled atmosphere research* (pp. 239–244). Elsevier.

Savvas, D., Ntatsi, G., Tsiridis, T. N., & Kefalas, P. (2017). Deep water culture for lettuce production: Influence of plant density and solution exchange rate. *IOP Conference Series: Earth and Environmental Science*, *110*(1), 012032.

Sepúlveda-Vera, E., Muñoz, V., Rubia, A., & Moreno-Caselles, J. (2019). Hydroponic culture of lettuce using biochar-amended coir as growth medium. *Scientifica*, *2019*.

Sharma, A., Manpoong, C., Devadas, V. S., Kartha, B. D., Pandey, H., & Wangsu, M. (2022). Crop hydroponics, phyto-hydroponics, crop production, and factors affecting soilless culture. *ACS Agricultural Science & Technology*, *2*(6), 1134–1150.

Sharma, N., Acharya, S., Kumar, K., Singh, N., & Chaurasia, O. P. (2018). Hydroponics as an advanced technique for vegetable production: An overview. *Journal of Soil and Water Conservation*, *17*(4), 364–371.

Sharma, S., Lishika, B., Shahi, A., & Kaushal, S. (2023). Hydroponics: The potential to enhance sustainable food production in non-arable areas. *Current Journal of Applied Science and Technology*, *42*(39), 13–23.

Shetty, H. M., Pai, K., & Mallya, N. (2021). Fully automated hydroponics system for smart farming. *International Journal of Engineering and Manufacturing*, *4*, 33–41.

Sigognault Flochlay, A., Thomas, E., & Sparagano, O. (2017). Poultry red mite (Dermanyssus gallinae) infestation: A broad impact parasitological disease that still remains a significant challenge for the egg-laying industry in Europe. *Parasites & Vectors*, *10*, 1–6.

Sogoni, A., Jimoh, M. O., Kambizi, L., & Laubscher, C. P. (2021). The impact of salt stress on plant growth, mineral composition, and antioxidant activity in tetragonia decumbens mill.: An underutilized edible halophyte in south Africa. *Horticulturae*, *7*(6), 140.

Son, J. E., Kim, H. J., & Ahn, T. I. (2020). Hydroponic systems. In *Plant factory* (pp. 273–283). Academic Press.

Sonneveld, C., Hernandez, J., & Stigter, J. (2017). Irrigation and fertilization management in soilless cultivation. In J. G. Atherton & J. R. Chalker-Scott (Eds.), *Advances in postharvest technologies for minimizing food loss* (pp. 241–264). Woodhead Publishing.

Sousa, R. D., Bragança, L., da Silva, M. V., & Oliveira, R. S. (2024). Challenges and solutions for sustainable food systems: The potential of home hydroponics. *Sustainability*, *16*(2), 817.

Specht, K., Siebert, R., Hartmann, I., Freisinger, U. B., Sawicka, M., Werner, A., Thomaier, S., Henckel, D., Walk, H., & Dierich, A. (2014). Urban agriculture of the future: An overview of sustainability aspects of food production in and on buildings. *Agriculture and Human Values*, *31*, 33–51.

Suryaningprang, A., Suteja, J., Mulyaningrum, M., & Herlinawati, E. (2021). Hydroponic: Empowering local farmer knowhow to gain value added on agriculture commodity. *Budapest International Research and Critics Institute (BIRCI-Journal): Humanities and Social Sciences*, *4*(1), 787–796.

Treftz, C., & Omaye, S. T. (2016). Hydroponics: Potential for augmenting sustainable food production in non-arable regions. *Nutrition & Food Science*, *46*(5), 672–684.

Tshayingwe, A., Jimoh, M. O., Sogoni, A., Wilmot, C. M., & Laubscher, C. P. (2023). Light intensity and growth media influence growth, nutrition, and phytochemical content in trachyandra divaricata Kunth. *Agronomy*, *13*(1), 247.

Van Lenteren, J. C., & Nicot, P. C. (2020). Integrated pest management methods and considerations concerning implementation in greenhouses. In Gullino et al. (Eds.), *Integrated pest and disease management in greenhouse crops* (pp. 177–193). Springer Nature.

Van Wyk, B. E., & Gorelik, B. (2017). The history and ethnobotany of Cape herbal teas. *South African Journal of Botany*, *110*, 18–38.

Verner, D., Vellani, S., Goodman, E., & Love, D. C. (2022). Frontier agriculture: Climate-smart and water-saving agriculture technologies for livelihoods and food security. In *New forms of urban agriculture: An urban ecology perspective* (pp. 159–186). Springer Nature.

Viljoen, C. C., Jimoh, M. O., & Laubscher, C. P. (2021). Studies of vegetative growth, inflorescence development and eco-dormancy formation of abscission layers in Streptocarpus formosus (Gesneriaceae). *Horticulturae*, *7*(6), 120.

Vinci, G., & Rapa, M. (2019). Hydroponic cultivation: Life cycle assessment of substrate choice. *British Food Journal*, *121*(8), 1801–1812.

Wang, R., Isozaki, M., Iwasaki, Y., & Muramatsu, Y. (2022). Root-zone temperature effects on spinach biomass production using a nutrient film technique system. *HortScience*, *57*(4), 532–540.

Warren, E., Hawkesworth, S., & Knai, C. (2015). Investigating the association between urban agriculture and food security, dietary diversity, and nutritional status: A systematic literature review. *Food Policy*, *53*, 54–66.

Weber, C. F. (2016). Nutrient content of cabbage and lettuce microgreens grown on vermicompost and hydroponic growing pads. *Journal of Horticulture*, *3*(4), 1–5.

Weinzierl, R., & Henn, T. (2020). Botanical insecticides and insecticidal soaps. In *Handbook of integrated pest management for turf and ornamentals* (pp. 541–555). CRC Press.

Wright, B. D., Bausch, W. C., & Knott, W. M. (1988). A hydroponic system for microgravity plant experiments. *Transactions of the ASAE*, *31*(2), 440–0446.

Yadav, R. K., Chiranjeevi, P., & Patil, S. A. (2020). Integrated drip hydroponics-microbial fuel cell system for wastewater treatment and resource recovery. *Bioresource Technology Reports*, *9*, 100392.

Yatagai, F., Honma, M., Dohmae, N., & Ishioka, N. (2019). Biological effects of space environmental factors: A possible interaction between space radiation and microgravity. *Life Sciences in Space Research*, *20*, 113–123.

Zhang, H., Gao, Y., Liu, J., Lin, Z., Lee, C. T., Hashim, H., & Li, C. (2021). Recovery of nutrients from fish sludge as liquid fertilizer to enhance sustainability of aquaponics: A Review. *Chemical Engineering Transactions*, *83*, 55–60.

Zhang, Y., Zhang, Y. K., & Li, Z. (2022). A new and improved aquaponics system model for food production patterns for urban architecture. *Journal of Cleaner Production*, *342*, 130867.

Index

Note: Page numbers in *italics* indicate figures, and page numbers in **bold** indicate tables in the text

Ash method, and seed germination, 12

D

E

F

G

L

M

N

O